# Network Information Systems

## A Dynamical Systems Approach

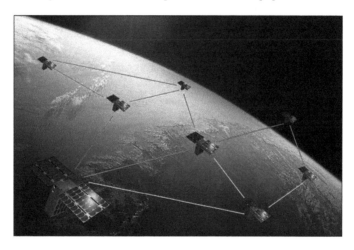

**Wassim M. Haddad**

The Georgia Institute of Technology, Atlanta, Georgia

**Qing Hui**

University of Nebraska-Lincoln, Lincoln, Nebraska

**Junsoo Lee**

University of South Carolina, Columbia, South Carolina

Society for Industrial and Applied Mathematics
Philadelphia

| | |
|---|---|
| *Publications Director* | Kivmars H. Bowling |
| *Executive Editor* | Elizabeth Greenspan |
| *Acquisitions Editor* | Elizabeth Greenspan |
| *Developmental Editor* | Rose Kolassiba |
| *Managing Editor* | Kelly Thomas |
| *Production Editor* | Lisa Briggeman |
| *Copyeditor* | Julia Cochrane |
| *Production Manager* | Donna Witzleben |
| *Production Coordinator* | Cally A. Shrader |
| *Compositor* | Scott Collins |
| *Graphic Designer* | Doug Smock |

**Library of Congress Control Number: 2023932635**

*Cover photo: National Aeronautics and Space Administration*

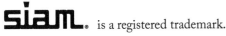

*To my students, who constantly inspire, challenge,*
*and impel me in my abiding pursuit of knowledge.*
W. M. H.

*To my wife Huijun Zhu, my daughter Iris Hui,*
*and my son Aiden Hui.*
Q. H.

*To my parents Jiwon Lee and Kwangyeon Kim, who made it*
*possible for me to pursue my passion for knowledge.*
J. L.

*Information entropy is the irreducible kernel in the universe and is more fundamental than matter itself, with information forming the very core of existence. For to produce change (motion) requires energy, whereas to direct this change requires information; id est, energy takes different forms, but these forms are determined by information.*

WASSIM M. HADDAD

# Contents

Preface                                                                          xiii

Chapter 1. Introduction                                                            1
  1.1    Network Dynamical Systems                                                  1
  1.2    A Brief Outline of the Monograph                                           4

Chapter 2. Distributed Nonlinear Control for Network Systems                       13
  2.1    Introduction                                                              13
  2.2    Notation and Definitions                                                  14
  2.3    The Consensus Problem in Dynamical Networks                               16
  2.4    Distributed Control for Parallel Formations                               33
  2.5    Distributed Control for Circular Formations                               43

Chapter 3. Thermodynamics-Based Control of Network Systems                         49
  3.1    Introduction                                                              49
  3.2    Energy, Entropy, and Thermal Equilibria                                   50
  3.3    Heat Transfer between Pairs of Subsystems                                 52
  3.4    Interconnection Structure Using Graphs                                    55
  3.5    Thermal Equilibria and Semistability                                      56
  3.6    Thermodynamics of Undirected Networks                                     63
  3.7    Thermodynamics of Directed Networks                                       66

Chapter 4. Finite Time Semistability, Consensus, and Formation
  Control for Nonlinear Dynamical Networks                                         77
  4.1    Introduction                                                              77
  4.2    Mathematical Preliminaries                                               79
  4.3    Lyapunov and Converse Lyapunov Theory for Semistability                   80
  4.4    Finite Time Semistability of Nonlinear Dynamical Systems                  87
  4.5    Homogeneity and Finite Time Semistability                                 92
  4.6    The Consensus Problem in Dynamical Networks                              101
  4.7    Distributed Control Algorithms for Finite Time Consensus
         and Parallel Formation                                                   104

**Chapter 5. Consensus Protocols for Network Systems with a Uniformly Continuous Quasi-Resetting Architecture**          **115**

| | | |
|---|---|---|
| 5.1 | Introduction | 115 |
| 5.2 | Notation, Definitions, and Graph-Theoretic Notions | 116 |
| 5.3 | A Quasi-Resetting Control Architecture for Approximating Resetting Controllers | 117 |
| 5.4 | Consensus Control Protocols | 121 |
| 5.5 | Stability Analysis | 123 |

**Chapter 6. Control Protocols for Dynamical Networks with Switching Communication Topologies**          **133**

| | | |
|---|---|---|
| 6.1 | Introduction | 133 |
| 6.2 | Mathematical Preliminaries | 134 |
| 6.3 | Semistability Theory for Differential Inclusions | 137 |
| 6.4 | Time-Varying Discontinuous Dynamical Systems | 150 |
| 6.5 | Lyapunov-Based Semistability Analysis for Time-Varying Discontinuous Dynamical Systems | 152 |
| 6.6 | Applications to Network Consensus with a Switching Topology | 163 |
| 6.7 | Application to Finite Time Rendezvous Problems | 173 |
| 6.8 | Discontinuous Consensus Protocols with Time-Varying Communication Links | 176 |

**Chapter 7. Robust Network Consensus Protocols**          **179**

| | | |
|---|---|---|
| 7.1 | Introduction | 179 |
| 7.2 | Mathematical Preliminaries | 179 |
| 7.3 | Semistability and Homogeneous Dynamical Systems | 183 |
| 7.4 | Robust Control Algorithms for Network Consensus Protocols | 184 |

**Chapter 8. Network Systems with Time Delays**          **197**

| | | |
|---|---|---|
| 8.1 | Introduction | 197 |
| 8.2 | Mathematical Preliminaries | 198 |
| 8.3 | Semistability and Equipartition of Linear Compartmental Systems with Time Delay | 201 |
| 8.4 | Semistability and Equipartition of Nonlinear Compartmental Systems with Time Delay | 204 |
| 8.5 | The Consensus Problem in Dynamical Networks | 213 |

**Chapter 9. Eulerian Swarming Networks and Continuum Thermodynamics**          **219**

| | | |
|---|---|---|
| 9.1 | Introduction | 219 |
| 9.2 | Mathematical Preliminaries | 221 |

9.3     A Thermodynamic Model for Large-Scale Swarms                           224
9.4     Boundary Semistable Control for Large-Scale Swarms                      231
9.5     Advection-Diffusion Network Models                                     238
9.6     Connections between Eulerian and Lagrangian Models for
        Information Consensus                                                  240

**Chapter 10. $\mathcal{H}_2$ Optimal Semistable Control for Network Systems     243**

10.1    Introduction                                                           243
10.2    $\mathcal{H}_2$ Semistability Theory                                    243
10.3    Optimal Semistable Stabilization                                        254
10.4    Optimal Fixed-Structure Control for Network Consensus                   257

**Chapter 11. Optimal Control for Linear and Nonlinear
        Semistabilization                                                       261**

11.1    Introduction                                                           261
11.2    Mathematical Preliminaries                                             262
11.3    Semistability Analysis of Nonlinear Systems                            273
11.4    Optimal Control for Semistabilization                                  277

**Chapter 12. Optimal Robust Consensus for Network Systems with
        Imperfect Information                                                   291**

12.1    Introduction                                                           291
12.2    Loss of Robustness with Imperfect Information                          293
12.3    Semistable Gaussian Linear-Quadratic Consensus                        298
12.4    Semistabilization and Information State Equipartitioning               304
12.5    Existence of Optimal Solutions for the Gaussian Linear-
        Quadratic Consensus Problem                                            327
12.6    A Numerical Algorithm for the Gaussian Linear-Quadratic
        Consensus Problem                                                      334

**Chapter 13. Mitigating the Effects of Sensor Uncertainties in
        Network Systems                                                         343**

13.1    Introduction                                                           343
13.2    Mathematical Preliminaries and Problem Formulation                     344
13.3    Adaptive Leader Following with Time-Invariant Sensor
        Uncertainties                                                          348
13.4    Adaptive Leader Following with Time-Varying Sensor
        Uncertainties                                                          352

**Chapter 14. Adaptive Estimation Using Network Identifiers                      367**

14.1    Introduction                                                           367
14.2    Mathematical Preliminaries and Problem Formulation                     368
14.3    Adaptive Estimation Problem                                            369
14.4    Adaptive Distributed Observers                                         371

14.5    Adaptive Consensus of Distributed Observers over
        Networks with Undirected Graph Topologies                       373
14.6    Extensions to Networks with Directed Graph Topologies           379

**Chapter 15.  Discrete-Time Network Systems                            387**

15.1    Introduction                                                    387
15.2    Mathematical Preliminaries                                      389
15.3    Semistability of Discrete Autonomous Systems                    391
15.4    Lyapunov and Converse Lyapunov Theorems for
        Semistability                                                   395
15.5    Finite Time Semistability of Discrete Autonomous Systems        400
15.6    Lyapunov and Converse Lyapunov Theorems for Finite
        Time Semistability                                              401
15.7    A Thermodynamic-Based Architecture for Asymptotic
        Network Consensus                                               412

**Chapter 16.  $\mathcal{H}_2$ Optimal Semistable Control for Discrete-Time
Network Systems                                                         423**

16.1    Introduction                                                    423
16.2    Discrete-Time $\mathcal{H}_2$ Semistability Theory              423
16.3    Optimal Semistable Stabilization                                435
16.4    Information Flow Models                                         437
16.5    Semistability of Information Flow Models                        439
16.6    Optimal Fixed-Structure Control of Network Consensus            445

**Chapter 17.  Optimal Network Resource Allocation                      449**

17.1    Introduction                                                    449
17.2    Resource Allocation Problem Formulation                         451
17.3    A Control-Theoretic Approach to Balanced Resource
        Allocation                                                      454
17.4    Semistable Linear-Quadratic Control Theory                      459
17.5    Characterization of Semistability via Semicontrollability       465
17.6    Optimization-Based Control Design                               473
17.7    A Numerical Algorithm for Optimal Resource Allocation           476

**Chapter 18.  Approximate Consensus for Network Systems with
Inaccurate Sensor Measurements                                         487**

18.1    Introduction                                                    487
18.2    Notation, Definitions, and Mathematical Preliminaries           488
18.3    Consensus Control Problem with Uncertain Interagent
        Location Measurements                                           489
18.4    Continuous-Time Consensus with a Connected Graph
        Topology                                                        491
18.5    Discrete-Time Consensus with a Connected Graph Topology         496

18.6    A Set-Valued Analysis Approach to Discrete-Time
        Consensus                                                     499

**Chapter 19. A Hybrid Thermodynamic Control Protocol for
Network Systems with Intermittent Information**                       **511**

19.1    Introduction                                                  511
19.2    Mathematical Preliminaries                                    512
19.3    A Hybrid Thermodynamic Consensus Control Architecture        515
19.4    Connections to Information Entropy and the Hybrid
        Second Law of Thermodynamics                                 521

**Chapter 20. Hybrid Control Protocols for Consensus, Parallel
Formations, and Collision Avoidance**                                **529**

20.1    Introduction                                                  529
20.2    Distributed Nonlinear Control Algorithms for Consensus       529
20.3    A Hybrid Control Architecture for Consensus                  531
20.4    Distributed Control for Parallel Formations                  544
20.5    Collision Avoidance via Hybrid Control                       548

**Chapter 21. Formation Control Protocols for Network Systems via
Hybrid Stabilization of Sets**                                       **555**

21.1    Introduction                                                  555
21.2    Hybrid Control and Impulsive Dynamical Systems               556
21.3    Hybrid Stabilization of Sets                                 562
21.4    Specialization to Linear Dynamical Systems                   566
21.5    Hybrid Control Design for Parallel and Rendezvous
        Formations                                                   569
21.6    Hybrid Control Design for Consensus in Multiagent
        Networks                                                     574
21.7    Hybrid Control Design for Cyclic Pursuit                     578

**Chapter 22. Conclusion**                                           **589**

**Bibliography**                                                     **591**

**Index**                                                           **619**

# Preface

Recent technological advances in communications and computation have spurred a broad interest in control of networks and control over networks. Network systems involve distributed decision-making for coordination of networks of dynamic agents and address a broad area of applications, including cooperative control of unmanned air vehicles, microsatellite clusters, mobile robotics, battle space management, congestion control in communication networks, intelligent vehicle/highway systems, large-scale manufacturing systems, and biological networks, to cite but a few examples. To address the problem of autonomy and complexity for control and coordination of network systems, in this monograph we look to system thermodynamics and dynamical systems theory for inspiration in developing innovative architectures for controlling network systems.

Specifically, we draw from the fundamental principles of dynamical systems theory and dynamical thermodynamics to develop a continuous-time, a discrete-time, and a hybrid dynamical system and control framework for linear and nonlinear large-scale network systems. In particular, thermodynamically inspired control algorithms are developed to address agent interactions, cooperative and noncooperative control, task assignments, and resource allocations. Our proposed framework extends the concepts of energy, entropy, and temperature to undirected and directed networks and uses continuous-time, discrete-time, and hybrid thermodynamic principles to design distributed control protocol algorithms for static and dynamic networked systems in the face of system uncertainty, exogenous disturbances, imperfect system network communication, and time delays.

More specifically, we use the zeroth, first, and second laws of thermodynamics to design distributed controllers that cause networked dynamical systems to emulate natural thermodynamic behavior involving inherent guarantees of robustness in network systems. In particular, the proposed controller architectures involve the exchange of state information between agents, guaranteeing that the closed-loop dynamical network is *semistable* (i.e., Lyapunov stable and convergent) to an equipartitioned equilibrium state representing a state of information consensus consistent with thermodynamic principles. Furthermore, novel control architectures for formation control, rendezvous, flocking, cyclic pursuit, and resource allocation problems are also presented. The proposed control architectures address model uncertainties, synchronism, system time delays, and switching network topologies for addressing robustness, information asynchrony between agents, message transmission and processing delays, communication link failures, and communication dropouts.

The underlying intention of this monograph is to present a general network systems framework using tools from algebraic graph theory, dynamical systems theory, and system thermodynamics. It is hoped that this monograph will help stimulate increased interaction between engineers, physicists, computer scientists, information scientists, and dynamical systems and control theorists.

## Acknowledgments

In some parts of the monograph we have relied on work we have done jointly with Jordan M. Berg, Sanjay P. Bhat, Makram Chahine, Vijaysekhar Chellaboina, Michael A. Dimitriou, Rafal Goebel, Andrea L'Afflitto, Sergey G. Nersesov, Taymour Sadikov, and Tansel Yucelen; it is a pleasure to acknowledge their contributions. The results reported in this monograph were obtained at the School of Aerospace Engineering, Georgia Institute of Technology, Atlanta. The research support provided by the Air Force Office of Scientific Research (AFOSR) over the years has been instrumental in allowing us to explore basic research topics that have led to some of the material in this monograph. We are indebted to AFOSR for its continued support.

*Wassim M. Haddad*

*Chapter One*

---

# Introduction

## 1.1 Network Dynamical Systems

Due to advances in embedded communication, computation, and control resources over the last several years, considerable research effort has been devoted to control of networks and control over networks [5, 80, 93, 101, 180, 183, 194, 206, 211, 220, 234, 238, 241, 258, 294, 296, 304, 305]. Network systems involve distributed decision-making for coordination of networks of dynamic agents involving information flow, enabling enhanced operational effectiveness via cooperative control in autonomous systems. These dynamical network systems cover a very broad spectrum of applications, including cooperative control of unmanned air vehicles (UAVs) and autonomous underwater vehicles (AUVs) for combat, surveillance, and reconnaissance [312]; distributed reconfigurable sensor networks for managing power levels of wireless networks [73]; air and ground transportation systems for air traffic control and payload transport and traffic management [290]; swarms of air and space vehicle formations for command and control between heterogeneous air and space assets [93, 305]; and congestion control in communication networks for routing the flow of information through a network [241].

To enable the applications for these complex large-scale network systems, cooperative control tasks such as formation control, rendezvous, flocking, cyclic pursuit, and consensus need to be developed [161, 167, 180, 203, 220, 234, 238, 289]. The formation control problem, which includes flocking and cyclic pursuit, wherein parallel and circular formations of vehicles are sought, typically requires *cohesion*, *separation*, and *alignment* constraints for individual agent steering, which describe how a given vehicle maneuvers based on the positions and velocities of nearby agents. Specifically, cohesion refers to a steering rule wherein a given vehicle attempts to move toward the average position of local vehicles, separation refers to collision avoidance with nearby vehicles, and alignment refers to velocity matching with nearby vehicles.

To realize these tasks, individual agents need to share information of the system objectives as well as the dynamical network. In particular, in many applications involving multiagent systems, groups of agents are required to agree on certain quantities of interest. Information consensus over dynamic information-exchange topologies guarantees agreement between agents for a given coordination task. Distributed consensus algorithms involve neighbor-to-neighbor interaction between agents, wherein agents update their information state based on the information states of the neighboring agents. A unique feature of the closed-loop dynamics under any control algorithm that achieves consensus in a dynamical network is the existence of a continuum of equilibria representing a state of consensus. Under such dynamics, the limiting consensus state achieved is not determined completely by the dynamics but depends on the initial system state as well.

In systems possessing a continuum of equilibria, *semistability*, and not asymptotic stability, is the relevant notion of stability [36, 38, 122]. Semistability is the property whereby every trajectory that starts in a neighborhood of a Lyapunov stable equilibrium converges to a (possibly different) Lyapunov stable equilibrium. Semistability thus implies Lyapunov stability and is implied by asymptotic stability. From a practical viewpoint, it is not sufficient to only guarantee that a network converges to a state of consensus since steady-state convergence is not sufficient to guarantee that small perturbations from the limiting state will lead to only small transient excursions from a state of consensus. It is also necessary to guarantee that the equilibrium states representing consensus are Lyapunov stable and, consequently, semistable.

Modern complex large-scale network systems can additionally give rise to systems which have nonsmooth dynamics. Examples include variable-structure systems where control inputs switch discontinuously between extreme input values to generate minimum-time or minimum-fuel system trajectories. Discontinuities can also be intentionally designed as part of the controller architecture to achieve hierarchical system stabilization [195, 196]. In particular, the complexity of modern controlled large-scale network systems is further exacerbated by the use of hierarchical embedded control subsystems within the feedback control system, that is, abstract decision-making units performing logical checks that identify system mode operation and specify the continuous-variable subcontroller to be activated.

Such systems typically possess a multiechelon hierarchical hybrid decentralized and distributed control architecture characterized by continuous-time dynamics at the lower levels of the hierarchy and discrete-time dynamics at the higher levels of the hierarchy. The lower-level units directly interact with the dynamical system to be controlled, while the higher-level units receive information from the lower-level units as inputs and provide (possibly discrete) output commands which serve to coordinate and reconcile the (sometimes competing) actions of the lower-level units [128].

The hierarchical controller organization reduces processor cost and controller complexity by breaking up the processing task into relatively small pieces and decomposing the fast and slow control functions. Typically, the higher-level units perform logical checks that determine system mode operation, whereas the lower-level units execute continuous-variable commands for a given system mode of operation. Due to their multiechelon hierarchical structure, hybrid dynamical systems are capable of simultaneously exhibiting continuous-time dynamics, discrete-time dynamics, logic commands, discrete events, and resetting events.

Such systems include switching network systems [50, 196, 247], nonsmooth impact systems [49, 51], biological network systems [192], sampled-data systems [138], discrete-event systems [246], intelligent vehicle/highway systems [216], constrained mechanical systems [49], flight control systems [298], and command and control for future battlefields, to cite but a few examples. The mathematical descriptions of many of these systems can be characterized by hybrid systems and impulsive differential equations with discontinuous right-hand sides [13, 14, 105, 106, 117, 128, 152, 192, 269].

To enable the autonomous operation of multiagent network systems, functional algorithms for agent coordination and control need to be developed. In particular, control algorithms need to address agent interactions, cooperative and noncooperative control, task assignments, and resource allocations. To realize these tasks, appropriate sensory and cognitive capabilities such as adaptation, learning, decision-making, and agreement (or consensus) on the agent and multiagent levels are required.

The common approach for addressing the autonomous operation of multiagent systems is to use distributed control algorithms involving neighbor-to-neighbor interaction between agents, wherein agents update their information state based on the information states of the neighboring agents. Since most multiagent network systems are highly interconnected and mutually interdependent, both physically and through a multitude of information and communication networks, these systems are characterized by high-dimensional, large-scale interconnected dynamical systems.

To develop distributed methods for control and coordination of autonomous multiagent systems, many researchers have looked to autonomous *swarm* systems appearing in nature for inspiration [194, 197, 226, 245, 258, 304]. These systems necessitate the development of relatively simple autonomous agents that are inherently distributed, self-organized, and truly scalable. Scalability follows from the fact that such systems do not involve centralized control and communication architectures. In addition, these systems should be inherently robust to individual agent failures, unplanned task assignment changes, and environmental changes.

In this monograph, we highlight the role that dynamical systems theory and system thermodynamics can play in the control of network systems. Specifically, we explore synergies between algebraic graph theory, dynamical systems theory, and dynamical thermodynamics in developing control methods for self-organizing teams of multiagent vehicles under sparse sensing. In self-organizing systems, the connection between the local subsystem interactions and the globally complex system behavior is often elusive. In nature, these systems are known as dissipative systems and consume energy and matter while maintaining their stable structure by dissipating entropy to the environment. A common phenomenon among these systems is that they evolve in accordance with the laws of (nonequilibrium) thermodynamics [118, 127]. Dynamical thermodynamics involves open interconnected dynamical systems that exchange matter and energy with their environment in accordance with the first law (conservation of energy) and the second law (nonconservation of entropy) of thermodynamics.

In light of the above, it seems both natural and appropriate to postulate the following paradigm for the nonlinear analysis and control law design of complex large-scale network systems and multiagent autonomous systems: develop a unified thermodynamic analysis and control systems design framework for continuous-time, discrete-time, and hybrid large-scale dynamical network systems and multiagent systems in the face of a specified level of modeling uncertainty. This paradigm provides a rigorous foundation for developing a unified network thermodynamic system analysis and synthesis framework for continuous-time and discrete-time systems, as well as systems that combine continuous-time and discrete-time dynamics possessing hybrid, hierarchical, and feedback structures. Correspondingly, the main goal of this monograph is to present analysis and control design tools for linear and nonlinear network systems and multiagent systems which support this paradigm.

## 1.2 A Brief Outline of the Monograph

The main objective of this monograph is to develop a general design framework for linear and nonlinear dynamical network systems, with an emphasis on a dynamical systems approach. The main contents of the monograph are as follows. In Chapter 2, we establish notation and definitions and develop a thermodynamic framework for addressing consensus problems for nonlinear multiagent dynamical systems with a fixed communication graph topology. Specifically, we present distributed nonlinear static and dynamic controller architectures for multiagent coordination. The proposed controller architectures are predicated on system thermodynamic notions resulting in controller architectures involving the exchange of information between agents that guarantee that the closed-loop dynamical network system is consistent with basic thermodynamic principles. Furthermore, novel control architectures for formation control and cyclic pursuit problems are also presented.

The zeroth and first laws of thermodynamics define the concepts of thermal equilibrium and thermal energy. The second law of thermodynamics determines whether a particular transfer of thermal energy can occur. Collectively, these fundamental laws of nature imply that a closed collection of thermodynamic subsystems will tend to thermal equilibrium. In Chapter 3, we generalize the concepts of energy, entropy, and temperature to undirected and directed networks of single integrators and demonstrate how thermodynamic principles can be applied to the design of distributed consensus control algorithms for networked dynamical systems.

In Chapter 4, we develop the theory for semistability and finite time semistability analysis and synthesis of systems having a continuum of equilibria. Semistability is the property whereby the solutions of a dynamical system converge to Lyapunov stable equilibrium points determined by the system initial conditions. More specifically, we merge the theories of semistability and finite time stability to develop a rigorous framework for finite time semistability. In particular, finite time semistability for a continuum of equilibria of continuous autonomous systems is established. Continuity of the settling-time function capturing the finite settling-time behavior of the dynamical system, as well as Lyapunov and converse Lyapunov theorems for semistability, are developed. In addition, necessary and sufficient conditions for finite time semistability of homogeneous systems are addressed by exploiting the fact that a homogeneous system is finite time semistable if and only if it is semistable and has a negative degree of homogeneity. Then we use these results to develop a general framework for designing semistable protocols in dynamical networks for achieving coordination tasks in finite time.

In Chapter 5, we develop a consensus protocol for networked multiagent systems using a resetting control architecture. Specifically, we design control protocols consisting of a delayed feedback quasi-resetting control law where controller resettings occur when the relative state measurements (i.e., distance) between an agent and its neighboring agents approach zero. In contrast to the impulsive resetting controllers introduced in Chapters 19–21, the proposed resetting is uniformly continuous, and, hence, our approach does not require any well-posedness assumptions imposed by impulsive resetting controllers. In addition, using a Lyapunov–Krasovskii functional, it is shown that the multiagent system reaches asymptotic state equipartitioning, where the system steady state is uniformly distributed over the system initial conditions. Finally, we develop $\mathcal{L}_\infty$ transient performance guarantees accounting for system overshoot and excessive control effort.

In Chapter 6, we develop semistability and finite time semistability for discontinuous dynamical systems. In particular, Lyapunov-based tests for strong and weak semistability as well as finite time semistability for autonomous and nonautonomous differential inclusions are established. Using these results, we then develop a framework for designing semistable and finite time semistable control protocols for dynamical networks with switching topologies. Specifically, we present distributed nonlinear static and dynamic output feedback controller architectures for multiagent network consensus and rendezvous problems with dynamically changing communication topologies.

Even though many consensus control protocol algorithms have been developed over the last several years in the literature, robustness properties of these algorithms involving nonlinear dynamics have been largely ignored. Robustness here refers to sensitivity of the control algorithm achieving semistability and consensus in the face of model uncertainty. In Chapter 7, we examine the robustness of several control algorithms for network consensus protocols with information model uncertainty of a specified structure. In particular, we develop sufficient conditions for robust stability of control protocol functions involving higher-order perturbation terms that scale in a consistent fashion with respect to a scaling operation on an underlying space, with the additional property that the protocol functions can be written as a sum of functions, each homogeneous with respect to a fixed scaling operation, that retain system semistability and consensus.

In many applications involving multiagent systems, groups of agents are required to agree on certain quantities of interest in the face of communication delays. In particular, it is important to develop consensus protocols for networks of dynamic agents with directed information flow, switching network topologies, and possible system time delays. In Chapter 8, we use dynamical system models to characterize dynamic algorithms for linear and nonlinear networks of dynamic agents in the presence of interagent communication delays that possess a continuum of semistable equilibria. In addition, we show that the steady-state distribution of the dynamic network is uniform, leading to system state equipartitioning or consensus. These results extend the results in the literature on consensus protocols for linear balanced networks to linear and nonlinear unbalanced networks with time delays.

In Chapter 9, we develop a thermodynamic framework for addressing consensus problems for Eulerian swarm models. Specifically, we present a distributed boundary controller architecture involving the exchange of information between uniformly distributed swarms over an $n$-dimensional (not necessarily Euclidian) space that guarantee that the closed-loop system is consistent with basic thermodynamic principles. In addition, we establish the existence of a unique continuously differentiable entropy functional for all equilibrium and nonequilibrium states of our thermodynamically consistent dynamical system. Information consensus and semistability are shown using the well-known Sobolev embedding theorems and the notion of generalized (or weak) solutions. Since the closed-loop system is guaranteed to satisfy basic thermodynamic principles, robustness to individual agent failures and unplanned individual agent behavior is automatically guaranteed.

In Chapter 10, we develop the $\mathcal{H}_2$ semistability theory for linear dynamical systems. Using this theory, we design $\mathcal{H}_2$ optimal semistable consensus controllers for linear dynamical networks. Unlike the standard $\mathcal{H}_2$ optimal control problem, a complicating feature of the $\mathcal{H}_2$ optimal semistable stabilization problem is that the closed-loop Lyapunov equation guaranteeing semistability can admit multiple solutions. An interesting feature of the proposed approach, however, is that a least squares solution over all possible semistabilizing solutions corresponds to the $\mathcal{H}_2$ optimal solution. It is shown that this least squares solution can be characterized by a linear matrix inequality (LMI) minimization problem. The proposed framework is then used to develop $\mathcal{H}_2$ optimal consensus protocols for continuous-time networks.

In Chapter 11, the classical Hamilton–Jacobi–Bellman optimal linear and nonlinear control problem is extended to address a weaker version of closed-loop asymptotic stability, namely, semistability, which is of paramount importance for consensus control of network dynamical systems. Specifically, we show that the optimal semistable state feedback controller can be solved using a form of the Hamilton–Jacobi–Bellman conditions that does not require the cost-to-go function to be sign definite. This result is then used to solve an optimal linear-quadratic network regulation problem using a Riccati equation approach.

In Chapter 12, we address an optimal robust equipartitioning problem for network systems in the face of imperfect state information using a semistable Gaussian linear-quadratic consensus control problem formulation. In particular, we consider network systems subject to Gaussian white noise disturbances, measurement noise, and a random distribution of the system initial conditions. More specifically, necessary and sufficient conditions for semistability and optimal information state equipartitioning are derived, and the existence of optimal solutions to this optimization problem is given for network systems with exogenous disturbances and information uncertainty. A globally convergent numerical algorithm is also presented to efficiently solve the proposed constrained optimization problem.

Many networked multiagent systems consist of interacting agents that locally exchange information, energy, or matter. Since these systems do not in general have a centralized architecture to monitor the activity of each agent, resilient distributed control system design for networked multiagent systems is essential in providing high system performance, reliability, and operation in the presence of system uncertainties. An important class of such system uncertainties that can significantly deteriorate the achievable closed-loop system performance is sensor uncertainties, which can arise due to low sensor quality, sensor failure, sensor bias, or detrimental environmental conditions. In Chapter 13, we develop a distributed *adaptive control* architecture for networked multiagent systems with undirected communication graph topologies to mitigate the effect of sensor uncertainties. Specifically, we consider agents having identical high-order, linear dynamics with agent interactions corrupted by unknown exogenous disturbances. We show that the proposed adaptive control architecture guarantees asymptotic stability of the closed-loop dynamical system when the exogenous disturbances are time invariant and uniform ultimate boundedness when the exogenous disturbances are time varying.

In Chapter 14, we depart from the network consensus control problem and present an *adaptive estimation* framework predicated on multiagent network identifiers with undirected and directed graph topologies. Specifically, the system state and plant parameters are identified online using $N$ agents implementing adaptive observers with an interagent communication architecture. The adaptive observer architecture includes an additive term which involves a penalty on the mismatch between the state and parameter estimates. The proposed architecture is shown to guarantee state and parameter estimate consensus. Furthermore, the proposed adaptive identifier architecture provides a measure of agreement of the state and parameter estimates that is independent of the network topology and guarantees that the deviation from the mean estimate for both the state and the parameter estimates converges to zero.

In Chapter 15, we develop semistability and finite time semistability analysis and synthesis of discrete-time dynamical systems having a continuum of equilibria. In particular, we build on the theories of semistability and finite time semistability for continuous-time dynamical systems developed in Chapters 2 and 4 to develop a rigorous framework for discrete semistability and discrete finite time semistability. Specifically, Lyapunov and converse Lyapunov theorems for semistability and finite time semistability are developed, and the regularity properties of the Lyapunov function establishing finite time semistability are shown to be related to the settling-time function capturing the finite settling-time behavior of the dynamical system. These results are then used to develop a general framework for designing semistable and finite time semistable consensus protocols for discrete dynamical networks for achieving multiagent coordination tasks asymptotically and in finite time. The proposed controller architectures involve the exchange of generalized energy state information between agents, guaranteeing that the closed-loop dynamical network is semistable to an equipartitioned equilibrium representing a state of consensus consistent with basic thermodynamic principles.

In Chapter 16, we build on the results of Chapter 10 to develop $\mathcal{H}_2$ semistability theory for linear discrete-time dynamical systems. Using this theory, we design $\mathcal{H}_2$ optimal semistable controllers for linear dynamical systems. As in the continuous-time semistable $\mathcal{H}_2$ optimal control problem, a complicating feature of the discrete-time $\mathcal{H}_2$ optimal semistable stabilization problem is that the closed-loop Lyapunov equation guaranteeing semistability can admit multiple solutions. As for the continuous-time problem, we show that a least squares solution over all possible semistabilizing solutions corresponds to the $\mathcal{H}_2$ optimal solution. Specifically, it is shown that this least squares solution can be characterized by an LMI minimization problem. The proposed framework is then used to develop $\mathcal{H}_2$ optimal semistable controllers for addressing the consensus control problem in networks of dynamic agents.

In Chapter 17, we present a control-theoretic framework to efficiently design balanced coordinated resource allocation algorithms in a network using semistabilization for discrete-time stochastic linear systems. Specifically, using the notion of semiobservability and a modified Lyapunov equation for capturing discrete-time semistability, necessary and sufficient conditions for an equivalent characterization of the proposed balanced coordinated resource allocation problem are derived. The proposed formulation provides a connection between the balanced coordinated network resource allocation problem and a discrete-time analogue of the optimal semistable Gaussian control problem developed in Chapter 12. We then convert the optimal semistable Gaussian control design problem into a constrained nonlinear optimization problem and develop numerical algorithms for the original balanced coordinated resource allocation problem using a class of randomized swarm optimization methods.

One of the main challenges in network consensus control is dealing with inaccurate sensor data. Specifically, for a group of mobile agents, the measurement of the exact location of the other agents relative to a particular agent is often inaccurate due to sensor measurement uncertainty or detrimental environmental conditions. In Chapter 18, we address the consensus problem for a group of agents with a connected, undirected, and time-invariant communication graph topology in the face of uncertain interagent measurement data. Using agent location uncertainty characterized by norm bounds centered at the neighboring agent's exact locations, we show that the agents reach an approximate consensus state and converge to a set centered at the centroid of the agents' initial locations. The diameter of the set is shown to be dependent on the graph Laplacian and the magnitude of the uncertainty norm bound. Furthermore, we show that if the network is all-to-all connected and the measurement uncertainty is characterized by a ball of radius $r$, then the diameter of the set to which the agents converge is $2r$. In addition, we also formulate our problem using set-valued analysis and develop a set-valued invariance principle to obtain set-valued consensus protocols.

In Chapter 19, we build on the results of Chapters 2 and 15 to develop a thermodynamically based framework for addressing consensus problems using a hybrid control protocol architecture with a dynamic communication topology wherein communication events are triggered via state-dependent resettings consistent with thermodynamic principles. The proposed hybrid controller architecture involves the exchange of intermittent state information between agents, guaranteeing that the closed-loop dynamical network is semistable to an equipartitioned equilibrium representing a state of information consensus consistent with basic thermodynamic principles.

In Chapter 20, we extend the results of Chapter 19 to develop a hybrid control framework that guarantees fast consensus for multiagent network systems. Specifically, we present several hybrid distributed controller architectures for multiagent coordination that guarantee improved transient performance of coordination tasks as compared to continuous-time or discrete-time control protocol architectures. The proposed controller architectures are predicated on a hybrid dynamic compensation structure involving the exchange of information between agents. A unique feature of the proposed framework is that the controller architectures are hybrid and can achieve finite time coordination, which leads to a significant improvement in transient closed-loop system performance. The overall closed-loop dynamics under these controller algorithms achieving consensus and parallel formation control possess discontinuous flows, since the controller architectures combine logical switchings with continuous dynamics.

In Chapter 21, we develop a hybrid control framework for addressing multiagent formation control protocols for general nonlinear dynamical systems using hybrid stabilization of sets. The proposed framework develops a novel class of fixed-order, energy-based hybrid dynamic controllers as a means of achieving cooperative control formations, which can include flocking, cyclic pursuit, rendezvous, and consensus control of multiagent systems. These dynamic controllers combine a logical switching architecture with the continuous system dynamics to guarantee that a system-generalized energy function whose zero level set characterizes a specified system formation is strictly decreasing across switchings. The proposed approach addresses general nonlinear dynamical systems and is not limited to systems involving single and double integrator dynamics for consensus and formation control or unicycle models for cyclic pursuit. Finally, we draw conclusions in Chapter 22.

*Chapter Two*

# Distributed Nonlinear Control for Network Systems

## 2.1 Introduction

Modern complex network dynamical systems are highly interconnected and mutually interdependent, both physically and through a multitude of information and communication constraints. Distributed decision-making for coordination of networks of dynamic agents involving information flow can be naturally captured by graph-theoretic notions. As discussed in Chapter 1, these dynamical network systems cover a very broad spectrum of applications, including cooperative control of unmanned air vehicles (UAVs) [312], autonomous underwater vehicles (AUVs) [281], distributed sensor networks [73], air and ground transportation systems [290], swarms of air and space vehicle formations [93, 305], congestion control in communication networks [241], routing the flow of packets through a network, and managing power levels as well as supply and logistic chains in network systems. Hence, it is not surprising that considerable research effort has been devoted to control of networks and control over networks in recent years [5, 80, 93, 101, 180, 183, 194, 206, 211, 220, 234, 238, 241, 258, 294, 304, 305].

A key application area within aerospace systems is cooperative control of vehicle formations using distributed and decentralized controller architectures. Distributed control refers to a control architecture wherein the control is distributed via multiple computational units that are interconnected through information and communication networks, whereas decentralized control refers to a control architecture wherein local decisions are based only on local information. Vehicle formations are typically dynamically decoupled; that is, the motion of a given agent or vehicle does not directly affect the motion of the other agents or vehicles. The multiagent system is coupled via the task which the agents or vehicles are required to perform.

In many applications involving multiagent systems, groups of agents are required to agree on certain quantities of interest. In particular, it is important to develop information consensus protocols for networks of dynamic agents wherein a unique feature of the closed-loop dynamics under any control algorithm that achieves consensus is the existence of a continuum of equilibria representing a state of equipartitioning or *consensus*. Under such dynamics, the limiting consensus state achieved is not determined completely by the dynamics but depends on the initial system state as well. For such systems possessing a continuum of equilibria, *semistability* [35, 36], not asymptotic stability, is the relevant notion of stability. As noted in Chapter 1, semistability is the property whereby every trajectory that starts in a neighborhood of a Lyapunov stable equilibrium converges to a (possibly different) Lyapunov stable equilibrium.

Alternatively, in other applications of multiagent systems, groups of agents are required to achieve and maintain a prescribed geometric shape. This *formation control* problem includes *flocking* [234, 294] and *cyclic pursuit* [220], wherein parallel and circular formations of vehicles are sought. For formation control of multiple vehicles, *cohesion*, *separation*, and *alignment* constraints are typically required for individual agent steering, which describes how a given vehicle maneuvers based on the positions and velocities of nearby agents. Specifically, cohesion refers to a steering rule wherein a given vehicle attempts to move toward the average position of local vehicles, separation refers to collision avoidance with nearby vehicles, and alignment refers to velocity matching with nearby vehicles.

Using graph-theoretic notions, in this chapter we develop a unified framework for addressing consensus, formation, and cyclic pursuit problems for multiagent dynamical systems. Specifically, we present continuous distributed and decentralized controller architectures for multiagent coordination. In contrast to virtually all of the existing results in the literature on control of networks, the majority of the proposed controllers are dynamic compensators. Some of the proposed controller architectures are predicated on the recently developed notion of system thermodynamics [127], resulting in thermodynamically consistent continuous controller architectures involving the exchange of information between agents that guarantee that the closed-loop dynamical network is consistent with basic thermodynamic principles. The proposed controllers also use undirected and directed graphs to accommodate for a full range of possible graph information topologies without limitations of bidirectional communication.

## 2.2  Notation and Definitions

In this section, we introduce notation and several definitions needed for developing many of the results of this monograph. In a definition or when a word is defined in the text, the concept defined is italicized. Italics in the running text are also used for emphasis. The definition of a word, phrase, or symbol is to be understood as an "if and only if" statement. Lowercase letters such as $x$ denote vectors, uppercase letters such as $A$ denote matrices, uppercase script letters such as $\mathcal{S}$ denote sets, and lowercase Greek letters such as $\alpha$ denote scalars; however, there are a few exceptions to this convention.

The notation $\mathcal{S}_1 \subset \mathcal{S}_2$ means that $\mathcal{S}_1$ is a proper subset of $\mathcal{S}_2$, whereas $\mathcal{S}_1 \subseteq \mathcal{S}_2$ means that either $\mathcal{S}_1$ is a proper subset of $\mathcal{S}_2$ or $\mathcal{S}_1$ is equal to $\mathcal{S}_2$. Throughout the monograph we use two basic types of mathematical statements, namely, *existential* and *universal* statements. An existential statement has the following form: there exists $x \in \mathcal{X}$ such that a certain condition $C$ is satisfied; a universal statement has the following form: condition $C$ holds for all $x \in \mathcal{X}$. For universal statements we often omit the words "for all" and write the following: condition $C$ holds, $x \in \mathcal{X}$.

The notation used in this monograph is fairly standard. Specifically, $\mathbb{R}$ (respectively, $\mathbb{C}$) denotes the set of real (respectively, complex) numbers, $\overline{\mathbb{Z}}_+$ denotes the set of nonnegative integers, $\mathbb{Z}_+$ denotes the set of positive integers, $\mathbb{R}^n$ (respectively, $\mathbb{C}^n$) denotes the set of $n \times 1$ real (respectively, complex) column vectors, $\mathbb{R}^{n \times m}$ (respectively, $\mathbb{C}^{n \times m}$) denotes the set of real (respectively, complex) $n \times m$ matrices, $\mathbb{S}^n$ denotes the set of $n \times n$ symmetric matrices, $\mathbb{N}^n$ (respectively, $\mathbb{P}^n$) denotes the set of $n \times n$ nonnegative definite (respectively, positive definite) matrices, $(\cdot)^{\mathrm{T}}$ denotes transpose, $(\cdot)^+$ denotes the Moore–Penrose generalized inverse, $(\cdot)^{\#}$ denotes the group generalized inverse, $(\cdot)^{\mathrm{D}}$ denotes the Drazin inverse, $\otimes$ denotes Kronecker product, $\oplus$ denotes Kronecker sum, $I_n$ or $I$ denotes the $n \times n$ identity matrix, and $\mathbf{e}$ denotes the ones vector of order $n$, that is, $\mathbf{e} = [1, \ldots, 1]^{\mathrm{T}}$. Furthermore, $\mathcal{L}_2$ denotes the space of square-integrable Lebesgue measurable functions on $[0, \infty)$, and $\mathcal{L}_\infty$ denotes the space of bounded Lebesgue measurable functions on $[0, \infty)$. Finally, we denote the boundary, the interior, and the closure of the set $\mathcal{S}$ by $\partial \mathcal{S}$, $\overset{\circ}{\mathcal{S}}$, and $\overline{\mathcal{S}}$, respectively.

We write $\| \cdot \|$ for the Euclidean vector norm; $\mathcal{R}(A)$ and $\mathcal{N}(A)$ for the range space and the null space of a matrix $A$, respectively; $\mathrm{spec}(A)$ for the spectrum of the square matrix $A$ including multiplicity; $\alpha(A)$ for the spectral abscissa of $A$ (i.e., $\alpha(A) = \max\{\mathrm{Re}\,\lambda : \lambda \in \mathrm{spec}(A)\}$); $\rho(A)$ for the spectral radius of $A$ (i.e., $\rho(A) = \max\{|\lambda| : \lambda \in \mathrm{spec}(A)\}$); and $\mathrm{ind}(A)$ for the index of $A$ (i.e., the size of the largest Jordan block of $A$ associated with $\lambda = 0$, where $\lambda \in \mathrm{spec}(A)$). For a matrix $A \in \mathbb{R}^{p \times q}$, $\mathrm{row}_i(A)$ and $\mathrm{col}_j(A)$ denote the $i$th row and the $j$th column of $A$, respectively, and $\sigma_{\min}(A)$ (respectively, $\sigma_{\max}(A)$) denotes the minimum (respectively, maximum) singular value of $A$.

Furthermore, we write $V'(x)$ for the Fréchet derivative of $V$ at $x$; $\mathcal{B}_\varepsilon(x)$, $x \in \mathbb{R}^n$, $\varepsilon > 0$, for the *open ball centered* at $x$ with *radius* $\varepsilon$ in the Euclidean norm; $\lambda_{\min}(A)$ (respectively, $\lambda_{\max}(A)$) for the minimum (respectively, maximum) eigenvalue of the Hermitian matrix $A$, $A \geq 0$ (respectively, $A > 0$) to denote the fact that the Hermitian matrix $A$ is nonnegative (respectively, positive) definite; inf to denote infimum (i.e., the greatest lower bound); sup to denote supremum (i.e., the least upper bound); and $x(t) \to \mathcal{M}$ as $t \to \infty$ to denote that $x(t)$ approaches the set $\mathcal{M}$ (i.e., for each $\varepsilon > 0$ there exists $T > 0$ such that $\mathrm{dist}(x(t), \mathcal{M}) < \varepsilon$ for all $t > T$, where $\mathrm{dist}(p, \mathcal{M}) \triangleq \inf_{x \in \mathcal{M}} \|p - x\|$). Finally, the notions of openness, convergence, continuity, and compactness that we use throughout the monograph refer to the topology generated on $\mathbb{R}^n$ by the norm $\| \cdot \|$.

The following definition is needed for several of the results in the monograph.

**Definition 2.1.** A continuous function $\gamma : [0, a) \to [0, \infty)$, where $a \in (0, \infty]$, is of *class* $\mathcal{K}$ if it is strictly increasing and $\gamma(0) = 0$. A continuous function $\gamma : [0, \infty) \to [0, \infty)$ is of *class* $\mathcal{K}_\infty$ if it is strictly increasing, $\gamma(0) = 0$, and $\gamma(s) \to \infty$ as $s \to \infty$. A continuous function $\gamma : [0, \infty) \to [0, \infty)$ is of *class* $\mathcal{L}$ if it is strictly decreasing and $\gamma(s) \to 0$ as $s \to \infty$. Finally, a continuous function $\gamma : [0, a) \times [0, \infty) \to [0, \infty)$ is of *class* $\mathcal{KL}$ if, for each fixed $s$, $\gamma(r, s)$ is of class $\mathcal{K}$ with respect to $r$ and, for each fixed $r$, $\gamma(r, s)$ is of class $\mathcal{L}$ with respect to $s$.

## 2.3 The Consensus Problem in Dynamical Networks

Information consensus problems appear frequently in coordination of multi-agent systems and involve finding a dynamic algorithm that enables a group of agents in a network to agree upon certain quantities of interest with undirected or directed information flow. In this monograph, we use undirected and directed graphs to represent a dynamical network and present solutions to the consensus problem for networks with both graph *topologies* (or information flow) [238].

Specifically, let $\mathfrak{G} = (\mathcal{V}, \mathcal{E}, \mathcal{A})$ be a weighted *directed graph* (or digraph) denoting the dynamical network (or dynamic graph), with the set of *nodes* (or vertices) $\mathcal{V} = \{v_1, \ldots, v_n\}$ involving a finite nonempty set denoting the agents; the set of *edges* $\mathcal{E} \subseteq \mathcal{V} \times \mathcal{V}$ involving a set of ordered pairs denoting the direction of information flow; and a weighted *adjacency* (or connectivity) matrix $\mathcal{A} \in \mathbb{R}^{n \times n}$ such that $\mathcal{A}_{(i,j)} = a_{ij} > 0$, $i, j = 1, \ldots, n$, if $(v_j, v_i) \in \mathcal{E}$, and $a_{ij} = 0$ otherwise. The edge $(i, j) \in \mathcal{E}$ denotes that agent $\mathcal{G}_j$ can obtain information from agent $\mathcal{G}_i$ but not necessarily vice versa. *Self edges* $(i, i) \in \mathcal{E}$ are allowed. The in-degree and out-degree of node $i$ are, respectively, defined as $\deg_{\text{in}}(i) \triangleq \sum_{j=1}^{n} a_{ji}$ and $\deg_{\text{out}}(i) \triangleq \sum_{j=1}^{n} a_{ij}$. We say that the node $i$ of a digraph $\mathcal{G}$ is *balanced* if and only if $\deg_{\text{in}}(i) = \deg_{\text{out}}(i)$. A graph $\mathfrak{G}$ is *balanced* if and only if all of its nodes are balanced, that is, $\sum_{j=1}^{n} a_{ij} = \sum_{j=1}^{n} a_{ji}$ for all $i = 1, \ldots, n$. Finally, we denote the *value* of the node $v_i$, $i = 1, \ldots, n$, at time $t$ by $x_i(t) \in \mathbb{R}$. The consensus problem involves the design of a dynamic algorithm that guarantees information state equipartition, that is, $\lim_{t \to \infty} x_i(t) = \alpha \in \mathbb{R}$ for $i = 1, \ldots, n$, where $\alpha$ depends on the system initial conditions.

The consensus problem involves the network system characterized by the dynamical system $\mathcal{G}$ given by

$$\dot{x}_i(t) = \sum_{j=1, j \neq i}^{q} \phi_{ij}(x_i(t), x_j(t)), \quad x_i(t_0) = x_{i0}, \quad t \geq t_0, \quad i = 1, \ldots, q,$$
(2.1)

where $\phi_{ij}(\cdot, \cdot)$, $i, j = 1, \ldots, q$, are locally Lipschitz continuous, or, in vector form,

$$\dot{x}(t) = f(x(t)), \quad x(t_0) = x_0, \quad t \geq t_0,$$
(2.2)

where $x(t) \triangleq [x_1(t), \ldots, x_q(t)]^{\mathrm{T}}$, $t \geq t_0$, and $f = [f_1, \ldots, f_q]^{\mathrm{T}} : \mathbb{R}^q \to \mathbb{R}^q$ is such that

$$f_i(x) = \sum_{j=1, j \neq i}^{q} \phi_{ij}(x_i, x_j).$$
(2.3)

Here, for each $i \in \{1, \ldots, q\}$, $x_i(t)$, $t \geq 0$, represents an *information state*, and $u_i(t) \triangleq f_i(x(t))$ is a distributed consensus control algorithm involving neighbor-to-neighbor interaction between agents. This nonlinear model is proposed in [127] and is called a *power balance equation*. Here, however, we address a more general model in that $\phi_{ij}(\cdot, \cdot)$ has no special structure, and $x$ need not be constrained to the nonnegative orthant of the state space.

For the statement of the main results of this section, the following definition is needed.

**Definition 2.2** (see [24]). A *directed graph* $\mathfrak{G}$ associated with the *adjacency matrix* $\mathcal{A} \in \mathbb{R}^{q \times q}$ has *vertices* $\{1, \ldots, q\}$ and an *arc* from vertex $i$ to vertex $j$, $i \neq j$, if $\mathcal{A}_{(j,i)} \neq 0$. A *graph* or *undirected graph* $\mathfrak{G}$ associated with the adjacency matrix $\mathcal{A} \in \mathbb{R}^{q \times q}$ is a directed graph for which the *arc set* is symmetric, that is, $\mathcal{A} = \mathcal{A}^{\mathrm{T}}$. We say that $\mathfrak{G}$ is *strongly connected* if, for every ordered pair of vertices $(i, j)$, $i \neq j$, there exists a *path* (i.e., sequence of arcs) leading from $i$ to $j$.

Recall that $\mathcal{A} \in \mathbb{R}^{q \times q}$ is *irreducible*; that is, there does not exist a permutation matrix such that $\mathcal{A}$ is cogredient to a lower block triangular matrix, if and only if $\mathfrak{G}$ is strongly connected (see Theorem 2.7 of [24]). Furthermore, note that for an undirected graph $\mathcal{A} = \mathcal{A}^{\mathrm{T}}$, and, hence, every undirected graph is balanced.

**Assumption 2.1.** The unweighted adjacency matrix $\mathcal{A} \in \mathbb{R}^{q \times q}$ associated with the multiagent dynamical system $\mathcal{G}$ given by (2.2) is defined by

$$\mathcal{A}_{(i,j)} = \begin{cases} 0 & \text{if } \phi_{ij}(x_i, x_j) \equiv 0, \\ 1, & \text{otherwise,} \end{cases} \qquad i \neq j, \quad i, j = 1, \ldots, q, \quad (2.4)$$

and

$$\mathcal{A}_{(i,i)} = -\sum_{k=1, k \neq i}^{q} \mathcal{A}_{(i,k)}, \quad i = j, \quad i = 1, \ldots, q, \qquad (2.5)$$

with rank $\mathcal{A} = q - 1$, and for $\mathcal{A}_{(i,j)} = 1$, $i \neq j$, $\phi_{ij}(x_i, x_j) = 0$ if and only if $x_i = x_j$.

**Assumption 2.2.** For $i, j = 1, \ldots, q$, $(x_i - x_j)\phi_{ij}(x_i, x_j) \leq 0$, $x_i, x_j \in \mathbb{R}$.

The condition that $\phi_{ij}(x_i, x_j) = 0$ if and only if $x_i = x_j$, $i \neq j$, implies that agents $\mathcal{G}_i$ and $\mathcal{G}_j$ are *connected* and, hence, can share information; alternatively, $\phi_{ij}(x_i, x_j) \equiv 0$ implies that agents $\mathcal{G}_i$ and $\mathcal{G}_j$ are *disconnected* and, hence, cannot share information.

Assumption 2.1 implies that if the information or energies in the connected agents $\mathcal{G}_i$ and $\mathcal{G}_j$ are equal, then information or energy exchange between these agents is not possible. This statement is reminiscent of the *zeroth law of thermodynamics*, which postulates that temperature equality is a necessary and sufficient condition for thermal equilibrium. Furthermore, if $\mathcal{A} = \mathcal{A}^{\mathrm{T}}$ and rank $\mathcal{A} = q - 1$, then it follows that the adjacency matrix $\mathcal{A}$ is irreducible, which implies that for every pair of agents $\mathcal{G}_i$ and $\mathcal{G}_j$, $i \neq j$, of $\mathcal{G}$ there exists a sequence of information connectors (information arcs) of $\mathcal{G}$ that connect $\mathcal{G}_i$ and $\mathcal{G}_j$.

Assumption 2.2 implies that energy or information flows from more energetic or information-rich agents to less energetic or information-poor agents and is reminiscent of the *second law of thermodynamics*, which states that heat (energy in transition) must flow in the direction of lower temperatures. For further details, see [127].

For the statement of the next result, let $\mathbf{e} \in \mathbb{R}^q$ denote the ones vector of order $q$, that is, $\mathbf{e} \triangleq [1, \ldots, 1]^\mathrm{T}$, and let $\mathbf{e}_i \in \mathbb{R}^n$ denote the elementary vector of order $n$ with one in the $i$th location and zeros elsewhere. Note that for a directed graph $\mathfrak{G}$, $\mathcal{A}\mathbf{e} = \mathcal{A}^\mathrm{T}\mathbf{e}$ implies that $\mathfrak{G}$ is balanced.

**Proposition 2.1.** Consider the multiagent dynamical system (2.2), and assume that Assumptions 2.1 and 2.2 hold. Then $f_i(x) = 0$ for all $i = 1, \ldots, q$ if and only if $x_1 = \cdots = x_q$. Furthermore, $\alpha\mathbf{e}$, $\alpha \in \mathbb{R}$, is an equilibrium state of (2.2).

**Proof.** If $x_i = x_j$ for all $(i, j) \in \mathcal{E}$, then $f_i(x) = 0$ for all $i = 1, \ldots, q$ is immediate from Assumption 2.1. Next, we show that $f_i(x) = 0$ for all $i = 1, \ldots, q$ implies that $x_1 = \cdots = x_q$. If $f_i(x) = 0$ for all $i = 1, \ldots, q$, then it follows from Assumption 2.2 that

$$0 = \sum_{i=1}^{q} x_i f_i(x) = \sum_{i=1}^{q}\sum_{j=1}^{q} x_i \phi_{ij}(x) = \sum_{i=1}^{q-1}\sum_{j=i+1}^{q} (x_i - x_j)\phi_{ij}(x) \leq 0,$$

where we have used the fact that $\phi_{ij}(x) = -\phi_{ji}(x)$ for all $i, j = 1, \ldots, q$. Hence, $(x_i - x_j)\phi_{ij}(x) = 0$ for all $i, j = 1, \ldots, q$. Now, the result follows from Assumption 2.1.

Alternatively, the proof can be shown using graph-theoretic concepts. Specifically, if $x_i = x_j$ for all $(i, j) \in \mathcal{E}$, then $f_i(x) = 0$ for all $i = 1, \ldots, q$ is immediate from Assumption 2.1. Next, we show that $f_i(x) = 0$ for all $i = 1, \ldots, q$ implies that $x_1 = \cdots = x_q$. If the values of all nodes are equal, then the result is immediate. Hence, assume that there exists a node $i^*$ such that $x_{i^*} \geq x_j$ for all $j \neq i^*$, $j \in \{1, \ldots, q\}$. If $(i, j) \in \mathcal{E}$, then we define a *neighbor* of node $i$ to be node $j$, and vice versa.

Define the initial node set $\mathcal{J}^{(0)} \triangleq \{i^*\}$, and denote the indices of all the first neighbors of node $i^*$ by $\mathcal{J}^{(1)} = \mathcal{N}_{i^*}$. Then $f_{i^*}(x) = 0$ implies that $\sum_{j \in \mathcal{N}_{i^*}} \phi_{i^*j}(x_{i^*}, x_j) = 0$. Since $x_j \leq x_{i^*}$ for all $j \in \mathcal{N}_{i^*}$ and, by Assumption 2.2, $\phi_{ij}(z_i, z_j) \leq 0$ for all $z_i \geq z_j$, it follows that $x_{i^*} = x_j$ for all the first neighbors $j \in \mathcal{J}^{(1)}$. Next, we define the $k$th neighbor of node $i^*$ and show that the value of node $i^*$ is equal to the values of all $k$th neighbors of node $i^*$ for $k = 1, \ldots, q-1$. The set of $k$th neighbors of node $i^*$ is defined by

$$\mathcal{J}^{(k)} \triangleq \mathcal{J}^{(k-1)} \cup \mathcal{N}_{\mathcal{J}^{(k-1)}}, \quad k \geq 1, \quad \mathcal{J}^{(0)} = \{i^*\}, \tag{2.6}$$

where $\mathcal{N}_{\mathcal{J}}$ denotes the set of neighbors of the node set $\mathcal{J} \subseteq \mathcal{V}$. By definition, $\{i^*\} \subset \mathcal{J}^{(k)} \subseteq \mathcal{V}$ for all $k \geq 1$ and $\mathcal{J}^{(k)}$ is a monotonically increasing sequence of node sets in the sense of set inclusions.

Next, we show that $\mathcal{J}^{(q-1)} = \mathcal{V}$. Suppose, *ad absurdum*, $\mathcal{V}\backslash\mathcal{J}^{(q-1)} \neq \emptyset$. Then, by definition, there exists one node $m \in \{1,\ldots,q\}$ disconnected from all the other nodes. Hence, $\mathcal{A}_{(m,i)} = \mathcal{A}_{(i,m)} = 0$, $i = 1,\ldots,q$, which implies that the adjacency matrix $\mathcal{A}$ has a row and a column of zeros. Without loss of generality, assume that $\mathcal{A}$ has the form

$$\mathcal{A} = \begin{bmatrix} \mathcal{A}_{\mathrm{s}} & 0_{(q-1)\times 1} \\ 0_{1\times(q-1)} & 0 \end{bmatrix},$$

where $\mathcal{A}_{\mathrm{s}} \in \mathbb{R}^{(q-1)\times(q-1)}$ denotes the adjacency matrix for the new undirected graph $\mathbb{G}$, which excludes node $m$ from the undirected graph $\mathfrak{G}$. In this case, since rank $\mathcal{A}_{\mathrm{s}} \leq q-2$, it follows that rank $\mathcal{A} < q-1$, which contradicts Assumption 2.1.

Using mathematical induction, we show that the values of all the nodes in $\mathcal{J}^{(k)}$ are equal for $k \geq 1$. This statement holds for $k = 1$. Assuming that the values of all the nodes in $\mathcal{J}^{(k)}$ are equal to the value of node $i^*$, we show that the values of all the nodes in $\mathcal{J}^{(k+1)}$ are equal to the value of node $i^*$ as well. Note that since $\mathfrak{G}$ is strongly connected, $\mathcal{N}_i \neq \emptyset$ for all $i \in \mathcal{V}$. If $\mathcal{N}_i \cap (\mathcal{J}^{(k+1)}\backslash\mathcal{J}^{(k)}) = \emptyset$ for all $i$, then it follows that $\mathcal{J}^{(k+1)} = \mathcal{J}^{(k)}$, and, hence, the statement holds. Thus, it suffices to show that $x_i = x_{i^*}$ for an arbitrary node $i \in \mathcal{J}^{(k)}$ with $\mathcal{N}_i \cap (\mathcal{J}^{(k+1)}\backslash\mathcal{J}^{(k)}) \neq \emptyset$. For node $i$, note that $\sum_{j\in\mathcal{N}_i} \phi_{ij}(x_i,x_j) = 0$. Furthermore, note that $\mathcal{N}_i = (\mathcal{N}_i \cap \mathcal{J}^{(k)}) \cup (\mathcal{N}_i \cap (\mathcal{V}\backslash\mathcal{J}^{(k)}))$, $\mathcal{V}\backslash\mathcal{J}^{(k)} = \mathcal{V}\backslash\mathcal{J}^{(k+1)} \cup (\mathcal{J}^{(k+1)}\backslash\mathcal{J}^{(k)})$, and $\mathcal{J}^{(k)} \subseteq \mathcal{V}$ for all $k$, and $\mathcal{J}^{(k+1)}$ contains the set of first neighbors of node $i$, or $\mathcal{N}_i \subseteq \mathcal{J}^{(k+1)}$. Then it follows that $\mathcal{N}_i \cap (\mathcal{V}\backslash\mathcal{J}^{(k)}) = \mathcal{N}_i \cap (\mathcal{J}^{(k+1)}\backslash\mathcal{J}^{(k)})$ and

$$\sum_{j\in\mathcal{N}_i\cap\mathcal{J}^{(k)}} \phi_{ij}(x_i,x_j) + \sum_{j\in\mathcal{N}_i\cap(\mathcal{J}^{(k+1)}\backslash\mathcal{J}^{(k)})} \phi_{ij}(x_i,x_j) = 0. \qquad (2.7)$$

Since $x_j = x_i$ for all nodes $j \in \mathcal{N}_i \cap \mathcal{J}^{(k)} \subseteq \mathcal{J}^{(k)}$, it follows that

$$\sum_{j\in\mathcal{N}_i\cap\mathcal{J}^{(k)}} \phi_{ij}(x_i,x_j) = 0,$$

and, hence, $\sum_{j\in\mathcal{N}_i\cap(\mathcal{J}^{(k+1)}\backslash\mathcal{J}^{(k)})} \phi_{ij}(x_i,x_j) = 0$. However, since $x_{i^*} = x_i \geq x_j$ for all $i \in \mathcal{J}^{(k)}$ and $j \in \mathcal{V}\backslash\mathcal{J}^{(k)}$, it follows that the values of all nodes in $\mathcal{N}_i \cap (\mathcal{J}^{(k+1)}\backslash\mathcal{J}^{(k)})$ are equal to $x_{i^*}$. Hence, the values of all nodes $i$ in the node set $\bigcup_{i\in\mathcal{J}^{(k)}} \mathcal{N}_i \cap (\mathcal{J}^{(k+1)}\backslash\mathcal{J}^{(k)}) = \mathcal{J}^{(k+1)} \cap (\mathcal{J}^{(k+1)}\backslash\mathcal{J}^{(k)}) = \mathcal{J}^{(k+1)}\backslash\mathcal{J}^{(k)}$ are equal to $x_{i^*}$; that is, the values of all the nodes in $\mathcal{J}^{(k+1)}$ are equal. Combining this result with the fact that $\mathcal{J}^{(q-1)} = \mathcal{V}$, it follows that the values of all the nodes in $\mathcal{V}$ are equal.

The second assertion is a direct consequence of the first assertion. $\square$

**Definition 2.3** (see [36]). Let $\mathcal{S} \subseteq \mathbb{R}^q$ be a positively invariant set with respect to the system (2.2). An equilibrium point $x \in \mathcal{S}$ of the system (2.2) is *semistable* with respect to $\mathcal{S}$ if it is Lyapunov stable and there exists an open subset $\mathcal{U}$ of $\mathcal{S}$ containing $x$ such that for all initial conditions in $\mathcal{U}$, the trajectory of (2.2) converges to a Lyapunov stable equilibrium point; that is, $\lim_{t \to \infty} s(t, x) = y$, where $y \in \mathcal{S}$ is a Lyapunov stable equilibrium point of (2.2) and $x \in \mathcal{U}$.

The following lemma is needed for the main result of this section.

**Lemma 2.1.** Let $A \in \mathbb{R}^{q \times q}$ and $A_{\mathrm{d}i} \in \mathbb{R}^{q \times q}$, $i = 1, \ldots, n_{\mathrm{d}}$, be given by

$$
A_{(i,j)} = \begin{cases} -\sum_{k=1, k \neq i}^{q} a_{ik}, & i = j, \\ 0, & i \neq j, \end{cases}
$$
$$
A_{\mathrm{d}(i,j)} = \begin{cases} 0, & i = j, \\ a_{ij}, & i \neq j, \end{cases} \quad i, j = 1, \ldots, q, \tag{2.8}
$$

or

$$
A_{(i,j)} = \begin{cases} -\sum_{k=1, k \neq i}^{q} a_{ki}, & i = j, \\ 0, & i \neq j, \end{cases}
$$
$$
A_{\mathrm{d}(i,j)} = \begin{cases} 0, & i = j, \\ a_{ij}, & i \neq j, \end{cases} \quad i, j = 1, \ldots, q, \tag{2.9}
$$

where $A_{\mathrm{d}} \triangleq \sum_{i=1}^{n_{\mathrm{d}}} A_{\mathrm{d}i}$, $a_{ij} \geq 0$, $i, j = 1, \ldots, q$, $i \neq j$. Assume that $\sum_{k=1, k \neq i}^{q} a_{ik} = \sum_{k=1, k \neq i}^{q} a_{ki}$ for each $i = 1, \ldots, q$. Then there exist nonnegative definite matrices $Q_i \in \mathbb{R}^{q \times q}$, $i = 1, \ldots, n_{\mathrm{d}}$, such that

$$
2A + \sum_{i=1}^{n_{\mathrm{d}}} (Q_i + A_{\mathrm{d}i}^{\mathrm{T}} Q_i^{\#} A_{\mathrm{d}i}) \leq 0. \tag{2.10}
$$

**Proof.** For each $i \in \{1, \ldots, n_{\mathrm{d}}\}$, let $Q_i$ be the diagonal matrix defined by

$$
Q_{i(l,l)} \triangleq \sum_{m=1, l \neq m}^{q} A_{\mathrm{d}i(l,m)}, \quad l = 1, \ldots, q, \tag{2.11}
$$

and note that $A + \sum_{i=1}^{n_{\mathrm{d}}} Q_i = 0$, $(A_{\mathrm{d}i} - Q_i)\mathbf{e} = 0$, and $Q_i Q_i^{\#} A_{\mathrm{d}i} = A_{\mathrm{d}i}$, $i = 1, \ldots, n_{\mathrm{d}}$. Hence, $M\mathbf{e} = 0$, where

$$
M \triangleq \begin{bmatrix} 2A + \sum_{i=1}^{n_{\mathrm{d}}} Q_i & A_{\mathrm{d}1}^{\mathrm{T}} & A_{\mathrm{d}2}^{\mathrm{T}} & \cdots & A_{\mathrm{d}n_{\mathrm{d}}}^{\mathrm{T}} \\ A_{\mathrm{d}1} & -Q_1 & 0 & \cdots & 0 \\ \vdots & \vdots & \vdots & \vdots & \vdots \\ A_{\mathrm{d}n_{\mathrm{d}}} & 0 & 0 & \cdots & -Q_{n_{\mathrm{d}}} \end{bmatrix}. \tag{2.12}
$$

Now, note that $M = M^{\mathrm{T}}$ and $M_{(i,j)} \geq 0$, $i, j = 1, \ldots, q$, $i \neq j$. Hence, by (ii) of Theorem 3.2 of [121], $M$ is *semistable*, that is, $\mathrm{Re}\,\lambda < 0$, or $\lambda = 0$ and $\lambda$ is semisimple,[1] where $\lambda \in \mathrm{spec}(A)$. Thus, $M \leq 0$, and since $Q_i Q_i^{\#} A_{\mathrm{d}i} = A_{\mathrm{d}i}$, $i = 1, \ldots, n_{\mathrm{d}}$, it follows from Proposition 8.2.3 of [26] that $M \leq 0$ if and only if (2.10) holds.

Alternatively, if $A \in \mathbb{R}^{q \times q}$ and $A_{\mathrm{d}i} \in \mathbb{R}^{q \times q}$, $i = 1, \ldots, n_{\mathrm{d}}$, are given by (2.9), then let $Q_i$ be the diagonal matrix defined by

$$Q_{i(l,l)} \triangleq \sum_{m=1, l\neq m}^{q} A_{\mathrm{d}i(m,l)}, \quad l = 1, \ldots, q. \tag{2.13}$$

The result now follows using similar arguments as above. $\qquad\square$

**Theorem 2.1.** Consider the multiagent dynamical system (2.2), and assume that Assumptions 2.1 and 2.2 hold.

(i) Assume that $\phi_{ij}(x_i, x_j) = -\phi_{ji}(x_j, x_i)$ for all $i, j = 1, \ldots, q$, $i \neq j$. Then, for every $\alpha \in \mathbb{R}$, $\alpha \mathbf{e}$ is a semistable equilibrium state of (2.2). Furthermore, $x(t) \to \frac{1}{q}\mathbf{e}\mathbf{e}^{\mathrm{T}}x(t_0)$ as $t \to \infty$ and $\frac{1}{q}\mathbf{e}\mathbf{e}^{\mathrm{T}}x(t_0)$ is a semistable equilibrium state.

(ii) Let $\phi_{ij}(x_i, x_j) = \mathcal{A}_{(i,j)}[\sigma(x_j) - \sigma(x_i)]$ for all $i, j = 1, \ldots, q$, $i \neq j$, where $\sigma(0) = 0$ and $\sigma(\cdot)$ is strictly increasing. Assume that $\mathcal{A}^{\mathrm{T}}\mathbf{e} = 0$. Then, for every $\alpha \in \mathbb{R}$, $\alpha \mathbf{e}$ is a semistable equilibrium state of (2.2). Furthermore, $x(t) \to \frac{1}{q}\mathbf{e}\mathbf{e}^{\mathrm{T}}x(t_0)$ as $t \to \infty$ and $\frac{1}{q}\mathbf{e}\mathbf{e}^{\mathrm{T}}x(t_0)$ is a semistable equilibrium state.

**Proof.** (i) It follows from Proposition 2.1 that $\alpha \mathbf{e} \in \mathbb{R}^q$, $\alpha \in \mathbb{R}$, is an equilibrium state of (2.2). To show Lyapunov stability of the equilibrium state $\alpha \mathbf{e}$, consider $V(x) = \frac{1}{2}(x - \alpha \mathbf{e})^{\mathrm{T}}(x - \alpha \mathbf{e})$ as a Lyapunov function candidate. Now, since $\phi_{ij}(x) = -\phi_{ji}(x)$, $x \in \mathbb{R}^q$, $i \neq j$, $i, j = 1, \ldots, q$, and $\mathbf{e}^{\mathrm{T}}f(x) = 0$, $x \in \mathbb{R}^q$, it follows from Assumption 2.2 that

$$\dot{V}(x) = (x - \alpha \mathbf{e})^{\mathrm{T}}\dot{x}$$
$$= (x - \alpha \mathbf{e})^{\mathrm{T}}f(x)$$
$$= x^{\mathrm{T}}f(x)$$
$$= \sum_{i=1}^{q} x_i \left[ \sum_{j=1, j\neq i}^{q} \phi_{ij}(x) \right]$$

---

[1]$\lambda \in \mathbb{C}$ is a semisimple eigenvalue of the matrix $A$ if and only if its algebraic multiplicity is equal to its geometric multiplicity [27, p. 322].

$$= \sum_{i=1}^{q} \sum_{j=i+1}^{q} (x_i - x_j)\phi_{ij}(x)$$

$$= \sum_{i=1}^{q} \sum_{j \in \mathcal{K}_i} (x_i - x_j)\phi_{ij}(x)$$

$$\leq 0, \quad x \in \mathbb{R}^q, \tag{2.14}$$

where $\mathcal{K}_i \triangleq \mathcal{N}_i \setminus \cup_{l=1}^{i-1}\{l\}$ and $\mathcal{N}_i \triangleq \{j \in \{1, \ldots, q\} : \phi_{ij}(x) = 0$ if and only if $x_i = x_j\}$, $i = 1, \ldots, q$, which establishes Lyapunov stability of the equilibrium state $\alpha\mathbf{e}$.

To show that $\alpha\mathbf{e}$ is semistable, let

$$\mathcal{R} \triangleq \{x \in \mathbb{R}^q : \dot{V}(x) = 0\}$$
$$= \{x \in \mathbb{R}^q : (x_i - x_j)\phi_{ij}(x) = 0, \ i = 1, \ldots, q, \ j \in \mathcal{K}_i\}.$$

Now, by Assumption 2.1 the directed graph associated with the adjacency matrix $\mathcal{A}$ for the large-scale dynamical system $\mathcal{G}$ is strongly connected, which implies that $\mathcal{R} = \{x \in \mathbb{R}^q : x_1 = \cdots = x_q\}$. Since the set $\mathcal{R}$ consists of the equilibrium states of (2.2), it follows that the largest invariant set $\mathcal{M}$ contained in $\mathcal{R}$ is given by $\mathcal{M} = \mathcal{R}$. Hence, it follows from the Krasovskii–LaSalle invariant set theorem [122, p. 147] that, for every initial condition $x(t_0) \in \mathbb{R}^q$, $x(t) \to \mathcal{M}$ as $t \to \infty$, and, hence, $\alpha\mathbf{e}$ is a semistable equilibrium state of (2.2).

Next, note that since $\mathbf{e}^{\mathrm{T}}x(t) = \mathbf{e}^{\mathrm{T}}x(t_0)$ and $x(t) \to \mathcal{M}$ as $t \to \infty$, it follows that $x(t) \to \frac{1}{q}\mathbf{e}\mathbf{e}^{\mathrm{T}}x(t_0)$ as $t \to \infty$. Hence, with $\alpha = \frac{1}{q}\mathbf{e}^{\mathrm{T}}x(t_0)$, $\alpha\mathbf{e} = \frac{1}{q}\mathbf{e}\mathbf{e}^{\mathrm{T}}x(t_0)$ is a semistable equilibrium state of (2.2).

(ii) It follows from Lemma 2.1 that there exists $Q_i$, $i = 1, \ldots, q$, such that (2.10) holds with $Q_i$ given by (2.11), and $A$ and $A_{\mathrm{d}i}$, $i = 1, \ldots, q$, are given by (2.8). Next, consider the nonnegative function given by

$$V(x) = 2 \sum_{i=1}^{q} \int_{0}^{x_i} \sigma(\theta)\mathrm{d}\theta. \tag{2.15}$$

Since $\sigma(\cdot)$ is a strictly increasing function, it follows from the mean value theorem that

$$V(x) = 2 \sum_{i=1}^{q} \sigma(\delta_i x_i)x_i \geq 2 \sum_{i=1}^{q} \sigma(\delta_i x_i)\delta_i x_i > 0$$

for all $x \neq 0$, where $0 < \delta_i < 1$, and, hence, there exists a class $\mathcal{K}$ function $\alpha(\cdot)$ such that $V(x) \geq \alpha(\|x\|)$. Now, the derivative of $V(x)$ along the trajectories of (2.2) is given by

$$\dot{V}(x) = 2\hat{\sigma}^{\mathrm{T}}(x)A\hat{\sigma}(x) + 2\sum_{i=1}^{q}\hat{\sigma}^{\mathrm{T}}(x)A_{\mathrm{d}i}\hat{\sigma}(x)$$

$$\leq -\sum_{i=1}^{q}[\hat{\sigma}^{\mathrm{T}}(x)Q_i\hat{\sigma}(x) - 2\hat{\sigma}^{\mathrm{T}}(x)A_{\mathrm{d}i}\hat{\sigma}(x) + \hat{\sigma}^{\mathrm{T}}(x)A_{\mathrm{d}i}^{\mathrm{T}}Q_i^{\#}A_{\mathrm{d}i}\hat{\sigma}(x)]$$

$$= -\sum_{i=1}^{q}[-Q_i\hat{\sigma}(x) + A_{\mathrm{d}i}\hat{\sigma}(x)]^{\mathrm{T}}Q_i^{\#}[-Q_i\hat{\sigma}(x) + A_{\mathrm{d}i}\hat{\sigma}(x)]$$

$$\leq 0, \quad x \in \mathbb{R}^q, \tag{2.16}$$

where $\hat{\sigma} : \mathbb{R}^q \to \mathbb{R}^q$ is given by $\hat{\sigma}(x) \triangleq [\sigma(x_1), \ldots, \sigma(x_q)]^{\mathrm{T}}$.

Next, let

$$\mathcal{R} \triangleq \{x \in \mathbb{R}^q : -Q_i\hat{\sigma}(x) + A_{\mathrm{d}i}\hat{\sigma}(x) = 0, i = 1, \ldots, q\}.$$

Then it follows from the Krasovskii–LaSalle invariant set theorem [122, p. 147] that $x(t) \to \mathcal{M}$ as $t \to \infty$, where $\mathcal{M}$ denotes the largest invariant set contained in $\mathcal{R}$. Now, since $A + \sum_{i=1}^{q}Q_i = 0$, it follows that

$$\mathcal{R} \subseteq \hat{\mathcal{R}} \triangleq \left\{x \in \mathbb{R}^q : A\hat{\sigma}(x) + \sum_{i=1}^{q}A_{\mathrm{d}i}\hat{\sigma}(x) = 0\right\}.$$

Hence, since $\mathrm{rank}(A + \sum_{i=1}^{q}A_{\mathrm{d}i}) = q - 1$ and $(A + \sum_{i=1}^{q}A_{\mathrm{d}i})\mathbf{e} = 0$, it follows that the largest invariant set $\hat{\mathcal{M}}$ contained in $\hat{\mathcal{R}}$ is given by $\hat{\mathcal{M}} = \{x \in \mathbb{R}^q : x = \alpha\mathbf{e}, \alpha \in \mathbb{R}\}$. Furthermore, since $\hat{\mathcal{M}} \subseteq \mathcal{R} \subseteq \hat{\mathcal{R}}$, it follows that $\mathcal{M} = \hat{\mathcal{M}}$.

Finally, to show the Lyapunov stability of $\alpha\mathbf{e}$, $\alpha \in \mathbb{R}$, consider the Lyapunov function candidate

$$\tilde{V}(x) = 2\sum_{i=1}^{q}\int_{\alpha}^{x_i}[\sigma(\theta) - \sigma(\alpha)]\mathrm{d}\theta.$$

Note that $\tilde{V}(\alpha\mathbf{e}) = 0$ and, by the mean value theorem, there exist $\delta_i \in (0, 1)$, $i = 1, \ldots, q$, such that

$$\tilde{V}(x) = 2\sum_{i=1}^{q}[\sigma(\alpha + \delta_i(x_i - \alpha)) - \sigma(\alpha)](x_i - \alpha)$$

$$\geq 2\sum_{i=1}^{q}[\sigma(\alpha + \delta_i(x_i - \alpha)) - \sigma(\alpha)]\delta_i(x_i - \alpha)$$

$$> 0, \quad x \neq \alpha\mathbf{e}.$$

The Lyapunov derivative of $\tilde{V}(x)$ along the trajectories of (2.2) is given by

$$\dot{\tilde{V}}(x) = 2\sum_{i=1}^{q}[\sigma(x_i) - \sigma(\alpha)]\dot{x}_i$$

$$= 2\sum_{i=1}^{q}[\sigma(x_i) - \sigma(\alpha)]f_i(x)$$

$$= 2\sum_{i=1}^{q}[\sigma(x_i) - \sigma(\alpha)]\left[\sum_{j=1,j\neq i}^{q}\phi_{ij}(x_i,x_j)\right]$$

$$= 2\sum_{i=1}^{q}[\sigma(x_i) - \sigma(\alpha)]\left[\sum_{j=1,j\neq i}^{q}\mathcal{A}_{(i,j)}(\sigma(x_j) - \sigma(x_i))\right]$$

$$= 2\sum_{i=1}^{q}[\sigma(x_i) - \sigma(\alpha)]\left[\sum_{j=1,j\neq i}^{q}\mathcal{A}_{(i,j)}[(\sigma(x_j) - \sigma(\alpha)) - (\sigma(x_i) - \sigma(\alpha))]\right].$$

Let $a_{ij} = \mathcal{A}_{(i,j)}$, $i,j = 1,\ldots,q$, and $A$ and $A_{\mathrm{d}l}$ be given by (2.9), with $l = 1,\ldots,q$, that is, $n_{\mathrm{d}} = q$. Then it follows from Lemma 2.1 that

$$\dot{\tilde{V}}(x) = 2\sum_{i=1}^{q}\sum_{j=1,j\neq i}^{q}(-a_{ij})[\sigma(x_i) - \sigma(\alpha)]^2$$

$$+ 2\sum_{i=1}^{q}\sum_{j=1,j\neq i}^{q}a_{ij}[\sigma(x_i) - \sigma(\alpha)][\sigma(x_j) - \sigma(\alpha)]$$

$$= 2[\hat{\sigma}(x) - \hat{\sigma}(\alpha\mathbf{e})]^{\mathrm{T}}A[\hat{\sigma}(x) - \hat{\sigma}(\alpha\mathbf{e})]$$

$$+ 2\sum_{i=1}^{q}[\hat{\sigma}(x) - \hat{\sigma}(\alpha\mathbf{e})]^{\mathrm{T}}A_{\mathrm{d}i}[\hat{\sigma}(x) - \hat{\sigma}(\alpha\mathbf{e})]$$

$$\leq -\sum_{i=1}^{q}\Big([\hat{\sigma}(x) - \hat{\sigma}(\alpha\mathbf{e})]^{\mathrm{T}}Q_i[\hat{\sigma}(x) - \hat{\sigma}(\alpha\mathbf{e})]$$

$$- 2[\hat{\sigma}(x) - \hat{\sigma}(\alpha\mathbf{e})]^{\mathrm{T}}A_{\mathrm{d}i}[\hat{\sigma}(x) - \hat{\sigma}(\alpha\mathbf{e})]$$

$$+ [\hat{\sigma}(x) - \hat{\sigma}(\alpha\mathbf{e})]^{\mathrm{T}}A_{\mathrm{d}i}^{\mathrm{T}}Q_i^{\#}A_{\mathrm{d}i}[\hat{\sigma}(x) - \hat{\sigma}(\alpha\mathbf{e})]\Big)$$

$$= -\sum_{i=1}^{q}\Big(-Q_i[\hat{\sigma}(x) - \hat{\sigma}(\alpha\mathbf{e})] + A_{\mathrm{d}i}[\hat{\sigma}(x) - \hat{\sigma}(\alpha\mathbf{e})]\Big)^{\mathrm{T}}Q_i^{\#}$$

$$\cdot \Big(-Q_i[\hat{\sigma}(x) - \hat{\sigma}(\alpha\mathbf{e})] + A_{\mathrm{d}i}[\hat{\sigma}(x) - \hat{\sigma}(\alpha\mathbf{e})]\Big)$$

$$\leq 0, \quad x \in \mathbb{R}^q.$$

Hence, for every $\alpha \in \mathbb{R}$, $\alpha\mathbf{e}$ is a Lyapunov stable equilibrium point of (2.2). Thus, $\alpha\mathbf{e}$ is a semistable equilibrium point of (2.2).                                    □

Theorem 2.1 implies that the steady-state value of the state of each agent $\mathcal{G}_i$ of the multiagent dynamical system $\mathcal{G}$ is equal; that is, the steady-state value of the multiagent dynamical system $\mathcal{G}$ given by

$$x_\infty = \frac{1}{q}\mathbf{e}\mathbf{e}^\mathrm{T} x(t_0) = \left[\frac{1}{q}\sum_{i=1}^{q} x_i(t_0)\right]\mathbf{e} \qquad (2.17)$$

is uniformly distributed over all multiagents of $\mathcal{G}$. This phenomenon is known as *equipartition of energy* [127] in system thermodynamics and *information consensus* or *protocol agreement* [238] in cooperative network systems.

Next, we generalize (ii) of Theorem 2.1 to the case where (2.3) has the nonlinear structure

$$\phi_{ij}(x_i, x_j) = a_{ij}(x_j) - a_{ji}(x_i), \quad i, j = 1, \ldots, q, \quad i \neq j, \qquad (2.18)$$

where $a_{ij} : \mathbb{R} \to \mathbb{R}$, $i, j = 1, \ldots, q$, $i \neq j$, are such that $a_{ij}(0) = 0$ and $a_{ij}(\cdot)$, $i, j = 1, \ldots, q$, $i \neq j$, is strictly increasing. For the next result define

$$f_{\mathrm{c}i}(x_i) \triangleq - \sum_{j=1,j\neq i}^{q} a_{ji}(x_i); \quad f_{\mathrm{d}i}(x) \triangleq \mathbf{e}_i \sum_{j=1}^{q} a_{ij}(x_j), \quad i = 1, \ldots, q; \qquad (2.19)$$

and $f_{\mathrm{c}}(x) \triangleq [f_{\mathrm{c}1}(x_1), \ldots, f_{\mathrm{c}q}(x_q)]^\mathrm{T}$.

**Theorem 2.2.** Consider the multiagent dynamical system given by (2.2), where $\phi_{ij}(x_i, x_j)$, $i, j = 1, \ldots, q$, $i \neq j$, is given by (2.18) and $f_{\mathrm{c}i}(\cdot)$, $i = 1, \ldots, q$, is strictly decreasing. Assume that $\mathbf{e}^\mathrm{T}[f_{\mathrm{c}}(x) + \sum_{i=1}^{q} f_{\mathrm{d}i}(x)] = 0$, $x \in \mathbb{R}^q$, and $f_{\mathrm{c}}(x) + \sum_{i=1}^{q} f_{\mathrm{d}i}(x) = 0$ if and only if $x = \alpha\mathbf{e}$ for every $\alpha \in \mathbb{R}$. Furthermore, assume there exist nonnegative diagonal matrices $P_i \in \mathbb{R}^{q \times q}$, $i = 1, \ldots, q$, such that every $P_i$ has only one nonzero diagonal entry and this entry is in the $(i, i)$th position, $P \triangleq \sum_{i=1}^{q} P_i$,

$$P_i^{\#} P_i f_{\mathrm{d}i}(x) = f_{\mathrm{d}i}(x), \quad x \in \mathbb{R}^q, \quad i = 1, \ldots, q, \qquad (2.20)$$

$$\sum_{i=1}^{q} [f_{\mathrm{d}i}(x) - f_{\mathrm{d}i}(\alpha\mathbf{e})]^\mathrm{T} P_i [f_{\mathrm{d}i}(x) - f_{\mathrm{d}i}(\alpha\mathbf{e})]$$
$$\leq [f_{\mathrm{c}}(x) - f_{\mathrm{c}}(\alpha\mathbf{e})]^\mathrm{T} P[f_{\mathrm{c}}(x) - f_{\mathrm{c}}(\alpha\mathbf{e})]^\mathrm{T}, \quad \alpha \in \mathbb{R}, \quad x \in \mathbb{R}^q. \qquad (2.21)$$

Then, for every $\alpha \in \mathbb{R}$, $\alpha\mathbf{e}$ is a semistable equilibrium state of (2.2). Furthermore, $x(t) \to \frac{1}{q}\mathbf{e}\mathbf{e}^\mathrm{T} x(t_0)$ as $t \to \infty$ and $\frac{1}{q}\mathbf{e}\mathbf{e}^\mathrm{T} x(t_0)$ is a semistable equilibrium state.

**Proof.** Consider the nonnegative function given by

$$V(x) = -2\sum_{i=1}^{q} \int_{0}^{x_i} P_{(i,i)} f_{ci}(\theta) \mathrm{d}\theta. \tag{2.22}$$

Since $f_{ci}(\cdot)$, $i = 1, \ldots, q$, is a strictly decreasing function, it follows that

$$V(x) = 2\sum_{i=1}^{q} P_{(i,i)}[-f_{ci}(\delta_i x_i)]x_i$$

$$\geq 2\sum_{i=1}^{q} P_{(i,i)}[-f_{ci}(\delta_i x_i)]\delta_i x_i$$

$$> 0$$

for all $x_i \neq 0$, where $0 < \delta_i < 1$, and, hence, there exists a class $\mathcal{K}$ function $\alpha(\cdot)$ such that $V(x) \geq \alpha(\|x\|)$. Now, note that the derivative of $V(x)$ along the trajectories of (2.2) is given by

$$\dot{V}(x) = -2f_{\mathrm{c}}^{\mathrm{T}}(x)Pf_{\mathrm{c}}(x) - 2\sum_{i=1}^{q} f_{\mathrm{c}}^{\mathrm{T}}(x)Pf_{\mathrm{d}i}(x)$$

$$\leq -f_{\mathrm{c}}^{\mathrm{T}}(x)Pf_{\mathrm{c}}(x) - 2\sum_{i=1}^{q} f_{\mathrm{c}}^{\mathrm{T}}(x)PP_i^{\#}P_i f_{\mathrm{d}i}(x)$$

$$-\sum_{i=1}^{q} f_{\mathrm{d}i}(x)P_i P_i^{\#} P_i f_{\mathrm{d}i}(x)$$

$$= -\sum_{i=1}^{q}[Pf_{\mathrm{c}}(x) + P_i f_{\mathrm{d}i}(x)]^{\mathrm{T}} P_i^{\#}[Pf_{\mathrm{c}}(x) + P_i f_{\mathrm{d}i}(x)]$$

$$\leq 0, \quad x \in \mathbb{R}^q, \tag{2.23}$$

where the first inequality in (2.23) follows from (2.20) and (2.21), and the last equality in (2.23) follows from the fact that

$$f_{\mathrm{c}}^{\mathrm{T}}(x)Pf_{\mathrm{c}}(x) = \sum_{i=1}^{q} f_{\mathrm{c}}^{\mathrm{T}}(x)PP_i^{\#}Pf_{\mathrm{c}}(x), \quad x \in \mathbb{R}^q.$$

Next, let

$$\mathcal{R} \triangleq \{x \in \mathbb{R}^q : Pf_{\mathrm{c}}(x) + P_i f_{\mathrm{d}i}(x) = 0, i = 1, \ldots, q\}.$$

Then it follows from the Krasovskii–LaSalle invariant set theorem [122, p. 147] that $x(t) \rightarrow \mathcal{M}$ as $t \rightarrow \infty$, where $\mathcal{M}$ denotes the largest invariant set contained in $\mathcal{R}$. Now, since $\mathbf{e}^{\mathrm{T}}(f_{\mathrm{c}}(x) + \sum_{i=1}^{q} f_{\mathrm{d}i}(x)) = 0, x \in \mathbb{R}^q$, it follows that

$$\mathcal{R} \subseteq \hat{\mathcal{R}} \triangleq \left\{ x \in \mathbb{R}^q : f_{\mathrm{c}}(x) + \sum_{i=1}^{q} f_{\mathrm{d}i}(x) = 0 \right\}$$
$$= \{ x \in \mathbb{R}^q : x = \alpha \mathbf{e}, \ \alpha \in \mathbb{R} \}, \tag{2.24}$$

which implies that $x(t) \to \hat{\mathcal{R}}$ as $t \to \infty$.

Finally, to show the Lyapunov stability of $\alpha \mathbf{e}$, $\alpha \in \mathbb{R}$, consider the Lyapunov function candidate

$$V(x) = -2 \sum_{i=1}^{q} \int_{\alpha}^{x_i} P_{(i,i)}(f_{\mathrm{c}i}(\theta) - f_{\mathrm{c}i}(\alpha)) \mathrm{d}\theta.$$

Note that $V(\alpha \mathbf{e}) = 0$ and, by the mean value theorem, there exist $\delta_i \in (0,1)$, $i = 1, \ldots, q$, such that

$$V(x) = 2 \sum_{i=1}^{q} P_{(i,i)}[f_{\mathrm{c}i}(\alpha) - f_{\mathrm{c}i}(\alpha + \delta_i(x_i - \alpha))](x_i - \alpha)$$
$$\geq 2 \sum_{i=1}^{q} P_{(i,i)}[f_{\mathrm{c}i}(\alpha) - f_{\mathrm{c}i}(\alpha + \delta_i(x_i - \alpha))]\delta_i(x_i - \alpha)$$
$$> 0, \quad x \neq \alpha \mathbf{e}.$$

The Lyapunov derivative of $V(x)$ along the trajectories of (2.2) is given by

$$\dot{V}(x) = -2 \sum_{i=1}^{q} P_{(i,i)}(f_{\mathrm{c}i}(x_i) - f_{\mathrm{c}i}(\alpha))\dot{x}_i$$
$$= -2 \sum_{i=1}^{q} P_{(i,i)}(f_{\mathrm{c}i}(x_i) - f_{\mathrm{c}i}(\alpha))f_i(x)$$
$$= -2 \sum_{i=1}^{q} P_{(i,i)}(f_{\mathrm{c}i}(x_i) - f_{\mathrm{c}i}(\alpha)) \left[ \sum_{j=1, j \neq i}^{q} \phi_{ij}(x_i, x_j) \right]$$
$$= -2 \sum_{i=1}^{q} P_{(i,i)}(f_{\mathrm{c}i}(x_i) - f_{\mathrm{c}i}(\alpha)) \left[ \sum_{j=1, j \neq i}^{q} (a_{ij}(x_j) - a_{ji}(x_i)) \right]$$
$$= -2 \sum_{i=1}^{q} P_{(i,i)}(f_{\mathrm{c}i}(x_i) - f_{\mathrm{c}i}(\alpha)) \left[ \sum_{j=1, j \neq i}^{q} a_{ij}(x_j) - \sum_{j=1, j \neq i}^{q} a_{ji}(x_i) \right]$$

$$= -2\sum_{i=1}^{q} P_{(i,i)}(f_{ci}(x_i) - f_{ci}(\alpha)) \left[ \sum_{j=1,j\neq i}^{q} a_{ij}(x_j) + f_{ci}(x_i) \right].$$

Since for every $\alpha \in \mathbb{R}$,

$$0 = \sum_{j=1,j\neq i}^{q} a_{ij}(\alpha) - \sum_{j=1,j\neq i}^{q} a_{ji}(\alpha) = \sum_{j=1,j\neq i}^{q} a_{ij}(\alpha) - f_{ci}(\alpha),$$

it follows that

$$\dot{V}(x) = -2\sum_{i=1}^{q} P_{(i,i)}(f_{ci}(x_i) - f_{ci}(\alpha))$$

$$\cdot \left[ \sum_{j=1,j\neq i}^{q} a_{ij}(x_j) - \sum_{j=1,j\neq i}^{q} a_{ij}(\alpha) + f_{ci}(x_i) - f_{ci}(\alpha) \right]$$

$$= -2\sum_{i=1}^{q} P_{(i,i)}(f_{ci}(x_i) - f_{ci}(\alpha)) \left[ \sum_{j=1,j\neq i}^{q} a_{ij}(x_j) - \sum_{j=1,j\neq i}^{q} a_{ij}(\alpha) \right]$$

$$- 2\sum_{i=1}^{q} P_{(i,i)}(f_{ci}(x_i) - f_{ci}(\alpha))^2$$

$$= -2\sum_{i=1}^{q} [f_c(x) - f_c(\alpha\mathbf{e})]^{\mathrm{T}} P[f_{di}(x) - f_{di}(\alpha\mathbf{e})]$$

$$- 2[f_c(x) - f_c(\alpha\mathbf{e})]^{\mathrm{T}} P[f_c(x) - f_c(\alpha\mathbf{e})]$$

$$\leq -[f_c(x) - f_c(\alpha\mathbf{e})]^{\mathrm{T}} P[f_c(x) - f_c(\alpha\mathbf{e})]$$

$$- 2\sum_{i=1}^{q} [f_c(x) - f_c(\alpha\mathbf{e})]^{\mathrm{T}} PP_i^{\#} P_i[f_{di}(x) - f_{di}(\alpha\mathbf{e})]$$

$$- \sum_{i=1}^{q} [f_{di}(x) - f_{di}(\alpha\mathbf{e})]^{\mathrm{T}} P_i P_i^{\#} P_i[f_{di}(x) - f_{di}(\alpha\mathbf{e})]$$

$$= -\sum_{i=1}^{q} [P(f_c(x) - f_c(\alpha\mathbf{e})) + P_i(f_{di}(x) - f_{di}(\alpha\mathbf{e}))]^{\mathrm{T}} P_i^{\#}$$

$$\cdot [P(f_c(x) - f_c(\alpha\mathbf{e})) + P_i(f_{di}(x) - f_{di}(\alpha\mathbf{e}))]$$

$$\leq 0, \quad x \in \mathbb{R}^q.$$

Hence, for every $\alpha \in \mathbb{R}$, $\alpha\mathbf{e}$ is a Lyapunov stable equilibrium point of (2.2). Thus, $\alpha\mathbf{e}$ is a semistable equilibrium point of (2.2). $\square$

Finally, we specialize Theorem 2.2 to the case where

$$\phi_{ij}(x_i, x_j) = a_{ij}\sigma(x_j) - a_{ji}\sigma(x_i) \tag{2.25}$$

and $\sigma : \mathbb{R} \to \mathbb{R}$ is such that $\sigma(u) = 0$ if and only if $u = 0$, $a_{ij} \in \mathbb{R}$, $i, j = 1, \ldots, q$, $i \neq j$. In this case, (2.2) can be rewritten as

$$\dot{x}(t) = A\hat{\sigma}(x(t)) + \sum_{i=1}^{q} A_{\mathrm{d}i}\hat{\sigma}(x(t)), \quad x(0) = x_0, \quad t \geq 0, \tag{2.26}$$

where $\hat{\sigma} : \mathbb{R}^q \to \mathbb{R}^q$ is given by $\hat{\sigma}(x) \triangleq [\sigma(x_1), \ldots, \sigma(x_q)]^{\mathrm{T}}$ and $A$ and $A_{\mathrm{d}i}$, $i = 1, \ldots, q$, are given by (2.9).

**Theorem 2.3.** Consider the multiagent dynamical system given by (2.26), where $\sigma : \mathbb{R} \to \mathbb{R}$ is such that $\sigma(0) = 0$ and $\sigma(\cdot)$ is strictly increasing. Assume that $(A + \sum_{i=1}^{q} A_{\mathrm{d}i})^{\mathrm{T}}\mathbf{e} = (A + \sum_{i=1}^{q} A_{\mathrm{d}i})\mathbf{e} = 0$ and $\mathrm{rank}(A + \sum_{i=1}^{q} A_{\mathrm{d}i}) = q - 1$. Then, for every $\alpha \in \mathbb{R}$, $\alpha\mathbf{e}$ is a semistable equilibrium point of (2.2). Furthermore, $x(t) \to \frac{1}{q}\mathbf{e}\mathbf{e}^{\mathrm{T}}x(t_0)$ as $t \to \infty$ and $\frac{1}{q}\mathbf{e}\mathbf{e}^{\mathrm{T}}x(t_0)$ is a semistable equilibrium state.

**Proof.** It follows from Lemma 2.1 that there exist $Q_i$, $i = 1, \ldots, q$, such that (2.10) holds with $Q_i$ given by (2.13). Now, since

$$A = -\sum_{i=1}^{q} Q_i = -\sum_{i=1}^{q} P_i^{\#} = -P^{-1},$$

where $P = \sum_{i=1}^{q} P_i$, it follows from (2.10) that, for all $x \in \mathbb{R}^q$,

$$0 \geq 2[\hat{\sigma}(x) - \hat{\sigma}(\alpha\mathbf{e})]^{\mathrm{T}} A[\hat{\sigma}(x) - \hat{\sigma}(\alpha\mathbf{e})]$$
$$+ 2[\hat{\sigma}(x) - \hat{\sigma}(\alpha\mathbf{e})]^{\mathrm{T}} \sum_{i=1}^{q} (Q_i + A_{\mathrm{d}i}^{\mathrm{T}}Q_i^{\#}A_{\mathrm{d}i})[\hat{\sigma}(x) - \hat{\sigma}(\alpha\mathbf{e})]$$
$$= -[f_{\mathrm{c}}(x) - f_{\mathrm{c}}(\alpha\mathbf{e})]^{\mathrm{T}} P[f_{\mathrm{c}}(x) - f_{\mathrm{c}}(\alpha\mathbf{e})]^{\mathrm{T}}$$
$$+ \sum_{i=1}^{q} [f_{\mathrm{d}i}(x) - f_{\mathrm{d}i}(\alpha\mathbf{e})]^{\mathrm{T}} P_i[f_{\mathrm{d}i}(x) - f_{\mathrm{d}i}(\alpha\mathbf{e})],$$

where $f_{\mathrm{c}}(x) = A\hat{\sigma}(x)$ and $f_{\mathrm{d}i}(x) = A_{\mathrm{d}i}\hat{\sigma}(x)$, $i = 1, \ldots, q$, $x \in \mathbb{R}^q$. Furthermore, since $P_i^{\#}P_iA_{\mathrm{d}i} = A_{\mathrm{d}i}$, $i = 1, \ldots, q$, it follows that $P_i^{\#}P_if_{\mathrm{d}i}(x) = f_{\mathrm{d}i}(x)$, $i = 1, \ldots, q$, $x \in \mathbb{R}^q$. Now, the result is an immediate consequence of Theorem 2.2 by noting that $\mathbf{e}^{\mathrm{T}}[f_{\mathrm{c}}(x) + \sum_{i=1}^{q} f_{\mathrm{d}i}(x)] = 0$ and $f_{\mathrm{c}}(x) + \sum_{i=1}^{q} f_{\mathrm{d}i}(x) = 0$ if and only if $x = \alpha\mathbf{e}$ for every $\alpha \in \mathbb{R}$. $\square$

Theorems 2.2 and 2.3 can be extended to address linear and nonlinear dynamical networks with multiple time delays. For details, see Chapter 8.

Note that the results of this section provide a generalization to Theorems 4 and 5 of [238], which establish information consensus protocols for the special structure $\phi_{ij}(x_i, x_j) = a_{ij}(x_i - x_j)$, $i, j = 1, \ldots, q$, $i \neq j$.

Next, we provide explicit connections of the consensus control architecture presented in (i) of Theorem 2.1 to a recently developed theory of thermodynamics [118]. To develop these connections, the following definition of entropy is needed.

**Definition 2.4.** For the distributed consensus protocol $\mathcal{G}$ given by (2.2), a function $\mathcal{S} : \mathbb{R}^q \to \mathbb{R}$ satisfying

$$\mathcal{S}(x(t_2)) \geq \mathcal{S}(x(t_1)), \quad t_2 \geq t_1 \geq t_0, \tag{2.27}$$

is called an *entropy* of $\mathcal{G}$.

The next theorem gives an explicit expression for the entropy function of the closed-loop, continuous-time multiagent dynamical system $\mathcal{G}$ given by (2.2).

**Theorem 2.4.** Consider the closed-loop, multiagent dynamical system $\mathcal{G}$ given by (2.2), and assume that Assumptions 2.1 and 2.2 hold. Then the function $\mathcal{S} : \mathbb{R}^q \to \mathbb{R}$ given by

$$\mathcal{S}(x) = \mathbf{e}^{\mathrm{T}} \mathbf{log}_e(c\,\mathbf{e} + x) - q \log_e c, \quad x \in \mathbb{R}^q, \tag{2.28}$$

is an entropy function of $\mathcal{G}$, where $\mathbf{log}_e(c\,\mathbf{e} + x)$ denotes the vector natural logarithm given by $[\log_e(c + x_1), \ldots, \log_e(c + x_q)]^{\mathrm{T}}$ and $c > \|x\|_\infty$.

**Proof.** Since $\phi_{ij}(x_i, x_j) = -\phi_{ji}(x_j, x_i)$, $i \neq j$, $i, j = 1, \ldots, q$, and $c > \|x\|_\infty$, it follows that

$$
\begin{aligned}
\dot{\mathcal{S}}(x(t)) &= \sum_{i=1}^{q} \frac{\dot{x}_i(t)}{c + x_i(t)} \\
&= \sum_{i=1}^{q} \sum_{j=1, j \neq i}^{q} \frac{\phi_{cij}(x_i(t), x_j(t))}{c + x_i(t)} \\
&= \sum_{i=1}^{q} \sum_{j=i+1}^{q} \left( \frac{\phi_{cij}(x_i(t), x_j(t))}{c + x_i(t)} - \frac{\phi_{cij}(x_i(t), x_j(t))}{c + x_j(t)} \right) \\
&= \sum_{i=1}^{q-1} \sum_{j=i+1}^{q} \frac{\phi_{cij}(x_i(t), x_j(t))[x_j(t) - x_i(t)]}{(c + x_i(t))(c + x_j(t))} \\
&\geq 0, \quad t \geq t_0.
\end{aligned}
\tag{2.29}
$$

Now, integrating (2.29) over $[t_1, t_2]$ yields (2.27). $\qquad \square$

Note that it follows from (2.29) that the entropy function given by (2.28) satisfies (2.27) as an equality for an equilibrium (equipartitioned) process and as a strict inequality for a nonequilibrium (nonequipartitioned) process. The entropy expression given by (2.28) is identical in form to the Boltzmann entropy for statistical thermodynamics and the Shannon entropy characterizing the amount of information [118]. In addition, note that $\mathcal{S}(x)$ given by (2.28) achieves a maximum when all the information states $x_i$, $i = 1, \ldots, q$, are equal [118]. Inequality (2.27) is a generalization of Clausius's inequality for equilibrium and nonequilibrium thermodynamics as well as reversible and irreversible thermodynamics as applied to adiabatically isolated continuous-time thermodynamic systems. For details, see [118].

**Example 2.1.** Consider the multiagent system comprising the controlled longitudinal motion of seven Boeing 747 airplanes [55] linearized at an altitude of 40 kft and a velocity of 774 ft/sec given by

$$\dot{z}_i(t) = A z_i(t) + B \delta_i(t), \quad z_i(0) = z_{i_0}, \quad i = 1, \ldots, 7, \quad t \geq 0, \qquad (2.30)$$

where $z_i(t) = [v_{x_i}(t), v_{z_i}(t), q_i(t), \theta_{e_i}(t)]^{\mathrm{T}} \in \mathbb{R}^4$, $t \geq 0$, is state vector of aircraft $i \in \{1, \ldots, 7\}$, with $v_{x_i}(t)$, $t \geq 0$, representing the $x$-body-axis component of the velocity of the airplane center of mass with respect to the reference axes (in ft/sec); $v_{z_i}(t)$, $t \geq 0$, representing the $z$-body-axis component of the velocity of the airplane center of mass with respect to the reference axes (in ft/sec); $q_i(t)$, $t \geq 0$, representing the $y$-body-axis component of the angular velocity of the airplane (pitch rate) with respect to the reference axes (in crad/sec); $\theta_{e_i}(t)$, $t \geq 0$, representing the pitch Euler angle of the airplane body axes with respect to the reference axes (in crad); $\delta(t)$, $t \geq 0$, representing the elevator control input (in crad); and

$$A = \begin{bmatrix} -0.003 & 0.039 & 0 & -0.332 \\ -0.065 & -0.319 & 7.74 & 0 \\ 0.020 & -0.101 & -0.429 & 0 \\ 0 & 0 & 1 & 0 \end{bmatrix}, \quad B = \begin{bmatrix} 0.010 \\ -0.180 \\ -1.16 \\ 0 \end{bmatrix}.$$

We propose a two-level control hierarchy composed of a lower-level controller for command following and a higher-level consensus controller for pitch rate consensus with the communication topology shown in Figure 2.1. To address the lower-level controller design, let $x_i(t)$, $i = 1, \ldots, 7$, $t \geq 0$, denote an information command generated by (2.1) (i.e., the guidance command), and let $s_i(t)$, $i = 1, \ldots, 7$, $t \geq 0$, denote the integrator state satisfying

$$\dot{s}_i(t) = E z_i(t) - x_i(t), \quad s_i(0) = s_{i_0}, \quad i = 1, \ldots, 7, \quad t \geq 0, \qquad (2.31)$$

where $E = [0, 0, 1, 0]$.

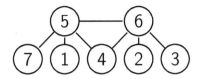

Figure 2.1 Aircraft communication topology.

Now, defining the augmented state

$$\hat{z}_i(t) \triangleq [z_i^{\mathrm{T}}(t), s_i(t)]^{\mathrm{T}} \in \mathbb{R}^5,$$

(2.30) and (2.31) give

$$\dot{\hat{z}}_i(t) = \hat{A}\hat{z}_i(t) + \hat{B}_1\delta_i(t) + \hat{B}_2 x_i(t), \quad \hat{z}_i(0) = \hat{z}_{i_0}, \quad i = 1, \dots, 7, \quad t \geq 0, \tag{2.32}$$

where

$$\hat{A} \triangleq \begin{bmatrix} A & 0 \\ E & 0 \end{bmatrix}, \quad \hat{B}_1 \triangleq \begin{bmatrix} B \\ 0 \end{bmatrix}, \quad \hat{B}_2 \triangleq \begin{bmatrix} 0 \\ -1 \end{bmatrix}. \tag{2.33}$$

Furthermore, let the elevator control input be given by

$$\delta_i(t) = -K\hat{z}_i(t), \quad i = 1, \dots, 7,$$

where

$$K = [-0.0157, 0.0831, -4.7557, -0.1400, -9.8603],$$

which is designed based on an optimal linear-quadratic regulator (LQR).

For the higher-level consensus controller design, we use (i) of Theorem 2.1 with function $\phi_{ij}(x_i(t), x_j(t)) = (x_j(t) - x_i(t))^{\frac{1}{3}}$, $i, j \in \{1, \dots, 7\}$, $i \neq j$, to generate the information state $x(t)$, $t \geq 0$. For our simulation, we set $x_1(0) = q_1(0) = 0$, $x_2(0) = q_2(0) = 16$, $x_3(0) = q_3(0) = 8$, $x_4(0) = q_4(0) = 5$, $x_5(0) = q_5(0) = 16$, $x_6(0) = q_6(0) = 19$, and $x_7(0) = q_7(0) = 20$, with all other initial conditions set to zero. Figure 2.2 shows the information command signals and pitch rate of each aircraft versus time for the thermodynamic control protocol with $\phi_{ij}(x_i(t), x_j(t)) = (x_j(t) - x_i(t))^{\frac{1}{3}}$, $i, j \in \{1, \dots, 7\}$, $i \neq j$. Finally, Figure 2.3 shows the total agent entropies versus time. Since every component of the information state $x(t)$, $t \geq 0$, is nonnegative, here we set $c = 1$ so that $q \log_e c = 0$. $\triangle$

## 2.4 Distributed Control for Parallel Formations

In this section, we develop a distributed controller for mobile agents to achieve parallel flocking formations [183, 234, 294, 295]. Specifically, consider $q$ mobile agents given by

$$\ddot{x}_i(t) = u_i(t), \quad x_i(0) = x_{i0}, \quad \dot{x}_i(0) = \dot{x}_{i0}, \quad t \geq 0, \tag{2.34}$$

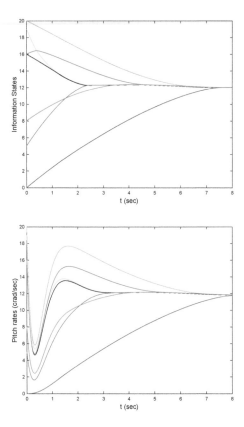

Figure 2.2 Closed-loop information command signal $x_i(t)$ (top) and pitch rate $q_i(t)$ (bottom) trajectories with the higher-level consensus protocol given in (i) of Theorem 2.1; $x_1(t), q_1(t)$ in blue; $x_2(t), q_2(t)$ in red; $x_3(t), q_3(t)$ in green; $x_4(t), q_4(t)$ in magenta; $x_5(t), q_5(t)$ in black; $x_6(t), q_6(t)$ in yellow; and $x_7(t), q_7(t)$ in cyan.

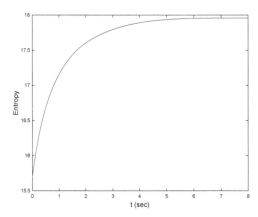

Figure 2.3 Total agent entropy versus time (with $c = 1$ in (2.28)) for control protocol.

where, for $i = 1, \ldots, q$ and $t \geq 0$, $x_i(t) \in \mathbb{R}^n$ is the position vector of the $i$th agent, $\dot{x}_i(t) \in \mathbb{R}^n$ is the velocity vector of the $i$th agent, and $u_i(t) \in \mathbb{R}^n$ is the control input of the $i$th agent. The control aim is to design a feedback control law so that parallel formation is achieved, wherein the agents $\mathcal{G}_i$ are collectively required to maintain a prescribed geometric shape with $\dot{x}_1 = \cdots = \dot{x}_q = c$, where $c \in \mathbb{R}^n$ is a constant vector, and the relative position between any two mobile agents is asymptotically stabilized to a constant value.

Here, we consider multiagent systems involving *nonholonomic dynamics*. Recall that a mechanical system described by the generalized coordinates $x \in \mathcal{X}$, where $\mathcal{X}$ is a smooth $n$-dimensional configuration manifold, subject to $m$ smooth constraints $A(x)\dot{x} = 0$, is *nonholonomic* if $A(\cdot)$ is not integrable. An equivalent description of such systems can be constructed by using a basis $G(x)$ of a distribution that annihilates $A(x)$ to describe allowable velocities $\dot{x} \in T_x\mathcal{X}$ as $\dot{x} = G(x)u$, where $T_x\mathcal{X}$ denotes the tangent space of $\mathcal{X}$ at $x$. This representation includes a wide range of multiagent models described by an element of the Lie group[2] $G$ of rigid motions in $\mathbb{R}^2$ or $\mathbb{R}^3$, denoted by $SE(2)$ or $SE(3)$.

Consider $q$ mobile agents, with each mobile agent described by the *driftless controllable nonholonomic system*

$$\dot{x}_{\mathrm{p}i}(t) = X_1(x_{\mathrm{p}i}(t))u_{1,i}(t) + \cdots + X_{m+1}(x_{\mathrm{p}i}(t))u_{m+1,i}(t),$$
$$x_{\mathrm{p}i}(0) = x_{\mathrm{p}i0}, \quad t \geq 0, \tag{2.35}$$

where, for each $i \in \{1, \ldots, q\}$ and $t \geq 0$, $x_{\mathrm{p}i}(t) \triangleq [x_i^{\mathrm{T}}(t), z_i^{\mathrm{T}}(t)]^{\mathrm{T}}$, $x_i(t) \in \mathbb{R}^n$, $z_i(t) \in \mathbb{R}^m$; $u_{l,i}(t) \in \mathbb{R}$ denotes the control input, $l = 1, \ldots, m$; $X_1 = [g^{\mathrm{T}}(z_i), 0_{1 \times m}]^{\mathrm{T}} \in \mathbb{R}^{m+n}$, $g : \mathbb{R}^m \to \mathbb{R}^n$, is smooth; and $X_j = \mathbf{e}_{n+j-1}$, $j = 2, \ldots, m+1$, where $\mathbf{e}_{n+j-1} \in \mathbb{R}^{m+n}$ is a vector whose $(n + j - 1)$th component is one and remaining components are zero. The aim here is to design control inputs $u_{j,i}$ so that all agents maintain a parallel formation.

Consider the dynamic controller given by

$$\dot{x}_{\mathrm{c}i}(t) = -\sum_{j=1, j \neq i}^{q} \mathcal{A}_{(i,j)}(x_{\mathrm{c}i}(t) - x_{\mathrm{c}j}(t)) + g^{\mathrm{T}}(z_i(t))\zeta_i(t), \tag{2.36}$$

$$\dot{z}_{\mathrm{c}i}(t) = -\sum_{j=1, j \neq i}^{q} \mathcal{A}_{(i,j)}(z_{\mathrm{c}i}(t) - z_{\mathrm{c}j}(t)) - \sum_{j=1, j \neq i}^{q} \mathcal{A}_{(i,j)}(z_i(t) - z_j(t)),$$

---

[2]A *group* is a set that is closed under an associative binary operation with respect to which there exists a unique identity element within the set and every element in the set has an inverse. A *Lie group* is a topological group that can be given an analytic structure such that the group operation and inversion are analytic. $SE(n)$ is a special Euclidean group in $n$ dimensions consisting of a rotation and a translation of a vector.

$$x_{ci}(0) = x_{ci0}, \quad z_{ci}(0) = z_{ci0}, \quad t \geq 0, \quad i = 1, \ldots, q, \quad j \in \mathcal{K}_i,$$

(2.37)

$$u_{1,i}(t) = -x_{ci}(t),$$

(2.38)

$$\begin{bmatrix} u_{2,i}(t) \\ \vdots \\ u_{m+1,i}(t) \end{bmatrix} = - \sum_{j=1,j\neq i}^{q} \mathcal{A}_{(j,i)}(z_{cj}(t) - z_{ci}(t)),$$

(2.39)

where

$$\zeta_i(t) \triangleq \sum_{j=1,j\neq i}^{q} \nabla_{x_i}(U_{ij}(x_i(t), x_j(t)) + U_{ji}(x_j(t), x_i(t))),$$

(2.40)

$\mathcal{K}_i \triangleq \{i_1, \ldots, i_{|\mathcal{K}_i|}\}$ denotes the indices of all the other agents which have a communication link with the $i$th agent, $|\mathcal{K}_i|$ denotes the cardinality of the set $\mathcal{K}_i$, $U_{ij}(x_i, x_j)$ is a *generalized potential function* satisfying $U_{ij}(x_i, x_j) \geq 0$, $U_{ij}(x_i, x_j) \equiv 0$ for $\mathcal{A}_{(i,j)} = 0$,

$$\nabla_{x_i} U_{ij}(x_i, x_j) + \nabla_{x_j} U_{ij}(x_i, x_j) = 0,$$

(2.41)

and

$$\sum_{i=1}^{q} \sum_{j=1,j\neq i}^{q} \nabla_{x_i}(U_{ij}(x_i, x_j) + U_{ji}(x_j, x_i)) = 0,$$

(2.42)

where $\nabla_{x_i} U_{ij}(x_i, x_j) \triangleq (\frac{\partial U_{ij}}{\partial x_i}(x_i, x_j))^{\mathrm{T}}$ is a gradient vector. The function $U_{ij}(x_i, x_j)$ is similar to the generalized potential function used in Theorem 20.4 and the artificial potential functions addressed in [234, 259, 294]. However, unlike the results presented in [259, 294], $U_{ij}(x_i, x_j)$ cannot be used to avoid collision between the agents. The collision avoidance problem will be addressed in Section 20.5.

**Theorem 2.5.** Consider the closed-loop system $\tilde{\mathcal{G}}$ consisting of $q$ mobile agents with dynamics (2.35) and the controller (2.36)–(2.39). Assume that Assumptions 2.1 and 2.2 hold and $\mathcal{A} = \mathcal{A}^{\mathrm{T}}$. Let $\mathcal{D} \subseteq \mathbb{R}^{(2m+n+1)q}$ be a compact positively invariant set with respect to the closed-loop system $\tilde{\mathcal{G}}$. Then there exists $\mathcal{D}_0 \subseteq \mathcal{D}$ such that if all the system initial conditions are in $\mathcal{D}_0$, then the state of the closed-loop system $\tilde{\mathcal{G}}$ approaches the largest invariant set $\mathcal{M}$ contained in

$$\bigcap_{i=1}^{q} \{x_{ci} = \alpha, \ z_i = \beta, \ z_{ci} = \gamma, \ \dot{x}_i = g(\beta)\alpha, \ g^{\mathrm{T}}(\beta)\zeta_i = 0\},$$

where $\alpha \in \mathbb{R}$, $\beta \in \mathbb{R}^m$, and $\gamma \in \mathbb{R}^m$. If, in addition, $\mathcal{D}_0 = \mathcal{D} = \mathbb{R}^{(2m+n+1)q}$, then the result is global.

**Proof.** Consider the nonnegative function given by

$$V(x, z, x_c, z_c) = \sum_{i=1}^{q} \sum_{j=1, j \neq i}^{q} U_{ij}(x_i, x_j) + \frac{1}{2}\|z\|_2^2 + \frac{1}{2}\|x_c\|_2^2 + \frac{1}{2}\|z_c\|_2^2, \quad (2.43)$$

where $x \triangleq [x_1^T, \ldots, x_q^T]^T$, $z \triangleq [z_1^T, \ldots, z_q^T]^T$, $x_c \triangleq [x_{c1}, \ldots, x_{cq}]^T$, and $z_c \triangleq [z_{c1}^T, \ldots, z_{cq}^T]^T$, $i = 1, \ldots, q$. The derivative of $V(x, z, x_c, z_c)$ along the trajectories of the closed-loop system (2.35)–(2.39) is given by

$$\dot{V}(x, z, x_c, z_c) = -\sum_{i=1}^{q-1} \sum_{j=i+1}^{q} \mathcal{A}_{(i,j)}(z_{ci} - z_{cj})^T(z_{ci} - z_{cj})$$

$$-\sum_{i=1}^{q-1} \sum_{j=i+1}^{q} \mathcal{A}_{(i,j)}(x_{ci} - x_{cj})^2$$

$$\leq 0, \quad (x, z, x_c, z_c) \in \mathcal{D}. \quad (2.44)$$

Next, let $\mathcal{R} \triangleq \{(x, z, x_c, z_c) \in \mathcal{D} : \dot{V}(x, z, x_c, z_c) = 0\}$. Since Assumptions 2.1 and 2.2 hold, it follows that

$$\mathcal{R} = \{(x, z, x_c, z_c) \in \mathcal{D} : z_{c1} = \cdots = z_{cq}, \, x_{c1} = \cdots = x_{cq}\}.$$

Let $\mathcal{M}$ denote the largest invariant set contained in $\mathcal{R}$. Now, it follows from Proposition 2.1 that on $\mathcal{M}$, $z_{ci} = \gamma$, $i = 1, \ldots, q$, where $\gamma \in \mathbb{R}^m$.

Next, it follows from (2.37) and (2.39) that $\dot{z}_i = 0$ and

$$\sum_{j=1, j \neq i}^{q} \mathcal{A}_{(i,j)}(z_i - z_j) = 0, \quad i = 1, \ldots, q.$$

Hence, it follows from Proposition 2.1 that $z_i = \beta$, where $\beta \in \mathbb{R}^m$, $i = 1, \ldots, m$. In this case, $\dot{x}_i - \dot{x}_j = g(z_j)x_{cj} - g(z_i)x_{ci} = 0$, and, hence,

$$\frac{d}{dt}\|x_c\|_2^2 = -2\sum_{i=1}^{q} \sum_{j=1, j \neq i}^{q} (\dot{x}_i - \dot{x}_j)^T \nabla_{x_i} U_{ij}(x_i, x_j) = 0,$$

which implies that $x_{ci} = \alpha$, $i = 1, \ldots, q$, where $\alpha \in \mathbb{R}$ is a constant. Now, it follows from the Krasovskii–LaSalle invariant set theorem [122, p. 147] that, for every system initial condition in $\mathcal{D}_0$, $(x(t), z(t), x_c(t), z_c(t)) \to \mathcal{M}$ as $t \to \infty$. The global result follows using standard arguments. $\square$

The controller (2.36)–(2.39) is a dynamic compensator which differs from the conventional controllers used in the literature for parallel formation control [183, 234, 294, 295]. The proposed controller is also a global controller. Moreover, note that the final configuration of the multiagent systems satisfies the constraint

$$g(\beta) \perp \text{span} \{\zeta_1, \ldots, \zeta_q\}, \tag{2.45}$$

which implies that the velocity vector $g(\beta)$ is perpendicular to the tangent space span$\{\zeta_1, \ldots, \zeta_q\}$. This property is key to designing distributed controllers for parallel formations. Finally, if $U_{ij}(\cdot, \cdot)$ is radially unbounded, then the assumption of the existence of a compact positively invariant set $\mathcal{D}$ can be relaxed by taking the level set $\{(x, z, x_c, z_c) \in \mathbb{R}^{(2m+n+1)q} : V(x, z, x_c, z_c) \leq c\}$ as the invariant set, where $c > 0$ is a sufficiently large constant.

**Example 2.2.** In this example, we consider the parallel formation problem with the multiagent dynamical system given by (2.36)–(2.39). Let $q = 10$ and $\mathcal{A}_{(i,i+1)} = \mathcal{A}_{(i+1,i)} = \mathcal{A}_{(i,i+2)} = \mathcal{A}_{(i+2,i)} = 1$, $i = 1, \ldots, 8$. Every other entry in $\mathcal{A}$ is zero. We choose $U_{ij}(x_i, x_j) = 1 - \exp(1 - \|x_i - x_j\|_2^2)$, $i, j = 1, \ldots, 10$, $i \neq j$. A group of 10 agents is initialized with random initial positions $(x, y)$ in the range of $[-10, 20] \times [-10, 10]$. The initial states of the controller (2.36) and (2.37) are chosen randomly within $[-10, 10]$.

Figure 2.4 illustrates parallel formations using the dynamic controller (2.36)–(2.40). Figures 2.5 and 2.6 show the time histories of the velocities and orientations for the parallel formation design.                    $\triangle$

Next, we extend the above kinematic result to dynamic models. Specifically, consider $q$ mobile agents, with each mobile agent described by the dynamical system

$$\dot{x}_i(t) = g(y_i(t)) z_i(t), \quad x_i(0) = x_{i0}, \quad t \geq 0, \tag{2.46}$$

$$\dot{y}_i(t) = w_i(t), \quad y_i(0) = y_{i0}, \tag{2.47}$$

$$\dot{z}_i(t) = u_i(t), \quad z_i(0) = z_{i0}, \tag{2.48}$$

$$\dot{w}_i(t) = v_i(t), \quad w_i(0) = w_{i0}, \tag{2.49}$$

where, for each $i \in \{1, \ldots, q\}$ and $t \geq 0$, $x_i(t) \in \mathbb{R}^n$, $y_i(t) \triangleq [y_{1i}(t), \ldots, y_{mi}(t)]^{\mathrm{T}} \in \mathbb{R}^m$, $z_i(t) \in \mathbb{R}$, $w_i(t) \triangleq [w_{1i}(t), \ldots, w_{mi}(t)]^{\mathrm{T}} \in \mathbb{R}^m$, $u_i(t) \in \mathbb{R}$, $v_i(t) \triangleq [v_{1i}(t), \ldots, v_{mi}(t)]^{\mathrm{T}} \in \mathbb{R}^m$, and $g : \mathbb{R}^m \to \mathbb{R}^n$ is a smooth function. The goal here is to design control inputs $u_i(t)$ and $v_i(t)$ so that all agents maintain a parallel formation. To address this problem, consider the

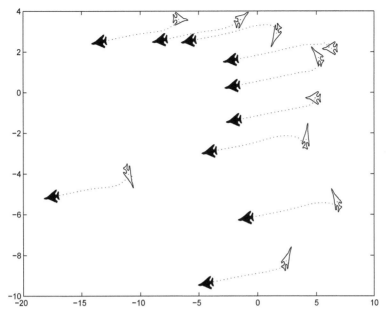

Figure 2.4 Parallel formation design.

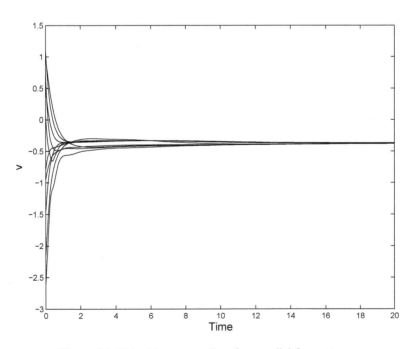

Figure 2.5 Velocities versus time for parallel formations.

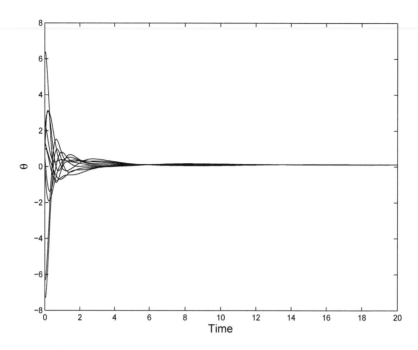

Figure 2.6 Orientations versus time for parallel formations.

controller architecture given by

$$\dot{x}_{ci}(t) = -\sum_{j=1,j\neq i}^{q} \mathcal{A}_{(i,j)}(x_{ci}(t) - x_{cj}(t)) - \sum_{j=1,j\neq i}^{q} \mathcal{A}_{(i,j)}(z_i(t) - z_j(t)),$$

$$x_{ci}(0) = x_{ci0}, \quad t \geq 0, \qquad (2.50)$$

$$u_i(t) = -\sum_{j=1,j\neq i}^{q} \mathcal{A}_{(j,i)}(x_{cj}(t) - x_{ci}(t)) - g^{\mathrm{T}}(y_i(t))\zeta_i(t), \qquad (2.51)$$

$$v_i(t) = -\sum_{j=1,j\neq i}^{q} \mathcal{A}_{(i,j)}(w_i(t) - w_j(t))$$

$$-\sum_{j=1,j\neq i}^{q} \mathcal{A}_{(i,j)}(y_i(t) - y_j(t)) - \sigma_{kk}(w_k(t)), \qquad (2.52)$$

where $k \in \{1, \ldots, q\}$; $\sigma_{kk}(\cdot)$ is such that $\sigma_{kk}(x)x > 0$ for $x \in \mathbb{R}$, $x \neq 0$; and $\zeta_i(t)$, $i, j = 1, \ldots, q$, $i \neq j$, is given by (2.40).

**Theorem 2.6.** Consider the closed-loop system $\tilde{\mathcal{G}}$ consisting of $q$ mobile agents with dynamics (2.46)–(2.49) and controller (2.50)–(2.52). Assume that Assumptions 2.1 and 2.2 hold and $\mathcal{A} = \mathcal{A}^{\mathrm{T}}$. Let $\mathcal{D}$ be a compact positively invariant set with respect to the closed-loop system $\tilde{\mathcal{G}}$. Then there exists $\mathcal{D}_0 \subseteq \mathcal{D}$ such that if all the system initial conditions are in $\mathcal{D}_0$, then the state of the closed-loop system $\tilde{\mathcal{G}}$ approaches the largest invariant set $\mathcal{M}$ contained in

$$\bigcap_{i=1}^{q} \{y_i = \alpha,\ z_i = \beta,\ w_i = 0,\ \dot{x}_i = g(\alpha)\beta,\ x_{\mathrm{c}i} = \gamma,\ g^{\mathrm{T}}(\alpha)\zeta_i = 0\},$$

where $\alpha \in \mathbb{R}^m$, $\beta \in \mathbb{R}$, and $\gamma \in \mathbb{R}$. If, in addition, $\mathcal{D} = \mathbb{R}^{(n+2m+2)q}$, then the result is global.

**Proof.** Consider the nonnegative function given by

$$V(x, y, z, w, x_{\mathrm{c}}) = \sum_{i=1}^{q} \sum_{j=1, j\neq i}^{q} U_{ij}(x_i, x_j) + \frac{1}{4} \sum_{i=1}^{q} \sum_{j=1, j\neq i}^{q} \mathcal{A}_{(i,j)} \|y_i - y_j\|_2^2$$
$$+ \frac{1}{2}\|x_{\mathrm{c}}\|_2^2 + \frac{1}{2}\|z\|_2^2 + \frac{1}{2}\|w\|_2^2, \tag{2.53}$$

where

$$x \triangleq [x_1^{\mathrm{T}}, \ldots, x_q^{\mathrm{T}}]^{\mathrm{T}},\ y \triangleq [y_1^{\mathrm{T}}, \ldots, y_q^{\mathrm{T}}]^{\mathrm{T}},\ z \triangleq [z_1, \ldots, z_q]^{\mathrm{T}},\ w \triangleq [w_1^{\mathrm{T}}, \ldots, w_q^{\mathrm{T}}]^{\mathrm{T}},$$

and

$$x_{\mathrm{c}} \triangleq [x_{\mathrm{c}1}, \ldots, x_{\mathrm{c}q}]^{\mathrm{T}}, \quad i = 1, \ldots, q.$$

The derivative of $V(x, y, z, w, x_{\mathrm{c}})$ along the trajectories of the closed-loop system (2.46)–(2.52) is given by

$$\dot{V}(x, y, z, w, x_{\mathrm{c}}) = -\sum_{i=1}^{q-1} \sum_{j=i+1}^{q} \mathcal{A}_{(i,j)}(x_{\mathrm{c}i} - x_{\mathrm{c}j})^2$$
$$-\sum_{i=1}^{q-1} \sum_{j=i+1}^{q} \mathcal{A}_{(i,j)}(w_i - w_j)^{\mathrm{T}}(w_i - w_j)$$
$$-\sum_{k\in\{1,\ldots,q\}} \sigma_{kk}(w_k)w_k$$
$$\leq 0, \quad (x, y, z, w, x_{\mathrm{c}}) \in \mathcal{D}. \tag{2.54}$$

Next, let $\mathcal{R} \triangleq \{(x, y, z, w, x_{\mathrm{c}}) \in \mathcal{D} : \dot{V}(x, y, z, w, x_{\mathrm{c}}) = 0\}$. Since Assumptions 2.1 and 2.2 hold, it follows from Proposition 2.1 that

$$\mathcal{R} = \{(x, y, z, w, x_{\mathrm{c}}) \in \mathcal{D} : w_1 = \cdots = w_q = 0,\ x_{\mathrm{c}1} = \cdots = x_{\mathrm{c}q},\ i = 1, \ldots, q\}.$$

Let $\mathcal{M}$ denote the largest invariant set contained in $\mathcal{R}$. Then it follows that on $\mathcal{M}$, $\dot{y}_i = 0$, $i = 1, \ldots, q$. In this case, it follows from (2.52) that $\sum_{j=1, j\neq i}^{q} \mathcal{A}_{(i,j)}(y_i - y_j) = 0$ for all $i = 1, \ldots, q$. Now, it follows from Proposition 2.1 that $y_i = \alpha$, $i = 1, \ldots, q$, where $\alpha \in \mathbb{R}^m$. In addition, since $\sum_{i=1}^{q} \dot{x}_{ci} = 0$ and $x_{c1} = \cdots = x_{cq}$ on $\mathcal{M}$, it follows that $x_{ci} = \gamma$, $i = 1, \ldots, q$, where $\gamma \in \mathbb{R}$, and, hence, it follows from (2.50) that $\sum_{j=1, j \neq i}^{q} \mathcal{A}_{(i,j)}(z_i - z_j) = 0$.

Next, it follows from Proposition 2.1 that $z_1 = \cdots = z_q$ and, hence, $\dot{x}_i - \dot{x}_j = G(y_i)z_i - G(y_j)z_j = 0$. Now,

$$\frac{\mathrm{d}}{\mathrm{d}t}\|z\|_2^2 = -2\sum_{i=1}^{q}\sum_{j=1, j \neq i}^{q}(\dot{x}_i - \dot{x}_j)^{\mathrm{T}}\nabla_{x_i}U_{ij}(x_i, x_j) = 0,$$

which implies that $z_i = \beta$, $i = 1, \ldots, q$, where $\beta \in \mathbb{R}$. Finally, it follows from the Krasovskii–LaSalle invariant set theorem [122, p. 147] that, for every system initial condition in $\mathcal{D}_0$, $(x(t), y(t), z(t), w(t), x_c(t)) \to \mathcal{M}$ as $t \to \infty$. The global result follows using standard arguments. $\square$

It is important to note that Theorems 2.5 and 2.6 can be easily generalized to hybrid protocols using the results developed in Section 20.3 by replacing (2.36), (2.37), and (2.50) with the corresponding hybrid consensus protocol architectures given in Section 20.3.

Finally, we note that the above results can be extended to the case where $\mathfrak{G}$ in Assumption 2.1 is a directed graph. For simplicity of exposition, we only consider $q$ multiagent dynamical systems given by (2.34). For this problem, consider the static controller given by

$$u_i(t) = \sum_{j=1, j \neq i}^{q} \mathcal{A}_{(i,j)}(\dot{x}_j(t) - \dot{x}_i(t))$$

$$- \sum_{j=1, j \neq i}^{q} \mathcal{A}_{(i,j)}[x_i(t) - x_j(t) - (d_i - d_j)], \qquad (2.55)$$

where $d_i \in \mathbb{R}^n$, $i = 1, \ldots, q$.

**Theorem 2.7.** Consider the closed-loop system $\tilde{\mathcal{G}}$ consisting of $q$ mobile agents with dynamics (2.34) and the controller (2.55). Assume that Assumptions 2.1 and 2.2 hold and $\mathcal{A}^{\mathrm{T}}\mathbf{e} = 0$. Then parallel formation is achieved for (2.34) by using the distributed feedback control law (2.55).

**Proof.** The proof follows by using the nonnegative function

$$V(\tilde{x}, \dot{\tilde{x}}, \tilde{x}_c) = \frac{1}{2}\sum_{i=1}^{q}\|x_i - d_i\|_2^2 + \frac{1}{2}\sum_{i=1}^{q}\|\dot{x}_i\|_2^2 \qquad (2.56)$$

and, hence, is omitted. $\square$

## 2.5 Distributed Control for Circular Formations

In this section, we develop dynamic compensators for achieving circular formations [183] involving cyclic pursuit [220]. The proposed controllers have a leaderless distributed architecture which is more robust than static, leader-based control designs [183, 220]. Consider the $q$ mobile autonomous agents in a plane described by the *unicycle model* given by

$$\dot{x}_i(t) = v_i(t) \cos \theta_i(t), \quad x_i(0) = x_{i0}, \quad t \geq 0, \tag{2.57}$$

$$\dot{y}_i(t) = v_i(t) \sin \theta_i(t), \quad y_i(0) = y_{i0}, \tag{2.58}$$

$$\dot{\theta}_i(t) = w_i(t), \quad \theta_i(0) = \theta_{i0}, \tag{2.59}$$

where, for each $i \in \{1, \ldots, q\}$, $[x_i, y_i]^{\mathrm{T}} \in \mathbb{R}^2$ denotes the position vector of the $i$th agent, $\theta_i \in \mathbb{R}$ denotes the orientation of the $i$th agent, $v_i \in \mathbb{R}$ denotes the velocity of the $i$th agent, $w_i \in \mathbb{R}$ denotes the angular velocity of the $i$th agent, and $u_i = [v_i, w_i]^{\mathrm{T}}$ is the control input of the $i$th agent.

For our first result, we assume that the graph $\mathfrak{G}$ of the communication topology for the mobile agents is a *complete graph*; that is, $\mathfrak{G}$ contains all possible edges, and, hence, every agent is connected to every other agent. The control aim is to design $u_i$ so that a circular formation is achieved; that is, the system state asymptotically converges to an invariant manifold characterized by the following constraints:

$$v_i = c_1, \quad w_i = c_2, \quad \sum_{i=1}^{q} \sin \theta_i = 0, \quad \sum_{i=1}^{q} \cos \theta_i = 0, \tag{2.60}$$

$$\sum_{j=1}^{q} (x_i - x_j) \cos \theta_i + (y_i - y_j) \sin \theta_i = 0, \quad i = 1, \ldots, q, \tag{2.61}$$

where $c_1 \in \mathbb{R}$ and $c_2 \in \mathbb{R}$.

Consider the dynamic controller architecture given by

$$\dot{x}_{\mathrm{c}i}(t) = - \sum_{j=1, j \neq i}^{q} (x_{\mathrm{c}i}(t) - x_{\mathrm{c}j}(t))$$

$$+ \sum_{j=1, j \neq i}^{q} [(x_i(t) - x_j(t)) \cos \theta_i(t) + (y_i(t) - y_j(t)) \sin \theta_i(t)],$$

$$x_{\mathrm{c}i}(0) = x_{\mathrm{c}i0}, \quad i = 1, \ldots, q, \quad t \geq 0, \tag{2.62}$$

$$\dot{z}_{\mathrm{c}i}(t) = - \sum_{j=1, j \neq i}^{q} (z_{\mathrm{c}i}(t) - z_{\mathrm{c}j}(t))$$

$$+ \sum_{j=1,j\neq i}^{q} \left( \theta_i(t) - \theta_j(t) - \frac{2(j-i)}{q}\pi \right),$$

$$z_{ci}(0) = z_{ci0}, \quad i = 1, \ldots, q, \quad t \geq 0, \tag{2.63}$$

$$v_i(t) = -x_{ci}(t)$$

$$- \sum_{j=1,j\neq i}^{q} [(x_i(t) - x_j(t))\cos\theta_i(t) + (y_i(t) - y_j(t))\sin\theta_i(t)], \tag{2.64}$$

$$\omega_i(t) = -z_{ci}(t), \tag{2.65}$$

where $x_{ci} \in \mathbb{R}$ and $z_{ci} \in \mathbb{R}$, $i = 1, \ldots, q$.

**Theorem 2.8.** Consider the closed-loop system $\tilde{\mathcal{G}}$ consisting of $q$ mobile agents in a plane with dynamics (2.57)–(2.59) and controller (2.62)–(2.65). Assume that the graph $\mathfrak{G}$ of the communication topology for the mobile agents is complete. Let $\mathcal{D} \subseteq \mathbb{R}^{5q}$ be a compact positively invariant set with respect to the closed-loop system $\tilde{\mathcal{G}}$. Then there exists $\mathcal{D}_0 \subseteq \mathcal{D}$ such that if all the system initial conditions are in $\mathcal{D}_0$, then the state of the closed-loop system $\tilde{\mathcal{G}}$ approaches the largest invariant set $\mathcal{M}$ contained in a manifold characterized by (2.60) and (2.61). If, in addition, $\mathcal{D}_0 = \mathcal{D} = \mathbb{R}^{5q}$, then the result is global.

**Proof.** Consider the nonnegative function

$$V(x, y, \theta, x_c, z_c) = \frac{1}{4}\sum_{i=1}^{q}\sum_{j=1,j\neq i}^{q}[(x_i - x_j)^2 + (y_i - y_j)^2] + \frac{1}{2}\|x_c\|_2^2$$

$$+ \frac{1}{4}\sum_{i=1}^{q}\sum_{j=1,j\neq i}^{q}\left(\theta_i - \theta_j - \frac{2(j-i)}{q}\pi\right)^2 + \frac{1}{2}\|z_c\|_2^2, \tag{2.66}$$

where $x \triangleq [x_1, \ldots, x_q]^{\mathrm{T}}$, $y \triangleq [y_1, \ldots, y_q]^{\mathrm{T}}$, $\theta \triangleq [\theta_1, \ldots, \theta_q]^{\mathrm{T}}$, $x_c \triangleq [x_{c1}, \ldots, x_{cq}]^{\mathrm{T}}$, and $z_c \triangleq [z_{c1}, \ldots, z_{cq}]^{\mathrm{T}}$. The derivative of $V(x, y, \theta, x_c, z_c)$ along the trajectories of the closed-loop system (2.57)–(2.59) and (2.62)–(2.65) is given by

$$\dot{V}(x, y, \theta, x_c, z_c) = -\sum_{i=1}^{q-1}\sum_{j=i+1}^{q}(x_{ci} - x_{cj})^2 - \sum_{i=1}^{q-1}\sum_{j=i+1}^{q}(z_{ci} - z_{cj})^2$$

$$- \left[\sum_{j=1,j\neq i}^{q}(x_i - x_j)\cos\theta_i + (y_i - y_j)\sin\theta_i\right]^2$$

$$\leq 0, \quad (x, y, \theta, x_c, z_c) \in \mathcal{D}. \tag{2.67}$$

Next, let $\mathcal{R} \triangleq \{(x, y, \theta, x_c, z_c) \in \mathcal{D} : \dot{V}(x, y, \theta, x_c, z_c) = 0\}$, and note that

$$\mathcal{R} = \left\{ (x, y, \theta, x_c, z_c) \in \mathcal{D} : x_{c1} = \cdots = x_{cq}, z_{c1} = \cdots = z_{cq}, \right.$$

$$\left. \sum_{j=1, j \neq i}^{q} (x_i - x_j) \cos \theta_i + (y_i - y_j) \sin \theta_i = 0, i = 1, \ldots, q \right\}.$$

Let $\mathcal{M}$ denote the largest invariant set contained in $\mathcal{R}$, and note that

$$\sum_{i=1}^{q} \sum_{j=1, j \neq i}^{q} \left( \theta_i - \theta_j - \frac{2(j-i)}{q} \pi \right) = 0.$$

Then it follows from (2.63) that on $\mathcal{M}$, $z_{ci} = c_1$, and, hence, $\omega_i = -c_1$, where $c_1 \in \mathbb{R}$, $i = 1, \ldots, q$. Now, it follows from (2.63) that

$$\sum_{j=1, j \neq i}^{q} \left( \theta_i - \theta_j - \frac{2(j-i)}{q} \pi \right) = 0$$

for each $i = 1, \ldots, q$. Hence, it follows from Proposition 2.1 that $\theta_i - \theta_j = \frac{2(j-i)}{q} \pi$ for each $i = 1, \ldots, q$, which further implies that $\sum_{i=1}^{q} \sin \theta_i = 0$ and $\sum_{i=1}^{q} \cos \theta_i = 0$.

Next, since $x_{c1} = \cdots = x_{cq}$ and

$$\sum_{j=1, j \neq i}^{q} (x_i - x_j) \cos \theta_i + (y_i - y_j) \sin \theta_i = 0$$

for each $i = 1, \ldots, q$ on $\mathcal{M}$, it follows from (2.62) that on $\mathcal{M}$, $x_{ci} = c_2$, $i = 1, \ldots, q$, where $c_2 \in \mathbb{R}$, and, hence, it follows from (2.64) that on $\mathcal{M}$, $v_i = -c_2$, $i = 1, \ldots, q$. Finally, it follows from the Krasovskii–LaSalle invariant set theorem [122, p. 147] that, for every system initial condition in $\mathcal{D}_0$, $(x(t), y(t), \theta(t), x_c(t), z_c(t)) \to \mathcal{M}$ as $t \to \infty$. The global result follows using standard arguments. $\qquad \square$

Next, we generalize the above result by relaxing the completeness condition on the graph $\mathfrak{G}$ of the communication topology for the mobile agents (2.57)–(2.59). Specifically, we assume that the graph $\mathfrak{G}$ satisfies Assumptions 2.1 and 2.2, $\mathcal{A}^T e = 0$, and assume that the circular center is fixed and is given by $(R_x, R_y)$. In this case, the invariant manifold is given by

$$v_i = c_1, \quad \omega_i = c_2, \quad (x_i - R_x) \cos \theta_i + (y_i - R_y) \sin \theta_i = 0,$$
$$i = 1, \ldots, q. \quad (2.68)$$

Consider the dynamic controller given by

$$\dot{x}_{ci}(t) = - \sum_{j=1,j\neq i}^{q} \mathcal{A}_{(i,j)}(x_{ci}(t) - x_{cj}(t)) + (x_i(t) - R_x)\cos\theta_i(t)$$
$$+ (y_i(t) - R_y)\sin\theta_i(t), \quad x_{ci}(0) = x_{ci0}, \quad t \geq 0, \tag{2.69}$$

$$\dot{z}_{ci}(t) = - \sum_{j=1,j\neq i}^{q} \mathcal{A}_{(i,j)}(z_{ci}(t) - z_{cj}(t))$$
$$+ \sum_{j=1,j\neq i}^{q} \mathcal{A}_{(i,j)}\left(\theta_i(t) - \theta_j(t) - \frac{2(j-i)}{q}\pi\right), \quad z_{ci}(0) = z_{ci0}, \tag{2.70}$$

$$v_i(t) = -x_{ci}(t) - [(x_i(t) - R_x)\cos\theta_i(t) + (y_i(t) - R_y)\sin\theta_i(t)], \tag{2.71}$$

$$\omega_i(t) = -z_{ci}(t), \tag{2.72}$$

where $x_{ci} \in \mathbb{R}$ and $z_{ci} \in \mathbb{R}$, $i = 1, \ldots, q$.

**Theorem 2.9.** Consider the closed-loop system $\tilde{\mathcal{G}}$ consisting of $q$ mobile agents in a plane with dynamics (2.57)–(2.59) and controller (2.69)–(2.72). Assume that the graph $\mathfrak{G}$ of the communication topology for the mobile agents (2.57)–(2.59) satisfies Assumptions 2.1 and 2.2, and assume that $\mathcal{A}^{\mathrm{T}}\mathbf{e} = 0$. Let $\mathcal{D} \subseteq \mathbb{R}^{5q}$ be a compact positively invariant set with respect to the closed-loop system $\tilde{\mathcal{G}}$. Then there exists $\mathcal{D}_0 \subseteq \mathcal{D}$ such that if all of the system initial conditions are in $\mathcal{D}_0$, then the state of the closed-loop system $\tilde{\mathcal{G}}$ approaches the largest invariant set $\mathcal{M}$ contained in a manifold characterized by (2.68). If, in addition, $\mathcal{D}_0 = \mathcal{D} = \mathbb{R}^{5q}$, then the result is global.

**Proof.** The proof is similar to the proof of Theorem 2.8 by considering the nonnegative function

$$V(x, y, \theta, x_c, z_c) = \frac{1}{2}\sum_{i=1}^{q}[(x_i - R_x)^2 + (y_i - R_y)^2]$$
$$+ \frac{1}{4}\sum_{i=1}^{q}\sum_{j=1,j\neq i}^{q}\mathcal{A}_{(i,j)}\left(\theta_i - \theta_j - \frac{2(j-i)}{q}\pi\right)^2$$
$$+ \frac{1}{2}\|x_c\|_2^2 + \frac{1}{2}\|z_c\|_2^2 \tag{2.73}$$

and, hence, is omitted. $\qquad\qquad\qquad\qquad\qquad\qquad\qquad\qquad\qquad\square$

**Example 2.3.** In this example, we consider the cyclic pursuit problem with the multiagent dynamical system given by (2.57)–(2.59). Let $q = 10$ and $\mathcal{A}_{(i,i+1)} = \mathcal{A}_{(i+1,i)} = \mathcal{A}_{(i,i+2)} = \mathcal{A}_{(i+2,i)} = 1$, $i = 1, \ldots, 8$. Every other entry in $\mathcal{A}$ is zero. Here, we choose $R_x = R_y = 0$. A group of 10 agents is initialized with random initial positions $(x, y)$ in the range of $[-10, 20] \times [-10, 10]$. The initial states of the controller (2.69) and (2.70) are chosen randomly within $[-10, 10]$.

Figure 2.7 illustrates circular formations using the dynamic controller (2.69)–(2.72). Figures 2.8 and 2.9 show the time histories of the velocities and orientations for circular formation design. $\triangle$

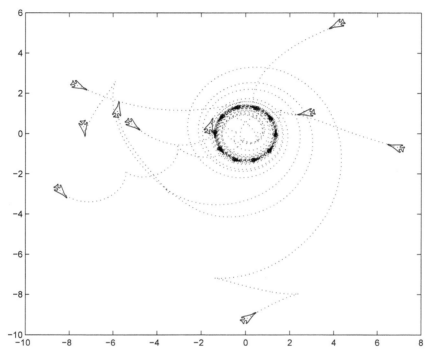

Figure 2.7 Circular formation design.

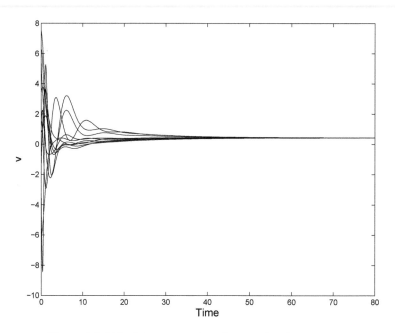

Figure 2.8 Velocities versus time for circular formations.

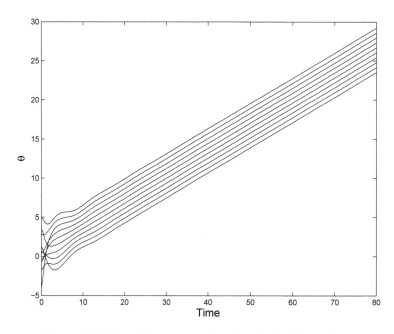

Figure 2.9 Orientations versus time for circular formations.

*Chapter Three*

---

# Thermodynamics-Based Control of Network Systems

## 3.1 Introduction

For a network of interconnected dynamical systems, it is often desired that some property of each subsystem approach a single common value across the network. For example, in a group of autonomous vehicles, this property might be a common heading angle or a shared communication frequency. As discussed in Chapter 2, designing a controller that ensures that a common value will be found is called the *consensus control problem* [224]. Achieving consensus with distributed controllers that can access only local information is called the *distributed consensus control problem*. Related topics include rendezvous, synchronization, flocking, and cyclic pursuit [224]. These topics arise in a broad variety of important applications, including cooperative control of unmanned air vehicles (UAVs), microsatellite clusters, mobile robots, and congestion control in communication networks, and will be addressed in later chapters.

A sizable body of work has emerged in recent years that addresses the distributed consensus problem using the tools of algebraic graph theory [56, 71, 93, 224, 236, 238]. In this chapter, we present an alternative perspective on the distributed consensus problem, based on *system thermodynamics*, a framework that unifies the foundational disciplines of thermodynamics and dynamical systems theory. System thermodynamics has been applied to achieve the formulation of classical thermodynamics in a dynamical systems setting [127]. System thermodynamics has also been used to apply thermodynamic principles to the analysis, design, and control of dynamic systems [125, 131, 133, 134].

To illustrate the relevance of thermodynamics to consensus control, consider conductive heat flow in a homogeneous, isotropic, thermally insulated body. Heat flows within the body according to Fourier's law of heat conduction, which states that the rate of heat flow $q$ through an area $A$ is proportional to the temperature gradient $\nabla T$. The constant of proportionality is an intrinsic material property called the *thermal conductivity* and is denoted by $\kappa$. Hence, $q = -\kappa A \nabla T$. One consequence of the second law of thermodynamics is that $\kappa$ cannot be negative; another is that heat will flow until the temperature of the body is uniform.

If the system is perturbed by local addition or removal of heat, then the temperature distribution will respond by stabilizing at a new uniform value. That is, the natural flow of heat under Fourier's law robustly maintains a globally stable "temperature consensus" in response to system disturbances and system uncertainty. Just as intuitive notions of energy and dissipation can guide controller design using Lyapunov- or passivity-based methods [122], so too can the laws of thermodynamics be abstracted and generalized to guide controller design for consensus control problems in networked systems [162].

This chapter develops a system thermodynamic framework for the distributed consensus control problem on static, finite-dimensional, undirected, and directed networks of first-order systems. This is a somewhat restricted class of problems, which has received extensive attention using other methods. However, system thermodynamics proves extremely effective in designing controllers for such systems, and the intuition provided by the thermodynamic analogies points the way toward extensions to a broader class of problems. The main objective of this chapter is to further develop the connection between fundamental thermodynamic principles and the control of network systems alluded to in Chapter 2.

## 3.2 Energy, Entropy, and Thermal Equilibria

In this section, we present definitions of internal energy, entropy, and temperature suitable for the analysis of a restricted class of thermodynamic systems. Specifically, we consider a network of $n$ lumped thermal masses interconnected by links along which heat can flow from one subsystem to another. We refer to an individual subsystem as $\mathcal{X}_i$, or simply subsystem $i$, and denote the composite system by $\mathcal{X} = \cup_{i=1}^{n} \mathcal{X}_i$.

*Internal energy*, or simply *energy*, may be thought of as the potential of a system to perform work on other systems and on the environment. Energy may be transferred between systems by mass transfer, heat transfer, or work. Subsequently, in this presentation, energy is transferred only as heat. We denote the internal energy of subsystem $i$ by $U_i$ and the vector of subsystem energies by $\boldsymbol{U} = [U_1, \ldots, U_n]^{\mathrm{T}}$. The total internal energy of the system is the sum of the subsystem energies and is given by

$$U_{\mathcal{X}} = \sum_{i=1}^{n} U_i.$$

An *isolated* system does not interchange energy with any other system or the environment. However, energy can be exchanged between the subsystems making up an isolated system.

*Entropy* is a measure of how well energy is distributed throughout a system. We denote the entropies of subsystem $i$ by $S_i$ and the vector of subsystem entropies by $\boldsymbol{S} = [S_1, \ldots, S_n]^{\mathrm{T}}$. The system entropy is the sum of the subsystem entropies and is given by

$$S_{\mathcal{X}} = \sum_{i=1}^{n} S_i.$$

A higher system entropy indicates a more uniform distribution of energy among subsystems; the subsystem entropies themselves have no meaning in isolation. For a fixed amount of total energy, system entropy is maximal when the energy is *equipartitioned*, that is, when every energy storage mode has equal energy. For details, see [127].

In classical thermodynamics, *temperature* is well defined only for a system or subsystem in equilibrium. Subsystem $i$ is assumed to be in internal thermal equilibrium, with corresponding temperature $T_i \geq 0$. We denote the vector of subsystem temperatures by $\boldsymbol{T} = [T_1, \ldots, T_n]^{\mathrm{T}}$. Any addition or removal of heat is assumed to be sufficiently slow that each subsystem remains in internal equilibrium. Since subsystems are generally not in equilibrium with each other, there is generally no meaning of "system temperature." When the subsystems are all at a single common temperature $\bar{T}$, then we say the system is in thermal equilibrium with temperature $\bar{T}$. We write $\bar{\boldsymbol{T}}$ for the temperature vector of a system in thermal equilibrium, that is, $\bar{\boldsymbol{T}} = \mathbf{e}\bar{T}$, where $\mathbf{e} \triangleq [1, \ldots, 1]^{\mathrm{T}}$ is the $n$-dimensional ones vector. Here we distinguish the notion of *thermal equilibrium*, meaning the condition of an isolated system at a uniform temperature, from the notion of *dynamic equilibrium*, meaning the condition of a dynamic system with time rate of change equal to zero.

Energy, entropy, and temperature are related by the *fundamental thermodynamic relationship* [48, 188]. In the absence of mechanical work, this relationship can be written as [127]

$$T_i = \left( \frac{\partial S_i}{\partial U_i} \right)^{-1}. \tag{3.1}$$

We will define entropy as a function of energy, that is, $S_i(U_i)$ and $S_{\mathcal{X}}(\boldsymbol{U})$. It will also be convenient to write the entropies as functions of temperature, using $U_i(T_i)$. Thus, we define $\boldsymbol{U}(\boldsymbol{T}) \triangleq [U_1(T_1), \ldots, U_n(T_n)]^{\mathrm{T}}$, $\tilde{S}_i(T_i) \triangleq S_i(U_i(T_i))$, and $\tilde{S}_{\mathcal{X}}(\boldsymbol{T}) \triangleq S_{\mathcal{X}}(\boldsymbol{U}(\boldsymbol{T})) = \sum_{i=1}^{n} \tilde{S}_i(T_i)$.

Let $\mathrm{d}Q_i$ be an infinitesimal amount of heat received by subsystem $i$. Since energy is assumed to be transferred only as heat, $\mathrm{d}U_i = \mathrm{d}Q_i$. The infinitesimal change in entropy that accompanies this heat addition is

$$\mathrm{d}S_i = \left( \frac{\partial S_i}{\partial U_i} \right) \mathrm{d}U_i = \frac{\mathrm{d}Q_i}{T_i}, \qquad (3.2)$$

where subsystem $\mathcal{X}_i$ is in equilibrium at temperature $T_i$. In terms of heat flow rates, (3.2) can be written as

$$\frac{\mathrm{d}S_i}{\mathrm{d}t} = \frac{1}{T_i} \frac{\mathrm{d}Q_i}{\mathrm{d}t} = \frac{q_i}{T_i}, \qquad (3.3)$$

where $q_i \triangleq \dot{Q}_i$. Here, the time rate of change is assumed to be sufficiently slow that the system remains in a slowly varying state of equilibrium. This state is sometimes referred to as a *quasi-equilibrium* state [118].

In this chapter, we use the following versions of the zeroth, first, and second laws of thermodynamics. For further details on the laws of thermodynamics, see [118].

**Zeroth Law of Thermodynamics.** If two subsystems are individually in thermal equilibrium with a third subsystem, then the two subsystems are also in thermal equilibrium with each other.

**First Law of Thermodynamics.** The increase in the internal energy of a subsystem is equal to the heat supplied to the subsystem.[3] The internal energy of an isolated system is constant.

**Second Law of Thermodynamics.** The entropy of an isolated system does not decrease.

## 3.3 Heat Transfer between Pairs of Subsystems

We begin by considering heat transfer between a pair of subsystems. Let $\mathrm{d}Q_{ij}$ denote the heat transferred from subsystem $j$ to subsystem $i$, and let $q_{ij} \triangleq \dot{Q}_{ij}$ denote the rate of flow of heat from subsystem $j$ to subsystem $i$. We denote the corresponding change (respectively, rate of change) in

---

[3]This version of the first law holds only when work performed by the system on the environment, and work done by the environment on the system, are assumed to be zero.

internal energy and entropy as $\mathrm{d}U_{ij}$ and $\mathrm{d}S_{ij}$ (respectively, $\dot{U}_{ij}$ and $\dot{S}_{ij}$). It is intuitive to think of two subsystems being linked pairwise by a link that conducts heat but does not store it. This notion is equivalent to pairwise energy conservation, that is, $q_{ji} = -q_{ij}$. If each subsystem is in quasi equilibrium, then the second law of thermodynamics implies

$$\dot{S}_{\mathcal{X}} = \sum_{i=1}^{n} \dot{S}_i = \sum_{i=1}^{n}\sum_{j=1}^{n} \dot{S}_{ij} \geq 0.$$

Consider first the case where heat transfer is restricted to be between a single pair of subsystems, namely, subsystem $i$ and subsystem $j$. This is the case when the system consists of only two subsystems, or when only two subsystems are physically connected. Then the total entropy change is given by

$$\mathrm{d}S_{\mathcal{X}} = \mathrm{d}S_i + \mathrm{d}S_j = \frac{\mathrm{d}Q_{ij}}{T_i} - \frac{\mathrm{d}Q_{ij}}{T_j} = \left(\frac{T_j - T_i}{T_i T_j}\right)\mathrm{d}Q_{ij}$$

or, equivalently, in terms of time rate of change,

$$\dot{S}_{\mathcal{X}} = \left(\frac{T_j - T_i}{T_i T_j}\right)q_{ij}.$$

By the second law of thermodynamics, $\mathrm{d}S_{\mathcal{X}} \geq 0$ or $\dot{S}_{\mathcal{X}} \geq 0$, implying $\mathrm{sign}(\mathrm{d}Q_{ij}) = \mathrm{sign}(T_j - T_i)$ or $\mathrm{sign}(q_{ij}) = \mathrm{sign}(T_j - T_i)$, where $\mathrm{sign}(\sigma) \triangleq \sigma/|\sigma|$, $\sigma \neq 0$, and $\mathrm{sign}(0) \triangleq 0$. Thus, for heat exchange between subsystem pairs, the second law of thermodynamics implies that if any heat is transferred, then it must move from the subsystem with the higher temperature to the subsystem with the lower temperature.

**Definition 3.1.** The heat transfer law between a pair of subsystems $i$ and $j$ is of *symmetric Fourier type* if it has the form

$$q_{ij} = \alpha_{ij}(T_j - T_i), \tag{3.4}$$

where $\alpha_{ij}(\cdot)$ is a function satisfying (i) the *sector bound condition*,

$$\delta_1 \leq \frac{\alpha_{ij}(\xi)}{\xi} \leq \delta_2, \quad \xi \neq 0, \tag{3.5}$$

with $\alpha(0) = 0$ and $0 < \delta_1 \leq \delta_2$, and (ii) the *pairwise symmetry condition*,

$$\alpha_{ji}(\xi) = -\alpha_{ij}(-\xi). \tag{3.6}$$

The linear form of Fourier's law, in which $q_{ij} = k_{ij}(T_j - T_i)$ and $q_{ji} = k_{ij}(T_i - T_j)$, is a heat transfer law of symmetric Fourier type. For the heat transfer law between two subsystems to be of symmetric Fourier type, the subsystems must be connected; the "heat transfer law" for an unconnected pair of subsystems, $q_{ij} = q_{ji} = 0$, violates the first inequality in the sector bound (3.5).

If the heat transfer law between a pair of subsystems $i$ and $j$ is of symmetric Fourier type, then the energy transfer between those subsystems automatically satisfies the first law of thermodynamics and the second law of thermodynamics. To see that the first law of thermodynamics is satisfied, note that the rate of change of the total internal energy of the subsystem pair is given by

$$
\begin{aligned}
\dot{U}_{ij} + \dot{U}_{ji} &= \alpha_{ij}(T_j - T_i) + \alpha_{ji}(T_i - T_j) \\
&= \alpha_{ij}(T_j - T_i) - \alpha_{ij}(T_j - T_i) \\
&= 0,
\end{aligned}
\tag{3.7}
$$

where the second equality follows from the symmetry condition (3.6).

To see that the second law of thermodynamics is satisfied, note that the rate of change of the total entropy of the subsystem pair is given by

$$
\begin{aligned}
\dot{S}_{ij} + \dot{S}_{ji} &= \frac{\alpha_{ij}(T_j - T_i)}{T_i} + \frac{\alpha_{ji}(T_i - T_j)}{T_j} \\
&= \left( \frac{1}{T_i} - \frac{1}{T_j} \right) \alpha_{ij}(T_j - T_i) \\
&= \frac{(T_j - T_i)}{T_i T_j} \alpha_{ij}(T_j - T_i) \\
&\geq 0,
\end{aligned}
\tag{3.8}
$$

where the second equality follows from the symmetry condition (3.6) and the inequality follows from the sector bound condition (3.5). The sector bound condition also implies that equality holds in (3.8) if and only if $T_i = T_j$.

If every pair of subsystems in a system either is disconnected or obeys a heat transfer law of symmetric Fourier type, then the first law of thermodynamics and the second law of thermodynamics will be satisfied at the system level. To see that the first law of thermodynamics is satisfied, consider the rate of change of the total internal energy of the system given by

$$
\dot{U}_{\mathcal{X}} = \sum_{i=1}^{n} \sum_{j=1}^{n} \dot{U}_{ij} = \sum_{i=1}^{n} \sum_{j=i+1}^{n} \left( \dot{U}_{ij} + \dot{U}_{ji} \right) = 0,
\tag{3.9}
$$

where the second summand is zero due to (3.7) if the subsystem pair $(i, j)$ is connected, as well as to the fact that $\dot{U}_{ij} = -\dot{U}_{ji} = 0$ if the subsystem pair $(i, j)$ is disconnected.

To see that the second law of thermodynamics is satisfied, consider the rate of change of the total entropy of the system given by

$$\dot{S}_{\mathcal{X}} = \sum_{i=1}^{n} \sum_{j=1}^{n} \dot{S}_{ij}$$

$$= \sum_{i=1}^{n} \sum_{j=i+1}^{n} \left( \dot{S}_{ij} + \dot{S}_{ji} \right)$$

$$\geq 0, \tag{3.10}$$

where the second summand is nonnegative due to (3.8) if the subsystem pair $(i, j)$ is connected, as well as to the fact that $\dot{S}_{ij} = \dot{S}_{ji} = 0$ if the subsystem pair $(i, j)$ is disconnected. Equality in (3.10) holds if and only if every connected pair of subsystems is at the same temperature.

## 3.4 Interconnection Structure Using Graphs

The interconnection of the thermal subsystems strongly influences the equilibrium properties of the system. For example, if the system consists of two or more disjoint sets of subsystems, then these decoupled components will generally not be in thermal equilibrium. We use the framework of graph theory to describe the interconnection structure of the subsystems [15, 102]. Every subsystem is a vertex of a directed graph $\mathfrak{G}$, with *vertices* $\mathcal{V}_{\mathfrak{G}}$ and *directed edges*, or *arcs*, $\mathcal{E}_{\mathfrak{G}}$. The total number of vertices in $\mathfrak{G}$ is denoted by $n_{\mathfrak{G}}$. Arcs are written as ordered pairs $(j, i)$. The arc $(j, i)$ is said to *initiate* at $j$ and *terminate* at $i$. Nodes $j$ and $i$ are called the *tail* and *head*, respectively, of the arc $(j, i)$. Loops are explicitly forbidden in the graphs considered here; that is, there are no arcs of the form $(i, i)$.

The total number of arcs in $\mathfrak{G}$ terminating at node $i$ is the *in-degree* of $i$, denoted by $d_{\mathfrak{G}}^{-}(i)$, and the total number of arcs in $\mathfrak{G}$ initiating at node $i$ is the *out-degree* of $i$ in $\mathfrak{G}$, denoted by $d_{\mathfrak{G}}^{+}(i)$. If node $j$ is the tail of an arc that terminates at $i$, we say that $j$ is a *direct predecessor* of $i$, and if node $j$ is the head of an arc that initiates at $i$, we say that $j$ is a *direct successor* of $i$ in $\mathfrak{G}$. We denote the set of all direct predecessors of $i$ in $\mathfrak{G}$ by $\mathcal{P}_{\mathfrak{G}}(i)$, and we denote the set of all direct successors of $i$ in $\mathfrak{G}$ by $\mathcal{S}_{\mathfrak{G}}(i)$. That is, $\mathcal{P}_{\mathfrak{G}}(i) \triangleq \{j : (j, i) \in \mathcal{E}_{\mathfrak{G}}\}$ and $\mathcal{S}_{\mathfrak{G}}(i) \triangleq \{j : (i, j) \in \mathcal{E}_{\mathfrak{G}}\}$.

A *strong path* in $\mathfrak{G}$ is an ordered sequence of arcs from $\mathcal{E}_{\mathfrak{G}}$ such that the head of any arc is the tail of the next. A strong path in $\mathfrak{G}$ can also be considered as a directed subgraph of $\mathfrak{G}$. A *strong cycle* is a strong path that begins and ends at the same vertex. Every strong cycle $\mathcal{C}$ satisfies $d_{\mathcal{C}}^{-}(i) = d_{\mathcal{C}}^{+}(i)$ for $i \in \mathcal{V}_{C}$. Vertices may appear in a strong cycle more than once; that is, $d_{\mathcal{C}}^{-}(i) = d_{\mathcal{C}}^{+}(i) \geq 1$ for $i \in \mathcal{V}_{C}$. If no vertex appears more than once, then the cycle is a *simple strong cycle*. For a simple strong cycle, $d_{\mathcal{C}}^{-}(i) = d_{\mathcal{C}}^{+}(i) = 1$ for $i \in \mathcal{V}_C$. A graph is *strongly connected* if a strong path exists from any vertex to any other vertex.

A property of a strongly connected graph is that it must contain a strong cycle that passes through every vertex in the graph at least once; we call such a cycle a *complete strong cycle*. A complete strong cycle need not include every edge. A strongly connected graph may contain more than one complete strong cycle. A complete strong cycle that contains each vertex exactly once is called a *simple complete strong cycle*. Every simple complete strong cycle satisfies $d_{\mathcal{C}}^{-}(i) = d_{\mathcal{C}}^{+}(i) = 1$ for $i \in \mathcal{V}_G$. The nodes of a simple complete strong cycle $\mathcal{C}$ can be renumbered so that $\mathcal{P}_{\mathcal{C}}(i) = \{i - 1\}$ and $\mathcal{S}_{\mathcal{C}}(i - 1) = \{i\}$ for $i = 1, \ldots, n_{\mathfrak{G}}$, where, for notational convenience, node zero is identified with node $n_{\mathfrak{G}}$.

The adjacency matrix in this chapter is denoted by $G \triangleq [g_{ij}]$, where $g_{ij} = 1$ if there is an edge initiating at $j$ and terminating at $i$, that is, if $(j, i) \in \mathcal{E}_{\mathfrak{G}}$. Otherwise, $g_{ij} = 0$. By assumption, $[g_{ii}] = 0$. Though all edges are directed, we say that the system graph is *undirected* if $g_{ji} = g_{ij}$, that is, if $G$ is symmetric. Otherwise, the system graph is said to be *directed*.

Heat transfer on an undirected graph can be considered pairwise on edges $(i, j)$ and $(j, i)$. Thus, the first law of thermodynamics and the second law of thermodynamics are automatically satisfied for a heat transfer law of symmetric Fourier type on an undirected graph. A linear heat flow law $q_{ij} = k_{ij}(T_j - T_i)$ is associated with the *weighted adjacency matrix* $K = [k_{ij}]$, which we also call the *thermal conductance matrix*. The symmetry condition implies that $K = K^{\mathrm{T}}$ for the linear form of Fourier's law.

## 3.5 Thermal Equilibria and Semistability

Consider a network of $n$ subsystems, each with thermal mass $M_i$, temperature $T_i$, energy $U_i$, and entropy $S_i$. Let the interconnection structure be defined by the graph $\mathfrak{G}$. The system is in thermal equilibrium if and only if $\boldsymbol{T} = \bar{\boldsymbol{T}} = \mathbf{e}\bar{T}$ for every $\bar{T} > 0$. In the terminology of dynamical systems theory, every thermal equilibrium is a *nonisolated* equilibrium point, since every thermal equilibrium with a slightly perturbed uniform temperature will also be an equilibrium point in the dynamical systems sense. Thermal systems have the property that, after a small perturbation, the system will return to thermal equilibrium, though typically at a slightly different temperature. This property is desirable for the distributed consensus control problem.

As depicted in Chapter 1, in terms of concepts from dynamical systems theory, neither Lyapunov stability nor asymptotic stability captures this behavior. The relevant concept is that of semistability. A detailed discussion of semistability is given in Chapter 4. In this chapter, we require the following definitions for an equilibrium point $\bar{x}$ of an autonomous dynamical system

$$\dot{x}(t) = f(x(t)), \quad x(0) = x_0, \quad t \geq 0, \tag{3.11}$$

where $f : \mathcal{D} \subseteq \mathbb{R}^n \to \mathbb{R}^n$. We denote the solution to this system with initial condition $x(0) = x_0 \in \mathcal{D}$ by $x(t; x_0)$.

**Definition 3.2.** An equilibrium point $\bar{x} \in \mathcal{D}$ of (3.11) is *Lyapunov stable* if, for every $\varepsilon > 0$, there exists $\delta_1(\varepsilon) > 0$ such that $\|x_0 - \bar{x}\| < \delta_1(\varepsilon)$ implies $\|x(t; x_0) - \bar{x}\| < \varepsilon$ for all $x_0 \in \mathcal{D}$ and $t \geq 0$.

**Definition 3.3.** An equilibrium point $\bar{x} \in \mathcal{D}$ of (3.11) is *semistable* if it is Lyapunov stable and if there exists $\delta_2 > 0$ such that, for all $x_0 \in \mathcal{D}$ satisfying $\|x_0 - \bar{x}\| < \delta_2$, $x(t; x_0)$ converges to a Lyapunov stable equilibrium point in $\mathcal{D}$, which need not be $\bar{x}$. A set of equilibrium points is semistable if every point in the set is semistable.

**Definition 3.4.** An equilibrium point $\bar{x} \in \mathcal{D}$ of (3.11) is *asymptotically stable* if it is Lyapunov stable and if there exists $\delta_3 > 0$ such that $\|x_0 - \bar{x}\| < \delta_3$ implies $x(t; x_0) \to \bar{x}$ as $t \to \infty$. An equilibrium point $\bar{x}$ is *asymptotically stable on a set* $\mathcal{W} \subset \mathcal{D}$ if $\bar{x}$ is asymptotically stable and $x_0 \in \mathcal{W}$ implies $x(t; x_0) \to \bar{x}$ as $t \to \infty$.

Clearly, semistability is a stronger property than Lyapunov stability but a weaker property than asymptotic stability. Nonisolated equilibrium points can be semistable but not asymptotically stable. For a *set* of nonisolated equilibria, asymptotic stability of the set and semistability of the set are independent properties [122]. Figure 3.1(a) shows a Lyapunov stable nonisolated equilibrium point. A trajectory starting nearby is guaranteed to remain nearby, but it could oscillate forever without converging. Figure 3.1(b) shows a semistable nonisolated equilibrium point. The trajectory converges to a nearby Lyapunov stable equilibrium point.

We now proceed to analyze the stability of the thermal equilibria of a thermodynamic system. First, we define the entropy of the $i$th subsystem by

$$S_i(U_i) \triangleq M_i \ln(U_i). \tag{3.12}$$

Using (3.1), it follows that

$$T_i = \left(\frac{\partial S_i}{\partial U_i}\right)^{-1} = \frac{U_i}{M_i}, \tag{3.13}$$

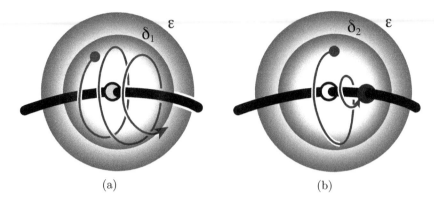

(a)                                                                      (b)

Figure 3.1 (a) Lyapunov stable nonisolated equilibrium point (hollow). The perturbed
trajectory need not converge to a new equilibrium. (b) Semistable nonisolated equilibrium point (hollow). Semistability guarantees convergence of the
perturbed trajectory to a nearby equilibrium point (filled) and is a stronger
property than Lyapunov stability.

which gives the familiar equation $U_i = M_i T_i$ relating the energy and temperature of a lumped thermal mass. With this relationship, the subsystem
and system entropies can be written directly as a function of temperature,
namely,

$$\tilde{S}_i(T_i) \triangleq S_i(M_i T_i) = M_i \ln(M_i T_i), \quad i = 1, \ldots, n,$$

and

$$\tilde{S}_{\mathcal{X}}(\boldsymbol{T}) \triangleq S_{\mathcal{X}}(M_1 T_1, \ldots, M_n T_n) = \sum_{i=1}^{n} \tilde{S}_i(T_i).$$

We refer to $S_{\mathcal{X}}(\boldsymbol{U})$ or $\tilde{S}_{\mathcal{X}}(\boldsymbol{T})$ as the *total entropy function*. Other forms
of the entropy function may be chosen, as in [118] and Chapter 2; however,
the features of those functions will not be required here. One consequence of
choosing (3.12) is that it does not satisfy the third law of thermodynamics
(Nernst's theorem), which states that the entropy is zero when the absolute
temperature is zero [127]. This is not a significant drawback for the purposes
of this chapter.

We first consider isolated systems, for which the total system energy
is conserved. The set of feasible subsystem temperatures corresponding to
a constant system energy $U_0$ is given by

$$\mathcal{T}(U_0) \triangleq \left\{ \boldsymbol{T} : T_i \geq 0, \quad i = 1, \ldots, n, \quad \text{and} \quad \sum_{i=1}^{n} M_i T_i = U_0 \right\}. \quad (3.14)$$

The entropy function has the following key property for isolated systems.

**Theorem 3.1.** The total entropy function $\tilde{S}_{\mathcal{X}}(\boldsymbol{T})$, with subsystem entropies $\tilde{S}_i(T_i) = M_i \ln(M_i T_i)$, $i = 1, \ldots, n$, restricted to $\mathcal{T}(U_0)$, has a unique maximum at $\boldsymbol{T}^* = \mathbf{e}T^*$, where $T^* \triangleq U_0/M$ and $M \triangleq \sum_{i=1}^n M_i$. Equivalently, the total entropy function $S_{\mathcal{X}}(\boldsymbol{U})$, restricted to the set

$$\mathcal{U} \triangleq \left\{ \boldsymbol{U} : U_i \geq 0, \quad i = 1, \ldots, n, \quad \text{and} \quad \sum_{i=1}^n U_i = U_0 \right\},$$

has a unique maximum at $\boldsymbol{U}^* = [U_1^*, U_2^*, \ldots, U_n^*]^{\mathrm{T}}$, where $U_i^* \triangleq M_i T^*$.

**Proof.** The entropy of the $i$th subsystem is given by

$$\begin{aligned}
\tilde{S}_i(T_i) &= M_i \ln(M_i T_i) \\
&= M_i \ln(M_i T_i) - M_i \ln(M_i T^*) + M_i \ln(M_i T^*) \\
&= M_i \ln(T_i/T^*) + M_i \ln(M_i T^*).
\end{aligned}$$

The corresponding total system entropy is given by

$$\tilde{S}_{\mathcal{X}}(\boldsymbol{T}) = \ln \left( \frac{T_1^{M_1} T_2^{M_2} \cdots T_n^{M_n}}{T^{*M_1 + M_2 + \cdots + M_n}} \right) + S_{\mathcal{X}}^*,$$

where $S_{\mathcal{X}}^* \triangleq \sum_{i=1}^n M_i \ln(M_i T^*)$ is the value of the entropy function at thermal equilibrium.

Now, writing

$$\left( \frac{T_1^{M_1} T_2^{M_2} \cdots T_n^{M_n}}{T^{*M_1 + M_2 + \cdots + M_n}} \right) = \left( \frac{T_1^{\mu_1} T_2^{\mu_2} \cdots T_n^{\mu_n}}{T^*} \right)^M,$$

where $\mu_i \triangleq M_i/M$, and noting that

$$T^* = U_0/M = (1/M) \sum_{i=1}^n M_i T_i = \sum_{i=1}^n \mu_i T_i,$$

it follows that

$$\frac{T_1^{\mu_1} T_2^{\mu_2} \cdots T_n^{\mu_n}}{T^*} = \frac{T_1^{\mu_1} T_2^{\mu_2} \cdots T_n^{\mu_n}}{\mu_1 T_1 + \mu_2 T_2 + \cdots + \mu_n T_n} \leq 1,$$

where the inequality follows from the generalized power mean inequality [143], which also provides that equality holds if and only if all of the $T_i$ are equal, that is, if and only if $\boldsymbol{T}$ is of the form $\mathbf{e}T$. However, since $T \sum_{i=1}^n M_i = TM = U_0$, the only $\boldsymbol{T}$ of this form that satisfies the energy constraint is $\mathbf{e}T^*$. Hence, $\tilde{S}_{\mathcal{X}}(\boldsymbol{T})$ achieves a maximum value of $S_{\mathcal{X}}^*$ at $\boldsymbol{T} = \boldsymbol{T}^*$.

The form of the theorem with energy as the independent variable follows from the equality of the two forms of the entropy function $S_{\mathcal{X}}(\boldsymbol{U}) =$

$\tilde{S}_{\mathcal{X}}(\boldsymbol{T})$, which implies that $S_{\mathcal{X}}(\boldsymbol{U})$ must have a unique maximum at $\boldsymbol{U}^* = [U_1^*, U_2^*, \ldots, U_n^*]^{\mathrm{T}}$, where $U_i^* = M_i T^*$, $i = 1, \ldots, n$.    □

One might surmise from Theorem 3.1 that the maximum entropy does not occur when energy is equipartitioned, since the $\bar{U}_i$'s are not equal but are instead weighted by the thermal masses $M_i$. However, energy equipartition refers to uniform distribution of energy over all storage modes, and the subsystems with higher thermal mass have a larger number of storage modes. Maximum entropy corresponds to energy equipartition in this sense.

For positive system energy $U_{\mathcal{X}}$, we denote the temperature vector corresponding to the maximum entropy point by

$$\boldsymbol{T}^*(U_{\mathcal{X}}) \triangleq (U_{\mathcal{X}}/M)\mathbf{e}. \tag{3.15}$$

Figure 3.2 depicts $\mathcal{T}(U_{\mathcal{X}})$, which is the set of feasible subsystem temperatures corresponding to system energy $U_{\mathcal{X}}$. We denote the set of feasible temperatures for any positive total system energy, which is the union of $\mathcal{T}(U_{\mathcal{X}})$ for all $U_{\mathcal{X}} \geq 0$, by $\mathcal{T}$. Figure 3.2 also depicts the set $\mathcal{T}^* \triangleq \cup_{U_{\mathcal{X}}>0} \boldsymbol{T}^*(U_{\mathcal{X}})$, that is, the set of all $\boldsymbol{T}^*(U_{\mathcal{X}})$ corresponding to a positive total system energy. Finally, the boundary of the feasible set $\mathcal{T}(U_{\mathcal{X}})$ is given by

$$\partial\mathcal{T}(U_{\mathcal{X}}) = \bigcup_{j=1}^{n} \left\{ \boldsymbol{T} \in \mathcal{T}(U_{\mathcal{X}}) \, : \, T_j = 0 \quad \text{and} \quad \sum_{i=1}^{n} \mu_i T_i = U_{\mathcal{X}}/M \right\}.$$

Since every point on $\partial\mathcal{T}(U_{\mathcal{X}})$ contains at least one $T_i = 0$, it follows that $\lim_{T \to \partial\mathcal{T}} \tilde{S}_{\mathcal{X}}(T) = -\infty$.

The following three properties characterize thermal equilibrium in isolated thermodynamic systems.

**Property 1.** Total system energy is conserved; that is, $U_{\mathcal{X}}(t) = U_0 \triangleq U_{\mathcal{X}}(0)$.

**Property 2.** $\boldsymbol{T}^*(U_0)$ is the unique equilibrium point of the system in $\mathcal{T}(U_0)$. $\boldsymbol{T}^*(U_0)$ is asymptotically stable in $\mathcal{T}(U_0)$.

**Property 3.** Every $\boldsymbol{T}^*(U_0)$ is a nonisolated equilibrium point in $\mathcal{T}$. The set $\mathcal{T}^*$ is semistable. Every trajectory starting in $\mathcal{T}$ converges to a point in $\mathcal{T}^*$.

Next, we state our main theorem for thermodynamic systems satisfying a heat transfer law of symmetric Fourier type.

**Theorem 3.2.** Consider a network of interconnected thermal masses

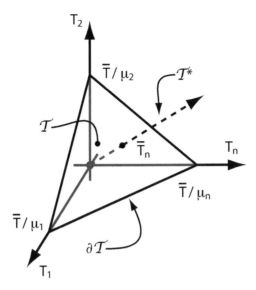

Figure 3.2   The subset of feasible temperatures $\mathcal{T}(U_\mathcal{X})$ corresponding to total system energy $U_\mathcal{X}$.

with a strongly connected and undirected system graph $\mathfrak{G}$. If the energy flow between the subsystems of $\mathfrak{G}$ is governed by a heat transfer law of symmetric Fourier type, then Properties 1–3 hold.

**Proof.** First, recall that a heat transfer law of symmetric Fourier type conserves energy if the network is undirected, and, hence, Property 1 holds. Next, note that the system of thermal masses with an interconnection law of symmetric Fourier type has dynamics

$$\dot{T}_i = \frac{1}{M_i} \sum_{j \in \mathcal{P}_\mathfrak{G}(i)} \alpha_{ij}(T_j - T_i), \quad i = 1, \ldots, n. \tag{3.16}$$

To show Property 2, first note that every $\boldsymbol{T} = \mathbf{e}T$ is an equilibrium point of (3.16). Furthermore, these points are the *only* equilibrium points. To see this, consider any $\boldsymbol{T}$ not of the form $\mathbf{e}T$, and note that there are a finite number of subsystems. Hence, there must exist at least one subsystem with maximum temperature $T_{\max}$. Since the network is strongly connected, at least one of the subsystems at $T_{\max}$, say $T_M$, must have at least one neighbor with a lower temperature. Thus, $\dot{T}_M < 0$, and, hence, this system is not in equilibrium. Finally, since $\boldsymbol{T}^*(U_0) = (U_0/M)\mathbf{e}$ is the only point of the form $\mathbf{e}T$ in $\mathcal{T}(U_0)$, this is the unique equilibrium point.

To show asymptotic stability on $\mathcal{T}(U_0)$, let $\tilde{S}^* = \tilde{S}(\boldsymbol{T}^*(U_0))$. By Theorem 3.1, $\tilde{S}^*$ is a unique maximum for the entropy function, and, hence,

$\tilde{S}(\boldsymbol{T}) - \tilde{S}^* \leq 0$, with equality holding if and only if $\boldsymbol{T} = \boldsymbol{T}^*(U_0)$. The function

$$V_1(\boldsymbol{T}) = (1/2)(\tilde{S}(\boldsymbol{T}) - \tilde{S}^*)^2$$

is zero at $\boldsymbol{T}^*(U_0)$ and positive at every other point in $\mathcal{T}(U_0)$. Furthermore, $\dot{V}_1(\boldsymbol{T})$ is zero at $\boldsymbol{T}^*(U_0)$ and negative at every other point in $\mathcal{T}(U_0)$. All that remains to show asymptotic stability of $\boldsymbol{T}^*(U_0)$ is to shift the origin of $V_1(\boldsymbol{T})$ to $\boldsymbol{T}^*(U_0)$ and show that the set $\mathcal{T}(U_0)$ is forward invariant.

To see this, define

$$V(\boldsymbol{T}') \triangleq (1/2)(\tilde{S}(\boldsymbol{T}' + \boldsymbol{T}^*(U_0)) - \tilde{S}^*)^2$$

and note that $V(\boldsymbol{T}')$ is positive definite on $\mathcal{T}(U_0)$, and further note that $\dot{V}(\boldsymbol{T}')$ is negative definite on $\mathcal{T}(U_0)$. To show forward invariance of the set $\mathcal{T}(U_0)$, consider the portion of the boundary of $\mathcal{T}(U_0)$ corresponding to $T_i = 0$, and take the inner product between the vector $\dot{\boldsymbol{T}}$ and the unit vector $\hat{\mathbf{u}}_i$ in the direction of $T_i$. This inner product is equal to the $i$th component of $\dot{\boldsymbol{T}}$ evaluated at the boundary point; that is,

$$\dot{\boldsymbol{T}}^{\mathrm{T}} \hat{\mathbf{u}}_i = \frac{1}{M_i} \sum_{j \in \mathcal{N}_{\mathfrak{G}}(i)} \alpha_{ij}(T_j - T_i) \Bigg|_{T_i = 0} = \frac{1}{M_i} \sum_{j \in \mathcal{N}_{\mathfrak{G}}(i)} \alpha_{ij}(T_j). \tag{3.17}$$

Now, by the sector bound condition (3.5), $\alpha_{ij}(T_j)/T_j > 0$, and, hence, since $T_j \geq 0$ for $j = 1, \dots, n$, with at least one $T_j > 0$, it follows that $\dot{\boldsymbol{T}}^{\mathrm{T}} \hat{\mathbf{u}}_i > 0$. This implies that the trajectories of the system point into $\mathcal{T}(U_0)$ along the boundary $\partial \mathcal{T}(U_0)$, and, hence, $\mathcal{T}(U_0)$ is an invariant set of (3.16). Thus, we conclude [306] that $\boldsymbol{T}^*(U_0)$ is asymptotically stable in $\mathcal{T}(U_0)$, with region of attraction equal to all of $\mathcal{T}(U_0)$. This completes the proof of Property 2.

To show the first part of Property 3, consider the equilibrium point $\boldsymbol{T}^*(U_0)$ and note that for every $\tau$ that does not result in negative temperatures, every point of the form $\mathbf{e}(T^*(U_0) + \tau)$ is also an equilibrium point of (3.16). Thus, every neighborhood of $\boldsymbol{T}^*(U_0)$ in $\mathcal{T}$ contains another equilibrium point, and, hence, $\boldsymbol{T}^*(U_0)$ is nonisolated.

To show that $\boldsymbol{T}^*(U_0)$ is semistable, we first show that $\boldsymbol{T}^*(U_0)$ is Lyapunov stable. To do this, consider the open ball of radius $\delta_1$ in $\mathcal{T}$ centered on $\boldsymbol{T}^*(U_0)$. Consider any initial point $\boldsymbol{T}'$ in this ball, and denote its energy by $U'$. Since energy is conserved by a heat transfer law of symmetric Fourier type, the resulting trajectory will converge to equilibrium point $\boldsymbol{T}^*(U') = \mathbf{e}(U'/M)$. Note that by (3.13), $U'$ satisfies $|U' - U_0| \leq \bar{M}\delta_1$, where

$\bar{M} \triangleq \max_{i=1,\ldots,n} M_i$. Therefore,

$$\|\boldsymbol{T}^*(U') - \boldsymbol{T}^*(U_0)\| = \|\mathbf{e}(U' - U_0)/M\| = \sqrt{n}|U' - U_0|/M \leq \sqrt{n}\delta_1\bar{\mu},$$

where $\bar{\mu} = \bar{M}/M$.

Next, given any $\varepsilon > 0$, choose $\delta_1 = \varepsilon/(2\sqrt{n}\bar{\mu})$. Now, since $\boldsymbol{T}^*(U')$ is asymptotically stable in $\mathcal{T}(U')$, for every $\varepsilon > 0$, choose $\delta_2 > 0$ such that $\|\boldsymbol{T} - \boldsymbol{T}^*(U')\| < \varepsilon/2$ for all $t > 0$. Then, given any $\varepsilon > 0$, choose $\delta_1$ and $\delta_2$ as above, and require that $\|\boldsymbol{T}' - \boldsymbol{T}(U_0)\| < \min(\delta_1, \delta_2)$, ensuring that

$$\|\boldsymbol{T}(t) - \boldsymbol{T}(U_0)\| \leq \|\boldsymbol{T}(t) - \boldsymbol{T}(U')\| + \|\boldsymbol{T}(U') - \boldsymbol{T}(U_0)\| < \varepsilon.$$

This shows that $\boldsymbol{T}^*(U_0)$ is Lyapunov stable. Semistability of $\boldsymbol{T}^*(U_0)$ now follows immediately from the asymptotic stability of $\boldsymbol{T}^*(U')$ in $\mathcal{T}(U')$ and the Lyapunov stability of $\boldsymbol{T}^*(U')$ in $\mathcal{T}$.

Finally, note that, since the system energy is conserved for every trajectory starting in $\mathcal{T}$, for every initial point in $\mathcal{T}$, the system will converge to a point in $\mathcal{T}^*$, which proves that $\mathcal{T}^*$ is semistable. $\qquad\square$

## 3.6 Thermodynamics of Undirected Networks

By generalizing the notions of temperature, energy, and entropy, system thermodynamics guides the design of distributed consensus controllers that cause networked dynamic systems to emulate natural thermodynamic behavior. On an undirected graph, Theorem 3.2 can be used with minor modifications to show that every control law of symmetric Fourier type achieves energy consensus. In general, it is information, not energy, that flows between subsystems. This is an important distinction, especially when treating directed networks. In the generalized case, energy flows into or out of a subsystem based on the temperatures of its neighbors, but that energy does not actually come from the neighbors, as it would in a natural thermodynamic system.

Consider a network of $n$ single integrators given by

$$\dot{x}_i(t) = u_i(t), \quad x_i(0) = x_{i0}, \quad t \geq 0, \quad i = 1, \ldots, n, \qquad (3.18)$$

where the network has a connectivity structure specified by the *undirected* graph $\mathfrak{G}$. Defining the state vector $x \triangleq [x_1, \ldots, x_n]^{\mathrm{T}}$ and input vector $u \triangleq [u_1, \ldots, u_n]^{\mathrm{T}}$, the system dynamics (3.18) can be written as

$$\dot{x}(t) = u(t), \quad x(0) = x_0, \quad t \geq 0. \qquad (3.19)$$

This system might be obtained by input-output linearization of a network of relative degree-one nonlinear systems. Alternatively, it might arise from a network of feedback-passivated nonlinear systems, in which case the subsystem storage functions can serve as energy functions.

The energy of a subsystem depends only on the state of that subsystem. Thus, we define a continuously differentiable, nonnegative internal energy function $U_i(x_i)$ for each subsystem of (3.18) and let

$$S_i(U_i) \triangleq M_i \ln(U_i), \tag{3.20}$$

$$T_i(U_i) \triangleq U_i/M_i, \tag{3.21}$$

$$\tilde{S}_i(T_i) \triangleq M_i \ln(M_i T_i). \tag{3.22}$$

Furthermore, let $U_{\mathcal{X}}(x) \triangleq \sum_{i=1}^{n} U_i(x_i)$, $S_{\mathcal{X}}(\boldsymbol{U}) \triangleq \sum_{i=1}^{n} S_i(U_i)$, and $\tilde{S}_{\mathcal{X}}(\boldsymbol{T})$ $\triangleq \sum_{i=1}^{n} \tilde{S}_i(T_i)$.

To obtain a controller that emulates the desirable stability properties of a thermodynamic system in thermal equilibrium, we specify that the "heat flow" between pairs of subsystems be governed by a law of symmetric Fourier type. Therefore, we take the time derivative of (3.21) and require that $\dot{U}_i$ be given by the sum of pairwise heat transfer laws of the form (3.4). That is, we require that $\dot{U}_i$ be given by

$$\dot{U}_i = \sum_{j \in \mathcal{P}_{\oplus}(i)} \alpha_{ij}(T_j - T_i), \quad i = 1, \dots, n, \tag{3.23}$$

where $\alpha_{ij}(\cdot)$ is of symmetric Fourier type. Hence, the evolution of the generalized temperature functions for each subsystem is given by

$$\dot{T}_i = \dot{U}_i/M_i = \frac{1}{M_i} \sum_{j \in \mathcal{P}_{\oplus}(i)} \alpha_{ij}(T_j - T_i), \quad i = 1, \dots, n. \tag{3.24}$$

Note that (3.24) is identical to (3.16), so the generalized subsystem temperatures will behave as physical temperatures in a thermodynamic system. It remains to express the control law (3.23) in terms of the original system variables. Expanding the left-hand side of (3.23) gives

$$\dot{U}_i(x_i) = U_i'(x_i)\dot{x}_i = U_i'(x_i)u_i, \quad i = 1, \dots, n. \tag{3.25}$$

Now, (3.23) gives

$$U_i'(x_i)u_i = \sum_{j \in \mathcal{P}_{\oplus}(i)} \alpha_{ij}(T_j - T_i), \quad i = 1, \dots, n, \tag{3.26}$$

which can be solved for $u_i$ to obtain the desired feedback control law

$$u_i(x) = \frac{1}{U_i'(x_i)} \sum_{j \in \mathcal{P}_{\oplus}(i)} \alpha_{ij}(T_j(U_j(x_j)) - T_i(U_i(x_i))), \quad i = 1, \dots, n. \tag{3.27}$$

This control is the basis for our main result on undirected graphs.

**Theorem 3.3.** Consider a network of single integrators with a strongly connected, undirected system graph $\mathfrak{G}$. For each subsystem of $\mathfrak{G}$, define a continuously differentiable nonnegative generalized energy function $U_i(x_i)$, with corresponding generalized entropy $S_i = M_i \ln(U_i)$ and generalized temperature $T_i = U_i/M_i$. If $U_i'(x_i) \neq 0$ for all $x_i$, $i = 1, \ldots, n$, then, under the action of every decentralized control law (3.27), where $\alpha_{ij}(\cdot)$ is of symmetric Fourier type, Properties 1–3 hold.

**Proof.** The proof is identical to that of Theorem 3.2 and, hence, is omitted.                                                                                                        □

Choosing $U_i(x_i)$ to be strictly increasing for $x_i > 0$ guarantees that $U_i'(x_i) \neq 0$ for $x_i > 0$, $i = 1, \ldots, n$. Note that it is not required that the subsystem energy functions be uniform. As an example for the case where $n = 3$, the mixed set of functions $U_1(x_1) = x_1$, $U_2(x_2) = e^{x_2}$, and $U_3(x_3) = x_3^2$ is a valid choice for the subsystem energies.

**Example 3.1.** Consider a network of four single integrators with the undirected graph structure shown in Figure 3.3. We apply a linear control law of symmetric Fourier type to this system for two choices of subsystem energies and temperatures. Specifically, for case 1 we let $U_i(x_i) = x_i$ and $T_i = U_i$, and for case 2 we let $U_i(x_i) = (1/2)x_i^2$ and $T_i = U_i$. Note that for both cases, $M_i = 1$ for all $i$. For the heat transfer law, we choose $\alpha_{ij}(\xi) = g_{ij}\xi$ for both cases, where $g_{ij}$ is the $(i,j)$th entry of the graph connectivity matrix.

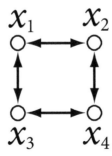

Figure 3.3 Simple four-subsystem network.

For the first simulation, we take

$$u_i(x) = \sum_{j=1}^{4} g_{ij}(x_j - x_i), \quad i = 1, \dots, 4,$$

and

$$\dot{S}_{\mathcal{X}}(x) = \sum_{i=1}^{4} u_i(x)/x_i.$$

This control can be seen to be of the form $u = -Lx$, where $L = D - K$ is the graph Laplacian and $D$ is the weighted degree matrix. This controller is known to solve the distributed consensus problem on an undirected or balanced graph [93, 236, 238]. The response is shown in Figure 3.4(a–c).

For the second simulation, we take

$$u_i(x) = \sum_{j=1}^{4} -(g_{ij}/2x_i)(x_j^2 - x_i^2), \quad i = 1, \dots, 4,$$

and

$$\dot{S}_{\mathcal{X}}(x) = \sum_{i=1}^{4} (2u_i(x)/x_i).$$

The response is shown in Figure 3.5(a–c).                                                                        △

## 3.7 Thermodynamics of Directed Networks

In Section 3.2, we used the property that energy transport is bidirectional; that is, the heat flowing from subsystem $j$ to subsystem $i$ must be equal and opposite to the heat flowing from subsystem $i$ to subsystem $j$. This is a natural assumption for thermodynamic systems. In Section 3.6, we assumed symmetry of information flow for networked dynamic systems; that is, it was assumed that if temperature $T_j$ is known to subsystem $i$, then temperature $T_i$ is known to subsystem $j$. However, many important network consensus problems are posed for *directed* networks, where such symmetry is not guaranteed.

In this section, we once again consider (3.18), which involves a network of $n$ single integrators. Now, however, the network topology is assumed to be directed. As before, we assume that the graph is strongly connected. From the point of view of the analysis, on a directed graph it is not possible to show compliance with the first law of thermodynamics and the second law of thermodynamics by considering edges pairwise. Instead it is necessary to use

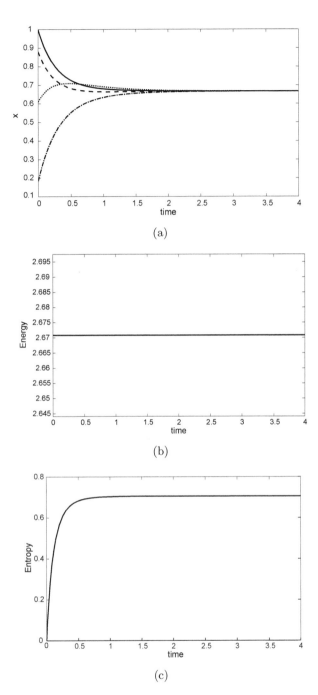

(a)

(b)

(c)

Figure 3.4 (a–c) Response of case 1 showing consensus reached with energy conserved and entropy strictly increasing. (a) Consensus variables $x_i$. (b) Energy $U_{\mathcal{X}}$. (c) Entropy $S_{\mathcal{X}}$.

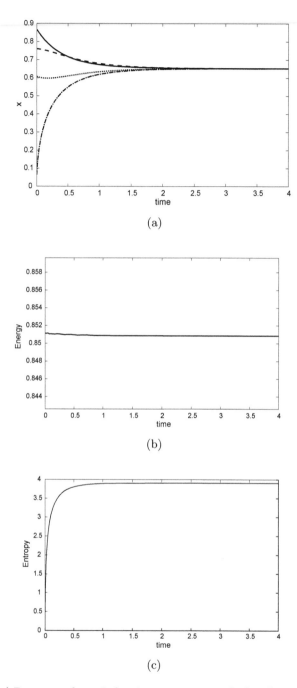

Figure 3.5 (a–c) Response of case 2 showing consensus reached with energy conserved and entropy strictly increasing. (a) Consensus variables $(1/2)x_i^2$. (b) Energy $U_\mathcal{X}$. (c) Entropy $S_\mathcal{X}$.

the global properties of the graph. The assumption of strong connectedness implies the existence of a complete strong cycle. We construct a distributed control law using this cycle and show that the result satisfies the first law of thermodynamics and the second law of thermodynamics for the system.

Our controller for a strongly connected directed network will again be based on the laws of heat flow. However, due to the asymmetry of information flow in the directed network, we use an *asymmetric* form of (3.4), where the pairwise symmetry condition (3.6) is removed.

**Definition 3.5.** The heat transfer law between a pair of subsystems $i$ and $j$ is of *general Fourier type* if it has the form

$$q_{ij} = \alpha_{ij}(T_j - T_i) \tag{3.28}$$

for $j \in \mathcal{P}_{\mathfrak{G}}(i)$, $i = 1, \ldots, n$, where $\alpha_{ij}(\cdot)$ is a function satisfying the sector bound condition

$$\delta_1 \leq \frac{\alpha_{ij}(\xi)}{\xi} \leq \delta_2, \quad \xi \neq 0, \tag{3.29}$$

with $\alpha_{ij}(0) = 0$ and $0 < \delta_1 \leq \delta_2$.

To develop a thermodynamics-based consensus controller for the dynamical system (3.18) on a directed network, we let all subsystem thermal masses be equal with value $m$. Let $U_i(x_i)$ denote the energy of the $i$th subsystem, and let the corresponding subsystem entropy and temperature be given by

$$S_i(U_i) \triangleq m \ln(U_i), \tag{3.30}$$

$$T_i(U_i) \triangleq U_i/m, \tag{3.31}$$

$$\tilde{S}_i(T_i) \triangleq m \ln(mT_i). \tag{3.32}$$

The following lemma is needed for our main result on directed networks.

**Lemma 3.1.** For $n$ positive values $T_1, \ldots, T_n$ such that $T_0 = T_n$,

$$\sum_{i=1}^{n} \frac{T_{i-1}}{T_i} \geq n,$$

with equality holding if and only if $T_1 = T_2 = \cdots = T_n$.

**Proof.** Let $f : \mathbb{R}^n \to \mathbb{R}$ be defined by

$$f(T_1, T_2, \ldots, T_n) \triangleq \frac{T_n}{T_1} + \sum_{i=2}^{n} \frac{T_{i-1}}{T_i} - n, \tag{3.33}$$

and note that $f(T, T, \ldots, T) = 0$ and

$$f(\lambda T_1, \lambda T_2, \ldots, \lambda T_n) = f(T_1, T_2, \ldots, T_n)$$

for every real $\lambda \neq 0$. Since all the $T_i$'s are assumed positive, without loss of generality define a reduced set of coordinates by $\mu_i \triangleq T_i/T_n, i = 1, \ldots, n-1$, where every $\mu_i > 0$. Now, $T_n/T_1 = 1/\mu_1$; $T_{n-1}/T_n = \mu_{n-1}$; and $T_{i-1}/T_i = \mu_{i-1}/\mu_i, i = 2, \ldots, n-2$. Therefore, the function (3.33) can be written as

$$f(T_1, T_2, \ldots, T_n) = \tilde{f}(\mu_1, \mu_2, \ldots, \mu_{n-1}) = \frac{1}{\mu_1} + \sum_{i=2}^{n-1} \frac{\mu_{i-1}}{\mu_i} + \mu_{n-1} - n.$$

Next, note that

$$\frac{\partial \tilde{f}}{\partial \mu_i} = -\frac{\mu_{i-1}}{\mu_i^2} + \frac{1}{\mu_{i+1}}, \quad i = 2, \ldots, n-2, \tag{3.34}$$

$$\frac{\partial \tilde{f}}{\partial \mu_1} = -\frac{1}{\mu_1^2} + \frac{1}{\mu_2}, \tag{3.35}$$

$$\frac{\partial \tilde{f}}{\partial \mu_{n-1}} = -\frac{\mu_{n-2}}{\mu_{n-1}^2} + 1. \tag{3.36}$$

Setting (3.35) to zero gives $\mu_2 = \mu_1^2$, and setting (3.34) to zero gives $\mu_{i+1} = \mu_i^2/\mu_{i-1}$. Proceeding by mathematical induction, we obtain the relationship $\mu_i = \mu_1^i$ for $i = 1, \ldots, n-1$. Setting (3.36) to zero yields $\mu_{n-1}^2 = \mu_{n-2}$. Substituting $\mu_i = \mu_1^i$ yields $\mu_1^{2n-2} = \mu_1^{n-2}$, which implies $\mu_1 = 1$. Thus, we obtain $\bar{\mu}_i = 1, i = 1, \ldots, n-1$, where the overbar indicates an extremal value. It follows that $\tilde{f}(\mu_1, \mu_2, \ldots, \mu_{n-1})$ has a unique critical point at $\bar{\mu} = (1, 1, \ldots, 1)$, and, hence, $f(T_1, T_2, \ldots, T_n)$ has a critical point at $T_1 = T_2 = \cdots = T_n$.

Furthermore, note that

$$\frac{\partial^2 \tilde{f}}{\partial \mu_i \partial \mu_{i-1}} = -\frac{1}{\mu_i^2}, \quad \frac{\partial^2 \tilde{f}}{\partial \mu_i^2} = -\frac{2\mu_{i-1}}{\mu_i^3}, \quad \frac{\partial^2 \tilde{f}}{\partial \mu_i \partial \mu_{i+1}} = -\frac{1}{\mu_{i+1}^2}, \quad i = 2, \ldots, n-2,$$

$$\frac{\partial^2 \tilde{f}}{\partial \mu_1^2} = \frac{2}{\mu_1^3}, \quad \frac{\partial^2 \tilde{f}}{\partial \mu_1 \partial \mu_2} = -\frac{1}{\mu_2^2},$$

$$\frac{\partial^2 \tilde{f}}{\partial \mu_{n-1} \partial \mu_{n-2}} = -\frac{1}{\mu_{n-1}^2}, \quad \frac{\partial^2 \tilde{f}}{\partial \mu_{n-1}^2} = \frac{2\mu_{n-2}}{\mu_{n-1}^3}.$$

Evaluating $\frac{\partial^2 \tilde{f}}{\partial \mu^2}$ at $\mu = \bar{\mu}$ yields

$$\left. \frac{\partial^2 \tilde{f}}{\partial \mu^2} \right|_{\mu=\bar{\mu}} = \begin{bmatrix} 2 & -1 & 0 & 0 & \cdots & 0 & 0 & 0 \\ -1 & 2 & -1 & 0 & \cdots & 0 & 0 & 0 \\ \vdots & \vdots & \vdots & \vdots & \ddots & \vdots & \vdots & \vdots \\ 0 & 0 & 0 & 0 & \cdots & -1 & 2 & -1 \\ 0 & 0 & 0 & 0 & \cdots & 0 & -1 & 2 \end{bmatrix}. \tag{3.37}$$

A tridiagonal symmetric matrix with main diagonal given by $(a_1, \ldots, a_n)$ and first super- and subdiagonals given by $(b_1, \ldots, b_{n-1})$ is positive definite if $b_i^2 \leq (1/4)a_i a_{i+1}(1 + \frac{\pi^2}{1+4\pi^2})$ for $i = 1, \ldots, n-1$ [4]. Therefore, (3.37) is positive definite, and, hence, the critical point $\bar{\mu} = (1, 1, \ldots, 1)$ is a unique minimum of $\tilde{f}(\mu_1, \mu_2, \ldots, \mu_{n-1})$ in the region where $\mu_i > 0$, $i = 1, \ldots, n-1$. Furthermore, note that

$$\sum_{i=1}^{n} \frac{T_{i-1}}{T_i} - n \geq 0,$$

with equality holding if and only if $T_1 = T_2 = \cdots = T_n$. $\qquad \square$

The controller is first illustrated for a network that admits a simple complete strong cycle $\mathcal{C}$, that is, a complete strong cycle that passes through every vertex of the graph exactly once. As discussed in Section 3.4, the vertices of $\mathcal{C}$ may be numbered so that

$$\mathcal{C} = \{(1, 2), (2, 3), \ldots, (n-1, n), (n, 1)\}.$$

We take the derivative of both sides of (3.31) to obtain

$$\dot{T}_i = \frac{1}{m} U_i'(x_i) u_i$$

and let the control for subsystem $i$ be given by

$$u_i(x) = \frac{\alpha}{U_i'(x_i)} \left[ T_{i-1}(U_{i-1}(x_{i-1})) - T_i(U_i(x_i)) \right], \quad i = 1, \ldots, n, \tag{3.38}$$

where $\alpha > 0$ is now a constant for all connected subsystems, and, for notational convenience, we identify index zero with subsystem $n$. The "heat transfer law" underlying this controller is given by

$$\dot{U}_i = q_i = \alpha(T_{i-1} - T_i), \quad i = 1, \ldots, n,$$

where $\alpha$ is now a multiplicative positive constant rather than a function.

This heat transfer law is of general Fourier type but not of symmetric Fourier type. Note that, in contrast to the controller (3.27) for the undirected graph, which depended on every direct predecessor of $i$, that is, on every node in $\mathcal{P}_{\mathfrak{G}}(i)$, the controller (3.38) depends on *one and only one* direct predecessor of $i$. While $i$ may have multiple direct predecessors in $\mathfrak{G}$, only one is the direct predecessor to $i$ in the complete strong cycle $\mathcal{C}$.

Conservation of total system energy follows immediately from

$$\dot{U}_{\mathcal{X}} = \alpha \sum_{i=1}^{n} (T_{i-1} - T_i) = 0.$$

The subsystem entropy functions satisfy $\dot{S}_i = \dot{U}_i/T_i$; that is,

$$\dot{S}_i(T_i) = \frac{\alpha(T_{i-1} - T_i)}{T_i} = \alpha \left( \frac{T_{i-1}}{T_i} - 1 \right), \quad i = 1, \ldots, n,$$

and

$$\dot{S}_{\mathcal{X}}(\boldsymbol{T}) = \alpha \sum_{i=1}^{n} \left( \frac{T_{i-1}}{T_i} - 1 \right) = \alpha \left[ \left( \sum_{i=1}^{n} \frac{T_{i-1}}{T_i} \right) - n \right].$$

It follows from Lemma 3.1 that $\dot{S}_{\mathcal{X}}(\boldsymbol{T}) \geq 0$ for all $\boldsymbol{T} \in \mathcal{T}$, with equality holding if and only if $\boldsymbol{T} = \bar{\boldsymbol{T}} = e\bar{T}$. Note that in general a strongly connected graph can have many distinct simple complete strong cycles. Different choices for $\mathcal{C}$ will result in different controllers; however, every such controller will have the properties just derived.

Not all networks admit a simple complete strong cycle. In some cases the shortest complete strong cycle $\mathcal{C}$ may traverse one or more edges and pass through one or more vertices multiple times. In such a case, we choose any complete strong cycle $\mathcal{C}$ and let $\nu_{ji}$ be the number of times the edge $(j, i)$ is traversed in $\mathcal{C}$. Now, let the distributed control law for subsystem $i$ be given by

$$u_i(x) = \frac{\alpha}{U_i'(x_i)} \sum_{j \in \mathcal{P}_{\mathcal{C}}(i)} \nu_{ji} \left[ T_j(U_j(x_j)) - T_i(U_i(x_i)) \right], \qquad (3.39)$$

where $\alpha > 0$ is a constant. This corresponds to the heat transfer law

$$\dot{U}_i = q_i = \alpha \sum_{j \in \mathcal{P}_{\mathcal{C}}(i)} \nu_{ji} \left( T_j - T_i \right), \quad i = 1, \ldots, n, \qquad (3.40)$$

which is of general Fourier type but not of symmetric Fourier type.

Next, we state our main result for general directed networks. Note that there can be several distinct complete strong cycles of $\mathfrak{G}$, any of which can be used to construct the distributed control law (3.39).

**Theorem 3.4.** Consider a network $\mathfrak{G}$ of strongly connected single integrators. For each subsystem of $\mathfrak{G}$, define a continuously differentiable, nonnegative generalized energy function $U_i(x_i)$, and let the subsystem entropies and subsystem temperatures be given by (3.30)–(3.32). If $U_i'(x_i) \neq 0$ for all $x_i, i = 1, \ldots, n$, then, under the action of the decentralized control (3.39), Properties 1–3 hold.

**Proof.** First, note that the time rate of change of the total system energy is given by

$$\dot{U}_{\mathcal{X}} = \sum_{i=1}^{n} \dot{U}_i$$

$$= \alpha \sum_{i=1}^{n} \sum_{j \in \mathcal{P}_{\mathcal{C}}(i)} \nu_{ji}(T_j - T_i)$$

$$= \alpha \sum_{i=1}^{n} \left( \sum_{j \in \mathcal{S}_{\mathcal{C}}(i)} \nu_{ij} - \sum_{j \in \mathcal{P}_{\mathcal{C}}(i)} \nu_{ji} \right) T_i$$

$$= \alpha \sum_{i=1}^{n} \left( d_{\mathcal{C}}^{+}(i) - d_{\mathcal{C}}^{-}(i) \right) T_i$$

$$= 0,$$

which shows that $U_{\mathcal{X}}(t) = U_0$, and, hence, Property 1 holds.

To show that the system entropy is nondecreasing under (3.39), we again proceed by renumbering the nodes to match their order in the complete strong cycle. This time, however, we do not simply permute the indices, but we also expand the total number of nodes. Specifically, we assign sequential indices to the ordered nodes in the complete strong cycle $\mathcal{C}$ and give a *new index* to each vertex every time it is traversed. Thus, the complete strong cycle $\mathcal{C}$ is converted to a simple complete strong cycle $\mathcal{C}'$ when expressed in the expanded set of indices. The total number of nodes $n'$ in the expanded cycle is given by

$$n' = \sum_{i=1}^{n} \sum_{j \in \mathcal{P}_{\mathcal{C}}(i)} \nu_{ji}, \tag{3.41}$$

where the summation (3.41) gives the number of arcs contained in $\mathcal{C}$, including multiplicity.

Next, the time rate of change of the system entropy is given by

$$\dot{\tilde{S}}_{\mathcal{X}}(\boldsymbol{T}) = \sum_{i=1}^{n} \dot{\tilde{S}}_i$$

$$= \alpha \sum_{i=1}^{n} \sum_{j \in \mathcal{P}_{\mathcal{C}}(i)} \frac{\nu_{ji}}{T_i} (T_j - T_i)$$

$$= \alpha \sum_{i=1}^{n} \sum_{j \in \mathcal{P}_{\mathcal{C}}(i)} \nu_{ji} \left( \frac{T_j}{T_i} - 1 \right), \tag{3.42}$$

where the summation in (3.42) is over each arc in the cycle $\mathcal{C}$, with the factor $\nu_{ji}$ accounting for multiplicity. By construction, the summation over all arcs of $\mathcal{C}$ with multiplicity is equal to the summation over all distinct arcs in $\mathcal{C}'$.

Now, denoting the vector of subsystem temperatures referenced using the expanded indices as $\boldsymbol{T}' = [T_1', T_2', T_3', \ldots, T_{n'}']^{\mathrm{T}}$, and denoting the representation of the system with these temperatures by $\mathcal{X}'$, it follows that

$$\dot{\tilde{S}}_{\mathcal{X}}(\boldsymbol{T}) = \alpha \sum_{i=1}^{n} \sum_{j \in \mathcal{P}_{\mathcal{C}}(i)} \nu_{ji} \left( \frac{T_j}{T_i} - 1 \right)$$

$$= \alpha \sum_{k=1}^{n'} \left( \frac{T_{k-1}'}{T_k'} - 1 \right)$$

$$\triangleq \dot{\tilde{S}}_{\mathcal{X}'}(\boldsymbol{T}'),$$

where, for notational convenience, $T_0'$ is identified with $T_{n'}'$. It now follows from Lemma 3.1 that $\dot{\tilde{S}}_{\mathcal{X}'}(\boldsymbol{T}') \geq 0$, where equality holds if and only if $T_1' = T_2' = T_3' = \cdots = T_{n'}'$. Thus, $\dot{\tilde{S}}_{\mathcal{X}}(\boldsymbol{T}) \geq 0$, where equality holds if and only if $T_1 = T_2 = T_3 = \cdots = T_n$. The fact that some of the reindexed temperatures are constrained to be equal implies that the function $f$ defined in the proof of Lemma 3.1 is constrained to a submanifold in $\mathbb{R}^{n'-1}$. However, this restriction does not change the conclusions of the lemma.

Finally, let $\tilde{S}^* = \tilde{S}(\boldsymbol{T}^*(U_0))$, and define

$$V(\boldsymbol{T}') \triangleq (1/2)(\tilde{S}(\boldsymbol{T}' + \boldsymbol{T}^*(U_0)) - \tilde{S}^*)^2.$$

The rest of the proof is now identical to the proof of Theorem 3.2 and, hence, is omitted. $\qquad \square$

**Example 3.2.** Consider the four-agent network shown in Figure 3.6. The dynamics of each agent are given by

$$\dot{x}_i(t) = u_i(t), \quad x_i(0) = x_{i0}, \quad t \geq 0, \quad i = 1, \ldots, 4.$$

This network does not admit a simple complete strong cycle; however, it does admit the complete strong cycle given by

$$\mathcal{C} = \{(1,2), (2,4), (4,3), (3,2), (2,4), (4,1)\}.$$

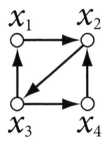

Figure 3.6 Directed network of four agents without a simple complete strong cycle.

Now, it follows from Theorem 3.4 that the controls

$$u_1 = \alpha \left[\partial U_1 / \partial x_1\right]^{-1} (T_4 - T_1),$$
$$u_2 = \alpha \left[\partial U_2 / \partial x_2\right]^{-1} \left[(T_1 - T_2) + (T_3 - T_2)\right],$$
$$u_3 = \alpha \left[\partial U_3 / \partial x_3\right]^{-1} (T_4 - T_3),$$
$$u_4 = 2\alpha \left[\partial U_4 / \partial x_4\right]^{-1} (T_2 - T_4)$$

achieve network consensus with energy conservation. Figure 3.7 shows the performance of this controller for three different choices of subsystem energy function. $\triangle$

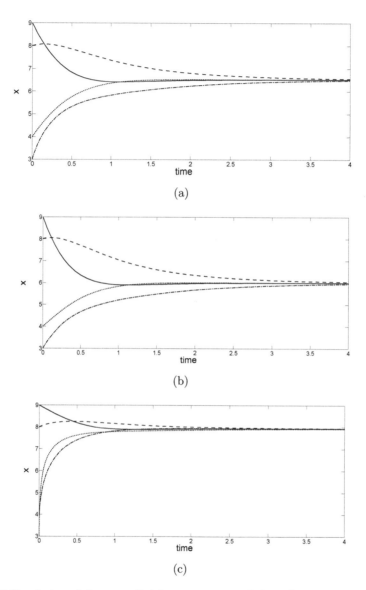

Figure 3.7 Simulation of the controlled four-agent network for different choices of subsystem energy function. (a) $U_i = x_i^2/2$. (b) $U_i = x_i$. (c) $U_i = e^{x_i}$.

## Chapter Four

---

# Finite Time Semistability, Consensus, and Formation Control for Nonlinear Dynamical Networks

## 4.1 Introduction

In a recent series of papers [35, 36], the authors developed a unified stability analysis framework for systems having a continuum of equilibria. Since every neighborhood of a nonisolated equilibrium contains another equilibrium, a nonisolated equilibrium cannot be asymptotically stable. Hence, asymptotic stability is not the appropriate notion of stability for systems having a continuum of equilibria. Two notions that are of particular relevance to such systems are convergence and semistability.

Convergence is the property whereby every system solution converges to a limit point that may depend on the system initial conditions. Semistability is the additional requirement that all solutions converge to limit points that are Lyapunov stable. Semistability for an equilibrium thus implies Lyapunov stability and is implied by asymptotic stability. It is important to note that semistability is not merely equivalent to asymptotic stability of the set of equilibria. Indeed, it is possible for a trajectory to converge to the set of equilibria without converging to any one equilibrium point, as examples in [36] show.

The dependence of the limiting state on the initial state is seen in numerous dynamical systems, including compartmental systems [179], which arise in chemical kinetics and biomedical, environmental, economic, power, and thermodynamic systems [118, 127]. For these systems, every trajectory that starts in a neighborhood of a Lyapunov stable equilibrium converges to a (possibly different) Lyapunov stable equilibrium, and, hence, these systems are semistable. Semistability is especially pertinent to networks of dynamic agents which exhibit convergence to a state of consensus in which the agents agree on certain quantities of interest. Semistability was first introduced in [58] for linear systems, and it was applied to matrix second-order systems in [28]. References [35, 36] consider semistability of nonlinear systems and give several stability results for systems having a continuum of equilibria based on nontangency and arc length of trajectories, respectively.

In addition to semistability, it is desirable that a dynamical system that exhibits semistability also possess the property that trajectories that converge to a Lyapunov stable system state must do so in finite time rather than merely asymptotically. Finite time convergence to an isolated Lyapunov stable equilibrium, that is, finite time stability, was rigorously studied in [34], although finite time stabilization of second-order systems was considered earlier in [32, 139]. More recently, researchers have considered finite time stabilization of higher-order systems [146], as well as finite time stabilization using output feedback [147]. Alternatively, discontinuous finite time stabilizing feedback controllers have been developed in the literature [100, 265, 266]. However, in practical implementation, discontinuous feedback controllers can lead to chattering behavior due to system uncertainty or measurement noise and, hence, may excite unmodeled high-frequency system dynamics.

In this chapter, we merge the theories of semistability and finite time stability developed in [34, 35, 36] to develop a rigorous framework for finite time semistability. In Section 4.3, we extend the theory of semistability given in [35, 36] by presenting new Lyapunov theorems as well as the first converse Lyapunov theorem for semistability, which holds with a smooth (i.e., infinitely differentiable) Lyapunov function. Next, in Section 4.4, we establish finite time semistability theory. We present the notions of finite time convergence and finite time semistability for nonlinear dynamical systems and develop several sufficient Lyapunov stability theorems for finite time semistability. Following [37], we exploit homogeneity as a means for verifying finite time convergence in Section 4.5. Our main result in this direction asserts that a homogeneous system is finite time semistable if and only if it is semistable and has a negative degree of homogeneity. This main result depends on a converse Lyapunov result for homogeneous semistable systems, which we develop.

While our converse result resembles a related result for asymptotically stable systems given in [37, 263], the proof of our result is rendered more difficult by the fact that our result does not hold under the notions of homogeneity considered in [37, 263]. More specifically, while previous treatments of homogeneity involved Euler vector fields representing asymptotically stable dynamics, our results involve homogeneity with respect to a semi-Euler vector field representing a semistable system having the same equilibria as the dynamics of interest. Consequently, our theory precludes the use of dilations commonly used in the literature on homogeneous systems (such as [263]) and requires us to adopt a more geometric description of homogeneity (see [37] and references therein).

Next, in Section 4.6, we use the main results of the chapter to develop a general, thermodynamically motivated framework for designing semistable protocols in dynamical networks for achieving coordination tasks in finite time. Distributed decision-making for coordination of networks of dynamic agents involving information flow can be naturally captured by graph-theoretic notions. As discussed in Chapter 1, these dynamical network systems cover a very broad spectrum of applications, and, hence, it is not surprising that considerable research effort has been devoted to control of networks and control over networks in recent years [93, 180, 220, 234, 238]. However, with the notable exception of [69], finite time coordination has not been addressed in the literature.

In many applications involving multiagent systems, groups of agents are required to agree on certain quantities of interest. In such applications, it is important to develop information consensus protocols for networks of dynamic agents. As shown in Chapter 2, an essential feature of the closed-loop dynamics under any control algorithm that achieves consensus in a dynamical network is the existence of a continuum of equilibria representing a state of consensus. Under such dynamics, the limiting consensus state achieved is not determined completely by the dynamics but depends on the initial system state.

From a practical viewpoint, it is not sufficient to guarantee only that a network converges to a state of consensus, since steady-state convergence is not sufficient to guarantee that small perturbations from the limiting state will lead to only small transient excursions from a state of consensus. It is also necessary to guarantee that the equilibrium states representing consensus are Lyapunov stable and, consequently, semistable. Hence, in Section 4.7, we use the results from Sections 4.4–4.6 to develop a unified distributed control framework based on finite time semistability for addressing the consensus and formation control problem in networks of agents.

## 4.2 Mathematical Preliminaries

In this chapter, we consider nonlinear dynamical systems of the form

$$\dot{x}(t) = f(x(t)), \quad x(0) = x_0, \quad t \in \mathcal{I}_{x_0}, \tag{4.1}$$

where $x(t) \in \mathcal{D} \subseteq \mathbb{R}^n$, $t \in \mathcal{I}_{x_0}$, is the system state vector; $\mathcal{D}$ is an open set; $f : \mathcal{D} \to \mathbb{R}^n$ is continuous on $\mathcal{D}$; $f^{-1}(0) \triangleq \{x \in \mathcal{D} : f(x) = 0\}$ is nonempty; and $\mathcal{I}_{x_0} = [0, \tau_{x_0})$, $0 \leq \tau_{x_0} \leq \infty$, is the maximal interval of existence for the solution $x(\cdot)$ of (4.1). A continuously differentiable function $x : \mathcal{I}_{x_0} \to \mathcal{D}$ is said to be a *solution* of (4.1) on the interval $\mathcal{I}_{x_0} \subset \mathbb{R}$ if $x$ satisfies (4.1) for all $t \in \mathcal{I}_{x_0}$. The continuity of $f$ implies that, for every $x_0 \in \mathcal{D}$, there exist $\tau_0 < 0 < \tau_1$ and a solution $x(\cdot)$ of (4.1) defined on $(\tau_0, \tau_1)$ such that $x(0) = x_0$. A solution $x$ is said to be *right maximally defined* if $x$ cannot be extended on the right (either uniquely or nonuniquely) to a solution of (4.1). Here, we assume that for every initial condition $x_0 \in \mathcal{D}$, (4.1) has a unique right maximally defined solution and this unique solution is defined on $[0, \infty)$.

Under these assumptions, the solutions of (4.1) define a continuous *global semiflow* on $\mathcal{D}$; that is, $s : [0, \infty) \times \mathcal{D} \to \mathcal{D}$ is a jointly continuous function satisfying the *consistency property* $s(0, x) = x$ and the *semigroup property* $s(t, s(\tau, x)) = s(t + \tau, x)$ for every $x \in \mathcal{D}$ and $t, \tau \in [0, \infty)$. Furthermore, we assume that for every initial condition $x_0 \in \mathcal{D} \backslash f^{-1}(0)$, (4.1) has a local unique solution for negative time. Given $t \in [0, \infty)$, we denote the *flow* $s(t, \cdot) : \mathcal{D} \to \mathcal{D}$ of (4.1) by $s_t(x_0)$ or $s_t$. Likewise, given $x \in \mathcal{D}$, we denote the *solution curve* or *trajectory* $s(\cdot, x) : [0, \infty) \to \mathcal{D}$ of (4.1) by $s^x(t)$ or $s^x$. Finally, the image of $\mathcal{U} \subset \mathcal{D}$ under the flow $s_t$ is defined as $s_t(\mathcal{U}) \triangleq \{y : y = s_t(x_0) \text{ for some } x_0 \in \mathcal{U}\}$.

A set $\mathcal{M} \subseteq \mathbb{R}^n$ is *positively invariant* if $s_t(\mathcal{M}) \subseteq \mathcal{M}$ for all $t \geq 0$. The set $\mathcal{M}$ is *negatively invariant* if, for every $z \in \mathcal{M}$ and every $t \geq 0$, there exists $x \in \mathcal{M}$ such that $s(t, x) = z$ and $s(\tau, x) \in \mathcal{M}$ for all $\tau \in [0, t]$. The set $\mathcal{M}$ is *invariant* if $s_t(\mathcal{M}) = \mathcal{M}$, $t \geq 0$. Note that a set is invariant if and only if it is positively and negatively invariant. Finally, a set $\mathcal{E} \subseteq \mathbb{R}^n$ is *connected* if and only if every pair of open sets $\mathcal{U}_i \subseteq \mathbb{R}^n$, $i = 1, 2$, satisfying $\mathcal{E} \subseteq \mathcal{U}_1 \cup \mathcal{U}_2$ and $\mathcal{U}_i \cap \mathcal{E} \neq \varnothing$, $i = 1, 2$, has a nonempty intersection. A *connected component* of the set $\mathcal{E} \subseteq \mathbb{R}^n$ is a connected subset of $\mathcal{E}$ that is not properly contained in any connected subset of $\mathcal{E}$.

## 4.3 Lyapunov and Converse Lyapunov Theory for Semistability

In this section, we develop necessary and sufficient conditions for semistability. In order to develop necessary and sufficient conditions for finite time semistability, we first need to establish a converse Lyapunov theorem for semistability. This extends some of the results in [12, 190, 221, 263, 311]. Converse Lyapunov theorems were extensively studied in [190, 221].

In particular, Massera [221] proved a converse Lyapunov theorem under the assumption that the vector field $f$ is locally Lipschitz continuous. For locally Lipschitz continuous vector fields, it has been shown that asymptotic stability implies the existence of a smooth (i.e., infinitely differentiable) Lyapunov function. Kurzweil [190] proved the existence of smooth Lyapunov functions for asymptotic stability under the assumption of $f$ only being continuous. Unlike asymptotic stability, Lyapunov stability for autonomous dynamical systems does not imply the existence of a continuous Lyapunov function. However, semistability does imply the existence of a smooth Lyapunov function. Before stating this result, we first present several definitions and a key proposition.

**Definition 4.1** (see [36]). An equilibrium point $x \in \mathcal{D}$ of (4.1) is *Lyapunov stable* if for every open subset $\mathcal{N}_\varepsilon$ of $\mathcal{D}$ containing $x$, there exists an open subset $\mathcal{N}_\delta$ of $\mathcal{D}$ containing $x$ such that $s_t(\mathcal{N}_\delta) \subset \mathcal{N}_\varepsilon$ for all $t \geq 0$. An equilibrium point $x \in \mathcal{D}$ of (4.1) is *semistable* if it is Lyapunov stable and there exists an open subset $\mathcal{U}$ of $\mathcal{D}$ containing $x$ such that for all initial conditions in $\mathcal{U}$, the trajectory of (4.1) converges to a Lyapunov stable equilibrium point; that is, $\lim_{t\to\infty} s(t,x) = y$, where $y \in \mathcal{D}$ is a Lyapunov stable equilibrium point of (4.1) and $x \in \mathcal{U}$. If, in addition, $\mathcal{U} = \mathcal{D} = \mathbb{R}^n$, then the equilibrium point $x \in \mathcal{D}$ of (4.1) is a *globally semistable equilibrium*. The system (4.1) is said to be *Lyapunov stable* if every equilibrium point of (4.1) is Lyapunov stable. The system (4.1) is said to be *semistable* if every equilibrium point of (4.1) is semistable. Finally, (4.1) is said to be *globally semistable* if every equilibrium of (4.1) is globally semistable.

**Definition 4.2.** The *domain of semistability* is the set of points $x_0 \in \mathcal{D}$ such that if $x(t)$ is a solution to (4.1) with $x(0) = x_0$, $t \geq 0$, then $x(t)$ converges to a Lyapunov stable equilibrium point in $\mathcal{D}$.

Note that if (4.1) is semistable, then its domain of semistability contains the set of equilibria in its interior.

Next, we present alternative equivalent characterizations of semistability of (4.1).

**Proposition 4.1.** Consider the nonlinear dynamical system (4.1). Then the following statements are equivalent.

(i) The system (4.1) is semistable.

(ii) For each $x_e \in f^{-1}(0)$, there exist class $\mathcal{K}$ and $\mathcal{L}$ functions $\alpha(\cdot)$ and $\beta(\cdot)$, respectively, and $\delta = \delta(x_e) > 0$, such that if $\|x_0 - x_e\| < \delta$, then $\|x(t) - x_e\| \leq \alpha(\|x_0 - x_e\|)$, $t \geq 0$, and $\mathrm{dist}(x(t), f^{-1}(0)) \leq \beta(t)$, $t \geq 0$.

(iii) For each $x_e \in f^{-1}(0)$, there exist class $\mathcal{K}$ functions $\alpha_1(\cdot)$ and $\alpha_2(\cdot)$, a class $\mathcal{L}$ function $\beta(\cdot)$, and $\delta = \delta(x_e) > 0$ such that if $\|x_0 - x_e\| < \delta$, then $\mathrm{dist}(x(t), f^{-1}(0)) \leq \alpha_1(\|x(t) - x_e\|)\beta(t) \leq \alpha_2(\|x_0 - x_e\|)\beta(t)$, $t \geq 0$.

**Proof.** To show that (i) implies (ii), suppose that (4.1) is semistable, and let $x_e \in f^{-1}(0)$. It follows from Lemma 4.5 of [186] that there exists $\delta = \delta(x_e) > 0$ and a class $\mathcal{K}$ function $\alpha(\cdot)$ such that if $\|x_0 - x_e\| \leq \delta$, then $\|x(t) - x_e\| \leq \alpha(\|x_0 - x_e\|)$, $t \geq 0$. Without loss of generality, we can assume that $\delta$ is such that $\overline{\mathcal{B}_\delta(x_e)}$ is contained in the domain of semistability of (4.1). Hence, for every $x_0 \in \overline{\mathcal{B}_\delta(x_e)}$, $\lim_{t \to \infty} x(t) = x^* \in f^{-1}(0)$ and, consequently, $\lim_{t \to \infty} \text{dist}(x(t), f^{-1}(0)) = 0$.

For each $\varepsilon > 0$ and $x_0 \in \overline{\mathcal{B}_\delta(x_e)}$, define $T_{x_0}(\varepsilon)$ to be the infimum of $T$ with the property that $\text{dist}(x(t), f^{-1}(0)) < \varepsilon$ for all $t \geq T$; that is, $T_{x_0}(\varepsilon) \triangleq \inf\{T : \text{dist}(x(t), f^{-1}(0)) < \varepsilon, t \geq T\}$. For each $x_0 \in \overline{\mathcal{B}_\delta(x_e)}$, the function $T_{x_0}(\varepsilon)$ is nonnegative and nonincreasing in $\varepsilon$, and $T_{x_0}(\varepsilon) = 0$ for sufficiently large $\varepsilon$.

Next, let $T(\varepsilon) \triangleq \sup\{T_{x_0}(\varepsilon) : x_0 \in \overline{\mathcal{B}_\delta(x_e)}\}$. We claim that $T$ is well defined. To show this, consider $\varepsilon > 0$ and $x_0 \in \overline{\mathcal{B}_\delta(x_e)}$. Since $\text{dist}(s(t, x_0), f^{-1}(0)) < \varepsilon$ for every $t > T_{x_0}(\varepsilon)$, it follows from the continuity of $s$ that, for every $\eta > 0$, there exists an open neighborhood $\mathcal{U}$ of $x_0$ such that $\text{dist}(s(t, z), f^{-1}(0)) < \varepsilon$ for every $z \in \mathcal{U}$. Hence, $\lim \sup_{z \to x_0} T_z(\varepsilon) \leq T_{x_0}(\varepsilon)$, implying that the function $x_0 \mapsto T_{x_0}(\varepsilon)$ is upper semicontinuous at the arbitrarily chosen point $x_0$ and hence on $\overline{\mathcal{B}_\delta(x_e)}$. Since an upper semicontinuous function defined on a compact set achieves its supremum, it follows that $T(\varepsilon)$ is well defined. The function $T(\cdot)$ is the pointwise supremum of a collection of nonnegative and nonincreasing functions and is hence nonnegative and nonincreasing. Moreover, $T(\varepsilon) = 0$ for every $\varepsilon > \max\{\alpha(\|x_0 - x_e\|) : x_0 \in \overline{\mathcal{B}_\delta(x_e)}\}$.

Let

$$\psi(\varepsilon) \triangleq \frac{2}{\varepsilon} \int_{\varepsilon/2}^{\varepsilon} T(\sigma) \mathrm{d}\sigma + \frac{1}{\varepsilon} \geq T(\varepsilon) + \frac{1}{\varepsilon}.$$

The function $\psi(\varepsilon)$ is positive, continuous, and strictly decreasing, and $\psi(\varepsilon) \to 0$ as $\varepsilon \to \infty$. Choose $\beta(\cdot) = \psi^{-1}(\cdot)$. Then $\beta(\cdot)$ is positive, continuous, and strictly decreasing, and $\beta(\sigma) \to 0$ as $\sigma \to \infty$. Furthermore, $T(\beta(\sigma)) < \psi(\beta(\sigma)) = \sigma$. Hence, $\text{dist}(x(t), f^{-1}(0)) \leq \beta(t)$, $t \geq 0$.

To show that (ii) implies (iii), suppose that (ii) holds, and let $x_e \in f^{-1}(0)$. Then it follows from Lemma 4.5 of [186] that $x_e$ is Lyapunov stable. Choosing $x_0$ sufficiently close to $x_e$, it follows from the inequality $\|x(t) - x_e\| \leq \alpha(\|x_0 - x_e\|)$, $t \geq 0$, that trajectories of (4.1) starting sufficiently close to $x_e$ are bounded, and, hence, the positive limit set of (4.1) is nonempty. Since $\lim_{t \to \infty} \text{dist}(x(t), f^{-1}(0)) = 0$, it follows that the positive limit set is contained in $f^{-1}(0)$.

Now, since every point in $f^{-1}(0)$ is Lyapunov stable, it follows from Proposition 5.4 of [36] that $\lim_{t\to\infty} x(t) = x^*$, where $x^* \in f^{-1}(0)$ is Lyapunov stable. If $x^* = x_e$, then it follows using similar arguments as above that there exists a class $\mathcal{L}$ function $\hat{\beta}(\cdot)$ such that

$$\mathrm{dist}(x(t), f^{-1}(0)) \leq \|x(t) - x_e\| \leq \hat{\beta}(t)$$

for every $x_0$ satisfying $\|x_0 - x_e\| < \delta$ and $t \geq 0$. Hence,

$$\mathrm{dist}(x(t), f^{-1}(0)) \leq \sqrt{\|x(t) - x_e\|}\sqrt{\hat{\beta}(t)}, \quad t \geq 0.$$

Next, consider the case where $x^* \neq x_e$, and let $\alpha_1(\cdot)$ be a class $\mathcal{K}$ function. In this case, note that

$$\lim_{t\to\infty} \mathrm{dist}(x(t), f^{-1}(0))/\alpha_1(\|x(t) - x_e\|) = 0,$$

and, hence, it follows using similar arguments as above that there exists a class $\mathcal{L}$ function $\beta(\cdot)$ such that

$$\mathrm{dist}(x(t), f^{-1}(0)) \leq \alpha_1(\|x(t) - x_e\|)\beta(t), \quad t \geq 0.$$

Finally, note that $\alpha_1 \circ \alpha$ is of class $\mathcal{K}$ (by Lemma 4.2 of [186]), and, hence, (iii) follows immediately.

To show that (iii) implies (i), suppose that (iii) holds, and let $x_e \in f^{-1}(0)$. Then it follows that $\alpha_1(\|x(t) - x_e\|) \leq \alpha_2(\|x(0) - x_e\|)$, $t \geq 0$; that is, $\|x(t) - x_e\| \leq \alpha(\|x(0) - x_e\|)$, where $t \geq 0$ and $\alpha = \alpha_1^{-1} \circ \alpha_2$ is of class $\mathcal{K}$ (by Lemma 4.2 of [186]). It now follows from Lemma 4.5 of [186] that $x_e$ is Lyapunov stable. Since $x_e$ was chosen arbitrarily, it follows that every equilibrium point is Lyapunov stable. Furthermore, $\lim_{t\to\infty} \mathrm{dist}(x(t), f^{-1}(0)) = 0$. Choosing $x_0$ sufficiently close to $x_e$, it follows from the inequality $\|x(t) - x_e\| \leq \alpha(\|x_0 - x_e\|)$, $t \geq 0$, that trajectories of (4.1) starting sufficiently close to $x_e$ are bounded, and, hence, the positive limit set of (4.1) is nonempty. Since every point in $f^{-1}(0)$ is Lyapunov stable, it follows from Proposition 5.4 of [36] that $\lim_{t\to\infty} x(t) = x^*$, where $x^* \in f^{-1}(0)$ is Lyapunov stable. Hence, by definition, (4.1) is semistable. □

Given a continuous function $V : \mathcal{D} \to \mathbb{R}$, the *upper right Dini derivative* of $V$ along the solution of (4.1) is defined by

$$\dot{V}(s(t, x)) \triangleq \limsup_{h\to 0^+} \frac{1}{h}[V(s(t+h, x)) - V(s(t, x))].$$

It is easy to see that $\dot{V}(x_e) = 0$ for every $x_e \in f^{-1}(0)$. Finally, if $V(\cdot)$ is continuously differentiable, then $\dot{V}(x) = V'(x)f(x)$.

Next, we present a sufficient condition for semistability.

**Theorem 4.1.** Consider the system (4.1). Let $\mathcal{U}$ be an open neighborhood of $f^{-1}(0)$, and assume that there exists a continuously differentiable function $V : \mathcal{U} \to \mathbb{R}$ such that $V'(x)f(x) < 0$, $x \in \mathcal{U} \backslash f^{-1}(0)$. If (4.1) is Lyapunov stable, then (4.1) is semistable.

**Proof.** Since (4.1) is Lyapunov stable by assumption, for every $z \in f^{-1}(0)$, there exists an open neighborhood $\mathcal{V}_z$ of $z$ such that $s([0, \infty) \times \mathcal{V}_z)$ is bounded and contained in $\mathcal{U}$. The set $\mathcal{V} \triangleq \bigcup_{z \in f^{-1}(0)} \mathcal{V}_z$ is an open neighborhood of $f^{-1}(0)$ contained in $\mathcal{U}$. Consider $x \in \mathcal{V}$ so that there exists $z \in f^{-1}(0)$ such that $x \in \mathcal{V}_z$ and $s(t, x) \in \mathcal{U}$, $t \geq 0$. Since $s([0, \infty) \times \mathcal{V}_z)$ is bounded, it follows that the positive limit set of $x$ is nonempty and invariant. Furthermore, it follows from the assumption that $\dot{V}(s(t, x)) \leq 0$, $t \geq 0$, and, hence, it follows from the Krasovskii–LaSalle invariant set theorem [147, p. 122] that $s(t, x) \to \mathcal{M}$ as $t \to \infty$, where $\mathcal{M}$ is the largest invariant set contained in the set $\mathcal{R} = \{y \in \mathcal{U} : V'(y)f(y) = 0\}$. Note that $\mathcal{R} = f^{-1}(0)$ is invariant, and, hence, $\mathcal{M} = \mathcal{R}$, which implies that $\lim_{t \to \infty} \text{dist}(s(t, x), f^{-1}(0)) = 0$. Finally, since every point in $f^{-1}(0)$ is Lyapunov stable, it follows from Proposition 5.4 of [36] that $\lim_{t \to \infty} s(t, x) = x^*$, where $x^* \in f^{-1}(0)$ is Lyapunov stable. Hence, by definition, (4.1) is semistable.                                                                                       $\square$

Next, we provide a converse Lyapunov theorem for semistability.

**Theorem 4.2.** Consider the system (4.1). Suppose that (4.1) is semistable with domain of semistability $\mathcal{D}_0$. Then there exist a smooth nonnegative function $V : \mathcal{D}_0 \to \overline{\mathbb{R}}_+$ and a class $\mathcal{K}_\infty$ function $\alpha(\cdot)$ such that the following conditions hold.

(i) $V(x) = 0$, $x \in f^{-1}(0)$.

(ii) $V(x) \geq \alpha(\text{dist}(x, f^{-1}(0)))$, $x \in \mathcal{D}_0$.

(iii) $V'(x)f(x) < 0$, $x \in \mathcal{D}_0 \backslash f^{-1}(0)$.

**Proof.** For any given solution $x(t)$ of (4.1), the change of time variable from $t$ to $\tau = \int_0^t (1 + \|f(x(s))\|) \mathrm{d}s$ results in the dynamical system

$$\frac{\mathrm{d}\bar{x}}{\mathrm{d}\tau} = \frac{f(\bar{x}(\tau))}{1 + \|f(\bar{x}(\tau))\|}, \quad \bar{x}(0) = x_0, \quad \tau \geq 0, \tag{4.2}$$

where $\bar{x}(\tau) = x(t)$. With a slight abuse of notation, let $\bar{s}(t, x)$, $t \geq 0$, denote the solution of (4.2) starting from $x \in \mathcal{D}_0$. Note that (4.2) implies that $\|\bar{s}(t, x) - \bar{s}(\tau, x)\| \leq |t - \tau|$, $x \in \mathcal{D}_0$, $t, \tau \geq 0$.

Next, define the function $U : \mathcal{D}_0 \to \overline{\mathbb{R}}_+$ by

$$U(x) \triangleq \sup_{t \geq 0} \left\{ \frac{1 + 2t}{1 + t} \text{dist}(\bar{s}(t, x), f^{-1}(0)) \right\}, \quad x \in \mathcal{D}_0. \tag{4.3}$$

Note that $U(\cdot)$ is well defined since (4.2) is semistable. Clearly, (i) holds with $V(\cdot)$ replaced by $U(\cdot)$. Furthermore, since $U(x) \geq \text{dist}(x, f^{-1}(0))$, $x \in \mathcal{D}_0$, it follows that (ii) holds with $V(\cdot)$ replaced by $U(\cdot)$.

To show that $U(\cdot)$ is continuous on $\mathcal{D}_0 \backslash f^{-1}(0)$, define $T : \mathcal{D}_0 \backslash f^{-1}(0) \to [0, \infty)$ by

$$T(z) \triangleq \inf\{h : \text{dist}(\bar{s}(t, z), f^{-1}(0)) < \text{dist}(z, f^{-1}(0))/2 \text{ for all } t \geq h > 0\},$$

and define $\mathcal{W}_\varepsilon \triangleq \{x \in \mathcal{D}_0 : \text{dist}(x, f^{-1}(0)) < \varepsilon\}$. Note that $\mathcal{W}_\varepsilon \supset f^{-1}(0)$ is open. Consider $z \in \mathcal{D}_0 \backslash f^{-1}(0)$, define $\lambda \triangleq \text{dist}(z, f^{-1}(0)) > 0$, and let $x_e \triangleq \lim_{t \to \infty} \bar{s}(t, z)$. Since $x_e$ is Lyapunov stable, it follows that there exists an open neighborhood $\mathcal{V}$ of $x_e$ such that all solutions of (4.2) in $\mathcal{V}$ remain in $\mathcal{W}_{\lambda/2}$. Since $x_e$ is semistable, it follows that there exists $h > 0$ such that $\bar{s}(h, z) \in \mathcal{V}$. Consequently, $\bar{s}(h + t, z) \in \mathcal{W}_{\lambda/2}$ for all $t \geq 0$, and, hence, it follows that $T(z)$ is well defined.

Next, by continuity of solutions of (4.2) on compact time intervals, it follows that there exists a neighborhood $\mathcal{U}$ of $z$ such that $\mathcal{U} \cap f^{-1}(0) = \emptyset$ and $\bar{s}(T(z), y) \in \mathcal{V}$ for all $y \in \mathcal{U}$. Now, it follows from the choice of $\mathcal{V}$ that $\bar{s}(T(z) + t, y) \in \mathcal{W}_{\lambda/2}$ for all $t \geq 0$ and $y \in \mathcal{U}$. Then, for every $t > T(z)$ and $y \in \mathcal{U}$,

$$[(1 + 2t)/(1 + t)]\text{dist}(\bar{s}(t, y), f^{-1}(0)) \leq 2\text{dist}(\bar{s}(t, y), f^{-1}(0)) \leq \lambda.$$

Therefore, for every $y \in \mathcal{U}$,

$$U(z) - U(y) = \sup_{t \geq 0} \left\{ \frac{1 + 2t}{1 + t} \text{dist}(\bar{s}(t, z), f^{-1}(0)) \right\}$$

$$- \sup_{t \geq 0} \left\{ \frac{1 + 2t}{1 + t} \text{dist}(\bar{s}(t, y), f^{-1}(0)) \right\}$$

$$= \sup_{0 \leq t \leq T(z)} \left\{ \frac{1 + 2t}{1 + t} \text{dist}(\bar{s}(t, z), f^{-1}(0)) \right\}$$

$$- \sup_{0 \leq t \leq T(z)} \left\{ \frac{1 + 2t}{1 + t} \text{dist}(\bar{s}(t, y), f^{-1}(0)) \right\}. \tag{4.4}$$

Hence,

$$|U(z) - U(y)|$$

$$\leq \sup_{0 \leq t \leq T(z)} \left| \frac{1 + 2t}{1 + t} \left( \text{dist}(\bar{s}(t, z), f^{-1}(0)) - \text{dist}(\bar{s}(t, y), f^{-1}(0)) \right) \right|$$

$$\leq 2 \sup_{0 \leq t \leq T(z)} \left| \operatorname{dist}(\bar{s}(t, z), f^{-1}(0)) - \operatorname{dist}(\bar{s}(t, y), f^{-1}(0)) \right|$$

$$\leq 2 \sup_{0 \leq t \leq T(z)} \operatorname{dist}(\bar{s}(t, z), \bar{s}(t, y)), \quad z \in \mathcal{D}_0 \backslash f^{-1}(0), \quad y \in \mathcal{U}. \quad (4.5)$$

Now, it follows from continuous dependence of solutions $\bar{s}(\cdot, \cdot)$ on system initial conditions (Theorem 3.4 of Chapter I of [141]) and (4.5) that $U(\cdot)$ is continuous at $z$. Furthermore, it follows from (4.5) that, for every sufficiently small $h > 0$,

$$|U(\bar{s}(h, z)) - U(z)| \leq 2 \sup_{0 \leq t \leq T(z)} \|\bar{s}(t, \bar{s}(h, z)) - \bar{s}(t, z)\|$$

$$= 2 \sup_{0 \leq t \leq T(z)} \|\bar{s}(t + h, z) - \bar{s}(t, z)\| \leq 2h,$$

which implies that $|\dot{U}(z)| \leq 2$. Since $z \in \mathcal{D}_0 \backslash f^{-1}(0)$ was chosen arbitrarily, it follows that $U(\cdot)$ is continuous, $|\dot{U}(\cdot)| \leq 2$, and $T(\cdot)$ is well defined on $\mathcal{D}_0 \backslash f^{-1}(0)$.

To show that $U(\cdot)$ is continuous on $f^{-1}(0)$, consider $x_e \in f^{-1}(0)$. Let $\{x_n\}_{n=1}^{\infty}$ be a sequence in $\mathcal{D}_0 \backslash f^{-1}(0)$ that converges to $x_e$. Since $x_e$ is Lyapunov stable, it follows from Lemma 4.5 of [186] that $x(t) \equiv x_e$ is the unique solution to (4.2) with $x_0 = x_e$. By continuous dependence of solutions $\bar{s}(\cdot, \cdot)$ on system initial conditions (Theorem 3.4 of Chapter I of [141]), $\bar{s}(t, x_n) \to \bar{s}(t, x_e) = x_e$ as $n \to \infty$, $t \geq 0$.

Let $\varepsilon > 0$, and note that it follows from (ii) of Proposition 4.1 that there exists $\delta = \delta(x_e) > 0$ such that, for every solution of (4.2) in $\mathcal{B}_\delta(x_e)$, there exists $\hat{T} = \hat{T}(x_e, \varepsilon) > 0$ such that $\bar{s}_t(\mathcal{B}_\delta(x_e)) \subset \mathcal{W}_\varepsilon$ for all $t \geq \hat{T}$. Next, note that there exists a positive integer $N_1$ such that $x_n \in \mathcal{B}_\delta(x_e)$ for all $n \geq N_1$. Now, it follows from (4.3) that

$$U(x_n) \leq 2 \sup_{0 \leq t \leq \hat{T}} \operatorname{dist}(\bar{s}(t, x_n), f^{-1}(0)) + 2\varepsilon, \quad n \geq N_1. \quad (4.6)$$

Furthermore, it follows from Lemma 3.1 of Chapter I of [141] that $\bar{s}(\cdot, x_n)$ converges to $\bar{s}(\cdot, x_e)$ uniformly on $[0, \hat{T}]$. Hence,

$$\lim_{n \to \infty} \sup_{0 \leq t \leq \hat{T}} \operatorname{dist}(\bar{s}(t, x_n), f^{-1}(0)) = \sup_{0 \leq t \leq \hat{T}} \operatorname{dist}\left( \lim_{n \to \infty} \bar{s}(t, x_n), f^{-1}(0) \right)$$

$$= \sup_{0 \leq t \leq \hat{T}} \operatorname{dist}(x_e, f^{-1}(0)) = 0,$$

which implies that there exists a positive integer $N_2 = N_2(x_e, \varepsilon) \geq N_1$ such that

$$\sup_{0 \leq t \leq \hat{T}} \operatorname{dist}(\bar{s}(t, x_n), f^{-1}(0)) < \varepsilon$$

for all $n \geq N_2$. Combining (4.6) with the above result yields $U(x_n) < 4\varepsilon$ for all $n \geq N_2$, which implies that $\lim_{n\to\infty} U(x_n) = 0 = U(x_e)$.

Next, we show that $U(\bar{x}(\tau))$ is strictly decreasing along the solution of (4.2) on $\mathcal{D}\backslash f^{-1}(0)$. Note that for every $x \in \mathcal{D}_0\backslash f^{-1}(0)$ and $0 < h \leq 1/2$ such that $\bar{s}(h, x) \in \mathcal{D}_0\backslash f^{-1}(0)$, it follows from the arguments preceding (4.4) that, for sufficiently small $h$, the supremum in the definition of $U(\bar{s}(h, x))$ is reached at some time $\hat{t}$ such that $0 \leq \hat{t} \leq T(x)$. Hence,

$$
\begin{aligned}
U(\bar{s}(h, x)) &= \operatorname{dist}(\bar{s}(\hat{t} + h, x), f^{-1}(0))\frac{1 + 2\hat{t}}{1 + \hat{t}} \\
&= \operatorname{dist}(\bar{s}(\hat{t} + h, x), f^{-1}(0))\frac{1 + 2\hat{t} + 2h}{1 + \hat{t} + h}\left[1 - \frac{h}{(1 + 2\hat{t} + 2h)(1 + \hat{t})}\right] \\
&\leq U(x)\left[1 - \frac{h}{2(1 + T(x))^2}\right],
\end{aligned}
\tag{4.7}
$$

which implies that $\dot{U}(x) \leq -\frac{1}{2}U(x)(1 + T(x))^{-2} < 0$, $x \in \mathcal{D}_0\backslash f^{-1}(0)$, and, hence, (iii) holds with $V(\cdot)$ replaced by $U(\cdot)$. The function $U(\cdot)$ now satisfies all of the conditions of the theorem except for smoothness.

To obtain smoothness, note that since $|\dot{U}(x)| \leq 2$ for every $x \in \mathcal{D}_0$, it follows that $\dot{U}(x)$ satisfies a boundedness condition in the sense of Wilson [311]. By Theorem 2.5 of [311], there exists a smooth function $W : \mathcal{D}_0\backslash f^{-1}(0) \to \mathbb{R}$ satisfying

$$
|W(x) - U(x)| < \frac{1}{4}U(x)(1 + T(x))^{-2} < \frac{1}{2}U(x), \quad x \in \mathcal{D}_0\backslash f^{-1}(0),
$$

and

$$
\dot{W}(x) \leq -\frac{1}{4}U(x)(1 + T(x))^{-2} < 0, \quad x \in \mathcal{D}_0\backslash f^{-1}(0).
$$

Next, we extend $W(\cdot)$ to all of $\mathcal{D}_0$ by taking $W(z) = 0$ for $z \in f^{-1}(0)$. Now, $W(\cdot)$ is a continuous Lyapunov function which is smooth on $\mathcal{D}_0\backslash f^{-1}(0)$. Taking $V(x) = W(x)e^{-(W(x))^{-2}}$ and noting that

$$
W(x) > \frac{1}{2}U(x) > \frac{1}{2}\operatorname{dist}(x, f^{-1}(0)), \quad x \in \mathcal{D}_0\backslash f^{-1}(0),
$$

so that $V(\cdot)$ satisfies (ii) with $\alpha(r) \triangleq (r/2)e^{-4/r^2}$, we obtain the desired smooth Lyapunov function. $\square$

## 4.4 Finite Time Semistability of Nonlinear Dynamical Systems

In this section, we establish the notion of finite time semistability and develop sufficient Lyapunov stability theorems for finite time semistability.

**Definition 4.3.** An equilibrium point $x_e \in f^{-1}(0)$ of (4.1) is said to be *finite time semistable* if there exist an open neighborhood $\mathcal{U} \subseteq \mathcal{D}$ of $x_e$ and a function $T : \mathcal{U}\backslash f^{-1}(0) \to (0, \infty)$, called the *settling-time function*, such that the following statements hold.

(i) For every $x \in \mathcal{U}\backslash f^{-1}(0)$, $s(t, x) \in \mathcal{U}\backslash f^{-1}(0)$ for all $t \in [0, T(x))$ and $\lim_{t \to T(x)} s(t, x)$ exists and is contained in $\mathcal{U} \cap f^{-1}(0)$.

(ii) $x_e$ is semistable.

An equilibrium point $x_e \in f^{-1}(0)$ of (4.1) is said to be *globally finite time semistable* if it is finite time semistable with $\mathcal{D} = \mathcal{U} = \mathbb{R}^n$. The system (4.1) is said to be *finite time semistable* if every equilibrium point in $f^{-1}(0)$ is finite time semistable. Finally, (4.1) is said to be *globally finite time semistable* if every equilibrium point in $f^{-1}(0)$ is globally finite time semistable.

It is easy to see from Definition 4.3 that, for all $x \in \mathcal{U}$, $T(x) = \inf\{t \in \overline{\mathbb{R}}_+ : f(s(t, x)) = 0\}$, where $T(\mathcal{U} \cap f^{-1}(0)) = \{0\}$.

**Lemma 4.1.** Suppose that (4.1) is finite time semistable. Let $x_e \in f^{-1}(0)$ be an equilibrium point of (4.1), and let $\mathcal{U} \subseteq \mathcal{D}$ be as in Definition 4.3. Furthermore, let $T : \mathcal{U} \to \overline{\mathbb{R}}_+$ be the settling-time function. Then $T$ is continuous on $\mathcal{U}$ if and only if $T$ is continuous at each $z_e \in \mathcal{U} \cap f^{-1}(0)$.

**Proof.** The proof is similar to the proof of Proposition 2.4 given in [34] and, hence, is omitted. $\square$

Next, we introduce a new definition which is weaker than finite time semistability and is needed for the next result.

**Definition 4.4.** The system (4.1) is said to be *finite time convergent to* $\mathcal{M} \subseteq f^{-1}(0)$ for $\mathcal{D}_0 \subseteq \mathcal{D}$ if for every $x_0 \in \mathcal{D}_0$, there exists a finite time $T = T(x_0) > 0$ such that $x(t) \in \mathcal{M}$ for all $t \geq T$.

The next result gives a sufficient condition for characterizing finite time convergence.

**Proposition 4.2.** Let $\mathcal{D}_0 \subseteq \mathcal{D}$ be positively invariant and $\mathcal{M} \subseteq f^{-1}(0)$. Assume that there exists a continuous function $V : \mathcal{D}_0 \to \mathbb{R}$ such that $\dot{V}(\cdot)$ is defined everywhere on $\mathcal{D}_0$; $V(x) = 0$ if and only if $x \in \mathcal{M} \subset \mathcal{D}_0$; and

$$-c_1|V(x)|^\alpha \leq \dot{V}(x) \leq -c_2|V(x)|^\alpha, \quad x \in \mathcal{D}_0\backslash\mathcal{M}, \tag{4.8}$$

where $c_1 \geq c_2 > 0$ and $0 < \alpha < 1$. Then (4.1) is finite time convergent to

$\mathcal{M}$ for $\{x \in \mathcal{D}_0 : V(x) \geq 0\}$. Alternatively, if $V$ is nonnegative and

$$\dot{V}(x) \leq -c_3(V(x))^\alpha, \quad x \in \mathcal{D}_0 \backslash \mathcal{M}, \tag{4.9}$$

where $c_3 > 0$, then (4.1) is finite time convergent to $\mathcal{M}$ for $\mathcal{D}_0$.

**Proof.** Note that (4.8) is also true for $x \in \mathcal{M}$. Applying the comparison lemma (Theorems 4.1 and 4.2 of [319]) to (4.8) yields $\mu(t, V(x), c_1) \leq V(s(t,x)) \leq \mu(t, V(x), c_2)$, $x \in \{z \in \mathcal{D}_0 : V(z) \geq 0\}$, where $\mu$ is given by

$$\mu(t,z,c) \triangleq \begin{cases} (|z|^{1-\alpha} - c(1-\alpha)t)^{\frac{1}{1-\alpha}}, & 0 \leq t < \frac{|z|^{1-\alpha}}{c(1-\alpha)}, \quad \alpha < 1, \\ 0, & t \geq \frac{|z|^{1-\alpha}}{c(1-\alpha)}, \quad \alpha < 1. \end{cases} \tag{4.10}$$

Hence, $V(s(t,x)) = 0$ for $t \geq \frac{|V(x)|^{1-\alpha}}{c_2(1-\alpha)}$, which implies that $s(t,x) \in \mathcal{M}$ for $t \geq \frac{|V(x)|^{1-\alpha}}{c_2(1-\alpha)}$. The conclusion follows. The second part of the conclusion can be proved similarly. $\qquad\square$

The next result establishes a relationship between finite time convergence and finite time semistability.

**Theorem 4.3.** Assume that there exists a continuous nonnegative function $V : \mathcal{D} \to \overline{\mathbb{R}}_+$ such that $\dot{V}(\cdot)$ is defined everywhere on $\mathcal{D}$; $V^{-1}(0) = f^{-1}(0)$; and there exists an open neighborhood $\mathcal{U} \subseteq \mathcal{D}$ such that $\mathcal{U} \cap f^{-1}(0)$ is nonempty and

$$\dot{V}(x) \leq w(V(x)), \quad x \in \mathcal{U} \backslash f^{-1}(0), \tag{4.11}$$

where $w : \mathbb{R} \to \mathbb{R}$ is continuous, $w(0) = 0$, and

$$\dot{z}(t) = w(z(t)), \quad z(0) = z_0 \in \mathbb{R}, \quad t \geq 0, \tag{4.12}$$

has a unique solution in forward time. If (4.12) is finite time convergent to the origin for $\overline{\mathbb{R}}_+$ and every point in $\mathcal{U} \cap f^{-1}(0)$ is a Lyapunov stable equilibrium point of (4.1), then every point in $\mathcal{U} \cap f^{-1}(0)$ is finite time semistable. Moreover, the settling-time function of (4.1) is continuous on an open neighborhood of $\mathcal{U} \cap f^{-1}(0)$. Finally, if $\mathcal{U} = \mathcal{D}$, then (4.1) is finite time semistable.

**Proof.** Consider $x_e \in \mathcal{U} \cap f^{-1}(0)$. Since $x(t) \equiv x_e$ is Lyapunov stable, it follows that there exists an open positively invariant set $\mathcal{S} \subseteq \mathcal{U}$ containing $x_e$. Next, it follows from (4.11) that

$$\dot{V}(s(t,x)) \leq w(V(s(t,x))), \quad x \in \mathcal{S}, \quad t \geq 0. \tag{4.13}$$

Now, applying the comparison lemma (Theorem 4.1 of [319]) to the inequality (4.13) with the comparison system (4.12) yields

$$V(s(t,x)) \le \psi(t, V(x)), \quad t \ge 0, \quad x \in \mathcal{S}, \tag{4.14}$$

where $\psi : [0, \infty) \times \mathbb{R} \to \mathbb{R}$ is the global semiflow of (4.12). Since (4.12) is finite time convergent to the origin for $\overline{\mathbb{R}}_+$, it follows from (4.14) and the nonnegativity of $V(\cdot)$ that

$$V(s(t,x)) = 0, \quad t \ge \hat{T}(V(x)), \quad x \in \mathcal{S}, \tag{4.15}$$

where $\hat{T}(\cdot)$ denotes the settling-time function of (4.12).

Next, since $s(0,x) = x$, $s(\cdot, \cdot)$ is jointly continuous and $V(s(t,x)) = 0$ is equivalent to $f(s(t,x)) = 0$ on $\mathcal{S}$, it follows that $\inf\{t \in \overline{\mathbb{R}}_+ : f(s(t,x)) = 0\} > 0$ for $x \in \mathcal{S} \backslash f^{-1}(0)$. Furthermore, it follows from (4.15) that $\inf\{t \in \overline{\mathbb{R}}_+ : f(s(t,x)) = 0\} < \infty$ for $x \in \mathcal{S}$. Define $T : \mathcal{S} \backslash f^{-1}(0) \to \overline{\mathbb{R}}_+$ by $T(x) = \inf\{t \in \overline{\mathbb{R}}_+ : f(s(t,x)) = 0\}$. Then it follows that every point in $\mathcal{S} \cap f^{-1}(0)$ is finite time semistable and $T$ is the settling-time function on $\mathcal{S}$. Furthermore, it follows from (4.15) that $T(x) \le \hat{T}(V(x))$, $x \in \mathcal{S}$. Since the settling-time function of a one-dimensional finite time stable system is continuous at the equilibrium, it follows that $T$ is continuous at each point in $\mathcal{S} \cap f^{-1}(0)$. Since $x_\mathrm{e} \in \mathcal{U} \cap f^{-1}(0)$ was chosen arbitrarily, it follows that every point in $\mathcal{U} \cap f^{-1}(0)$ is finite time semistable, while Lemma 4.1 implies that $T$ is continuous on an open neighborhood of $\mathcal{U} \cap f^{-1}(0)$.

The last statement follows by noting that, if $\mathcal{U} = \mathcal{D}$, then $\mathcal{U}$ is positively invariant by our assumptions on (4.1), and, hence, the preceding arguments hold with $\mathcal{S} = \mathcal{U}$.    $\square$

**Example 4.1.** Consider the nonlinear dynamical system given by

$$\dot{x}_1(t) = (1 - x_1^2(t) - x_2^2(t))^{\frac{1}{3}}(x_1(t) - x_2(t)), \quad x_1(0) = x_{10}, \quad t \ge 0, \tag{4.16}$$

$$\dot{x}_2(t) = (1 - x_1^2(t) - x_2^2(t))^{\frac{1}{3}}(x_1(t) + x_2(t)), \quad x_2(0) = x_{20}, \tag{4.17}$$

where $x_1 \in \mathbb{R}$ and $x_2 \in \mathbb{R}$. For this system, we show that all the points in $\mathcal{S}^1 \triangleq \{(x_1, x_2) \in \mathbb{R}^2 : x_1^2 + x_2^2 = 1\}$ are finite time semistable. To see this, consider $V(x) = \frac{1}{4}(x_1^2 + x_2^2 - 1)^2$. Let $0 < c < 1$ and $\mathcal{U} = \{(x_1, x_2) \in \mathbb{R}^n : x_1^2 + x_2^2 > c\}$. Then

$$\dot{V}(x) = -(x_1^2 + x_2^2)|x_1^2 + x_2^2 - 1|^{\frac{4}{3}} \le -2^{\frac{4}{3}} c (V(x))^{\frac{2}{3}}, \quad (x_1, x_2) \in \mathcal{U}.$$

Next, we show that every point in $\mathcal{S}^1$ is Lyapunov stable. This can be shown by using the nontangency-based Lyapunov tests developed in [36].

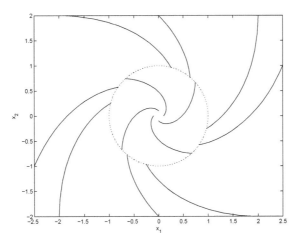

Figure 4.1 Phase portrait for Example 4.1.

In particular, it follows from Example 4.2 of [36] that for every $x \in \mathcal{S}^1$, $f$ is nontangent to $\mathcal{S}^1$. Now, it follows from Corollary 7.2 of [36] that every point in $\mathcal{S}^1$ is Lyapunov stable. Hence, with $c_3 = c2^{\frac{4}{3}}$, $\alpha = \frac{2}{3}$, and $w(x) = -c_3 \mathrm{sign}(x)|x|^\alpha$, it follows from the second conclusion of Proposition 4.2 and Theorem 4.3 that every point in $\mathcal{S}^1$ is finite time semistable. Figure 4.1 shows the phase portrait of (4.16) and (4.17).                     $\triangle$

**Theorem 4.4.** Assume that there exists a continuous nonnegative function $V : \mathcal{D} \to \overline{\mathbb{R}}_+$ such that $\dot{V}(\cdot)$ is defined everywhere on $\mathcal{D}$, $V^{-1}(0) = f^{-1}(0)$, and there exists an open neighborhood $\mathcal{U} \subseteq \mathcal{D}$ such that $\mathcal{U} \cap f^{-1}(0)$ is nonempty and (4.9) holds for all $x \in \mathcal{U} \backslash f^{-1}(0)$. Furthermore, assume that there exists a continuous nonnegative function $W : \mathcal{U} \to \overline{\mathbb{R}}_+$ such that $\dot{W}(\cdot)$ is defined everywhere on $\mathcal{U}$; $W^{-1}(0) = \mathcal{U} \cap f^{-1}(0)$; and

$$\|f(x)\| \le -c_0 \dot{W}(x), \quad x \in \mathcal{U} \backslash f^{-1}(0), \tag{4.18}$$

where $c_0 > 0$. Then every point in $\mathcal{U} \cap f^{-1}(0)$ is finite time semistable.

**Proof.** For any $x_{\mathrm{e}} \in \mathcal{U} \cap f^{-1}(0)$, since $W(x) \ge 0 = W(x_{\mathrm{e}})$ for all $x \in \mathcal{U}$, it follows from (i) of Theorem 5.2 of [35] that $x_{\mathrm{e}}$ is a Lyapunov stable equilibrium and, hence, every point in $\mathcal{U} \cap f^{-1}(0)$ is Lyapunov stable. Now, it follows from the second conclusion of Proposition 4.2 and Theorem 4.3, with $w(x) = -c_3 \mathrm{sign}(x)|x|^\alpha$, that every point in $\mathcal{U} \cap f^{-1}(0)$ is finite time semistable.                     $\square$

**Example 4.2.** Consider the dynamical system given by (4.16) and (4.17). Let $V(x) = \frac{1}{4}(x_1^2 + x_2^2 - 1)^2$ and $\hat{V}(x) = \frac{1}{2}(\sqrt{x_1^2 + x_2^2} - 1)^2$. It follows from Example 4.1 that $\dot{V}(x) \le -2^{\frac{4}{3}}c_1(V(x))^{\frac{2}{3}}$ for all $x \in \mathcal{U}$, where

$\mathcal{U}$ is as in Example 4.1. Since $\|f(x)\| = |x_1^2 + x_2^2 - 1|^{\frac{1}{3}}\sqrt{x_1^2 + x_2^2}$ and $\dot{\hat{V}}(x) = (\sqrt{x_1^2 + x_2^2} - 1)\sqrt{x_1^2 + x_2^2}(1 - x_1^2 - x_2^2)^{\frac{1}{3}}$ for all $x \in \mathcal{U}$, it follows that $\|f(x)\| = -(2\hat{V}(x))^{-\frac{1}{2}}\dot{\hat{V}}(x)$ for all $x \in \mathcal{U}\backslash\mathcal{S}^1$. Now, taking $W(x) = (2\hat{V}(x))^{\frac{1}{2}}$ yields $\|f(x)\| = -\dot{W}(x)$ for all $x \in \mathcal{U}\backslash\mathcal{S}^1$. Hence, it follows from Theorem 4.4 that every point in $\mathcal{S}^1$ is finite time semistable.    △

## 4.5 Homogeneity and Finite Time Semistability

In this section, we develop necessary and sufficient conditions for finite time semistability of homogeneous dynamical systems. In what follows, we will need to consider a complete vector field $\nu$ on $\mathbb{R}^n$ such that the solutions of the differential equation $\dot{y}(t) = \nu(y(t))$ define a continuous *global flow* $\psi : \mathbb{R} \times \mathbb{R}^n \to \mathbb{R}^n$ on $\mathbb{R}^n$, where $\nu^{-1}(0) = f^{-1}(0)$. For each $\tau \in \mathbb{R}$, the map $\psi_\tau(\cdot) = \psi(\tau, \cdot)$ is a homeomorphism and $\psi_\tau^{-1} = \psi_{-\tau}$. We define a function $V : \mathbb{R}^n \to \mathbb{R}$ to be *homogeneous of degree* $l \in \mathbb{R}$ *with respect to* $\nu$ if and only if $(V \circ \psi_\tau)(x) = e^{l\tau}V(x)$, $\tau \in \mathbb{R}$, $x \in \mathbb{R}^n$. Our assumptions imply that every connected component of $\mathbb{R}^n\backslash f^{-1}(0)$ is invariant under $\nu$.

The *Lie derivative* of a continuous function $V : \mathbb{R}^n \to \mathbb{R}$ with respect to $\nu$ is given by $L_\nu V(x) \triangleq \lim_{t \to 0^+} \frac{1}{t}[V(\psi(t, x)) - V(x)]$ whenever the limit on the right-hand side exists. If $V$ is a continuous homogeneous function of degree $l > 0$, then $L_\nu V$ is defined everywhere and satisfies $L_\nu V = lV$. We assume that the vector field $\nu$ is a *semi-Euler vector field*; that is, the dynamical system

$$\dot{y}(t) = -\nu(y(t)), \quad y(0) = y_0, \quad t \geq 0, \tag{4.19}$$

is globally semistable. Thus, for each $x \in \mathbb{R}^n$, $\lim_{\tau \to \infty} \psi(-\tau, x) = x^* \in \nu^{-1}(0)$, and for each $x_e \in \nu^{-1}(0)$, there exists $z \in \mathbb{R}^n$ such that $x_e = \lim_{\tau \to \infty} \psi(-\tau, z)$. Finally, we say that the vector field $f$ is *homogeneous of degree* $k \in \mathbb{R}$ *with respect to* $\nu$ if and only if $\nu^{-1}(0) = f^{-1}(0)$ and, for every $t \in \overline{\mathbb{R}}_+$ and $\tau \in \mathbb{R}$,

$$s_t \circ \psi_\tau = \psi_\tau \circ s_{e^{k\tau}t}, \tag{4.20}$$

where $\circ$ denotes the composition operator. Note that if $V : \mathbb{R}^n \to \mathbb{R}$ is a homogeneous function of degree $l$ such that $L_f V(x)$ is defined everywhere, then $L_f V(x)$ is a homogeneous function of degree $l + k$.

Finally, note that if $\nu$ and $f$ are continuously differentiable in a neighborhood of $x \in \mathbb{R}^n$, then (4.20) holds at $x$ for sufficiently small $t$ and $\tau$ if and only if $[\nu, f](x) = kf(x)$ in a neighborhood of $x \in \mathbb{R}^n$, where the Lie bracket $[\nu, f]$ of $\nu$ and $f$ can be computed by using $[\nu, f] = \frac{\partial f}{\partial x}\nu - \frac{\partial \nu}{\partial x}f$.

The following lemmas are needed for the main results of this section.

**Lemma 4.2.** Consider the dynamical system (4.19). Let $\mathcal{D}_{\mathrm{c}} \subset \mathbb{R}^n$ be a compact set satisfying $\mathcal{D}_{\mathrm{c}} \cap \nu^{-1}(0) = \varnothing$. Then, for every open set $\mathcal{U}$ satisfying $\mathcal{U} \supset \nu^{-1}(0)$, there exist $\tau_1, \tau_2 > 0$ such that $\psi_{-t}(\mathcal{D}_{\mathrm{c}}) \subset \mathcal{U}$ for all $t > \tau_1$ and $\psi_\tau(\mathcal{D}_{\mathrm{c}}) \cap \mathcal{U} = \varnothing$ for all $\tau > \tau_2$.

**Proof.** Let $\mathcal{U}$ be an open neighborhood of $\nu^{-1}(0)$. Since every $z \in \nu^{-1}(0)$ is Lyapunov stable under $\nu$, it follows that there exists an open neighborhood $\mathcal{V}_z$ containing $z$ such that $\psi_{-t}(\mathcal{V}_z) \subseteq \mathcal{U}$ for all $t \geq 0$. Hence, $\mathcal{V} \triangleq \bigcup_{z \in \nu^{-1}(0)} \mathcal{V}_z$ is open and $\psi_{-t}(\mathcal{V}) \subseteq \mathcal{U}$ for all $t \geq 0$.

Next, consider the collection of nested sets $\{\mathcal{D}_t\}_{t>0}$, where

$$\mathcal{D}_t = \{x \in \mathcal{D}_{\mathrm{c}} : \psi_h(x) \notin \mathcal{V}, h \in [-t, 0]\}$$

$$= \mathcal{D}_{\mathrm{c}} \cap \left( \mathbb{R}^n \setminus \left( \bigcup_{h \in [-t, 0]} \psi_h^{-1}(\mathcal{V}) \right) \right), \quad t > 0.$$

For each $t > 0$, $\mathcal{D}_t$ is a compact set. Therefore, if $\mathcal{D}_t$ is nonempty for each $t > 0$, then there exists $x \in \bigcap_{t>0} \mathcal{D}_t$; that is, there exists $x \in \mathcal{D}_{\mathrm{c}}$ such that $\psi_{-t}(x) \notin \mathcal{V}$ for all $t > 0$, which contradicts the fact that the domain of semistability of (4.19) is $\mathbb{R}^n$. Hence, there exists $\tau > 0$ such that $\mathcal{D}_\tau = \varnothing$; that is, $\mathcal{D}_{\mathrm{c}} \subset \bigcup_{h \in [-\tau, 0]} \psi_h^{-1}(\mathcal{V})$. Therefore, for every $t > \tau$,

$$\psi_{-t}(\mathcal{D}_{\mathrm{c}}) \subset \bigcup_{h \in [-\tau, 0]} \psi_{-t}(\psi_h^{-1}(\mathcal{V})) = \bigcup_{h \in [-\tau, 0]} \psi_{-t-h}(\mathcal{V}) \subseteq \mathcal{U}.$$

The second conclusion follows using similar arguments as above. $\square$

**Lemma 4.3.** Suppose that $f : \mathbb{R}^n \to \mathbb{R}^n$ is homogeneous of degree $k \in \mathbb{R}$ with respect to $\nu$ and (4.1) is (locally) semistable. Then the domain of semistability of (4.1) is $\mathbb{R}^n$.

**Proof.** Let $\mathcal{A} \subseteq \mathbb{R}^n$ be the domain of semistability and $x \in \mathbb{R}^n$. Note that $\mathcal{A}$ is an open neighborhood of $\nu^{-1}(0)$. Since every point in $\nu^{-1}(0)$ is a globally semistable equilibrium under $-\nu$, there exists $\tau > 0$ such that $z = \psi_{-\tau}(x) \in \mathcal{A}$. Then it follows from (4.20) that $s(t, x) = s(t, \psi_\tau(z)) = \psi_\tau(s(e^{k\tau}t, z))$. Since $\lim_{t \to \infty} s(t, z) = x^* \in f^{-1}(0)$, it follows that

$$\lim_{t \to \infty} s(t, x) = \lim_{t \to \infty} \psi_\tau \left( s(e^{k\tau}t, z) \right) = \psi_\tau \left( \lim_{t \to \infty} s(e^{k\tau}t, z) \right) = \psi_\tau(x^*) = x^*,$$

which implies that $x \in \mathcal{A}$. Since $x \in \mathbb{R}^n$ is arbitrary, $\mathcal{A} = \mathbb{R}^n$. $\square$

**Theorem 4.5.** Suppose that $f : \mathbb{R}^n \to \mathbb{R}^n$ is homogeneous of degree $k \in \mathbb{R}$ with respect to $\nu$ and (4.1) is semistable. Then for every

$l > \max\{-k, 0\}$, there exists a continuous nonnegative function $V : \mathbb{R}^n \to \overline{\mathbb{R}}_+$ that is homogeneous of degree $l$ with respect to $\nu$; is continuously differentiable on $\mathbb{R}^n \backslash f^{-1}(0)$; and satisfies $V^{-1}(0) = f^{-1}(0)$, $V'(x)f(x) < 0$, $x \in \mathbb{R}^n \backslash f^{-1}(0)$, and for each $x_e \in f^{-1}(0)$ and each bounded open neighborhood $\mathcal{D}_0$ containing $x_e$, there exist $c_1 = c_1(\mathcal{D}_0) \geq c_2 = c_2(\mathcal{D}_0) > 0$ such that

$$-c_1[V(x)]^{\frac{l+k}{l}} \leq V'(x)f(x) \leq -c_2[V(x)]^{\frac{l+k}{l}}, \quad x \in \mathcal{D}_0. \qquad (4.21)$$

**Proof.** Choose $l > \max\{-k, 0\}$. First, we prove that there exists a continuous Lyapunov function $V$ on $\mathbb{R}^n$ that is homogeneous of degree $l$ with respect to $\nu$, is continuously differentiable on $\mathbb{R}^n \backslash f^{-1}(0)$, and satisfies $V'(x)f(x) < 0$ for $x \in \mathbb{R}^n \backslash f^{-1}(0)$. Choose any nondecreasing smooth function $g : \overline{\mathbb{R}}_+ \to [0, 1]$ such that $g(s) = 0$ for $s \leq a$, $g(s) = 1$ for $s \geq b$, and $g'(s) > 0$ on $(a, b)$, where $0 < a < b$ are constants. It follows from Theorem 4.2 and Lemma 4.3 that there exists a continuously differentiable Lyapunov function $U(\cdot)$ on $\mathbb{R}^n$ satisfying all of the properties in Theorem 4.2.

Next, define

$$V(x) \triangleq \int_{-\infty}^{+\infty} e^{-l\tau} g(U(\psi(\tau, x))) \mathrm{d}\tau, \quad x \in \mathbb{R}^n. \qquad (4.22)$$

Let $\mathcal{U}$ be a bounded open set satisfying $\overline{\mathcal{U}} \cap f^{-1}(0) = \varnothing$. Since every point in $\nu^{-1}(0)$ is a globally semistable equilibrium point under $-\nu$, it follows that for each $x \in \overline{\mathcal{U}}$, $\lim_{\tau \to +\infty} U(\psi(\tau, x)) = +\infty$ and $\lim_{\tau \to +\infty} U(\psi(-\tau, x)) = 0$. Now, it follows from Lemma 4.2 that there exist time instants $\tau_1 < \tau_2$ such that for each $x \in \overline{\mathcal{U}}$, $U(\psi(\tau, x)) \leq a$ for all $\tau \leq \tau_1$ and $U(\psi(\tau, x)) \geq b$ for all $\tau \geq \tau_2$. Hence,

$$V(x) = \int_{\tau_1}^{\tau_2} e^{-l\tau} g(U(\psi(\tau, x))) \mathrm{d}\tau + \frac{e^{-l\tau_2}}{l}, \quad x \in \mathcal{U}, \qquad (4.23)$$

which implies that $V$ is well defined, positive, and continuously differentiable on $\mathcal{U}$.

Next, since $U(\cdot)$ satisfies (i) and (ii) of Theorem 4.2, it follows from (4.22) and (4.23) that $V^{-1}(0) = f^{-1}(0)$. Since for any $\sigma \in \mathbb{R}$ and $x \in \mathbb{R}^n$,

$$V(\psi(\sigma, x)) = \int_{-\infty}^{+\infty} e^{-l\tau} g(U(\psi(\tau + \sigma, x))) \mathrm{d}\tau = e^{l\sigma} V(x), \qquad (4.24)$$

by definition, $V$ is homogeneous of degree $l$. In addition, it follows from (4.20) and (4.23) that

$$V'(x)f(x) = \int_{\tau_1}^{\tau_2} e^{-l\tau} g'(U(\psi(\tau, x))) \frac{\mathrm{d}}{\mathrm{d}t} U(s(e^{-k\tau}t, \psi(\tau, x))) \Big|_{t=0} \mathrm{d}\tau$$

$$= \int_{\tau_1}^{\tau_2} e^{-(l+k)\tau} g'(U(\psi(\tau,x)))U'(\psi(\tau,x))f(\psi(\tau,x))\mathrm{d}\tau$$
$$< 0, \quad x \in \mathcal{U}, \tag{4.25}$$

which implies that $V'f$ is negative and continuous on $\mathcal{U}$. Now, since $\mathcal{U}$ is arbitrary, it follows that $V$ is well defined and continuously differentiable and $V'f$ is negative and continuous on $\mathbb{R}^n \backslash f^{-1}(0)$.

Next, to show continuity at points in $f^{-1}(0)$, we define $T : \mathbb{R}^n \backslash f^{-1}(0) \to \mathbb{R}$ by $T(x) = \sup\{t \in \mathbb{R} : U(\psi(\tau,x)) \le a \text{ for all } \tau \le t\}$ and note that the continuity of $U$ implies that $U(\psi(T(x),x)) = a$ for all $x \in \mathbb{R}^n \backslash f^{-1}(0)$. Let $x_e \in f^{-1}(0)$, and consider a sequence $\{x_k\}_{k=1}^{\infty}$ in $\mathbb{R}^n \backslash f^{-1}(0)$ converging to $x_e$. We claim that the sequence $\{T(x_k)\}_{k=1}^{\infty}$ has no bounded subsequence, so that $\lim_{k\to\infty} T(x_k) = \infty$.

To prove our claim by contradiction, suppose that $\{T(x_{k_i})\}_{i=1}^{\infty}$ is a bounded subsequence. Without loss of generality, we may assume that the sequence $\{T(x_{k_i})\}_{i=1}^{\infty}$ converges to $h \in \mathbb{R}$. Then, by joint continuity of $\psi$,

$$\lim_{i\to\infty} \psi(T(x_{k_i}),x_{k_i}) = \psi(h,x_e) = x_e,$$

so that $\lim_{i\to\infty} U(\psi(T(x_{k_i}),x_{k_i})) = U(x_e) = 0$. However, this contradicts our observation above that $U(\psi(T(x),x)) = a$ for all $x \in \mathbb{R}^n \backslash f^{-1}(0)$. The contradiction leads us to conclude that $\lim_{k\to\infty} T(x_k) = \infty$. Now, for each $k = 1, 2, \ldots$, it follows that

$$V(x_k) = \int_{T(x_k)}^{\infty} e^{-l\tau} g(U(\psi(\tau,x_k)))\mathrm{d}\tau \le \int_{T(x_k)}^{\infty} e^{-l\tau}\mathrm{d}\tau = l^{-1}e^{-lT(x_k)},$$

so that $\lim_{k\to\infty} V(x_k) = 0 = V(x_e)$. Since $x_e$ was chosen arbitrarily, it follows that $V$ is continuous at every $x_e \in f^{-1}(0)$.

To show that $V$ possesses the last property, let $x_e \in f^{-1}(0)$, and choose a bounded open neighborhood $\mathcal{D}_0$ of $x_e$. Let $\mathcal{Q} = \psi(\mathbb{R}_+ \times \mathcal{D}_0)$. For every $\varepsilon > 0$, define $\mathcal{Q}_\varepsilon = \mathcal{Q} \cap V^{-1}(\varepsilon)$. For every $\varepsilon > 0$, define the continuous map $\tau_\varepsilon : \mathbb{R}^n \backslash f^{-1}(0) \to \mathbb{R}$ by $\tau_\varepsilon(x) \triangleq l^{-1}\ln(\varepsilon/V(x))$, and note that, for every $x \in \mathbb{R}^n \backslash f^{-1}(0)$, $\psi(t,x) \in V^{-1}(\varepsilon)$ if and only if $t = \tau_\varepsilon(x)$. Next, define $\beta_\varepsilon : \mathbb{R}^n \backslash f^{-1}(0) \to \mathbb{R}^n$ by $\beta_\varepsilon \triangleq \psi(\tau_\varepsilon(x),x)$. Note that, for every $\varepsilon > 0$, $\beta_\varepsilon$ is continuous, and $\beta_\varepsilon(x) \in V^{-1}(\varepsilon)$ for every $x \in \mathbb{R}^n \backslash f^{-1}(0)$.

Consider $\varepsilon > 0$. $\mathcal{Q}_\varepsilon$ is the union of the images of connected components of $\mathcal{D}_0 \backslash f^{-1}(0)$ under the continuous map $\beta_\varepsilon$. Since every connected component of $\mathbb{R}^n \backslash f^{-1}(0)$ is invariant under $\nu$, it follows that the image of each connected component $\mathcal{U}$ of $\mathbb{R}^n \backslash f^{-1}(0)$ under $\beta_\varepsilon$ is contained in $\mathcal{U}$ itself. In particular, the images of connected components of $\mathcal{D}_0 \backslash f^{-1}(0)$ under $\beta_\varepsilon$ are all disjoint. Thus, each connected component of $\mathcal{Q}_\varepsilon$ is the image of exactly one connected component of $\mathcal{D}_0 \backslash f^{-1}(0)$ under $\beta_\varepsilon$. Finally, if $\varepsilon$ is small enough so that $V^{-1}(\varepsilon) \cap \mathcal{D}_0$ is nonempty, then $V^{-1}(\varepsilon) \cap \mathcal{D}_0 \subseteq \mathcal{Q}_\varepsilon$, and, hence, every connected component of $\mathcal{Q}_\varepsilon$ has a nonempty intersection with $\mathcal{D}_0 \backslash f^{-1}(0)$.

We claim that $\mathcal{Q}_\varepsilon$ is bounded for every $\varepsilon > 0$. It is easy to verify that, for every $\varepsilon_1, \varepsilon_2 \in (0, \infty)$, $\mathcal{Q}_{\varepsilon_2} = \psi_h(\mathcal{Q}_{\varepsilon_1})$ with $h = l^{-1} \ln(\varepsilon_2 / \varepsilon_1)$. Hence, it suffices to prove that there exists $\varepsilon > 0$ such that $\mathcal{Q}_\varepsilon$ is bounded. To arrive at a contradiction, suppose, *ad absurdum*, that $\mathcal{Q}_\varepsilon$ is unbounded for every $\varepsilon > 0$. Choose a bounded open neighborhood $\mathcal{V}$ of $\overline{\mathcal{D}}_0$ and a sequence $\{\varepsilon_i\}_{i=1}^\infty$ in $(0, \infty)$ converging to zero. By our assumption, for every $i = 1, 2, \ldots$, at least one connected component of $\mathcal{Q}_{\varepsilon_i}$ must contain a point in $\mathbb{R}^n \backslash \mathcal{V}$. On the other hand, for sufficiently large $i$, every connected component of $\mathcal{Q}_{\varepsilon_i}$ has a nonempty intersection with $\mathcal{D}_0 \subset \mathcal{V}$. It follows that $\mathcal{Q}_{\varepsilon_i}$ has a nonempty intersection with the boundary of $\mathcal{V}$ for every sufficiently large $i$. Hence, there exists a sequence $\{x_i\}_{i=1}^\infty$ in $\mathcal{D}_0$ and a sequence $\{t_i\}_{i=1}^\infty$ in $(0, \infty)$ such that $y_i \triangleq \psi_{t_i}(x_i) \in V^{-1}(\varepsilon_i) \cap \partial \mathcal{V}$ for every $i = 1, 2, \ldots$.

Since $\mathcal{V}$ is bounded, we can assume that the sequence $\{y_i\}_{i=1}^\infty$ converges to $y \in \partial \mathcal{V}$. Continuity implies that $V(y) = \lim_{i \to \infty} V(y_i) = \lim_{i \to \infty} \varepsilon_i = 0$. Since $V^{-1}(0) = f^{-1}(0) = \nu^{-1}(0)$, it follows that $y$ is Lyapunov stable under $-\nu$. Since $y \notin \overline{\mathcal{D}}_0$, there exists an open neighborhood $\mathcal{U}$ of $y$ such that $\mathcal{U} \cap \mathcal{D}_0 = \emptyset$. The sequence $\{y_i\}_{i=1}^\infty$ converges to $y$, while $\psi_{-t_i}(y_i) = x_i \in \mathcal{D}_0 \subset \mathbb{R}^n \backslash \mathcal{U}$, which contradicts Lyapunov stability. This contradiction implies that there exists $\varepsilon > 0$ such that $\mathcal{Q}_\varepsilon$ is bounded. It now follows that $\mathcal{Q}_\varepsilon$ is bounded for every $\varepsilon > 0$.

Finally, consider $x \in \mathcal{D}_0 \backslash f^{-1}(0)$. Choose $\varepsilon > 0$, and note that $\psi_{\tau_\varepsilon(x)}(x) \in \mathcal{Q}_\varepsilon$. Furthermore, note that $V'(x)f(x) < 0$ for all $x \in \mathbb{R}^n \backslash f^{-1}(0)$, $V'(x)f(x)$ is continuous on $\mathbb{R}^n \backslash f^{-1}(0)$, and $\overline{\mathcal{Q}}_\varepsilon \cap f^{-1}(0) = \emptyset$. Then, by homogeneity, $V(\psi_{\tau_\varepsilon(x)}(x)) = \varepsilon$, and, hence,

$$\min_{z \in \overline{\mathcal{Q}}_\varepsilon} V'(z)f(z) \leq V'(\psi_{\tau_\varepsilon(x)}(x))f(\psi_{\tau_\varepsilon(x)}(x)) \leq \max_{z \in \overline{\mathcal{Q}}_\varepsilon} V'(z)f(z). \quad (4.26)$$

Since $V'(\psi_{\tau_\varepsilon(x)}(x))f(\psi_{\tau_\varepsilon(x)}(x))$ is homogeneous of degree $l + k$, it follows that

$$V'(\psi_{\tau_\varepsilon(x)}(x))f(\psi_{\tau_\varepsilon(x)}(x)) = e^{(l+k)\tau_\varepsilon(x)} V'(x)f(x) = \varepsilon^{\frac{l+k}{l}} V(x)^{-\frac{l+k}{l}} V'(x)f(x).$$

Let $c_1 \triangleq -\varepsilon^{-\frac{l+k}{l}} \min_{z \in \overline{\mathcal{Q}}_\varepsilon} V'(z)f(z)$ and $c_2 \triangleq -\varepsilon^{-\frac{l+k}{l}} \max_{z \in \overline{\mathcal{Q}}_\varepsilon} V'(z)f(z)$. Note that $c_1$ and $c_2$ are positive and well defined since $\overline{\mathcal{Q}}_\varepsilon$ is compact. Hence, the theorem is proved. $\qquad \square$

The following result represents the main application of homogeneity [37] to finite time semistability and finite time stabilization.

**Theorem 4.6.** Suppose $f$ is homogeneous of degree $k \in \mathbb{R}$ with respect to $\nu$. Then (4.1) is finite time semistable if and only if (4.1) is semistable and $k < 0$. In addition, if (4.1) is finite time semistable, then the settling-time function $T(\cdot)$ is homogeneous of degree $-k$ with respect to $\nu$ and $T(\cdot)$ is continuous on $\mathbb{R}^n$.

**Proof.** Since finite time semistability implies semistability, it suffices to prove that if (4.1) is semistable, then (4.1) is finite time semistable if and only if $k < 0$. Suppose that (4.1) is finite time semistable, and let $l > \max\{-k, 0\}$. Then for each $x_{\mathrm{e}} \in f^{-1}(0)$, it follows from Theorem 4.5 that there exist a bounded, open, and positively invariant set $\mathcal{S}$ containing $x_{\mathrm{e}}$ and a continuous nonnegative function $V : \mathcal{S} \to \overline{\mathbb{R}}_+$ that is homogeneous of degree $l + k$ and is such that $V'(x)f(x)$ is continuous, negative on $\mathcal{S} \backslash f^{-1}(0)$, and homogeneous of degree $l + k$, and (4.21) holds.

Now, suppose, *ad absurdum*, that $k \geq 0$ and $x \in \mathcal{S} \backslash f^{-1}(0)$. Then applying the comparison lemma (Theorem 4.2 in [319]) to the first inequality in (4.21) yields $V(s(t, x)) \geq \pi(t, V(x))$, where $\pi$ is given by

$$
\pi(t, x) = \begin{cases} \mathrm{sign}(x) \left( \frac{1}{|x|^{\alpha-1}} + c_1(\alpha - 1)t \right)^{-\frac{1}{\alpha-1}}, & \alpha > 1, \\ e^{-c_1 t} x, & \alpha = 1, \end{cases} \tag{4.27}
$$

and where $\mathrm{sign}(x) \triangleq x/|x|$, $x \neq 0$, and $\mathrm{sign}(0) \triangleq 0$, with $\alpha = l + k/l \geq 1$. Since, in this case, $\pi(t, V(x)) > 0$ for all $t \geq 0$, we have $s(t, x) \notin \mathcal{S} \cap f^{-1}(0)$ for every $t \geq 0$; that is, $x_{\mathrm{e}}$ is not a finite time semistable equilibrium under $f$, which is a contradiction. Hence, $k < 0$.

Conversely, if $k < 0$, pick $x_{\mathrm{e}} \in f^{-1}(0)$. Choose an open neighborhood $\mathcal{D}_0$ of $x_{\mathrm{e}}$ such that (4.22) holds. Next, $\mathcal{S}_{x_{\mathrm{e}}}$ is chosen to be a bounded, positively invariant neighborhood of $x_{\mathrm{e}}$ contained in $\mathcal{D}_0$. Then it follows from Theorem 4.5 that there exists a continuous nonnegative function $V(\cdot)$ such that (4.21) holds on $\mathcal{S}_{x_{\mathrm{e}}}$. Now, with $c = c_2 > 0$, $0 < \alpha = 1 + k/l < 1$, $\mathcal{D}_0 = \mathcal{S}_{x_{\mathrm{e}}}$, and $w(x) = -c\,\mathrm{sign}(x)|x|^\alpha$, it follows from Proposition 4.2 and Theorem 4.3 that $x_{\mathrm{e}}$ is finite time semistable on $\mathcal{S}_{x_{\mathrm{e}}}$.

Define $\mathcal{S} \triangleq \bigcup_{x_{\mathrm{e}} \in f^{-1}(0)} \mathcal{S}_{x_{\mathrm{e}}}$. Then $\mathcal{S}$ is an open neighborhood of $f^{-1}(0)$ such that every solution in $\mathcal{S}$ converges in finite time to a Lyapunov stable equilibrium. Hence, (4.1) is finite time semistable. Lemma 4.3 then implies that (4.1) is globally finite time semistable and $T(\cdot)$ is defined on $\mathbb{R}^n$. By Proposition 4.2 with $\mathcal{D}_0 = \mathcal{S}_{x_{\mathrm{e}}}$ and Theorem 4.3, it follows that $T(\cdot)$ is continuous on $\mathcal{S}_{x_{\mathrm{e}}}$. Next, since $x_{\mathrm{e}} \in f^{-1}(0)$ was chosen arbitrarily, it follows from Lemma 4.1 that $T(\cdot)$ is continuous on $\mathbb{R}^n$.

Finally, let $x \in \mathbb{R}^n$, and note that since every point in $\nu^{-1}(0) = f^{-1}(0)$ is a globally semistable equilibrium under $-\nu$, there exists $\tau > 0$ such that $z \triangleq \psi_{-\tau}(x) \in \mathcal{S}$. Then it follows from (4.20) that $s(t, x) = s(t, \psi_\tau(z)) = \psi_\tau(s(e^{k\tau}t, z))$, and, hence, $f(s(t, x)) = 0$ if and only if $f(s(e^{k\tau}t, z)) = 0$. Now, it follows that for $x \in \mathcal{S}$, $T(\psi_{-\tau}(x)) = T(z) = e^{k\tau}T(x)$. By definition, it follows that $T(\cdot)$ is homogeneous of degree $-k$ with respect to $\nu$. $\qquad\square$

In order to use Theorem 4.6 to prove finite time semistability of a homogeneous system, *a priori* information of semistability for the system is needed, which is not easy to obtain. To overcome this, we need to develop some sufficient conditions to establish finite time semistability. Recall that a function $V : \mathbb{R}^n \to \mathbb{R}$ is said to be *weakly proper* if and only if for every $c \in \mathbb{R}$, every connected component of the set $\{x \in \mathbb{R}^n : V(x) \leq c\} = V^{-1}((-\infty, c])$ is compact [36].

**Proposition 4.3.** Assume that $f$ is homogeneous of degree $k < 0$ with respect to $\nu$. Furthermore, assume that there exists a weakly proper, continuous function $V : \mathbb{R}^n \to \mathbb{R}$ such that $\dot{V}$ is defined on $\mathbb{R}^n$ and satisfies $\dot{V}(x) \leq 0$ for all $x \in \mathbb{R}^n$. If every point in the largest invariant subset $\mathcal{N}$ of $\dot{V}^{-1}(0)$ is a Lyapunov stable equilibrium point of (4.1), then (4.1) is finite time semistable.

**Proof.** Since $V(\cdot)$ is weakly proper, it follows from Proposition 3.1 of [36] that the positive orbit $s^x([0, \infty))$ of $x \in \mathbb{R}^n$ is bounded in $\mathbb{R}^n$. Since every solution is bounded, it follows from the hypotheses on $V(\cdot)$ that for every $x \in \mathbb{R}^n$, the omega limit set $\omega(x)$ is nonempty and contained in the largest invariant subset $\mathcal{N}$ of $\dot{V}^{-1}(0)$. Since every point in $\mathcal{N}$ is a Lyapunov stable equilibrium point, it follows from Proposition 5.4 of [36] that the omega limit set $\omega(x)$ contains a single point for every $x \in \mathbb{R}^n$. And since $\lim_{t \to \infty} s(t, x) \in \mathcal{N}$ is Lyapunov stable for every $x \in \mathbb{R}^n$, by definition, the system (4.1) is semistable. Hence, it follows from Theorem 4.6 that (4.1) is finite time semistable. $\qquad\square$

**Example 4.3.** Consider the nonlinear dynamical system given by

$$\dot{x}_1(t) = (x_2(t) - x_1(t))^{\frac{1}{3}} + (x_3(t) - x_1(t))^{\frac{1}{3}}, \quad x_1(0) = x_{10}, \quad t \geq 0, \quad (4.28)$$

$$\dot{x}_2(t) = (x_1(t) - x_2(t))^{\frac{1}{3}} + (x_3(t) - x_2(t))^{\frac{1}{3}}, \quad x_2(0) = x_{20}, \quad (4.29)$$

$$\dot{x}_3(t) = (x_1(t) - x_3(t))^{\frac{1}{3}} + (x_2(t) - x_3(t))^{\frac{1}{3}}, \quad x_3(0) = x_{30}, \quad (4.30)$$

where $x_i \in \mathbb{R}$, $i = 1, 2, 3$. For each $a \in \mathbb{R}$, $x_1 = x_2 = x_3 = a$ is the equilibrium point of (4.28)–(4.30). We show that all of the equilibrium points in (4.28)–(4.30) are finite time semistable. Note that the vector field

$f$ of (4.28)–(4.30) is homogeneous of degree $-2$ with respect to the semi-Euler vector field

$$\nu(x) = (2x_1 - x_2 - x_3)\frac{\partial}{\partial x_1} + (2x_2 - x_1 - x_3)\frac{\partial}{\partial x_2} + (2x_3 - x_1 - x_2)\frac{\partial}{\partial x_3}.$$

The differential operator notation in $\nu(x)$ is standard differential geometric notation used to write coordinate expressions for vector fields. This notation is based on the fact that there is a one-to-one correspondence between first-order linear differential operators on real-valued functions and vector fields.

Next, consider $V(x) = \frac{1}{2}x_1^2 + \frac{1}{2}x_2^2 + \frac{1}{2}x_3^2$. Then $\dot{V}(x(t)) \leq 0$, $t \geq 0$, and $\mathcal{N} = \{x \in \mathbb{R}^4 : x_1 = x_2 = x_3 = a\}$. Now, it follows from the Lyapunov function candidate

$$V(x - a\mathbf{e}) = \frac{1}{2}(x_1 - a)^2 + \frac{1}{2}(x_2 - a)^2 + \frac{1}{2}(x_3 - a)^2$$

that

$$\dot{V}(x - a\mathbf{e}) = -(x_1 - x_2)^{\frac{4}{3}} - (x_2 - x_3)^{\frac{4}{3}} - (x_3 - x_1)^{\frac{4}{3}} \leq 0,$$

which implies that every point in $\mathcal{N}$ is a Lyapunov stable equilibrium point of (4.28)–(4.30). Hence, it follows from Proposition 4.3 that the system (4.28)–(4.30) is finite time semistable. In fact, $x_1(t) = x_2(t) = x_3(t) = \frac{1}{3}(x_{10} + x_{20} + x_{30})$ for $t \geq T(x_0)$. Figure 4.2 shows the state trajectories versus time.                                                                    △

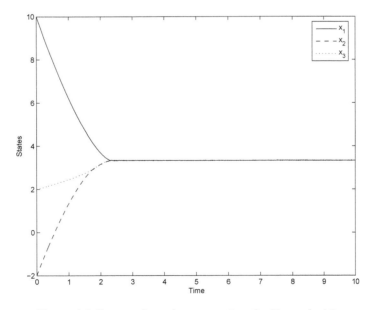

Figure 4.2 State trajectories versus time for Example 4.3.

Note that in Proposition 4.3 Lyapunov stability is needed for finite time semistability. However, finding the corresponding Lyapunov function can be a difficult task. To overcome this drawback, we use the nontangency-based approach [36] to guarantee finite time semistability by testing a condition on the vector field $f$, which avoids proving Lyapunov stability. Before we state this result, we need some new notation and definitions, which can be found in [36].

Given a set $\mathcal{E} \subseteq \mathbb{R}^n$, let $\mathrm{co}\,\mathcal{E}$ denote the union of the convex hulls of the connected components of $\mathcal{E}$, and let $\mathrm{coco}\,\mathcal{E}$ denote the cone generated by $\mathrm{co}\,\mathcal{E}$. Given $x \in \mathbb{R}^n$, the *direction cone* $\mathcal{F}_x$ of $f$ at $x$ relative to $\mathbb{R}^n$ is the intersection of all sets of the form $\overline{\mathrm{coco}\,(f(\mathcal{U})\backslash\{0\})}$, where $\mathcal{U} \subseteq \mathbb{R}^n$ is an open neighborhood of $x$. Let $z \in \mathcal{E} \subseteq \mathbb{R}^n$. A vector $v \in \mathbb{R}^n$ is *tangent* to $\mathcal{E}$ at $z \in \mathcal{E}$ if and only if there exist a sequence $\{z_i\}_{i=1}^\infty$ in $\mathcal{E}$ converging to $z$ and a sequence $\{h_i\}_{i=1}^\infty$ of positive real numbers converging to zero such that $\lim_{i\to\infty} \frac{1}{h_i}(z_i - z) = v$. The *tangent cone* to $\mathcal{E}$ at $z$ is the closed cone $T_z\mathcal{E}$ of all vectors tangent to $\mathcal{E}$ at $z$. Finally, the vector field $f$ is *nontangent* to the set $\mathcal{E}$ at the point $z \in \mathcal{E}$ if and only if $T_z\mathcal{E} \cap \mathcal{F}_z \subseteq \{0\}$.

**Proposition 4.4.** Assume that $f$ is homogeneous of degree $k < 0$ with respect to $\nu$. Furthermore, assume that there exists a weakly proper, continuous function $V : \mathbb{R}^n \to \mathbb{R}$ such that $\dot{V}$ is defined on $\mathbb{R}^n$ and satisfies $V(x) \geq 0$, $x \in \mathbb{R}^n$; $V(z) = 0$ for $z \in f^{-1}(0)$; and $\dot{V}(x) \leq 0$ for all $x \in \mathbb{R}^n$. For every $z \in f^{-1}(0)$, let $\mathcal{N}_z$ denote the largest negatively invariant connected subset of $\dot{V}^{-1}(0)$ containing $z$. If $f$ is nontangent to $\mathcal{N}_z$ at the point $z \in f^{-1}(0)$, then (4.1) is finite time semistable.

**Proof.** Since $V(x) \geq 0 = V(z)$ and $\dot{V}(x) \leq 0 = \dot{V}(z)$ for all $x \in \mathbb{R}$ and $z \in f^{-1}(0)$, with all the given conditions, it follows from (ii) of Theorem 7.1 of [36] that $x$ is Lyapunov stable. Now, it follows from Proposition 4.3 that (4.1) is finite time semistable. $\qquad\square$

**Example 4.4.** Consider the dynamical system given by

$$\dot{x}_1(t) = (x_3(t) - x_4(t))^{\frac{1}{3}}, \quad x_1(0) = x_{10}, \quad t \geq 0, \tag{4.31}$$

$$\dot{x}_2(t) = (x_4(t) - x_3(t))^{\frac{1}{3}}, \quad x_1(0) = x_{10}, \tag{4.32}$$

$$\dot{x}_3(t) = \mathrm{sign}(x_4(t) - x_3(t))(x_4(t) - x_3(t))^{\frac{2}{3}} + x_2(t) - x_1(t), \quad x_1(0) = x_{10}, \tag{4.33}$$

$$\dot{x}_4(t) = \mathrm{sign}(x_3(t) - x_4(t))(x_3(t) - x_4(t))^{\frac{2}{3}} + x_1(t) - x_2(t), \quad x_1(0) = x_{10}, \tag{4.34}$$

where $x_i \in \mathbb{R}$, $i = 1, 2, 3, 4$. For each $a, b \in \mathbb{R}$, $x_1 = x_2 = a$ and $x_3 = x_4 = b$ are the equilibrium points of (4.31)–(4.34). We show that all the equilibrium points in (4.31)–(4.34) are finite time semistable. Note that the vector field $f$ of (4.31)–(4.34) is homogeneous of degree $-2$ with respect to the semi-Euler vector field

$$\nu(x) = 2(x_1 - x_2)\frac{\partial}{\partial x_1} + 2(x_2 - x_1)\frac{\partial}{\partial x_2}$$
$$+ 3(x_3 - x_4)\frac{\partial}{\partial x_3} + 3(x_4 - x_3)\frac{\partial}{\partial x_4}.$$

Now, consider the function

$$V(x) = \frac{1}{2}(x_1 - x_2)^2 + \frac{3}{4}(x_3 - x_4)^{\frac{4}{3}},$$

and note that

$$\dot{V}(x) = -2|x_3 - x_4| \leq 0.$$

Let

$$\mathcal{R} \triangleq \{x \in \mathbb{R}^4 : \dot{V}(x) = 0\} = \{x \in \mathbb{R}^4 : x_3 = x_4\},$$

and let $\mathcal{N}$ denote the largest negatively invariant set contained in $\mathcal{R}$. On $\mathcal{N}$, it follows from (4.31)–(4.34) that $\dot{x}_1 = \dot{x}_2 = 0$, $\dot{x}_3 = \dot{x}_4 = 0$, and $x_1 = x_2$. Hence, $\mathcal{N} = \{x \in \mathbb{R}^4 : x_1 = x_2 = a, x_3 = x_4 = b\}$, $a, b \in \mathbb{R}$, which implies that $\mathcal{N}$ is the set of equilibrium points.

Next, we show that $f$ for (4.31)–(4.34) is nontangent to $\mathcal{N}$ at the point $z \in \mathcal{N}$. To see this, note that the tangent cone $T_z\mathcal{N}$ to the equilibrium set $\mathcal{N}$ is orthogonal to the vectors $\mathbf{u}_1 \triangleq [1, -1, 0, 0]^T$ and $\mathbf{u}_2 \triangleq [0, 0, 1, -1]^T$. On the other hand, since $f(z) \in \text{span}\{\mathbf{u}_1, \mathbf{u}_2\}$ for all $z \in \mathbb{R}^4$, it follows that the direction cone $\mathcal{F}$ of $f$ at $z \in \mathcal{N}$ relative to $\mathbb{R}^4$ satisfies $\mathcal{F}_z \subseteq \text{span}\{\mathbf{u}_1, \mathbf{u}_2\}$. Hence, $T_z\mathcal{N} \cap \mathcal{F}_z = \{0\}$, which implies that the vector field $f$ is nontangent to the set of equilibria $\mathcal{N}$ at the point $z \in \mathcal{N}$. Note that for every $z \in \mathcal{N}$, the set $\mathcal{N}_z$ required by Proposition 4.4 is contained in $\mathcal{N}$.

Since nontangency to $\mathcal{N}$ implies nontangency to $\mathcal{N}_z$ at the point $z \in \mathcal{N}$, it follows from Proposition 4.4 that the system (4.31)–(4.34) is finite time semistable. In particular, $x_1(t) = x_2(t) = \frac{1}{2}(x_{10} + x_{20})$ and $x_3(t) = x_4(t) = \frac{1}{2}(x_{30} + x_{40})$ for $t \geq T(x_0)$. Figure 4.3 shows the state trajectories versus time. $\triangle$

## 4.6 The Consensus Problem in Dynamical Networks

In this section, we revisit the nonlinear consensus problem introduced in Chapter 2. Specifically, we recall that $\mathfrak{G} = (\mathcal{V}, \mathcal{E}, \mathcal{A})$ is a directed graph (or

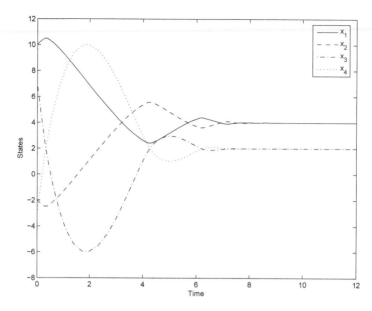

Figure 4.3 State trajectories versus time for Example 4.4.

digraph) denoting the dynamical network (or dynamic graph) with the set of nodes (or vertices) $\mathcal{V} = \{1, \ldots, q\}$ involving a finite nonempty set denoting the agents; the set of edges $\mathcal{E} \subseteq \mathcal{V} \times \mathcal{V}$ involving a set of ordered pairs denoting the direction of information flow; and an adjacency (or connectivity) matrix $\mathcal{A} \in \mathbb{R}^{q \times q}$ such that $\mathcal{A}_{(i,j)} = 1$, $i, j = 1, \ldots, q$, if $(j, i) \in \mathcal{E}$, while $\mathcal{A}_{(i,j)} = 0$ if $(j, i) \notin \mathcal{E}$. The edge $(j, i) \in \mathcal{E}$ denotes that agent $j$ can obtain information from agent $i$, but not necessarily vice versa. Moreover, we assume that $\mathcal{A}_{(i,i)} = 0$ for all $i \in \mathcal{V}$. A graph or undirected graph $\mathfrak{G}$ associated with the adjacency matrix $\mathcal{A} \in \mathbb{R}^{q \times q}$ is a directed graph for which the arc set is symmetric; that is, $\mathcal{A} = \mathcal{A}^{\mathrm{T}}$.

Weighted graphs can also be considered here; however, since this extension does not alter any of the conceptual results in the chapter, we do not consider it for simplicity of exposition. Finally, we denote the *value* of the node $i \in \{1, \ldots, q\}$ at time $t$ by $x_i(t) \in \mathbb{R}$.

Recall that the consensus problem involves the network system characterized by the dynamical system $\mathcal{G}$ given by

$$\dot{x}_i(t) = \sum_{j=1, j \neq i}^{q} \phi_{ij}(x_i(t), x_j(t)), \quad x_i(t_0) = x_{i0}, \quad t \geq t_0, \quad i = 1, \ldots, q,$$
(4.35)

or, in vector form,

$$\dot{x}(t) = f(x(t)), \quad x(t_0) = x_0, \quad t \geq t_0,$$
(4.36)

where $x(t) \triangleq [x_1(t), \ldots, x_q(t)]^{\mathrm{T}}$, $t \geq t_0$, and $f = [f_1, \ldots, f_q]^{\mathrm{T}} : \mathbb{R}^q \to \mathbb{R}^q$ is such that $f_i(x) = \sum_{j=1, j \neq i}^q \phi_{ij}(x_i, x_j)$.

For the statement of the main results of this section, recall the following definition and assumptions from Chapter 2.

**Definition 4.5** (see [24]). A directed graph $\mathfrak{G}$ is *strongly connected* if for any ordered pair of vertices $(i, j)$, $i \neq j$, there exists a *path* (i.e., a sequence of arcs) leading from $i$ to $j$.

**Assumption 4.1.** The *connectivity matrix*[4] $\mathcal{C} \in \mathbb{R}^{q \times q}$ associated with the multiagent dynamical system $\mathcal{G}$ is defined by

$$\mathcal{C}_{(i,j)} \triangleq \begin{cases} 0 & \text{if } \phi_{ij}(x_i, x_j) \equiv 0, \\ 1, & \text{otherwise,} \end{cases} \quad i \neq j, \quad i, j = 1, \ldots, q, \quad (4.37)$$

and $\mathcal{C}_{(i,i)} = -\sum_{k=1, k \neq i}^q \mathcal{C}_{(i,k)}$, $i = 1, \ldots, q$, with $\operatorname{rank} \mathcal{C} = q - 1$, and for $\mathcal{C}_{(i,j)} = 1$, $i \neq j$, $\phi_{ij}(x_i, x_j) = 0$ if and only if $x_i = x_j$.

**Assumption 4.2.** For $i, j = 1, \ldots, q$, $(x_i - x_j)\phi_{ij}(x_i, x_j) \leq 0$, $x_i, x_j \in \mathbb{R}$.

As discussed in Chapter 2, Assumptions 4.1 and 4.2 are a restatement of the zeroth law and second law of thermodynamics as applied to the network consensus control problem.

**Proposition 4.5.** Consider the multiagent dynamical system (4.36), and assume that Assumptions 4.1 and 4.2 hold. Then $f_i(x) = 0$ for all $i = 1, \ldots, q$ if and only if $x_1 = \cdots = x_q$. Furthermore, $\alpha e$, $\alpha \in \mathbb{R}$, is an equilibrium state of (4.36).

**Proof.** This is a restatement of Proposition 2.1. $\square$

**Theorem 4.7.** Consider the multiagent dynamical system (4.36), and assume that Assumptions 4.1 and 4.2 hold. Furthermore, assume that $\phi_{ij}(x_i, x_j) = -\phi_{ji}(x_j, x_i)$ for all $i, j = 1, \ldots, q$, $i \neq j$. Then, for every $\alpha \in \mathbb{R}$, $\alpha e$ is a semistable equilibrium state of (4.36). Furthermore, $x(t) \to \frac{1}{q} e e^{\mathrm{T}} x(t_0)$ as $t \to \infty$, and $\frac{1}{q} e e^{\mathrm{T}} x(t_0)$ is a semistable equilibrium state.

**Proof.** This is a restatement of Proposition 2.1 and (i) of Theorem 2.1. $\square$

---

[4]The negative of the connectivity matrix, that is, $-\mathcal{C}$, is known as the Laplacian of the directed graph $\mathfrak{G}$ in the literature.

Theorem 4.7 implies that the steady-state value of the information state in each agent $\mathcal{G}_i$ of the multiagent dynamical system $\mathcal{G}$ is equal; that is, the steady-state value of the multiagent dynamical system $\mathcal{G}$ given by

$$x_\infty = \frac{1}{q} \mathbf{e} \mathbf{e}^{\mathrm{T}} x(t_0) = \left[ \frac{1}{q} \sum_{i=1}^{q} x_i(t_0) \right] \mathbf{e}$$

is uniformly distributed over all multiagents of $\mathcal{G}$.

## 4.7 Distributed Control Algorithms for Finite Time Consensus and Parallel Formation

In this section, we combine the thermodynamically motivated information consensus framework for multiagent dynamic networks developed in Section 4.6 with the finite time semistability and homogeneity theory developed in Sections 4.3–4.5 to design distributed finite time consensus protocols for cooperative network systems. Specifically, consider $q$ continuous-time integrator agents with dynamics

$$\dot{x}_i(t) = u_i(t), \quad x_i(0) = x_{i0}, \quad t \geq 0, \tag{4.38}$$

where for each $i \in \{1, \ldots, q\}$, $x_i(t) \in \mathbb{R}$ denotes the information state and $u_i(t) \in \mathbb{R}$ denotes the information control input for all $t \geq 0$. The general consensus protocol is given by

$$u_i(t) = \sum_{j=1, j \neq i}^{q} \phi_{ij}(x_i(t), x_j(t)), \tag{4.39}$$

where $\phi_{ij}(\cdot, \cdot)$ satisfies Assumptions 4.1 and 4.2 and $\phi_{ij}(x_i, x_j) = -\phi_{ji}(x_j, x_i)$ for all $i, j = 1, \ldots, q$, $i \neq j$. Note that (4.38) and (4.39) describe an interconnected network where information states are updated using a distributed controller involving neighbor-to-neighbor interaction between agents.

**Theorem 4.8.** Consider the closed-loop multiagent system $\mathcal{G}$ given by (4.38) and (4.39). Assume that Assumptions 4.1 and 4.2 hold and $\phi_{ij}(x_i, x_j) = -\phi_{ji}(x_j, x_i)$ for all $i, j = 1, \ldots, q$, $i \neq j$. Furthermore, assume that the vector field $f$ of the closed-loop system (4.38) and (4.39) is homogeneous of degree $k \in \mathbb{R}$ with respect to

$$\nu(x) = -\sum_{i=1}^{q} \left[ \sum_{j=1, j \neq i}^{q} \mu_{ij}(x_i, x_j) \right] \frac{\partial}{\partial x_i},$$

where $x \triangleq [x_1, \ldots, x_q]^{\mathrm{T}} \in \mathbb{R}^q$ and $\mu_{ij}(\cdot, \cdot)$ satisfies Assumption 4.2; $\mu_{ij}(x_i, x_j) = -\mu_{ji}(x_j, x_i)$; and $\mu_{ij}(x_i, x_j) = 0$ if and only if $x_i = x_j$ for all $i, j = 1, \ldots, q$,

$i \neq j$. Then, for every $x_e \in \mathbb{R}$, $x_e \mathbf{e}$ is a finite time semistable equilibrium state of $\mathcal{G}$ if and only if $k < 0$. Furthermore, if $k < 0$, then $x(t) = \frac{1}{q}\mathbf{e}\mathbf{e}^T x(0)$ for all $t \geq T(x(0))$ and $\frac{1}{q}\mathbf{e}\mathbf{e}^T x(0)$ is a finite time semistable equilibrium state, where $T(x(0)) \geq 0$.

**Proof.** Suppose that $k < 0$. It follows from Theorem 4.7 that $x_e \mathbf{e} \in \mathbb{R}^q$, $x_e \in \mathbb{R}$, is a semistable equilibrium state of the closed-loop homogeneous system (4.38) and (4.39). Furthermore, $x(t) \to \frac{1}{q}\mathbf{e}\mathbf{e}^T x(0)$ as $t \to \infty$, and $\frac{1}{q}\mathbf{e}\mathbf{e}^T x(0)$ is a semistable equilibrium state. Next, it can be shown using similar arguments as in the proof of Theorem 4.7 that (4.19) is globally semistable with

$$\nu(x) = -\sum_{i=1}^{q} \left[ \sum_{j=1, j \neq i}^{q} \mu_{ij}(x_i, x_j) \right] \frac{\partial}{\partial x_i}.$$

Now, it follows from Theorem 4.6 that $x_e \mathbf{e}$ is a finite time semistable equilibrium state by noting that the vector field $\sum_{j=1, j\neq i}^{q} \phi_{ij}(x_i, x_j)$ is homogeneous of degree $k < 0$ with respect to the semi-Euler vector field

$$\nu(x) = -\sum_{i=1}^{q} \left[ \sum_{j=1, j \neq i}^{q} \mu_{ij}(x_i, x_j) \right] \frac{\partial}{\partial x_i}.$$

Hence, with $x_e = \frac{1}{q}\mathbf{e}^T x(0)$, $x_e \mathbf{e} = \frac{1}{q}\mathbf{e}\mathbf{e}^T x(0)$ is a finite time semistable equilibrium state. The converse follows as a direct consequence of Theorem 4.6. $\square$

The following corollary to Theorem 4.8 gives a concrete form for

$$\phi_{ij}(x_i, x_j), \quad i, j = 1, \ldots, q, \quad i \neq j.$$

**Corollary 4.1.** Consider the closed-loop multiagent system $\mathcal{G}$ given by (4.38) and (4.39) with

$$\phi_{ij}(x_i, x_j) = \mathcal{C}_{(i,j)} \operatorname{sign}(x_j - x_i)|x_j - x_i|^{\alpha}, \tag{4.40}$$

where $\alpha > 0$ and $\mathcal{C}_{(i,j)}$ is as in (4.37) with $\mathcal{C} = \mathcal{C}^T$. Assume that Assumptions 4.1 and 4.2 hold. Then, for every $x_e \in \mathbb{R}$, $x_e \mathbf{e}$ is a finite time semistable equilibrium state of $\mathcal{G}$ if and only if $\alpha < 1$. Furthermore, if $\alpha < 1$, then $x(t) = \frac{1}{q}\mathbf{e}\mathbf{e}^T x(0)$ for all $t \geq T(x(0))$ and $\frac{1}{q}\mathbf{e}\mathbf{e}^T x(0)$ is a finite time semistable equilibrium state, where $T(x(0)) \geq 0$.

**Proof.** The Lie bracket of $\nu(x) = -\sum_{i=1}^{q} \left[ \sum_{j=1, j\neq i}^{q} (x_j - x_i) \right] \frac{\partial}{\partial x_i}$ and the vector field $f$ of the closed-loop system (4.38) and (4.39) with (4.40) is given by

$$[\nu, f] = \left[ \sum_{i=1}^{q} \frac{\partial f_1}{\partial x_i}\nu_i - \frac{\partial \nu_1}{\partial x_i}f_i, \ldots, \sum_{i=1}^{q} \frac{\partial f_q}{\partial x_i}\nu_i - \frac{\partial \nu_q}{\partial x_i}f_i \right]^{\mathrm{T}}.$$

Since, for each $i, j = 1, \ldots, q$,

$$\frac{\partial f_j}{\partial x_i}\nu_i - \frac{\partial \nu_j}{\partial x_i}f_i$$

$$= \begin{cases} \mathcal{C}_{(j,i)}\alpha|x_i - x_j|^{\alpha-1}\left[\sum_{s=1,s\neq i}^{q}(x_i - x_s)\right] \\ \quad + \sum_{k=1,k\neq i}^{q}\mathcal{C}_{(i,k)}\mathrm{sign}(x_k - x_i)|x_k - x_i|^{\alpha}, & i \neq j, \\ \left[\sum_{k=1,k\neq j}^{q}\mathcal{C}_{(j,k)}\alpha|x_k - x_j|^{\alpha-1}\right]\left[\sum_{s=1,s\neq j}^{q}(x_s - x_j)\right] \\ \quad -(q-1)\sum_{k=1,k\neq j}^{q}\mathcal{C}_{(j,k)}\mathrm{sign}(x_k - x_j)|x_k - x_j|^{\alpha}, & i = j, \end{cases}$$

and noting that $\mathcal{C}_{(i,j)} = \mathcal{C}_{(j,i)}$, $i, j = 1, \ldots, q$, $i \neq j$, it follows that for each $j = 1, \ldots, q$,

$$\sum_{i=1}^{q}\frac{\partial f_j}{\partial x_i}\nu_i - \frac{\partial \nu_j}{\partial x_i}f_i$$

$$= \frac{\partial f_j}{\partial x_j}\nu_j - \frac{\partial \nu_j}{\partial x_j}f_j + \sum_{i=1,i\neq j}^{q}\frac{\partial f_j}{\partial x_i}\nu_i - \frac{\partial \nu_j}{\partial x_i}f_i$$

$$= \alpha \sum_{k=1,k\neq j}^{q}\mathcal{C}_{(j,k)}\mathrm{sign}(x_k - x_j)|x_k - x_j|^{\alpha}$$

$$+ \sum_{k=1,k\neq j}^{q}\sum_{s=1,s\neq j,k}^{q}\mathcal{C}_{(j,k)}\alpha|x_k - x_j|^{\alpha-1}(x_s - x_j)$$

$$- (q-1)\sum_{k=1,k\neq j}^{q}\mathcal{C}_{(j,k)}\mathrm{sign}(x_k - x_j)|x_k - x_j|^{\alpha}$$

$$+ \alpha \sum_{i=1,i\neq j}^{q}\mathcal{C}_{(j,i)}\mathrm{sign}(x_i - x_j)|x_i - x_j|^{\alpha}$$

$$+ \sum_{i=1,i\neq j}^{q}\sum_{s=1,s\neq i,j}^{q}\mathcal{C}_{(j,i)}\alpha|x_i - x_j|^{\alpha-1}(x_i - x_s)$$

$$+ \sum_{i=1}^{q}\sum_{k=1,k\neq i}^{q}\mathcal{C}_{(i,k)}\mathrm{sign}(x_k - x_i)|x_k - x_i|^{\alpha}$$

$$- \sum_{k=1,k\neq j}^{q}\mathcal{C}_{(j,k)}\mathrm{sign}(x_k - x_j)|x_k - x_j|^{\alpha}$$

$$= q(\alpha - 1) \sum_{i=1, i \neq j}^{q} \mathcal{C}_{(j,i)} \text{sign}(x_i - x_j) |x_i - x_j|^{\alpha}$$

$$= q(\alpha - 1) f_j, \tag{4.41}$$

which implies that the vector field $f$ is homogeneous of degree $k = q(\alpha - 1)$ with respect to the semi-Euler vector field

$$\nu(x) = - \sum_{i=1}^{q} \left[ \sum_{j=1, j \neq i}^{q} (x_j - x_i) \right] \frac{\partial}{\partial x_i}.$$

Now, the result is a direct consequence of Theorem 4.8.        □

Note that Example 4.3 serves as a special case of Corollary 4.1. More important, note that the proposed protocol (4.40) is different from the protocols given in [69, 73] since (4.40) is a distributed *continuous* protocol and is not based on a nonsmooth gradient flow. Furthermore, this protocol does not satisfy the conditions of Theorem 4 or Theorem 5 of [69]. It is also important to note that the proposed protocol can achieve superior performance over the protocols given in [69] since the closed-loop system generated by (4.40) results in continuous closed-loop vector fields as opposed to discontinuous closed-loop vector fields based on nonsmooth gradient flows, which can lead to chattering behavior. In addition, the proposed protocol tends to have a faster settling time. Finally, a key advantage of continuous (but non-Lipschitzian) closed-loop systems over Lipschitzian closed-loop systems is that continuous finite time controllers tend to have better robustness and disturbance rejection properties [32, 34].

A majority of the results in the literature consider only *static* consensus protocols. A natural question regarding (4.38) is how to design finite time *dynamic* compensators to achieve network consensus. This question is important because dynamic controllers can be used to design finite time consensus protocols for multiagent coordination via output feedback. To begin to address this question, we consider $q$ continuous-time integrator agents given by (4.38) and the dynamic compensators given by

$$\dot{x}_{ci}(t) = \sum_{j=1, j \neq i}^{q} \phi_{ij}(x_{ci}(t), x_{cj}(t)) + \sum_{j=1, j \neq i}^{q} \eta_{ij}(x_i(t), x_j(t)),$$

$$x_{ci}(0) = x_{ci0}, \quad t \geq 0, \tag{4.42}$$

$$u_i(t) = - \sum_{j=1, j \neq i}^{q} \mu_{ij}(x_{ci}(t), x_{cj}(t)), \tag{4.43}$$

where $\phi_{ij}(\cdot,\cdot)$, $\eta_{ij}(\cdot,\cdot)$, and $\mu_{ij}(\cdot,\cdot)$, $i,j = 1,\ldots,q$, $i \neq j$, satisfy Assumptions 4.1 and 4.2. Furthermore, $\phi_{ij}(\cdot,\cdot)$, $\eta_{ij}(\cdot,\cdot)$, and $\mu_{ij}(\cdot,\cdot)$ are chosen such that the vector field of the closed-loop system (4.38), (4.42), and (4.43) is homogeneous with respect to given semi-Euler vector fields. Recall that if the closed-loop system is semistable and homogeneous of degree $k < 0$ with respect to a given semi-Euler vector field, then the closed-loop system is finite time semistable.

As an example, consider

$$\phi_{ij}(x_{\mathrm{c}i}, x_{\mathrm{c}j}) = \mathcal{C}_{(i,j)}\mathrm{sign}(x_{\mathrm{c}j} - x_{\mathrm{c}i})|x_{\mathrm{c}j} - x_{\mathrm{c}i}|^{\frac{1+\alpha}{2}},$$
$$\mu_{ij}(x_{\mathrm{c}i}, x_{\mathrm{c}j}) = \mathcal{C}_{(i,j)}\mathrm{sign}(x_{\mathrm{c}j} - x_{\mathrm{c}i})|x_{\mathrm{c}j} - x_{\mathrm{c}i}|^{\alpha},$$

and

$$\eta_{ij}(x_i, x_j) = \mathcal{C}_{(i,j)}(x_j - x_i)$$

for $0 < \alpha < 1$ and $i,j = 1,\ldots,q$, $i \neq j$. Note that the dynamic compensator (4.42) has a similar structure to (4.36) with additional input supply. Thus, the proposed controller architecture can be viewed as an *interconnection of thermodynamic controllers*; for details see [127]. Finally, note that Example 4.4 is a special case of the closed-loop system given by (4.38), (4.42), and (4.43) with $\phi_{ij}(\cdot,\cdot)$, $\eta_{ij}(\cdot,\cdot)$, and $\mu_{ij}(\cdot,\cdot)$, $i,j = 1,\ldots,q$, $i \neq j$, as specified above.

**Theorem 4.9.** Consider the closed-loop system given by (4.38), (4.42), and (4.43) with $\phi_{ij}(\cdot,\cdot)$, $\eta_{ij}(\cdot,\cdot)$, and $\mu_{ij}(\cdot,\cdot)$, $i,j = 1,\ldots,q$, $i \neq j$, as specified above. Assume that Assumptions 4.1 and 4.2 hold and $\mathcal{C} = \mathcal{C}^{\mathrm{T}}$. Then, for every $a \in \mathbb{R}$ and $b \in \mathbb{R}$, $(x(t), x_{\mathrm{c}}(t)) \equiv (a\mathbf{e}, b\mathbf{e})$ is a finite time semistable equilibrium state of (4.38), (4.42), and (4.43). Furthermore, $x(t) = \frac{1}{q}\mathbf{e}\mathbf{e}^{\mathrm{T}}x(0)$ and $x_{\mathrm{c}}(t) = \frac{1}{q}\mathbf{e}\mathbf{e}^{\mathrm{T}}x_{\mathrm{c}}(0)$ for all $t \geq T(x(0), x_{\mathrm{c}}(0))$, and $(\frac{1}{q}\mathbf{e}\mathbf{e}^{\mathrm{T}}x(0), \frac{1}{q}\mathbf{e}\mathbf{e}^{\mathrm{T}}x_{\mathrm{c}}(0))$ is a finite time semistable equilibrium state.

**Proof.** Let $\lambda > 0$. Using similar arguments as in the proof of Corollary 4.1, it can be shown that the closed-loop system given by (4.38), (4.42), and (4.43) is homogeneous of degree $k = q\lambda\frac{\alpha-1}{1+\alpha} < 0$ with respect to the semi-Euler vector field

$$\nu(x, x_{\mathrm{c}}) = -\lambda \sum_{i=1}^{q}\left[\sum_{j=1,j\neq i}^{q}(x_j - x_i)\right]\frac{\partial}{\partial x_i}$$
$$-\frac{2\lambda}{1+\alpha}\sum_{i=1}^{q}\left[\sum_{j=1,j\neq i}^{q}(x_{\mathrm{c}j} - x_{\mathrm{c}i})\right]\frac{\partial}{\partial x_{\mathrm{c}i}}.$$

Next, note that, for every $a, b \in \mathbb{R}$, $x(t) \equiv a\mathbf{e}$ and $x_c(t) \equiv b\mathbf{e}$ are the equilibrium points for the closed-loop system. Consider the nonnegative function given by

$$V(\tilde{x}) = \frac{1}{4} \sum_{i=1}^{q} \sum_{j=1, j\neq i}^{q} \mathcal{C}_{(i,j)}(x_i - x_j)^2$$

$$+ \frac{1}{2 + 2\alpha} \sum_{i=1}^{q} \sum_{j=1, j\neq i}^{q} \mathcal{C}_{(i,j)} |x_{ci} - x_{cj}|^{1+\alpha}, \qquad (4.44)$$

where $\tilde{x} \triangleq [x^{\mathrm{T}}, x_c^{\mathrm{T}}]^{\mathrm{T}} \in \mathbb{R}^{2q}$. In this case, the derivative of $V(\cdot)$ along the trajectories of the closed-loop system is given by

$$\dot{V}(\tilde{x}) = -2 \sum_{i=1}^{q} \sum_{j=i+1}^{q-1} \mu_{ij}(x_{ci}, x_{cj}) \phi_{ij}(x_{ci}, x_{cj}) \leq 0, \quad \tilde{x} \in \mathbb{R}^{2q}.$$

Let

$$\mathcal{R} \triangleq \{\tilde{x} \in \mathbb{R}^{2q} : \dot{V}(\tilde{x}) = 0\} = \{\tilde{x} \in \mathbb{R}^{2q} : x_{c1} = \cdots = x_{cq}\},$$

and let $\mathcal{N}$ denote the largest negatively invariant set of $\mathcal{R}$. On $\mathcal{N}$, it follows from (4.38), (4.42), and (4.43) that $\dot{x}_i = 0$, $\dot{x}_{ci} = 0$, and $x_1 = \cdots = x_q$, $i = 1, \ldots, q$. Hence, $\mathcal{N} = \{\tilde{x} \in \mathbb{R}^{2q} : x = a\mathbf{e}, x_c = b\mathbf{e}\}$, $a, b \in \mathbb{R}$, which implies that $\mathcal{N}$ is the set of equilibrium points.

Since the graph $\mathfrak{G}$ of the closed-loop system is strongly connected, assume, without loss of generality, that $\mathcal{C}_{(i,i+1)} = \mathcal{C}_{(q,1)} = 1$, where $i = 1, \ldots, q-1$. Now, for $q = 2$, it was shown in Example 4.4 that the vector field $f$ of the closed-loop system given by (4.38), (4.42), and (4.43) is nontangent to $\mathcal{N}$ at a point $\tilde{x} \in \mathcal{N}$. Next, we show that, for $q \geq 3$, the vector field $f$ of the closed-loop system given by (4.38), (4.42), and (4.43) is nontangent to $\mathcal{N}$ at a point $\tilde{x} \in \mathcal{N}$.

To see this, note that the tangent cone $T_{\tilde{x}}\mathcal{N}$ to the equilibrium set $\mathcal{N}$ is orthogonal to the $2q$ vectors

$$\mathbf{u}_i \triangleq [0_{1\times(i-1)}, \mathcal{C}_{(i,i+1)}, -\mathcal{C}_{(i,i+1)}, 0_{1\times(2q-i-1)}]^{\mathrm{T}} \in \mathbb{R}^{2q},$$

$$\mathbf{u}_q \triangleq [-\mathcal{C}_{(q,1)}, 0_{1\times(q-2)}, \mathcal{C}_{(q,1)}, 0_{1\times q}]^{\mathrm{T}} \in \mathbb{R}^{2q},$$

$$\mathbf{v}_i \triangleq [0_{1\times(q+i-1)}, -\mathcal{C}_{(i,i+1)}, \mathcal{C}_{(i,i+1)}, 0_{1\times(q-i-1)}]^{\mathrm{T}} \in \mathbb{R}^{2q},$$

$$\mathbf{v}_q \triangleq [0_{1\times q}, \mathcal{C}_{(q,1)}, 0_{1\times(q-2)}, -\mathcal{C}_{(q,1)}]^{\mathrm{T}} \in \mathbb{R}^{2q},$$

$i = 1, \ldots, q-1$, $q \geq 3$. Alternatively, since

$$f(\tilde{x}) \in \mathrm{span}\{\mathbf{u}_1, \ldots, \mathbf{u}_q, \mathbf{v}_1, \ldots, \mathbf{v}_q\}$$

for all $\tilde{x} \in \mathbb{R}^{2q}$, it follows that the direction cone $\mathcal{F}_{\tilde{x}}$ of $f$ at $\tilde{x} \in \mathcal{N}$ relative to $\mathbb{R}^{2q}$ satisfies $\mathcal{F}_{\tilde{x}} \subseteq \mathrm{span}\{\mathbf{u}_1, \ldots, \mathbf{u}_q, \mathbf{v}_1, \ldots, \mathbf{v}_q\}$. Hence, $T_{\tilde{x}}\mathcal{N} \cap \mathcal{F}_{\tilde{x}} = \{0\}$, which implies that the vector field $f$ is nontangent to the set of equilibria $\mathcal{N}$ at the point $\tilde{x} \in \mathcal{N}$. Note that for every $z \in \mathcal{N}$, the set $\mathcal{N}_z$ required by Proposition 4.4 is contained in $\mathcal{N}$. Since nontangency to $\mathcal{N}$ implies nontangency to $\mathcal{N}_z$ at the point $z \in \mathcal{N}$, it follows from Proposition 4.4 that the closed-loop system (4.38), (4.42), and (4.43) is finite time semistable. $\square$

Finally, we apply the developed theory to design finite time distributed controllers for parallel formations [183] such as flocking [234]. Specifically, consider $q$ continuous-time double integrator agents with dynamics

$$\ddot{x}_i(t) = u_i(t), \quad x_i(0) = x_{i0}, \quad \dot{x}_i(0) = \dot{x}_{i0}, \quad t \geq 0, \qquad (4.45)$$

where, for each $i \in \{1, \ldots, q\}$, $x_i(t) = [x_{1i}(t), x_{2i}(t), x_{3i}(t)]^{\mathrm{T}} \in \mathbb{R}^3$ denotes the position, $\dot{x}_i(t) = [\dot{x}_{1i}(t), \dot{x}_{2i}(t), \dot{x}_{3i}(t)]^{\mathrm{T}} \in \mathbb{R}^3$ denotes the velocity, and $u_i(t) = [u_{1i}(t), u_{2i}(t), u_{3i}(t)]^{\mathrm{T}} \in \mathbb{R}^3$ is the control input. We seek a continuous distributed feedback control law $u_i$ involving transmission of both $x_i$ and $\dot{x}_i$ between agents so that *finite time parallel formation* is achieved; that is, the velocity $\dot{x}_i$ reaches a constant vector in finite time for all $i = 1, \ldots, q$, and the relative position between two agents reaches a constant value in finite time.

**Theorem 4.10.** Consider the dynamical system given by (4.45). Then finite time parallel formation for (4.45) is achieved under the distributed feedback control law given by the static controller

$$u_{ri} = \sum_{j=1,j\neq i}^{q} \phi_{rij}(\dot{x}_{ri}, \dot{x}_{rj}) - \sum_{j=1,j\neq i}^{q} \mathcal{C}_{(i,j)}\mathrm{sign}(\psi_\alpha(x_{ri}, x_{rj}))\big|\psi_\alpha(x_{ri}, x_{rj})\big|^{\frac{\alpha}{2-\alpha}},$$

$$(4.46)$$

where $0 < \alpha < 1$, $\phi_{rij}(\dot{x}_{ri}, \dot{x}_{rj}) = \mathcal{C}_{(i,j)}\mathrm{sign}(\dot{x}_{rj} - \dot{x}_{ri})|\dot{x}_{rj} - \dot{x}_{ri}|^\alpha$ satisfies Assumptions 4.1 and 4.2, $\mathcal{C}_{(i,j)}$ is as in (4.37) with $\mathcal{C} = \mathcal{C}^{\mathrm{T}}$, $\psi_\alpha(x_{ri}, x_{rj}) \triangleq x_{ri} - x_{rj} - d_{rij}$, and $d_{rij} = -d_{rji} \in \mathbb{R}$, $i, j = 1, \ldots, q$, $i \neq j$, $r = 1, 2, 3$.

**Proof.** For the distributed control law (4.46), let $z_{rij} \triangleq \psi_\alpha(x_{ri}, x_{rj})$, $i, j = 1, \ldots, q$, $i \neq j$, $r = 1, 2, 3$, and consider the augmented closed-loop system

$$\dot{z}_{rij}(t) = \dot{x}_{ri}(t) - \dot{x}_{rj}(t), \quad z_{rij}(0) = z_{rij0}, \quad t \geq 0, \quad i, j = 1, \ldots, q,$$
$$i \neq j, \quad r = 1, 2, 3, \qquad (4.47)$$
$$\ddot{x}_{ri}(t) = \sum_{j=1,j\neq i}^{q} \phi_{rij}(\dot{x}_{ri}(t), \dot{x}_{rj}(t)) - \sum_{j=1,j\neq i}^{q} \mathcal{C}_{(i,j)}\mathrm{sign}(z_{rij}(t))|z_{rij}(t)|^{\frac{\alpha}{2-\alpha}},$$

$$\dot{x}_{ri}(0) = \dot{x}_{ri0}. \tag{4.48}$$

It can be shown using similar arguments as in the proof of Corollary 4.1 that the closed-loop system given by (4.47) and (4.48) is homogeneous of degree $k = q(\alpha - 1) < 0$ with respect to the semi-Euler vector field

$$\nu_r = -\sum_{i=1}^{q} \left[ \sum_{j=1,j\neq i}^{q} (\dot{x}_{rj} - \dot{x}_{ri}) \right] \frac{\partial}{\partial \dot{x}_{ri}} + q(2-\alpha) \sum_{i=1}^{q} \sum_{j=1,j\neq i}^{q} z_{rij} \frac{\partial}{\partial z_{rij}}.$$

Next, consider the nonnegative function

$$V_r(z_r, \dot{x}_{(r)}) = \frac{1}{2} \sum_{i=1}^{q} \dot{x}_{ri}^2 + \frac{2-\alpha}{4} \sum_{i=1}^{q} \sum_{j=1,j\neq i}^{q} \mathcal{C}_{(i,j)} |z_{rij}|^{\frac{2}{2-\alpha}}, \tag{4.49}$$

where $z_r \triangleq [z_{r12}, z_{r13}, \ldots, z_{r1q}, z_{r21}, z_{r23}, \ldots, z_{r2q}, \ldots, z_{rq(q-1)}] \in \mathbb{R}^{q^2-q}$ and $x_{(r)} \triangleq [x_{r1}, \ldots, x_{rq}]^{\mathrm{T}} \in \mathbb{R}^q$, $r = 1, 2, 3$. In this case, the derivative of $V_r(\cdot)$ along the trajectories of the closed-loop system is given by

$$
\begin{aligned}
\dot{V}_r(z_r, \dot{x}_{(r)}) =& \sum_{i=1}^{q} \dot{x}_{ri} \sum_{j=1,j\neq i}^{q} \phi_{rij}(\dot{x}_{ri}, \dot{x}_{rj}) \\
& - \sum_{i=1}^{q} \dot{x}_{ri} \sum_{j=1,j\neq i}^{q} \mathcal{C}_{(i,j)} \mathrm{sign}(\psi_\alpha(x_{ri}, x_{rj})) \\
& \cdot |\psi_\alpha(x_{ri}, x_{rj})|^{\frac{\alpha}{2-\alpha}} + \frac{1}{2} \sum_{i=1}^{q} \sum_{j=1,j\neq i}^{q} \mathcal{C}_{(i,j)} \mathrm{sign}(\psi_\alpha(x_{ri}, x_{rj})) \\
& \cdot |\psi_\alpha(x_{ri}, x_{rj})|^{\frac{\alpha}{2-\alpha}} (\dot{x}_{ri} - \dot{x}_{rj}) \\
=& \sum_{i=1}^{q-1} \sum_{j=i+1}^{q} (\dot{x}_{ri} - \dot{x}_{rj}) \phi_{rij}(\dot{x}_{ri}, \dot{x}_{rj}) \\
\leq& 0, \quad (z_r, \dot{x}_{(r)}) \in \mathbb{R}^{q^2-q} \times \mathbb{R}^q. \tag{4.50}
\end{aligned}
$$

Now, let

$$
\begin{aligned}
\mathcal{R}_r \triangleq& \{ (z_r, \dot{x}_{(r)}) \in \mathbb{R}^{q^2} : \dot{V}_r(z_r, \dot{x}_{(r)}) = 0 \} \\
=& \left\{ (z_r, \dot{x}_{(r)}) \in \mathbb{R}^{q^2} : \sum_{i=1}^{q-1} \sum_{j=i+1}^{q} (\dot{x}_{ri} - \dot{x}_{rj}) \phi_{rij}(\dot{x}_{ri}, \dot{x}_{rj}) = 0, \right. \\
& \left. \quad i = 1, \ldots, q-1 \right\}, \quad r = 1, 2, 3,
\end{aligned}
$$

and note that, by assumption, $\mathcal{R}_r = \{(z_r, \dot{x}_{(r)}) \in \mathbb{R}^{q^2} : \dot{x}_{r1} = \cdots = \dot{x}_{rq}\}$. Furthermore, since $\dot{x}_{r1} = \cdots = \dot{x}_{rq}$, it follows that $\dot{z}_{rij} = 0$, $i, j = 1, \ldots, q$, $i \neq j$, $r = 1, 2, 3$. Let $\mathcal{M}_r$ denote the largest invariant set contained in $\mathcal{R}_r$. On $\mathcal{M}_r$,

$$\frac{\mathrm{d}}{\mathrm{d}t}|z_{rij}|^{\frac{2}{2-\alpha}} = \frac{2}{2-\alpha}\mathrm{sign}(z_{rij})|z_{rij}|^{\frac{\alpha}{2-\alpha}}\dot{z}_{rij} = 0,$$

and, hence,

$$\frac{1}{2}\frac{\mathrm{d}}{\mathrm{d}t}\sum_{i=1}^{q}\dot{x}_{ri}^2 = \dot{V}_r - \frac{2-\alpha}{4}\sum_{i=1}^{q}\sum_{j=1,j\neq i}^{q}\mathcal{C}_{(i,j)}\frac{\mathrm{d}}{\mathrm{d}t}|z_{rij}|^{\frac{2}{2-\alpha}} = 0,$$

which implies that $\dot{x}_{r1} = \cdots = \dot{x}_{rq} = c$, where $c \in \mathbb{R}$. Finally, since $\sum_{j=1,j\neq i}^{q}\mathcal{C}_{(i,j)}\mathrm{sign}(z_{rij})|z_{rij}|^{\frac{\alpha}{2-\alpha}} = 0$ on $\mathcal{M}_r$ and, for each $i \in \{1, \ldots, q\}$, $z_{rij} = -z_{rji}$ and $\dot{z}_{rij} = 0$, it follows from Proposition 4.5 that $z_{rij} = 0$, $k = 1, 2, 3$.

To show Lyapunov stability of $\dot{x}_{(r)}(t) \equiv c\mathbf{e}$ and $z_r(t) \equiv 0$, consider the shifted Lyapunov function candidate

$$\tilde{V}_r(z_r, \dot{x}_{(r)}) = \frac{1}{2}\sum_{i=1}^{q}(\dot{x}_{ri} - c)^2 + \frac{2-\alpha}{4}\sum_{i=1}^{q}\sum_{j=1,j\neq i}^{q}\mathcal{C}_{(i,j)}|z_{rij}|^{\frac{2}{2-\alpha}}, \qquad (4.51)$$

where $r = 1, 2, 3$. The rest of the proof now follows using identical arguments as above and invoking Proposition 4.3 with

$$\nu_r(x_r, z_r) = -\sum_{i=1}^{q}\left[\sum_{j=1,j\neq i}^{q}(\dot{x}_{rj} - \dot{x}_{ri})\right]\frac{\partial}{\partial\dot{x}_{ri}} + q(2-\alpha)\sum_{i=1}^{q}\sum_{j=1,j\neq i}^{q}z_{rij}\frac{\partial}{\partial z_{rij}}$$

for the closed-loop system given by (4.47) and (4.48) for showing finite time parallel formation. $\qquad \square$

**Example 4.5.** To illustrate the efficacy of the controller proposed in Theorem 4.10 for finite time parallel formation control, let $q = 3$, $r = 1$, $\alpha = \frac{1}{3}$, $d_{112} = 2$, $d_{123} = 1$, and $d_{131} = 3$. The initial conditions are given by $x_{1i}(0) = [-3, 2, 5]^{\mathrm{T}}$ and $\dot{x}_{1i}(0) = [0.5, 1, 2]^{\mathrm{T}}$, $i = 1, 2, 3$. Figures 4.4 and 4.5 show the positions and the velocities versus time, respectively, where $v_{1i} \triangleq \dot{x}_{1i}$, $i = 1, 2, 3$. $\qquad \triangle$

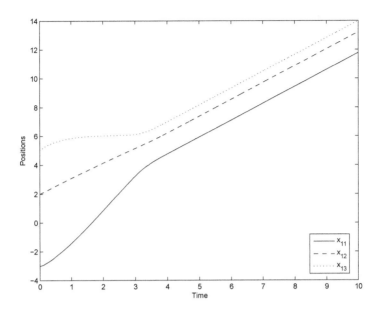

Figure 4.4  Positions versus time for finite time parallel formation.

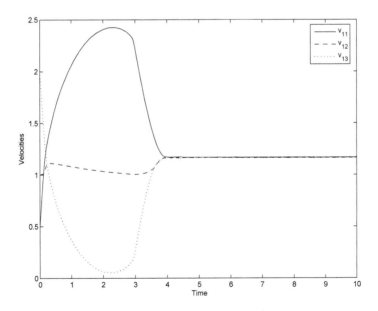

Figure 4.5  Velocities versus time for finite time parallel formation.

*Chapter Five*

---

# Consensus Protocols for Network Systems with a Uniformly Continuous Quasi-Resetting Architecture

## 5.1 Introduction

Networked multiagent systems consist of a group of agents that locally sense their environment, communicate with one another, and process information in order to achieve a given set of system objectives. Since they have widespread applications in physics, biology, social sciences, economics, and engineering, it is not surprising that the last decade has witnessed an increased interest in these systems (see, e.g., [224, 236, 256] and references therein).

For a multivehicle aerospace network of interconnected systems, it is often desired that some property of each vehicle approach a single common value across the network. For example, in a group of autonomous aerospace vehicles, this property might be a common heading angle or a shared communication frequency. As discussed in Chapters 2 and 4, designing a controller that ensures a set of system objectives is called the consensus control problem [224, 236, 256]. Consensus control protocols employ a distributed controller architecture wherein local information is accessed and processed.

In this chapter, we present a novel network consensus control protocol using an approximate resetting architecture. The notion of resetting in feedback control design was originally introduced in [65]. Specifically, a nonlinear integral controller architecture was proposed, where the integrator resets its output to zero whenever its input is zero. Since resetting controllers are nonlinear, closed-loop system performance is not constrained to the Bode integral limitation theorem [149] and, hence, can be used to overcome fundamental performance limitations of linear controllers for achieving fast system response without excessive overshoot and control effort [19, 20, 115, 231].

In Section 5.2 of this chapter, we establish some additional definitions and notation and review some basic results from graph theory which provide the mathematical foundation for designing quasi-resetting controllers for multiagent systems. In Section 5.3, we highlight our proposed approximate resetting controller approach using a simple first-order integrator model. In Section 5.4, we generalize the ideas presented in Section 5.3 to develop continuous approximate resetting controllers for networked multiagent systems. This controller framework leads to a new consensus protocol architecture consisting of a delayed feedback control law with uniformly continuous quasi resettings occurring when the relative state measurements (i.e., distance) between an agent and its neighboring agents approach zero. Furthermore, we show that the proposed framework does not require any well-posedness assumptions or the time regularization that is typically imposed by hybrid resetting controllers (see Chapters 19–21).

We then turn our attention to stability and convergence in Section 5.5. Specifically, using a Lyapunov–Krasovskii functional, we show that the multiagent system reaches asymptotic agreement, wherein the system steady state is uniformly distributed over the system initial conditions, preserving the centroid of the network. In addition, we develop $\mathcal{L}_\infty$ consensus performance guarantees while accounting for system overshoot constraints and excessive control effort.

## 5.2 Notation, Definitions, and Graph-Theoretic Notions

In this section, we establish some additional notation and definitions and recall some basic results from graph theory [102, 224]. We write $\lambda_i(A)$ for the $i$th eigenvalue of the square matrix $A$, $\mathrm{diag}(a)$ for the diagonal matrix with the entries $a \in \mathbb{R}^n$ on its diagonal, and $[A]_{ij}$ for the $(i,j)$th entry of the matrix $A$. Furthermore, for a signal $x(t) \in \mathbb{R}^n$, $t \geq 0$, the extended $\mathcal{L}_{\infty e}$ norm and the $\mathcal{L}_\infty$ norm [185, Section 5.5] are defined, respectively, as $\|x\|_{\mathcal{L}_{\infty e}} \triangleq \max_{1 \leq i \leq n}(\sup_{0 \leq t \leq T} |x_i|)$ and $\|x\|_{\mathcal{L}_\infty} \triangleq \max_{1 \leq i \leq n}(\sup_{t \geq 0} |x_i|)$, where $x_i(t)$ denotes the $i$th component of $x(t)$ and $T > 0$.

As noted in Chapter 3, graphs are broadly adopted to encode interactions between groups of agents. An *undirected* graph $\mathfrak{G}$ is defined by a set $\mathcal{V}_\mathfrak{G} = \{1, \ldots, n\}$ of *nodes* and a set $\mathcal{E}_\mathfrak{G} \subset \mathcal{V}_\mathfrak{G} \times \mathcal{V}_\mathfrak{G}$ of *edges*. If $(i,j) \in \mathcal{E}_\mathfrak{G}$, then the nodes $i$ and $j$ are *neighbors* and the neighboring relation is indicated by $i \sim j$. The *degree* of a node is given by the number of its neighbors. If $d_i$ is the degree of node $i$, then the *degree* matrix of a graph $\mathfrak{G}$ denoted by $\mathcal{D}(\mathfrak{G}) \in \mathbb{R}^{n \times n}$ is given by $\mathcal{D}(\mathfrak{G}) \triangleq \mathrm{diag}(d)$, where $d = [d_1, \ldots, d_n]^\mathrm{T}$.

A *path* $i_0 i_1 \cdots i_L$ is a finite sequence of nodes such that $i_{k-1} \sim i_k$, $k = 1, \ldots, L$, and recall that a graph $\mathfrak{G}$ is connected if there exists a path

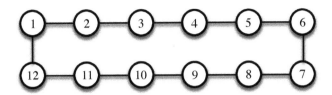

Figure 5.1 A networked multiagent system represented by an undirected graph.

between any pair of distinct nodes. The adjacency matrix of a graph $\mathfrak{G}$ in this chapter is denoted by $\mathcal{A}(\mathfrak{G}) \in \mathbb{R}^{n \times n}$, where

$$
[\mathcal{A}(\mathfrak{G})]_{ij} \triangleq \begin{cases} 1 & \text{if } (i,j) \in \mathcal{E}_{\mathfrak{G}}, \\ 0, & \text{otherwise.} \end{cases} \tag{5.1}
$$

The *Laplacian matrix* of a graph $\mathcal{L}(\mathfrak{G}) \in \mathbb{N}^n \cap \mathbb{S}^n$ plays a central role in graph theory for networked multiagent systems and is given by $\mathcal{L}(\mathfrak{G}) \triangleq \mathcal{D}(\mathfrak{G}) - \mathcal{A}(\mathfrak{G})$, where the eigenvalues of the Laplacian for a connected undirected graph can be ordered as

$$
0 = \lambda_{\min}(\mathcal{L}(\mathfrak{G})) \triangleq \lambda_1(\mathcal{L}(\mathfrak{G})) < \lambda_2(\mathcal{L}(\mathfrak{G})) \le \cdots \le \lambda_{\max}(\mathcal{L}(\mathfrak{G})) \triangleq \lambda_n(\mathcal{L}(\mathfrak{G})). \tag{5.2}
$$

It can be easily shown that $\mathbf{e}_n$ is the eigenvector corresponding to the zero eigenvalue $\lambda_{\min}(\mathcal{L}(\mathfrak{G}))$ of $\mathcal{L}(\mathfrak{G})$, and, hence, $\mathcal{L}(\mathfrak{G})\mathbf{e}_n = 0_n$ holds.

As an example, consider the networked multiagent system shown in Figure 5.1. For this system, the graph Laplacian $\mathcal{L}(\mathfrak{G})$ has the form

$$
\mathcal{L}(\mathfrak{G}) \triangleq \begin{bmatrix} 2 & -1 & 0 & 0 & 0 & \dots \\ -1 & 2 & -1 & 0 & 0 & \dots \\ 0 & -1 & 2 & -1 & 0 & \dots \\ \vdots & \vdots & \vdots & \vdots & \vdots & \ddots \end{bmatrix} \in \mathbb{N}^{12} \cap \mathbb{S}^{12}. \tag{5.3}
$$

## 5.3 A Quasi-Resetting Control Architecture for Approximating Resetting Controllers

In this section, we highlight the salient features of our quasi-resetting control framework for approximating resetting hybrid controllers using a simple first-order integrator model. Specifically, to elucidate our proposed approach, consider the scalar dynamical system given by

$$
\dot{x}(t) = u(t), \quad x(0) = 0, \quad t \ge 0, \tag{5.4}
$$

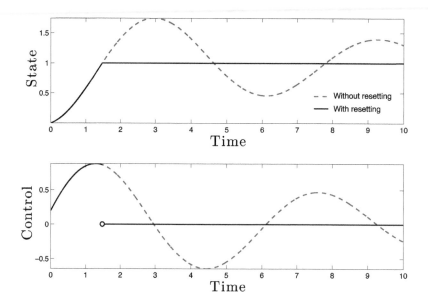

Figure 5.2 Tracking performance for a constant command $c = 1$ using the control law given by (5.5)–(5.7) with and without resetting ($k_1 = 0.2$ and $k_2 = 1$).

where $x(t) \in \mathbb{R}$, $t \geq 0$, is the system state and the control law $u(t)$, $t \geq 0$, is given by

$$u(t) = -k_1\big(x(t) - c\big) - v(t), \tag{5.5}$$
$$\dot{v}(t) = k_2\big(x(t) - c\big), \quad v(0) = 0, \quad x(t) - c \neq 0, \tag{5.6}$$
$$v(t^+) = 0, \quad x(t) - c = 0, \tag{5.7}$$

where $c \in \mathbb{R}$ is a constant command; $v(t) \in \mathbb{R}$, $t \geq 0$, is an integrator state; $v(t^+) \triangleq \lim_{\varepsilon \to 0} v(t + \varepsilon)$; and $k_i > 0$, $i = 1, 2$, are the design parameters. The tracking performance based on the control law (5.5)–(5.7) for $c = 1$ is shown in Figure 5.2 without and with resetting. In particular, this simple example illustrates the desired effect of the resetting action of the integrator and how it yields satisfactory tracking performance in finite time.

The stability properties of the above simple example can be shown by considering the Lyapunov function candidate $V(e, v) = e^2 + k_2^{-1} v^2$, where $e(t) \triangleq x(t) - c$. Note that $V(0, 0) = 0$ and $V(e, v) > 0$ for all $(e, v) \neq 0$. Since

$$\dot{V}\big(e(t), v(t)\big) = -2k_1 e^2(t) \leq 0, \quad \big(e(t), v(t)\big) \notin \mathcal{Z},$$

Table 5.1 Algorithm for the resetting controller in (5.6) and (5.7).

```
if |x − c| ≤ ε;  v = 0;
else;  v = v + Δt[k₃(x − c)];  end.
```

and

$$\Delta V\big(e(t), v(t)\big) \triangleq V\big(e(t^+), v(t^+)\big) - V\big(e(t), v(t)\big)$$
$$= -k_2^{-1} v^2(t)$$
$$\leq 0, \quad \big(e(t), v(t)\big) \in \mathcal{Z},$$

where $\mathcal{Z} \triangleq \{(e, v) \in \mathbb{R} \times \mathbb{R} : e = 0\}$ is the *resetting set*, it follows from Theorem 19.1 in Chapter 19 that $\big(e(t), v(t)\big) \to \mathcal{M} \triangleq \{(0, 0)\}$ as $t \to \infty$, where $\mathcal{M}$ is the largest invariant set contained in

$$\mathcal{R} \triangleq \{\big(e(t), v(t)\big) \in \mathbb{R} \times \mathbb{R} : \big(e(t), v(t)\big) \notin \mathcal{Z}, \dot{V}\big(e(t), v(t)\big) = 0\}$$
$$\cup \{\big(e(t), v(t)\big) \in \mathbb{R} \times \mathbb{R} : \big(e(t), v(t)\big) \in \mathcal{Z}, \Delta V\big(e(t), v(t)\big) = 0\}.$$

In order to apply Theorem 19.1, we require that (i) if $(e(t), v(t)) \in \mathcal{Z}$, then $\big(e(t^+), v(t^+)\big) \notin \mathcal{Z}$, and (ii) if at time $t$ the trajectory $\big(e(t), v(t)\big)$ belongs to the closure of $\mathcal{Z}$ but not $\mathcal{Z}$, then there exists $\varepsilon > 0$ such that for all $0 < \delta < \varepsilon$, $\big(e(t + \delta), v(t + \delta)\big) \notin \mathcal{Z}$.

Assumptions (i) and (ii) guarantee that for a particular system initial condition, the resetting system times are well defined and distinct; see Chapter 19 and [128]. Even though these assumptions can be satisfied for the preceding example by redefining the resetting set as in, for example, [16], for certain problem formulations this may prove to be a difficult task. Furthermore, even if assumptions (i) and (ii) hold, the closed-loop system with a resetting controller can exhibit Zeno solutions, wherein solutions exhibit infinitely many resettings in finite time [128]. To circumvent this problem, a time regularization approach can be used [16, 231], wherein resetting is avoided if a minimum dwell time between the resetting times is enforced. However, in general, it is desirable to eliminate such assumptions since they tend to require specialized hybrid controller architectures [125, 135].

In practice, implementation of a resetting controller may not be feasible. To see this, consider the algorithm given in Table 5.1 for the resetting controller (5.6) and (5.7), where $\varepsilon > 0$ is a small constant and $\Delta t$ is the sampling time. Here, we approximate (5.6) by resorting to a first-order Euler integration method and implement (5.7) by using `if-else` logic. Even though $\Delta t \to 0$ results in the exact implementation of (5.6) and (5.7),[5] this implementation is not possible in practice.

---

[5]Note that to implement (5.7) exactly we need $\Delta t \to 0$, which implies $\varepsilon \to 0$.

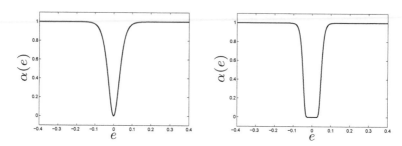

Figure 5.3 Function $\alpha(x(t) - c)$ in (5.9) for $\xi = 25$ and $\theta = 1$ (left) and $\xi = 50$ and $\theta = 25$ (right).

Table 5.2 Algorithm for the quasi-resetting controller in (5.8).

$$v = \left[1 - \mathrm{sech}^2\left(\xi(x - c)\right)\right]^{\theta} v + k_3\left(x - c\right)$$

To address the implementation issues discussed above, consider the delayed feedback control law given by (5.5) with, in place of (5.6) and (5.7),

$$v(t) = \alpha\left(x(t) - c\right) v(t - \tau) + k_3\left(x(t) - c\right), \tag{5.8}$$

where

$$\alpha\left(x(t) - c\right) \triangleq \left[1 - \mathrm{sech}^2\left(\xi[x(t) - c]\right)\right]^{\theta}; \tag{5.9}$$

$\xi > 0$, $\theta > 0$, $k_3 > 0$ are design parameters; and $\tau > 0$ is a time-delay design parameter (see Figure 5.3). In the case where $x(t) - c$ is sufficiently bounded away from zero, it follows from (5.9) that $\alpha(x(t) - c) \approx 1$, and, hence, (5.8) behaves as an integrator since $v(t) \approx v(t - \tau) + k_3(x(t) - c)$. Alternatively, if $x(t) - c = 0$, then $\alpha(x(t) - c) = 0$, and, hence, $v(t) = 0$, so that approximate resetting occurs.

In contrast to standard impulsive resetting controllers [125, 135], this approximate resetting control architecture involves a quasi-impulsive uniformly continuous switch. From an implementation perspective, the time-delay design parameter $\tau$ can be chosen to be sufficiently small so that $\tau = \Delta t$; see Table 5.2 for the implementation of (5.8).

Figure 5.4 revisits the preceding example and shows the tracking performance for $c = 1$ with the delayed feedback control law given by (5.5) and (5.8) for two different values of $\xi$. For both values of $\xi$, the controller achieves satisfactory tracking performance. Furthermore, note that for $\xi = 250$, the proposed approximate resetting controller (5.8) recovers the tracking performance shown in Figure 5.2 for the resetting controller (5.6) and (5.7).

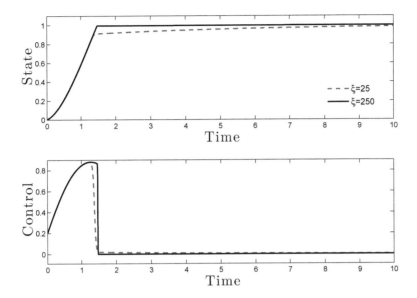

Figure 5.4 Tracking performance for a constant command $c = 1$ using the delayed control law given by (5.5) and (5.8) with $\xi = 25$ and $\xi = 250$ ($\tau = 0.001$, $k_3 = 0.001$, and $\theta = 1$).

## 5.4 Consensus Control Protocols

In this section, we generalize the ideas presented in Section 5.3 to develop continuous approximate resetting controllers for networked multiagent systems. Specifically, we consider a system of $n$ agents exchanging information using local measurements, with $\mathfrak{G}$ defining a connected undirected graph topology and with nodes and edges representing agents and interagent information exchange links, respectively.[6] Specifically, let $x_i(t) \in \mathbb{R}^m$ denote the state of node $i$ at time $t \geq 0$ whose dynamics is described by the single integrator dynamics

$$\dot{x}_i(t) = u_i(t), \quad x_i(0) = x_{i0}, \quad i = 1, \dots, n, \tag{5.10}$$

where $u_i(t) \in \mathbb{R}^N$, $t \geq 0$, is the control input of node $i$. Assuming that agent $i$ has access to the relative state information with respect to its neighbors, a standard solution of the consensus problem can be achieved by applying

---

[6]In this chapter, we assume that the network is static, and, hence, agent evolution will not cause edges to appear or disappear in the network. Thus, the proposed architecture does not address information link failures and communication dropouts. This extension is addressed in Chapter 6.

the standard control protocol [224]

$$u_i(t) = -k_1 \sum_{i \sim j} \big( x_i(t) - x_j(t) \big), \tag{5.11}$$

where $k_1 > 0$ is a design parameter. Here, (5.10), in conjunction with (5.11), can be represented by the graph Laplacian dynamics

$$\dot{x}(t) = -k_1 \mathcal{L}(\mathfrak{G}) \otimes I_m \, x(t), \quad x(0) = x_0, \quad t \geq 0, \tag{5.12}$$

where $x(t) = [x_1^{\mathrm{T}}(t), \dots, x_n^{\mathrm{T}}(t)]^{\mathrm{T}} \in \mathbb{R}^{mn}$ denotes the aggregated state vector of the multiagent system.

Although our results can be directly extended to the case of (5.12), for simplicity of exposition, we will focus on individual agent states evolving in $\mathbb{R}$ (i.e., $N = 1$). In this case, (5.12) becomes

$$\dot{x}(t) = -k_1 \mathcal{L}(\mathfrak{G}) x(t), \quad x(0) = x_0, \quad t \geq 0. \tag{5.13}$$

Since the undirected graph $\mathfrak{G}$ is connected, it can be shown that $x(t) \to \frac{1}{n} \mathbf{e}_n \mathbf{e}_n^{\mathrm{T}} x_0$ as $t \to \infty$ [224]. This shows that agents reach asymptotic agreement to a steady-state value composed of an average of the system initial conditions, that is, *centroid* convergence of the network.

Next, we generalize the control protocol given by (5.11) by developing a continuous quasi-resetting control protocol for networked multiagent systems. Specifically, let

$$u_i(t) = -k_1 \sum_{i \sim j} \big( x_i(t) - x_j(t) \big) - \sum_{i \sim j} \big( v_i(t) - v_j(t) \big), \tag{5.14}$$

$$v_i(t) = \alpha_i \left( \sum_{i \sim j} \big( x_i(t) - x_j(t) \big) \right) v_i(t - \tau) + k_2 \sum_{i \sim j} \big( x_i(t) - x_j(t) \big), \tag{5.15}$$

where

$$\alpha_i \left( \sum_{i \sim j} \big( x_i(t) - x_j(t) \big) \right) \triangleq \left[ 1 - \operatorname{sech}^2 \left( \xi \sum_{i \sim j} \big( x_i(t) - x_j(t) \big) \right) \right]^{\theta}; \tag{5.16}$$

$\xi > 0$, $\theta > 0$, $v_i(t) \in \mathbb{R}$, $k_1 > 0$, and $k_2 > 0$ are design parameters; and $\tau > 0$ is a time-delay design parameter.

Note that in the case where $[\mathcal{L}(\mathfrak{G})x]_i \triangleq \sum_{i \sim j} \big( x_i(t) - x_j(t) \big)$ is sufficiently bounded away from zero, then $\alpha_i(\cdot) \approx 1$, and, hence, (5.15) serves as an integrator since

$$v_i(t) \approx v_i(t - \tau) + k_2 \sum_{i \sim j} \big(x_i(t) - x_j(t)\big). \tag{5.17}$$

Alternatively, if $[\mathcal{L}(\mathfrak{G})x]_i = 0$, that is, the relative state measurements between an agent $i$ and its neighboring agents vanish, then $\alpha_i(\cdot) = 0$, and, hence, $v_i(t) = 0$, so that resetting occurs.

The proposed uniformly continuous quasi-resetting control protocol (5.14) can be related to standard impulsive resetting controllers (see Chapter 20). To see this, let $\xi \to \infty$ and $\theta \to \infty$ in (5.16). In this case, $\alpha_i(\cdot) = 1$ and $\alpha_i(\cdot) = 0$ when $[\mathcal{L}(\mathfrak{G})x]_i \neq 0$ and $[\mathcal{L}(\mathfrak{G})x]_i = 0$, respectively. For $[\mathcal{L}(\mathfrak{G})]_i \neq 0$ (i.e., $\alpha_i(\cdot) = 1$), it follows from (5.15) that

$$v_i(t) = v_i(t - \tau) + k_2 \sum_{i \sim j} \big(x_i(t) - x_j(t)\big) \tag{5.18}$$

or, equivalently,

$$\big[v_i(t) - v_i(t - \tau)\big]/\tau = \hat{k}_2 \sum_{i \sim j} \big(x_i(t) - x_j(t)\big), \tag{5.19}$$

where $\hat{k}_2 \triangleq k_2/\tau$. Now, letting $\tau \to 0$ yields

$$\dot{v}_i(t) = \hat{k}_2 \sum_{i \sim j} \big(x_i(t) - x_j(t)\big), \tag{5.20}$$

where $\dot{v}_i(t) = \lim_{\tau \to 0} \big[v_i(t) - v_i(t - \tau)\big]/\tau$.

Alternatively, for $[\mathcal{L}(\mathfrak{G})x]_i = 0$ (i.e., $\alpha_i(\cdot) = 0$), (5.15) can be viewed as $v_i(t^+) = 0$ in light of the conditions on $\xi$, $\theta$, and $\tau$. That is, for $\xi \to \infty$, $\theta \to \infty$, and $\tau \to 0$, the standard impulsive reset controller[7] is given by

$$\dot{v}_i(t) = \hat{k}_2 \sum_{i \sim j} \big(x_i(t) - x_j(t)\big), \quad [\mathcal{L}(\mathfrak{G})x]_i \neq 0, \tag{5.21}$$

$$v(t^+) = 0, \quad [\mathcal{L}(\mathfrak{G})x]_i = 0. \tag{5.22}$$

As noted in Section 5.3, using (5.21) and (5.22) instead of (5.15) and (5.16) requires well-posedness assumptions and time regularization to guarantee that the system resetting times are distinct and well defined, as well as to guarantee the absence of Zeno solutions.

## 5.5 Stability Analysis

In this section, we establish stability properties of the proposed consensus control protocol with quasi resetting given by (5.14) and (5.15) for $t \geq 0$.

---

[7] The impulsive resetting controller given by (5.21) and (5.22) can be viewed as a multiagent system version of the Clegg integrator discussed in [65].

Let $x(t) = [x_1(t), \ldots, x_n(t)]^{\mathrm{T}} \in \mathbb{R}^n$ and $v(t) = [v_1(t), \ldots, v_n(t)]^{\mathrm{T}} \in \mathbb{R}^n$ for $t \geq 0$. Then (5.10), in conjunction with (5.14) and (5.15), can be written as

$$\dot{x}(t) = -k_1 \mathcal{L}(\mathfrak{G}) x(t) - \mathcal{L}(\mathfrak{G}) v(t), \quad x(0) = x_0, \quad t \geq 0, \tag{5.23}$$

$$v(t) = D(\mathfrak{G}) v(t - \tau) + k_2 \mathcal{L}(\mathfrak{G}) x(t), \tag{5.24}$$

where $D(\mathfrak{G}) \in \mathbb{P}^n \cap \mathbb{S}^n$ is given by

$$D(\mathfrak{G}) \triangleq \mathrm{diag}\big(\alpha(\cdot)\big), \quad \alpha(\cdot) = \big[\alpha_1(\cdot), \ldots, \alpha_n(\cdot)\big]^{\mathrm{T}}. \tag{5.25}$$

Note that, for every $\kappa > 0$, $I_n - D^\kappa(\mathfrak{G}) \in \mathbb{N}^n \cap \mathbb{S}^n$.

The following key lemmas are necessary for the main results of this section. For the following results, we consider the transformation given by

$$x(t) = \mathrm{ave}\big(x(t)\big)\mathbf{e}_n + \delta(t), \tag{5.26}$$

where $\mathrm{ave}\big(x(t)\big) \triangleq \frac{1}{n} \mathbf{e}_n^{\mathrm{T}} x(t)$ and $\delta(t) \in \mathbb{R}^n$, $t \geq 0$.

**Lemma 5.1.** Consider the dynamical system given by

$$\dot{x}(t) = -k_1 \mathcal{L}(\mathfrak{G}) x(t) - \mathcal{L}(\mathfrak{G}) \rho(t), \quad x(0) = x_0, \quad t \geq 0, \tag{5.27}$$

where $\rho(t) \in \mathbb{R}^n$, $t \geq 0$, is an arbitrary vector. Then $\mathrm{ave}\big(x(t)\big) = \frac{1}{n} \mathbf{e}_n^{\mathrm{T}} x_0$ for all $t \geq 0$.

**Proof.** Differentiating $\mathrm{ave}\big(x(t)\big)$ with respect to time yields

$$\frac{\mathrm{d}}{\mathrm{d}t} \mathrm{ave}\big(x(t)\big) = \frac{1}{n} \mathbf{e}_n^{\mathrm{T}} \dot{x}(t) = \frac{1}{n} \mathbf{e}_n^{\mathrm{T}} \Big[ -k_1 \mathcal{L}(\mathfrak{G}) x(t) - \mathcal{L}(\mathfrak{G}) \rho(t) \Big] = 0_n, \tag{5.28}$$

where in (5.28) we used the fact that $\mathcal{L}(\mathfrak{G}) \mathbf{e}_n = 0_n$ and $\mathcal{L}(\mathfrak{G}) = \mathcal{L}^{\mathrm{T}}(\mathfrak{G})$. Hence, $\mathrm{ave}\big(x(t)\big) = \mathrm{ave}\big(x(0)\big)$, $t \geq 0$, which proves the result. $\qquad \square$

**Lemma 5.2.** Consider the dynamical system given by (5.23) with $x(t)$ given by (5.26). If $\delta(t) \to 0$ as $t \to \infty$, then $x(t) \to \frac{1}{n} \mathbf{e}_n \mathbf{e}_n^{\mathrm{T}} x_0$ as $t \to \infty$.

**Proof.** It follows from (5.26) that $\delta(t) \to 0$ as $t \to \infty$ implies that $x(t) \to \mathrm{ave}\big(x(t)\big)\mathbf{e}_n$ as $t \to \infty$. Furthermore, since the form of (5.23) is identical to (5.27), it follows from Lemma 5.1 that $\mathrm{ave}\big(x(t)\big) = \frac{1}{n} \mathbf{e}_n^{\mathrm{T}} x_0$, $t \geq 0$, and, hence, $x(t) \to \frac{1}{n} \mathbf{e}_n \mathbf{e}_n^{\mathrm{T}} x_0$ as $t \to \infty$. $\qquad \square$

Note that if $\lim_{t \to \infty} \delta(t) = 0$, then asymptotic agreement with the system steady state uniformly distributed over the system initial conditions preserving the centroid of the network is guaranteed.

An alternative consensus protocol using a quasi-resetting architecture can be constructed by considering

$$u_i(t) = -k_1 \sum_{i \sim j} \big(x_i(t) - x_j(t)\big) - v_i(t) \tag{5.29}$$

and (5.15). In this case, we have

$$\dot{x}(t) = -k_1 \mathcal{L}(\mathfrak{G})x(t) - v(t), \quad x(0) = x_0, \quad t \geq 0, \tag{5.30}$$

and (5.24). However, since the form of (5.30) is different than (5.27), $\mathrm{ave}\big(x(t)\big) \neq \mathrm{ave}\big(x(0)\big)$, $t \geq 0$; that is, centroid convergence of the network is no longer guaranteed at steady state. Note that for $\alpha_i(\cdot) = 1$ and $\tau \to 0$, (5.29) and (5.15) can be viewed as a delayed version of the proportional-integral protocols proposed in [297, 321]. To see this, let $\hat{k}_2 \triangleq k_2/\tau$, and note that the integral action

$$\dot{v}_i(t) = \hat{k}_2 \sum_{i \sim j} \big(x_i(t) - x_j(t)\big) \tag{5.31}$$

can be recovered, where $\dot{v}_i(t) = \lim_{\tau \to 0} \big[v_i(t) - v_i(t - \tau)\big]/\tau$.

Lemma 5.2 suggests that we need $\lim_{t \to \infty} \delta(t) = 0$ when solving the consensus problem. Thus, we use the transformation given by (5.26) to write (5.23) and (5.24) as

$$\dot{\delta}(t) = -k_1 \mathcal{L}(\mathfrak{G})\delta(t) - \mathcal{L}(\mathfrak{G})v(t), \quad \delta(0) = \delta_0, \quad t \geq 0, \tag{5.32}$$
$$v(t) = D(\mathfrak{G})v(t - \tau) + k_2 \mathcal{L}(\mathfrak{G})\delta(t). \tag{5.33}$$

The following theorem is necessary for the main results of this chapter.

**Theorem 5.1.** Consider the networked multiagent system given by (5.10), where agents exchange information using local measurements and with $\mathfrak{G}$ defining a connected undirected graph topology. Furthermore, consider the consensus protocol given by (5.14), (5.15), and (5.16). Then

$$\lim_{t \to \infty} \big(\delta(t), v(t)\big) = (0, 0).$$

**Proof.** Consider the Lyapunov–Krasovskii functional candidate given by

$$V\big(\delta, v\big) = \delta^{\mathrm{T}}\delta + k_2^{-1} \int_{-\tau}^{0} v^{\mathrm{T}}(\mu)v(\mu)\mathrm{d}\mu, \tag{5.34}$$

and note that $V(0, 0) = 0$ and $V\big(\delta, v\big) > 0$ for all $\big(\delta, v\big) \neq (0, 0)$. Differentiating $\delta^{\mathrm{T}}\delta$ along the system trajectories of (5.32) yields

$$\frac{\mathrm{d}}{\mathrm{d}t}\delta^{\mathrm{T}}(t)\delta(t)$$

$$= 2\delta^{\mathrm{T}}(t)\left[-k_1\mathcal{L}(\mathfrak{G})\delta(t) - \mathcal{L}(\mathfrak{G})v(t)\right]$$

$$= -2k_1\delta^{\mathrm{T}}(t)\mathcal{L}(\mathfrak{G})\delta(t) - 2\delta^{\mathrm{T}}(t)\mathcal{L}(\mathfrak{G})\left[D(\mathfrak{G})v(t-\tau) + k_2\mathcal{L}(\mathfrak{G})\delta(t)\right]$$

$$= -2k_1\delta^{\mathrm{T}}(t)\mathcal{L}(\mathfrak{G})\delta(t) - 2\delta^{\mathrm{T}}(t)\mathcal{L}(\mathfrak{G})D(\mathfrak{G})v(t-\tau)$$

$$\quad - 2k_2\delta^{\mathrm{T}}(t)\mathcal{L}^2(\mathfrak{G})\delta(t). \tag{5.35}$$

Furthermore, differentiating $\int_{-\tau}^{0} v^{\mathrm{T}}(\mu)v(\mu)\mathrm{d}\mu$ with respect to time yields

$$\frac{\mathrm{d}}{\mathrm{d}t}\int_{t-\tau}^{t} v^{\mathrm{T}}(\mu)v(\mu)\mathrm{d}\mu$$

$$= v^{\mathrm{T}}(t)v(t) - v^{\mathrm{T}}(t-\tau)v(t-\tau)$$

$$= -v^{\mathrm{T}}(t-\tau)\left[I_n - D^2(\mathfrak{G})\right]v(t-\tau) + 2k_2\delta^{\mathrm{T}}(t)\mathcal{L}(\mathfrak{G})D(\mathfrak{G})v(t-\tau)$$

$$\quad + k_2^2\delta^{\mathrm{T}}(t)\mathcal{L}^2(\mathfrak{G})\delta(t)$$

$$\leq 2k_2\delta^{\mathrm{T}}(t)\mathcal{L}(\mathfrak{G})D(\mathfrak{G})v(t-\tau) + k_2^2\delta^{\mathrm{T}}(t)\mathcal{L}^2(\mathfrak{G})\delta(t), \quad t \geq 0, \tag{5.36}$$

where we used the fact that $I_n - D^2(\mathfrak{G}) \in \mathbb{N}^n \cap \mathbb{S}^n$. Next, it follows from (5.34), (5.35), and (5.36) that

$$\dot{V}\big(\delta(t),v(t)\big) \leq -2k_1\delta^{\mathrm{T}}(t)\mathcal{L}(\mathfrak{G})\delta(t) - 2k_2\delta^{\mathrm{T}}(t)\mathcal{L}^2(\mathfrak{G})\delta(t) \leq 0, \quad t \geq 0. \tag{5.37}$$

Now, let

$$\mathcal{R} \triangleq \left\{\big(\delta(t),v(t)\big)\in \mathbb{R}^n \times \mathbb{R}^n : \dot{V}\big(\delta(t),v(t)\big)= 0\right\},$$

and let $\mathcal{M}$ be the largest invariant set contained in $\mathcal{R}$. Note that, in this case, since $\mathcal{L}(\mathfrak{G})\delta(t) = 0_n$, it follows that $\sum_{i \sim j}\big(\delta_i(t) - \delta_j(t)\big)= 0$ for all $(i,j) \in \mathcal{E}_{\mathfrak{G}}$. Using similar arguments as in [237, Thm. 3], it follows from the connectivity of the graph $\mathfrak{G}$ that $\delta_i(t) = \delta_j(t)$ for all $i, j \in \mathcal{V}_{\mathfrak{G}}$. Furthermore, by (5.26), $\delta(t) = \big(I_n - \frac{1}{n}\mathbf{e}_n\mathbf{e}_n^{\mathrm{T}}\big)x(t)$, which implies that $\sum_{i=1}^{n}\delta_i(t) = 0$. Finally, in this case, since $D(\mathfrak{G}) = 0_{n\times n}$, it follows from (5.33) that $v(t) = 0$. Hence, $\big(\delta(t),v(t)\big) \to \mathcal{M} = \{(0,0)\}$ as $t \to \infty$.  $\square$

The following results, highlighting asymptotic agreement of the networked multiagent system and worst-case transient performance guarantees, respectively, are now immediate. For the first result, recall the definition of semistability given in Chapter 4.

**Theorem 5.2.** Consider the networked multiagent system given by (5.10), where agents exchange information using local measurements and with $\mathfrak{G}$ defining a connected undirected graph topology. Furthermore, consider the consensus protocol given by (5.14), (5.15), and (5.16). Then

$\lim_{t\to\infty} x(t) = \frac{1}{n}\mathbf{e}_n\mathbf{e}_n^{\mathrm{T}}x_0$, and $\frac{1}{n}\mathbf{e}_n\mathbf{e}_n^{\mathrm{T}}x_0$ is a semistable equilibrium state of (5.10) with $u_i(t)$ given by (5.14).

**Proof.** The proof is a direct consequence of Lemma 5.2 and Theorem 5.1 using the Lyapunov–Krasovskii functional given by $V(x,v) = x^{\mathrm{T}}x + k_2^{-1}\int_{-\tau}^{0} v^{\mathrm{T}}(\mu)v(\mu)\mathrm{d}\mu$. $\qquad\square$

**Theorem 5.3.** Consider the networked multiagent system given by (5.10), where agents exchange information using local measurements and with $\mathfrak{G}$ defining a connected undirected graph topology. Furthermore, consider the consensus protocol given by (5.14), (5.15), and (5.16). Then

$$\left\| x - \frac{1}{n}\mathbf{e}_n\mathbf{e}_n^{\mathrm{T}}x_0 \right\|_{\mathcal{L}_\infty} \leq \sqrt{V\left(\left[I_n - \frac{1}{n}\mathbf{e}_n\mathbf{e}_n^{\mathrm{T}}\right]x_0, v(0)\right)}, \qquad (5.38)$$

where $V(\cdot)$ is given by (5.34).

**Proof.** It follows from (5.37) that $\dot{V}\big(\delta(t), v(t)\big) \leq 0$ for all $t \geq 0$, and, hence,

$$V\big(\delta(t), v(t)\big) \leq V\big(\delta(0), v(0)\big), \quad t \geq 0. \qquad (5.39)$$

Now, using the fact that $V\big(\delta(t), v(t)\big) \geq \|\delta(t)\|_2^2$, $t \geq 0$, (5.39) yields

$$\|\delta(t)\|_2 \leq \sqrt{V(\delta(0), v(0))}, \quad t \geq 0. \qquad (5.40)$$

Since $\|\cdot\|_\infty \leq \|\cdot\|_2$ and every vector norm $\|\cdot\| : \mathbb{R}^n \to \mathbb{R}$ is uniformly continuous on $\mathbb{R}^n$, (5.40) yields

$$\|\delta\|_{\mathcal{L}_{\infty e}} \leq \sqrt{V(\delta(0), v(0))}. \qquad (5.41)$$

Now, (5.38) is a direct consequence of (5.41) since (5.41) holds uniformly in $T > 0$; $\delta(t) = x(t) - \mathrm{ave}\big(x(t)\big)\mathbf{e}_n$; and $\mathrm{ave}\big(x(t)\big) = \frac{1}{n}\mathbf{e}_n^{\mathrm{T}}x_0$, $t \geq 0$. $\qquad\square$

Note that it follows from (5.26) and (5.38) that the worst-case transient performance bound for an agent $i$ is given by

$$\|x_i\|_{\mathcal{L}_\infty} \leq \frac{1}{n}\|\mathbf{e}_n\mathbf{e}_n^{\mathrm{T}}x_0\|_2 + \sqrt{V\left(\left[I_n - \frac{1}{n}\mathbf{e}_n\mathbf{e}_n^{\mathrm{T}}\right]x_0, v(0)\right)}. \qquad (5.42)$$

Next, we consider a networked multiagent system represented by the undirected graph shown in Figure 5.1 for our first two examples and then consider three networked Boeing 747 airplanes on an undirected line graph for our third example. Our aim is to compare the performance of the standard consensus protocol given by (5.11) with the proposed consensus protocol (5.14)–(5.16).

**Example 5.1.** For the undirected graph shown in Figure 5.1, let

$$x_0 = \begin{bmatrix} 6, -5, 4, -3, 2, -1, 1, -2, 3, -4, 5, -6 \end{bmatrix}^{\mathrm{T}}. \qquad (5.43)$$

Furthermore, let $k_1 = 7.5$ for the standard consensus protocol given by (5.11), and let $k_1 = 2.5$, $k_2 = 0.001$, $\tau = 0.0001$, $\xi = 100$, and $\theta = 1$ for the proposed consensus protocol given by (5.14), (5.15), and (5.16). Figures 5.5 and 5.6 show the results for the standard and proposed consensus protocols, respectively. Note that even though both protocols achieve similar performance in terms of settling time, the control effort of the latter protocol is significantly less in magnitude than that of the former protocol.          $\triangle$

**Example 5.2.** For the undirected graph shown in Figure 5.1, let the (random) initial condition be given by

$$x_0 = \begin{bmatrix} -5.33, 4.67, 1.75, -0.15, 13.9, -7.83, -0.42, \\ 9.02, 2.49, 0.21, -3.67, 16.15 \end{bmatrix}^{\mathrm{T}}. \qquad (5.44)$$

Furthermore, let $k_1 = 9$ for the standard consensus protocol given by (5.11), and let $k_1 = 4.5$, $k_2 = 0.003$, $\tau = 0.0001$, $\xi = 100$, and $\theta = 1$ for the proposed consensus protocol given by (5.14), (5.15), and (5.16). Figures 5.7 and 5.8 show the results for the standard and proposed consensus protocols, respectively. Once again, both protocols achieve similar performance in terms of settling time; however, the control effort of the proposed protocol is less in magnitude than that of the standard protocol.          $\triangle$

**Example 5.3.** In this example, we use the proposed architecture for

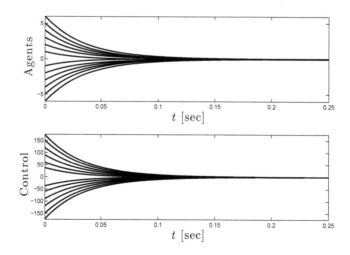

Figure 5.5 Agent and control responses for the standard consensus protocol given by (5.11) with $k_1 = 7.5$ (Example 5.1).

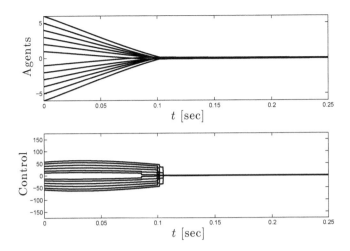

Figure 5.6  Agent and control responses for the proposed consensus protocol with resetting given by (5.14), (5.15), and (5.16) with $k_1 = 2.5$, $k_2 = 0.001$, $\tau = 0.0001$, $\xi = 100$, and $\theta = 1$ (Example 5.1).

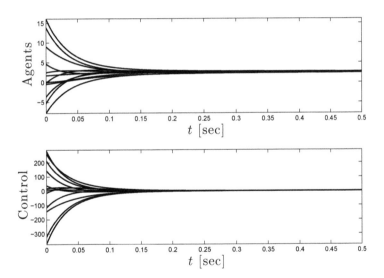

Figure 5.7  Agent and control responses for the standard consensus protocol given by (5.11) with $k_1 = 9$ (Example 5.2).

pitch rate consensus of commercial airplanes. Specifically, we consider the multiagent system representing the controlled longitudinal motion of three Boeing 747 airplanes addressed in Example 2.1 linearized at an altitude of 40 kft and a velocity of 774 ft/sec given by

$$\dot{z}_i(t) = Az_i(t) + B\delta_i(t), \quad z_i(0) = z_{i_0}, \quad i = 1, 2, 3, \quad t \geq 0, \qquad (5.45)$$

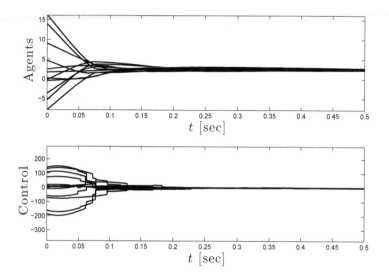

Figure 5.8 Agent and control responses for the proposed consensus protocol with resetting given by (5.14), (5.15), and (5.16) with $k_1 = 4.5$, $k_2 = 0.003$, $\tau = 0.0001$, $\xi = 100$, and $\theta = 1$ (Example 5.2).

where the state $z_i(t) = [v_{x_i}(t), v_{z_i}(t), q_i(t), \theta_{e_i}(t)]^{\mathrm{T}} \in \mathbb{R}^4$, $t \geq 0$, of agent $i$, $i = 1, 2, 3$; control input $\delta_i(t)$; and system matrices $A$ and $B$ are as defined in Example 2.1.

As in Example 2.1, we propose a two-level control hierarchy composed of a lower-level controller for command following and a higher-level controller for pitch rate consensus of the three airplanes given by (5.45). To address lower-level controller design, let $x_i(t)$, $i = 1, 2, 3$, $t \geq 0$, be a command generated by (5.10) (i.e., the guidance command), and let $s_i(t)$, $i = 1, 2, 3$, $t \geq 0$, denote the integrator state satisfying

$$\dot{s}_i(t) = Ez_i(t) - x_i(t), \quad s_i(0) = s_{i_0}, \quad i = 1, 2, 3, \quad t \geq 0, \tag{5.46}$$

where $E = [0, 0, 1, 0]$.

Now, defining the augmented state

$$\hat{z}_i(t) \triangleq [z_i^{\mathrm{T}}(t), s_i(t)]^{\mathrm{T}},$$

(5.45) and (5.46) give

$$\dot{\hat{z}}_i(t) = \hat{A}\hat{z}_i(t) + \hat{B}_1\delta_i(t) + \hat{B}_2 x_i(t), \quad \hat{z}_i(0) = \hat{z}_{i_0}, \quad i = 1, 2, 3, \quad t \geq 0, \tag{5.47}$$

where $\hat{A}$, $\hat{B}_1$, and $\hat{B}_2$ are given by (2.33). Furthermore, let the elevator

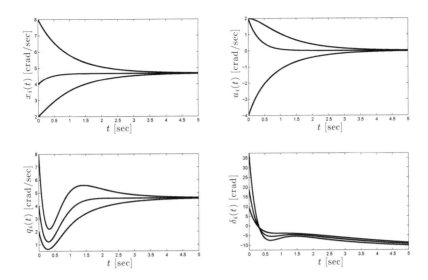

Figure 5.9 Agent guidance state ($x_i(t)$, $t \geq 0$), guidance input ($u_i(t)$, $t \geq 0$), pitch rate ($q_i(t)$, $t \geq 0$), and elevator control ($\delta_i(t)$, $t \geq 0$) responses for the standard consensus protocol given by (5.11) with $k_1 = 1$ (Example 5.3).

control input be given by

$$\delta_i(t) = -K\hat{z}_i(t), \quad i = 1, 2, 3,$$

where

$$K = [-0.0157, 0.0831, -4.7557, -0.1400, -9.8603],$$

which is designed based on an optimal LQR.

For the higher-level controller design, we first use the standard consensus protocol given by (5.11) and then use the proposed consensus protocol given by (5.14), (5.15), and (5.16) to generate an $x_i(t)$, $t \geq 0$, that has a direct effect on the lower-level controller design to achieve pitch rate consensus. Figures 5.9 and 5.10 present the results for all initial conditions set to zero and $x_1(0) = 8$, $x_2(0) = 4$, and $x_3(0) = 2$. It can be seen from the figures that the aircraft reach a pitch rate consensus faster with less control effort ($u_i(t)$, $t \geq 0$) with the proposed consensus protocol shown in Figure 5.10 than with the standard consensus protocol shown in Figure 5.9.    $\triangle$

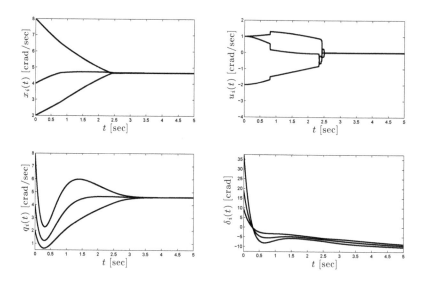

Figure 5.10  Agent guidance state $(x_i(t),\ t \geq 0)$, guidance input $(u_i(t),\ t \geq 0)$, pitch rate $(q_i(t),\ t \geq 0)$, and elevator control $(\delta_i(t),\ t \geq 0)$ responses for the proposed consensus protocol with resetting given by (5.14), (5.15), and (5.16) with $k_1 = 0.5$, $k_2 = 0.00025$, $\tau = 0.001$, $\xi = 250$, and $\theta = 1$ (Example 5.3).

*Chapter Six*

# Control Protocols for Dynamical Networks with Switching Communication Topologies

## 6.1 Introduction

Since communication links among multiagent systems are often unreliable due to multipath effects and exogenous disturbances, the information exchange topologies in network systems are often dynamic. In particular, link failures or creations in network multiagent systems result in switchings of the communication topology. This is the case, for example, if information between agents is exchanged by means of line-of-sight sensors that experience periodic communication dropouts due to agent motion. Variation in network topology is introduced through control input discontinuities, which in turn give rise to discontinuous dynamical systems. In this case, the vector field defining the dynamical system is a discontinuous function of the state, and, hence, system stability can be analyzed using nonsmooth Lyapunov theory, involving concepts such as weak and strong stability notions, differential inclusions, and generalized gradients of locally Lipschitz continuous functions and proximal subdifferentials of lower semicontinuous functions [70].

To address agreement problems in switching networks with state-dependent topologies, in this chapter we extend the theory of semistability to discontinuous time-invariant and time-varying dynamical systems. In particular, we develop sufficient conditions to guarantee weak and strong invariance of Filippov solutions. Moreover, we present Lyapunov-based tests for strong and weak semistability for autonomous and nonautonomous differential inclusions. In addition, we develop sufficient conditions for finite time semistability of autonomous discontinuous dynamical systems. Achieving agreement in finite time allows the dynamical network to use exact information in addressing other system tasks. Furthermore, using our consensus algorithms we develop a coordination framework for multivehicle rendezvous problems.

The contents of the chapter are as follows. In Section 6.2, we review some basic results on differential inclusions which provide the mathematical foundation for designing consensus protocols for network systems with dynamically changing communication topologies. In Sections 6.3–6.5, we develop a Lyapunov-based semistability and finite time semistability theory for time-invariant and time-varying differential inclusions. In addition, we develop new Lyapunov-based results for semistability that do not make assumptions of sign definiteness on the Lyapunov functions. Instead, our results extend the results of [36] to discontinuous systems and use the notion of nontangency between the discontinuous vector field and weakly invariant or weakly negatively invariant subsets of the level or sublevel sets of the Lyapunov function.

It is important to note that our stability results are different from the results in the literature [64, 82] since the Lipschitz conditions in [64, 82] are not valid for the autonomous differential inclusions considered in this chapter. In Sections 6.6–6.8, we use the results of Sections 6.3 and 6.5 to develop information protocols for consensus, parallel formation, and rendezvous problems for multiagent networks with dynamically changing communication topologies.

## 6.2 Mathematical Preliminaries

Consider the differential equation given by

$$\dot{x}(t) = f(x(t)), \quad x(0) = x_0, \quad t \geq 0, \tag{6.1}$$

where $f : \mathbb{R}^q \to \mathbb{R}^q$ is Lebesgue measurable[8] and locally essentially bounded with respect to $x$ [96, 97]; that is, $f$ is bounded on a bounded neighborhood of every point $x$, excluding sets of measure zero, and admits an equilibrium point $x_e \in \mathbb{R}^q$; that is, $f(x_e) = 0$.

An absolutely continuous function[9] $x : [0, \tau] \to \mathbb{R}^q$ is said to be a *Filippov solution* [96, 97] of (6.1) on the interval $[0, \tau]$ with initial condition $x(0) = x_0$ if $x(t)$ satisfies

$$\dot{x}(t) \in \mathcal{K}[f](x(t)), \quad \text{a.a. } t \in [0, \tau], \tag{6.2}$$

---

[8]A function $f : \mathcal{D} \to \mathbb{R}^q$ is *Lebesgue measurable* if the inverse image of $f$ on every open (or Borel) set is Lebesgue measurable.

[9]A function $x : [t_0, t_1] \to \mathbb{R}$ is absolutely continuous on $[t_0, t_1]$ if and only if, for every $\varepsilon > 0$, there exists $\delta > 0$ such that, for each finite collection $\{(a_1, b_1), \ldots, (a_n, b_n)\}$ of disjoint open intervals with $\sum_{i=1}^n (b_i - a_i) < \delta$, $\sum_{i=1}^n |x(b_i) - x(a_i)| < \varepsilon$. Equivalently, $x(\cdot)$ is absolutely continuous if and only if there exists a Lebesgue integrable function $\kappa : [t_0, t_1] \to \mathbb{R}$ such that $x(t) = x(t_0) + \int_{t_0}^{t_1} \kappa(s)\mathrm{d}s$, $t \in [t_0, t_1]$. Note that every absolutely continuous function is continuous; the converse, however, is not true.

where the *Filippov set-valued map* $\mathcal{K}[f] : \mathbb{R}^q \to \mathcal{B}(\mathbb{R}^q)$ is defined by

$$\mathcal{K}[f](x) \triangleq \bigcap_{\delta > 0} \bigcap_{\mu(\mathcal{S})=0} \overline{\text{co}} \{ f(\mathcal{B}_\delta(x) \backslash \mathcal{S}) \}, \quad x \in \mathbb{R}^q, \tag{6.3}$$

where $\mathcal{B}(\mathbb{R}^q)$ denotes the collection of all subsets of $\mathbb{R}^q$, $\mu(\cdot)$ denotes the Lebesgue measure in $\mathbb{R}^q$, "$\overline{\text{co}}$" denotes the convex closure, and $\bigcap_{\mu(\mathcal{S})=0}$ denotes the intersection over all sets $\mathcal{S}$ of Lebesgue measure zero.[10]

Note that $\mathcal{K}[f] : \mathbb{R}^q \to \mathcal{B}(\mathbb{R}^q)$ is a map that assigns sets to points. Dynamical systems of the form given by (6.2) are called *differential inclusions* in the literature [10], and, for each state $x \in \mathbb{R}^q$, they specify a *set* of possible evolutions rather than a single one. It follows from (1) of Theorem 1 of [240] that there exists a set $\mathcal{N}_f \subset \mathbb{R}^q$ of measure zero such that, for every set $\mathcal{W} \subset \mathbb{R}^q$ of measure zero,

$$\mathcal{K}[f](x) = \overline{\text{co}} \left\{ \lim_{i \to \infty} f(x_i) : x_i \to x, \, x_i \notin \mathcal{N}_f \cup \mathcal{W} \right\}. \tag{6.4}$$

Since the Filippov set-valued map given by (6.3) is upper semicontinuous with nonempty, convex, and compact values, and $\mathcal{K}[f]$ is also locally bounded [97, p. 85], it follows that Filippov solutions to (6.1) exist [97, Thm. 1, p. 77]. Recall that the solution $t \mapsto x(t)$ to (6.1) is a *right maximal solution* if it cannot be extended forward in time. We assume that all right maximal Filippov solutions to (6.1) exist on $[0, \infty)$, and, hence, we assume that (6.1) is *forward complete*.

We say that a set $\mathcal{M}$ is *weakly positively invariant* (respectively, *strongly positively invariant*) with respect to (6.1) if, for every $x_0 \in \mathcal{M}$, $\mathcal{M}$ contains a right maximal solution (respectively, all right maximal solutions) of (6.1) [11, 267]. The set $\mathcal{M} \subseteq \mathbb{R}^q$ is *weakly negatively invariant* if, for every $x \in \mathcal{N}$ and $t \geq 0$, there exist $z \in \mathcal{N}$ and a Filippov solution $\psi(\cdot)$ to (6.1) with $\psi(0) = z$ such that $\psi(t) = x$ and $\psi(\tau) \in \mathcal{N}$ for all $\tau \in [0, t]$. Finally, the set $\mathcal{M} \subseteq \mathbb{R}^q$ is *weakly invariant* if $\mathcal{M}$ is weakly positively invariant as well as weakly negatively invariant.

An equilibrium point of (6.1) is a point $x_e \in \mathbb{R}^n$ such that $0 \in \mathcal{K}[f](x_e)$. It is easy to see that $x_e$ is an equilibrium point of (6.1) if and only if the constant function $x(\cdot) = x_e$ is a Filippov solution of (6.1). We denote the set of equilibrium points of (6.1) by $\mathcal{E}$. Since the set-valued map $\mathcal{K}[f]$ is upper semicontinuous, it follows that $\mathcal{E}$ is closed. To develop Lyapunov theory

---

[10]Alternatively, we can consider Krasovskii solutions of (6.1) wherein the possible misbehavior of the derivative of the state on null measure sets is not ignored; that is, $\mathcal{K}[f](x)$ is replaced with $\mathcal{K}[f](x) = \bigcap_{\delta > 0} \overline{\text{co}} \{ f(\mathcal{B}_\delta(x)) \}$ and $f$ is assumed to be locally bounded.

for nonsmooth dynamical systems of the form given by (6.1), we need to introduce the notion of generalized derivatives and gradients. Here we focus on Clarke generalized derivatives and gradients [63].

**Definition 6.1** (see [11, 63]). Let $V : \mathbb{R}^q \to \mathbb{R}$ be a locally Lipschitz continuous function. The *Clarke upper generalized derivative* of $V(x)$ at $x$ in the direction of $v$ is defined by

$$V^o(x, v) \triangleq \limsup_{y \to x, h \to 0^+} \frac{V(y + hv) - V(y)}{h}. \tag{6.5}$$

The *Clarke generalized gradient* $\partial V : \mathbb{R}^q \to \mathcal{B}(\mathbb{R}^q)$ of $V(x)$ at $x$ is the set

$$\partial V(x) \triangleq \mathrm{co} \left\{ \lim_{i \to \infty} \nabla V(x_i) : x_i \to x, \, x_i \notin \mathcal{N} \cup \mathcal{S} \right\}, \tag{6.6}$$

where "co" denotes the convex hull, $\nabla$ denotes the nabla operator, $\mathcal{N}$ is the set of measure zero of points where $\nabla V$ does not exist, and $\mathcal{S}$ is an arbitrary set of measure zero in $\mathbb{R}^q$.

Note that (6.5) always exists. Furthermore, note that it follows from Theorem 2.5.1 of [63] that (6.6) is well defined and consists of all convex combinations of all the possible limits of the gradient at neighboring points where $V$ is differentiable.

In order to state the main results of this chapter, we need some additional notation and definitions. Given a locally Lipschitz continuous function $V : \mathbb{R}^q \to \mathbb{R}$, the *set-valued Lie derivative* $\mathcal{L}_f V : \mathbb{R}^q \to \mathcal{B}(\mathbb{R})$ of $V$ with respect to (6.1) [11, 73] is defined as

$$\mathcal{L}_f V(x) \triangleq \left\{ a \in \mathbb{R} : \text{there exists } v \in \mathcal{K}[f](x) \text{ such that} \right.$$
$$\left. p^{\mathrm{T}} v = a \text{ for all } p \in \partial V(x) \right\}. \tag{6.7}$$

If $V(x)$ is continuously differentiable at $x$, then $\mathcal{L}_f V(x) = \{\nabla V(x) \cdot v, v \in \mathcal{K}[f](x)\}$. In the case where $\mathcal{L}_f V(x)$ is nonempty, we use $\max \mathcal{L}_f V(x)$ to denote the largest element of $\mathcal{L}_f V(x)$. Finally, recall that a function $V : \mathbb{R}^q \to \mathbb{R}$ is *regular* at $x \in \mathbb{R}^q$ [63, Def. 2.3.4] if, for all $v \in \mathbb{R}^q$, the usual right directional derivative $V'_+(x, v) \triangleq \lim_{h \to 0^+} \frac{1}{h}[V(x + hv) - V(x)]$ exists and $V'_+(x, v) = V^o(x, v)$. $V$ is called *regular* on $\mathbb{R}^q$ if it is regular at every $x \in \mathbb{R}^q$.

The next definition introduces the notion of semistability for discontinuous dynamical systems.

**Definition 6.2.** Let $\mathcal{D} \subseteq \mathbb{R}^q$ be an open strongly positively invariant set with respect to (6.1). An equilibrium point $z \in \mathcal{D}$ of (6.1) is *Lyapunov stable* if, for every $\varepsilon > 0$, there exists $\delta = \delta(\varepsilon) > 0$ such that, for every initial condition $x_0 \in \mathcal{B}_\delta(z)$ and every Filippov solution $x(t)$ with the initial condition $x(0) = x_0$, $x(t) \in \mathcal{B}_\varepsilon(z)$ for all $t \geq 0$. An equilibrium point $z \in \mathcal{D}$ of (6.1) is *semistable* if $z$ is Lyapunov stable and there exists an open subset $\mathcal{D}_0$ of $\mathcal{D}$ containing $z$ such that, for all initial conditions in $\mathcal{D}_0$, the Filippov solutions of (6.1) converge to a Lyapunov stable equilibrium point. The system (6.1) is *semistable* with respect to $\mathcal{D}$ if every Filippov solution with initial condition in $\mathcal{D}$ converges to a Lyapunov stable equilibrium. Finally, (6.1) is said to be globally semistable if (6.1) is semistable with respect to $\mathbb{R}^q$.

Next, we introduce the definition of finite time semistability of (6.1).

**Definition 6.3.** Let $\mathcal{D} \subseteq \mathbb{R}^q$ be an open strongly positively invariant set with respect to (6.1). An equilibrium point $x_e \in \mathcal{E}$ of (6.1) is said to be *finite time semistable* if there exist an open neighborhood $\mathcal{U} \subseteq \mathcal{D}$ of $x_e$ and a function $T : \mathcal{U}\backslash\mathcal{E} \to (0, \infty)$, called the *settling-time function*, such that the following statements hold.

(i) For every $x \in \mathcal{U}\backslash\mathcal{E}$ and every Filippov solution $\psi(t)$ of (6.1) with $\psi(0) = x$, $\psi(t) \in \mathcal{U}\backslash\mathcal{E}$ for all $t \in [0, T(x))$, and $\lim_{t \to T(x)} \psi(t)$ exists and is contained in $\mathcal{U} \cap \mathcal{E}$.

(ii) $x_e$ is semistable.

An equilibrium point $x_e \in \mathcal{E}$ of (6.1) is said to be *globally finite time semistable* if it is finite time semistable with $\mathcal{D} = \mathcal{U} = \mathbb{R}^q$. The system (6.1) is said to be *finite time semistable* if every equilibrium point in $\mathcal{E}$ is finite time semistable. Finally, (6.1) is said to be *globally finite time semistable* if every equilibrium point in $\mathcal{E}$ is globally finite time semistable.

Given an absolutely continuous curve $\gamma : [0, \infty) \to \mathbb{R}^q$, *the positive limit set of $\gamma$* is the set $\Omega(\gamma)$ of points $y \in \mathbb{R}^q$ for which there exists an increasing divergent sequence $\{t_i\}_{i=1}^{\infty}$ satisfying $\lim_{i \to \infty} \gamma(t_i) = y$. We denote the positive limit set of a Filippov solution $\psi(\cdot)$ of (6.1) by $\Omega(\psi)$. The positive limit set of a bounded Filippov solution of (6.1) is nonempty and weakly invariant with respect to (6.1) [97, Lem. 4, p. 130].

## 6.3 Semistability Theory for Differential Inclusions

In this section, we develop Lyapunov-based semistability theory for discontinuous dynamical systems of the form given by (6.1). The following proposition is needed for the main results of this section.

**Proposition 6.1.** Let $\mathcal{D} \subseteq \mathbb{R}^q$ be an open strongly positively invariant set with respect to (6.1), and let $\psi(\cdot)$ be a Filippov solution of (6.2) with $\psi(0) \in \mathcal{D}$. If $z \in \Omega(\psi) \cap \mathcal{D}$ is a Lyapunov stable equilibrium point, then $z = \lim_{t \to \infty} \psi(t)$ and $\Omega(\psi) = \{z\}$.

**Proof.** Suppose that $z \in \Omega(\psi) \cap \mathcal{D}$ is Lyapunov stable, and let $\varepsilon > 0$. Since $z$ is Lyapunov stable, there exists $\delta = \delta(\varepsilon) > 0$ such that, for every $y \in \mathcal{B}_\delta(z)$ and every Filippov solution $\eta(\cdot)$ of (6.2) satisfying $\eta(0) = y$, $\eta(t) \in \mathcal{B}_\varepsilon(z)$ for all $t \geq 0$. Now, since $z \in \Omega(\psi)$, it follows that there exists a divergent sequence $\{t_i\}_{i=1}^\infty$ in $[0, \infty)$ such that $\lim_{i \to \infty} \psi(t_i) = z$, and, hence, there exists $k \geq 1$ such that $\psi(t_k) \in \mathcal{B}_\delta(z)$. It now follows from our construction of $\delta$ that $\psi(t) \in \mathcal{B}_\varepsilon(z)$ for all $t \geq t_k$. Since $\varepsilon$ was chosen arbitrarily, it follows that $z = \lim_{t \to \infty} \psi(t)$. Thus, $\lim_{n \to \infty} \psi(t_n) = z$ for every divergent sequence $\{t_n\}_{n=1}^\infty$, and, hence, $\Omega(\psi) = \{z\}$. $\square$

Next, we present sufficient conditions for semistability of (6.1). Here, we adopt the convention that $\max \emptyset = -\infty$.

**Theorem 6.1.** Let $\mathcal{D} \subseteq \mathbb{R}^q$ be an open strongly positively invariant set with respect to (6.1), and let $V : \mathcal{D} \to \mathbb{R}$ be locally Lipschitz continuous and regular on $\mathcal{D}$. Assume that, for each $x \in \mathcal{D}$ and each Filippov solution $\psi(\cdot)$ satisfying $\psi(0) = x$, there exists a compact subset of $\mathcal{D}$ containing $\psi(t)$ for all $t \geq 0$. Furthermore, assume that $\max \mathcal{L}_f V(x) \leq 0$ for almost all $x \in \mathcal{D}$ such that $\mathcal{L}_f V(x) \neq \emptyset$. Finally, define

$$\mathcal{Z} \triangleq \{x \in \mathcal{D} : 0 \in \mathcal{L}_f V(x)\}. \tag{6.8}$$

If every point in the largest weakly positively invariant subset $\mathcal{M}$ of $\overline{\mathcal{Z}} \cap \mathcal{D}$ is a Lyapunov stable equilibrium point, then (6.1) is semistable with respect to $\mathcal{D}$.

**Proof.** Let $x \in \mathcal{D}$, $\psi(\cdot)$ be a Filippov solution to (6.1) with $\psi(0) = x$, and let $\Omega(\psi)$ be the positive limit set of $\psi$. First, we show that $\Omega(\psi) \subseteq \overline{\mathcal{Z}}$. Since either $\max \mathcal{L}_f V(x) \leq 0$ or $\mathcal{L}_f V(x) = \emptyset$ for almost all $x \in \mathcal{D}$, it follows from Lemma 1 of [11] that $\frac{\mathrm{d}}{\mathrm{d}t} V(\psi(t))$ exists and is contained in $\mathcal{L}_f V(\psi(t))$ for almost every $t \geq 0$. Now, by assumption,

$$V(\psi(t)) - V(\psi(\tau)) = \int_\tau^t \frac{\mathrm{d}}{\mathrm{d}t} V(\psi(s)) \mathrm{d}s \leq 0, \quad t \geq \tau,$$

and, hence, $V(\psi(t)) \leq V(\psi(\tau))$, $t \geq \tau$, which implies that $V(\psi(t))$ is a nonincreasing function of time.

The continuity of $V$ and the boundedness of $\psi$ imply that $V(\psi(\cdot))$ is bounded. Hence, $\gamma_x \triangleq \lim_{t \to \infty} V(\psi(t))$ exists. Next, consider $p \in \Omega(\psi)$.

Then there exists an increasing unbounded sequence $\{t_n\}_{n=1}^{\infty}$ in $[0, \infty)$ such that $\psi(t_n) \to p$ as $n \to \infty$. Since $V$ is continuous on $\mathcal{D}$, it follows that

$$V(p) = V\left(\lim_{n\to\infty} \psi(t_n)\right) = \lim_{n\to\infty} V(\psi(t_n)) = \gamma_x,$$

and, hence, $V(p) = \gamma_x$ for $p \in \Omega(\psi)$. In other words, $\Omega(\psi)$ is contained in a level set of $V$.

Let $y \in \Omega(\psi)$. Since $\Omega(\psi)$ is weakly positively invariant, there exists a Filippov solution $\hat{\psi}(\cdot)$ of (6.1) such that $\hat{\psi}(0) = y$ and $\hat{\psi}(t) \in \Omega(\psi)$ for all $t \geq 0$. Since $V(\Omega(\psi)) = \{V(y)\}$, $\frac{d}{dt}V(\hat{\psi}(t)) = 0$, and, hence, it follows from Lemma 1 of [11] that $0 \in \mathcal{L}_f V(\hat{\psi}(t))$; that is, $\hat{\psi}(t) \in \mathcal{Z}$ for almost all $t \in [0, \hat{t}]$. In particular, $y \in \mathcal{Z}$. Since $y \in \Omega(\psi)$ was chosen arbitrarily, it follows that $\Omega(\psi) \subseteq \overline{\mathcal{Z}}$.

Next, since $\Omega(\psi)$ is weakly positively invariant, it follows that $\Omega(\psi) \subseteq \mathcal{M}$. Moreover, since every point in $\mathcal{M}$ is a Lyapunov stable equilibrium point of (6.1), it follows from Proposition 6.1 that $\Omega(\psi)$ contains a single point and $\lim_{t\to\infty} \psi(t)$ is a Lyapunov stable equilibrium. Now, since $x \in \mathcal{D}$ was chosen arbitrarily, it follows from Definition 6.2 that (6.1) is semistable with respect to $\mathcal{D}$.    $\square$

The following corollary to Theorem 6.1 provides sufficient conditions for *finite time* semistability of (6.1).

**Corollary 6.1.** Let $\mathcal{D} \subseteq \mathbb{R}^q$ be an open strongly positively invariant set with respect to (6.1), and let $V : \mathcal{D} \to \mathbb{R}$ be locally Lipschitz continuous and regular on $\mathcal{D}$. Assume that $\max \mathcal{L}_f V(x) < 0$ for almost all $x \in \mathcal{D}\backslash\mathcal{E}$ such that $\mathcal{L}_f V(x) \neq \emptyset$. If every equilibrium in $\mathcal{D}$ is Lyapunov stable, then every equilibrium in $\mathcal{D}$ is semistable. If, in addition, $\max \mathcal{L}_f V(x) \leq -\varepsilon < 0$ for almost every $x \in \mathcal{D}\backslash\mathcal{E}$ such that $\mathcal{L}_f V(x) \neq \emptyset$, then (6.1) is finite time semistable.

**Proof.** To prove the first statement, suppose every equilibrium in $\mathcal{D}$, that is, every point in $\mathcal{E}\cap\mathcal{D}$, is Lyapunov stable. By Lyapunov stability, there exists an open set $\mathcal{D}'$ containing $\mathcal{E} \cap \mathcal{D}$ such that $\mathcal{D}'$ is strongly positively invariant with respect to (6.1) and each Filippov solution having initial condition in $\mathcal{D}'$ is bounded.

Let $\mathcal{M}$ denote the largest weakly positively invariant subset of the set $\mathcal{Z}' \triangleq \{x \in \mathcal{D}' : 0 \in \mathcal{L}_f V(x)\}$. Note that $0 \in \mathcal{L}_f V(x)$ for every $x \in \mathcal{E}$. Since $\mathcal{E} \cap \mathcal{D}$ is weakly positively invariant and contained in $\mathcal{D}'$, it follows that $\mathcal{E} \cap \mathcal{D} \subseteq \mathcal{M}$. Since either $\max \mathcal{L}_f V(x) < 0$ or $\mathcal{L}_f V(x) = \emptyset$ for almost all $x \in \mathcal{D}\backslash\mathcal{E}$, it follows that $\mathcal{Z}' \subseteq \mathcal{E}$. Hence, it follows that $\mathcal{M} = \mathcal{E} \cap \mathcal{D}$.

Theorem 6.1 now implies that (6.1) is semistable with respect to $\mathcal{D}'$. Since $\mathcal{E} \cap \mathcal{D} = \mathcal{E} \cap \mathcal{D}'$, it follows that every equilibrium in $\mathcal{D}$ is semistable.

If, in addition, $\max \mathcal{L}_f V(x) \leq -\varepsilon < 0$ for almost every $x \in \mathcal{D} \backslash \mathcal{E}$ such that $\mathcal{L}_f V(x) \neq \emptyset$, then it follows from Proposition 2.8 of [73] that every Filippov solution originating in $\mathcal{D}'$ reaches $\mathcal{Z}'$ in finite time. Thus, it follows from Definition 6.3 that (6.1) is finite time semistable.                    $\square$

**Example 6.1.** Consider the nonlinear switched dynamical system on $\mathcal{D} = \mathbb{R}^2$ given by

$$\dot{x}_1(t) = f_{\sigma(t)}(x_2(t)) - g_{\sigma(t)}(x_1(t)), \quad x_1(0) = x_{10}, \quad t \geq 0, \quad \sigma(t) \in \mathcal{S}, \quad (6.9)$$
$$\dot{x}_2(t) = g_{\sigma(t)}(x_1(t)) - f_{\sigma(t)}(x_2(t)), \quad x_2(0) = x_{20}, \quad (6.10)$$

where $x_1, x_2 \in \mathbb{R}$; $\sigma : [0, \infty) \to \mathcal{S}$ is a piecewise constant switching signal; $\mathcal{S}$ is a finite index set for every $\sigma \in \mathcal{S}$; $f_\sigma(\cdot)$ and $g_\sigma(\cdot)$ are Lipschitz continuous; $f_\sigma(x_2) - g_\sigma(x_1) = 0$ if and only if $x_1 = x_2$; and $(x_1 - x_2)(f_\sigma(x_2) - g_\sigma(x_1)) \leq 0$, $x_1, x_2 \in \mathbb{R}$. Note that $f^{-1}(0) = \{(x_1, x_2) \in \mathbb{R}^2 : x_1 = x_2 = \alpha, \alpha \in \mathbb{R}\}$.

To show that (6.9)–(6.10) is semistable, consider the Lyapunov function candidate

$$V(x_1 - \alpha, x_2 - \alpha) = \frac{1}{2}(x_1 - \alpha)^2 + \frac{1}{2}(x_2 - \alpha)^2,$$

where $\alpha \in \mathbb{R}$. Now, it follows that

$$\begin{aligned}
\dot{V}(x_1 - \alpha, x_2 - \alpha) &= (x_1 - \alpha)[f_\sigma(x_2) - g_\sigma(x_1)] + (x_2 - \alpha)[g_\sigma(x_1) - f_\sigma(x_2)] \\
&= x_1[f_\sigma(x_2) - g_\sigma(x_1)] + x_2[g_\sigma(x_1) - f_\sigma(x_2)] \\
&= (x_1 - x_2)[f_\sigma(x_2) - g_\sigma(x_1)] \\
&\leq 0, \quad (x_1, x_2) \in \mathbb{R} \times \mathbb{R}, \quad (6.11)
\end{aligned}$$

which, by Theorem 1 of [11], implies that $x_1 = x_2 = \alpha$ is Lyapunov stable for all $\alpha \in \mathbb{R}$.

Next, we rewrite (6.9) and (6.10) in the form of the differential inclusion (6.2), where $x \triangleq [x_1, x_2]^{\mathrm{T}} \in \mathbb{R}^2$ and $f(x) \triangleq [f_\sigma(x_2) - g_\sigma(x_1), g_\sigma(x_1) - f_\sigma(x_2)]^{\mathrm{T}}$. Let $v_x$ be an arbitrary element of $\mathcal{K}[f](x)$, and note that the Clarke upper generalized derivative of $V(x) = \frac{1}{2}x_1^2 + \frac{1}{2}x_2^2$ along a vector $v_x \in \mathcal{K}[f](x)$ is given by $V^o(x, v_x) = x^{\mathrm{T}} v_x$. Furthermore, note that the set $\mathcal{D}_c \triangleq \{x \in \mathbb{R}^2 : V(x) \leq c\}$, where $c > 0$, is a compact set. Next, consider $\max V^o(x, v_x) \triangleq \max_{v_x \in \mathcal{K}[f]}\{x^{\mathrm{T}} v_x\}$. It follows from Theorem 1 of [240] and (6.11) that

$$x^{\mathrm{T}} \mathcal{K}[f](x) = \mathcal{K}[x^{\mathrm{T}} f](x) = \mathcal{K}\left[(x_1 - x_2)(f_\sigma(x_2) - g_\sigma(x_1))\right](x),$$

and, hence, by definition of $\mathcal{K}[f](x)$, it follows that

$$\max V^o(x, v_x) = \max \overline{\mathrm{co}}\{(x_1 - x_2)(f_\sigma(x_2) - g_\sigma(x_1))\}.$$

Note that since, by (6.11), $(x_1 - x_2)(f_\sigma(x_2) - g_\sigma(x_1)) \leq 0$, $x \in \mathbb{R}^2$, it follows that $\max V^o(x, v_x)$ cannot be positive, and, hence, the largest value that $\max V^o(x, v_x)$ can achieve is zero.

Finally, let

$$\mathcal{R} \triangleq \{(x_1, x_2) \in \mathbb{R}^2 : (x_1 - x_2)(f_\sigma(x_2) - g_\sigma(x_1)) = 0\}$$
$$= \{(x_1, x_2) \in \mathbb{R}^2 : x_1 = x_2 = \alpha, \alpha \in \mathbb{R}\}.$$

Since $\mathcal{R}$ consists of equilibrium points, it follows that $\mathcal{M} = \mathcal{R}$. Note that $\max \mathcal{L}_f V(x) \leq \max V^o(x, v_x)$ for each $x \in \mathbb{R}^2$ [11]. Hence, it follows from Theorem 6.1 that $x_1 = x_2 = \alpha$ is semistable for all $\alpha \in \mathbb{R}$.       $\triangle$

**Example 6.2.** Consider the discontinuous dynamical system on $\mathcal{D} = \mathbb{R}^2$ given by

$$\dot{x}_1(t) = \mathrm{sign}(x_2(t) - x_1(t)), \quad x_1(0) = x_{10}, \quad t \geq 0, \qquad (6.12)$$
$$\dot{x}_2(t) = \mathrm{sign}(x_1(t) - x_2(t)), \quad x_2(0) = x_{20}, \qquad (6.13)$$

where $x_1, x_2 \in \mathbb{R}$; $\mathrm{sign}(x) \triangleq x/|x|$ for $x \neq 0$; and $\mathrm{sign}(0) \triangleq 0$. Let $f(x_1, x_2) \triangleq [\mathrm{sign}(x_2 - x_1), \mathrm{sign}(x_1 - x_2)]^{\mathrm{T}}$. Consider $V(x_1, x_2) = \frac{1}{2}(x_1 - \alpha)^2 + \frac{1}{2}(x_2 - \alpha)^2$, where $\alpha \in \mathbb{R}$. Since $V(x_1, x_2)$ is differentiable at $x = (x_1, x_2)$, it follows that $\mathcal{L}_f V(x_1, x_2) = [x_1 - \alpha, x_2 - \alpha] \mathcal{K}[f](x_1, x_2)$.

Now, it follows from Theorem 1 of [240] that

$$\begin{aligned}
[x_1 - \alpha, x_2 - \alpha]\mathcal{K}[f](x) &= \mathcal{K}[[x_1 - \alpha, x_2 - \alpha]f](x) \\
&= \mathcal{K}[-(x_1 - x_2)\mathrm{sign}(x_1 - x_2)](x) \\
&= -(x_1 - x_2)\mathcal{K}[\mathrm{sign}(x_1 - x_2)](x) \\
&= -(x_1 - x_2)\mathrm{SGN}(x_1 - x_2) \\
&= -|x_1 - x_2|, \quad (x_1, x_2) \in \mathbb{R}^2, \qquad (6.14)
\end{aligned}$$

where $\mathrm{SGN}(\cdot)$ is defined by [240, 279]

$$\mathrm{SGN}(x) \triangleq \begin{cases} -1, & x < 0, \\ [-1, 1], & x = 0, \\ 1, & x > 0. \end{cases} \qquad (6.15)$$

Hence, $\max \mathcal{L}_f V(x_1, x_2) \leq 0$ for almost all $(x_1, x_2) \in \mathbb{R}^2$. Now, it follows from Theorem 2 of [11] that $(x_1, x_2) = (\alpha, \alpha)$ is Lyapunov stable.

Next, note that $0 \in \mathcal{L}_f V(x_1, x_2)$ if and only if $x_1 = x_2$, and, hence, $\mathcal{Z} = \{(x_1, x_2) \in \mathbb{R}^2 : x_1 = x_2\}$. Since $\mathcal{Z}$ is weakly positively invariant and every

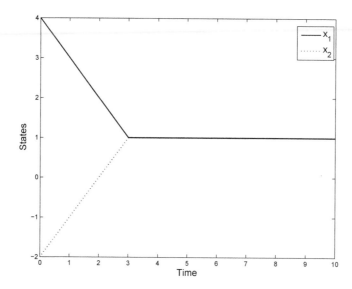

Figure 6.1 State trajectories versus time for Example 6.2.

point in $\mathcal{Z}$ is a Lyapunov stable equilibrium, it follows from Theorem 6.1 that the system (6.12) and (6.13) is semistable.

Finally, we show that the system (6.12) and (6.13) is finite time semistable. To see this, consider the nonnegative function $U(x_1, x_2) = |x_1 - x_2|$. Note that

$$\partial U(x_1, x_2) = \begin{cases} \{\text{sign}(x_1 - x_2)\} \times \{\text{sign}(x_2 - x_1)\}, & x_1 \neq x_2, \\ [-1, 1] \times [-1, 1], & x_1 = x_2. \end{cases} \quad (6.16)$$

Hence, it follows that

$$\mathcal{L}_f U(x_1, x_2) = \begin{cases} \{-2\}, & x_1 \neq x_2, \\ \{0\}, & x_1 = x_2, \end{cases} \quad (6.17)$$

which implies that $\max \mathcal{L}_f U(x_1, x_2) = -2 < 0$ for almost all $(x_1, x_2) \in \mathbb{R}^2 \backslash \mathcal{Z}$. Now, it follows from Corollary 6.1 that the system (6.12) and (6.13) is globally finite time semistable. Figure 6.1 shows the solutions of (6.12) and (6.13) for $x_{10} = 4$ and $x_{20} = -2$. △

Note that Theorem 6.1 and Corollary 6.1 require verifying Lyapunov stability for concluding semistability and finite time semistability, respectively. However, finding the corresponding Lyapunov function can be a difficult task. To overcome this drawback, we extend the nontangency-based approach of [36] to discontinuous dynamical systems in order to guarantee semistability and finite time semistability by testing a condition on the vector field $f$ which avoids proving Lyapunov stability. Before stating our result, we introduce some notation and definitions as well as extended versions of some results from [36].

A set $\mathcal{E} \subseteq \mathbb{R}^q$ is *connected* if and only if every pair of open sets $\mathcal{U}_i \subseteq \mathbb{R}^q$, $i = 1, 2$, satisfying $\mathcal{E} \subseteq \mathcal{U}_1 \cup \mathcal{U}_2$ and $\mathcal{U}_i \cap \mathcal{E} \neq \emptyset$, $i = 1, 2$, has a nonempty intersection. A *connected component* of the set $\mathcal{E} \subseteq \mathbb{R}^q$ is a connected subset of $\mathcal{E}$ that is not properly contained in any connected subset of $\mathcal{E}$. Given a set $\mathcal{E} \subseteq \mathbb{R}^q$, let $\mathrm{coco}\,\mathcal{E}$ denote the convex cone generated by $\mathcal{E}$.

**Definition 6.4.** Given $x \in \mathbb{R}^q$, the *direction cone* $\mathcal{F}_x$ of $f$ at $x$ is the intersection of closed convex cones of the form $\bigcap_{\mu(\mathcal{S})=0} \mathrm{coco}\{f(\mathcal{U}\backslash\mathcal{S})\}$, where $\mathcal{U} \subseteq \mathbb{R}^q$ is an open neighborhood of $x$. Let $\mathcal{E} \subseteq \mathbb{R}^q$. A vector $v \in \mathbb{R}^q$ is *tangent* to $\mathcal{E}$ at $z \in \mathcal{E}$ if there exist a sequence $\{z_i\}_{i=1}^{\infty}$ in $\mathcal{E}$ converging to $z$ and a sequence $\{h_i\}_{i=1}^{\infty}$ of positive real numbers converging to zero such that $\lim_{i\to\infty} \frac{1}{h_i}(z_i - z) = v$. The *tangent cone* to $\mathcal{E}$ at $z$ is the closed cone $T_z\mathcal{E}$ of all vectors tangent to $\mathcal{E}$ at $z$. Finally, the vector field $f$ is *nontangent* to the set $\mathcal{E}$ at the point $z \in \mathcal{E}$ if $T_z\mathcal{E} \cap \mathcal{F}_z \subseteq \{0\}$.

**Definition 6.5.** Given a point $x \in \mathbb{R}^q$ and a bounded open neighborhood $\mathcal{U} \subset \mathbb{R}^q$ of $x$, the *restricted prolongation* of $x$ with respect to $\mathcal{U}$ is the set $\mathcal{R}_x^{\mathcal{U}} \subseteq \overline{\mathcal{U}}$ of all subsequential limits of sequences of the form $\{\psi_i(t_i)\}_{i=1}^{\infty}$, where $\{t_i\}_{i=1}^{\infty}$ is a sequence in $[0, \infty)$; $\psi_i(\cdot)$ is a Filippov solution to (6.1) with $\psi_i(0) = x_i$, $i = 1, 2, \ldots$; and $\{x_i\}_{i=1}^{\infty}$ is a sequence in $\mathcal{U}$ converging to $x$ such that the set $\{z \in \mathbb{R}^q : z = \psi_i(t), t \in [0, t_i]\}$ is contained in $\overline{\mathcal{U}}$ for every $i = 1, 2, \ldots$.

**Proposition 6.2.** Let $\mathcal{D} \subseteq \mathbb{R}^q$ be an open strongly positively invariant set with respect to (6.1). Furthermore, let $x \in \mathcal{D}$, and let $\mathcal{U} \subseteq \mathcal{D}$ be a bounded open neighborhood of $x$. Then $\mathcal{R}_x^{\mathcal{U}}$ is connected. Moreover, if $x$ is an equilibrium point of (6.1), then $\mathcal{R}_x^{\mathcal{U}}$ is weakly negatively invariant.

**Proof.** The proof of connectedness is similar to the proof of the first part of Proposition 6.1 of [36] and, hence, is omitted. To prove weak negative invariance, suppose that $x \in \mathcal{D}$ is an equilibrium point of (6.1), and consider $z \in \mathcal{R}_x^{\mathcal{U}}$. Then there exist a sequence $\{t_i\}_{i=1}^{\infty}$ in $[0, \infty)$; a sequence $\{x_i\}_{i=1}^{\infty}$ in $\mathcal{D}$ converging to $x$; and a sequence $\{\psi_i(\cdot)\}_{i=1}^{\infty}$ of Filippov solutions of (6.1) such that $\lim_{i\to\infty} \psi_i(t_i) = z$ and, for every $i$, $\psi_i(0) = x_i$ and $\psi_i(h) \in \overline{\mathcal{U}}$ for every $h \in [0, t_i]$.

Now, let $t \geq 0$. First, assume $z = x$. Then $\psi \equiv x$ is a Filippov solution of (6.1) such that $\psi(0) = x$, $\psi(t) = z$, and $\psi(\tau) \in \mathcal{R}_x^{\mathcal{U}}$ for all $\tau \in [0, t]$. Next consider the case $z \neq x$. First, suppose that the sequence $\{t_i\}_{i=1}^{\infty}$ has a subsequence $\{t_{i_k}\}_{k=1}^{\infty}$ in $[0, t]$. By choosing a subsequence if necessary, we may assume that the subsequence $\{t_{i_k}\}_{k=1}^{\infty}$ converges to $T$. Necessarily,

$T \leq t$. By Lemma 1 in [97, p. 87], a subsequence of the sequence $\{\psi_{i_k}\}_{k=1}^{\infty}$ converges uniformly on compact subsets of $(0, T)$ to a Filippov solution $\psi$ of (6.1). Moreover, the solution $\psi$ satisfies $\psi(0) = x$ and $\psi(T) = z$. For each $s \in [0, T]$, $\psi(s)$ is a subsequential limit of the sequence $\{\psi_{i_k}(s)\}_{k=1}^{\infty}$ and, hence, is contained in $\mathcal{R}_x^{\mathcal{U}}$. It is now easy to verify that the function $\beta : [0, t] \to \mathcal{D}$ defined by

$$
\begin{aligned}
\beta(s) &= x, & 0 \leq s \leq t - T, \\
&= \psi(s - t + T), & t - T < s \leq t,
\end{aligned}
$$

is a Filippov solution of (6.1) satisfying $\beta(0) = x$, $\beta(t) = z$, and $\beta(s) \in \mathcal{R}_x^{\mathcal{U}}$ for all $s \in [0, t]$.

Next, suppose that the sequence $\{t_i\}_{i=1}^{\infty}$ has no subsequence in $[0, t]$. Then there exists $N > 0$ such that $t_i > t$ for all $i \geq N$. For each $i$, define $\beta_i : [0, t] \to \mathcal{D}$ by $\beta(s) = \psi_{i+N}(t_{i+N} - t + s)$. Clearly, each $\beta_i$ is a Filippov solution of (6.1). Moreover, the sequence $\{\beta_i(t)\}_{i=1}^{\infty}$ converges to $z$. Let $y \in \mathcal{D}$ be a subsequential limit of the bounded sequence $\{\beta_i(0)\}_{i=1}^{\infty}$. By definition, $y \in \mathcal{R}_x^{\mathcal{U}}$. By Lemma 1 in [97, p. 87], a subsequence of $\{\beta_i\}_{i=1}^{\infty}$ converges uniformly on compact subsets of $(0, t)$ to a Filippov solution $\beta$ of (6.1). Moreover, we may choose the subsequence such that $\beta(0) = y$ and $\beta(t) = z$. Finally, for each $s \in [0, t]$, $\beta(s)$ is a subsequential limit of the sequence $\{\beta_i(s)\}_{i=1}^{\infty}$ and, hence, is in $\mathcal{R}_x^{\mathcal{U}}$. We have thus shown that there exists a Filippov solution $\beta$ defined on $[0, t]$ such that $\beta(s) \in \mathcal{R}_x^{\mathcal{U}}$ for all $s \in [0, t]$ and $\beta(t) = z$. Since $t \geq 0$ and $z \in \mathcal{R}_x^{\mathcal{U}}$ were chosen to be arbitrary, it follows that $\mathcal{R}_x^{\mathcal{U}}$ is weakly negatively invariant. $\square$

The following two lemmas and proposition extend related results from [36] and are needed for the main result of this section.

**Lemma 6.1.** Let $\mathcal{D} \subseteq \mathbb{R}^q$ be an open strongly positively invariant set with respect to (6.1), and let $V : \mathcal{D} \to \mathbb{R}$ be locally Lipschitz continuous and regular on $\mathcal{D}$. Assume that $V(x) \geq 0$ for all $x \in \mathcal{D}$, $V(z) = 0$ for all $z \in \mathcal{E}$, and $\max \mathcal{L}_f V(x) \leq 0$ for almost every $x \in \mathcal{D}$ such that $\mathcal{L}_f V(x) \neq \emptyset$. For every $z \in \mathcal{E}$, let $\mathcal{N}_z$ denote the largest weakly negatively invariant connected subset of $\overline{\mathcal{Z}} \cap \mathcal{D}$ containing $z$, where $\mathcal{Z}$ is given by (6.8). Then, for every $x \in \mathcal{E}$ and every bounded open neighborhood $\mathcal{V} \subset \mathcal{D}$ of $x$, $\mathcal{R}_x^{\mathcal{V}} \subseteq \mathcal{N}_x$.

**Proof.** Let $x \in \mathcal{E}$, and let $\mathcal{V} \subset \mathcal{D}$ be a bounded open neighborhood of $x$. Consider $z \in \mathcal{R}_x^{\mathcal{V}}$. Let $\{x_i\}_{i=1}^{\infty}$ be a sequence in $\mathcal{V}$ converging to $x$, and let $\{t_i\}_{i=1}^{\infty}$ be a sequence in $[0, \infty)$ such that the sequence $\{\psi_i(t_i)\}_{i=1}^{\infty}$ converges to $z$ and, for every $i$, $\psi_i(\tau) \in \overline{\mathcal{V}} \subset \mathcal{D}$ for every $\tau \in [0, t_i]$, where $\psi_i(\cdot)$ is a Filippov solution to (6.1) with $\psi_i(0) = x_i$. Since either $\max \mathcal{L}_f V(y) \leq 0$ or $\mathcal{L}_f V(y) = \emptyset$ for almost every $y \in \mathcal{D}$, it follows from Lemma 1 of [11] that

$\frac{\mathrm{d}}{\mathrm{d}t}V(\psi(t))$ exists and is contained in $\mathcal{L}_f V(\psi(t))$ for almost all $t \in [0, \tau]$, where $\psi(\cdot)$ is a Filippov solution to (6.1) with $\psi(0) = y$. Now, by assumption,

$$V(\psi(\tau)) - V(y) = \int_0^\tau \frac{\mathrm{d}}{\mathrm{d}t} V(\psi(s)) \mathrm{d}s \leq 0, \quad \tau \geq 0,$$

and, hence, $V(\psi(\tau)) \leq V(y)$ for $y \in \mathcal{D}$ and $\tau \geq 0$.

Next, note that $V(z) = \lim_{i\to\infty} V(\psi_i(t_i)) \leq \lim_{i\to\infty} V(x_i) = V(x)$, and, hence, $V(z) \leq V(x)$. Since $V(z) \geq 0$ and $V(x) = 0$ by assumption, it follows that $V(z) = V(x) = 0$. Hence, $\mathcal{R}_x^{\mathcal{V}} \subseteq V^{-1}(0) \cap \overline{\mathcal{V}} \subset V^{-1}(0)$. By Proposition 6.2, $\mathcal{R}_x^{\mathcal{V}}$ is weakly negatively invariant and connected and $x \in \mathcal{R}_x^{\mathcal{V}}$. Hence, $\mathcal{R}_x^{\mathcal{V}} \subseteq \mathcal{M}_x$, where $\mathcal{M}_x$ denotes the largest weakly negatively invariant connected subset of $V^{-1}(0)$ containing $x$.

Finally, we show that $\mathcal{M}_x \subseteq \mathcal{N}_x$. Let $z \in \mathcal{M}_x$ and $t > 0$. By weak negative invariance, there exist $w \in \mathcal{M}_x$ and a Filippov solution $\psi(\cdot)$ to (6.1) satisfying $\psi(0) = w$ such that $\psi(t) = z$ and $\psi(\tau) \in \mathcal{M}_x \subseteq V^{-1}(0)$ for all $\tau \in [0, t]$. Thus, $V(\psi(\tau)) = V(x) = 0$ for every $\tau \in [0, t]$, and, hence, by Lemma 1 of [11], $0 \in \mathcal{L}_f V(\psi(\tau))$ for almost every $\tau \in [0, t]$; that is, $\psi(\tau) \in \mathcal{Z}$ for almost every $\tau \in [0, t]$. It immediately follows that $z \in \overline{\mathcal{Z}}$ and, hence, $\mathcal{M}_x \subseteq \overline{\mathcal{Z}}$. Since $\mathcal{M}_x$ is weakly negatively invariant and connected, contains $x$, and is contained in $\mathcal{U}$, it follows that $\mathcal{M}_x \subseteq \mathcal{N}_x$. Hence, $\mathcal{R}_x^{\mathcal{V}} \subseteq \mathcal{M}_x \subseteq \mathcal{N}_x$. $\qquad\square$

**Lemma 6.2.** Let $\mathcal{D} \subseteq \mathbb{R}^q$ be an open strongly positively invariant set with respect to (6.1). Furthermore, let $x \in \mathcal{D}$, and let $\{x_i\}_{i=1}^\infty$ be a sequence in $\mathcal{D}$ converging to $x$. Let $\mathcal{I}_i \subseteq [0, \infty)$, $i = 1, 2, \ldots$, be intervals containing zero, and let $\mathcal{B} \subseteq \mathcal{D}$ be the set of all subsequential limits contained in $\mathcal{D}$ of sequences of the form $\{\psi_i(\tau_i)\}_{i=1}^\infty$, where, for each $i$, $\tau_i \in \mathcal{I}_i$ and $\psi_i : \mathcal{I}_i \to \mathcal{D}$ is a Filippov solution of (6.1) satisfying $\psi_i(0) = x_i$. Then $\mathcal{B} = \{x\}$ if and only if $f$ is nontangent to $\mathcal{B}$ at $x$.

**Proof.** First, we note that $x \in \mathcal{B}$ since $x = \lim_{i\to\infty} \psi_i(0)$. Necessity now follows by noting that if $\mathcal{B} = \{x\}$, then $T_x \mathcal{B} = \{0\}$ and, hence, $T_x \mathcal{B} \cap \mathcal{F}_x \subseteq \{0\}$.

To prove sufficiency, suppose that $z_0 \in \mathcal{B}$, $z_0 \neq x$. Let $\{\mathcal{U}_k\}_{k=1}^\infty$ be a nested sequence of bounded open neighborhoods of $x$ in $\mathcal{D}$ such that $\overline{\mathcal{U}_{k+1}} \subset \mathcal{U}_k$ and $x_k \in \mathcal{U}_k$ for every $k = 1, 2, \ldots$, $\bigcap_k \mathcal{U}_k = \{x\}$, and $z_0 \notin \mathcal{U}_1$. Since $z_0 \in \mathcal{B}$, there exists a sequence $\{\tau_i\}_{i=1}^\infty$ such that $\tau_i \in \mathcal{I}_i$ for every $i$ and $\lim_{i\to\infty} \psi_i(\tau_i) = z_0 \notin \mathcal{U}_1$. The continuity of Filippov solutions implies that, for every $k$, there exists a sequence $\{h_j^k\}_{j=k}^\infty$ in $[0, \infty)$ such that, for every $j \geq k$, $h_j^k \in \mathcal{I}_j$, $h_j^k \leq \tau_j$, $\psi_j(\tau) \in \mathcal{U}_k$ for every $\tau \in [0, h_j^k)$, and $\psi_j(h_j^k) \in$

$\partial \mathcal{U}_k$. For each $k$, let $z_k \in \partial \mathcal{U}_k$ be a subsequential limit of the bounded sequence $\{\psi_j(h_j^k)\}_{j=k}^\infty$. Then, for every $k$, it follows that $z_k \in \mathcal{B}$, $z_k \neq x$, and $\lim_{k \to \infty} z_k = x$. Now, consider a subsequential limit $v$ of the bounded sequence $\{\|z_k - x\|^{-1}(z_k - x)\}$. Clearly, $v \in T_x \mathcal{B}$. Also $\|v\| = 1$ so that $v \neq 0$. We claim that $v \in \mathcal{F}_x$.

Let $\mathcal{V} \subseteq \mathcal{D}$ be an open neighborhood of $x$, and consider $\varepsilon > 0$. By construction, there exists $k$ such that $\left\| v - \|z_k - x\|^{-1}(z_k - x) \right\| < \varepsilon/3$. Moreover, since $\bigcap_i \mathcal{U}_i = \{x\}$, we can assume that $\mathcal{U}_k \subseteq \mathcal{V}$. Since $z_k$ belongs to the boundary of an open neighborhood of $x$, $\delta \triangleq \|z_k - x\| > 0$. Since $z_k = \lim_{i \to \infty} \psi_i(h_i^k)$ and $x = \lim_{i \to \infty} x_i$, there exists $i$ such that $x_i \in \mathcal{V}$, $\|x - x_i\| < \varepsilon \delta/3$, and $\|z_k - \psi_i(h_i^k)\| < \varepsilon \delta/3$. Let $\mathcal{S} \subset \mathcal{D}$ be a zero measure set. Then $\mathcal{K}[f](\psi_i(\tau)) \subseteq \mathrm{co}\{f(\mathcal{V} \backslash \mathcal{S})\}$ for all $\tau \in [0, h_i^k]$, so that $\dot{\psi}_i(\tau) \in \mathrm{co}\{f(\mathcal{V} \backslash \mathcal{S})\}$ for almost every $\tau \in [0, h_i^k]$. Therefore, it follows from Theorem I.6.13 of [310, p. 145] that

$$w \triangleq \psi_i(h_i^k) - x_i = \int_0^{h_i^k} \dot{\psi}_i(\tau) \mathrm{d}\tau$$

is contained in the convex cone generated by $\mathrm{co}\{f(\mathcal{V} \backslash \mathcal{S})\}$. Since $\mathcal{S}$ was chosen to be an arbitrary zero measure set, it follows that

$$w \in \bigcap_{\mu(\mathcal{S})=0} \mathrm{coco}\{f(\mathcal{V} \backslash \mathcal{S})\}.$$

Now,

$$\left\| v - \delta^{-1} w \right\| = \left\| v - \delta^{-1}(z_k - x) - \delta^{-1}(\psi(h_i^k, x_i) - z_k) - \delta^{-1}(x - x_i) \right\|$$
$$\leq \left\| v - \|z_k - x\|^{-1}(z_k - x) \right\|$$
$$\quad + \delta^{-1}\|\psi(h_i^k, x_i) - z_k\| + \delta^{-1}\|x - x_i\|$$
$$< \varepsilon.$$

We have thus shown that, for every $\varepsilon > 0$, there exists

$$w \in \bigcap_{\mu(\mathcal{S})=0} \mathrm{coco}\{f(\mathcal{V} \backslash \mathcal{S})\}$$

and $\delta > 0$ such that $w \neq 0$ and $\|v - \delta^{-1} w\| < \varepsilon$. It follows that $v$ is contained in the closed convex cone $\overline{\bigcap_{\mu(\mathcal{S})=0} \mathrm{coco}\{f(\mathcal{V} \backslash \mathcal{S})\}}$. Since $\mathcal{V}$ was chosen to be an arbitrary open neighborhood of $x$, it follows that $v$ is contained in $\mathcal{F}_x$. Thus, if $\mathcal{B} \neq \{x\}$, then there exists $v \in \mathbb{R}^q$ such that $v \neq 0$ and $v \in T_x \mathcal{B} \cap \mathcal{F}_x$; that is, $f$ is not nontangent to $\mathcal{B}$ at $x$. Sufficiency now follows. $\qquad \square$

**Proposition 6.3.** Let $\mathcal{D} \subseteq \mathbb{R}^q$ be an open strongly positively invariant set with respect to (6.1). Furthermore, let $x \in \mathcal{D}$, and let $\mathcal{U} \subseteq \mathcal{D}$ be a bounded open neighborhood of $x$. If the vector field $f$ of (6.1) is nontangent to $\mathcal{R}_x^{\mathcal{U}}$ at $x$, then the point $x$ is a Lyapunov stable equilibrium of (6.1).

**Proof.** Since $f$ is nontangent to $\mathcal{R}_x^{\mathcal{U}}$ at $x$, by definition, it follows that $T_x \mathcal{R}_x^{\mathcal{U}} \cap \mathcal{F}_x \subseteq \{0\}$. Let $z \in \mathcal{R}_x^{\mathcal{U}}$. Then there exist a sequence $\{x_i\}_{i=1}^{\infty}$ converging to $x$, a sequence $\{t_i\}_{i=1}^{\infty}$ in $[0, \infty)$, and a sequence $\{\psi_i\}_{i=1}^{\infty}$ of Filippov solutions of (6.2) such that $\psi_i(0) = x_i$ and $\psi([0, t_i]) \subseteq \overline{\mathcal{U}}$ for every $i = 1, 2, \ldots$, and $\lim_{i \to \infty} \psi_i(t_i) = z$.

First, suppose that the sequence $\{t_i\}_{i=1}^{\infty}$ converges to zero. Then it follows from Theorem 11 of [96] that there exists a Filippov solution $\hat{\psi}(\cdot)$ to (6.1) with $\hat{\psi}(0) = x$ such that $\lim_{i \to \infty} \psi_i(t_i) = \hat{\psi}(0) = x$. Next, suppose that the sequence $\{t_i\}_{i=1}^{\infty}$ does not converge to zero. Then there exists a subsequence $\{t_{i_k}\}_{k=1}^{\infty}$ of the sequence $\{t_i\}_{i=1}^{\infty}$ such that $\liminf_{k \to \infty} t_{i_k} > 0$. Let $\mathcal{I}_k \triangleq [0, t_{i_k}]$ for each $k$, and let $\mathcal{B} \subseteq \overline{\mathcal{U}}$ denote the set of all subsequential limits of sequences of the form $\{\psi_{i_k}(\tau_k)\}_{k=1}^{\infty}$, where $\tau_k \in \mathcal{I}_k$ for every $k$. By construction, $z \in \mathcal{B}$ and $\mathcal{B} \subseteq \mathcal{R}_x^{\mathcal{U}}$. Hence, $T_x \mathcal{B} \cap \mathcal{F}_x \subseteq T_x \mathcal{R}_x^{\mathcal{U}} \cap \mathcal{F}_x \subseteq \{0\}$; that is, $f$ is nontangent to $\mathcal{B}$ at $x$. Now, it follows from Lemma 6.2 that $\mathcal{B} = \{x\}$. Hence, $z = x$. Since $z \in \mathcal{R}_x^{\mathcal{U}}$ is arbitrary, it follows that $\mathcal{R}_x^{\mathcal{U}} = \{x\}$.

Suppose, *ad absurdum*, that $x$ is not a Lyapunov stable equilibrium. Then there exist a bounded open neighborhood $\mathcal{V} \subseteq \mathcal{U}$ of $x$, a sequence $\{x_i\}_{i=1}^{\infty}$ in $\mathcal{V}$ converging to $x$, a sequence $\{\psi_i\}_{i=1}^{\infty}$ of Filippov solutions to (6.2), and a sequence $\{t_i\}_{i=1}^{\infty}$ in $[0, \infty)$ such that $\psi_i(x_i) = x_i$ and $\psi_i(t_i) \in \partial \mathcal{V}$ for every $i$. Without loss of generality, we can assume that the sequence $\{t_i\}_{i=1}^{\infty}$ is chosen such that, for every $i$, $\psi_i(h) \in \mathcal{V}$ for all $h \in [0, t_i)$. Now, every subsequential limit of the bounded sequence $\{\psi_i(t_i)\}_{i=1}^{\infty}$ is distinct from $x$ by construction and is contained in $\mathcal{R}_x^{\mathcal{U}}$ by definition, which implies that $\mathcal{R}_x^{\mathcal{U}} \setminus \{x\} \neq \emptyset$. This contradicts our earlier conclusion that $\mathcal{R}_x^{\mathcal{U}} = \{x\}$. Hence, $x$ is Lyapunov stable. $\qquad \square$

The following theorem gives sufficient conditions for semistability using nontangency of the vector field $f$.

**Theorem 6.2.** Let $\mathcal{D} \subseteq \mathbb{R}^q$ be an open strongly positively invariant set with respect to (6.1), and let $V : \mathcal{D} \to \mathbb{R}$ be locally Lipschitz continuous and regular on $\mathcal{D}$. Assume that $V(x) \geq 0$ for all $x \in \mathcal{D}$, $V(z) = 0$ for all $z \in \mathcal{E} \cap \mathcal{D}$, and $\max \mathcal{L}_f V(x) \leq 0$ for almost every $x \in \mathcal{D}$ such that $\mathcal{L}_f V(x) \neq \emptyset$. Furthermore, for every $z \in \mathcal{E}$, let $\mathcal{N}_z$ denote the largest weakly negatively invariant connected subset of $\overline{\mathcal{Z}} \cap \mathcal{D}$ containing $z$, where $\mathcal{Z}$ is given by (6.8). If $f$ is nontangent to $\mathcal{N}_z$ at every $z \in \mathcal{E}$, then every equilibrium in $\mathcal{D}$ is semistable.

**Proof.** Let $\mathcal{V} \subset \mathcal{D}$ be a bounded open neighborhood of $x \in \mathcal{E} \cap \mathcal{D}$. Since $f$ is nontangent to $\mathcal{N}_x$ at the point $x \in \mathcal{E} \cap \mathcal{V}$, it follows that $T_x \mathcal{N}_x \cap \mathcal{F}_x \subseteq \{0\}$. Next, we show that $f$ is nontangent to $\mathcal{R}_x^{\mathcal{V}}$ at the point $x$. It follows from Lemma 6.1 that $\mathcal{R}_x^{\mathcal{V}} \subseteq \mathcal{N}_x$. Hence, $T_x \mathcal{R}_x^{\mathcal{V}} \cap \mathcal{F}_x \subseteq T_x \mathcal{N}_x \cap \mathcal{F}_x \subseteq \{0\}$; that is, $T_x \mathcal{R}_x^{\mathcal{V}} \cap \mathcal{F}_x \subseteq \{0\}$. By definition, $f$ is nontangent to $\mathcal{R}_x^{\mathcal{V}}$ at the point $x$. Now, it follows from Proposition 6.3 that $x$ is a Lyapunov stable equilibrium. Since $x \in \mathcal{E} \cap \mathcal{D}$ was chosen arbitrarily, it follows that every equilibrium of (6.1) in $\mathcal{D}$ is Lyapunov stable.

By Lyapunov stability of $x$, it follows that there exists a strongly positively invariant neighborhood $\mathcal{U} \subset \mathcal{V}$ of $x$ that is open and bounded and such that $\overline{\mathcal{U}} \subset \mathcal{V}$. Consider $z \in \mathcal{U}$, and let $\psi(\cdot)$ be a Filippov solution of (6.1) with $\psi(0) = z$. Then $\psi(\cdot)$ is bounded in $\mathcal{D}$. Hence, it follows from [97, p. 129] and Theorem 3 of [11] that $\Omega(\psi) \subseteq \overline{\mathcal{U}}$ is nonempty and contained in $\overline{\mathcal{Z}}$.

Let $w \in \Omega(\psi)$. The invariance and connectedness of $\Omega(\psi)$ imply that $\Omega(\psi) \subseteq \mathcal{N}_w$. Hence, $T_w \Omega(\psi) \cap \mathcal{F}_w \subseteq T_w \mathcal{N}_w \cap \mathcal{F}_w \subseteq \{0\}$. Now, it follows from Lemma 6.2 (see the proof of Proposition 5.2 of [36]) that $\lim_{t \to \infty} \psi(t)$ exists. Since $z \in \mathcal{U}$ was chosen arbitrarily, it follows that every Filippov solution in $\mathcal{U}$ converges to a limit. The strong invariance of $\mathcal{U}$ implies that the limit of every Filippov solution in $\mathcal{U}$ is contained in $\overline{\mathcal{U}}$. Since every equilibrium in $\overline{\mathcal{U}} \subset \mathcal{V}$ is Lyapunov stable, it follows from Theorem 6.1 that $x$ is semistable. Finally, since $x \in \mathcal{E} \cap \mathcal{D}$ was chosen arbitrarily, it follows that every equilibrium in $\mathcal{D}$ is semistable. $\qquad\square$

**Example 6.3.** Consider the discontinuous dynamical system on $\mathcal{D} = \mathbb{R}^4$ given by

$$\dot{x}_1(t) = \text{sign}(x_3(t) - x_4(t)), \quad x_1(0) = x_{10}, \quad t \geq 0, \tag{6.18}$$

$$\dot{x}_2(t) = \text{sign}(x_4(t) - x_3(t)), \quad x_2(0) = x_{20}, \tag{6.19}$$

$$\dot{x}_3(t) = \text{sign}(x_4(t) - x_3(t)) + \text{sign}(x_2(t) - x_1(t)), \quad x_3(0) = x_{30}, \tag{6.20}$$

$$\dot{x}_4(t) = \text{sign}(x_3(t) - x_4(t)) + \text{sign}(x_1(t) - x_2(t)), \quad x_4(0) = x_{40}, \tag{6.21}$$

where $x_1, x_2, x_3, x_4 \in \mathbb{R}$. Let $f : \mathbb{R}^4 \to \mathbb{R}^4$ denote the vector field of (6.18)–(6.21) and $x \triangleq [x_1, x_2, x_3, x_4] \in \mathbb{R}^4$, and consider the function

$$V(x) = |x_1 - x_2| + |x_3 - x_4|.$$

Next, note that

$$
\partial V(x) = \begin{cases}
\{\mathrm{sign}(x_1 - x_2)\} \times \{\mathrm{sign}(x_2 - x_1)\} \\
\quad \times \{\mathrm{sign}(x_3 - x_4)\} \times \{\mathrm{sign}(x_4 - x_3)\}, \quad x_1 \neq x_2, \, x_3 \neq x_4, \\
[-1, 1] \times [-1, 1] \times \{\mathrm{sign}(x_3 - x_4)\} \times \{\mathrm{sign}(x_4 - x_3)\}, \\
\qquad\qquad x_1 = x_2, \, x_3 \neq x_4, \\
\{\mathrm{sign}(x_1 - x_2)\} \times \{\mathrm{sign}(x_2 - x_1)\} \times [-1, 1] \times [-1, 1], \\
\qquad\qquad x_1 \neq x_2, \, x_3 = x_4, \\
\overline{\mathrm{co}}\{(1, 1), (-1, 1), (-1, -1), (1, -1)\}, \quad x_1 = x_2, \, x_3 = x_4,
\end{cases}
$$

and, hence,

$$
\mathcal{L}_f V(x) = \begin{cases}
\{-2\}, & x_1 \neq x_2, \, x_3 \neq x_4, \\
\varnothing, & x_1 = x_2, \, x_3 \neq x_4, \\
\varnothing, & x_1 \neq x_2, \, x_3 = x_4, \\
\{0\}, & x_1 = x_2, \, x_3 = x_4,
\end{cases}
\tag{6.22}
$$

which implies that $\max \mathcal{L}_f V(x) \leq 0$ for almost every $x \in \mathbb{R}^4$ such that $\mathcal{L}_f V(x) \neq \varnothing$. Consequently, $\mathcal{Z} = \{x \in \mathbb{R}^4 : x_1 = x_2, x_3 = x_4\}$. Let $\mathcal{N}$ denote the largest weakly negatively invariant subset contained in $\mathcal{Z}$. On $\mathcal{N}$, it follows from (6.18)–(6.21) that $\dot{x}_1 = \dot{x}_2 = 0$ and $\dot{x}_3 = \dot{x}_4 = 0$. Hence, $\mathcal{N} = \{x \in \mathbb{R}^4 : x_1 = x_2 = a, x_3 = x_4 = b\}$, $a, b \in \mathbb{R}$, which implies that $\mathcal{N}$ is the set of equilibrium points.

Next, we show that $f$ for (6.18)–(6.21) is nontangent to $\mathcal{N}$ at the point $z \in \mathcal{N}$. To see this, note that the tangent cone $T_z\mathcal{N}$ to the equilibrium set $\mathcal{N}$ is orthogonal to the vectors $\mathbf{u}_1 \triangleq [1, -1, 0, 0]^\mathrm{T}$ and $\mathbf{u}_2 \triangleq [0, 0, 1, -1]^\mathrm{T}$. On the other hand, since $f(z) \in \mathrm{span}\{\mathbf{u}_1, \mathbf{u}_2\}$ for all $z \in \mathbb{R}^4$, it follows that $f(\mathcal{V}) \subseteq \mathrm{span}\{\mathbf{u}_1, \mathbf{u}_2\}$ for every subset $\mathcal{V} \subseteq \mathbb{R}^4$. Consequently, the direction cone $\mathcal{F}_z$ of $f$ at $z \in \mathcal{N}$ relative to $\mathbb{R}^4$ satisfies $\mathcal{F}_z \subseteq \mathrm{span}\{\mathbf{u}_1, \mathbf{u}_2\}$. Hence, $T_z\mathcal{N} \cap \mathcal{F}_z = \{0\}$, which implies that the vector field $f$ is nontangent to the set of equilibria $\mathcal{N}$ at the point $z \in \mathcal{N}$. Note that for every $z \in \mathcal{N}$, the set $\mathcal{N}_z$ required by Theorem 6.2 is contained in $\mathcal{N}$. Since nontangency to $\mathcal{N}$ implies nontangency to $\mathcal{N}_z$ at the point $z \in \mathcal{N}$, it follows from Theorem 6.2 that every equilibrium point of (6.18)–(6.21) in $\mathbb{R}^4$ is semistable.

Finally, note that either $\max \mathcal{L}_f V(x) \leq -2 < 0$ or $\mathcal{L}_f V(x) = \varnothing$ for almost all $x \in \mathbb{R}^4 \backslash \mathcal{Z}$, so it follows from Corollary 6.1 that (6.18)–(6.21) is globally finite time semistable. Figure 6.2 shows the solutions of (6.18)–(6.21) for $x_{10} = 4$, $x_{20} = -2$, $x_{30} = 1$, and $x_{40} = -3$. $\qquad \triangle$

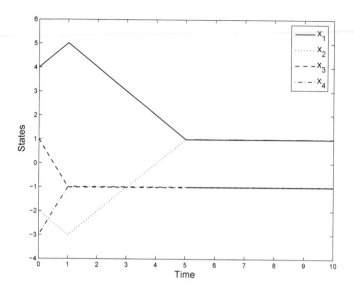

Figure 6.2 State trajectories versus time for Example 6.3.

## 6.4 Time-Varying Discontinuous Dynamical Systems

In this and the next section, we consider time-varying differential equations given by

$$\dot{x}(t) = f(t, x(t)), \quad x(t_0) = x_0, \quad t \geq t_0, \qquad (6.23)$$

where $t \in \mathbb{R}$, $x(t) \in \mathbb{R}^q$, and $f : \mathbb{R} \times \mathbb{R}^q \to \mathbb{R}^q$ is Lebesgue measurable and locally essentially bounded [96, 97]. We assume that the equilibrium set $\mathcal{E} \triangleq \{x \in \mathbb{R}^q : f(t, x) = 0 \text{ for all } t \in \mathbb{R}\}$ is closed. An absolutely continuous function $x : [t_0, \tau] \to \mathbb{R}^q$ is said to be a *Filippov solution* [96, 97] of (6.23) on the interval $[t_0, \tau]$ with initial condition $x(t_0) = x_0$ if $x(t)$ satisfies

$$\dot{x}(t) \in \mathcal{K}[f](t, x(t)), \quad \text{a. a.} \quad t \in [t_0, \tau], \qquad (6.24)$$

where the Filippov set-valued map $\mathcal{K}[f] : [0, \infty) \times \mathbb{R}^q \to \mathcal{B}(\mathbb{R}^q)$ is defined by

$$\mathcal{K}[f](t, x) \triangleq \bigcap_{\delta > 0} \bigcap_{\mu(\mathcal{S}) = 0} \overline{\text{co}} \{f(t, \mathcal{B}_\delta(x) \backslash \mathcal{S})\}, \quad (t, x) \in [t_0, \infty) \times \mathbb{R}^q. \qquad (6.25)$$

Note that it follows from [70] that there exists a set $\mathcal{N}_f \subset \mathbb{R}^q$ of measure zero such that

$$\mathcal{K}[f](t, x) = \overline{\text{co}} \left\{ \lim_{i \to \infty} f(t, x_i) : x_i \to x, \ x_i \notin \mathcal{N}_f \cup \mathcal{W} \right\}, \qquad (6.26)$$

where $\mathcal{W} \subset \mathbb{R}^q$ is an arbitrary set of measure zero. Since the Filippov set-valued map given by (6.25) is upper semicontinuous with nonempty, convex, and compact values, and it is also locally bounded, it follows that Filippov solutions to (6.23) exist [97].

Let $\mathcal{S}$ be a given closed subset of $\mathbb{R}^q$. Then the pair $(\mathcal{S}, \mathcal{K}[f](t, x))$ is called *weakly invariant* (respectively, *strongly invariant*) if for all initial conditions $(t_0, x_0)$ with $x_0 \in \mathcal{S}$, $\mathcal{S}$ contains a Filippov solution (respectively, all Filippov solutions) $x(\cdot)$ of (6.2) on $[t_0, \infty)$ satisfying $x(t_0) = x_0$. Recall that an equilibrium point $x_e \in \mathcal{E}$ of (6.23) is an equilibrium point of (6.24) if and only if $0 \in \mathcal{K}[f](t, x_e)$ for all $t \in [0, \infty)$.

An equilibrium point $x_e \in \mathcal{E}$ of (6.23) is *Lyapunov stable* if for every $t_0 \in \mathbb{R}$ and every $\varepsilon > 0$, there exists $\delta = \delta(t_0, \varepsilon) > 0$ such that for every $\|x_0 - x_e\| \leq \delta$, the Filippov solutions $x(t)$, $t \geq t_0$, with the initial condition $x(t_0) = x_0$ satisfy $\|x(t) - x_e\| < \varepsilon$ for all $t \geq t_0$. An equilibrium point $x_e \in \mathcal{E}$ of (6.23) is *uniformly Lyapunov stable* if for every $\varepsilon > 0$, there exists $\delta = \delta(\varepsilon) > 0$ such that for every $\|x_0 - x_e\| \leq \delta$, the Filippov solutions $x(t)$, $t \geq t_0$, with the initial condition $x(t_0) = x_0$ satisfy $\|x(t) - x_e\| < \varepsilon$ for all $t \geq t_0$ and for all $t_0 \in \mathbb{R}$.

The following definitions are needed.

**Definition 6.6.** (i) An equilibrium point $x_e \in \mathcal{E}$ of (6.23) is *weakly semistable* (respectively, *semistable*) if for every $t_0 \in \mathbb{R}$, $x_e$ is Lyapunov stable and there exists $\delta = \delta(t_0) > 0$ such that for every $\|x_0 - x_e\| \leq \delta$, a Filippov solution (respectively, every Filippov solution) $x(t)$, $t \geq t_0$, with the initial condition $x(t_0) = x_0$ satisfies $\lim_{t \to \infty} x(t) = z$ and $z \in \mathcal{E}$ is a Lyapunov stable equilibrium point. The system (6.23) is *weakly semistable* (respectively, *semistable*) if all of the equilibrium points of (6.23) are weakly semistable (respectively, semistable).

(ii) An equilibrium point $x_e \in \mathcal{E}$ of (6.23) is *uniformly weakly semistable* (respectively, *uniformly semistable*) if $x_e$ is uniformly Lyapunov stable and there exists $\delta > 0$ such that for every $\|x_0 - x_e\| \leq \delta$, a Filippov solution (respectively, every Filippov solution) $x(t)$, $t \geq t_0$, with the initial condition $x(t_0) = x_0$ satisfies $\lim_{t \to \infty} x(t) = z$ uniformly in $t_0 \in \mathbb{R}$; that is, for every $\varepsilon > 0$, there exists $T = T(\varepsilon) > 0$ such that $\|x(t)\| < \varepsilon$ for every $t \geq t_0 + T(\varepsilon)$ and every $x_0 \in \mathbb{R}^q$, and $z \in \mathcal{E}$ is a uniformly Lyapunov stable equilibrium point. The system (6.23) is *uniformly weakly semistable* (respectively, *uniformly semistable*) if all of the equilibrium points of (6.23) are uniformly weakly semistable (respectively, uniformly semistable).

**Definition 6.7** (see [64]). Let $\mathcal{S}$ be a closed subset of $\mathbb{R}^q$. Given $u \notin \mathcal{S}$, let $x \in \mathcal{S}$ be such that $\|x - u\| = \inf_{s \in \mathcal{S}} \|s - u\|$. Then $x$ is called a *projection* of $u$ onto $\mathcal{S}$. The set of all such projections is denoted by $\text{proj}(u, \mathcal{S})$. The vector $u - x$ (and all of its nonnegative multiples) defines a *proximal normal direction* to $\mathcal{S}$ at $x$. The set of all vectors constructed in this way (for fixed $x$, by varying $u$) is called the *proximal normal cone* to $\mathcal{S}$ at $x$ and is denoted by $\mathcal{N}_{\mathcal{S}}^{P}(x)$.

**Definition 6.8** (see [97]). The *contingent set* denoted by $\text{Cont}(t_0, x_0)$ is the set of all limit points of the sequences $\frac{x_i(t_i) - x_0}{t_i - t_0}$ as $t_i \to t_0$, where $x_i(\cdot)$ is a Filippov solution to (6.23) on $[t_0, t_i]$ satisfying $x_i(t_0) = x_0$, $i = 1, 2, \ldots$.

## 6.5 Lyapunov-Based Semistability Analysis for Time-Varying Discontinuous Dynamical Systems

In this section, we develop a Lyapunov-based semistability theory for time-varying discontinuous dynamical systems of the form given by (6.23). The following lemmas are needed for the main results of this section.

**Lemma 6.3.** Let $\mathcal{S}$ be a closed subset of $\mathbb{R}^q$. Assume that there exists $M > 0$ such that for every $(t, x) \in \mathbb{R}^{q+1}$ and almost every $v \in \mathcal{K}[f](t, x)$, $\|v\| \leq M$. If $(\mathcal{S}, \mathcal{K}[f](t, x))$ is weakly invariant, then $\mathcal{K}[f](t, x) \cap \text{Cont}(t, x) \neq \emptyset$ for every $x \in \mathcal{S}$ and $t \geq t_0$.

**Proof.** Since $(\mathcal{S}, \mathcal{K}[f](t, x))$ is weakly invariant, it follows that for every $x_0 \in \mathcal{S}$ there exists a Filippov solution $x(\cdot)$ to (6.23) on $[t_0, \infty)$ such that $x(t) \in \mathcal{S}$ for all $t \geq t_0$, where $x(t_0) = x_0$. Hence, for a sequence $\{t_n\}_{n=1}^{\infty}$ satisfying $\lim_{n \to \infty} t_n = t_0$, it follows that there exist Filippov solutions $x_n(\cdot)$ to (6.23) on $[t_0, t_n]$ such that $x_n(t_n) \in \mathcal{S}$ with $x_n(t_0) = x_0$. Since $\|v\| \leq M$ for every $v \in \mathcal{K}[f](t, x)$, it follows that $\|\dot{x}_n(t)\| \leq M$ almost everywhere $t \geq t_0$, where $\dot{x}_n(t) \in \mathcal{K}[f](t, x_n(t))$. Note that

$$x_n(t_n) - x_0 = \int_{t_0}^{t_n} \dot{x}_n(t) \, \mathrm{d}t.$$

Then it follows that $\|x_n(t_n) - x_0\| \leq M(t_n - t_0)$ for all $n = 1, 2, \ldots$. Hence, we can take a subsequence $\{t_{n_i}\}_{i=1}^{\infty}$ satisfying $\frac{x_{n_i}(t_{n_i}) - x_0}{t_{n_i} - t_0} \to \nu$ as $n_i \to \infty$ for some $\nu$. Note that $\nu \in \text{Cont}(t_0, x_0)$ by definition. Next, we show that $\nu \in \mathcal{K}[f](t_0, x_0)$.

For a given $\delta > 0$ and all sufficiently large $n_i$, it follows that the set $\{x_{n_i}(t) : t_0 \leq t \leq t_{n_i}\}$ is contained in $\mathcal{B}_{\delta}(x_0)$. Furthermore, for a given $\varepsilon > 0$ and sufficiently small $\delta$, it follows from Theorem 1 of [97, p. 87] that for $x \in \mathcal{B}_{\delta}(x_0)$ and $|t - t_0| < \sigma$, $\sigma > 0$,

$$\mathcal{K}[f](t, x) \subset \mathcal{K}[f](t_0, x_0) + \varepsilon \mathcal{B},$$

where

$$A + \varepsilon \mathcal{B} \triangleq \{y : y \in \mathcal{B}_\varepsilon(x), x \in A\}.$$

Hence, for sufficiently large $n_i$, it follows from Theorem 1 of [97, p. 70] that

$$\frac{x_{n_i}(t_{n_i}) - x_0}{t_{n_i} - t_0} \in \text{Cont}(t_0, x_0) \subset \mathcal{K}[f](t, x) \subset \mathcal{K}[f](t_0, x_0) + \varepsilon \mathcal{B},$$

which implies that $\nu \in \mathcal{K}[f](t_0, x_0) + \varepsilon \overline{\mathcal{B}}$, where $A + \varepsilon \overline{\mathcal{B}} \triangleq \{y : y \in \overline{\mathcal{B}}_\varepsilon(x), x \in A\}$. Since $\varepsilon$ was chosen arbitrarily, it follows that $\nu \in \mathcal{K}[f](t_0, x_0)$. $\square$

For the next result, we write $\langle \cdot, \cdot \rangle$ for the inner product in a Hilbert space.

**Lemma 6.4.** Let $\mathcal{S}$ be a closed subset of $\mathbb{R}^q$, and consider $(t, x) \in [t_0, t_0 + a] \times \overline{\mathcal{B}}_b(x_0)$ for (6.24). Assume that for every $(t, z) \in [t_0, t_0 + d] \times \overline{\mathcal{B}}_b(x_0)$ there exists $w \in \text{proj}(z, \mathcal{S})$ such that $\langle f(t, z), z - w \rangle \leq 0$, where $d = \min\{a, \frac{b}{m}\}$ and $m = \sup_{(t,x) \in [t_0, t_0+a] \times \overline{\mathcal{B}}_b(x_0)} \|\mathcal{K}[f](t, x)\|$. Then $\text{dist}(x(t), \mathcal{S}) \leq \text{dist}(x(t_0), \mathcal{S})$ for every $t \in [t_0, t_0 + d]$, where $x(\cdot)$ is a Filippov solution of (6.24) on $[t_0, t_0 + d]$ with $x(t_0) = x_0$.

**Proof.** First, it follows from Lemma 15 of [97, p. 66] that $m < \infty$. For $k = 1, 2, \ldots$, let $h_k = d/k$ and $t_{ki} = t_0 + i h_k$, $i = 0, 1, \ldots, k$. Next, construct an approximate solution $x_k(t)$ to (6.23) as follows. Let $x_k(t_{k0}) = x_0$. If for some $i \geq 0$ the value $x_k(t_{ki}) = x_{ki}$ is defined and $\|x_{ki} - x_0\| \leq m(t_{ki} - t_0)$, then define $x_k(t)$, $t_{ki} < t \leq t_{k,i+1}$, by

$$x_k(t) \triangleq x_{ki} + \int_{t_{ki}}^t f(s, x_{ki}) ds.$$

Hence, $x_k(t)$ is constructed successively on the intervals $[t_{ki}, t_{k,i+1}]$, $i = 0, 1, \ldots, k - 1$. Furthermore, it follows that $\|x_k(t) - x_0\| \leq m(t - t_0)$, $t_{ki} < t \leq t_{k,i+1}$. Since

$$\dot{x}_k(t) = f(t, x_{ki}) \in \mathcal{K}[f](t, x_{ki}),$$

it follows that $\|\dot{x}_k(t)\| \leq m$ for almost all $t \geq t_0$. Hence, the functions $\{x_k(t)\}_{k=1}^\infty$ are uniformly bounded and equicontinuous. By the Arzelà–Ascoli theorem [61, p. 180] and Lemma 1 of [97, p. 76], there exists a subsequence of $x_k(t)$ uniformly converging to $x(t)$, where $x(\cdot)$ is a Filippov solution of (6.24) with $x(t_0) = x_0$.

Next, it follows that for each $i = 0, 1, \ldots, k$, there exists a point $w_{ki} \in \text{proj}(x_{ki}, \mathcal{S})$ such that $\langle f(t, x_{ki}), x_{ki} - w_{ki} \rangle \leq 0$, $t_{ki} < t \leq t_{k,i+1}$. Hence,

$$(\text{dist}(x_{k1}, \mathcal{S}))^2 \leq \|x_{k1} - w_{k0}\|^2$$

$$= \|x_{k1} - x_{k0}\|^2 + \|x_{k0} - w_{k0}\|^2 + 2\langle x_{k1} - x_{k0}, x_{k0} - w_{k0}\rangle$$

$$\leq m^2(t_{k1} - t_0)^2 + (\text{dist}(x_0, \mathcal{S}))^2 + 2\int_{t_0}^{t_1} \langle f(t, x_0), x_0 - w_{k0}\rangle \mathrm{d}t$$

$$\leq m^2(t_{k1} - t_0)^2 + (\text{dist}(x_0, \mathcal{S}))^2. \tag{6.27}$$

Similarly,

$$(\text{dist}(x_{ki}, \mathcal{S}))^2 \leq (\text{dist}(x_{k,i-1}, \mathcal{S}))^2 + m^2(t_{ki} - t_{k,i-1})^2.$$

Thus,

$$(\text{dist}(x_{ki}, \mathcal{S}))^2 \leq (\text{dist}(x_0, \mathcal{S}))^2 + m^2 \sum_{r=1}^{i}(t_{kr} - t_{k,r-1})^2$$

$$\leq (\text{dist}(x_0, \mathcal{S}))^2 + m^2 h_k d. \tag{6.28}$$

Let $\{x_{n_k}(t)\}_{k=1}^{\infty}$ be a subsequence of $x_k(t)$ uniformly converging to $x(t)$. Note that $h_{n_k} \to 0$ as $n_k \to \infty$. Hence, taking the limit on both sides of (6.28) yields $\text{dist}(x(t), \mathcal{S}) \leq \text{dist}(x(t_0), \mathcal{S})$ for every $t \in [t_0, t_0 + d]$. $\qquad\square$

Next, we present necessary and sufficient conditions for characterizing weak invariance. It is important to note that our results are different from the results in [64, 82] since the Lipschitz conditions in [64, 82] do not hold for the nonautonomous differential inclusion discussed in this section; see Examples 6.4 and 6.5 below. A similar observation holds for Proposition 6.7 below.

**Proposition 6.4.** Let $\mathcal{S}$ be a closed subset of $\mathbb{R}^q$. Assume that there exists $M > 0$ such that for every $(t, x) \in \mathbb{R}^{q+1}$ and almost every $v \in \mathcal{K}[f](t, x)$, $\|v\| \leq M$. Then $(\mathcal{S}, \mathcal{K}[f](t, x))$ is weakly invariant if and only if, for every $\zeta \in \mathcal{N}_{\mathcal{S}}^P(x)$,

$$\min_{v \in \mathcal{K}[f](t,x)} \langle \zeta, v\rangle \leq 0, \quad t \in \mathbb{R}, \quad x \in \mathcal{S}. \tag{6.29}$$

**Proof.** To show necessity, define the function $f_\mathrm{P}$ as follows. For every $x \in \mathbb{R}^n$ and $t \in \mathbb{R}$, choose any $w = w(x) \in \text{proj}(x, \mathcal{S})$ and let $v \in \mathcal{K}[f](t, w)$ minimize the function $v \mapsto \langle v, x - w\rangle$ over $\mathcal{K}[f](t, w)$. Set $f_\mathrm{P}(t, x) = v$, $x \in \mathbb{R}^n, t \in \mathbb{R}$. Since $x - w \in \mathcal{N}_{\mathcal{S}}^P(w)$, it follows from (6.29) that $\langle f_\mathrm{P}(t, x), x - w\rangle \leq 0$. Note that $\|f_\mathrm{P}(t, x)\| = \|v\| \leq M$, $x \in \mathbb{R}^n$, $t \in \mathbb{R}$. Hence, by taking $t_0 = 0$, $a = 1$, and $b = M$ in Lemma 6.4, it follows that the Filippov solutions $x(\cdot)$ to $\dot{x}(t) = f_\mathrm{P}(t, x(t))$ with $x(0) = x_0$ on $[0, 1]$ satisfy $\text{dist}(x(t), \mathcal{S}) \leq \text{dist}(x_0, \mathcal{S})$, which implies that if $x_0 \in \mathcal{S}$, then $x(t) \in \mathcal{S}$ for all $t \in [0, 1]$. We can extend $x(\cdot)$ to $[0, \infty)$ by considering the interval $[n, n+1]$ successively for $n = 1, 2, \ldots$.

To complete the proof, we need to show that $x(\cdot)$ is a Filippov solution to (6.24). Define the Filippov set-valued map $\mathcal{K}_\mathcal{S}[f](t, x)$ by

$$\mathcal{K}_\mathcal{S}[f](t, x) \triangleq \mathrm{co}\{\mathcal{K}[f](t, w) : w \in \mathrm{proj}(x, \mathcal{S})\}.$$

We claim that $\mathcal{K}_\mathcal{S}[f](t, x) = \mathcal{K}[f](t, x)$ for $x \in \mathcal{S}$. To see this, note that if $x \in \mathcal{S}$, then $w = x \in \mathcal{S}$. Hence, it follows from the definition of differential inclusions that

$$\mathcal{K}_\mathcal{S}[f](t, x) = \mathrm{co}\{\mathcal{K}[f](t, x) : x \in \mathcal{S}\} = \mathcal{K}[f](t, x).$$

Next, since $f_\mathrm{P} \in \mathcal{K}_\mathcal{S}[f]$, it follows that $\mathcal{K}[f_\mathrm{P}] \subseteq \mathcal{K}_\mathcal{S}[f]$. By definition, the Filippov solution $x(\cdot)$ of

$$\dot{x}(t) = f_\mathrm{P}(t, x(t))$$

satisfies

$$\dot{x}(t) \in \mathcal{K}_\mathcal{S}[f](t, x(t))$$

almost everywhere on $[0, 1]$ with $x(0) = x_0$. Since $x(t) \in \mathcal{S}$ on $[0, 1]$ and $\mathcal{K}_\mathcal{S}[f](t, x) = \mathcal{K}[f](t, x)$ for $x \in \mathcal{S}$, it follows that $x(\cdot)$ is a Filippov solution to (6.24).

To show sufficiency, suppose that $(\mathcal{S}, \mathcal{K}[f])$ is weakly invariant. Then it follows from Lemma 6.3 that $\mathcal{K}[f](t, x) \cap \mathrm{Cont}(t, x) \neq \emptyset$ for every $x \in \mathcal{S}$ and $t \geq t_0$. Next, we show that $\mathrm{Cont}(t, x) \subseteq \mathcal{H}_\mathcal{S}(x) \triangleq \{\eta \in \mathbb{R}^n : \langle \zeta, \eta \rangle \leq 0, \zeta \in \mathcal{N}_\mathcal{S}^P(x)\}$ for $x \in \mathcal{S}$. To see this, choose $\nu \in \mathrm{Cont}(t_0, x_0)$. Then it follows that $\nu = \lim_{i \to \infty} \frac{x_i(t_i) - x_0}{t_i - t_0}$, where $t_i \to t_0$ as $i \to \infty$ and $t_i > t_0$. Let $\zeta \in \mathcal{N}_\mathcal{S}^P(x_0)$. Then $\langle \zeta, x_i(t_i) - x_0 \rangle = \langle w - x_0, x_i(t_i) - x_0 \rangle$, where $w \in \mathrm{proj}(x_0, \mathcal{S})$. Since $\|w - x_0\| \leq \|x_i(t_i) - x_0\|$, it follows from the Cauchy–Schwarz inequality that

$$\langle w - x_0, x_i(t_i) - x_0 \rangle \leq \|w - x_0\| \|x_i(t_i) - x_0\| \leq \|x_i(t_i) - x_0\|^2.$$

Hence, $\langle \zeta, \frac{x_i(t_i) - x_0}{t_i - t_0} \rangle \leq \|x_i(t_i) - x_0\| \|\frac{x_i(t_i) - x_0}{t_i - t_0}\|$.

Finally, note that since $\lim_{i \to \infty} x_i(t_i) = x_0$, it follows that

$$\langle \zeta, \nu \rangle = \left\langle \zeta, \lim_{i \to \infty} \frac{x_i(t_i) - x_0}{t_i - t_0} \right\rangle$$
$$= \lim_{i \to \infty} \left\langle \zeta, \frac{x_i(t_i) - x_0}{t_i - t_0} \right\rangle$$

$$\leq \lim_{i \to \infty} \|x_i(t_i) - x_0\| \left\| \frac{x_i(t_i) - x_0}{t_i - t_0} \right\|$$

$$= 0\|\nu\|$$

$$= 0,$$

which implies that $\nu \in \mathcal{H}_{\mathcal{S}}(x_0)$. This shows that for every $\zeta \in \mathcal{N}^P_{\mathcal{S}}(x)$, $\langle \zeta, v \rangle \leq 0$, where $v \in \mathcal{K}[f](t,x) \cap \mathrm{Cont}(t,x)$, which implies that (6.29) holds. $\qquad\square$

The following propositions are needed for the main results of this section. For the first proposition, recall that the *epigraph* of a function $f : \mathcal{X} \to \mathbb{R}$ is defined by the $\alpha$-sublevel set $\mathrm{Ep}(f) \triangleq \{(x, \alpha) \in \mathcal{X} \times \mathbb{R} : f(x) \leq \alpha\}$ [262, p. 23].

**Proposition 6.5.** Assume that there exists $M > 0$ such that for every $(t,x) \in \mathbb{R}^{q+1}$ and almost every $v \in \mathcal{K}[f](t,x)$, $\|v\| \leq M$. Furthermore, assume that there exist a continuously differentiable function $V(\cdot)$ and a continuous function $W(\cdot)$ such that the following statements hold.

(i) $\alpha(\|x\|) \leq V(x) \leq \beta(\|x\|)$, $x \in \mathbb{R}^q$, where $\alpha(\cdot)$ and $\beta(\cdot)$ are class $\mathcal{K}_\infty$ functions.

(ii) $\min_{v \in \mathcal{K}[f](t,x)} \langle \nabla V(x), v \rangle \leq -W(x)$ for all $x \in \mathbb{R}^q$ and $t \in \mathbb{R}$, where $W(x) \geq 0$ for all $x \in \mathbb{R}^q$.

Then $(V^{-1}([0,c]), \mathcal{K}[f](t,x))$ is weakly invariant, and, for every $x_0 \in \mathbb{R}^q$, there exists a Filippov solution $x(\cdot)$ to (6.23) on $[t_0, \infty)$ with $x(t_0) = x_0$ such that $x(t) \to W^{-1}(0)$ as $t \to \infty$, where $c > 0$.

**Proof.** Since $V(\cdot)$ is continuously differentiable, it follows from Proposition 2 of [10, p. 32] that $\{\nabla V(x)\} = \partial V(x)$, $x \in \mathbb{R}^q$. Thus, it follows from (ii) that $\min_{v \in \mathcal{K}[f](t,x)} \langle p, v \rangle \leq 0$, $p \in \partial V(x)$, $x \in \mathbb{R}^q$. Consider the epigraph of $V(\cdot)$ defined by

$$\mathrm{Ep}(V) \triangleq \{(x, z) \in \mathbb{R}^q \times \mathbb{R} : V(x) \leq z\}.$$

Note that $\mathrm{Ep}(V)$ is closed. Let $(\zeta, \lambda) \in \mathbb{R}^q \times \mathbb{R}$ belong to $\mathcal{N}^P_{\mathrm{Ep}(V)}(x, z)$ for some $(x, z) \in \mathrm{Ep}(V)$. We show that for $(\zeta, \lambda) \in \mathcal{N}^P_{\mathrm{Ep}(V)}(x, z)$, there exists $v \in \mathcal{K}[f](t,x)$ such that $\langle \zeta, v \rangle \leq 0$.

First, we show that $\lambda \leq 0$. Let $y$ be in the domain of $V$ and $(y^*, 0) \in \mathcal{N}^P_{\mathrm{Ep}(V)}(y, V(y))$ with $y^* \neq 0$. Without loss of generality, assume $\|y^*\| = 1$.

Then there exists $(x, V(y)) \notin \mathrm{Ep}(V)$ such that

$$\|(x, V(y)) - (y, V(y))\| = \inf_{(s,V(s)) \in \mathrm{Ep}(V)} \|(x, V(s)) - (s, V(s))\|$$

and $(x - y)/\|x - y\| = y^*$, where $(y, V(y)) \in \mathrm{Ep}(V)$. By Proposition 2.1 of [253], we can assume, without loss of generality, that

$$(y^*, 0) \in \partial \mathrm{dist}((x, V(y)), \mathrm{Ep}(V)).$$

Note that, for every $(\hat{x}, V(\hat{y}))$, it follows from the definition of an epigraph that

$$\mathrm{dist}((\hat{x}, V(\hat{y})), \mathrm{Ep}(V)) \leq \mathrm{dist}((\hat{x}, V(\hat{y}) - t), \mathrm{Ep}(V))$$

for every $t > 0$. Suppose that there exists $(\hat{x}, V(\hat{y}))$ arbitrarily close to $(x, V(y))$ and $t > 0$ arbitrarily small so that

$$\mathrm{dist}((\hat{x}, V(\hat{y})), \mathrm{Ep}(V)) < \mathrm{dist}((\hat{x}, V(\hat{y}) - t), \mathrm{Ep}(V)).$$

Then it follows from Theorem 1.4 of [253] that there exists

$$(\zeta, \lambda) \in \partial \mathrm{dist}((\bar{x}, V(\bar{y})), \mathrm{Ep}(V)),$$

where $(\bar{x}, V(\bar{y}))$ is arbitrarily close to $(x, V(y))$ such that $\langle (\zeta, \lambda), (\hat{x}, V(\hat{y}) - t) - (\hat{x}, V(\hat{y})) \rangle > 0$, which implies that $\lambda < 0$. For the case where

$$\mathrm{dist}((\hat{x}, V(\hat{y})), \mathrm{Ep}(V)) = \mathrm{dist}((\hat{x}, V(\hat{y}) - t), \mathrm{Ep}(V)), \quad t > 0,$$

it follows that $\langle (\zeta, \lambda), (\hat{x}, V(\hat{y}) - t) - (\hat{x}, V(\hat{y})) \rangle = 0$, which implies that $\lambda = 0$. Hence, $\lambda \leq 0$.

If $\lambda < 0$, then $(\zeta/(-\lambda), -1) \in \mathcal{N}_{\mathrm{Ep}(V)}^{P}(x, z)$, which implies that $-\zeta/\lambda \in \partial V(x)$. Now, it follows from (ii) that there exists $v \in \mathcal{K}[f](t, x)$ such that $\langle (-\zeta/\lambda), v \rangle \leq 0$ and, hence, $\langle \zeta, v \rangle \leq 0$. Alternatively, if $\lambda = 0$, then $(\zeta, 0) \in \mathcal{N}_{\mathrm{Ep}(V)}^{P}(x, V(x))$. Now, it follows from Theorem 2.4 of [253] that there exist sequences $\{(\zeta_i, -\varepsilon_i)\}_{i=1}^{\infty}$, with $\varepsilon_i > 0$, and $\{x_i\}_{i=1}^{\infty}$ such that

$$\lim_{i \to \infty} (\zeta_i, -\varepsilon_i) = (\zeta, 0), \quad (\zeta_i, -\varepsilon_i) \in \mathcal{N}_{\mathrm{Ep}(V)}^{P}(x_i, V(x_i)),$$

and $\lim_{i \to \infty} x_i = x$. Using the above result for the case where $\lambda < 0$, it follows that there exists $v_i \in \mathcal{K}[f](t, x_i)$ such that $\langle \zeta_i, v_i \rangle \leq 0$. By assumption, the sequence $\{v_i\}_{i=1}^{\infty}$ is uniformly bounded. Hence, there exists a subsequence $\{n_i\}_{i=1}^{\infty}$ such that $\{v_{n_i}\}_{i=1}^{\infty}$ converges to the limit $v$. Furthermore, $v \in \mathcal{K}[f](t, x)$ since $\mathcal{K}[f]$ is upper semicontinuous. Thus, $\langle \zeta, v \rangle \leq 0$.

Since for $(\zeta, \lambda) \in \mathcal{N}_{\mathrm{Ep}(V)}^{P}(x, z)$ there exists $v \in \mathcal{K}[f](t, x)$ such that $\langle \zeta, v \rangle \leq 0$, it follows from Proposition 6.4 that the pair $(\mathrm{Ep}(V), \mathcal{K}[f] \times \{0\})$ is weakly invariant, and, hence, for every $x_0 \in \mathbb{R}^q$, there exists a Filippov solution $x(\cdot)$ to (6.23) on $[t_0, \infty)$ with $x(t_0) = x_0$ such that $V(x(t)) \le V(x_0)$ for all $t \ge t_0$, which implies that $(V^{-1}([0, c]), \mathcal{K}[f])$ is weakly invariant.

To show the second assertion, define a function $U : \mathbb{R}^q \times \mathbb{R} \to \mathbb{R}$ by $U(x, y) \triangleq V(x) + y$ and a set-valued map $\mathcal{F}(t, x, y) \triangleq \mathcal{K}[f](t, x) \times \{y : y = W(x)\}$. We claim that, for every $\alpha \in \mathbb{R}^q$, there exists a Filippov solution $z = (x, y)$ to the differential inclusion $\dot{z} \in \mathcal{F}(t, z)$ almost everywhere on $[t_0, \infty)$, with $x(t_0) = \alpha$ and $y(t_0) = 0$, such that $U(x(t), y(t)) \le U(\alpha, 0)$ for all $t \ge t_0$. Let $(\zeta, \eta) \in \partial U(x, y)$. Then $\zeta \in \partial V(x)$ and $\eta = 1$. Since $\langle v, \zeta \rangle \le -W(x)$ for some $v \in \mathcal{K}[f](t, x)$, it follows that $\langle v, \zeta \rangle + W(x) \le 0$ or, equivalently, $\langle (v, W(x)), (\zeta, 1) \rangle \le 0$.

Using similar arguments as above, it can be shown that the pair $(\mathrm{Ep}(U), \mathcal{F} \times \{0\})$ is weakly invariant, which implies that, for every $\alpha \in \mathbb{R}^q$, there exists a Filippov solution $(x, y)$ to $\dot{z} \in \mathcal{F}(t, z)$ almost everywhere on $[t_0, \infty)$, with $x(t_0) = \alpha$ and $y(t_0) = 0$, such that $U(x(t), y(t)) \le U(\alpha, 0)$ for all $t \ge t_0$. Note that $U(x(t), y(t)) \le U(\alpha, 0)$ for $t \ge t_0$ implies that

$$V(x(t)) + \int_{t_0}^{t} W(x(\tau)) \mathrm{d}\tau \le V(\alpha),$$

where $x(\cdot)$ is a Filippov solution to (6.23). Hence, $V(x(t))$ and $\int_{t_0}^{t} W(x(\tau)) \mathrm{d}\tau$ are bounded for almost all $t \ge t_0$. Furthermore, note that $\dot{x}(t)$ is uniformly bounded for almost all $t \ge t_0$. Now, using similar arguments as in the proof of Theorem 8.4 of [186], it can be shown that $x(t) \to W^{-1}(0)$ as $t \to \infty$. $\square$

For the next result, we write $x(t) \rightrightarrows \mathcal{M}$ as $t \to \infty$ to mean that $x(t)$ approaches the set $\mathcal{M}$ uniformly in the initial time $t_0 \in \mathbb{R}$.

**Proposition 6.6.** Consider the time-varying discontinuous dynamical system (6.23). Assume that every point in $\mathcal{E}$ is Lyapunov stable. Furthermore, assume that, for a given $x_0 \in \mathbb{R}^q$, there exists a Filippov solution to (6.23) satisfying $x(t) \to \mathcal{E}$ as $t \to \infty$. Then $x(t) \to z$ as $t \to \infty$, where $z \in \mathcal{E}$. Alternatively, assume that every point in $\mathcal{E}$ is uniformly Lyapunov stable and, for given $x_0 \in \mathbb{R}^q$, there exists a Filippov solution to (6.23) satisfying $x(t) \rightrightarrows \mathcal{E}$ as $t \to \infty$. Then $x(t) \rightrightarrows z$ as $t \to \infty$, where $z \in \mathcal{E}$.

**Proof.** The proof is similar to the proof of Proposition 6.1 and, hence, is omitted. $\square$

Next, we present sufficient conditions for weak semistability and uniform weak semistability for (6.23).

**Theorem 6.3.** Assume that there exists $M > 0$ such that for almost every $v \in \mathcal{K}[f](t, x)$, $\|v\| \leq M$. Furthermore, assume that there exist a continuously differentiable function $V(\cdot)$ and a continuous function $W(\cdot)$ such that (i) and (ii) of Proposition 6.5 hold and $\mathcal{E} \subseteq W^{-1}(0)$. If every point in $W^{-1}(0)$ is a Lyapunov stable equilibrium of (6.23), then (6.23) is weakly semistable. Alternatively, if every point in $W^{-1}(0)$ is a uniformly Lyapunov stable equilibrium of (6.23), then (6.23) is uniformly weakly semistable.

**Proof.** It follows from Proposition 6.5 that there exists a Filippov solution $x(\cdot)$ to (6.23) such that $x(t) \to W^{-1}(0)$ as $t \to \infty$. Since every point in $W^{-1}(0)$ is a Lyapunov stable equilibrium of (6.23), it follows that $W^{-1}(0) \subseteq \mathcal{E}$. Furthermore, since, by assumption, $\mathcal{E} \subseteq W^{-1}(0)$, it follows that $W^{-1}(0) = \mathcal{E}$. Hence, $x(t) \to \mathcal{E}$ as $t \to \infty$, and every point in $\mathcal{E}$ is Lyapunov stable. Now, it follows from Proposition 6.6 that $x(t) \to z$ as $t \to \infty$, where $z \in \mathcal{E}$. By definition, (6.23) is weakly semistable.

To show the second assertion, note that since $\dot{x}(t)$ is uniformly bounded, it follows using similar arguments as in the proof of Proposition 6.5 that $x(t) \rightrightarrows W^{-1}(0)$ as $t \to \infty$. Now, using similar arguments as above, it can be shown that (6.23) is uniformly weakly semistable. $\qquad \square$

If all the conditions in Theorem 6.3 are satisfied and (6.23) has a unique Filippov solution, then it follows from Theorem 6.3 that (6.23) is semistable. Sufficient conditions for guaranteeing uniqueness of Filippov solutions can be found in [70, 97].

**Example 6.4.** Consider the time-varying discontinuous dynamical system given by

$$\dot{x}_1(t) = \frac{1 + 2t^2}{1 + t^2} \operatorname{sign}(x_2(t) - x_1(t)), \quad x_1(t_0) = x_{10}, \quad t \geq t_0, \qquad (6.30)$$

$$\dot{x}_2(t) = \frac{1 + 2t^2}{1 + t^2} \operatorname{sign}(x_1(t) - x_2(t)), \quad x_2(t_0) = x_{20}, \qquad (6.31)$$

where $x_1, x_2 \in \mathbb{R}$. Note that, for $x = [x_1, x_2]^{\mathrm{T}}$,

$$\mathcal{K}[f](t, x) = \begin{cases} \left\{\frac{1+2t^2}{1+t^2}\right\} \times \left\{-\frac{1+2t^2}{1+t^2}\right\}, & x_2 > x_1, \\ \left[-\frac{1+2t^2}{1+t^2}, \frac{1+2t^2}{1+t^2}\right] \times \left[-\frac{1+2t^2}{1+t^2}, \frac{1+2t^2}{1+t^2}\right], & x_1 = x_2, \quad t \geq t_0. \\ \left\{-\frac{1+2t^2}{1+t^2}\right\} \times \left\{\frac{1+2t^2}{1+t^2}\right\}, & x_1 > x_2, \end{cases}$$

$$(6.32)$$

Clearly, $\|v\| \leq 2\sqrt{2}$ for almost all $v \in \mathcal{K}[f](t, x)$.

Next, consider the function

$$V(x_1, x_2) = \frac{1}{2}(x_1 - \alpha)^2 + \frac{1}{2}(x_2 - \alpha)^2,$$

where $\alpha \in \mathbb{R}$. Then it follows from the time-dependent version of Theorem 1 of [240] that

$$
\begin{aligned}
[x_1 - \alpha, x_2 - \alpha]^{\mathrm{T}} \mathcal{K}[f](t, x) &= \mathcal{K}[[x_1 - \alpha, x_2 - \alpha]^{\mathrm{T}} f](t, x) \\
&= \mathcal{K}\left[ -\frac{1 + 2t^2}{1 + t^2}(x_1 - x_2)\mathrm{sign}(x_1 - x_2) \right](t, x) \\
&= -\frac{1 + 2t^2}{1 + t^2}(x_1 - x_2)\mathcal{K}[\mathrm{sign}(x_1 - x_2)](x) \\
&= -\frac{1 + 2t^2}{1 + t^2}(x_1 - x_2)\mathrm{SGN}(x_1 - x_2) \\
&= -\frac{1 + 2t^2}{1 + t^2}|x_1 - x_2|, \quad t \in \mathbb{R}, \quad (x_1, x_2) \in \mathbb{R}^2,
\end{aligned}
$$

(6.33)

which further implies that $\langle \nabla V(x_1, x_2), v \rangle \leq -|x_1 - x_2|$ for every $v \in \mathcal{K}[f](t, x)$. Now, it follows from Theorem 1 of [97, p. 153] that $x_1 = x_2 = \alpha$ is Lyapunov stable. In fact, it can be shown that $x_1 = x_2 = \alpha$ is uniformly Lyapunov stable.

Next, let $W(x_1, x_2) = |x_1 - x_2|$ and note that $W^{-1}(0) = \{(x_1, x_2) \in \mathbb{R}^2 : x_1 = x_2\} = \mathcal{E}$. Now, it follows from Theorem 6.3 that (6.30)–(6.31) is weakly semistable. Moreover, it can be shown that (6.30)–(6.31) is uniformly weakly semistable. Figure 6.3 shows the solutions of (6.30)–(6.31) for $x_{10} = 4$, $x_{20} = -2$, and $t_0 = 0, 1, 2, 3$. $\triangle$

The next proposition characterizes strong invariance of (6.23).

**Proposition 6.7.** Consider the time-varying discontinuous dynamical system (6.23). Let $\mathcal{S}$ be a closed subset of $\mathbb{R}^q$, and assume that there exists $M > 0$ such that for every $(t, x) \in \mathbb{R}^{q+1}$, $\|f(t, x)\| \leq M$ for almost all $t \in \mathbb{R}$ and $x \in \mathbb{R}^q$. Then $(\mathcal{S}, \mathcal{K}[f](t, x))$ is strongly invariant if and only if, for every $\zeta \in \mathcal{N}_{\mathcal{S}}^P(x)$ and $x \in \mathcal{S}$,

$$\max_{v \in \mathcal{K}[f](t, x)} \langle \zeta, v \rangle \leq 0, \quad t \in \mathbb{R}, \quad x \in \mathcal{S}. \tag{6.34}$$

**Proof.** First, note that it follows from $\|f(t, x)\| \leq M$ for almost all $t \in \mathbb{R}$ and $x \in \mathbb{R}^q$, and from (6.26), that, for almost every $v \in \mathcal{K}[f](t, x)$, $\|v\| \leq M$.

To show necessity, let $x_0 \in \mathcal{S}$, and define the Filippov set-valued func-

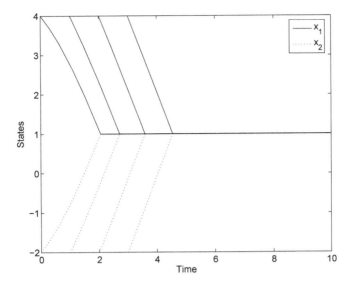

Figure 6.3 State trajectories versus time for Example 6.4.

tion $G$ by

$$G(t,x) \triangleq \{v \in \mathcal{K}[f](t,x) : \langle \zeta, v \rangle \leq 0, \zeta \in \mathcal{N}_{\mathcal{S}}^{P}(x)\}, \quad (t,x) \in [t_0, \infty) \times \mathcal{S}. \tag{6.35}$$

Note that the pair $(\mathcal{S}, G)$ is weakly invariant. Then it follows that there exists a Filippov solution $y(\cdot)$ to the differential inclusion given by

$$\dot{y}(t) \in G(t, y(t)), \quad y(t_0) = x_0, \quad \text{a. a.} \quad t \geq t_0, \tag{6.36}$$

such that $y(t) \in \mathcal{S}$ for all $t \geq t_0$. Note that $G(t,x) = \mathcal{K}[f](t,x)$ provided that (6.34) holds and $y(t_0) = x_0$. Now, it follows from Theorem 1 of [97, p. 87] that, for $\varepsilon > 0$, $\|x(t) - y(t)\| \leq \varepsilon$ for all $t \in [t_0, \tau]$, where $x(\cdot)$ denotes any Filippov solution of (6.23) with $x(t_0) = x_0$. If $\text{dist}(y(t), \partial \mathcal{S}) > 0$ for all $t \geq t_0$, then by taking $\varepsilon < \text{dist}(y(t), \partial \mathcal{S})$ it follows that $x(t) \in \mathcal{S}$ for all $t \geq t_0$.

Alternatively, consider the case where $\text{dist}(y(t), \partial \mathcal{S}) = 0$. In this case, we claim that $x(t) \in \mathcal{S}$ for all $t \geq t_0$. To see this, suppose, *ad absurdum*, that there exists a time instant $t^*$ such that $x(t^*) \in \partial \mathcal{S}$ and $x(t) \notin \mathcal{S}$ for $t^* < t \leq t^* + \delta$. Then it follows that $\langle \dot{x}(t^*), \zeta^* \rangle > 0$ for $\zeta^* \in \mathcal{N}_{\mathcal{S}}^{P}(x(t^*))$. Note that $\dot{x}(t^*) \in \mathcal{K}[f](t^*, x(t^*))$. Hence, $\langle v^*, \zeta^* \rangle > 0$ for some $v^* \in \mathcal{K}[f](t^*, x(t^*))$, which contradicts (6.34). Thus, for $\text{dist}(y(t), \partial \mathcal{S}) = 0$, $x(t) \in \mathcal{S}$ for all $t \geq t_0$. Thus, $(\mathcal{S}, \mathcal{K}[f](t,x))$ is strongly invariant.

To show sufficiency, consider any $\tilde{x} \in \mathcal{S}$. Let $\tilde{v} \in \mathcal{K}[f](t, \tilde{x})$ be given. Define the set-valued function $\mathcal{F}(t,x) \triangleq \{g(t,x)\}$, where $g(t,x)$ is such that

$\|g(t,x) - \tilde{v}\| = \inf_{\mu \in \mathcal{K}[f](t,x)} \|\mu - \tilde{v}\|$ for some fixed $t \in \mathbb{R}$. Note that $g(t,\tilde{x}) = \tilde{v}$. Next, since $(\mathcal{S}, \mathcal{F})$ is strongly invariant, it follows that $(\mathcal{S}, \mathcal{F})$ is weakly invariant, and, hence, by Proposition 6.4, $\langle \tilde{\zeta}, \tilde{v} \rangle \leq 0$ for any $\tilde{\zeta} \in \mathcal{N}_{\mathcal{S}}^{P}(\tilde{x})$. Since $\tilde{v}$ is arbitrary in $\mathcal{K}[f](t,\tilde{x})$, it follows that (6.34) holds.                          $\square$

Finally, we present sufficient conditions for semistability and uniform semistability for (6.23).

**Theorem 6.4.** Assume that there exists $M > 0$ such that for almost every $(t,x) \in \mathbb{R}^{q+1}$, $\|f(t,x)\| \leq M$. Furthermore, assume that there exist a continuously differentiable function $V(\cdot)$ and a continuous function $W(\cdot)$ such that (i) of Proposition 6.5 holds, $\mathcal{E} \subseteq W^{-1}(0)$, and

$$\max_{v \in \mathcal{K}[f](t,x)} \langle \nabla V(x), v \rangle \leq -W(x) \qquad (6.37)$$

for every $x \in \mathcal{S}$ and $t \in \mathbb{R}$. If every point in $W^{-1}(0)$ is a Lyapunov stable equilibrium of (6.23), then (6.23) is semistable. Alternatively, if every point in $W^{-1}(0)$ is a uniformly Lyapunov stable equilibrium of (6.23), then (6.23) is uniformly semistable.

**Proof.** Using similar arguments as in the proof of Proposition 6.5 and Proposition 6.7, it can be shown that every Filippov solution $x(\cdot)$ of (6.23) satisfies $x(t) \to W^{-1}(0)$ as $t \to \infty$. Since every point in $W^{-1}(0)$ is a Lyapunov stable equilibrium of (6.23), it follows that $W^{-1}(0) \subseteq \mathcal{E}$. Since, by assumption, $\mathcal{E} \subseteq W^{-1}(0)$, it follows that $W^{-1}(0) = \mathcal{E}$. Hence, $x(t) \to \mathcal{E}$ as $t \to \infty$, and every point in $\mathcal{E}$ is Lyapunov stable. Now, it follows from Proposition 6.6 that $x(t) \to z$ as $t \to \infty$, where $z \in \mathcal{E}$. By definition, (6.23) is semistable.

To prove the second assertion, note that since $\dot{x}(t)$ is uniformly bounded for almost all $t \geq t_0$, it follows using similar arguments as in the proof of Proposition 6.5 that $x(t) \rightrightarrows W^{-1}(0)$ as $t \to \infty$. Now, using similar arguments as above, it can be shown that (6.23) is uniformly semistable.                          $\square$

**Example 6.5.** Consider the time-varying discontinuous dynamical system given by

$$\dot{x}_1(t) = (2 - \cos t)\mathrm{sign}(x_2(t) - x_1(t)), \quad x_1(t_0) = x_{10}, \quad t \geq t_0, \quad (6.38)$$
$$\dot{x}_2(t) = (2 - \cos t)\mathrm{sign}(x_1(t) - x_2(t)), \quad x_2(t_0) = x_{20}, \qquad (6.39)$$

where $x_1, x_2 \in \mathbb{R}$. Clearly, $\|f(t,x)\| \leq 3\sqrt{2}$ for almost all $t \geq t_0$ and $x \in \mathbb{R}^2$. Next, consider

$$V(x_1, x_2) = \frac{1}{2}(x_1 - \alpha)^2 + \frac{1}{2}(x_2 - \alpha)^2,$$

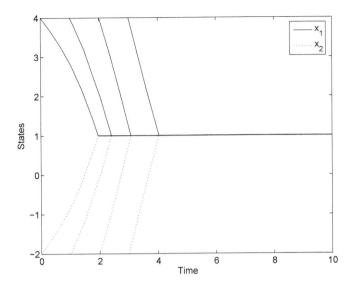

Figure 6.4 State trajectories versus time for Example 6.5.

where $\alpha \in \mathbb{R}$. Then it follows from the time-dependent version of Theorem 1 of [240] that

$$
\begin{aligned}
[x_1 - \alpha, x_2 - \alpha]^{\mathrm{T}} \mathcal{K}[f](t, x) \\
&= \mathcal{K}[[x_1 - \alpha, x_2 - \alpha]^{\mathrm{T}} f](t, x) \\
&= \mathcal{K}\left[-(2 - \cos t)(x_1 - x_2)\mathrm{sign}(x_1 - x_2)\right](t, x) \\
&= -(2 - \cos t)(x_1 - x_2)\mathcal{K}[\mathrm{sign}(x_1 - x_2)](x) \\
&= -(2 - \cos t)(x_1 - x_2)\mathrm{SGN}(x_1 - x_2) \\
&= -(2 - \cos t)|x_1 - x_2|, \quad t \in \mathbb{R}, \quad (x_1, x_2) \in \mathbb{R}^2, \quad (6.40)
\end{aligned}
$$

which implies that $\langle \nabla V(x_1, x_2), v \rangle \leq -|x_1 - x_2|$ for every $v \in \mathcal{K}[f](t, x)$.

Now, it follows from Theorem 1 of [97, p. 153] that $x_1 = x_2 = \alpha$ is Lyapunov stable. In fact, it can be shown that $x_1 = x_2 = \alpha$ is uniformly Lyapunov stable. Next, let $W(x_1, x_2) = |x_1 - x_2|$, and note that $W^{-1}(0) = \{(x_1, x_2) \in \mathbb{R}^2 : x_1 = x_2\} = \mathcal{E}$. Now, it follows from Theorem 6.4 that (6.38)–(6.39) is semistable. Moreover, it can be shown that (6.38)–(6.39) is uniformly semistable. Figure 6.4 shows the solutions of (6.38)–(6.39) for $x_{10} = 4$, $x_{20} = -2$, and $t_0 = 0, 1, 2, 3$. $\qquad \triangle$

## 6.6 Applications to Network Consensus with a Switching Topology

As noted in Section 6.1, communication links among multiagent systems are often unreliable due to multipath effects and exogenous disturbances leading to dynamic information exchange topologies. In this section, we use the semistability theory developed in Section 6.3 to develop switched consensus protocols to achieve agreement over a network with switching topology. Specifically, we consider $q$ mobile agents with dynamics $\mathcal{G}_i$ given by

$$\dot{x}_i(t) = u_i(t), \quad x_i(0) = x_{i0}, \quad t \geq 0, \tag{6.41}$$

where for each $i \in \{1, \ldots, q\}$, $x_i(t) \in \mathbb{R}$ denotes the information state and $u_i(t) \in \mathbb{R}$ denotes the information control input for all $t \geq 0$. The nonlinear consensus protocol is given by

$$u_i(t) = \sum_{j=1, j \neq i}^{q} \phi_{ij}(x_i(t), x_j(t)), \quad i = 1, \ldots, q, \tag{6.42}$$

where $\phi_{ij}(\cdot, \cdot)$, $i, j = 1, \ldots, q$, are Lebesgue measurable and locally essentially bounded.

Note that (6.41) and (6.42) describe an interconnected network $\mathcal{G}$ with a graph topology $\mathfrak{G} = (\mathcal{V}, \mathcal{E}, \mathcal{A})$, where $\mathcal{V} = \{1, \ldots, q\}$ denotes the set of nodes (or vertices) involving a finite nonempty set denoting the agents; $\mathcal{E} \subseteq \mathcal{V} \times \mathcal{V}$ denotes the set of edges involving a set of ordered pairs denoting the direction of information flow; and $\mathcal{A}$ denotes an adjacency matrix such that $\mathcal{A}_{(i,j)} = 1$, $i, j = 1, \ldots, q$, if $(j, i) \in \mathcal{E}$, and zero otherwise. Furthermore, note that it follows from (6.41) and (6.42) that information states are updated using a distributed nonlinear controller involving neighbor-to-neighbor interaction between agents.

The following assumptions are a restatement of Assumptions 4.1 and 4.2 and are needed for the main results of this section.

**Assumption 6.1.** The *connectivity matrix* $\mathcal{C} \in \mathbb{R}^{q \times q}$ associated with the multiagent dynamical system (6.41) and (6.42) is defined by

$$\mathcal{C}_{(i,j)} \triangleq \begin{cases} 0 & \text{if } \phi_{ij}(x_i, x_j) \equiv 0, \\ 1, & \text{otherwise,} \end{cases} \quad i \neq j, \quad i, j = 1, \ldots, q, \tag{6.43}$$

and $\mathcal{C}_{(i,i)} \triangleq -\sum_{k=1, k \neq i}^{q} \mathcal{C}_{(i,k)}$, $i = 1, \ldots, q$, with $\operatorname{rank} \mathcal{C} = q - 1$, and for $\mathcal{C}_{(i,j)} = 1$, $i \neq j$, $\phi_{ij}(x_i, x_j) = 0$ if and only if $x_i = x_j$.

**Assumption 6.2.** For $i, j = 1, \ldots, q$, $(x_i - x_j)\phi_{ij}(x_i, x_j) \leq 0$, $x_i, x_j \in \mathbb{R}$.

To address the network consensus problem with a switching topology, consider the switched controller $\mathcal{G}_{si}$ given by

$$u_i(t) = \sum_{j=1,j\neq i}^{q} \phi_{ij}^{\sigma(t)}(x_i(t), x_j(t)), \quad i = 1, \ldots, q, \tag{6.44}$$

where $\sigma : [0, \infty) \to \mathcal{S}$ is a piecewise constant switching signal, $\mathcal{S}$ is a finite index set, and $\phi_{ij}^{\sigma} : \mathbb{R} \times \mathbb{R} \to \mathbb{R}$ is Lebesgue measurable and locally essentially bounded and satisfies Assumptions 6.1 and 6.2 for every $\sigma \in \mathcal{S}$. We denote by $t_i$, $i = 1, 2, \ldots$, the consecutive discontinuities of $\sigma$. Furthermore, we assume that $\mathcal{C} = \mathcal{C}^{\mathrm{T}}$ in Assumption 6.1, where $\mathcal{C} = \mathcal{C}(t)$, $t \geq 0$. The assumption $\mathcal{C} = \mathcal{C}^{\mathrm{T}}$ implies that the underlying dynamic graph for the multiagent system $\mathcal{G}$ given by (6.41) and (6.42) is undirected. For details, see [162].

**Theorem 6.5.** Consider the closed-loop system $\tilde{\mathcal{G}}$ given by the multiagent dynamical system (6.41) and the switched controller (6.44). Assume that Assumptions 6.1 and 6.2 hold for every $\sigma \in \mathcal{S}$. Furthermore, assume that $\mathcal{C} = \mathcal{C}^{\mathrm{T}}$, where $\mathcal{C} = \mathcal{C}(t)$, $t \geq 0$, in Assumption 6.1. Then, for every $\alpha \in \mathbb{R}$, $x_1 = \cdots = x_q = \alpha$ is a semistable state of $\tilde{\mathcal{G}}$. Furthermore, $x_i(t) \to \frac{1}{q}\sum_{i=1}^{q} x_{i0}$, and $\frac{1}{q}\sum_{i=1}^{q} x_{i0}$ is a semistable equilibrium state.

**Proof.** Consider the Lyapunov function candidate

$$V(x) = \frac{1}{2}(x - \alpha\mathbf{e})^{\mathrm{T}}(x - \alpha\mathbf{e}), \tag{6.45}$$

where $x \triangleq [x_1, \ldots, x_q]^{\mathrm{T}} \in \mathbb{R}^q$ and $\alpha \in \mathbb{R}$. Next, we rewrite the closed-loop system (6.41) and (6.44) as the differential inclusion (6.2). For every $v \in \mathcal{K}[f](x)$, let $V^o(x, v) \triangleq x^{\mathrm{T}}v$ and $\max V^o(x, v) \triangleq \max_{v \in \mathcal{K}[f]}\{x^{\mathrm{T}}v\}$. Now, it follows from Theorem 1 of [240] that

$$x^{\mathrm{T}}\mathcal{K}[f](x) = \mathcal{K}[x^{\mathrm{T}}f](x)$$

$$= \mathcal{K}\left[\sum_{i=1}^{q} x_i \left(\sum_{j=1,j\neq i}^{q} \phi_{ij}^{\sigma}(x_i, x_j)\right)\right](x)$$

$$= \mathcal{K}\left[\sum_{i=1}^{q-1}\sum_{j=i+1}^{q} (x_i - x_j)\phi_{ij}^{\sigma}(x_i, x_j)\right](x), \quad x \in \mathbb{R}^q, \tag{6.46}$$

and, hence, by the definition of differential inclusions, it follows that

$$\max V^o(x, v) = \max \overline{\mathrm{co}}\left\{\sum_{i=1}^{q-1}\sum_{j=i+1}^{q} (x_i - x_j)\phi_{ij}^{\sigma}(x_i, x_j)\right\}.$$

Next, note that since, by Assumption 6.2,

$$\sum_{i=1}^{q-1}\sum_{j=i+1}^{q}(x_i - x_j)\phi_{ij}^{\sigma}(x_i, x_j) \leq 0, \quad x_i \in \mathbb{R},$$

it follows that $\max V^o(x, v)$ cannot be positive, and, hence, the largest value that $\max V^o(x, v)$ can achieve is zero, which establishes Lyapunov stability of $x \equiv \alpha\mathbf{e}$.

Finally, note that $0 \in \mathcal{L}_f V(x)$ if and only if

$$\sum_{i=1}^{q-1}\sum_{j=i+1}^{q}(x_i - x_j)\phi_{ij}^{\sigma}(x_i, x_j) = 0,$$

and, hence,

$$\mathcal{Z} \triangleq \left\{ x \in \mathbb{R}^q : \sum_{i=1}^{q-1}\sum_{j=i+1}^{q}(x_i - x_j)\phi_{ij}^{\sigma}(x_i, x_j) = 0 \right\}.$$

Now, it follows from Proposition 4.5 that $\mathcal{Z} = \{x \in \mathbb{R}^q : x_1 = \cdots = x_q\}$. Since $\mathcal{Z}$ consists of equilibrium points, it follows that $\mathcal{M} = \mathcal{Z}$. Hence, it follows from Theorem 6.1 that $x = \alpha\mathbf{e}$ is semistable for all $\alpha \in \mathbb{R}$. $\qquad\square$

Note that Example 6.1 serves as a special case of Theorem 6.5.

**Example 6.6.** To further illustrate Theorem 6.5, we consider a four-agent system with consensus protocol (6.44) of the form

$$\phi_{ij}^{\sigma}(x_i, x_j) = \mathcal{C}_{\sigma(i,j)}\tanh(x_j - x_i), \quad i, j = 1, 2, 3, 4, \quad i \neq j.$$

The index set is $\mathcal{S} = \{1, 2\}$, $t_i = i$ denotes the switching time instant, and the communication topology is switched between the following two connectivity matrices:

$$\mathcal{C}_1 = \begin{bmatrix} -2 & 1 & 0 & 1 \\ 1 & -2 & 1 & 0 \\ 0 & 1 & -2 & 1 \\ 1 & 0 & 1 & -2 \end{bmatrix}, \quad \mathcal{C}_2 = \begin{bmatrix} -1 & 1 & 0 & 0 \\ 1 & -2 & 0 & 1 \\ 0 & 0 & -1 & 1 \\ 0 & 1 & 1 & -2 \end{bmatrix}. \quad (6.47)$$

For $t_i \leq s < t_{i+1}$, $\sigma(s) = 1$ for even $i$ and $\sigma(s) = 2$ for odd $i$. Figure 6.5 shows the states of the closed-loop system (6.41) and (6.44) with a switching topology as characterized by (6.47). $\qquad\triangle$

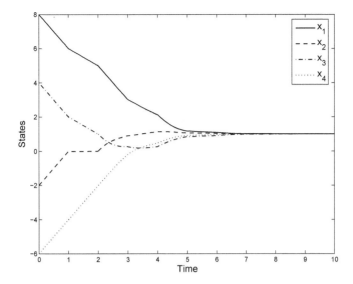

Figure 6.5 State trajectories for Example 6.6.

Next, we extend Theorem 6.5 to the discontinuous controllers $\mathcal{G}_{ni}$ of the form

$$u_i = \sum_{j=1, j \neq i}^{q} \mathcal{C}_{(i,j)} \operatorname{sign}(x_j - x_i), \quad i = 1, \ldots, q. \tag{6.48}$$

It is important to note that the consensus protocol (6.48) is a logic-based, distributed decision-making protocol. Although a similar consensus protocol based on nonsmooth gradient flows is proposed in [69], the key difference between (6.48) and the one in [69] is that (6.48) is a *distributed* protocol, while the consensus protocol in [69] is a *centralized* protocol.

In Chapter 4, we showed that the *continuous* consensus protocol given by

$$u_i = \sum_{j=1, j \neq i}^{q} \mathcal{C}_{(i,j)} \operatorname{sign}(x_j - x_i)|x_j - x_i|^{\alpha}, \quad i = 1, \ldots, q, \tag{6.49}$$

achieves finite time consensus for $0 < \alpha < 1$. Next, we show that (6.49) also achieves finite time consensus for $\alpha = 0$. Note that in this case, (6.49) reduces to (6.48). Furthermore, note that Example 6.2 is a special case of the closed-loop system given by (6.41) and (6.48).

**Theorem 6.6.** Consider the closed-loop system $\tilde{\mathcal{G}}$ given by the multiagent dynamical system (6.41) and the discontinuous controller (6.48). Assume that Assumptions 6.1 and 6.2 hold. Furthermore, assume that

$\mathcal{C} = \mathcal{C}^{\mathrm{T}}$ in Assumption 6.1. Then, for every $\alpha \in \mathbb{R}$, $x_1 = \cdots = x_q = \alpha$ is a finite time semistable state of $\tilde{\mathcal{G}}$. Furthermore, $x_i(t) = \frac{1}{q} \sum_{i=1}^{q} x_{i0}$ for $t \geq T(x_{10}, \ldots, x_{q0})$, and $\frac{1}{q} \sum_{i=1}^{q} x_{i0}$ is a semistable equilibrium state.

**Proof.** Consider the Lyapunov function candidate (6.45). Since $V(x)$ is differentiable at $x$, it follows that $\mathcal{L}_f V(x) = (x - \alpha \mathbf{e})^{\mathrm{T}} \mathcal{K}[f](x)$. Now, it follows from Theorem 1 of [240] that

$$
\begin{aligned}
(x - \alpha \mathbf{e})^{\mathrm{T}} \mathcal{K}[f](x) &= \mathcal{K}[(x - \alpha \mathbf{e})^{\mathrm{T}} f](x) \\
&= \mathcal{K}[x^{\mathrm{T}} f](x) \\
&= \mathcal{K} \left[ \sum_{i=1}^{q} x_i \sum_{j=1, j \neq i}^{q} \mathcal{C}_{(i,j)} \mathrm{sign}(x_j - x_i) \right](x) \\
&= \mathcal{K} \left[ -\sum_{i=1}^{q} \sum_{j=1, j \neq i}^{q} \mathcal{C}_{(i,j)} (x_i - x_j) \mathrm{sign}(x_i - x_j) \right](x) \\
&\subseteq -\sum_{i=1}^{q} \sum_{j=1, j \neq i}^{q} \mathcal{C}_{(i,j)} (x_i - x_j) \mathcal{K}[\mathrm{sign}(x_i - x_j)](x) \\
&= -\sum_{i=1}^{q} \sum_{j=1, j \neq i}^{q} \mathcal{C}_{(i,j)} (x_i - x_j) \mathrm{SGN}(x_i - x_j) \\
&= -\sum_{i=1}^{q} \sum_{j=1, j \neq i}^{q} \mathcal{C}_{(i,j)} |x_i - x_j|, \quad x \in \mathbb{R}^q, \qquad (6.50)
\end{aligned}
$$

which implies that $\max \mathcal{L}_f V(x) \leq 0$ for almost all $x \in \mathbb{R}^q$. Hence, it follows from Theorem 2 of [11] that $x_1 = \cdots = x_q = \alpha$ is Lyapunov stable.

Next, note that since

$$
\begin{aligned}
\mathcal{L}_f V(x) &= \mathcal{K} \left[ -\sum_{i=1}^{q} \sum_{j=1, j \neq i}^{q} \mathcal{C}_{(i,j)} (x_i - x_j) \mathrm{sign}(x_i - x_j) \right](x) \\
&= \mathcal{K} \left[ -\sum_{i=1}^{q} \sum_{j=1, j \neq i}^{q} \mathcal{C}_{(i,j)} |x_i - x_j| \right](x),
\end{aligned}
$$

it follows that $0 \in \mathcal{L}_f V(x)$ if and only if $x_1 = \cdots = x_q$, and, hence, $\mathcal{Z} = \{x \in \mathbb{R}^q : x_1 = \cdots = x_q\}$. Since the largest weakly positively invariant subset $\mathcal{M}$ of $\mathcal{Z}$ is given by $\mathcal{M} = \{x \in \mathbb{R}^q : x_1 = \cdots = x_q = \alpha, \alpha \in \mathbb{R}\}$, it follows from Theorem 6.1 that $\tilde{\mathcal{G}}$ is semistable.

Finally, we show that $\tilde{\mathcal{G}}$ is finite time semistable. To see this, consider the nonnegative function

$$U(x) = \frac{1}{2} \sum_{i=1}^{q} \sum_{j=1, j \neq i}^{q} \mathcal{C}_{(i,j)} |x_i - x_j|.$$

In this case, using similar arguments as in Example 6.2, it follows that

$$\mathcal{L}_f U(x) = \begin{cases} \left\{ -2 \sum_{i=1}^{q} \sum_{j=1, j \neq i}^{q} \mathcal{C}_{(i,j)} \right\}, & x_i \neq x_j, \ i, j = 1, \dots, q, \ i \neq j, \\ \varnothing, & x_k = x_l \text{ for some} \\ & k, l \in \{1, \dots, q\}, \ k \neq l, \\ \{0\}, & x_1 = \dots = x_q, \end{cases}$$

which implies that either $\max \mathcal{L}_f U(x) \leq -2 \sum_{i=1}^{q} \sum_{j=1, j \neq i}^{q} \mathcal{C}_{(i,j)} < 0$ or $\mathcal{L}_f U(x) = \varnothing$ for almost all $x \in \mathbb{R}^q \backslash \mathcal{Z}$. Hence, it follows from Corollary 6.1 that $\tilde{\mathcal{G}}$ is globally finite time semistable. $\qquad \square$

Next, we design discontinuous *dynamic* consensus protocols for (6.41). In contrast to the static controllers addressed in [180] and [238], the proposed controller is a dynamic compensator. This controller architecture allows us to design finite time consensus protocols via quantized feedback in a dynamical network. Specifically, consider the $q$ mobile agents with dynamics $\mathcal{G}_i$ given by (6.41). Furthermore, consider the discontinuous dynamic compensators $\mathcal{G}_{ci}$ given by

$$\dot{x}_{ci}(t) = \sum_{j=1, j \neq i}^{q} \mathcal{C}_{(i,j)} \text{sign}(x_{cj}(t) - x_{ci}(t)) + \sum_{j=1, j \neq i}^{q} \mathcal{C}_{(i,j)} \text{sign}(x_i(t) - x_j(t)),$$
$$x_{ci}(0) = x_{ci0}, \quad t \geq 0, \qquad (6.51)$$

$$u_i(t) = \sum_{j=1, j \neq i}^{q} \mathcal{C}_{(j,i)} \text{sign}(x_{cj}(t) - x_{ci}(t)), \qquad (6.52)$$

where $x_{ci}(t) \in \mathbb{R}$, $t \geq 0$, and $i = 1, \dots, q$.

**Theorem 6.7.** Consider the closed-loop system $\tilde{\mathcal{G}}$ given by the multiagent dynamical system (6.41) and the discontinuous dynamic controller (6.51)–(6.52). Assume that Assumption 6.1 holds and $\mathcal{C} = \mathcal{C}^{\text{T}}$. Then, for every $\alpha \in \mathbb{R}$ and $\beta \in \mathbb{R}$, $x_1 = \dots = x_q = \alpha$ and $x_{c1} = \dots = x_{cq} = \beta$ is a finite time semistable state of $\tilde{\mathcal{G}}$. Furthermore, $x_i(t) = \frac{1}{q} \sum_{i=1}^{q} x_{i0}$ and $x_{ci}(t) = \frac{1}{q} \sum_{i=1}^{q} x_{ci0}$ for all $t \geq T(x_{10}, \dots, x_{q0}, x_{c10}, \dots, x_{cq0})$, and

$$\left( \frac{1}{q} \sum_{i=1}^{q} x_{i0}, \frac{1}{q} \sum_{i=1}^{q} x_{ci0} \right)$$

is a semistable equilibrium state.

**Proof.** Note that for every $a, b \in \mathbb{R}$, $x(t) \equiv a\mathbf{e}$ and $x_c(t) \equiv b\mathbf{e}$ are the equilibrium points for the closed-loop system $\tilde{\mathcal{G}}$. Consider the nonnegative function given by

$$V(\tilde{x}) = \frac{1}{2} \sum_{i=1}^{q} \sum_{j=1, j \neq i}^{q} \mathcal{C}_{(i,j)} |x_i - x_j| + \frac{1}{2} \sum_{i=1}^{q} \sum_{j=1, j \neq i}^{q} \mathcal{C}_{(i,j)} |x_{ci} - x_{cj}|, \quad (6.53)$$

where $\tilde{x} \triangleq [x^{\mathrm{T}}, x_c^{\mathrm{T}}]^{\mathrm{T}} \in \mathbb{R}^{2q}$. In this case, using similar arguments as in Example 6.3, it follows that

$$\mathcal{L}_f V(\tilde{x})$$
$$= \begin{cases} \left\{ -2\sum_{i=1}^{q} \sum_{j=1, j \neq i}^{q} \mathcal{C}_{(i,j)} \right\}, & x_i \neq x_j, \ x_{ci} \neq x_{cj}, \ i, j = 1, \ldots, q, \ i \neq j, \\ \emptyset, & x_k = x_l \text{ or } x_{ck} = x_{cl} \\ & \quad \text{for some } k, l \in \{1, \ldots, q\}, \ k \neq l, \\ \{0\}, & x_1 = \cdots = x_q, \ x_{c1} = \cdots = x_{cq}, \end{cases}$$
$$(6.54)$$

which implies that either $\max \mathcal{L}_f V(\tilde{x}) \leq 0$ or $\mathcal{L}_f V(\tilde{x}) = \emptyset$ for almost all $\tilde{x} \in \mathbb{R}^{2q}$.

Next, define

$$\mathcal{Z} \triangleq \{\tilde{x} \in \mathbb{R}^{2q} : x_1 = \cdots = x_q, x_{c1} = \cdots = x_{cq}\},$$

and let $\mathcal{N}$ denote the largest negatively invariant set of $\mathcal{Z}$. On $\mathcal{N}$, it follows from (6.41), (6.51), and (6.52) that $\dot{x}_i = 0$ and $\dot{x}_{ci} = 0$, $i = 1, \ldots, q$. Hence, $\mathcal{N} = \{\tilde{x} \in \mathbb{R}^{2q} : x = a\mathbf{e}, x_c = b\mathbf{e}\}$, $a, b \in \mathbb{R}$, which implies that $\mathcal{N}$ is the set of equilibrium points.

Since the connectivity matrix $\mathcal{C}$ of the closed-loop system is irreducible, assume, without loss of generality, that $\mathcal{C}_{(i,i+1)} = \mathcal{C}_{(q,1)} = 1$, where $i = 1, \ldots, q-1$. Now, for $q = 2$, it was shown in Example 6.3 that the vector field $f$ of the closed-loop system given by (6.41), (6.51), and (6.52) is nontangent to $\mathcal{N}$ at a point $\tilde{x} \in \mathcal{N}$. Next, we show that for $q \geq 3$, the vector field $f$ of the closed-loop system given by (6.41), (6.51), and (6.52) is nontangent to $\mathcal{N}$ at a point $\tilde{x} \in \mathcal{N}$.

To see this, note that the tangent cone $T_{\tilde{x}} \mathcal{N}$ to the equilibrium set $\mathcal{N}$ is orthogonal to the $2q$ vectors

$$\mathbf{u}_i \triangleq [0_{1 \times (i-1)}, \mathcal{C}_{(i,i+1)}, -\mathcal{C}_{(i,i+1)}, 0_{1 \times (2q-i-1)}]^{\mathrm{T}} \in \mathbb{R}^{2q},$$
$$\mathbf{u}_q \triangleq [-\mathcal{C}_{(q,1)}, 0_{1 \times (q-2)}, \mathcal{C}_{(q,1)}, 0_{1 \times q}]^{\mathrm{T}} \in \mathbb{R}^{2q},$$
$$\mathbf{v}_i \triangleq [0_{1 \times (q+i-1)}, -\mathcal{C}_{(i,i+1)}, \mathcal{C}_{(i,i+1)}, 0_{1 \times (q-i-1)}]^{\mathrm{T}} \in \mathbb{R}^{2q},$$

and

$$\mathbf{v}_q \triangleq [0_{1 \times q}, \mathcal{C}_{(q,1)}, 0_{1 \times (q-2)}, -\mathcal{C}_{(q,1)}]^{\mathrm{T}} \in \mathbb{R}^{2q}, \quad i = 1, \ldots, q-1 \quad q \geq 3.$$

Since $f(\tilde{x}) \in \mathrm{span}\{\mathbf{u}_1, \ldots, \mathbf{u}_q, \mathbf{v}_1, \ldots, \mathbf{v}_q\}$ for all $\tilde{x} \in \mathbb{R}^{2q}$, it follows that $f(\mathcal{V}) \subseteq \mathrm{span}\{\mathbf{u}_1, \ldots, \mathbf{u}_q, \mathbf{v}_1, \ldots, \mathbf{v}_q\}$ for every subset $\mathcal{V} \subseteq \mathbb{R}^{2q}$. Consequently, the direction cone $\mathcal{F}_{\tilde{x}}$ of $f$ at $\tilde{x} \in \mathcal{N}$ relative to $\mathbb{R}^{2q}$ satisfies $\mathcal{F}_{\tilde{x}} \subseteq \mathrm{span}\{\mathbf{u}_1, \ldots, \mathbf{u}_q, \mathbf{v}_1, \ldots, \mathbf{v}_q\}$. Hence, $T_{\tilde{x}}\mathcal{N} \cap \mathcal{F}_{\tilde{x}} = \{0\}$, which implies that the vector field $f$ is nontangent to the set of equilibria $\mathcal{N}$ at the point $\tilde{x} \in \mathcal{N}$. Note that for every $z \in \mathcal{N}$, the set $\mathcal{N}_z$ required by Theorem 6.2 is contained in $\mathcal{N}$. Since nontangency to $\mathcal{N}$ implies nontangency to $\mathcal{N}_z$ at the point $z \in \mathcal{N}$, it follows from Theorem 6.2 that the closed-loop system $\tilde{\mathcal{G}}$ is semistable.

Finally, note that either $\max \mathcal{L}_f V(x) \leq -2 \sum_{i=1}^{q} \sum_{j=1, j \neq i}^{q} \mathcal{C}_{(i,j)} < 0$ or $\mathcal{L}_f V(\tilde{x}) = \varnothing$ for almost all $x \in \mathbb{R}^4 \backslash \mathcal{Z}$, and, hence, it follows from Corollary 6.1 that $\tilde{\mathcal{G}}$ is globally finite time semistable. $\qquad \square$

The dynamic compensator (6.51) and (6.52) is a state feedback controller. A natural question regarding (6.41) is how to design finite time consensus protocols for multiagent coordination via *output feedback*. To address this question, we consider $q$ continuous-time integrator agents given by (6.41) and with outputs given by

$$y_i = \sum_{j=1, j \neq i}^{q} \mathcal{C}_{(i,j)}(x_j - x_i), \quad i = 1, \ldots, q. \tag{6.55}$$

In addition, consider the dynamic output feedback compensator given by

$$\dot{x}_{ci}(t) = \sum_{j=1, j \neq i}^{q} \mathcal{C}_{(i,j)} \mathrm{sign}(x_{cj}(t) - x_{ci}(t)) + y_i(t),$$
$$x_{ci}(0) = x_{ci0}, \quad t \geq 0, \tag{6.56}$$

$$u_i(t) = \sum_{j=1, j \neq i}^{q} \mathcal{C}_{(j,i)} \mathrm{sign}(x_{cj}(t) - x_{ci}(t)), \tag{6.57}$$

where $x_{ci}(t) \in \mathbb{R}$, $t \geq 0$, and $i = 1, \ldots, q$.

**Theorem 6.8.** Consider the closed-loop system $\tilde{\mathcal{G}}$ given by the multiagent dynamical system (6.41) and the discontinuous dynamic output feedback controller (6.56) and (6.57) with (6.55). Assume that Assumption 6.1 holds and $\mathcal{C} = \mathcal{C}^{\mathrm{T}}$. Then, for every $\alpha \in \mathbb{R}$ and $\beta \in \mathbb{R}$, $x_1 = \cdots = x_q = \alpha$ and $x_{c1} = \cdots = x_{cq} = \beta$ is a finite time semistable state of $\tilde{\mathcal{G}}$. Furthermore, $x_i(t) = \frac{1}{q} \sum_{i=1}^{q} x_{i0}$ and $x_{ci}(t) = \frac{1}{q} \sum_{i=1}^{q} x_{ci0}$

for all $t \geq T(x_{10}, \ldots, x_{q0}, x_{c10}, \ldots, x_{cq0})$, and $(\frac{1}{q} \sum_{i=1}^{q} x_{i0}, \frac{1}{q} \sum_{i=1}^{q} x_{ci0})$ is a semistable equilibrium state.

**Proof.** The proof is similar to the proof of Theorem 6.7 with

$$V(\tilde{x}) = \frac{1}{2} \sum_{i=1}^{q} \sum_{j=1, j \neq i}^{q} \mathcal{C}_{(i,j)}(x_i - x_j)^2 + \frac{1}{2} \sum_{i=1}^{q} \sum_{j=1, j \neq i}^{q} \mathcal{C}_{(i,j)}|x_{ci} - x_{cj}|$$

and, hence, is omitted.                                                    □

**Example 6.7.** To illustrate Theorem 6.8, consider (6.41) with $q = 2$ and with the dynamic feedback compensator (6.56)–(6.57). Figure 6.6 shows the states of the closed-loop system (6.41), (6.55), (6.56), and (6.57).   △

As an application of the proposed finite time consensus protocol, we present a control algorithm for achieving *finite time parallel formation* [183]. Specifically, we consider $q$ mobile agents with dynamics $\mathcal{G}_i$ given by

$$\dot{x}_i(t) = v_i(t), \quad x_i(0) = x_{i0}, \quad t \geq 0, \tag{6.58}$$
$$\dot{v}_i(t) = u_i(t), \quad v_i(0) = v_{i0}, \tag{6.59}$$

where, for each $i \in \{1, \ldots, q\}$, $x_i(t) \in \mathbb{R}$ denotes the position, $v_i(t) \in \mathbb{R}$ denotes the velocity, and $u_i(t) \in \mathbb{R}$ denotes the control input for all $t \geq 0$. The control aim is to design a distributed feedback control law $u_i$ involving the transmission of both $x_i$ and $v_i$ between agents so that finite time parallel formation is achieved; that is, the velocity $v_i$ reaches a constant vector in

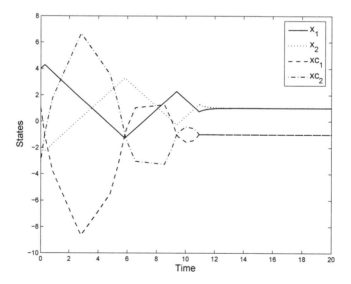

Figure 6.6 State trajectories for Example 6.7.

finite time for all $i = 1, \ldots, q$, and the relative position between two agents reaches a constant value in finite time.

**Theorem 6.9.** Consider the dynamical system given by (6.58) and (6.59). Then finite time parallel formation for (6.58) and (6.59) is achieved under the distributed feedback control law given by the static controller

$$u_i = \sum_{j=1, j\neq i}^{q} \mathcal{C}_{(i,j)}\text{sign}(v_j - v_i) - \sum_{j=1, j\neq i}^{q} \mathcal{C}_{(i,j)}\text{sign}(x_i - x_j - d_{ij}),$$

$$i = 1, \ldots, q, \qquad (6.60)$$

where $\mathcal{C}_{(i,j)}$ is as in (6.43), with $\mathcal{C} = \mathcal{C}^{\text{T}}$ and $d_{ij} = -d_{ji} \in \mathbb{R}$, $i, j = 1, \ldots, q$, $i \neq j$.

**Proof.** The proof is similar to the proof of Theorem 6.6 with

$$V(x, v) = \frac{1}{2}v^{\text{T}}v + \frac{1}{2}\sum_{i=1}^{q} \sum_{j=1, j\neq i}^{q} \mathcal{C}_{(i,j)}|x_i - x_j - d_{ij}|, \qquad (6.61)$$

where $v \triangleq [v_1, \ldots, v_q]^{\text{T}}$ and $x \triangleq [x_1, \ldots, x_q]^{\text{T}}$, and, hence, is omitted. $\qquad \square$

## 6.7 Application to Finite Time Rendezvous Problems

An important application of the network information consensus problem is the rendezvous problem. In particular, a cooperative rendezvous task requires that each agent determine the rendezvous time and location through team negotiation. In this section, we develop a coordination framework for finite time multiagent rendezvous. Here, we assume that each agent has a limited sensing region which is bounded within a cyclic area of radius $r$.

Specifically, we consider $q$ mobile agents with dynamics $\mathcal{G}_i$ given by (6.41) and with the switching controller $\mathcal{G}_{si}$ given by

$$u_i = \sum_{j=1, j\neq i}^{q} \mathcal{C}_{(i,j)}G(\|x_j - x_i\|)\text{sign}(x_j - x_i), \quad i = 1, \ldots, q, \qquad (6.62)$$

where

$$G(x) = \begin{cases} 1, & x < r, \\ \frac{1}{2}, & x = r, \\ 0, & x > r, \end{cases} \qquad (6.63)$$

and $r > 0$ is a constant.

**Theorem 6.10.** Consider the closed-loop system $\tilde{\mathcal{G}}$ given by the multiagent dynamical system (6.41) and the hybrid controller (6.62). Let $\mathcal{D} \triangleq \{x \in \mathbb{R}^q : \|x_i - x_j\| < r, i, j = 1, \ldots, q\}$, and assume that Assumptions 6.1 and 6.2 hold. Furthermore, assume that $\mathcal{C} = \mathcal{C}^{\mathrm{T}}$ in Assumption 6.1. Then, for every $\alpha \in \mathbb{R}$, $x_1 = \cdots = x_q = \alpha$ is a finite time semistable state of $\tilde{\mathcal{G}}$ with respect to $\mathcal{D}$. Furthermore, $x_i(t) = \frac{1}{q} \sum_{i=1}^{q} x_{i0}$ for $t \geq T(x_{10}, \ldots, x_{q0})$, and $\frac{1}{q} \sum_{i=1}^{q} x_{i0}$ is a semistable equilibrium state for $x_0 \in \mathcal{D}$.

**Proof.** The proof is similar to the proof of Theorem 6.7 and, hence, is omitted.                                                                    $\square$

Theorem 6.10 gives a coordination control architecture for multiagent rendezvous where a consensus manager is applied via a distributed consensus algorithm to guarantee that all agents achieve consensus on a rendezvous objective in finite time. Note that if the communication graph is initially connected, it remains connected throughout the closed-loop system level and rendezvous is achieved in finite time under the proposed hybrid control algorithm.

Next, we design a hybrid *dynamic* rendezvous protocol for (6.41). In contrast to the static controllers addressed in [204] and [74], the proposed controller is a dynamic compensator. This controller architecture allows us to design finite time rendezvous protocols via quantized feedback in a dynamical network. Specifically, consider the $q$ mobile agents with the dynamics $\mathcal{G}_i$ given by (6.41). Furthermore, consider the hybrid dynamic compensators $\mathcal{G}_{ci}$ given by

$$\dot{x}_{ci}(t) = \sum_{j=1, j \neq i}^{q} \mathcal{C}_{(i,j)} G(\|x_{ci} - x_{cj}\|) \mathrm{sign}(x_{cj}(t) - x_{ci}(t))$$

$$+ \sum_{j=1, j \neq i}^{q} \mathcal{C}_{(i,j)} G(\|x_i - x_j\|) \mathrm{sign}(x_i(t) - x_j(t)), \quad x_{ci}(0) = x_{ci0},$$

$$t \geq 0, \qquad (6.64)$$

$$u_i(t) = \sum_{j=1, j \neq i}^{q} \mathcal{C}_{(j,i)} G(\|x_i - x_j\|) \mathrm{sign}(x_{cj}(t) - x_{ci}(t)), \qquad (6.65)$$

where $x_{ci}(t) \in \mathbb{R}$, $t \geq 0$, and $i = 1, \ldots, q$.

**Theorem 6.11.** Consider the closed-loop system $\tilde{\mathcal{G}}$ given by the multiagent dynamical system (6.41) and the hybrid dynamic controller (6.64)–(6.65). Let $\tilde{\mathcal{D}} \triangleq \{(x, x_c) : \|x_i - x_j\| < r, \|x_{ci} - x_{cj}\| < r, i, j = 1, \ldots, q\}$, and assume that Assumption 6.1 holds and $\mathcal{C} = \mathcal{C}^{\mathrm{T}}$. Then, for every $\alpha \in \mathbb{R}$ and $\beta \in \mathbb{R}$, $x_1 = \cdots = x_q = \alpha$ and $x_{c1} = \cdots = x_{cq} = \beta$ is a finite time semistable state of $\tilde{\mathcal{G}}$ with respect to $\tilde{\mathcal{D}}$. Furthermore, $x_i(t) = \frac{1}{q}\sum_{i=1}^{q} x_{i0}$ and $x_{ci}(t) = \frac{1}{q}\sum_{i=1}^{q} x_{ci0}$ for all $t \geq T(x_{10}, \ldots, x_{q0}, x_{c10}, \ldots, x_{cq0})$, and $(\frac{1}{q}\sum_{i=1}^{q} x_{i0}, \frac{1}{q}\sum_{i=1}^{q} x_{ci0})$ is a semistable equilibrium state for $(x_0, x_{c0}) \in \tilde{\mathcal{D}}$.

**Proof.** The proof is similar to the proof of Theorem 6.7 and, hence, is omitted. $\square$

The dynamic compensator (6.64) and (6.65) is a state feedback controller. Next, we design a finite time rendezvous protocol for multiagent coordination via output feedback. Specifically, we consider the $q$ continuous-time integrator agents given by (6.41) with output $y_i$ given by

$$y_i = \sum_{j=1, j\neq i}^{q} \mathcal{C}_{(i,j)} G(\|x_i - x_j\|)(x_j - x_i), \quad i = 1, \ldots, q. \qquad (6.66)$$

In addition, consider the dynamic output feedback compensator given by

$$\dot{x}_{ci}(t) = \sum_{j=1, j\neq i}^{q} \mathcal{C}_{(i,j)} G(\|x_{ci} - x_{cj}\|)\mathrm{sign}(x_{cj}(t) - x_{ci}(t)) + y_i(t),$$
$$x_{ci}(0) = x_{ci0}, \quad t \geq 0, \qquad (6.67)$$

$$u_i(t) = \sum_{j=1, j\neq i}^{q} \mathcal{C}_{(j,i)} G(\|x_i - x_j\|)\mathrm{sign}(x_{cj}(t) - x_{ci}(t)), \qquad (6.68)$$

where $x_{ci}(t) \in \mathbb{R}$, $t \geq 0$, and $i = 1, \ldots, q$.

**Theorem 6.12.** Consider the closed-loop system $\tilde{\mathcal{G}}$ given by the multiagent dynamical system (6.41) and the hybrid dynamic controller (6.67)–(6.68). Let $\tilde{\mathcal{D}}$ be as in Theorem 6.11, and assume that Assumption 6.1 holds and $\mathcal{C} = \mathcal{C}^{\mathrm{T}}$. Then, for every $\alpha \in \mathbb{R}$ and $\beta \in \mathbb{R}$, $x_1 = \cdots = x_q = \alpha$ and $x_{c1} = \cdots = x_{cq} = \beta$ is a finite time semistable state of $\tilde{\mathcal{G}}$ with respect to $\tilde{\mathcal{D}}$. Furthermore, $x_i(t) = \frac{1}{q}\sum_{i=1}^{q} x_{i0}$ and $x_{ci}(t) = \frac{1}{q}\sum_{i=1}^{q} x_{ci0}$ for all $t \geq T(x_{10}, \ldots, x_{q0}, x_{c10}, \ldots, x_{cq0})$, and $(\frac{1}{q}\sum_{i=1}^{q} x_{i0}, \frac{1}{q}\sum_{i=1}^{q} x_{ci0})$ is a semistable equilibrium state for $(x_0, x_{c0}) \in \tilde{\mathcal{D}}$.

**Proof.** The proof is similar to the proof of Theorem 6.8 and, hence, is omitted. $\square$

## 6.8 Discontinuous Consensus Protocols with Time-Varying Communication Links

Finally, in this section we consider discontinuous consensus protocols $\mathcal{G}$ with time-dependent and state-dependent communication links given by

$$\dot{x}_i(t) = \sum_{j=1, j \neq i}^{q} \mathcal{C}_{(i,j)}(x_i(t), x_j(t)) a_{ij}(t, x_i(t), x_j(t)) \mathrm{sign}(x_j(t) - x_i(t)),$$

$$x_i(t_0) = x_{i0}, \quad t \geq t_0, \quad i = 1, \ldots, q, \qquad (6.69)$$

where $t \geq t_0$; $x_i(t) \in \mathbb{R}$; $a_{ij} : \mathbb{R}^3 \to \mathbb{R}$ satisfies $a_{ij}(t, x_i, x_j) = a_{ji}(t, x_j, x_i)$ and $m \leq a_{ij}(t, x_i, x_j) \leq M$, $a_{ij}(t, x_i, x_j) \not\equiv 0$, $i, j = 1, \ldots, q$, $i \neq j$; $0 < m < M$ is a constant; and $\mathcal{C}_{(i,j)} : \mathbb{R}^2 \to \mathbb{R}$ satisfies the following assumption.

**Assumption 6.3.** The connectivity matrix $\mathcal{C}(x) \in \mathbb{R}^{q \times q}$, $x \triangleq [x_1, \ldots, x_q]^{\mathrm{T}} \in \mathbb{R}^q$, associated with $\mathcal{G}$ is defined by

$$\mathcal{C}_{(i,j)}(x_i, x_j) \triangleq \begin{cases} 0 & \text{if } (j, i) \in \mathcal{E}, \\ 1, & \text{otherwise,} \end{cases} \quad i \neq j, \quad i, j = 1, \ldots, q, \quad (6.70)$$

and $\mathcal{C}_{(i,i)}(x_i, x_i) = -\sum_{k=1, k \neq i}^{q} \mathcal{C}_{(i,k)}(x_i, x_k)$, $i = 1, \ldots, q$, with $\mathrm{rank}\, \mathcal{C}(x) = q - 1$, $x \in \mathbb{R}^q$, and $\mathcal{C}(x) = \mathcal{C}^{\mathrm{T}}(x)$, $x \in \mathbb{R}^q$.

**Theorem 6.13.** Consider the time-varying discontinuous consensus protocol $\mathcal{G}$ given by (6.69). Assume that Assumption 6.3 holds. Then $\mathcal{G}$ is uniformly semistable and $x_i(t) \rightrightarrows \frac{1}{q} \sum_{i=1}^{q} x_{i0}$ as $t \to \infty$, $i = 1, \ldots, q$.

**Proof.** First, note that $\|f(t, x)\| \leq M(q-1)\sqrt{q}$ for almost all $t \geq t_0$ and $x \in \mathbb{R}^q$. Next, consider the Lyapunov function candidate (6.45), and note that

$$
\begin{aligned}
(x - \alpha \mathbf{e})^{\mathrm{T}} \mathcal{K}[f](t, x) &= \mathcal{K}[(x - \alpha \mathbf{e})^{\mathrm{T}} f](t, x) \\
&= \mathcal{K}[x^{\mathrm{T}} f](t, x) \\
&= \mathcal{K}\left[\sum_{i=1}^{q} x_i \sum_{j=1, j \neq i}^{q} \mathcal{C}_{(i,j)} a_{ij} \mathrm{sign}(x_i - x_j)\right](t, x) \\
&= \mathcal{K}\left[-\sum_{i=1}^{q} \sum_{j=1, j \neq i}^{q} \mathcal{C}_{(i,j)} a_{ij}(x_i - x_j) \mathrm{sign}(x_i - x_j)\right](t, x) \\
&\subseteq -\sum_{i=1}^{q} \sum_{j=1, j \neq i}^{q} \mathcal{C}_{(i,j)} a_{ij}(x_i - x_j) \mathcal{K}[\mathrm{sign}(x_i - x_j)](x)
\end{aligned}
$$

$$= -\sum_{i=1}^{q} \sum_{j=1, j\neq i}^{q} \mathcal{C}_{(i,j)} a_{ij} (x_i - x_j) \mathrm{SGN}(x_i - x_j)$$

$$= -\sum_{i=1}^{q} \sum_{j=1, j\neq i}^{q} \mathcal{C}_{(i,j)} a_{ij} |x_i - x_j|,$$

$$(t, x) \in [t_0, \infty) \times \mathbb{R}^q, \tag{6.71}$$

which implies that

$$\langle \nabla V(x), v \rangle \leq -\sum_{i=1}^{q} \sum_{j=1, j\neq i}^{q} m \mathcal{C}_{(i,j)} |x_i - x_j|, \quad v \in \mathcal{K}[f](t, x).$$

Now, it follows from Theorem 1 of [97, p. 153] that $x_1 = \cdots = x_q = \alpha$ is Lyapunov stable. In fact, it can be shown that $x_1 = \cdots = x_q = \alpha$ is uniformly Lyapunov stable.

Next, let

$$W(x) = \sum_{i=1}^{q} \sum_{j=1, j\neq i}^{q} m \mathcal{C}_{(i,j)} |x_i - x_j|,$$

and note that $W^{-1}(0) = \{x \in \mathbb{R}^q : x_1 = \cdots = x_q\} = \mathcal{E}$. Now, it follows from Theorem 6.4 that $\mathcal{G}$ is uniformly semistable. Finally, since $\sum_{i=1}^{q} \dot{x}_i(t) = 0$, $t \geq t_0$, it follows that $x_i(t) \rightrightarrows \frac{1}{q} \sum_{i=1}^{q} x_{i0}$ as $t \to \infty$, $i = 1, \ldots, q$. $\qquad \square$

Note that Example 6.5 serves as a special case of Theorem 6.13.

# Chapter Seven

---

# Robust Network Consensus Protocols

## 7.1 Introduction

Even though many consensus protocol algorithms have been developed over the last several years in the literature (see [162, 166, 180, 203, 220, 228, 234, 238, 289] and the numerous references therein), and some robustness issues have been considered [8, 41, 43, 86, 116, 227, 238], robustness properties of these algorithms involving nonlinear dynamics have been largely ignored. Robustness here refers to sensitivity of the control algorithm achieving semistability and consensus in the face of model uncertainty. In this chapter, we build on the results of [162, 166] as well as Chapters 2 and 4 to examine the robustness of several control algorithms for network consensus protocols with information model uncertainty of a specified structure.

In particular, we develop sufficient conditions for robust stability of control protocol functions involving higher-order perturbation terms that scale in a consistent fashion with respect to a scaling operation on an underlying space with the additional property that the protocol functions can be written as a sum of functions, each homogeneous with respect to a fixed scaling operation, that retain system semistability and consensus. In addition, control protocol functions containing higher-order perturbation terms involving a thermodynamic information structure are also explored. Unlike results in this chapter, [8, 207] do not consider the effect of higher-order perturbation terms appearing in the control functions. In this sense, our work complements the work reported in [8, 207].

## 7.2 Mathematical Preliminaries

In this chapter, we consider nonlinear dynamical systems of the form

$$\dot{x}(t) = f(x(t)), \quad x(0) = x_0, \quad t \in \mathcal{I}_{x_0}, \tag{7.1}$$

where $x(t) \in \mathcal{D} \subseteq \mathbb{R}^n$, $t \in \mathcal{I}_{x_0}$, is the system state vector; $\mathcal{D}$ is an open set; $f : \mathcal{D} \to \mathbb{R}^n$ is continuous on $\mathcal{D}$; $f^{-1}(0) \triangleq \{x \in \mathcal{D} : f(x) = 0\}$ is nonempty;

and $\mathcal{I}_{x_0} = [0, \tau_{x_0})$, $0 \le \tau_{x_0} \le \infty$, is the maximal interval of existence for the solution $x(\cdot)$ of (7.1). A continuously differentiable function $x : \mathcal{I}_{x_0} \to \mathcal{D}$ is said to be a *solution* of (7.1) on the interval $\mathcal{I}_{x_0} \subset \mathbb{R}$ if $x$ satisfies (7.1) for all $t \in \mathcal{I}_{x_0}$. The continuity of $f$ implies that, for every $x_0 \in \mathcal{D}$, there exist $\tau_0 < 0 < \tau_1$ and a solution $x(\cdot)$ of (7.1) defined on $(\tau_0, \tau_1)$ such that $x(0) = x_0$.

A solution $x$ is said to be *right maximally defined* if $x$ cannot be extended on the right (either uniquely or nonuniquely) to a solution of (7.1). Here, we assume that for every initial condition $x_0 \in \mathcal{D}$, (7.1) has a unique right maximally defined solution, and this unique solution is defined on $[0, \infty)$. Furthermore, we assume that $f(\cdot)$ is locally Lipschitz continuous on $\mathcal{D} \backslash f^{-1}(0)$. Note that the local Lipschitzness of $f(\cdot)$ on $\mathcal{D} \backslash f^{-1}(0)$ implies local uniqueness in forward and backward time for nonequilibrium initial states.

As noted in Chapter 4, under these assumptions on $f$, the solutions of (7.1) define a continuous global semiflow on $\mathcal{D}$; that is, $s : [0, \infty) \times \mathcal{D} \to \mathcal{D}$ is a jointly continuous function satisfying the consistency property $s(0, x) = x$ and the semigroup property $s(t, s(\tau, x)) = s(t + \tau, x)$ for every $x \in \mathcal{D}$ and $t, \tau \in [0, \infty)$. Given $t \in [0, \infty)$, we denote the flow $s(t, \cdot) : \mathcal{D} \to \mathcal{D}$ of (7.1) by $s_t(x_0)$ or $s_t$.

Given a continuous function $V : \mathcal{D} \to \mathbb{R}$, the *upper right Dini derivative* of $V$ along the solution of (7.1) is defined by

$$\dot{V}(s(t, x)) \triangleq \limsup_{h \to 0^+} \frac{1}{h}[V(s(t + h, x)) - V(s(t, x))]. \tag{7.2}$$

It is easy to see that $\dot{V}(x_e) = 0$ for every $x_e \in f^{-1}(0)$. In addition, note that $\dot{V}(x) = \dot{V}(s(0, x))$. Finally, if $V(\cdot)$ is continuously differentiable, then $\dot{V}(x) = V'(x)f(x)$.

In the following, we will need to consider a complete vector field $\nu$ on $\mathbb{R}^n$, that is, a vector field $\nu$ such that the solutions of the differential equation $\dot{y}(t) = \nu(y(t))$ define a continuous *global flow* $\psi : \mathbb{R} \times \mathbb{R}^n \to \mathbb{R}^n$ on $\mathbb{R}^n$, where $\nu^{-1}(0) = f^{-1}(0)$. For each $\tau \in \mathbb{R}$, the map $\psi_\tau(\cdot) = \psi(\tau, \cdot)$ is a homeomorphism and $\psi_\tau^{-1} = \psi_{-\tau}$. Our assumptions imply that every connected component of $\mathbb{R}^n \backslash f^{-1}(0)$ is invariant under $\nu$.

Recall that a function $V : \mathbb{R}^n \to \mathbb{R}$ is said to be *homogeneous of degree* $l \in \mathbb{R}$ *with respect to* $\nu$ if and only if

$$(V \circ \psi_\tau)(x) = e^{l\tau}V(x), \quad \tau \in \mathbb{R}, \quad x \in \mathbb{R}^n. \tag{7.3}$$

Note that if $l \ne 0$, then it follows from (7.3) that $V(x) = 0$ if $x \in \nu^{-1}(0)$.

The following proposition provides a useful comparison between positive definite homogeneous functions with respect to an equilibrium set.

**Proposition 7.1.** Assume that $V_1(\cdot)$ and $V_2(\cdot)$ are continuous real-valued functions on $\mathbb{R}^n$, homogeneous with respect to $\nu$ of degrees $l_1 > 0$ and $l_2 > 0$, respectively, and $V_1(\cdot)$ satisfies $V_1(x) > 0$ for $x \in \mathbb{R}^n \backslash \nu^{-1}(0)$. Then, for each $x_e \in \nu^{-1}(0)$ and each bounded open neighborhood $\mathcal{D}_0$ containing $x_e$, there exist $c_1 = c_1(\mathcal{D}_0) \in \mathbb{R}$ and $c_2 = c_2(\mathcal{D}_0) \in \mathbb{R}$, where $c_2 \geq c_1$, such that

$$c_1 (V_1(x))^{\frac{l_2}{l_1}} \leq V_2(x) \leq c_2 (V_1(x))^{\frac{l_2}{l_1}}, \quad x \in \mathcal{D}_0. \tag{7.4}$$

If, in addition, $V_2(x) < 0$ for $x \in \mathbb{R}^n \backslash \nu^{-1}(0)$, then $c_1$ and $c_2$ in (7.4) may be chosen to additionally satisfy $c_1 \leq c_2 < 0$.

**Proof.** Let $x_e \in \nu^{-1}(0)$, and choose a bounded open neighborhood $\mathcal{D}_0$ of $x_e$. Let $\mathcal{Q} = \psi(\mathbb{R}_+ \times \mathcal{D}_0)$. For every $\varepsilon > 0$, define $\mathcal{Q}_\varepsilon = \mathcal{Q} \cap V_1^{-1}(\varepsilon)$; define the continuous map $\tau_\varepsilon : \mathbb{R}^n \backslash \nu^{-1}(0) \to \mathbb{R}$ by $\tau_\varepsilon(x) \triangleq l^{-1} \ln(\varepsilon / V_1(x))$; and note that, for every $x \in \mathbb{R}^n \backslash \nu^{-1}(0)$, $\psi(t, x) \in V_1^{-1}(\varepsilon)$ if and only if $t = \tau_\varepsilon(x)$. Next, define $\beta_\varepsilon : \mathbb{R}^n \backslash \nu^{-1}(0) \to \mathbb{R}^n$ by $\beta_\varepsilon \triangleq \psi(\tau_\varepsilon(x), x)$. Note that, for every $\varepsilon > 0$, $\beta_\varepsilon$ is continuous, and $\beta_\varepsilon(x) \in V_1^{-1}(\varepsilon)$ for every $x \in \mathbb{R}^n \backslash \nu^{-1}(0)$.

Consider $\varepsilon > 0$. $\mathcal{Q}_\varepsilon$ is the union of the images of connected components of $\mathcal{D}_0 \backslash \nu^{-1}(0)$ under the continuous map $\beta_\varepsilon$. Since every connected component of $\mathbb{R}^n \backslash \nu^{-1}(0)$ is invariant under $-\nu$, it follows that the image of each connected component $\mathcal{U}$ of $\mathbb{R}^n \backslash \nu^{-1}(0)$ under $\beta_\varepsilon$ is contained in $\mathcal{U}$. In particular, the images of the connected components of $\mathcal{D}_0 \backslash \nu^{-1}(0)$ under $\beta_\varepsilon$ are all disjoint. Thus, each connected component of $\mathcal{Q}_\varepsilon$ is the image of exactly one connected component of $\mathcal{D}_0 \backslash \nu^{-1}(0)$ under $\beta_\varepsilon$. Finally, if $\varepsilon$ is small enough so that $V_1^{-1}(\varepsilon) \cap \mathcal{D}_0$ is nonempty, then $V_1^{-1}(\varepsilon) \cap \mathcal{D}_0 \subseteq \mathcal{Q}_\varepsilon$, and, hence, every connected component of $\mathcal{Q}_\varepsilon$ has a nonempty intersection with $\mathcal{D}_0 \backslash \nu^{-1}(0)$.

We claim that $\mathcal{Q}_\varepsilon$ is bounded for every $\varepsilon > 0$. It is easy to verify that, for every $\varepsilon_1, \varepsilon_2 \in (0, \infty)$, $\mathcal{Q}_{\varepsilon_2} = \psi_h(\mathcal{Q}_{\varepsilon_1})$ with $h = l^{-1} \ln(\varepsilon_2 / \varepsilon_1)$. Hence, it suffices to prove that there exists $\varepsilon > 0$ such that $\mathcal{Q}_\varepsilon$ is bounded. To arrive at a contradiction, suppose, *ad absurdum*, that $\mathcal{Q}_\varepsilon$ is unbounded for every $\varepsilon > 0$. Choose a bounded open neighborhood $\mathcal{V}$ of $\overline{\mathcal{D}_0}$ and a sequence $\{\varepsilon_i\}_{i=1}^{\infty}$ in $(0, \infty)$ converging to zero. By our assumption, for every $i = 1, 2, \ldots,$ at least one connected component of $\mathcal{Q}_{\varepsilon_i}$ must contain a point in $\mathbb{R}^n \backslash \mathcal{V}$.

On the other hand, for $i$ sufficiently large, every connected component of $\mathcal{Q}_{\varepsilon_i}$ has a nonempty intersection with $\mathcal{D}_0 \subset \mathcal{V}$. It follows that $\mathcal{Q}_{\varepsilon_i}$ has a nonempty intersection with the boundary of $\mathcal{V}$ for every $i$ sufficiently large. Hence, there exist a sequence $\{x_i\}_{i=1}^{\infty}$ in $\mathcal{D}_0$ and a sequence $\{t_i\}_{i=1}^{\infty}$ in $(0, \infty)$

such that $y_i \triangleq \psi_{t_i}(x_i) \in V_1^{-1}(\varepsilon_i) \cap \partial \mathcal{V}$ for every $i = 1, 2, \ldots$. Since $\mathcal{V}$ is bounded, we can assume that the sequence $\{y_i\}_{i=1}^{\infty}$ converges to $y \in \partial \mathcal{V}$. Continuity implies that $V_1(y) = \lim_{i \to \infty} V_1(y_i) = \lim_{i \to \infty} \varepsilon_i = 0$. Since $V_1^{-1}(0) = \nu^{-1}(0)$, it follows that $y$ is Lyapunov stable under $-\nu$. Since $y \notin \overline{\mathcal{D}}_0$, there exists an open neighborhood $\mathcal{W}$ of $y$ such that $\mathcal{W} \cap \mathcal{D}_0 = \emptyset$. The sequence $\{y_i\}_{i=1}^{\infty}$ converges to $y$, while $\psi_{-t_i}(y_i) = x_i \in \mathcal{D}_0 \subset \mathbb{R}^n \backslash \mathcal{W}$, which contradicts Lyapunov stability. This contradiction implies that there exists $\varepsilon > 0$ such that $\mathcal{Q}_\varepsilon$ is bounded. It now follows that $\mathcal{Q}_\varepsilon$ is bounded for every $\varepsilon > 0$.

Finally, consider $x \in \mathcal{D}_0 \backslash \nu^{-1}(0)$, choose $\varepsilon > 0$, and note that

$$\psi_{\tau_\varepsilon(x)}(x) \in \mathcal{Q}_\varepsilon.$$

Furthermore, note that $V_2(x)$ is continuous on $x \in \mathbb{R}^n \backslash \nu^{-1}(0)$ and $\overline{\mathcal{Q}}_\varepsilon \cap \nu^{-1}(0) = \emptyset$. Then, by homogeneity, $V_1(\psi_{\tau_\varepsilon(x)}(x)) = \varepsilon$, and, hence,

$$\min_{z \in \overline{\mathcal{Q}}_\varepsilon} V_2(z) \leq V_2(\psi_{\tau_\varepsilon(x)}(x)) \leq \max_{z \in \overline{\mathcal{Q}}_\varepsilon} V_2(z). \tag{7.5}$$

Since $V_2(\psi_{\tau_\varepsilon(x)}(x))$ is homogeneous of degree $l_2$, it follows that

$$V_2(\psi_{\tau_\varepsilon(x)}(x)) = e^{l_2 \tau_\varepsilon(x)} V_2(x) = \varepsilon^{-\frac{l_2}{l_1}} (V_1(x))^{-\frac{l_2}{l_1}} V_2(x).$$

Let $c_1 \triangleq \varepsilon^{\frac{l_2}{l_1}} \min_{z \in \overline{\mathcal{Q}}_\varepsilon} V_2(z)$ and $c_2 \triangleq \varepsilon^{\frac{l_2}{l_1}} \max_{z \in \overline{\mathcal{Q}}_\varepsilon} V_2(z)$. Note that $c_1$ and $c_2$ are well defined, and, hence, the first assertion is proved. Finally, if $V_2(x) < 0$ for $x \in \mathbb{R}^n \backslash \nu^{-1}(0)$, then it follows from the definitions of $c_1$ and $c_2$ that $c_1 \leq c_2 < 0$.                                                       $\square$

Recall that the *Lie derivative* of a continuous function $V : \mathbb{R}^n \to \mathbb{R}$ with respect to $\nu$ is given by

$$L_\nu V(x) \triangleq \lim_{t \to 0^+} \frac{1}{t} [V(\psi(t, x)) - V(x)] \tag{7.6}$$

whenever the limit on the right-hand side exists. If $V$ is a continuous homogeneous function of degree $l > 0$, then $L_\nu V$ is defined everywhere and satisfies $L_\nu V = lV$. We assume that the vector field $\nu$ is a *semi-Euler vector field*; that is, the dynamical system

$$\dot{y}(t) = -\nu(y(t)), \quad y(0) = y_0, \quad t \geq 0, \tag{7.7}$$

is globally semistable. Thus, for each $x \in \mathbb{R}^n$, $\lim_{\tau \to \infty} \psi(-\tau, x) = x^* \in \nu^{-1}(0)$, and for each $x_e \in \nu^{-1}(0)$, there exists $z \in \mathbb{R}^n$ such that $x_e = \lim_{\tau \to \infty} \psi(-\tau, z)$. If $\nu^{-1}(0) = \{0\}$, then the semi-Euler vector field becomes the *Euler vector field* given in [37]. Finally, recall that the vector field $f$ is *homogeneous of degree* $k \in \mathbb{R}$ *with respect to* $\nu$ if and only if $\nu^{-1}(0) = f^{-1}(0)$ and, for every $t \in \overline{\mathbb{R}}_+$ and $\tau \in \mathbb{R}$,

$$s_t \circ \psi_\tau = \psi_\tau \circ s_{e^{k\tau}t}. \tag{7.8}$$

As discussed in Chapter 4, if $V : \mathbb{R}^n \to \mathbb{R}$ is a homogeneous function of degree $l$ such that $L_f V(x)$ is defined everywhere, then $L_f V(x)$ is a homogeneous function of degree $l + k$. Finally, recall that if $\nu$ and $f$ are continuously differentiable in a neighborhood of $x \in \mathbb{R}^n$, then (7.8) holds at $x$ for sufficiently small $t$ and $\tau$ if and only if $[\nu, f](x) = kf(x)$ in a neighborhood of $x \in \mathbb{R}^n$, where the Lie bracket $[\nu, f]$ of $\nu$ and $f$ can be computed using $[\nu, f] = \frac{\partial f}{\partial x}\nu - \frac{\partial \nu}{\partial x}f$.

## 7.3 Semistability and Homogeneous Dynamical Systems

Homogeneity of dynamical systems is a property whereby system vector fields scale in relation to a scaling operation or *dilation* on the state space. In this section, we present a robustness result for a vector field that can be written as a sum of several vector fields, each of which is homogeneous with respect to a certain fixed dilation. First, however, we present a result that shows that a semistable homogeneous system admits a homogeneous Lyapunov function. This is a weaker version of Theorem 6.2 of [37], which considers asymptotically stable homogeneous systems.

**Theorem 7.1.** Suppose $f : \mathbb{R}^n \to \mathbb{R}^n$ is homogeneous of degree $k \in \mathbb{R}$ with respect to $\nu$ and (7.1) is semistable under $f$. Then, for every $l > \max\{-k, 0\}$, there exists a continuous nonnegative function $V : \mathbb{R}^n \to \overline{\mathbb{R}}_+$ that is homogeneous of degree $l$ with respect to $\nu$, and continuously differentiable on $\mathbb{R}^n \backslash f^{-1}(0)$, where $V^{-1}(0) = f^{-1}(0)$, and $V'(x)f(x) < 0$ for $x \in \mathbb{R}^n \backslash f^{-1}(0)$.

**Proof.** This is a restatement of Theorem 4.5. $\qquad \square$

Next, we state the main theorem of this section, which involves a robustness result of a vector field that can be written as a sum of several vector fields.

**Theorem 7.2.** Let $f = g_1 + \cdots + g_p$, where, for each $i = 1, \ldots, p$, the vector field $g_i$ is continuous and homogeneous of degree $m_i$ with respect to $\nu$, and $m_1 < m_2 < \cdots < m_p$. If every equilibrium point in $g_1^{-1}(0)$ is semistable under $g_1$ and Lyapunov stable under $f$, then every equilibrium point in $g_1^{-1}(0)$ is semistable under $f$.

**Proof.** Let every point in $g_1^{-1}(0)$ be a semistable equilibrium under $g_1$. Choose $l > \max\{-m_1, 0\}$. Then it follows from Theorem 7.1 that there exists a continuous homogeneous function $V : \mathbb{R}^n \to \mathbb{R}$ of degree $l$ such that $V(x) = 0$ for $x \in g_1^{-1}(0)$, $V(x) > 0$ for $x \in \mathbb{R}^n \backslash g_1^{-1}(0)$, and $L_{g_1}V$ satisfies $L_{g_1}V(x) = 0$ for $x \in g_1^{-1}(0)$ and $L_{g_1}V(x) < 0$ for $x \in \mathbb{R}^n \backslash g_1^{-1}(0)$. For each $i \in \{1, \ldots, p\}$, $L_{g_i}V$ is continuous and homogeneous of degree $l + m_i > 0$ with respect to $\nu$. Let $x_e \in g_1^{-1}(0)$ and $\mathcal{U}$ be a bounded neighborhood of $x_e$. Then it follows from Proposition 7.1 and Theorem 7.1 that there exist $c_1 > 0$, $c_2, \ldots, c_p \in \mathbb{R}$ such that

$$L_{g_i}V(x) \leq -c_i(V(x))^{\frac{l+m_i}{l}}, \quad x \in \mathcal{U}, \quad i = 1, \ldots, p. \qquad (7.9)$$

Hence, for every $x \in \mathcal{U}$,

$$L_f V(x) \leq -\sum_{i=1}^{p} c_i(V(x))^{\frac{l+m_i}{l}} = (V(x))^{\frac{l+m_1}{l}}(-c_1 + U(x)), \qquad (7.10)$$

where $U(x) \triangleq -\sum_{i=2}^{p} c_i(V(x))^{\frac{m_i-m_1}{l}}$.

Since $m_i - m_1 > 0$ for every $i \geq 2$, it follows that the function $U(\cdot)$, which takes the value zero on the set $g_1^{-1}(0) \cap \mathcal{U}$, is continuous. Hence, there exists an open neighborhood $\mathcal{V} \subseteq \mathcal{U}$ of $x_e$ such that $U(x) < c_1/2$ for all $x \in \mathcal{V}$. Now, it follows from (7.10) that

$$L_f V(x) \leq -\frac{c_1}{2}(V(x))^{\frac{l+m_1}{l}}, \quad x \in \mathcal{V}. \qquad (7.11)$$

Since $x_e$ is Lyapunov stable, it follows that one can find a bounded neighborhood $\mathcal{W}$ of $x_e$ such that solutions in $\mathcal{W}$ remain in $\mathcal{V}$. Take an initial condition in $\mathcal{W}$. Since the solution is bounded (remains in $\mathcal{U}$), it follows from the Krasovskii–LaSalle invariance theorem that this solution converges to its compact positive limit set in $f^{-1}(0)$. Since all points in $f^{-1}(0)$ are Lyapunov stable, it follows from Proposition 5.4 of [36] that the positive limit set is a singleton involving a Lyapunov stable equilibrium in $f^{-1}(0)$. Since $x_e$ was chosen arbitrarily, it follows that all equilibria in $g_1^{-1}(0)$ are semistable.                                                                     $\square$

## 7.4 Robust Control Algorithms for Network Consensus Protocols

In this section, we apply the results of [162] and the results of Section 7.3 to develop sufficient conditions for robust stability of protocol consensus for dynamical networks [217, 238, 315]. In particular, using the thermodynamically motivated information consensus framework for multiagent nonlinear systems that achieve semistability and consensus developed in Chapters 2 and 4, we develop sufficient conditions for robust stability of control protocol functions involving higher-order perturbation terms.

These higher-order terms involve control functions that scale in a consistent fashion with respect to a scaling operation on an underlying space with the additional property that the control functions can be written as a sum of homogeneous functions with respect to a fixed scaling operation. In addition, we develop control protocol functions containing higher-order perturbation terms involving thermodynamic information structures.

Consider $q$ continuous-time integrator agents with dynamics

$$\dot{x}_i(t) = u_i(t), \quad x_i(0) = x_{i0}, \quad t \geq 0, \tag{7.12}$$

where, for each $i \in \{1, \ldots, q\}$, $x_i(t) \in \mathbb{R}$ denotes the information state and $u_i(t) \in \mathbb{R}$ denotes the information control input for all $t \geq 0$. The consensus protocol is given by

$$u_i(t) = f_i(x(t)) = \sum_{j=1, j \neq i}^{q} \phi_{ij}(x_i(t), x_j(t)), \tag{7.13}$$

where $\phi_{ij}(\cdot, \cdot)$ satisfies the conditions in Theorem 7.3 below. Note that (7.12) and (7.13) describe an interconnected network where information states are updated using a distributed controller involving neighbor-to-neighbor interaction between agents. Hence, the consensus problem involves the trajectories of the dynamical network characterized by the multiagent dynamical system $\mathcal{G}$ given by

$$\dot{x}_i(t) = \sum_{j=1, j \neq i}^{q} \phi_{ij}(x_i(t), x_j(t)), \quad x_i(0) = x_{i0}, \quad t \geq 0, \quad i = 1, \ldots, q, \tag{7.14}$$

or, in vector form,

$$\dot{x}(t) = f(x(t)), \quad x(0) = x_0, \quad t \geq 0, \tag{7.15}$$

where $x(t) \triangleq [x_1(t), \ldots, x_q(t)]^{\mathrm{T}}$, $t \geq 0$, and $f = [f_1, \ldots, f_q]^{\mathrm{T}} : \mathcal{D} \to \mathbb{R}^q$ is such that

$$f_i(x) = \sum_{j=1, j \neq i}^{q} \phi_{ij}(x_i, x_j), \tag{7.16}$$

where $\mathcal{D} \subseteq \mathbb{R}^q$ is open. This nonlinear model is proposed in Chapter 2.

Next, define

$$\mathcal{C}_{(i,j)} \triangleq \begin{cases} 0 & \text{if } \phi_{ij}(x_i, x_j) \equiv 0, \\ 1, & \text{otherwise,} \end{cases} \quad i \neq j, \quad i, j = 1, \ldots, q, \tag{7.17}$$

$$\mathcal{C}_{(i,i)} = -\sum_{k=1,\,k\neq i}^{q} \mathcal{C}_{(i,k)}, \quad i = 1, \ldots, q. \tag{7.18}$$

**Theorem 7.3.** Consider the multiagent dynamical system (7.15), and assume that Assumptions 2.1 and 2.2 hold with $\mathcal{A}_{(i,j)} = \mathcal{C}_{(i,j)}$, $i, j = 1, \ldots, q$. Then the following statements hold.

(i) Assume that $\phi_{ij}(x_i, x_j) = -\phi_{ji}(x_j, x_i)$ for all $i, j = 1, \ldots, q$, $i \neq j$. Then, for every $\alpha \in \mathbb{R}$, $\alpha\mathbf{e}$ is a semistable equilibrium state of (7.15). Furthermore, $x(t) \to \frac{1}{q}\mathbf{e}\mathbf{e}^{\mathrm{T}}x_0$ as $t \to \infty$, and $\frac{1}{q}\mathbf{e}\mathbf{e}^{\mathrm{T}}x_0$ is a semistable equilibrium state.

(ii) Let $\phi_{ij}(x_i, x_j) = \mathcal{C}_{(i,j)}[\sigma(x_j) - \sigma(x_i)]$ for all $i, j = 1, \ldots, q$, $i \neq j$, where $\sigma(0) = 0$ and $\sigma(\cdot)$ is strictly increasing, and assume that $\mathcal{C}^{\mathrm{T}}\mathbf{e} = 0$. Then, for every $\alpha \in \mathbb{R}$, $\alpha\mathbf{e}$ is a semistable equilibrium state of (7.15). Furthermore, $x(t) \to \frac{1}{q}\mathbf{e}\mathbf{e}^{\mathrm{T}}x_0$ as $t \to \infty$, and $\frac{1}{q}\mathbf{e}\mathbf{e}^{\mathrm{T}}x_0$ is a semistable equilibrium state.

**Proof.** This is a restatement of Theorem 2.1 with $\mathcal{A}_{(i,j)} = \mathcal{C}_{(i,j)}$, $i, j = 1, \ldots, q$. $\square$

Note that the assumption

$$\phi_{ij}(x_i, x_j) = -\phi_{ji}(x_j, x_i), \quad i, j = 1, \ldots, q, \quad i \neq j,$$

in (i) of Theorem 7.3 implies that $\mathcal{C} = \mathcal{C}^{\mathrm{T}}$, and, hence, the underlying graph topology for the multiagent system $\mathcal{G}$ given by (7.12) and (7.13) is undirected. Furthermore, since $\phi_{ij}(x_i, x_j)$ is not restricted to a specified structure, the consensus protocol algorithm is not restricted to a particular reference. Alternatively, in (ii) of Theorem 7.3 the assumption $\mathcal{C}^{\mathrm{T}}\mathbf{e} = 0$ implies that the underlying directed graph of $\mathcal{G}$ is balanced. To see this, recall that for a directed graph $\mathfrak{G}$, $\mathcal{A}\mathbf{e} = \mathcal{A}^{\mathrm{T}}\mathbf{e}$ implies that $\mathfrak{G}$ is balanced. Since $\mathcal{C} = \mathcal{A} - \mathcal{N}$, where $\mathcal{A}$ denotes the normalized adjacency matrix and $\mathcal{N} \triangleq \mathrm{diag}\left[\sum_{j=1}^{q} \mathcal{A}_{(1,j)}, \ldots, \sum_{j=1}^{q} \mathcal{A}_{(q,j)}\right] \in \mathbb{R}^{q \times q}$, it follows that $\mathcal{A}\mathbf{e} = \mathcal{A}^{\mathrm{T}}\mathbf{e}$ if and only if $\mathcal{C}\mathbf{e} = \mathcal{C}^{\mathrm{T}}\mathbf{e}$. Hence, $\mathcal{C}^{\mathrm{T}}\mathbf{e} = 0$ implies that $\mathfrak{G}$ is balanced.

Theorem 7.3 implies that the steady-state value of the information state in each agent $\mathcal{G}_i$ of the multiagent dynamical system $\mathcal{G}$ is equal; that is, the steady-state value of the multiagent dynamical system $\mathcal{G}$ given by

$$x_\infty = \frac{1}{q}\mathbf{e}\mathbf{e}^{\mathrm{T}}x_0 = \left[\frac{1}{q}\sum_{i=1}^{q} x_{i0}\right]\mathbf{e} \tag{7.19}$$

is uniformly distributed over all multiagents of $\mathcal{G}$.

Next, consider (7.12) and (7.13), and assume that the vector field $f = [f_1, \ldots, f_q]$ is homogeneous of degree $k \in \mathbb{R}$ with respect to $\nu$. Finally, consider the generalized (or perturbed) consensus protocol architecture

$$\dot{z}_i(t) = \sum_{j=1, j \neq i}^{q} \phi_{ij}(z_i(t), z_j(t)) + \Delta_i(z), \quad z_i(0) = z_{i0},$$

$$i = 1, \ldots, q, \quad t \geq 0, \tag{7.20}$$

where $\Delta = [\Delta_1, \ldots, \Delta_q]^{\mathrm{T}} : \mathbb{R}^q \to \mathbb{R}$ is a continuous function such that $\Delta$ is homogeneous of degree $l \in \mathbb{R}$ with respect to $\nu$ and (7.20) possesses unique solutions in forward time for initial conditions in $\mathbb{R}^q \backslash \{\alpha \mathbf{e} : \alpha \in \mathbb{R}\}$.

**Theorem 7.4.** Consider the nominal consensus protocol (7.12)–(7.13) and the generalized consensus protocol (7.20). If $\{\alpha \mathbf{e} : \alpha \in \mathbb{R}\} = \Delta^{-1}(0)$, every equilibrium point in $\{\alpha \mathbf{e} : \alpha \in \mathbb{R}\}$ is a Lyapunov stable equilibrium of (7.20), and $k < l$, then every equilibrium point in $\{\alpha \mathbf{e} : \alpha \in \mathbb{R}\}$ is a semistable equilibrium of (7.12)–(7.13) and (7.20).

**Proof.** It follows from Proposition 2.1 that, for every $\alpha \in \mathbb{R}$, $\alpha \mathbf{e}$ is an equilibrium point of (7.12) and (7.13). Next, it follows from Theorem 7.3 that $\alpha \mathbf{e}$ is a semistable equilibrium state of (7.12) and (7.13). Now, the result is a direct consequence of Theorem 7.2.                     $\square$

As a special case of Theorem 7.4, consider the nominal linear consensus protocol given by

$$\dot{x}_i(t) = \sum_{j=1, j \neq i}^{q} \mathcal{C}_{(i,j)}[x_j(t) - x_i(t)], \quad x_i(0) = x_{i0}, \quad i = 1, \ldots, q, \quad t \geq 0,$$

$$\tag{7.21}$$

where, for each $i \in \{1, \ldots, q\}$, $x_i \in \mathbb{R}$, $\mathcal{C}$ satisfies Assumption 2.1, and $\mathcal{C}^{\mathrm{T}} = \mathcal{C}$. Next, consider the generalized consensus protocol given by

$$\dot{z}_i(t) = \sum_{j=1, j \neq i}^{q} \mathcal{C}_{(i,j)}[z_j(t) - z_i(t)] + \sum_{j=1, j \neq i}^{q} \delta_{ij}(z_j(t) - z_i(t)),$$

$$z_i(0) = z_{i0}, \quad i = 1, \ldots, q, \quad t \geq 0, \tag{7.22}$$

and assume that $\delta_{ij} : \mathbb{R} \to \mathbb{R}$ is continuously differentiable and satisfies $\delta_{ij} \equiv 0$ if $\mathcal{C}_{(i,j)} = 0$, $\delta_{ij}(\lambda z) = \lambda^{1+r} \delta_{ij}(z)$ for all $\lambda > 0$ and for some $r \geq 0$, and $\delta_{ij}(z) = -\delta_{ji}(-z)$ for $z \in \mathbb{R}$ and $i, j = 1, \ldots, q$, $i \neq j$. Finally, let $\Delta = [\Delta_1, \ldots, \Delta_q]^{\mathrm{T}}$, where $\Delta_i = \sum_{j=1, j \neq i}^{q} \delta_{ij}(z_j - z_i)$, $i = 1, \ldots, q$.

**Proposition 7.2.** For $i, j = 1, \ldots, q$, $i \neq j$, let $\delta_{ij} : \mathbb{R} \to \mathbb{R}$ be continuously differentiable such that $\delta_{ij} \equiv 0$ if $\mathcal{C}_{(i,j)} = 0$, $\delta_{ij}(\lambda z) = \lambda^{1+r}\delta_{ij}(z)$ for all $\lambda > 0$ and some $r \geq 0$, and $\delta_{ij}(z) = -\delta_{ji}(-z)$ for all $z \in \mathbb{R}$. Furthermore, let $\Delta = [\Delta_1, \ldots, \Delta_q]^{\mathrm{T}}$, where $\Delta_i = \sum_{j=1, j \neq i}^{q} \delta_{ij}(z_j - z_i)$, $i = 1, \ldots, q$. Then $\Delta$ is homogeneous of degree $qr$ with respect to the semi-Euler vector field

$$\nu(x) = -\sum_{i=1}^{q}\left[\sum_{j=1, j \neq i}^{q}(x_j - x_i)\right]\frac{\partial}{\partial x_i}.$$

**Proof.** First, note that the Lie bracket of

$$\nu(x) = -\sum_{i=1}^{q}\left[\sum_{j=1, j \neq i}^{q}(x_j - x_i)\right]\frac{\partial}{\partial x_i}$$

and the vector field $\Delta$ is given by

$$[\nu, \Delta] = \left[\sum_{i=1}^{q}\frac{\partial \Delta_1}{\partial x_i}\nu_i - \sum_{i=1}^{q}\frac{\partial \nu_1}{\partial x_i}\Delta_i, \ldots, \sum_{i=1}^{q}\frac{\partial \Delta_q}{\partial x_i}\nu_i - \sum_{i=1}^{q}\frac{\partial \nu_q}{\partial x_i}\Delta_i\right]^{\mathrm{T}}.$$

Now, it follows from (7.8) and the assumptions on $\delta_{ij}$ that $\Delta_i$, $i = 1, \ldots, q$, is homogeneous of degree $r$ with respect to the standard dilation of the form $\Delta_\lambda(x_1, \ldots, x_q) = (\lambda x_1, \ldots, \lambda x_q)$ or, equivalently, the Euler vector field $\tilde{\nu}(x) = x_1\frac{\partial}{\partial x_1} + \cdots + x_q\frac{\partial}{\partial x_q}$ [37]. Hence, $[\tilde{\nu}, \Delta_i] = r\Delta_i$, $i = 1, \ldots, q$, or, equivalently,

$$\sum_{i=1}^{q}\frac{\partial \Delta_j}{\partial x_i}x_i = (r+1)\Delta_j, \quad j = 1, \ldots, q. \tag{7.23}$$

Next, note that

$$\nu_i = -\sum_{j=1, j \neq i}^{q}(x_j - x_i) = qx_i - \sum_{j=1}^{q}x_j, \quad i = 1, \ldots, q, \tag{7.24}$$

and

$$\sum_{i=1}^{q}\frac{\partial \Delta_j}{\partial x_i} = \sum_{i=1}^{q}\sum_{s=1, s \neq j}^{q}\frac{\partial \delta_{js}(x_s - x_j)}{\partial x_i} = 0, \quad j = 1, \ldots, q. \tag{7.25}$$

Hence, it follows that

$$\sum_{i=1}^{q} \frac{\partial \Delta_j}{\partial x_i} \nu_i = \sum_{i=1}^{q} \frac{\partial \Delta_j}{\partial x_i} \left( q x_i - \sum_{j=1}^{q} x_j \right)$$

$$= q \sum_{i=1}^{q} \frac{\partial \Delta_j}{\partial x_i} x_i - \left( \sum_{i=1}^{q} \frac{\partial \Delta_j}{\partial x_i} \right) \left( \sum_{j=1}^{q} x_j \right)$$

$$= q(r+1)\Delta_j, \quad j = 1, \ldots, q. \tag{7.26}$$

Alternatively, note that

$$\sum_{i=1}^{q} \Delta_i = \sum_{i=1}^{q} \sum_{j=1, j \neq i}^{q} \delta_{ij}(x_j - x_i) = 0, \tag{7.27}$$

and, hence,

$$\sum_{i=1}^{q} \frac{\partial \nu_j}{\partial x_i} \Delta_i = (q-1)\Delta_j - \sum_{i=1, i \neq j}^{q} \Delta_i$$

$$= q\Delta_j - \sum_{i=1}^{q} \Delta_i$$

$$= q\Delta_j, \quad j = 1, \ldots, q. \tag{7.28}$$

Thus,

$$\sum_{i=1}^{q} \frac{\partial \Delta_j}{\partial x_i} \nu_i - \sum_{i=1}^{q} \frac{\partial \nu_j}{\partial x_i} \Delta_i = qr\Delta_j, \quad j = 1, \ldots, q, \tag{7.29}$$

or, equivalently, $[\nu, \Delta] = qr\Delta$, which implies that the vector field $\Delta$ is homogeneous of degree $qr$ with respect to the semi-Euler vector field $\nu(x) = -\sum_{i=1}^{q}[\sum_{j=1, j \neq i}^{q}(x_j - x_i)]\frac{\partial}{\partial x_i}$. □

**Corollary 7.1.** The vector field of (7.21) is homogeneous of degree $k = 0$ with respect to the semi-Euler vector field

$$\nu(x) = -\sum_{i=1}^{q} \left[ \sum_{j=1, j \neq i}^{q} (x_j - x_i) \right] \frac{\partial}{\partial x_i}.$$

**Proof.** The result is a direct consequence of Proposition 7.2 by setting $r = 0$. □

**Corollary 7.2.** Consider the linear nominal consensus protocol (7.21) and the generalized nonlinear consensus protocol (7.22). Then every equilibrium point in $\{\alpha \mathbf{e} : \alpha \in \mathbb{R}\}$ is a semistable equilibrium of (7.21) and (7.22). Furthermore, $z(t) \to \frac{1}{q}\mathbf{e}\mathbf{e}^{\mathrm{T}}z_0$ as $t \to \infty$, and $\frac{1}{q}\mathbf{e}\mathbf{e}^{\mathrm{T}}z_0$ is a semistable equilibrium state.

**Proof.** It follows from (i) of Theorem 7.3 that $\alpha \mathbf{e}$, $\alpha \in \mathbb{R}$, is a semistable equilibrium of (7.21). Next, it follows from Corollary 7.1 that the right-hand side of (7.21) is homogeneous of degree $k = 0$ with respect to the semi-Euler vector field

$$\nu(x) = -\sum_{i=1}^{q} \left[ \sum_{j=1,j\neq i}^{q} (x_j - x_i) \right] \frac{\partial}{\partial x_i}.$$

To show that every point in $\{\alpha \mathbf{e} : \alpha \in \mathbb{R}\}$ is a Lyapunov stable equilibrium of (7.22), consider the Lyapunov function candidate given by

$$V(z - \alpha \mathbf{e}) = \frac{1}{2}\|z - \alpha \mathbf{e}\|^2.$$

Then it follows that

$$\dot{V}(z - \alpha \mathbf{e})$$
$$= (z - \alpha \mathbf{e})^{\mathrm{T}}\dot{z}$$
$$= \sum_{i=1}^{q}(z_i - \alpha)\sum_{j=1,j\neq i}^{q}\mathcal{C}_{(i,j)}[z_j - z_i] + \sum_{i=1}^{q}(z_i - \alpha)\sum_{j=1,j\neq i}^{q}\delta_{ij}(z_j - z_i)$$
$$= -\sum_{i=1}^{q-1}\sum_{j=i+1}^{q}\mathcal{C}_{(i,j)}[z_i - z_j]^2 + \sum_{i=1}^{q-1}\sum_{j=i+1}^{q}(z_i - z_j)\delta_{ij}(z_j - z_i)$$
$$= -\sum_{i=1}^{q-1}\sum_{j=i+1}^{q}\mathcal{C}_{(i,j)}[z_i - z_j]^2 + \sum_{i=1}^{q-1}\sum_{j=i+1}^{q}\mathcal{C}_{(i,j)}[z_i - z_j]\delta_{ij}(z_j - z_i),$$

$$z \in \mathbb{R}^q. \quad (7.30)$$

Next, since, by homogeneity of $\delta_{ij}$, $\delta_{ij}(\cdot)$ is such that $\lim_{z\to 0}\delta_{ij}(z)/z = 0$, it follows that for every $\gamma > 0$, there exists $\varepsilon_{ij} > 0$ such that $|\delta_{ij}(z)| \leq \gamma|z|$ for all $|z| < \varepsilon_{ij}$. Hence,

$$\sum_{i=1}^{q-1}\sum_{j=i+1}^{q}\mathcal{C}_{(i,j)}[z_i - z_j]\delta_{ij}(z_j - z_i) \leq \sum_{i=1}^{q-1}\sum_{j=i+1}^{q}\gamma\mathcal{C}_{(i,j)}[z_i - z_j]^2,$$

$$|z_i - z_j| < \varepsilon_{ij}. \quad (7.31)$$

Now, choosing $\gamma \leq 1$, it follows from (7.30) and (7.31) that

$$\dot{V}(z - \alpha\mathbf{e}) \leq -\sum_{i=1}^{q-1}\sum_{j=i+1}^{q}(1-\gamma)\mathcal{C}_{(i,j)}[z_i - z_j]^2$$

$$\leq 0, \quad |z_i - z_j| < \varepsilon_{ij}, \tag{7.32}$$

which establishes Lyapunov stability of the equilibrium state $\alpha\mathbf{e}$. The result now follows from Theorem 7.4. $\qquad\square$

It is important to note that Corollary 7.2 still holds for the case where the generalized consensus protocol has the nonlinear form

$$\dot{z}(t) = \mathcal{C}z(t) + \sum_{i=1}^{p}g_i(z(t)), \quad z(0) = z_0, \quad t \geq 0, \tag{7.33}$$

where, for each $i \in \{1, \ldots, q\}$, $g_i(z)$ is homogeneous of degree $l_i > 0$ with respect to $\nu(x) = -\sum_{i=1}^{q}[\sum_{j=1, j\neq i}^{q}(x_j - x_i)]\frac{\partial}{\partial x_i}$ and $l_1 < \cdots < l_p$.

As an application of Corollary 7.2, consider the Kuramoto model [286] given by

$$\dot{x}_1(t) = \sin(x_2(t) - x_1(t)), \quad x_1(0) = x_{10}, \quad t \geq 0, \tag{7.34}$$
$$\dot{x}_2(t) = \sin(x_1(t) - x_2(t)), \quad x_2(0) = x_{20}. \tag{7.35}$$

Note that for sufficiently small $x$, $\sin x$ can be approximated by $x - x^3/3! + \cdots + (-1)^{p-1}x^{2p-1}/(2p-1)!$, where $p$ is a positive integer. The truncated system associated with (7.34) and (7.35) is given by

$$\dot{x}_1 = x_2 - x_1 - \frac{1}{3!}(x_2 - x_1)^3 + \cdots + \frac{(-1)^{p-1}}{(2p-1)!}(x_2 - x_1)^{2p-1}, \tag{7.36}$$

$$\dot{x}_2 = x_1 - x_2 - \frac{1}{3!}(x_1 - x_2)^3 + \cdots + \frac{(-1)^{p-1}}{(2p-1)!}(x_1 - x_2)^{2p-1} \tag{7.37}$$

or, equivalently,

$$\begin{bmatrix} \dot{x}_1 \\ \dot{x}_2 \end{bmatrix} = \begin{bmatrix} -1 & 1 \\ 1 & -1 \end{bmatrix}\begin{bmatrix} x_1 \\ x_2 \end{bmatrix} + \sum_{i=1}^{p-1}g_i(x_1, x_2), \tag{7.38}$$

where

$$g_i(x_1, x_2) \triangleq \frac{(-1)^i}{(2i+1)!}\begin{bmatrix} (x_2 - x_1)^{2i+1} \\ (x_1 - x_2)^{2i+1} \end{bmatrix}, \quad i = 1, \ldots, p-1. \tag{7.39}$$

It can be easily shown that all the conditions of Corollary 7.2 hold for (7.38). Hence, it follows from Corollary 7.2 that every equilibrium point in $\{\alpha[1,1]^{\mathrm{T}} : \alpha \in \mathbb{R}\}$ is a local semistable equilibrium of (7.36) and (7.37), which implies that the equilibrium set $\{\alpha[1,1]^{\mathrm{T}} : \alpha \in \mathbb{R}\}$ of (7.36) and (7.37) has the same stability properties as the linear nominal system

$$\begin{bmatrix} \dot{x}_1 \\ \dot{x}_2 \end{bmatrix} = \begin{bmatrix} -1 & 1 \\ 1 & -1 \end{bmatrix} \begin{bmatrix} x_1 \\ x_2 \end{bmatrix}. \tag{7.40}$$

It should be noted that while our analysis above holds for every $p$, it does not imply that the exact model (7.34) and (7.35) is semistable.

Note that Corollary 7.2 deals with the undirected graph $\mathfrak{G} = (\mathcal{V}, \mathcal{E}, \mathcal{A})$, where $\mathcal{A}$ is a symmetric adjacency matrix. Next, we consider the case where $\mathfrak{G}$ is a directed graph and the control protocol functions involving higher-order perturbation terms are not homogeneous. The following lemma is needed for the next result. First, however, recall that for a diagonal matrix $A \in \mathbb{R}^{q \times q}$, the Drazin inverse $A^{\mathrm{D}} \in \mathbb{R}^{q \times q}$ is given by $A^{\mathrm{D}}_{(i,i)} = 0$ if $A_{(i,i)} = 0$ and $A^{\mathrm{D}}_{(i,i)} = 1/A_{(i,i)}$ if $A_{(i,i)} \neq 0$, $i = 1, \ldots, q$ [26, p. 227].

**Lemma 7.1.** Suppose that $A \in \mathbb{R}^{q \times q}$ and $A_{\mathrm{d}} \in \mathbb{R}^{q \times q}$ satisfy

$$A_{(i,j)} = \begin{cases} C_{(i,i)}, & i = j, \\ 0, & i \neq j, \end{cases} \quad A_{\mathrm{d}(i,j)} = \begin{cases} 0, & i = j, \\ C_{(i,j)}, & i \neq j, \end{cases} \quad i, j = 1, \ldots, q. \tag{7.41}$$

Assume that $C^{\mathrm{T}}\mathbf{e} = 0$. Then, for every $A_{\mathrm{d}i}$, $i = 1, \ldots, n_{\mathrm{d}}$, such that $\sum_{i=1}^{n_{\mathrm{d}}} A_{\mathrm{d}i} = A_{\mathrm{d}}$, there exist nonnegative definite matrices $Q_i \in \mathbb{R}^{q \times q}$, $i = 1, \ldots, n_{\mathrm{d}}$, such that

$$2A + \sum_{i=1}^{n_{\mathrm{d}}} (Q_i + A_{\mathrm{d}i}^{\mathrm{T}} Q_i^{\mathrm{D}} A_{\mathrm{d}i}) \leq 0. \tag{7.42}$$

**Proof.** For each $i \in \{1, \ldots, n_{\mathrm{d}}\}$, let $Q_i$ be the diagonal matrix defined by

$$Q_{i(l,l)} \triangleq \sum_{m=1, m \neq l}^{q} A_{\mathrm{d}i(l,m)}, \quad l = 1, \ldots, q, \tag{7.43}$$

and note that $A + \sum_{i=1}^{n_{\mathrm{d}}} Q_i = 0$, $(A_{\mathrm{d}i} - Q_i)\mathbf{e} = 0$, and $Q_i Q_i^{\mathrm{D}} A_{\mathrm{d}i} = A_{\mathrm{d}i}$, $i = 1, \ldots, n_{\mathrm{d}}$. Hence, $M\mathbf{e} = 0$, where

$$M \triangleq \begin{bmatrix} 2A + \sum_{i=1}^{n_{\mathrm{d}}} Q_i & A_{\mathrm{d}1}^{\mathrm{T}} & A_{\mathrm{d}2}^{\mathrm{T}} & \cdots & A_{\mathrm{d}n_{\mathrm{d}}}^{\mathrm{T}} \\ A_{\mathrm{d}1} & -Q_1 & 0 & \cdots & 0 \\ \vdots & \vdots & \vdots & \vdots & \vdots \\ A_{\mathrm{d}n_{\mathrm{d}}} & 0 & 0 & \cdots & -Q_{n_{\mathrm{d}}} \end{bmatrix}. \tag{7.44}$$

Now, note that $M = M^T$ and $M_{(i,j)} \geq 0$, $i,j = 1,\ldots,q$, $i \neq j$. Hence, by (ii) of Theorem 3.2 in [121], $M$ is semistable, that is, $\mathrm{Re}\,\lambda < 0$, or $\lambda = 0$ and $\lambda$ is semisimple, where $\lambda \in \mathrm{spec}(A)$. Thus, $M \leq 0$, and since $Q_i Q_i^D A_{di} = A_{di}$, $i = 1,\ldots,n_d$, it follows from Proposition 8.2.3 of [26] that $M \leq 0$ if and only if (7.42) holds. $\qquad\square$

**Theorem 7.5.** Consider the linear nominal consensus protocol (7.21), where $\mathcal{C}$ satisfies Assumption 2.1 with $\mathcal{A}_{(i,j)} = \mathcal{C}_{(i,j)}$, $i,j = 1,\ldots,q$; $\mathcal{C}^T\mathbf{e} = 0$; and the generalized nonlinear consensus protocol given by

$$\dot{z}_i(t) = \sum_{j=1,j\neq i}^{q} \mathcal{C}_{(i,j)}[z_j(t) - z_i(t)] + \sum_{j=1,j\neq i}^{q} \mathcal{H}_{(i,j)}[\sigma(z_j(t)) - \sigma(z_i(t))],$$

$$z_i(0) = z_{i0}, \quad i = 1,\ldots,q, \quad t \geq 0, \qquad (7.45)$$

where $\sigma(\cdot)$ satisfies $\sigma(0) = 0$ and $\sigma : \mathbb{R} \to \mathbb{R}$ is strictly increasing, and the matrix $\mathcal{H} = [\mathcal{H}_{(i,j)}]$ satisfies Assumption 2.1; $\mathcal{H}^T\mathbf{e} = 0$; $\mathcal{H}_{(i,j)} = 0$ whenever $\mathcal{C}_{(i,j)} = 0$, $i,j = 1,\ldots,q$, $i \neq j$; and $\mathcal{H} = \mathcal{C} - \mathcal{L}$, where $\mathcal{L}^T = \mathcal{L} \in \mathbb{R}^{q\times q}$. Then every equilibrium point in $\{\alpha\mathbf{e} : \alpha \in \mathbb{R}\}$ is a semistable equilibrium of (7.21) and (7.45). Furthermore, $z(t) \to \frac{1}{q}\mathbf{e}\mathbf{e}^T z_0$ as $t \to \infty$, and $\frac{1}{q}\mathbf{e}\mathbf{e}^T z_0$ is a semistable equilibrium state.

**Proof.** It follows from (ii) of Theorem 7.3 that $\alpha\mathbf{e}$, $\alpha \in \mathbb{R}$, is a semistable equilibrium of (7.21). Next, note that (7.45) can be rewritten as

$$\dot{z}_i(t) = \sum_{j=1,j\neq i}^{q} \mathcal{H}_{(i,j)}[(z_j(t) + \sigma(z_j(t))) - (z_i(t) + \sigma(z_i(t)))]$$

$$+ \sum_{j=1,j\neq i}^{q} \mathcal{L}_{(i,j)}[z_j(t) - z_i(t)], \quad z_i(0) = z_{i0},$$

$$i = 1,\ldots,q, \quad t \geq 0. \qquad (7.46)$$

Define $\hat{\sigma} : \mathbb{R}^q \to \mathbb{R}^q$ by $\hat{\sigma}(z) \triangleq [\sigma(z_1),\ldots,\sigma(z_q)]^T$. Now, for $C \in \mathbb{R}^{q\times q}$ and $C_d \in \mathbb{R}^{q\times q}$ satisfying

$$C_{(i,j)} = \begin{cases} \mathcal{H}_{(i,i)}, & i = j, \\ 0, & i \neq j, \end{cases} \quad C_{d(i,j)} = \begin{cases} 0, & i = j, \\ \mathcal{H}_{(i,j)}, & i \neq j, \end{cases} \quad i,j = 1,\ldots,q,$$

$$(7.47)$$

it follows from Lemma 7.1 that, for every $C_{di}$, $i = 1,\ldots,n_d$, such that $\sum_{i=1}^{n_d} C_{di} = C_d$, there exist nonnegative definite matrices $Q_i \in \mathbb{R}^{q\times q}$, $i = 1,\ldots,q$, such that

$$2C + \sum_{i=1}^{q}(Q_i + C_{di}^T Q_i^D C_{di}) \leq 0. \qquad (7.48)$$

To show that every equilibrium point $\alpha\mathbf{e}$, $\alpha \in \mathbb{R}$, of (7.45) is Lyapunov stable, consider the Lyapunov function candidate given by

$$V(z - \alpha\mathbf{e}) = \|z - \alpha\mathbf{e}\|^2 + 2\sum_{i=1}^{q}\int_{\alpha}^{z_i}[\sigma(\theta) - \sigma(\alpha)]\mathrm{d}\theta. \qquad (7.49)$$

Now, the derivative of $V(z - \alpha\mathbf{e})$ along the trajectories of (7.45) is given by

$$
\begin{aligned}
&\dot{V}(z - \alpha\mathbf{e}) \\
&= 2[z - \alpha\mathbf{e} + \hat{\sigma}(z) - \hat{\sigma}(\alpha\mathbf{e})]^{\mathrm{T}}C[z - \alpha\mathbf{e} + \hat{\sigma}(z) - \hat{\sigma}(\alpha\mathbf{e})] \\
&\quad + 2\sum_{i=1}^{q}[z - \alpha\mathbf{e} + \hat{\sigma}(z) - \hat{\sigma}(\alpha\mathbf{e})]^{\mathrm{T}}C_{\mathrm{d}i}[z - \alpha\mathbf{e} + \hat{\sigma}(z) - \hat{\sigma}(\alpha\mathbf{e})] \\
&\quad + 2\sum_{i=1}^{q}[z_i - \alpha + \sigma(z_i) - \sigma(\alpha)]\sum_{j=1,j\neq i}^{q}\mathcal{L}_{(i,j)}(z_j - z_i) \\
&\leq -\sum_{i=1}^{q}[z - \alpha\mathbf{e} + \hat{\sigma}(z) - \hat{\sigma}(\alpha\mathbf{e})]^{\mathrm{T}}Q_i[z - \alpha\mathbf{e} + \hat{\sigma}(z) - \hat{\sigma}(\alpha\mathbf{e})] \\
&\quad + \sum_{i=1}^{q}2[z - \alpha\mathbf{e} + \hat{\sigma}(z) - \hat{\sigma}(\alpha\mathbf{e})]^{\mathrm{T}}C_{\mathrm{d}i}[z - \alpha\mathbf{e} + \hat{\sigma}(z) - \hat{\sigma}(\alpha\mathbf{e})] \\
&\quad - \sum_{i=1}^{q}[z - \alpha\mathbf{e} + \hat{\sigma}(z) - \hat{\sigma}(\alpha\mathbf{e})]^{\mathrm{T}}C_{\mathrm{d}i}^{\mathrm{T}}Q_i^{\mathrm{D}}C_{\mathrm{d}i}[z - \alpha\mathbf{e} + \hat{\sigma}(z) - \hat{\sigma}(\alpha\mathbf{e})] \\
&\quad - 2\sum_{i=1}^{q-1}\sum_{j=i+1}^{q}\mathcal{L}_{(i,j)}(z_i - z_j)[\sigma(z_i) - \sigma(z_j)] - 2\sum_{i=1}^{q-1}\sum_{j=i+1}^{q}\mathcal{L}_{(i,j)}(z_i - z_j)^2 \\
&= -\sum_{i=1}^{q}(-Q_i[z - \alpha\mathbf{e} + \hat{\sigma}(z) - \hat{\sigma}(\alpha\mathbf{e})] + C_{\mathrm{d}i}[z - \alpha\mathbf{e} + \hat{\sigma}(z) - \hat{\sigma}(\alpha\mathbf{e})])^{\mathrm{T}}Q_i^{\mathrm{D}} \\
&\quad \cdot (-Q_i[z - \alpha\mathbf{e} + \hat{\sigma}(z) - \hat{\sigma}(\alpha\mathbf{e})] + C_{\mathrm{d}i}[z - \alpha\mathbf{e} + \hat{\sigma}(z) - \hat{\sigma}(\alpha\mathbf{e})]) \\
&\quad - \sum_{i=1}^{q}[z - \alpha\mathbf{e} + \hat{\sigma}(z) - \hat{\sigma}(\alpha\mathbf{e})]^{\mathrm{T}}C_{\mathrm{d}i}^{\mathrm{T}}Q_i^{\mathrm{D}}C_{\mathrm{d}i}[z - \alpha\mathbf{e} + \hat{\sigma}(z) - \hat{\sigma}(\alpha\mathbf{e})] \\
&\quad - 2\sum_{i=1}^{q-1}\sum_{j=i+1}^{q}\mathcal{L}_{(i,j)}(z_i - z_j)[\sigma(z_i) - \sigma(z_j)] - 2\sum_{i=1}^{q-1}\sum_{j=i+1}^{q}\mathcal{L}_{(i,j)}(z_i - z_j)^2 \\
&\leq 0, \quad z \in \mathbb{R}^q, \qquad\qquad (7.50)
\end{aligned}
$$

which establishes the Lyapunov stability of $\alpha\mathbf{e}$.

Finally, let $\mathcal{R} \triangleq \{x \in \mathbb{R}^q : \dot{V}(x) = 0\}$ and

$$\tilde{\mathcal{R}} \triangleq \{x \in \mathbb{R}^q : -Q_i[x + \hat{\sigma}(x)] + C_{\mathrm{d}i}[x + \hat{\sigma}(x)] = 0, \ i = 1,\ldots,q\},$$

and note that $\mathcal{R} \subseteq \tilde{\mathcal{R}}$. Then it follows from the Krasovskii–LaSalle invariant set theorem [122, p. 147] that $x(t) \to \mathcal{M}$ as $t \to \infty$, where $\mathcal{M}$ denotes the largest invariant set contained in $\mathcal{R}$. Now, since $C + \sum_{i=1}^{q} Q_i = 0$, it follows that

$$\mathcal{R} \subseteq \tilde{\mathcal{R}} \subseteq \hat{\mathcal{R}} \triangleq \left\{ x \in \mathbb{R}^q : C\hat{\sigma}(x) + \sum_{i=1}^{q} C_{di}\hat{\sigma}(x) = 0 \right\}.$$

Hence, since $C + \sum_{i=1}^{q} C_{di} = \mathcal{H}$, rank $\mathcal{H} = q-1$, and $\mathcal{H}\mathbf{e} = 0$, it follows that the largest invariant set $\hat{\mathcal{M}}$ contained in $\hat{\mathcal{R}}$ is given by $\hat{\mathcal{M}} = \{x \in \mathbb{R}^q : x = \alpha\mathbf{e}, \alpha \in \mathbb{R}\}$. Furthermore, since $\hat{\mathcal{M}} \subseteq \mathcal{R} \subseteq \hat{\mathcal{R}}$, it follows that $\mathcal{M} = \hat{\mathcal{M}}$. Hence, using similar arguments as in the proof of the implication (iii) $\Rightarrow$ (i) of Proposition 4.1, it follows that every equilibrium point in $\{\alpha\mathbf{e} : \alpha \in \mathbb{R}\}$ is a semistable equilibrium of (7.21) and (7.45).                    $\square$

**Example 7.1.** Consider the generalized consensus protocol given by

$$\dot{x}_1(t) = x_2(t) - x_1(t) + x_3(t) - x_1(t) + a(\sigma(x_2(t)) - \sigma(x_1(t))),$$
$$x_1(0) = x_{10}, \quad t \geq 0, \qquad (7.51)$$
$$\dot{x}_2(t) = x_3(t) - x_2(t) + a(\sigma(x_3(t)) - \sigma(x_2(t))), \quad x_2(0) = x_{20}, \qquad (7.52)$$
$$\dot{x}_3(t) = x_4(t) - x_3(t) + x_1(t) - x_3(t) + a(\sigma(x_4(t)) - \sigma(x_3(t))),$$
$$x_3(0) = x_{30}, \qquad (7.53)$$
$$\dot{x}_4(t) = x_5(t) - x_4(t) + a(\sigma(x_5(t)) - \sigma(x_4(t))) \quad x_4(0) = x_{40}, \qquad (7.54)$$
$$\dot{x}_5(t) = x_1(t) - x_5(t) + a(\sigma(x_1(t)) - \sigma(x_5(t))), \quad x_5(0) = x_{50}, \qquad (7.55)$$

where $\sigma(x) = \text{sign}(x)|x|^{\alpha+1}$, $\text{sign}(x) \triangleq x/|x|$ for $x \neq 0$, $\text{sign}(0) \triangleq 0$, and $\alpha \geq 0$. Note that (7.51)–(7.55) can be rewritten in the form of (7.45) with

$$C = \begin{bmatrix} -2 & 1 & 1 & 0 & 0 \\ 0 & -1 & 1 & 0 & 0 \\ 1 & 0 & -2 & 1 & 0 \\ 0 & 0 & 0 & -1 & 1 \\ 1 & 0 & 0 & 0 & -1 \end{bmatrix},$$

$$\mathcal{H} = \begin{bmatrix} -1 & 1 & 0 & 0 & 0 \\ 0 & -1 & 1 & 0 & 0 \\ 0 & 0 & -1 & 1 & 0 \\ 0 & 0 & 0 & -1 & 1 \\ 1 & 0 & 0 & 0 & -1 \end{bmatrix},$$

$$\mathcal{L} = \mathcal{C} - \mathcal{H} = \begin{bmatrix} -1 & 0 & 1 & 0 & 0 \\ 0 & 0 & 0 & 0 & 0 \\ 1 & 0 & -1 & 0 & 0 \\ 0 & 0 & 0 & 0 & 0 \\ 0 & 0 & 0 & 0 & 0 \end{bmatrix}.$$

Then it follows from Theorem 7.5 that every point in $\{(x_1, x_2, x_3, x_4, x_5) \in \mathbb{R}^5 : x_1 = x_2 = x_3 = x_4 = x_5 = c, c \in \mathbb{R}\}$ is a semistable equilibrium state of (7.51)–(7.55) with $a > 0$ and $a = 0$. Let $[x_{10}, x_{20}, x_{30}, x_{40}, x_{50}]^T = [5, 3, -5, 3, -1]^T$, $a = 6$, and $\alpha = 2$. Figure 7.1 shows the state trajectories versus time.                                                                                   $\triangle$

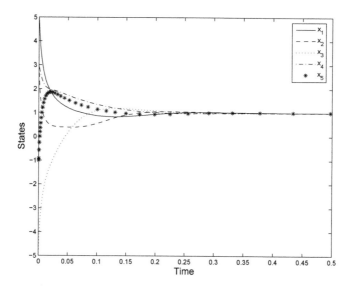

Figure 7.1 State trajectories versus time for (7.51)–(7.55).

*Chapter Eight*

---

# Network Systems with Time Delays

## 8.1 Introduction

Nonnegative and compartmental models [30, 91, 121, 178, 270] can also be used for modeling agreement problems in dynamical networks with directed graphs and switching topologies [237, 238]. Specifically, distributed decision-making for coordination of networks of dynamic agents involving information flow can be naturally captured by compartmental models. By properly formulating these systems in terms of subsystem interaction involving energy/mass transfer, the dynamical models of many of these systems can be derived from mass, energy, and information balance considerations that involve dynamic states whose values are nonnegative. Hence, it follows from physical considerations that the state trajectory of such systems remains in the nonnegative orthant of the state space for nonnegative initial conditions.

Since state information in large-scale networks can only be observed or relayed to controllers after a time delay, network system stability is affected by the presence of varying time lags throughout the network. In this chapter, we use compartmental dynamical system models to characterize dynamic algorithms for linear and nonlinear networks of dynamic agents in the presence of interagent communication delays that possess a continuum of semistable equilibria, that is, protocol algorithms that guarantee convergence to Lyapunov stable equilibria. In addition, we show that the steady-state distribution of the dynamic network is uniform, leading to system state equipartitioning or consensus.

To accurately describe the information in transit between the dynamic agents, it is necessary to include in any mathematical model of the system dynamics some information on the past system states. In this case, the state of the system at a given time involves a piece of the trajectories in the space of continuous functions defined on an interval in the nonnegative orthant of the state space. This of course leads to (infinite-dimensional) delay dynamical systems [142, 189].

## 8.2 Mathematical Preliminaries

In this section, we introduce notation, several definitions, and some key results concerning linear nonnegative dynamical systems with time delay [120, 130] that are necessary for developing some of the main results of this chapter. Specifically, for $x \in \mathbb{R}^n$ we write $x \geq\geq 0$ (respectively, $x >> 0$) to indicate that every component of $x$ is nonnegative (respectively, positive). In this case, we say that $x$ is *nonnegative* or *positive*, respectively. Likewise, $A \in \mathbb{R}^{n \times m}$ is *nonnegative*[11] or *positive* if every entry of $A$ is nonnegative or positive, respectively, which is written as $A \geq\geq 0$ or $A >> 0$, respectively. Let $\overline{\mathbb{R}}_+^n$ and $\mathbb{R}_+^n$ denote the nonnegative and positive orthants of $\mathbb{R}^n$; that is, if $x \in \mathbb{R}^n$, then $x \in \overline{\mathbb{R}}_+^n$ and $x \in \mathbb{R}_+^n$ are equivalent, respectively, to $x \geq\geq 0$ and $x >> 0$.

The following definition introduces the notion of a nonnegative (respectively, positive) function.

**Definition 8.1.** Let $T > 0$. A real function $x : [0, T] \to \mathbb{R}^n$ is a *nonnegative* (respectively, *positive*) *function* if $x(t) \geq\geq 0$ (respectively, $x(t) >> 0$) on the interval $[0, T]$.

The next definition introduces the notion of essentially nonnegative matrices and compartmental matrices.

**Definition 8.2** (see [24]). Let $A \in \mathbb{R}^{n \times n}$. $A$ is *essentially nonnegative* if $A_{(i,j)} \geq 0$, $i, j = 1, \ldots, n$, $i \neq j$. $A$ is *compartmental* if $A$ is essentially nonnegative and $A^{\mathrm{T}} \mathbf{e} \leq\leq 0$.

In the first part of this chapter, we consider linear time-delay dynamical systems $\mathcal{G}$ of the form

$$\dot{x}(t) = Ax(t) + \sum_{i=1}^{n_{\mathrm{d}}} A_{\mathrm{d}i} x(t - \tau_i), \quad x(\theta) = \eta(\theta), \quad -\bar{\tau} \leq \theta \leq 0, \quad t \geq 0, \quad (8.1)$$

where $x(t) \in \mathbb{R}^n$, $t \geq 0$, $A \in \mathbb{R}^{n \times n}$, $A_{\mathrm{d}i} \in \mathbb{R}^{n \times n}$, $\tau_i \in \mathbb{R}$, $i = 1, \ldots, n_{\mathrm{d}}$, $\bar{\tau} = \max_{i \in \{1, \ldots, n_{\mathrm{d}}\}} \tau_i$,

$$\eta(\cdot) \in \mathcal{C}_+ \triangleq \{\psi(\cdot) \in \mathcal{C}([-\bar{\tau}, 0], \mathbb{R}^n) : \psi(\theta) \geq\geq 0, \theta \in [-\bar{\tau}, 0]\}$$

is a continuous vector-valued function specifying the initial state of the sys-

---

[11]In this chapter, it is important to distinguish between a square nonnegative (respectively, positive) matrix and a nonnegative definite (respectively, positive definite) matrix.

tem, and $\mathcal{C}([-\bar{\tau}, 0], \mathbb{R}^n)$ denotes a Banach space[12] of continuous functions mapping the interval $[-\bar{\tau}, 0]$ into $\mathbb{R}^n$ with the topology of uniform convergence. Note that the state of (8.1) at time $t$ is the *piece of the trajectories* $x$ between $t - \tau$ and $t$ or, equivalently, the *element* $x_t$ in the space of continuous functions defined on the interval $[-\bar{\tau}, 0]$ and taking values in $\mathbb{R}^n$; that is, $x_t \in \mathcal{C}([-\bar{\tau}, 0], \mathbb{R}^n)$, where $x_t(\theta) \triangleq x(t + \theta)$, $\theta \in [-\bar{\tau}, 0]$.

Furthermore, since for a given time $t$ the piece of the trajectories $x_t$ is defined on $[-\bar{\tau}, 0]$, the uniform norm $\|x_t\| = \sup_{\theta \in [-\bar{\tau}, 0]} \|x(t+\theta)\|$, where $\|\cdot\|$ denotes the Euclidean vector norm, is used for the definitions of Lyapunov and asymptotic stability of (8.1). For further details, see [142, 189]. In addition, note that since $\eta(\cdot)$ is continuous, it follows from Theorem 2.1 of [142, p. 14] that there exists a unique solution $x(\eta)$ defined on $[-\bar{\tau}, \infty)$ that coincides with $\eta$ on $[-\bar{\tau}, 0]$ and satisfies (8.1) for all $t \geq 0$. Finally, recall that if the positive orbit $\gamma^+(\eta(\theta))$ of (8.1) is bounded, then $\gamma^+(\eta(\theta))$ is *precompact* [140]; that is, $\gamma^+(\eta(\theta))$ can be enclosed in the union of a finite number of $\varepsilon$-balls around elements of $\gamma^+(\eta(\theta))$.

Theorem 8.1 below gives necessary and sufficient conditions for asymptotic stability of the linear time-delay nonnegative dynamical system $\mathcal{G}$ given by (8.1). For this result, the following definition and proposition are needed.

**Definition 8.3.** The linear time-delay dynamical system given by (8.1) is *nonnegative* if for every $\eta(\cdot) \in \mathcal{C}_+$, the solution $x(t)$, $t \geq 0$, to (8.1) is nonnegative.

**Proposition 8.1** (see [120, 130]). The linear time-delay dynamical system $\mathcal{G}$ given by (8.1) is nonnegative if and only if $A \in \mathbb{R}^{n \times n}$ is essentially nonnegative and $A_{\mathrm{d}i} \in \mathbb{R}^{n \times n}$, $i = 1, \ldots, n_\mathrm{d}$, is nonnegative.

**Theorem 8.1** (see [120, 130]). Consider the linear time-delay dynamical system $\mathcal{G}$ given by (8.1), where $A \in \mathbb{R}^{n \times n}$ is essentially nonnegative and $A_{\mathrm{d}i} \in \mathbb{R}^{n \times n}$, $i = 1, \ldots, n_\mathrm{d}$, is nonnegative. If there exist $p, r \in \mathbb{R}^n$ such that $p >> 0$ and $r \geq\geq 0$ (respectively, $r >> 0$) satisfying

$$0 = \left( A + \sum_{i=1}^{n_\mathrm{d}} A_{\mathrm{d}i} \right)^{\mathrm{T}} p + r, \tag{8.2}$$

then $\mathcal{G}$ is Lyapunov (respectively, asymptotically) stable for all $\bar{\tau} \in [0, \infty)$. Conversely, if $\mathcal{G}$ is asymptotically stable for all $\bar{\tau} \in [0, \infty)$, then there exist

---

[12]A *Banach space* is a vector space of bounded functions defined on a compact set endowed with a Chebyshev norm, that is, the norm placed on bounded functions on a set $\mathcal{D}$ that assigns to each function the supremum of the moduli of the values of the function on $\mathcal{D}$. Equivalently, a Banach space is a complete normed space wherein every Cauchy sequence converges to an element in the space.

$p, r \in \mathbb{R}^n$ such that $p \gg 0$ and $r \gg 0$ satisfying (8.2).

Next, we consider a subclass of nonnegative systems, namely, compartmental systems. As noted in Section 8.1, compartmental dynamical systems are of major importance in network systems.

**Definition 8.4** (see [120, 130]). The linear time-delay dynamical system (8.1) is called a *compartmental dynamical system* if $A \in \mathbb{R}^{n \times n}$ is essentially nonnegative; $A_{di} \in \mathbb{R}^{n \times n}$, $i = 1, \ldots, n_d$, is nonnegative; and $A + \sum_{i=1}^{n_d} A_{di}$ is a compartmental matrix.

Note that the linear time-delay dynamical system (8.1) is compartmental if $A$ and $A_d \triangleq \sum_{i=1}^{n_d} A_{di}$ are given by

$$A_{(i,j)} = \begin{cases} -\sum_{k=1}^n a_{ki}, & i = j, \\ 0, & i \neq j, \end{cases} \qquad A_{d(i,j)} = \begin{cases} 0, & i = j, \\ a_{ij}, & i \neq j, \end{cases} \qquad (8.3)$$

where $a_{ii} \geq 0$, $i \in \{1, \ldots, n\}$, denotes the loss coefficients of the $i$th compartment and $a_{ij} \geq 0$, $i \neq j$, $i, j \in \{1, \ldots, n\}$, denotes the transfer coefficients from the $j$th compartment to the $i$th compartment.

The following results are necessary for developing some of the main results of this chapter.

**Proposition 8.2** (see [121]). Let $A \in \mathbb{R}^{n \times n}$ be essentially nonnegative, and assume there exists $p \in \mathbb{R}_+^n$ such that $A^{\mathrm{T}} p \leq\leq 0$. Then $A$ is semistable; that is, $\mathrm{Re}\, \lambda < 0$, or $\lambda = 0$ and $\lambda$ is semisimple, where $\lambda \in \mathrm{spec}(A)$.

**Corollary 8.1.** Let $A \in \mathbb{R}^{n \times n}$ be an essentially nonnegative matrix such that $A = A^{\mathrm{T}}$. If there exists $p \in \mathbb{R}_+^n$ such that $A^{\mathrm{T}} p \leq\leq 0$, then $A \leq 0$.

**Proof.** The proof is a direct consequence of Proposition 8.2 by noting that if $A$ is symmetric, then semistability implies that $A \leq 0$. $\square$

**Lemma 8.1.** Let $X \in \mathbb{R}^{n \times n}$ and $Z \in \mathbb{R}^{m \times m}$ be such that $X = X^{\mathrm{T}}$ and $Z = Z^{\mathrm{T}}$, and let $Y \in \mathbb{R}^{n \times m}$ be such that $Y = YZ^{\mathrm{D}}Z$. Then

$$M \triangleq \begin{bmatrix} X & Y \\ Y^{\mathrm{T}} & Z \end{bmatrix} \leq 0 \qquad (8.4)$$

if and only if $Z \leq 0$ and $X - YZ^{\mathrm{D}}Y^{\mathrm{T}} \leq 0$.

**Proof.** Define

$$T \triangleq \begin{bmatrix} I_n & -YZ^{\mathrm{D}} \\ 0 & I_m \end{bmatrix},$$

and note that $\det T \neq 0$. Now, noting that $TMT^{\mathrm{T}} \leq 0$ if and only if $M \leq 0$ and

$$
\begin{aligned}
TMT^{\mathrm{T}} &= \begin{bmatrix} I_n & -YZ^{\mathrm{D}} \\ 0 & I_m \end{bmatrix} \begin{bmatrix} X & Y \\ Y^{\mathrm{T}} & Z \end{bmatrix} \begin{bmatrix} I_n & 0 \\ -Z^{\mathrm{D}}Y^{\mathrm{T}} & I_m \end{bmatrix} \\
&= \begin{bmatrix} X - YZ^{\mathrm{D}}Y^{\mathrm{T}} & 0 \\ 0 & Z \end{bmatrix} \\
&\leq 0,
\end{aligned}
$$

the result follows immediately.                                         □

## 8.3 Semistability and Equipartition of Linear Compartmental Systems with Time Delay

In this section, we present sufficient conditions for semistability and system state equipartition for linear compartmental dynamical systems with time delay. Note that for addressing the stability of the zero solution of a time-delay nonnegative system, the usual stability definitions given in [142] need to be slightly modified. In particular, stability notions for nonnegative dynamical systems need to be defined with respect to relatively open subsets of $\overline{\mathbb{R}}_+^n$ containing the equilibrium solution $x_t \equiv 0$. For a similar definition, see [121]. In this case, standard Lyapunov–Krasovskii stability theorems for linear and nonlinear time-delay systems [142] can be used directly with the required sufficient conditions verified on $\overline{\mathbb{R}}_+^n$.

The following lemma is needed for the main theorem of this section.

**Lemma 8.2.** Let $A \in \mathbb{R}^{n \times n}$ and $A_{\mathrm{d}i} \in \mathbb{R}^{n \times n}$, $i = 1, \ldots, n_{\mathrm{d}}$, be given by (8.3). Assume that $(A + \sum_{i=1}^{n_{\mathrm{d}}} A_{\mathrm{d}i})\mathbf{e} = 0$. Then there exist nonnegative definite matrices $Q_i \in \mathbb{R}^{n \times n}$, $i = 1, \ldots, n_{\mathrm{d}}$, such that

$$
A + A^{\mathrm{T}} + \sum_{i=1}^{n_{\mathrm{d}}} (Q_i + A_{\mathrm{d}i}^{\mathrm{T}} Q_i^{\mathrm{D}} A_{\mathrm{d}i}) \leq 0. \tag{8.5}
$$

**Proof.** For each $i \in \{1, \ldots, n_{\mathrm{d}}\}$, let $Q_i$ be the diagonal matrix defined by

$$
Q_{i(l,l)} \triangleq \sum_{m=1, l \neq m}^{n_{\mathrm{d}}} A_{\mathrm{d}i(l,m)}, \tag{8.6}
$$

and note that it follows from (8.6) and the definition of the Drazin inverse that $(A_{\mathrm{d}i} - Q_i)\mathbf{e} = 0$ and $Q_i Q_i^{\mathrm{D}} A_{\mathrm{d}i} = A_{\mathrm{d}i}$, $i = 1, \ldots, n_{\mathrm{d}}$. Since $A$ and $Q_i$, $i = 1, \ldots, n_{\mathrm{d}}$, are diagonal and $(A + \sum_{i=1}^{n_{\mathrm{d}}} A_{\mathrm{d}i})\mathbf{e} = 0$, it follows that

$A + \sum_{i=1}^{n_\mathrm{d}} Q_i = 0$. Hence, $M\mathbf{e} = 0$, where

$$
M \triangleq
\begin{bmatrix}
A + A^\mathrm{T} + \sum_{i=1}^{n_\mathrm{d}} Q_i & A_{\mathrm{d}1}^\mathrm{T} & A_{\mathrm{d}2}^\mathrm{T} & \cdots & A_{\mathrm{d}n_\mathrm{d}}^\mathrm{T} \\
A_{\mathrm{d}1} & -Q_1 & 0 & \cdots & 0 \\
\vdots & \vdots & \vdots & \vdots & \vdots \\
A_{\mathrm{d}n_\mathrm{d}} & 0 & 0 & \cdots & -Q_{n_\mathrm{d}}
\end{bmatrix}. \tag{8.7}
$$

Now, it follows from Corollary 8.1 that $M \leq 0$, and since $Q_i Q_i^\mathrm{D} A_{\mathrm{d}i} = A_{\mathrm{d}i}$, $i = 1, \ldots, n_\mathrm{d}$, it follows from Lemma 8.1 that $M \leq 0$ if and only if (8.5) holds. $\qquad\square$

For the next result, recall that the equilibrium solution $x_t \equiv x_\mathrm{e}$ to (8.1) is semistable if and only if $x_\mathrm{e}$ is Lyapunov stable and $\lim_{t\to\infty} x(t)$ exists.

**Theorem 8.2.** Consider the linear time-delay dynamical system given by (8.1), where $A$ and $A_{\mathrm{d}i}$, $i = 1, \ldots, n_\mathrm{d}$, are given by (8.3). Assume that $(A + \sum_{i=1}^{n_\mathrm{d}} A_{\mathrm{d}i})^\mathrm{T} \mathbf{e} = (A + \sum_{i=1}^{n_\mathrm{d}} A_{\mathrm{d}i})\mathbf{e} = 0$ and $\mathrm{rank}(A + \sum_{i=1}^{n_\mathrm{d}} A_{\mathrm{d}i}) = n - 1$. Then, for every $\alpha \geq 0$, $\alpha\mathbf{e}$ is a semistable equilibrium point of (8.1). Furthermore, $x(t) \to \alpha^*\mathbf{e}$ as $t \to \infty$, where

$$
\alpha^* = \frac{\mathbf{e}^\mathrm{T} \eta(0) + \sum_{i=1}^{n_\mathrm{d}} \int_{-\tau_i}^0 \mathbf{e}^\mathrm{T} A_{\mathrm{d}i} \eta(\theta)\mathrm{d}\theta}{n + \sum_{i=1}^{n_\mathrm{d}} \tau_i \mathbf{e}^\mathrm{T} A_{\mathrm{d}i}\mathbf{e}}. \tag{8.8}
$$

**Proof.** It follows from Lemma 8.2 that there exist nonnegative matrices $Q_i$, $i = 1, \ldots, n_\mathrm{d}$, such that (8.5) holds. Now, consider the Lyapunov–Krasovskii functional $V : \mathcal{C}_+ \to \mathbb{R}$ given by

$$
V(\psi(\cdot)) = \psi^\mathrm{T}(0)\psi(0) + \sum_{i=1}^{n_\mathrm{d}} \int_{-\tau_i}^0 \psi^\mathrm{T}(\theta) A_{\mathrm{d}i}^\mathrm{T} Q_i^\mathrm{D} A_{\mathrm{d}i}\psi(\theta)\mathrm{d}\theta, \tag{8.9}
$$

and note that the directional derivative of $V(x_t)$ along the trajectories of (8.1) is given by

$$
\begin{aligned}
\dot{V}(x_t) =\ & 2x^\mathrm{T}(t)\dot{x}(t) + \sum_{i=1}^{n_\mathrm{d}} x^\mathrm{T}(t) A_{\mathrm{d}i}^\mathrm{T} Q_i^\mathrm{D} A_{\mathrm{d}i} x(t) \\
& - \sum_{i=1}^{n_\mathrm{d}} x^\mathrm{T}(t - \tau_i) A_{\mathrm{d}i}^\mathrm{T} Q_i^\mathrm{D} A_{\mathrm{d}i} x(t - \tau_i) \\
=\ & 2x^\mathrm{T}(t) A x(t) + 2x^\mathrm{T}(t) \sum_{i=1}^{n_\mathrm{d}} A_{\mathrm{d}i} x(t - \tau_i) + \sum_{i=1}^{n_\mathrm{d}} x^\mathrm{T}(t) A_{\mathrm{d}i}^\mathrm{T} Q_i^\mathrm{D} A_{\mathrm{d}i} x(t) \\
& - \sum_{i=1}^{n_\mathrm{d}} x^\mathrm{T}(t - \tau_i) A_{\mathrm{d}i}^\mathrm{T} Q_i^\mathrm{D} A_{\mathrm{d}i} x(t - \tau_i)
\end{aligned}
$$

$$\leq -\sum_{i=1}^{n_\mathrm{d}} [x^\mathrm{T}(t)Q_i x(t) - 2x^\mathrm{T}(t)A_{\mathrm{d}i}x(t-\tau_i)$$

$$+ x^\mathrm{T}(t-\tau_i)A_{\mathrm{d}i}^\mathrm{T} Q_i^\mathrm{D} A_{\mathrm{d}i}x(t-\tau_i)]$$

$$= -\sum_{i=1}^{n_\mathrm{d}} [-Q_i x(t) + A_{\mathrm{d}i}x(t-\tau_i)]^\mathrm{T} Q_i^\mathrm{D} [-Q_i x(t) + A_{\mathrm{d}i}x(t-\tau_i)]$$

$$\leq 0, \qquad t \geq 0. \tag{8.10}$$

Next, let

$$\mathcal{R} \triangleq \{\psi(\cdot) \in \mathcal{C}_+ : -Q_i\psi(0) + A_{\mathrm{d}i}\psi(-\tau_i) = 0, i = 1, \ldots, n_\mathrm{d}\},$$

and note that since the positive orbit $\gamma^+(\eta(\theta))$ of (8.1) is bounded, $\gamma^+(\eta(\theta))$ belongs to a compact subset of $\mathcal{C}_+$, and, hence, it follows from Theorem 3.2 of [142] that $x_t \to \mathcal{M}$, where $\mathcal{M}$ denotes the largest invariant set contained in $\mathcal{R}$. Now, since $A + \sum_{i=1}^{n_\mathrm{d}} Q_i = 0$, it follows that

$$\mathcal{R} \subset \hat{\mathcal{R}} \triangleq \left\{ \psi(\cdot) \in \mathcal{C}_+ : A\psi(0) + \sum_{i=1}^{n_\mathrm{d}} A_{\mathrm{d}i}\psi(-\tau_i) = 0 \right\}.$$

Hence, since $\mathrm{rank}(A + \sum_{i=1}^{n_\mathrm{d}} A_{\mathrm{d}i}) = n - 1$ and $(A + \sum_{i=1}^{n_\mathrm{d}} A_{\mathrm{d}i})\mathbf{e} = 0$, it follows that the largest invariant set $\hat{\mathcal{M}}$ contained in $\hat{\mathcal{R}}$ is given by

$$\hat{\mathcal{M}} = \{\psi \in \mathcal{C}_+ : \psi(\theta) = \alpha\mathbf{e}, \ \theta \in [-\bar{\tau}, 0], \ \alpha \geq 0\}.$$

Furthermore, since $\hat{\mathcal{M}} \subset \mathcal{R} \subset \hat{\mathcal{R}}$, it follows that $\mathcal{M} = \hat{\mathcal{M}}$.

Next, define the functional $E : \mathcal{C}_+ \to \mathbb{R}$ by

$$E(\psi(\cdot)) = \mathbf{e}^\mathrm{T}\psi(0) + \sum_{i=1}^{n_\mathrm{d}} \int_{-\tau_i}^{0} \mathbf{e}^\mathrm{T} A_{\mathrm{d}i}\psi(\theta)\mathrm{d}\theta, \tag{8.11}$$

and note that $\dot{E}(x_t) \equiv 0$ along the trajectories of (8.1). Thus, for all $t \geq 0$,

$$E(x_t) = E(\eta(\cdot)) = \mathbf{e}^\mathrm{T}\eta(0) + \sum_{i=1}^{n_\mathrm{d}} \int_{-\tau_i}^{0} \mathbf{e}^\mathrm{T} A_{\mathrm{d}i}\eta(\theta)\mathrm{d}\theta, \tag{8.12}$$

which implies that $x_t \to \mathcal{M} \cap \mathcal{E}$, where $\mathcal{E} \triangleq \{\psi(\cdot) \in \mathcal{C}_+ : E(\psi(\cdot)) = E(\eta(\cdot))\}$. Hence, since $\mathcal{M} \cap \mathcal{E} = \{\alpha^*\mathbf{e}\}$, it follows that $x(t) \to \alpha^*\mathbf{e}$, where $\alpha^*$ is given by (8.8).

Finally, Lyapunov stability of $\alpha\mathbf{e}$, $\alpha \geq 0$, follows by considering the Lyapunov–Krasovskii functional

$$V(\psi(\cdot)) = (\psi(0) - \alpha\mathbf{e})^{\mathrm{T}}(\psi(0) - \alpha\mathbf{e})$$

$$+ \sum_{i=1}^{n_{\mathrm{d}}} \int_{-\tau_i}^{0} (\psi(\theta) - \alpha\mathbf{e})^{\mathrm{T}} A_{\mathrm{d}i}^{\mathrm{T}} Q_i^{\mathrm{D}} A_{\mathrm{d}i}(\psi(\theta) - \alpha\mathbf{e})\mathrm{d}\theta$$

and noting that $V(\psi) \geq \|\psi(0) - \alpha\mathbf{e}\|_2^2$.                             $\square$

Note that if $n_{\mathrm{d}} = n^2 - n$, $A_{\mathrm{d}} = A_{\mathrm{d}}^{\mathrm{T}}$, and $(A + A_{\mathrm{d}})\mathbf{e} = 0$, then (8.1) can be rewritten as

$$\dot{x}_i(t) = - \sum_{j=1, j\neq i}^{n} a_{ij}[x_i(t) - x_j(t - \tau_{ij})], \quad x(\theta) = \eta(\theta), \quad -\bar{\tau} \leq \theta \leq 0, \quad t \geq 0,$$

(8.13)

where $i = 1, \ldots, n$ and $\tau_{ij} \in [0, \bar{\tau}]$, $i \neq j$, $i, j = 1, \ldots, n$, which implies that the rate of material or information transfer from the $i$th compartment to the $j$th compartment is proportional to the difference $x_j(t - \tau_{ij}) - x_i(t)$. Hence, the rate of material transfer is positive (respectively, negative) if $x_j(t - \tau_{ij}) > x_i(t)$ (respectively, $x_j(t - \tau_{ij}) < x_i(t)$). Equation (8.13) is an information flow balance equation that governs the information exchange among coupled subsystems and is completely analogous to the equations of thermal transfer with subsystem information playing the role of temperatures.

Furthermore, note that since $a_{ij} \geq 0$, $i \neq j$, $i, j = 1, \ldots, n$, information energy flows from more energetic (information-rich) subsystems to less energetic (information-poor) subsystems, which is consistent with the second law of thermodynamics requiring that heat (energy in transition) must flow in the direction of lower temperatures.

## 8.4 Semistability and Equipartition of Nonlinear Compartmental Systems with Time Delay

In this section, we extend the results of Section 8.3 to nonlinear compartmental systems with time delay. Specifically, we consider nonlinear time-delay dynamical systems $\mathcal{G}$ of the form

$$\dot{x}(t) = f(x(t)) + f_{\mathrm{d}}(x(t - \tau_1), \ldots, x(t - \tau_{n_{\mathrm{d}}})), \quad x(\theta) = \eta(\theta),$$
$$-\bar{\tau} \leq \theta \leq 0, \quad t \geq 0, \qquad (8.14)$$

where $x(t) \in \mathbb{R}^n$, $t \geq 0$; $f : \mathbb{R}^n \to \mathbb{R}^n$ is locally Lipschitz continuous and $f(0) = 0$; $f_{\mathrm{d}} : \mathbb{R}^n \times \cdots \times \mathbb{R}^n \to \mathbb{R}^n$ is locally Lipschitz continuous and $f_{\mathrm{d}}(0, \ldots, 0) = 0$; $\bar{\tau} = \max_{i \in \{1, \ldots, n_{\mathrm{d}}\}} \tau_i$, $\tau_i \geq 0$, $i = 1, \ldots, n_{\mathrm{d}}$; and $\eta(\cdot) \in \mathcal{C} = \mathcal{C}([-\bar{\tau}, 0], \mathbb{R}^n)$ is a continuous vector-valued function specifying the initial state of the system. Note that since $\eta(\cdot)$ is continuous, it follows from Theorem 2.3 of [142, p. 44] that there exists a unique solution $x(\eta)$ defined on $[-\bar{\tau}, \infty)$ that coincides with $\eta$ on $[-\bar{\tau}, 0]$ and satisfies (8.14) for all $t \geq 0$.

In addition, recall that if the positive orbit $\gamma^+(\eta(\theta))$ of (8.14) is bounded, then $\gamma^+(\eta(\theta))$ is precompact [140].

The following definitions generalize the notions of essential nonnegativity and nonnegativity to vector fields.

**Definition 8.5** (see [121]). Let $f = [f_1, \ldots, f_n]^T : \mathcal{D} \to \mathbb{R}^n$, where $\mathcal{D}$ is an open subset of $\mathbb{R}^n$ that contains $\overline{\mathbb{R}}_+^n$. Then $f$ is *essentially nonnegative* if $f_i(x) \geq 0$ for all $i = 1, \ldots, n$ and $x \in \overline{\mathbb{R}}_+^n$ such that $x_i = 0$, where $x_i$ denotes the $i$th element of $x$. $f$ is *compartmental* if $f$ is essentially nonnegative and $\mathbf{e}^T f(x) \leq 0$, $x \in \overline{\mathbb{R}}_+^n$.

**Definition 8.6** (see [123]). Let $f = [f_1, \ldots, f_n]^T : \mathcal{D} \to \mathbb{R}^n$, where $\mathcal{D}$ is an open subset of $\mathbb{R}^n$ that contains $\overline{\mathbb{R}}_+^n$. Then $f$ is *nonnegative* if $f_i(x) \geq 0$ for all $i = 1, \ldots, n$ and $x \in \overline{\mathbb{R}}_+^n$.

Note that if $f(x) = Ax$, where $A \in \mathbb{R}^{n \times n}$, then $f(\cdot)$ is essentially nonnegative if and only if $A$ is essentially nonnegative, and $f(\cdot)$ is nonnegative if and only if $A$ is nonnegative.

**Definition 8.7** (see [120]). The nonlinear time-delay dynamical system $\mathcal{G}$ given by (8.14) is *nonnegative* if for every $\eta(\cdot) \in \mathcal{C}_+$, where $\mathcal{C}_+ \triangleq \{\psi(\cdot) \in \mathcal{C} : \psi(\theta) \geq\geq 0, \ \theta \in [-\bar{\tau}, 0]\}$, the solution $x(t)$, $t \geq 0$, to (8.14) is nonnegative.

**Proposition 8.3** (see [120]). Consider the nonlinear time-delay dynamical system $\mathcal{G}$ given by (8.14). If $f(\cdot)$ is essentially nonnegative and $f_{\mathrm{d}}(\cdot)$ is nonnegative, then $\mathcal{G}$ is nonnegative.

For the remainder of this chapter, we assume that $f(\cdot)$ is essentially nonnegative and $f_{\mathrm{d}}(\cdot)$ is nonnegative so that, for every $\eta(\cdot) \in \mathcal{C}_+$, the nonlinear time-delay dynamical system $\mathcal{G}$ given by (8.14) is nonnegative.

Next, we consider a subclass of nonlinear nonnegative systems, namely, nonlinear compartmental systems.

**Definition 8.8.** The nonlinear time-delay dynamical system (8.14) is called a *compartmental dynamical system* if $F(\cdot)$ is compartmental, where $F(x) \triangleq f(x) + f_{\mathrm{d}}(x, x, \ldots, x)$.

Note that the nonlinear time-delay dynamical system is compartmental if $f(\cdot)$ and $f_{\mathrm{d}} = [f_{\mathrm{d}1}, \ldots, f_{\mathrm{d}n}]^{\mathrm{T}}$ are given by

$$f_i(x(t)) = -\sum_{j=1, j \neq i}^{n} a_{ji}(x(t)), \qquad (8.15)$$

$$f_{\mathrm{d}i}(x(t - \tau_1), \ldots, x(t - \tau_{n_{\mathrm{d}}})) = \sum_{j=1, j \neq i}^{n} a_{ij}(x(t - \tau_{ij})), \qquad (8.16)$$

where $a_{ii}(x(\cdot)) \geq 0$, $x(\cdot) \in \mathcal{C}_+$, $a_{ii}(0) = 0$, $i \in \{1, \ldots, n\}$, denotes the instantaneous rate of flow of material loss of the $i$th compartment; $a_{ij}(x(\cdot)) \geq 0$, $x(\cdot) \in \mathcal{C}_+$, $i \neq j$, $i, j \in \{1, \ldots, n\}$, denotes the instantaneous rate of material flow from the $j$th compartment to the $i$th compartment; $\tau_{ij}$, $i \neq j$, $i, j \in \{1, \ldots, n\}$, denotes the transfer time of material flow from the $j$th compartment to the $i$th compartment; and $a_{ii}(\cdot)$ and $a_{ij}(\cdot)$ are such that if $x_i = 0$, then $a_{ii}(x) = 0$ and $a_{ji}(x) = 0$ for all $i, j = 1, \ldots, n$, and $x \in \overline{\mathbb{R}}_+^n$. Note that the above constraints imply that $f(\cdot)$ is essentially nonnegative and $f_{\mathrm{d}}(\cdot)$ is nonnegative.

The next result generalizes Theorem 8.2 to nonlinear time-delay compartmental systems of the form

$$\dot{x}(t) = f(x(t)) + \sum_{i=1}^{n_{\mathrm{d}}} f_{\mathrm{d}i}(x(t - \tau_i)), \quad x(\theta) = \eta(\theta), \quad -\bar{\tau} \leq \theta \leq 0, \quad t \geq 0,$$

$$(8.17)$$

where $f : \overline{\mathbb{R}}_+^n \to \overline{\mathbb{R}}_+^n$ is given by $f(x) = [f_1(x_1), \ldots, f_n(x_n)]^{\mathrm{T}}$; $f(0) = 0$; $f_{\mathrm{d}i} : \overline{\mathbb{R}}_+^n \to \overline{\mathbb{R}}_+^n$, $i = 1, \ldots, n_{\mathrm{d}}$; and $f_{\mathrm{d}}(0) = 0$. Furthermore, we assume that $f_i(\cdot)$, $i = 1, \ldots, n$, are strictly decreasing functions.

**Theorem 8.3.** Consider the nonlinear time-delay dynamical system given by (8.17), where $f_i(\cdot)$, $i = 1, \ldots, n$, is strictly decreasing and $f_i(0) = 0$. Assume that $\mathbf{e}^{\mathrm{T}}[f(x) + \sum_{i=1}^{n_{\mathrm{d}}} f_{\mathrm{d}i}(x)] = 0$, $x \in \overline{\mathbb{R}}_+^n$, and $f(x) + \sum_{i=1}^{n_{\mathrm{d}}} f_{\mathrm{d}i}(x) = 0$ if and only if $x = \alpha \mathbf{e}$ for some $\alpha \geq 0$. Furthermore, assume that there exist nonnegative diagonal matrices $P_i \in \overline{\mathbb{R}}_+^{n \times n}$, $i = 1, \ldots, n_{\mathrm{d}}$, such that $P \triangleq \sum_{i=1}^{n_{\mathrm{d}}} P_i > 0$,

$$P_i^{\mathrm{D}} P_i f_{\mathrm{d}i}(x) = f_{\mathrm{d}i}(x), \quad x \in \overline{\mathbb{R}}_+^n, \quad i = 1, \ldots, n_{\mathrm{d}}, \qquad (8.18)$$

$$\sum_{i=1}^{n_{\mathrm{d}}} f_{\mathrm{d}i}^{\mathrm{T}}(x) P_i f_{\mathrm{d}i}(x) \leq f^{\mathrm{T}}(x) P f(x), \quad x \in \overline{\mathbb{R}}_+^n. \qquad (8.19)$$

Then, for every $\alpha \geq 0$, $\alpha \mathbf{e}$ is a semistable equilibrium point of (8.17). Furthermore, $x(t) \to \alpha^* e$ as $t \to \infty$, where $\alpha^*$ satisfies

$$n\alpha^* + \sum_{i=1}^{n_d} \tau_i \mathbf{e}^T f_{d_i}(\alpha^* \mathbf{e}) = \mathbf{e}^T \eta(0) + \sum_{i=1}^{n_d} \int_{-\tau_i}^{0} \mathbf{e}^T f_{d_i}(\eta(\theta)) d\theta. \qquad (8.20)$$

**Proof.** Consider the Lyapunov–Krasovskii functional $V : \mathcal{C}_+ \to \mathbb{R}$ given by

$$V(\psi(\cdot)) = -2\sum_{i=1}^{n} \int_{0}^{\psi_i(0)} P_{(i,i)} f_i(\zeta) d\zeta + \sum_{i=1}^{n_d} \int_{-\tau_i}^{0} f_{d_i}^T(\psi(\theta)) P_i f_{d_i}(\psi(\theta)) d\theta. \qquad (8.21)$$

Since $f_i(\cdot)$, $i = 1, \ldots, n$, is a strictly decreasing function, it follows that

$$V(\psi) \geq 2\sum_{i=1}^{n} P_{(i,i)} [-f_i(\delta_i \psi_i(0))] \psi_i(0) > 0, \quad \psi(0) \neq 0,$$

where $0 < \delta_i < 1$, and, hence, there exists a class $\mathcal{K}$ function $\alpha(\cdot)$ such that $V(\psi) \geq \alpha(\|\psi(0)\|)$. Now, note that the directional derivative of $V(x_t)$ along the trajectories of (8.17) is given by

$$\dot{V}(x_t)$$

$$= -2f^T(x(t))P\dot{x}(t) + \sum_{i=1}^{n_d} f_{d_i}^T(x(t))P_i f_{d_i}(x(t))$$

$$- \sum_{i=1}^{n_d} f_{d_i}^T(x(t - \tau_i)) P_i f_{d_i}(x(t - \tau_i))$$

$$= -2f^T(x(t))Pf(x(t)) - 2\sum_{i=1}^{n_d} f^T(x(t))P f_{d_i}(x(t - \tau_i))$$

$$+ \sum_{i=1}^{n_d} f_{d_i}^T(x(t))P_i f_{d_i}(x(t)) - \sum_{i=1}^{n_d} f_{d_i}^T(x(t - \tau_i)) P_i f_{d_i}(x(t - \tau_i))$$

$$\leq -f^T(x(t))Pf(x(t)) - 2\sum_{i=1}^{n_d} f^T(x(t))PP_i^D P_i f_{d_i}(x(t - \tau_i))$$

$$- \sum_{i=1}^{n_d} f_{d_i}^T(x(t - \tau_i)) P_i P_i^D P_i f_{d_i}(x(t - \tau_i))$$

$$= -\sum_{i=1}^{n_d} [Pf(x(t)) + P_i f_{d_i}(x(t - \tau_i))]^T P_i^D [Pf(x(t)) + P_i f_{d_i}(x(t - \tau_i))]$$

$$\leq 0, \quad t \geq 0, \qquad (8.22)$$

where the first inequality in (8.22) follows from (8.18) and (8.19) and the last equality in (8.22) follows from the fact that

$$f^{\mathrm{T}}(x)Pf(x) = \sum_{i=1}^{n_{\mathrm{d}}} f^{\mathrm{T}}(x)PP_i^{\mathrm{D}}Pf(x), \quad x \in \overline{\mathbb{R}}_+^n.$$

Next, let

$$\mathcal{R} \triangleq \{\psi(\cdot) \in \mathcal{C}_+ : Pf(\psi(0)) + P_i f_{\mathrm{d}i}(\psi(-\tau_i)) = 0, \quad i = 1, \ldots, n_{\mathrm{d}}\},$$

and note that since the positive orbit $\gamma^+(\eta(\theta))$ of (8.17) is bounded, $\gamma^+(\eta(\theta))$ belongs to a compact subset of $\mathcal{C}_+$, and, hence, it follows from Theorem 3.2 of [142] that $x_t \to \mathcal{M}$, where $\mathcal{M}$ denotes the largest invariant set (with respect to (8.17)) contained in $\mathcal{R}$. Now, since

$$\mathbf{e}^{\mathrm{T}}\left(f(x) + \sum_{i=1}^{n_{\mathrm{d}}} f_{\mathrm{d}i}(x)\right) = 0, \quad x \in \overline{\mathbb{R}}_+^n,$$

it follows that

$$\mathcal{R} \subset \hat{\mathcal{R}} \triangleq \left\{\psi(\cdot) \in \mathcal{C}_+ : f(\psi(0)) + \sum_{i=1}^{n_{\mathrm{d}}} f_{\mathrm{d}i}(\psi(-\tau_i)) = 0\right\}$$
$$= \{\psi(\cdot) \in \mathcal{C}_+ : \psi(\theta) = \alpha\mathbf{e}, \ t \in [-\bar{\tau}, 0], \ \alpha \geq 0\},$$

which implies that $x_t \to \hat{\mathcal{R}}$ as $t \to \infty$.

Next, define the functional $E : \mathcal{C}_+ \to \mathbb{R}$ by

$$E(\psi(\cdot)) = \mathbf{e}^{\mathrm{T}}\psi(0) + \sum_{i=1}^{n_{\mathrm{d}}} \int_{-\tau_i}^{0} \mathbf{e}^{\mathrm{T}} f_{\mathrm{d}i}(\psi(\theta))\mathrm{d}\theta, \tag{8.23}$$

and note that $\dot{E}(x_t) \equiv 0$ along the trajectories of (8.17). Thus, for all $t \geq 0$,

$$E(x_t) = E(\eta(\cdot)) = \mathbf{e}^{\mathrm{T}}\eta(0) + \sum_{i=1}^{n_{\mathrm{d}}} \int_{-\tau_i}^{0} \mathbf{e}^{\mathrm{T}} f_{\mathrm{d}i}(\eta(\theta))\mathrm{d}\theta, \tag{8.24}$$

which implies that $x_t \to \hat{\mathcal{R}} \cap \mathcal{E}$, where $\mathcal{E} \triangleq \{\psi(\cdot) \in \mathcal{C}_+ : E(\psi(\cdot)) = E(\eta(\cdot))\}$. Hence, $\hat{\mathcal{R}} \cap \mathcal{E} = \{\alpha^*\mathbf{e}\}$, and it follows that $x(t) \to \alpha^*\mathbf{e}$, where $\alpha^*$ satisfies (8.20).

Finally, Lyapunov stability of $\alpha\mathbf{e}$, $\alpha \geq 0$, follows by considering the Lyapunov–Krasovskii functional

$$V(\psi(\cdot)) = -2\sum_{i=1}^{n} \int_{\alpha}^{\psi_i(0)} P_{(i,i)}(f_i(\zeta) - f_i(\alpha))\mathrm{d}\zeta$$

$$+ \sum_{i=1}^{n_\mathrm{d}} \int_{-\tau_i}^{0} [f_{\mathrm{d}i}(\psi(\theta)) - f_{\mathrm{d}i}(\alpha\mathbf{e})]^\mathrm{T} P_i[f_{\mathrm{d}i}(\psi(\theta)) - f_{\mathrm{d}i}(\alpha\mathbf{e})]\mathrm{d}\theta$$

and noting that

$$V(\psi) \geq 2\sum_{i=1}^{n} P_{(i,i)}[f_i(\alpha) - f_i(\alpha + \delta_i(\psi_i(0) - \alpha))](\psi_i(0) - \alpha) > 0, \quad \psi_i(0) \neq \alpha,$$

where $0 < \delta_i < 1$. □

Theorem 8.3 establishes semistability and state equipartition for the special case of nonlinear compartmental systems of the form (8.15) and (8.16), where $f(\cdot)$ and $f_{\mathrm{d}i}(\cdot)$, $i = 1, \ldots, n$, satisfy (8.18) and (8.19). For general $n$-dimensional nonlinear compartmental systems with time delay and vector fields given by (8.17), it is not possible to guarantee semistability and state equipartition. However, semistability without state equipartition may be shown.

For example, consider the nonlinear time-delay compartmental dynamical system given by

$$\dot{x}_1(t) = -a_{21}(x_1(t)) + a_{12}(x_2(t - \tau_{12})), \quad x_1(\theta) = \eta_1(\theta),$$
$$-\bar{\tau} \leq \theta \leq 0, \quad t \geq 0, \qquad (8.25)$$
$$\dot{x}_2(t) = -a_{12}(x_2(t)) + a_{21}(x_1(t - \tau_{21})), \quad x_2(\theta) = \eta_2(\theta),$$
$$-\bar{\tau} \leq \theta \leq 0, \quad t \geq 0, \qquad (8.26)$$

where $x_1(t), x_2(t) \in \mathbb{R}$, $t \geq 0$; $a_{12} : \overline{\mathbb{R}}_+ \to \overline{\mathbb{R}}_+$ and $a_{21} : \overline{\mathbb{R}}_+ \to \overline{\mathbb{R}}_+$ satisfy $a_{12}(0) = a_{21}(0) = 0$ and $a_{12}(\cdot)$ and $a_{21}(\cdot)$ are strictly increasing; $\tau_{12}, \tau_{21} > 0$; $\bar{\tau} = \max\{\tau_{12}, \tau_{21}\}$; and $\eta_1(\cdot), \eta_2(\cdot) \in \mathcal{C}_+ = \mathcal{C}([-\bar{\tau}, 0], \overline{\mathbb{R}}_+)$. Note that (8.25) and (8.26) can have multiple equilibria with all the equilibria lying on the curve $a_{21}(u) = a_{12}(v)$, $u, v \geq 0$. It follows from the conditions on $a_{12}(\cdot)$ and $a_{21}(\cdot)$ that all system equilibria lie on the curve $y = a_{12}^{-1}(a_{21}(x))$ in the $(x, y)$ plane, where $a_{12}^{-1}(\cdot)$ denotes the inverse function of $a_{12}(\cdot)$.

Consider the functional $E : \mathcal{C}_+ \times \mathcal{C}_+ \to \mathbb{R}$ given by

$$E(\psi_1, \psi_2) = \psi_1(0) + \psi_2(0) + \int_{-\tau_{12}}^{0} a_{12}(\psi_2(\theta))\mathrm{d}\theta$$

$$+ \int_{-\tau_{21}}^{0} a_{21}(\psi_1(\theta))\mathrm{d}\theta. \qquad (8.27)$$

Now, it can be easily shown that the directional derivative of $E(\psi_1, \psi_2)$ along the trajectories of (8.25) and (8.26) is identically zero for all $t \geq 0$, which implies that, for all $t \geq 0$,

$$E(x_{1t}, x_{2t}) = E(\eta_1, \eta_2)$$
$$= \eta_1(0) + \eta_2(0) + \int_{-\tau_{12}}^{0} a_{12}(\eta_2(\theta))\mathrm{d}\theta + \int_{-\tau_{21}}^{0} a_{21}(\eta_1(\theta))\mathrm{d}\theta. \tag{8.28}$$

Next, consider the functional $V : \mathcal{C}_+ \times \mathcal{C}_+ \rightarrow \mathbb{R}$ given by

$$V(\psi_1, \psi_2) = 2 \int_0^{\psi_1(0)} a_{21}(\theta)\mathrm{d}\theta + 2 \int_0^{\psi_2(0)} a_{12}(\theta)\mathrm{d}\theta$$
$$+ \int_{-\tau_{12}}^{0} a_{12}^2(\psi_2(\theta))\mathrm{d}\theta + \int_{-\tau_{21}}^{0} a_{21}^2(\psi_1(\theta))\mathrm{d}\theta, \tag{8.29}$$

and note that the directional derivative of $V(\psi_1, \psi_2)$ along the trajectories of (8.25) and (8.26) is given by

$$\dot{V}(x_{1t}, x_{2t}) = -[a_{21}(x_1(t)) - a_{12}(x_2(t - \tau_{12}))]^2$$
$$- [a_{12}(x_2(t)) - a_{21}(x_1(t - \tau_{21}))]^2. \tag{8.30}$$

Now, using similar arguments as in the proof of Theorem 8.3, it follows that $(x_1(t), x_2(t)) \rightarrow (\alpha^*, a_{12}^{-1}(a_{21}(\alpha^*)))$ as $t \rightarrow \infty$, where $\alpha^*$ is the solution to the equation

$$\alpha^* + a_{12}^{-1}(a_{21}(\alpha^*)) + (\tau_{12} + \tau_{21})a_{21}(\alpha^*)$$
$$= \eta_1(0) + \eta_2(0) + \int_{-\tau_{12}}^{0} a_{12}(\eta_2(\theta))\mathrm{d}\theta + \int_{-\tau_{21}}^{0} a_{21}(\eta_1(\theta))\mathrm{d}\theta \tag{8.31}$$

and $(\alpha^*, a_{12}^{-1}(a_{21}(\alpha^*)))$ is a Lyapunov stable equilibrium state.

The above analysis shows that all two-dimensional nonlinear compartmental dynamical systems of the form (8.25) and (8.26) are semistable with system states reaching equilibria lying on the curve $y = a_{12}^{-1}(a_{21}(x))$ in the $(x, y)$ plane.

To demonstrate the utility of Theorem 8.3, we consider a nonlinear two-compartment time-delay dynamical system given by

$$\dot{x}_1(t) = -\sum_{i=1}^{n_{\mathrm{d}}} [a_i(x_1(t)) + a_i(x_2(t - \tau_i))], \quad x_1(\theta) = \eta_1(\theta),$$
$$-\bar{\tau} \leq \theta \leq 0, \ t \geq 0, \tag{8.32}$$

$$\dot{x}_2(t) = \sum_{i=1}^{n_{\mathrm{d}}}[a_i(x_1(t-\tau_i)) - a_i(x_2(t))], \quad x_2(\theta) = \eta_2(\theta),$$

$$-\bar{\tau} \le \theta \le 0, \tag{8.33}$$

where $a_i : \overline{\mathbb{R}}_+ \to \overline{\mathbb{R}}_+$, $i = 1, \dots, n_{\mathrm{d}}$, are such that, for every $i = 1, \dots, n_{\mathrm{d}}$,

$$[a_i(x_1) - a_i(x_2)](x_1 - x_2) > 0, \quad x_1 \ne x_2, \tag{8.34}$$

and $a_i(0) = 0$. If $x_1$ and $x_2$ represent system energies, then (8.32) and (8.33) capture energy flow balance between the two compartments, and (8.34) is consistent with the second law of thermodynamics; that is, energy flows from the more energetic compartment to the less energetic compartment [127].

Furthermore, since $a_i(0) = 0$, (8.34) implies that $a_i(\cdot)$, $i = 1, \dots, n_{\mathrm{d}}$, is strictly increasing. Now, note that (8.32) and (8.33) can be written in the form of (8.17) with

$$f(x) = \begin{bmatrix} -\sum_{i=1}^{n_{\mathrm{d}}} a_i(x_1) \\ -\sum_{i=1}^{n_{\mathrm{d}}} a_i(x_2) \end{bmatrix}, \quad f_{\mathrm{d}i}(x) = \begin{bmatrix} a_i(x_2) \\ a_i(x_1) \end{bmatrix}, \quad i = 1, \dots, n_{\mathrm{d}},$$

which implies that $f_j(x_j)$, $j = 1, 2$, are strictly decreasing. Next, with $P_i = I_n$, $i = 1, \dots, n_{\mathrm{d}}$, (8.18) and (8.19) are trivially satisfied, and, hence, it follows from Theorem 8.3 that $x_1(t) - x_2(t) \to 0$ as $t \to \infty$.

Next, we consider nonlinear compartmental time-delay dynamical systems of the form

$$\dot{x}_i(t) = -\sum_{j=1, j\ne i}^{n} a_{ji}(x_i(t)) + \sum_{j=1, j\ne i}^{n} a_{ij}(x_j(t-\tau_i)), \quad x(\theta) = \eta(\theta),$$

$$-\bar{\tau} \le \theta \le 0, \quad t \ge 0, \tag{8.35}$$

where $i = 1, \dots, n$, $a_{ij} : \overline{\mathbb{R}}_+ \to \overline{\mathbb{R}}_+$, $i \ne j$, $i, j \in \{1, \dots, n\}$, are such that $a_{ij}(0) = 0$ and $a_{ij}(\cdot)$, $i \ne j$, $i, j = 1, \dots, n$, is strictly increasing. Note that since each transfer coefficient $a_{ij}(\cdot)$ is a function of only $x_j$ and not $x$, the nonlinear compartmental system (8.35) is a *nonlinear donor-controlled compartmental system* [179]. In this case, (8.35) can be written in the form given by (8.17) with $n_{\mathrm{d}} = n$:

$$f_i(x_i) = -\sum_{j=1, j\ne i}^{n} a_{ji}(x_i), \quad f_{\mathrm{d}i}(x) = \mathbf{e}_i \sum_{j=1}^{n} a_{ij}(x_j), \quad i = 1, \dots, n. \tag{8.36}$$

Next, with $P_i = \mathbf{e}_i \mathbf{e}_i^{\mathrm{T}}$, $i = 1, \dots, n$, so that $P = I_n$, it follows that

(8.18) is trivially satisfied and (8.19) holds if and only if

$$\sum_{i=1}^{n}\left[\sum_{j=1,i\neq j}^{n} a_{ij}(x_j)\right]^2 \leq \sum_{i=1}^{n}\left[\sum_{j=1,i\neq j}^{n} a_{ji}(x_i)\right]^2, \quad x \in \overline{\mathbb{R}}_+^n. \tag{8.37}$$

In the case where $n = 2$, (8.37) is trivially satisfied, and, hence, it follows from Theorem 8.3 that $x_1(t) - x_2(t) \to 0$ as $t \to \infty$.

In general, (8.37) does not hold for arbitrary strictly increasing functions $a_{ij}(\cdot)$. However, if $a_{ij}(\cdot) = \sigma(\cdot)$, $i \neq j$, $i,j = 1,\ldots,n$, where $\sigma : \overline{\mathbb{R}}_+ \to \overline{\mathbb{R}}_+$ is such that $\sigma(0) = 0$ and is strictly increasing, (8.37) holds if and only if

$$\sum_{i=1}^{n}\left[\sum_{j=1,i\neq j}^{n} \sigma(x_j)\right]^2 \leq \sum_{i=1}^{n}\left[\sum_{j=1,i\neq j}^{n} \sigma(x_i)\right]^2, \quad x \in \mathbb{R}_+^n. \tag{8.38}$$

In this case, since

$$0 \geq (n-1)\sum_{i=1}^{n}\sigma^2(x_i) + (n-2)\sum_{i=1}^{n}\sum_{j=1,j\neq i}^{n}\sigma(x_i)\sigma(x_j) - (n-1)^2\sum_{i=1}^{n}\sigma^2(x_i)$$

$$= (n-2)\sum_{i=1}^{n}\sum_{j=1,j\neq i}^{n}(\sigma(x_i) - \sigma(x_j))^2,$$

(8.38) holds, and, hence, it follows from Theorem 8.3 that $x_i(t) - x_j(t) \to 0$ as $t \to \infty$, where $i \neq j$, $i,j = 1,\ldots,n$.

Next, we specialize Theorem 8.3 to nonlinear time-delay compartmental systems of the form

$$\dot{x}(t) = A\hat{\sigma}(x(t)) + \sum_{i=1}^{n_d} A_{di}\hat{\sigma}(x(t - \tau_i)), \quad x(\theta) = \eta(\theta),$$

$$-\bar{\tau} \leq \theta \leq 0, \quad t \geq 0, \tag{8.39}$$

where $\hat{\sigma} : \overline{\mathbb{R}}_+^n \to \overline{\mathbb{R}}_+^n$ is given by $\hat{\sigma}(x) = [\sigma(x_1), \sigma(x_2), \ldots, \sigma(x_n)]^T$, where $\sigma : \overline{\mathbb{R}}_+ \to \overline{\mathbb{R}}_+$ is such that $\sigma(u) = 0$ if and only if $u = 0$, and $A$ and $A_{di}$, $i = 1,\ldots,n_d$, are as given by (8.3).

**Theorem 8.4.** Consider (8.39) where $\sigma : \overline{\mathbb{R}}_+ \to \overline{\mathbb{R}}_+$ is such that $\sigma(0) = 0$ and $\sigma(\cdot)$ is strictly increasing. Assume that

$$\left(A + \sum_{i=1}^{n_d} A_{di}\right)^T \mathbf{e} = \left(A + \sum_{i=1}^{n_d} A_{di}\right)\mathbf{e} = 0$$

and

$$\text{rank}\left(A + \sum_{i=1}^{n_d} A_{di}\right) = n - 1.$$

Then, for every $\alpha \geq 0$, $\alpha\mathbf{e}$ is a semistable equilibrium point of (8.39). Furthermore, $x(t) \to \alpha^*\mathbf{e}$ as $t \to \infty$, where $\alpha^*$ satisfies

$$n\alpha^* + \sigma(\alpha^*)\sum_{i=1}^{n_d}\tau_i\mathbf{e}^T A_{di}\mathbf{e} = \mathbf{e}^T\eta(0) + \sum_{i=1}^{n_d}\int_{-\tau_i}^{0}\mathbf{e}^T A_{di}\hat{\sigma}(\eta(\theta))d\theta. \quad (8.40)$$

**Proof.** It follows from Lemma 8.2 that there exists $Q_i$, $i = 1, \ldots, n_d$, such that (8.5) holds with $Q_i$ given by (8.6). Now, since

$$A = -\sum_{i=1}^{n_d} Q_i = -\sum_{i=1}^{n_d} P_i^D = -P^{-1},$$

where $P = \sum_{i=1}^{n_d} P_i$, it follows from (8.5) that, for all $x \in \overline{\mathbb{R}}_+^n$,

$$0 \geq 2\hat{\sigma}^T(x)A\hat{\sigma}(x) + \hat{\sigma}^T(x)\sum_{i=1}^{n_d}(Q_i + A_{di}^T Q_i^D A_{di})\hat{\sigma}(x)$$

$$= -f^T(x)Pf(x) + \sum_{i=1}^{n_d} f_{di}^T(x)P_i f_{di}(x),$$

where $f(x) = A\hat{\sigma}(x)$ and $f_{di}(x) = A_{di}\hat{\sigma}(x)$, $i = 1, \ldots, n_d$, $x \in \overline{\mathbb{R}}_+^n$. Furthermore, since $P_i^D P_i A_{di} = A_{di}$, $i = 1, \ldots, n_d$, it follows that $P_i^D P_i f_{di}(x) = f_{di}(x)$, $i = 1, \ldots, n_d$, $x \in \overline{\mathbb{R}}_+^n$.

Now, the result is an immediate consequence of Theorem 8.3 by noting that $\mathbf{e}^T[f(x) + \sum_{i=1}^{n_d} f_{di}(x)] = 0$ and $f(x) + \sum_{i=1}^{n_d} f_{di}(x) = 0$ if and only if $x = \alpha\mathbf{e}$ for some $\alpha \geq 0$. $\qquad\square$

## 8.5 The Consensus Problem in Dynamical Networks

In this section, we apply the results of Sections 8.3 and 8.4 to the consensus problem in dynamical networks [217, 237, 238, 241, 315]. Recall that the consensus problem involves finding a dynamic algorithm that enables a group of agents in a network to agree upon certain quantities of interest with directed information flow subject to possible link failures and time delays. As in [238], we use directed graphs to represent a dynamical network and present solutions to the consensus problem for networks with *balanced* graph *topologies* (or information flow) [238] and unknown arbitrary time delays.

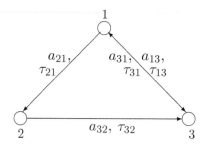

Figure 8.1 Three-state network system.

The consensus problem we consider involves the dynamical network characterized by the dynamical system

$$\dot{x}(t) = u(t), \quad x(0) = x_0, \quad t \geq 0, \tag{8.41}$$

where $x(t) \triangleq [x_1(t), \ldots, x_n(t)]^{\mathrm{T}}$ is the state of the network and $u(t) \triangleq [u_1(t), \ldots, u_m(t)]^{\mathrm{T}}$ is the input to the network with components $u_i(t)$ depending only on the states of the nodes $i$ and its neighbors. Specifically, the consensus problem deals with the design of an input $u(t)$ such that $x(t)$ converges to $\alpha \mathbf{e}$ as $t \to \infty$, where $\alpha \in \mathbb{R}$.

Due to the presence of directional constraints on information flow and system time delays, $u_i(t)$ is constrained to the feedback form

$$u_i(t) = f_i(x_i(t), x_{j_1}(t - \tau_{ij_1}), \ldots, x_{j_{m_i}}(t - \tau_{ij_{m_i}})),$$

where $\tau_{ij_k} > 0$, $j_k \in \mathcal{N}_i \triangleq \{j \in \{1, \ldots, n\} : (j, i) \in \mathcal{E}\}$, are unknown constant time delays between nodes $i$ and $j_k$. For notational convenience we additionally define the parameters $\tau_{ij} \triangleq 0$ if $(j, i) \notin \mathcal{E}$.

As an example, consider the network given in Figure 8.1, where $\mathcal{V} = \{1, 2, 3\}$ and $\mathcal{E} = \{(1, 2), (2, 3), (1, 3), (3, 1)\}$, with adjacency matrix $\mathcal{A}$ with entries $a_{13}$, $a_{21}$, $a_{31}$, and $a_{32} > 0$, and with the remaining entries being zeros. In this case, the input to the network is given by

$$u_1(t) = f_1(x_1(t), x_3(t - \tau_{13})),$$
$$u_2(t) = f_2(x_2(t), x_1(t - \tau_{21})),$$
$$u_3(t) = f_3(x_3(t), x_2(t - \tau_{32}), x_1(t - \tau_{31})),$$

so that, for $i = 1, 2, 3$, $\dot{x}_i(t)$ is dependent only on the states (values) of the nodes that are accessible by node $i$ and with $\tau_{ij}$ denoting the communication delay from node $j$ to node $i$.

Next, we apply Theorems 8.2 and 8.4 and present linear and nonlinear solutions for the consensus problem. Specifically, first we choose

$$f_i(x(t)) = -\sum_{j=1,i\neq j}^{n} a_{ji}x_i(t) + \sum_{j=1,i\neq j}^{n} a_{ij}x_j(t - \tau_{ij}), \quad i = 1,\ldots,n, \quad (8.42)$$

so that the *closed-loop* system is given by

$$\dot{x}_i(t) = -\sum_{j=1,i\neq j}^{n} a_{ji}x_i(t) + \sum_{j=1,i\neq j}^{n} a_{ij}x_j(t - \tau_{ij}), \quad x_i(\theta) = \eta_i(\theta),$$
$$-\bar{\tau} \leq \theta \leq 0, \quad t \geq 0, \quad (8.43)$$

for all $i = 1,\ldots,n$ or, equivalently,

$$\dot{x}(t) = Ax(t) + \sum_{l=1}^{n_{\mathrm{d}}} A_{\mathrm{d}l}x(t - \tau_l), \quad x(\theta) = \eta(\theta), \quad -\bar{\tau} \leq \theta \leq 0, \quad t \geq 0,$$
$$(8.44)$$

where $n_{\mathrm{d}} \triangleq n^2$; $A \in \mathbb{R}^{n\times n}$; and $A_{\mathrm{d}l} \in \mathbb{R}^{n\times n}$, $l = 1,\ldots,n_{\mathrm{d}}$, with

$$A = \mathrm{diag}\left[-\sum_{j=2}^{n} a_{j1}, \ldots, -\sum_{j=1}^{n-1} a_{jn}\right]; \quad (8.45)$$

$A_{\mathrm{d}((i-1)n+j)} = a_{ij}\mathbf{e}_i\mathbf{e}_j^{\mathrm{T}}$; and $\tau_{((i-1)n+j)} = \tau_{ij}$, $i,j = 1,\ldots,n$. Note that if $(j,i) \notin \mathcal{E}$, then $A_{\mathrm{d}((i-1)n+j)} = 0$, which implies that the algorithm is consistent with the directional constraints.

Furthermore, it can be easily shown that $(A + A_{\mathrm{d}})^{\mathrm{T}}\mathbf{e} = 0$, where $A_{\mathrm{d}} \triangleq \sum_{l=1}^{n_{\mathrm{d}}} A_{\mathrm{d}l}$, and $\mathrm{rank}(A + A_{\mathrm{d}}) = n - 1$ if and only if for every pair of nodes $(i,j) \in \mathcal{V}$ there exists a *path* from node $i$ to node $j$ [110]. Here, we assume that the adjacency matrix $\mathcal{A}$ is chosen such that $(A + A_{\mathrm{d}})\mathbf{e} = 0$ so that the linear time-delay closed-loop dynamical system (8.44) satisfies all of the conditions of Theorem 8.2. Hence, it follows from Theorem 8.2 that the dynamical network given by (8.44) solves the consensus problem; that is, $\lim_{t\to\infty}x_i(t) = \lim_{t\to\infty} x_j(t) = \alpha^*$, $i,j = 1,\ldots,n$, $i \neq j$, where $\alpha^*$ is given by (8.8).

Alternatively, it follows from Theorem 8.4 that the nonlinear dynamical network given by

$$\dot{x}(t) = A\hat{\sigma}(x(t)) + \sum_{i=1}^{n_{\mathrm{d}}} A_{\mathrm{d}i}\hat{\sigma}(x(t - \tau_i)), \quad x(\theta) = \eta(\theta), \quad -\bar{\tau} \leq \theta \leq 0, \quad t \geq 0,$$
$$(8.46)$$

also solves the nonlinear consensus problem where $\sigma(\cdot)$ and $\hat{\sigma}(\cdot)$ satisfy the conditions in Theorem 8.4. In this case, $\lim_{t\to\infty}x_i(t) = \lim_{t\to\infty} x_j(t) = \alpha^*$, $i,j = 1,\ldots,n$, $i \neq j$, where $\alpha^*$ is a solution to (8.40). Note that if $\sigma(\theta) = \theta$, then (8.46) specializes to (8.44). Although both (8.44) and (8.46) solve the same network consensus problem, the nonlinear function $\sigma(\cdot)$ within $\hat{\sigma}(\cdot)$ may be used to enhance the performance of the dynamic algorithm or satisfy other constraints. For example, choosing $\sigma(\theta) = \tanh(\theta)$, we can constrain bandwidth information from one agent to another.

**Example 8.1.** To illustrate the two algorithms given by (8.44) and (8.46), consider the network given by the graph shown in Figure 8.2 [238], where $a_{ij}$ and $\tau_{ij}$ denote the weight and the time delay for each edge shown. Here, we choose $a_{(i,j)} = 1$ if $(i,j) \in \mathcal{E}$ so that $(A + A_{\mathrm{d}})\mathbf{e} = 0$. In addition, it can be easily shown that $\mathrm{rank}(A + A_{\mathrm{d}}) = n - 1 = 9$.

With $x_0 = [1\ 2\ 3\ 4\ 5\ 6\ 7\ 8\ 9\ 10]^{\mathrm{T}}$, Figures 8.3 and 8.4 demonstrate the agreement between all nodes for the algorithms given by (8.44) and (8.46), respectively, with $\sigma(\theta) = \tanh(\theta)$ in (8.46). Finally, Figures 8.5 and 8.6 show the control input versus time for both linear and nonlinear consensus algorithms. Note that the maximum amplitude of the linear consensus algorithm is about six times that of the nonlinear consensus algorithm, and, as expected, the settling time of the nonlinear algorithm is longer than that of the linear algorithm.                                                                        $\triangle$

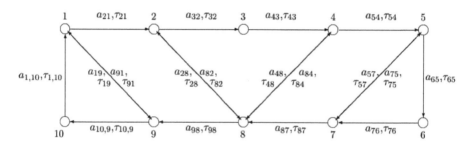

Figure 8.2 Balanced network.

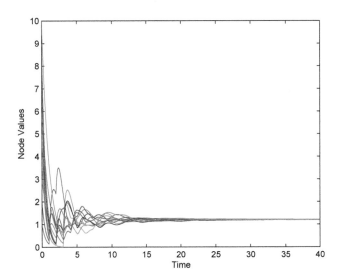

Figure 8.3 Linear consensus algorithm.

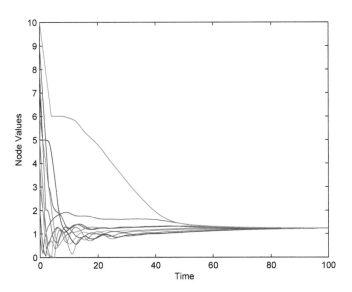

Figure 8.4 Nonlinear consensus algorithm.

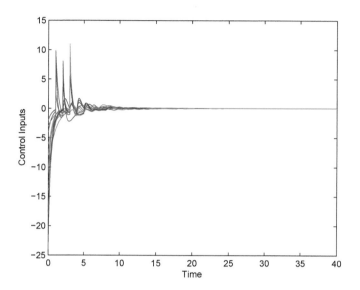

Figure 8.5  Linear consensus algorithm.

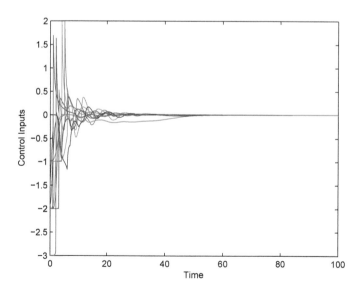

Figure 8.6  Nonlinear consensus algorithm.

*Chapter Nine*

---

# Eulerian Swarming Networks and Continuum Thermodynamics

## 9.1 Introduction

To develop distributed methods for control and coordination of autonomous multiagent systems, many researchers have looked to autonomous *swarm* systems appearing in nature for inspiration [194, 197, 226, 245, 258, 304]. In particular, biology has shown that many species, such as insect swarms, ungulate flocks, fish schools, ant colonies, and bacterial colonies, *self-organize* in nature [21, 57, 233, 243]. These biological aggregations give rise to remarkably complex global behaviors from simple local interactions between large numbers of relatively unintelligent agents without the need for centralized control. The spontaneous development (i.e., self-organization) of these autonomous biological systems and their spatiotemporal evolution to more complex states often appear without any external system interaction. In other words, structure morphing into coherent groups is internal to the system and results from local interactions among subsystem components that are independent of the physical nature of the individual components.

These local interactions often make up a simple set of rules that lead to remarkably complex global behaviors. *Complexity* here refers to the quality of a system wherein interacting subsystems self-organize to form hierarchical evolving structures exhibiting *emergent* system behaviors. Hence, a complex dynamical system is a system that is greater than the sum of its subsystems or parts. In addition, the spatially distributed sensing and actuation control architecture prevalent in such systems is inherently robust to individual subsystem (or agent) failures and unplanned behavior at the individual subsystem (or agent) level.

The connection between the local subsystem interactions and the globally complex system behavior is often elusive.[13] Complex dynamical systems

---

[13]This is true for nature in general and was most eloquently stated first by the ancient Greek philosopher Herakleitos (∼535–∼475 B.C.) in his 123rd fragment— Φύσις κρύπτεσθαι φιλεί (Nature loves to hide).

involving self-organizing components forming spatiotemporal evolving structures that exhibit a hierarchy of emergent system properties are not limited to biological aggregation systems. Such systems include, for example, nervous systems, immune systems, ecological systems, quantum particle systems, chemical reaction systems, economic systems, cellular systems, and galaxies, to cite but a few examples. These systems are known as *dissipative systems* [118, 127, 188] and consume energy and matter while maintaining their stable structure by dissipating entropy to the environment.

For example, as in biology,[14] in the physical universe billions of stars and galaxies interact to form self-organizing dissipative nonequilibrium structures [188, 251]. The fundamental common phenomenon among these systems is that they evolve in accordance to the laws of (nonequilibrium) thermodynamics, which are among the most firmly established laws of nature. System thermodynamics, in the sense of [118, 127], involves open interconnected dynamical systems that exchange matter and energy with their environment in accordance with the first law (conservation of energy) and the second law (nonconservation of entropy) of thermodynamics.

Self-organization can spontaneously occur in such systems by invoking the two fundamental axioms of the science of heat. Namely, (i) if the energies in the connected subsystems of an interconnected system are equal, then energy exchange between these subsystems is not possible, and (ii) energy flows from more energetic subsystems to less energetic subsystems. These axioms establish the existence of a system entropy function as well as *equipartition of energy* [118, 127] in system thermodynamics and *information consensus* [162] in cooperative networks, an *emergent* behavior in thermodynamic systems as well as swarm systems. Hence, in complex interconnected dynamical systems, self-organization is not a property of the system's parts but rather emerges as a result of the nonlinear subsystem interactions.

In light of the above discussion, engineering swarm systems necessitates the development of relatively simple autonomous agents that are inherently distributed, self-organized, and truly scalable. Scalability follows from the fact that such systems do not involve centralized control and communication architectures. In addition, engineered swarming systems should be inherently robust to individual agent failures, unplanned task assignment changes, and environmental changes. Mathematical models for large-scale swarms can involve Lagrangian or Eulerian models. In a Lagrangian model,

---

[14]All living systems are dissipative systems, but the converse is not necessarily true. Dissipative living systems involve pattern interactions by which life emerges. This nonlinear interaction between the subsystems making up a living system is characterized by *autopoiesis* (self-creation).

each agent is modeled as a particle governed by a difference or differential equation, whereas an Eulerian model describes the local energy or information flux for a distribution of swarms with an advection-diffusion (conservation) equation. The two formulations can be connected by a Fokker–Planck approximation relating jump distance distributions of individual agents to terms in the advection-diffusion equation [233].

As discussed in Chapter 1, in many applications involving multiagent systems, groups of agents are required to agree on certain quantities of interest. In particular, it is important to develop information consensus protocols for networks of dynamic agents wherein a unique feature of the closed-loop dynamics under any control algorithm that achieves consensus is the existence of a continuum of equilibria representing a state of equipartitioning or consensus. Under such dynamics, the limiting consensus state achieved is not determined completely by the dynamics but depends on the initial system state as well. For such systems possessing a continuum of equilibria, semistability [35, 36], not asymptotic stability, is the relevant notion of stability.

In this chapter, we develop distributed boundary control algorithms for addressing the consensus problem for an Eulerian swarm model. The proposed distributed boundary controller architectures are predicated on the recently developed notion of continuum thermodynamics [118, 127], resulting in controller architectures involving the exchange of information between uniformly distributed swarms over an $n$-dimensional (not necessarily Euclidian) space that guarantee that the closed-loop system is consistent with basic thermodynamic principles. For our thermodynamically consistent model, we further establish the existence of a unique continuously differentiable entropy functional for all equilibrium and nonequilibrium states of our system. Information consensus and semistability are shown using the well-known Sobolev embedding theorems and the notion of generalized (or weak) solutions. Finally, since the closed-loop system is guaranteed to satisfy basic thermodynamic principles, robustness to individual agent failures and unplanned individual agent behavior is automatically guaranteed.

## 9.2 Mathematical Preliminaries

In this chapter, we consider an Eulerian swarm model involving a nonlocal spatiotemporal distribution of swarm density. Specifically, consider the evolution equation for swarm aggregations defined over a compact connected set $\mathcal{V} \subset \mathbb{R}^n$ with a smooth boundary $\partial \mathcal{V}$ and volume $\mathrm{vol}\,\mathcal{V}$ characterized by the *conservation* equation [88, 118]

$$\frac{\partial u(x,t)}{\partial t} = -\nabla \cdot \phi(x, u(x,t), \nabla u(x,t)), \quad x \in \mathcal{V}, \quad t \geq t_0, \qquad (9.1)$$

$$u(x, t_0) = u_{t_0}(x) \in \mathcal{X}, \quad x \in \mathcal{V}, \quad \phi(x, u(x, t), \nabla u(x, t)) \cdot \mathbf{n}(x) \geq 0,$$
$$x \in \partial\mathcal{V}, \quad t \geq t_0, \tag{9.2}$$

where $u : \mathcal{V} \times [0, \infty) \to \overline{\mathbb{R}}_+$ denotes the density distribution at the point $x = [x_1, \ldots, x_n]^{\mathrm{T}} \in \mathcal{V}$ and time instant $t \geq t_0$, $\phi : \mathcal{V} \times [0, \infty) \times \mathbb{R}^n \to \mathbb{R}^n$ denotes a continuously differentiable *flux* function, $\nabla$ denotes the nabla operator, "·" denotes the dot product in $\mathbb{R}^n$, $\mathbf{n}^{\mathrm{T}}(x)$ denotes the outward normal vector to the boundary $\partial\mathcal{V}$ at $x \in \partial\mathcal{V}$, and $\mathcal{X}$ denotes a space of two-times continuously differentiable scalar functions defined on $\mathcal{V}$. Here, we assume that $\mathcal{V} = \{x \in \mathbb{R}^n : f(x) \leq 0\}$ and $\partial\mathcal{V} = \{x \in \mathbb{R}^n : f(x) = 0\}$, where $f : \mathbb{R}^n \to \mathbb{R}$ is a given continuously differentiable function, and, consequently, the outward normal vector to the boundary $\partial\mathcal{V}$ at $x \in \partial\mathcal{V}$ is given by $\mathbf{n}^{\mathrm{T}}(x) = \nabla f(x)$.

Equations (9.1) and (9.2) involve an information (or energy) flow equation for a uniformly distributed continuous system. Specifically, note that for a smooth, bounded region $\mathcal{V} \subset \mathbb{R}^n$, the integral $\int_{\mathcal{V}} u(x, t) \mathrm{d}\mathcal{V}$ denotes the total information (or energy) amount within $\mathcal{V}$ at time $t$. Hence, the rate of information change within $\mathcal{V}$ is governed by the flux function $\phi : \mathcal{V} \times \overline{\mathbb{R}}_+ \times \mathbb{R}^n \to \mathbb{R}^n$, which controls the rate of information transmission through the boundary $\partial\mathcal{V}$. Hence, for each time $t$,

$$\frac{\mathrm{d}}{\mathrm{d}t} \int_{\mathcal{V}} u(x, t) \mathrm{d}\mathcal{V} = -\int_{\partial\mathcal{V}} \phi(x, u(x, t), \nabla u(x, t)) \cdot \mathbf{n}(x) \mathrm{d}\mathcal{S}_{\mathcal{V}}, \tag{9.3}$$

where $\mathrm{d}\mathcal{S}_{\mathcal{V}}$ denotes an infinitesimal surface element of the boundary of the set $\mathcal{V}$. Using the divergence theorem, it follows from (9.3) that

$$\frac{\mathrm{d}}{\mathrm{d}t} \int_{\mathcal{V}} u(x, t) \mathrm{d}\mathcal{V} = -\int_{\partial\mathcal{V}} \phi(x, u(x, t), \nabla u(x, t)) \cdot \mathbf{n}(x) \mathrm{d}\mathcal{S}_{\mathcal{V}}$$
$$= -\int_{\mathcal{V}} \nabla \cdot \phi(x, u(x, t), \nabla u(x, t)) \mathrm{d}\mathcal{V}. \tag{9.4}$$

Since the region $\mathcal{V} \subset \mathbb{R}^n$ is arbitrary, it follows that the conservation equation over a unit volume within the continuum $\mathcal{V}$ involving the rate of information density change within the continuum is given by (9.1) and (9.2). The physical interpretation of (9.1) and (9.2) is straightforward. In particular, if $u(x, t)$ is an information (or energy) density at point $x \in \mathcal{V}$ and time $t \geq t_0$, then the conservation equation (9.1) describes the time evolution of the information (or energy) density $u(x, t)$ over the region $\mathcal{V}$, while the boundary condition in (9.2) involving the dot product implies that the information (or energy) of the system (9.1)–(9.2) can be either stored or transmitted but not supplied through the boundary of $\mathcal{V}$ from the environment.

We denote the information (or energy) distribution over the set $\mathcal{V}$ at time $t \geq t_0$ by $u_t \in \mathcal{X}$ so that for each $t \geq t_0$ the set of mappings generated by $u_t(x) \equiv u(x, t)$ for every $x \in \mathcal{V}$ gives the *flow* of (9.1) and (9.2). We assume that the function $\phi(\cdot, \cdot, \cdot)$ is continuously differentiable so that (9.1) and (9.2) admit a unique solution $u(x, t)$, $x \in \mathcal{V}$, $t \geq t_0$, and $u(\cdot, t) \in \mathcal{X}$, $t \geq t_0$, is continuously dependent on the initial information (or energy) distribution $u_{t_0}(x)$, $x \in \mathcal{V}$. It is well known, however, that nonlinear partial differential equations need not have smooth differentiable solutions (*classical solutions*), and one has to use the notion of Schwartz distributions, which provides a framework in which the information (or energy) density function $u(x, t)$ may be differentiated in a generalized sense infinitely often [88]. In this case, one has a well-defined notion of solutions that have jump discontinuities, which propagate as shock waves. Thus, one has to deal with *generalized* or *weak* solutions wherein uniqueness is lost. In this case, the *Clausius–Duhem* inequality is invoked for identifying the physically relevant (i.e., thermodynamically admissible) solution [75, 88].

If $u_{t_0}$ is a two-times continuously differentiable function with compact support and its derivative is sufficiently small on $[t_0, \infty)$, then the classical solution to (9.1) and (9.2) can break down at a finite time. As a consequence of this, one may only hope to find generalized (or weak) solutions to (9.1) and (9.2) over the semi-infinite interval $[t_0, \infty)$, that is, $\mathcal{L}_\infty$ functions[15] $u(\cdot, \cdot)$ that satisfy (9.1) in the sense of distributions, which provides a framework in which $u(\cdot, \cdot)$ may be differentiated in a general sense infinitely often.

It is important to note that we do *not* assume strict hyperbolicity of (9.1) and (9.2) since our interest in this chapter is to address semistability, and, hence, (9.1) and (9.2) cannot be hyperbolic. Thus, many results on the well-posedness of solutions of (9.1) and (9.2) developed in the literature are not applicable in this case. Furthermore, the linearization method also fails to provide any stability information due to nonhyperbolicity. Global well-posedness of smooth solutions of nonhyperbolic partial differential equations of the form (9.1) and (9.2) remains an open problem in mathematics [95]. Finally, the control aim here is to design a *boundary control* law so that the corresponding closed-loop system achieves semistability and *uniform information distribution* [127].

In this chapter, $\mathcal{L}_2$ denotes the space of square-integrable Lebesgue measurable functions on $\mathcal{V}$, and the $\mathcal{L}_2$ operator norm $\|\cdot\|_{\mathcal{L}_2}$ on $\mathcal{X}$ is used for

---

[15] $\mathcal{L}_\infty$ denotes the space of bounded Lebesgue measurable functions on $\mathcal{V}$ and provides the broadest framework for weak solutions. Alternatively, a natural function class for weak solutions is the space $\mathcal{BV}$ consisting of functions of bounded variation. Recall that a bounded measurable function $u(x, t)$ has locally bounded variation if its distributional derivatives are locally finite Radon measures.

the definitions of Lyapunov, semi-, and asymptotic stability. Furthermore, we introduce the Sobolev spaces[16]

$$\mathcal{W}_2^0(\mathcal{V}) \triangleq \{u_t : \mathcal{V} \to \mathbb{R} : u_t \in \mathrm{C}^0(\mathcal{V}) \cap \mathcal{L}_2(\mathcal{V})\}_{\mathrm{co}} \subset \mathcal{L}_2(\mathcal{V}), \tag{9.5}$$

$$\mathcal{W}_2^1(\mathcal{V}) \triangleq \{u_t : \mathcal{V} \to \mathbb{R} : u_t \in \mathrm{C}^1(\mathcal{V}) \cap \mathcal{L}_2(\mathcal{V}), (\nabla u_t)^{\mathrm{T}} \in \mathcal{L}_2(\mathcal{V})\}_{\mathrm{co}}, \tag{9.6}$$

where $\mathrm{C}^r(\mathcal{V})$ denotes a function space defined on $\mathcal{V}$ with $r$-continuous derivatives and $\{\cdot\}_{\mathrm{co}}$ denotes the completion[17] of $\{\cdot\}$ in $\mathcal{L}_2$ in the sense of [307], with norms

$$\|u_t\|_{\mathcal{W}_2^0} \triangleq \|u_t\|_{\mathcal{L}_2} = \left[\int_{\mathcal{V}} u_t^2(x) \mathrm{d}\mathcal{V}\right]^{\frac{1}{2}}, \tag{9.7}$$

$$\|u_t\|_{\mathcal{W}_2^1} \triangleq \left[\|u_t\|_{\mathcal{W}_2^0}^2 + D(u_t, u_t)\right]^{\frac{1}{2}} \tag{9.8}$$

defined on $\mathcal{W}_2^0(\mathcal{V})$ and $\mathcal{W}_2^1(\mathcal{V})$, respectively, where the gradient $\nabla u_t(x)$ in (9.8) is interpreted in the sense of a generalized gradient [307] and

$$D(u_t, u_t) \triangleq \int_{\mathcal{V}} \nabla u_t(x) \nabla^{\mathrm{T}} u_t(x) \mathrm{d}\mathcal{V}$$

is the *Dirichlet integral* of $u$ [98, p. 88].

Physically, the Dirichlet integral term represents the potential energy in $\mathcal{V}$ of the *electrostatic field* $-\nabla u$. Note that since the solutions to (9.1) and (9.2) are assumed to be two-times continuously differentiable functions on a compact set $\mathcal{V}$ and $\phi$ is continuously differentiable, it follows that $u_t(x)$, $t \geq t_0$, belongs to $\mathcal{W}_2^0(\mathcal{V})$ and $\mathcal{W}_2^1(\mathcal{V})$.

## 9.3 A Thermodynamic Model for Large-Scale Swarms

The nonlinear conservation equation (9.1)–(9.2) can exhibit a full range of nonlinear behavior, including bifurcations, limit cycles, and even chaos. To ensure a thermodynamically consistent information (or energy) flow model involving a diffusive (parabolic) character, additional assumptions are required. In this section, we develop a large-scale swarm model that is consistent with basic thermodynamic principles. First, however, we establish several key definitions and stability results for nonlinear infinite-dimensional

---

[16] A *Sobolev space* is a vector space of locally summable functions having weak derivatives of various orders and endowed with a norm formed from a combination of $\mathcal{L}_p$ norms.

[17] The space $\{\cdot\}$ defined as part of (9.6) is not complete with respect to the norm generated by the inner product (9.8). This space can be completed by adding the limit points of all Cauchy sequences in $\{\cdot\}$. In this way, $\{\cdot\}$ is embedded in the larger normed space $\{\cdot\}_{\mathrm{co}}$, which is complete. Of course, it follows from the Riesz–Fischer theorem [264, p. 125] that $\mathcal{L}_2$ is complete with respect to the norm generated by the inner product (9.7).

systems. Here, the state space is assumed to be a Banach space with fully nonlinear dynamics.

Let $\mathcal{B}$ be a Banach space with norm $\| \cdot \|_{\mathcal{B}}$. A *dynamical system* $\mathcal{G}$ on $\mathcal{B}$ is the triple $(\mathcal{B}, [t_0, \infty), s)$, where $s : [t_0, \infty) \times \mathcal{B} \to \mathcal{B}$ is such that the following axioms hold: (i) (*continuity*) $s(\cdot, \cdot)$ is jointly continuous, (ii) (*consistency*) $s(t_0, z_0) = z_0$ for all $t_0 \in \mathbb{R}$ and $z_0 \in \mathcal{B}$, and (iii) (*semigroup property*) $s(t + \tau, z_0) = s(\tau, s(t, z_0))$ for all $z_0 \in \mathcal{B}$ and $t, \tau \in [t_0, \infty)$. Given $t \in [0, \infty)$, we denote the *flow* $s(t, \cdot) : \mathcal{B} \to \mathcal{B}$ of $\mathcal{G}$ by $s_t(x_0)$ or $s_t$. Likewise, given $x \in \mathcal{B}$, we denote the *solution curve* or *trajectory* $s(\cdot, x) : [0, \infty) \to \mathcal{B}$ of $\mathcal{G}$ by $s^x(t)$ or $s^x$. The *positive limit set* of $x \in \mathcal{B}$ is the set $\omega(x)$ of points $z \in \mathcal{B}$ such that there exists an increasing sequence $\{t_i\}_{i=1}^{\infty}$ satisfying $s(t_i, x) \to z$ as $i \to \infty$. Finally, the image of $\mathcal{U} \subset \mathcal{B}$ under the flow $s_t$ is defined by $s_t(\mathcal{U}) \triangleq \{y : y = s_t(x_0) \text{ for some } x_0 \in \mathcal{U}\}$.

An *equilibrium point* of $\mathcal{G}$ is a point $z \in \mathcal{B}$ such that $s(t, z) = s(t_0, z)$ for all $t \geq t_0$. A set $\mathcal{M} \subseteq \mathcal{B}$ is *positively invariant* if $s_t(\mathcal{M}) \subseteq \mathcal{M}$ for all $t \geq 0$. The set $\mathcal{M}$ is *negatively invariant* if, for every $z \in \mathcal{M}$ and every $t \geq 0$, there exists $x \in \mathcal{M}$ such that $s(t, x) = z$ and $s(\tau, x) \in \mathcal{M}$ for all $\tau \in [0, t]$. The set $\mathcal{M}$ is *invariant* if $s_t(\mathcal{M}) = \mathcal{M}$, $t \geq 0$. Note that a set is invariant if and only if it is positively and negatively invariant.

**Definition 9.1.** Let $\mathcal{G}$ be a dynamical system on a Banach space $\mathcal{B}$ with norm $\| \cdot \|_{\mathcal{B}}$, and let $\mathcal{D}$ be a positively invariant set with respect to $\mathcal{G}$. An equilibrium point $x \in \mathcal{D}$ of $\mathcal{G}$ is *Lyapunov stable* if for every relatively open subset $\mathcal{N}_\varepsilon$ of $\mathcal{D}$ containing $x$, there exists a relatively open subset $\mathcal{N}_\delta$ of $\mathcal{D}$ containing $x$ such that $s_t(\mathcal{N}_\delta) \subseteq \mathcal{N}_\varepsilon$ for all $t \geq t_0$. An equilibrium point $x \in \mathcal{D}$ of $\mathcal{G}$ is *semistable* if it is Lyapunov stable and there exists a relatively open subset $\mathcal{U}$ of $\mathcal{D}$ containing $x$ such that for all initial conditions in $\mathcal{U}$, the trajectory $s(\cdot, \cdot)$ of $\mathcal{G}$ converges to a Lyapunov stable equilibrium point; that is, $\lim_{t \to \infty} s(t, z) = y$, where $y \in \mathcal{D}$ is a Lyapunov stable equilibrium point of $\mathcal{G}$ and $z \in \mathcal{U}$. Finally, an equilibrium point $x \in \mathcal{D}$ of $\mathcal{G}$ is *asymptotically stable* if it is Lyapunov stable and there exists a relatively open subset $\mathcal{U}$ of $\mathcal{D}$ containing $x$ such that $\lim_{t \to \infty} s(t, z) = x$ for all $z \in \mathcal{U}$.

The next result gives a sufficient condition to guarantee semistability of the equilibria of $\mathcal{G}$. For the statement of this result, let $\mathcal{B}$ and $\mathcal{C}$ be Banach spaces and recall that $\mathcal{B}$ is *compactly embedded* in $\mathcal{C}$ if $\mathcal{B} \subset \mathcal{C}$ and a unit ball in $\mathcal{B}$ belongs to a compact subset in $\mathcal{C}$. Furthermore, define

$$\dot{V}(z) \triangleq \lim_{h \to 0^+} \frac{1}{h}[V s(t_0 + h, z) - V(z)], \quad z \in \mathcal{B}, \tag{9.9}$$

for a given continuous function $V : \mathcal{B} \to \mathbb{R}$ and every $z \in \mathcal{B}$ such that the limit in (9.9) exists.

**Theorem 9.1.** Let $\mathcal{B}$ and $\mathcal{C}$ be Banach spaces such that $\mathcal{B}$ is compactly embedded in $\mathcal{C}$, and let $\mathcal{G}$ be a dynamical system defined in $\mathcal{B}$ and $\mathcal{C}$. Assume that there exist locally Lipschitz continuous functions $V_{\mathcal{B}} : \mathcal{B} \to \mathbb{R}$ and $V_{\mathcal{C}} : \mathcal{C} \to \mathbb{R}$ such that $V_{\mathcal{B}}(z) \geq 0$, $z \in \mathcal{B}_c$, and $V_{\mathcal{C}}(z) \geq 0$, $z \in \mathcal{C}_c$, where $\mathcal{B}_c = \{z \in \mathcal{B} : V_{\mathcal{B}}(z) < \eta\}$ and $\mathcal{C}_c = \{z \in \mathcal{C} : V_{\mathcal{C}}(z) < \eta\}$ for some $\eta > 0$ such that $\mathcal{B}_c \subset \mathcal{C}_c$. Furthermore, assume that $V_{\mathcal{B}}(s(t, z_0)) \leq V_{\mathcal{B}}(s(\tau, z_0))$ for all $t_0 \leq \tau \leq t$ and $z_0 \in \mathcal{B}_c$ and $V_{\mathcal{B}}(s(t, z_0)) \leq V_{\mathcal{B}}(s(\tau, z_0))$ for all $t_0 \leq \tau \leq t$ and $z_0 \in \mathcal{C}_c$. If $\mathcal{B}_c$ is bounded and every point in the largest invariant subset $\mathcal{M}$ contained in $\mathcal{R}$ given by $\mathcal{R} \triangleq \{z \in \overline{\mathcal{C}_c} : \dot{V}_{\mathcal{C}}(z) = 0\}$ is a Lyapunov stable equilibrium point of $\mathcal{G}$, then every equilibrium point in $\mathcal{M}$ is semistable.

**Proof.** First note that the assumptions on $V_{\mathcal{B}}$ imply that the trajectory $s(t, x)$ of $\mathcal{G}$ remains in $\mathcal{B}_c$ for all $x \in \mathcal{B}_c$ and $t \geq t_0$. Furthermore, since $\mathcal{B}$ is compactly embedded in $\mathcal{C}$, $s(t, x)$ is contained in a compact set of $\mathcal{C}_c$ for all $x \in \mathcal{B}_c$ and $t \geq t_0$. Now, it follows from Lemma 3 and Theorem 1 of [140] that, for every $x \in \mathcal{B}_c$, the positive limit set $\omega(x)$ of $x$ is nonempty and contained in the largest invariant subset $\mathcal{M}$ of $\mathcal{R}$. Since every point in $\mathcal{M}$ is a Lyapunov stable equilibrium point, it follows that every point in $\omega(x)$ is a Lyapunov stable equilibrium point.

Next, let $z \in \omega(x)$, and let $\mathcal{U}_\varepsilon$ be an open neighborhood of $z$. By Lyapunov stability of $z$, it follows that there exists a relatively open subset $\mathcal{U}_\delta$ containing $z$ such that $s_t(\mathcal{U}_\delta) \subseteq \mathcal{U}_\varepsilon$ for every $t \geq t_0$. Since $z \in \omega(x)$, it follows that there exists $h \geq 0$ such that $s(h, x) \in \mathcal{U}_\delta$. Thus, $s(t + h, x) = s_t(s(h, x)) \in s_t(\mathcal{U}_\delta) \subseteq \mathcal{U}_\varepsilon$ for every $t > t_0$. Hence, since $\mathcal{U}_\varepsilon$ was chosen arbitrarily, it follows that $z = \lim_{t \to \infty} s(t, x)$. Now, it follows that $\lim_{i \to \infty} s(t_i, x) \to z$ for every divergent sequence $\{t_i\}$, and, hence, $\omega(x) = \{z\}$. Finally, since $\lim_{t \to \infty} s(t, x) \in \mathcal{M}$ is Lyapunov stable for every $x \in \mathcal{B}_c$, it follows from the definition of semistability that every equilibrium point in $\mathcal{M}$ is semistable. $\qquad\square$

The following assumptions are needed for the main results of the chapter. For the statement of these assumptions, $\phi : \mathcal{V} \times \overline{\mathbb{R}}_+ \times \mathbb{R}^n \to \mathbb{R}^n$ denotes the system information (or energy) flow within the continuum $\mathcal{V}$; that is,

$$\phi(x, u(x, t), \nabla u(x, t))$$
$$= [\phi_1(x, u(x, t), \nabla u(x, t)), \ldots, \phi_n(x, u(x, t), \nabla u(x, t))]^{\mathrm{T}},$$

where $\phi_i(\cdot, \cdot, \cdot)$ denotes the information (or energy) flow through a unit area per unit time in the $x_i$ direction for all $i = 1, \ldots, n$ and $\nabla u(x, t) \triangleq [D_1 u(x, t), \ldots, D_n u(x, t)]$, $x \in \mathcal{D}$, $t \geq t_0$, denotes the gradient of $u(\cdot, t)$ with respect to the spatial variable $x$.

**Assumption 9.1.** For every $x \in \mathcal{V}$ and unit vector $\mathbf{u} \in \mathbb{R}^n$, $\phi(x, u_t(x)$,

$\nabla u_t(x)) \cdot \mathbf{u} = 0$ if and only if $\nabla u_t(x)\mathbf{u} = 0$.

**Assumption 9.2.** For every $x \in \mathcal{V}$ and unit vector $\mathbf{u} \in \mathbb{R}^n$, $\phi(x, u_t(x),$ $\nabla u_t(x)) \cdot \mathbf{u} > 0$ if and only if $\nabla u_t(x)\mathbf{u} < 0$, and $\phi(x, u_t(x), \nabla u_t(x)) \cdot \mathbf{u} < 0$ if and only if $\nabla u_t(x)\mathbf{u} > 0$.

Note that Assumptions 9.1 and 9.2 are the infinite-dimensional versions of Assumptions 2.1 and 2.2. Specifically, Assumption 9.1 implies that $\phi_i(x, u_t(x), \nabla u_t(x)) = 0$ if and only if $D_i u_t(x) = 0$, $x \in \mathcal{V}$, $i = 1, \ldots, n$, while Assumption 9.2 implies that $\phi_i(x, u_t(x), \nabla u_t(x))D_i u_t(x) \leq 0$, $x \in \mathcal{V}$, $i = 1, \ldots, n$, which further implies that $\nabla u_t(x)\phi(x, u_t(x), \nabla u_t(x)) \leq 0$, $x \in \mathcal{V}$. The physical interpretation of Assumption 9.1 is that if the flux function $\phi$ in a certain direction is zero, then information or energy density change in this direction is not possible. This statement is reminiscent of the zeroth law of thermodynamics, which postulates that temperature equality is a necessary and sufficient condition for thermal equilibrium.

Assumption 9.2 implies that information or energy flows from information-rich or more energetic regions to information-poor or less energetic regions and is reminiscent of the second law of thermodynamics, which states that heat (energy in transition) must flow in the direction of lower temperatures. For further details of these assumptions, see [118, 127].

The following proposition shows that the solution $u(x, t), x \in \mathcal{V}$, $t \geq t_0$, to (9.1) and (9.2) is nonnegative for all nonnegative initial information density distributions $u_{t_0}(x) \geq 0$, $x \in \mathcal{V}$.

**Proposition 9.1.** Consider the dynamical system $\mathcal{G}$ given by (9.1) and (9.2). Assume that Assumptions 9.1 and 9.2 hold. Furthermore, assume that if $u(\hat{x}, \hat{t}) = 0$ for some $\hat{x} \in \partial\mathcal{V}$ and $\hat{t} \geq t_0$, then $\phi(\hat{x}, u(\hat{x}, \hat{t}), \nabla u(\hat{x}, \hat{t})) = 0$. Then the solution $u(x, t)$, $x \in \mathcal{V}$, $t \geq t_0$, to (9.1)–(9.2) is nonnegative for all nonnegative initial density distributions $u_{t_0}(x) \geq 0$, $x \in \mathcal{V}$.

**Proof.** Note that if $u(\hat{x}, \hat{t}) = 0$ for some $\hat{x}$ in the interior of $\mathcal{V}$ and $\hat{t} \geq t_0$, then it follows from Assumption 9.2 that $\phi(y, u(y, \hat{t}), \nabla u(y, \hat{t}))$ is directed toward the point $\hat{x}$ for all points $y$ in a sufficiently small neighborhood of $\hat{x}$. This property, along with (9.1), implies that $\frac{\partial u(\hat{x}, \hat{t})}{\partial t} \geq 0$. Alternatively, if $u(\hat{x}, \hat{t}) = 0$ for some $\hat{x} \in \partial\mathcal{V}$ and $\hat{t} \geq t_0$, then it follows from (9.1) and Assumptions 9.1 and 9.2 that $\frac{\partial u(\hat{x}, \hat{t})}{\partial t} \geq 0$. Thus, the solution to (9.1) and (9.2) is nonnegative for all nonnegative initial density distributions. $\square$

Next, we show that a Clausius-type inequality holds for the Eulerian swarm model $\mathcal{G}$ given by (9.1) and (9.2). For this result, note that it follows from Assumption 9.1 that for $\phi(x, u(x,t), \nabla u(x,t)) \cdot \mathbf{n}(x) \equiv 0$, the function $u(x,t) = \alpha$, $x \in \mathcal{V}$, $t \geq t_0$, $\alpha \geq 0$, is the solution to (9.1) and (9.2) with $u_{t_0}(x) = \alpha$, $x \in \mathcal{V}$. Thus, we define an equilibrium process for the system $\mathcal{G}$ as a process where the trajectory of $\mathcal{G}$ moves along the equilibrium manifold

$$\mathcal{M}_e \triangleq \{u_t \in \mathcal{X} : u_t(x) = \alpha, \ x \in \mathcal{V}, \ \alpha \geq 0\},$$

that is, $u(x,t) = \alpha(t)$, $x \in \mathcal{V}$, $t \geq t_0$, for some $\mathcal{L}_\infty$ function $\alpha : [0, \infty) \to \overline{\mathbb{R}}_+$. A nonequilibrium process is a process that does not lie on $\mathcal{M}_e$.

The next result establishes a Clausius-type inequality for equilibrium and nonequilibrium states of the infinite-dimensional dynamical system $\mathcal{G}$.

**Proposition 9.2.** Consider the dynamical system $\mathcal{G}$ given by (9.1) and (9.2), and assume that Assumptions 9.1 and 9.2 hold. Then, for every initial energy density distribution $u_{t_0} \in \mathcal{X}$, $t_f \geq t_0$, such that $u_{t_f}(x) = u_{t_0}(x)$, $x \in \mathcal{V}$,

$$\int_{t_0}^{t_f} \left[ -\int_{\partial \mathcal{V}} \frac{\phi(x, u(x,t), \nabla u(x,t)) \cdot \mathbf{n}(x)}{c + u(x,t)} d\mathcal{S}_\mathcal{V} \right] dt \leq 0, \qquad (9.10)$$

where $c > 0$ and $u(x,t)$, $x \in \mathcal{V}$, $t \geq t_0$, is the solution to (9.1) and (9.2). Furthermore,

$$\int_{t_0}^{t_f} \left[ -\int_{\partial \mathcal{V}} \frac{\phi(x, u(x,t), \nabla u(x,t)) \cdot \mathbf{n}(x)}{c + u(x,t)} d\mathcal{S}_\mathcal{V} \right] dt = 0 \qquad (9.11)$$

if and only if there exists an $\mathcal{L}_\infty$ function $\alpha : [t_0, t_f] \to \overline{\mathbb{R}}_+$ such that $u(x,t) = \alpha(t)$, $x \in \mathcal{V}$, $t \in [t_0, t_f]$.

**Proof.** It follows from (9.1), the Green–Gauss theorem, and Assumption 9.2 that

$$\int_{t_0}^{t_f} \left[ -\int_{\partial \mathcal{V}} \frac{\phi(x, u(x,t), \nabla u(x,t)) \cdot \mathbf{n}(x)}{c + u(x,t)} d\mathcal{S}_\mathcal{V} \right] dt$$

$$= \int_{t_0}^{t_f} \int_{\mathcal{V}} \frac{\frac{\partial u(x,t)}{\partial t} + \nabla \cdot \phi(x, u(x,t), \nabla u(x,t))}{c + u(x,t)} d\mathcal{V}dt$$

$$- \int_{t_0}^{t_f} \int_{\partial \mathcal{V}} \frac{\phi(x, u(x,t), \nabla u(x,t)) \cdot \mathbf{n}(x)}{c + u(x,t)} d\mathcal{S}_\mathcal{V}dt$$

$$= \int_{\mathcal{V}} \log_e \left( \frac{c + u(x, t_f)}{c + u(x, t_0)} \right) d\mathcal{V}$$

$$+ \int_{t_0}^{t_f} \int_{\partial \mathcal{V}} \frac{\phi(x, u(x,t), \nabla u(x,t)) \cdot \mathbf{n}(x)}{c + u(x,t)} d\mathcal{S}_\mathcal{V}dt$$

$$+ \int_{t_0}^{t_f} \int_{\mathcal{V}} \frac{\nabla u(x,t)\phi(x, u(x,t), \nabla u(x,t))}{(c + u(x,t))^2} d\mathcal{V}dt$$

$$-\int_{t_0}^{t_f}\int_{\partial\mathcal{V}}\frac{\phi(x,u(x,t),\nabla u(x,t))\cdot\mathbf{n}(x)}{c+u(x,t)}\mathrm{d}\mathcal{S}_{\mathcal{V}}\mathrm{d}t$$

$$=\int_{t_0}^{t_f}\int_{\mathcal{V}}\frac{\nabla u(x,t)\phi(x,u(x,t),\nabla u(x,t))}{(c+u(x,t))^2}\mathrm{d}\mathcal{V}\mathrm{d}t$$

$$\leq 0, \tag{9.12}$$

which proves (9.10).

To show (9.11), note that it follows from (9.12), Assumption 9.1, and Assumption 9.2 that (9.11) holds if and only if $\nabla u(x,t) = 0$ for all $x \in \mathcal{V}$ and $t \in [t_0, t_f]$ or, equivalently, there exists an $\mathcal{L}_\infty$ function $\alpha : [t_0, t_f] \to \overline{\mathbb{R}}_+$ such that $u(x,t) = \alpha(t)$, $x \in \mathcal{V}$, $t \in [t_0, t_f]$.  $\square$

Inequality (9.10) is a generalization of Clausius's inequality for reversible and irreversible thermodynamics as applied to Eulerian swarm models and restricts the manner in which the system loses information over cyclic motions.

Next, we define an entropy functional for the continuum dynamical system $\mathcal{G}$.

**Definition 9.2.** For the dynamical system $\mathcal{G}$ given by (9.1) and (9.2), the functional $\mathcal{S} : \mathcal{X} \to \mathbb{R}$ satisfying

$$\mathcal{S}(u_{t_2}) \geq \mathcal{S}(u_{t_1}) + \int_{t_1}^{t_2} q(t)\mathrm{d}t \tag{9.13}$$

for all $t_2 \geq t_1 \geq t_0$, where

$$q(t) \triangleq -\int_{\partial\mathcal{V}}\frac{\phi(x,u(x,t),\nabla u(x,t))\cdot\mathbf{n}(x)}{c+u(x,t)}\mathrm{d}\mathcal{S}_{\mathcal{V}} \tag{9.14}$$

and $c > 0$, is called the *entropy* functional of $\mathcal{G}$.

In the next theorem, we present a unique, continuously differentiable entropy functional for the dynamical system $\mathcal{G}$. This result holds for equilibrium and nonequilibrium processes.

**Theorem 9.2.** Consider the dynamical system $\mathcal{G}$ given by (9.1) and (9.2), and assume that Assumptions 9.1 and 9.2 hold. Then the functional $\mathcal{S} : \mathcal{X} \to \mathbb{R}$ given by

$$\mathcal{S}(u_t) = \int_{\mathcal{V}}\log_e(c+u_t(x))\mathrm{d}\mathcal{V} - \mathrm{vol}\mathcal{V}\log_e c \tag{9.15}$$

is a unique (modulo a constant of integration), continuously differentiable entropy functional of $\mathcal{G}$. Furthermore, if $u_t \notin \mathcal{M}_e$, $t \geq t_0$, where $u_t = u(x,t)$ denotes the solution to (9.1) and (9.2) and $\mathcal{M}_e = \{u_t \in \mathcal{X} : u_t = \alpha, \alpha \geq 0\}$, then (9.15) satisfies

$$\mathcal{S}(u_{t_2}) > \mathcal{S}(u_{t_1}) + \int_{t_1}^{t_2} q(t)\mathrm{d}t. \tag{9.16}$$

**Proof.** It follows from the Green–Gauss theorem, Assumption 9.2, and (9.15) that

$$\begin{aligned}
\dot{\mathcal{S}}(u_t) &= \int_{\mathcal{V}} \frac{1}{c + u(x,t)} \frac{\partial u(x,t)}{\partial t}\, \mathrm{d}\mathcal{V} \\
&= \int_{\mathcal{V}} \frac{1}{c + u(x,t)} \left(-\nabla \cdot \phi(x, u(x,t), \nabla u(x,t))\right) \mathrm{d}\mathcal{V} \\
&= -\int_{\mathcal{V}} \frac{\nabla u(x,t)\phi(x, u(x,t), \nabla u(x,t))}{(c + u(x,t))^2}\mathrm{d}\mathcal{V} \\
&\quad - \int_{\partial\mathcal{V}} \frac{\phi(x, u(x,t), \nabla u(x,t)) \cdot \mathbf{n}(x)}{c + u(x,t)}\mathrm{d}\mathcal{S}_{\mathcal{V}} \\
&\geq q(t). \tag{9.17}
\end{aligned}$$

Now, integrating (9.17) over $[t_1, t_2]$ yields (9.13). Furthermore, if $u_t \notin \mathcal{M}_e$, $t \geq t_0$, then it follows from Assumption 9.1, Assumption 9.2, and (9.17) that (9.16) holds.

To show that (9.15) is a unique, continuously differentiable entropy functional of $\mathcal{G}$, let $\mathcal{S}(u_t)$ be a continuously differentiable entropy functional of $\mathcal{G}$ so that $\mathcal{S}(u_t)$ satisfies (9.13) or, equivalently,

$$\begin{aligned}
\dot{\mathcal{S}}(u_t) &\geq -\int_{\partial\mathcal{V}} \frac{\phi(x, u_t, \nabla u_t) \cdot \mathbf{n}(x)}{c + u_t}\mathrm{d}\mathcal{S}_{\mathcal{V}} \\
&= -\int_{\mathcal{V}} \nabla \cdot (\mu(u_t)S(x,t))\mathrm{d}\mathcal{V} \\
&= -\mu(u_t)S(x,t), \quad t \geq t_0, \tag{9.18}
\end{aligned}$$

where $\mu(u_t) \triangleq \frac{1}{c+u_t}$, $S(x,t) \triangleq \phi(x, u_t, \nabla u_t)$, $u_t$, $t \geq t_0$, denotes the solution to (9.1) and (9.2) and $\dot{\mathcal{S}}(u_t)$ denotes the time derivative of $\mathcal{S}(u_t)$ along the solution $u_t$, $t \geq t_0$. Hence, it follows from (9.18) that

$$\mathcal{S}'(u_t)[-\nabla \cdot S(x,t)] \geq -\mu(u_t)S(x,t), \quad u_t \in \overline{\mathbb{R}}_+, \quad x \in \mathcal{V}, \quad t \geq t_0; \tag{9.19}$$

that is,

$$\mathcal{S}'(u_t)\left[-S(x,t) - \int_{\mathcal{V}} \nabla^2 S(x,t)\mathrm{d}\mathcal{V}\right] \geq -\mu(u_t)S(x,t), \quad u_t \in \overline{\mathbb{R}}_+,$$
$$x \in \mathcal{V}, \quad t \geq t_0, \tag{9.20}$$

which implies that there exist continuous functions $\ell : \overline{\mathbb{R}}_+ \to \mathbb{R}^p$ and $\mathcal{W} : \overline{\mathbb{R}}_+ \to \mathbb{R}^{p \times q}$ such that

$$0 = \mathcal{S}'(u_t) \left[ -S(x,t) - \int_{\mathcal{V}} \nabla^2 S(x,t) d\mathcal{V} \right] + \mu(u_t) S(x,t)$$
$$- [\ell(u_t) + \mathcal{W}(u_t) S(x,t)]^{\mathrm{T}} [\ell(u_t) + \mathcal{W}(u_t) S(x,t)],$$
$$u_t \in \overline{\mathbb{R}}_+, \quad x \in \mathcal{V}, \quad t \geq t_0. \tag{9.21}$$

Now, equating coefficients of equal powers (of $S$), it follows that $\mathcal{W}(u_t) \equiv 0$, $\mathcal{S}'(u_t) = \mu(u_t)$, $u_t \in \overline{\mathbb{R}}_+$, and

$$0 = \mathcal{S}'(u_t) \int_{\mathcal{V}} \nabla^2 S(x,t) d\mathcal{V} + \ell^{\mathrm{T}}(u_t) \ell(u_t), \quad u_t \in \overline{\mathbb{R}}_+. \tag{9.22}$$

Hence, $\mathcal{S}(u_t) = \int_{\mathcal{V}} \log_e(c + u_t(x)) d\mathcal{V} - \mathrm{vol}\mathcal{V} \log_e c$, $u_t \in \overline{\mathbb{R}}_+$. Thus, (9.15) is a unique, continuously differentiable entropy functional for $\mathcal{G}$. $\qquad \square$

It follows from Theorem 9.2 that if no information flow is allowed into or out of $\mathcal{V}$ (i.e., the system is isolated), then $\mathcal{S}(u_{t_2}) \geq \mathcal{S}(u_{t_1})$, $t_2 \geq t_1$. This shows that for an adiabatically isolated system, the entropy of the final state is greater than or equal to the entropy of the initial state.

## 9.4 Boundary Semistable Control for Large-Scale Swarms

In this section, we develop a boundary controller that guarantees that the infinite-dimensional information flow model (9.1)–(9.2) has convergent flows to Lyapunov stable uniform equilibrium information density distributions determined by the system initial information density distribution. First, we show that if no information flow is allowed into or out of $\mathcal{V}$ (i.e., the boundary $\partial \mathcal{V}$ is insulated), then (9.1)–(9.2) is Lyapunov stable.

**Theorem 9.3.** Consider the dynamical system given by (9.1) and (9.2). Assume that Assumptions 9.1 and 9.2 hold. If

$$\phi(x, u(x,t), \nabla u(x,t)) \cdot \mathbf{n}(x) = 0, \quad x \in \partial \mathcal{V}, \quad t \geq t_0, \tag{9.23}$$

then $u(x,t) \equiv \alpha$, $\alpha \geq 0$, is Lyapunov stable.

**Proof.** It follows from Assumption 9.1 that $u(x,t) \equiv \alpha$, $\alpha \geq 0$, is an equilibrium state for (9.1) and (9.2). To show Lyapunov stability of the equilibrium state $u(x,t) \equiv \alpha$, consider the shifted Lyapunov functional candidate

$$V(u_t - \alpha) = \frac{1}{2} \int_{\mathcal{V}} (u_t(x) - \alpha)^2 d\mathcal{V} = \frac{1}{2} \|u_t - \alpha\|_{\mathcal{L}_2}^2. \tag{9.24}$$

Now, it follows from the Green–Gauss theorem and Assumptions 9.1 and 9.2 that

$$
\begin{aligned}
\dot{V}(u_t - \alpha) &= \int_{\mathcal{V}} (u(x,t) - \alpha)\frac{\partial u(x,t)}{\partial t}\mathrm{d}\mathcal{V} \\
&= -\int_{\mathcal{V}} u(x,t)\nabla \cdot \phi(x,u(x,t),\nabla u(x,t))\mathrm{d}\mathcal{V} \\
&\quad + \alpha\int_{\mathcal{V}} \nabla \cdot \phi(x,u(x,t),\nabla u(x,t))\mathrm{d}\mathcal{V} \\
&= \int_{\mathcal{V}} \nabla u(x,t)\phi(x,u(x,t),\nabla u(x,t))\mathrm{d}\mathcal{V} \\
&\quad - \int_{\partial\mathcal{V}} u(x,t)\phi(x,u(x,t),\nabla u(x,t)) \cdot \mathbf{n}(x)\mathrm{d}\mathcal{S}_{\mathcal{V}} \\
&\quad + \alpha\int_{\partial\mathcal{V}} \phi(x,u(x,t),\nabla u(x,t)) \cdot \mathbf{n}(x)\mathrm{d}\mathcal{S}_{\mathcal{V}} \\
&= \int_{\mathcal{V}} \nabla u(x,t)\phi(x,u(x,t),\nabla u(x,t))\mathrm{d}\mathcal{V} \\
&\leq 0, \quad u_t \in \mathcal{W}_2^0(\mathcal{V}),
\end{aligned}
\tag{9.25}
$$

which establishes Lyapunov stability of the equilibrium state $u(x,t) \equiv \alpha$. $\square$

Next, we show that the total $\mathcal{L}_2$ norm of the energy of (9.1) and (9.2) is nonincreasing.

**Proposition 9.3.** Consider the dynamical system given by (9.1) and (9.2). Assume that Assumptions 9.1 and 9.2 hold. If either $u(x,t) = 0$ for all $x \in \partial\mathcal{V}$ and $t \geq t_0$ or (9.23) holds, then $\|u_t\|_{\mathcal{W}_2^0} \leq \|u_\tau\|_{\mathcal{W}_2^0}$ for all $t_0 \leq \tau \leq t$.

**Proof.** Assume that $u(x,t) = 0$ for all $x \in \partial\mathcal{V}$ and $t \geq t_0$, and consider the functional

$$
V(u_t) = \|u_t\|_{\mathcal{W}_2^0}^2.
\tag{9.26}
$$

Now, it follows from the Green–Gauss theorem and Assumptions 9.1 and 9.2 that

$$
\begin{aligned}
\frac{1}{2}\dot{V}(u_t) &= \int_{\mathcal{V}} u(x,t)\frac{\partial u(x,t)}{\partial t}\mathrm{d}\mathcal{V} \\
&= -\int_{\mathcal{V}} u(x,t)\nabla \cdot \phi(x,u(x,t),\nabla u(x,t))\mathrm{d}\mathcal{V} \\
&= \int_{\mathcal{V}} \nabla u(x,t)\phi(x,u(x,t),\nabla u(x,t))\mathrm{d}\mathcal{V} \\
&\quad - \int_{\partial\mathcal{V}} u(x,t)\phi(x,u(x,t),\nabla u(x,t)) \cdot \mathbf{n}(x)\mathrm{d}\mathcal{S}_{\mathcal{V}} \\
&\leq 0, \quad u_t \in \mathcal{W}_2^0(\mathcal{V}),
\end{aligned}
\tag{9.27}
$$

which implies that $\|u_t\|_{\mathcal{W}_2^0} \leq \|u_\tau\|_{\mathcal{W}_2^0}$ for all $t_0 \leq \tau \leq t$.

Alternatively, if (9.23) holds, then

$$\frac{1}{2}\dot{V}(u_t) = \int_{\mathcal{V}} \nabla u(x,t)\phi(x, u(x,t), \nabla u(x,t))\mathrm{d}\mathcal{V} \leq 0, \quad u_t \in \mathcal{W}_2^0(\mathcal{V}), \quad (9.28)$$

which implies that $\|u_t\|_{\mathcal{W}_2^0} \leq \|u_\tau\|_{\mathcal{W}_2^0}$ for all $t_0 \leq \tau \leq t$.    $\square$

Next, we present necessary and sufficient conditions for semistability of the swarm aggregation model (9.1)–(9.2).

**Theorem 9.4.** Consider the dynamical system given by (9.1) and (9.2). Assume that Assumptions 9.1 and 9.2 hold, and assume that $D(u_t, u_t) \leq D(u_\tau, u_\tau)$ for all $t_0 \leq \tau \leq t$. Then, for every $\alpha \geq 0$, $u(x,t) \equiv \alpha$ is a semistable equilibrium state of (9.1) and (9.2) if and only if (9.23) holds. In this case, $u(x,t) \to \frac{1}{\mathrm{vol}\mathcal{V}} \int_{\mathcal{V}} u_{t_0}(x)\mathrm{d}\mathcal{V}$ as $t \to \infty$ for every initial condition $u_{t_0} \in \mathcal{W}_2^1(\mathcal{V})$ and every $x \in \mathcal{V}$; moreover, $\frac{1}{\mathrm{vol}\mathcal{V}} \int_{\mathcal{V}} u_{t_0}(x)\mathrm{d}\mathcal{V}$ is a semistable equilibrium state of (9.1) and (9.2).

**Proof.** Assume that (9.23) holds. Then it follows from Theorem 9.3 that $u(x,t) \equiv \alpha$, $\alpha \geq 0$, is Lyapunov stable. Next, to show semistability of this equilibrium state, consider the Lyapunov functionals (9.26) and

$$\mathcal{E}(u_t) = \|u_t\|_{\mathcal{W}_2^1}^2, \quad u_t \in \mathcal{W}_2^1(\mathcal{V}). \quad (9.29)$$

It follows from Proposition 9.3 that $V(u_t)$ is a nonincreasing functional of time for all $u_{t_0} \in \mathcal{W}_2^0(\mathcal{V})$. Furthermore, note that $\mathcal{E}(u_t) = V(u_t) + D(u_t, u_t)$. Hence, by assumption, $\mathcal{E}(u_t)$ is a nonincreasing functional of time for all $u_{t_0} \in \mathcal{W}_2^1(\mathcal{V})$.

Next, since the functionals $V(u_t)$ and $\mathcal{E}(u_t)$ are nonincreasing and bounded from below by zero, it follows that $V(u_t)$ and $\mathcal{E}(u_t)$ are bounded functionals for every $u_{t_0} \in \mathcal{W}_2^1(\mathcal{V})$. This implies that the positive orbit

$$\mathcal{O}_{u_{t_0}}^+ \triangleq \{u_t \in \mathcal{W}_2^1(\mathcal{V}) : u_t(x) = u(x,t), x \in \mathcal{V}, t \in [t_0, \infty)\}$$

of (9.1) and (9.2) is bounded in $\mathcal{W}_2^1(\mathcal{V})$ for all $u_{t_0} \in \mathcal{W}_2^1(\mathcal{V})$. Furthermore, it follows from Sobolev's embedding theorem [283, 307] that $\mathcal{W}_2^1(\mathcal{V})$ is compactly embedded in $\mathcal{W}_2^0(\mathcal{V})$, and, hence, $\mathcal{O}_{u_{t_0}}^+$ is contained in a compact subset of $\mathcal{W}_2^0(\mathcal{V})$.

Next, define the sets $\mathcal{D}_{\mathcal{W}_2^1} = \{u_t \in \mathcal{W}_2^1(\mathcal{V}) : \mathcal{E}(u_t) < \eta\}$ and $\mathcal{D}_{\mathcal{W}_2^0} = \{u_t \in \mathcal{W}_2^0(\mathcal{V}) : V(u_t) < \eta\}$ for some arbitrary $\eta > 0$. Note that $\mathcal{D}_{\mathcal{W}_2^1}$ and $\mathcal{D}_{\mathcal{W}_2^0}$ are invariant sets with respect to (9.1) and (9.2). Moreover, it follows from the definition of $\mathcal{E}(u_t)$ and $V(u_t)$ that $\mathcal{D}_{\mathcal{W}_2^1}$ and $\mathcal{D}_{\mathcal{W}_2^0}$ are bounded sets in $\mathcal{W}_2^1(\mathcal{V})$ and $\mathcal{W}_2^0(\mathcal{V})$, respectively, and $\mathcal{D}_{\mathcal{W}_2^1} \subset \mathcal{D}_{\mathcal{W}_2^0}$.

Next, let

$$\mathcal{R} \triangleq \{u_t \in \overline{\mathcal{D}}_{\mathcal{W}_2^0} : \dot{V}(u_t) = 0\}$$
$$= \{u_t \in \overline{\mathcal{D}}_{\mathcal{W}_2^0} : \nabla u_t(x)\phi(x, u_t(x), \nabla u_t(x)) = 0, \ x \in \mathcal{V}\}.$$

Now, it follows from Assumption 9.1 that

$$\mathcal{R} = \{u_t \in \overline{\mathcal{D}}_{\mathcal{W}_2^0} : \nabla u_t(x) = 0, \ x \in \mathcal{V}\}$$

or

$$\mathcal{R} = \left\{u_t \in \mathcal{W}_2^0(\mathcal{V}) : u_t(x) \equiv \sigma, \ 0 \leq \sigma \leq \sqrt{\frac{\eta}{\mathrm{vol}\mathcal{V}}}\right\};$$

that is, $\mathcal{R}$ is the set of uniform density distributions that are the equilibrium states of (9.1) and (9.2).

Since the set $\mathcal{R}$ consists of only the equilibrium states of (9.1) and (9.2), it follows that the largest invariant set $\mathcal{M}$ contained in $\mathcal{R}$ is given by $\mathcal{M} = \mathcal{R}$. Hence, noting that $\mathcal{M}$ belongs to the set of generalized (weak) solutions of (9.1) and (9.2) defined on $\mathcal{R}$, it follows from Theorem 9.1 that $u(x, t) \equiv \alpha$ is a semistable equilibrium state of (9.1) and (9.2). Moreover, since $\eta > 0$ can be arbitrarily large but finite and $\mathcal{E}(u_t)$ is radially unbounded, the previous statement holds for all $u_{t_0} \in \mathcal{W}_2^1(\mathcal{V})$. Next, note that since, by the divergence theorem,

$$\int_{\mathcal{V}} \frac{\partial u(x, t)}{\partial t} d\mathcal{V} = -\int_{\mathcal{V}} \nabla \cdot \phi(x, u(x, t), \nabla u(x, t)) d\mathcal{V}$$
$$= -\int_{\partial \mathcal{V}} \phi(x, u(x, t), \nabla u(x, t)) \cdot \mathbf{n}(x) d\mathcal{S}_{\mathcal{V}}$$
$$= 0, \tag{9.30}$$

it follows that $\int_{\mathcal{V}} u(x, t) d\mathcal{V} = \int_{\mathcal{V}} u_{t_0}(x) d\mathcal{V}$, $t \geq t_0$, which implies that

$$u(x, t) \to \frac{1}{\mathrm{vol}\mathcal{V}} \int_{\mathcal{V}} u_{t_0}(x) d\mathcal{V} \text{ as } t \to \infty.$$

Conversely, assume that, for every $\alpha \geq 0$, $u(x, t) \equiv \alpha$ is a semistable equilibrium state of (9.1) and (9.2). Suppose, *ad absurdum*, that there exists at least one point $x_\mathrm{p} \in \partial\mathcal{V}$ such that $\phi(x_\mathrm{p}, u_t(x_\mathrm{p}, \nabla u_t(x_\mathrm{p}))) \cdot \mathbf{n}(x_\mathrm{p}) > 0$. Consider the Lyapunov functional (9.26), and note that the Lyapunov derivative of $V(u_t)$ is given by (9.27). Let

$$\mathcal{R} \triangleq \{u_t \in \overline{\mathcal{D}}_{\mathcal{W}_2^0} : \dot{V}(u_t) = 0\}$$

$$= \{u_t \in \overline{\mathcal{D}}_{\mathcal{W}_2^0} : \nabla u_t(x)\phi(x, u_t(x), \nabla u_t(x)) = 0, \ x \in \mathcal{V}\}$$
$$\cap \{u_t \in \overline{\mathcal{D}}_{\mathcal{W}_2^0} : u(x,t)\phi(x, u_t(x), \nabla u_t(x)) \cdot \mathbf{n}(x) = 0, \ x \in \partial\mathcal{V}\}.$$

Now, since Assumption 9.1 holds, it follows that

$$\mathcal{R} = \{u_t \in \overline{\mathcal{D}}_{\mathcal{W}_2^0} : \nabla u_t(x) = 0, \ x \in \mathcal{V}\} \cap \{u_t \in \overline{\mathcal{D}}_{\mathcal{W}_2^0} : u_t(x_\mathrm{p}) = 0, \ x_\mathrm{p} \in \partial\mathcal{V}\}$$
$$= \{0\},$$

and the largest invariant set $\mathcal{M}$ contained in $\mathcal{R}$ is given by $\mathcal{M} = \{0\}$. By assumption, $\mathcal{E}(u_t)$ is a nonincreasing functional of time for all $u_{t_0} \in \mathcal{W}_2^1(\mathcal{V})$, and since $\mathcal{E}(u_t)$ is bounded from below by zero, the positive orbit $\mathcal{O}_{u_{t_0}}^+$ of (9.1) and (9.2) is bounded in $\mathcal{W}_2^1(\mathcal{V})$.

Hence, since $\mathcal{W}_2^1(\mathcal{V})$ is compactly embedded in $\mathcal{W}_2^0(\mathcal{V})$, it follows from Sobolev's embedding theorem [283, 307] that $\mathcal{O}_{u_{t_0}}^+$ is contained in a compact subset of $\mathcal{W}_2^0(\mathcal{V})$. Thus, it follows from Theorem 3 of [140] that for every initial density distribution $u_{t_0} \in \mathcal{D}_{\mathcal{W}_2^0}$, $u(x,t) \to \mathcal{M} = \{0\}$ as $t \to \infty$ with respect to the norm $\| \cdot \|_{\mathcal{W}_2^0}$, which shows asymptotic stability of the zero equilibrium state of (9.1) and (9.2). However, since asymptotic stability of (9.1) and (9.2) is equivalent to semistability of (9.1) and (9.2) if and only if the equilibrium state of (9.1) and (9.2) is zero, this contradicts the assumption that, for every $\alpha \geq 0$, $u(x,t) \equiv \alpha$ is an equilibrium state of (9.1) and (9.2). Hence, (9.23) holds. $\qquad\square$

Theorem 9.4 shows that the swarm aggregation model (9.1)–(9.2) with Assumptions 9.1 and 9.2 has convergent flows to Lyapunov stable uniform equilibrium information density distributions determined by the system initial information density distribution. As for lumped parameter systems (i.e., Lagrangian networks), this phenomenon is known as equipartition of energy [127] in continuum thermodynamics and *information consensus* or *protocol agreement* [162] in Eulerian swarming network systems.

**Corollary 9.1.** Consider the dynamical system $\mathcal{G}$ given by (9.1) and (9.2). Assume that Assumptions 9.1 and 9.2 hold, and assume that

$$\nabla^2 u_t(x)\nabla \cdot \phi(x, u_t(x), \nabla u_t(x)) \leq 0, \quad x \in \mathcal{V}, \quad u_t \in \mathcal{W}_2^1(\mathcal{V}), \qquad (9.31)$$

where $\nabla^2 \triangleq \nabla \cdot \nabla$ denotes the Laplace operator. Then, for every $\alpha \geq 0$, $u(x,t) \equiv \alpha$ is a semistable equilibrium state of (9.1) and (9.2) if and only if (9.23) holds. In this case, $u(x,t) \to \frac{1}{\mathrm{vol}\mathcal{V}} \int_{\mathcal{V}} u_{t_0}(x)\mathrm{d}\mathcal{V}$ as $t \to \infty$ for every initial condition $u_{t_0} \in \mathcal{W}_2^1(\mathcal{V})$ and every $x \in \mathcal{V}$; moreover, $\frac{1}{\mathrm{vol}\mathcal{V}} \int_{\mathcal{V}} u_{t_0}(x)\mathrm{d}\mathcal{V}$ is a semistable equilibrium state of (9.1) and (9.2).

**Proof.** The result is a direct consequence of Theorem 9.4 by showing that the Dirichlet integral $D(u_t, u_t)$ of $u_t$ is nonincreasing. To see this, note that it follows from the Green–Gauss theorem and (9.23) that

$$\frac{1}{2}\dot{D}(u_t, u_t) = \int_{\mathcal{V}} \nabla u(x,t) \frac{\partial}{\partial t}(\nabla u(x,t))^{\mathrm{T}} \mathrm{d}\mathcal{V}$$

$$= \int_{\partial\mathcal{V}} \frac{\partial u(x,t)}{\partial t} D_{\mathbf{n}(x)} u(x,t) \mathrm{d}\mathcal{S}_{\mathcal{V}}$$

$$+ \int_{\mathcal{V}} \nabla^2 u(x,t) \nabla \cdot \phi(x, u(x,t), \nabla u(x,t)) \mathrm{d}\mathcal{V}, \qquad (9.32)$$

where $D_{\mathbf{n}(x)} u(x,t) \triangleq \nabla u(x,t) \mathbf{n}(x)$ denotes the directional derivative of $u(x,t)$ along $\mathbf{n}(x)$ at $x \in \partial\mathcal{V}$. Next, it follows from (9.23) and Assumption 9.1, with $\mathbf{u} = \mathbf{n}(x)$, that $D_{\mathbf{n}(x)} u(x,t) = 0$, $x \in \partial\mathcal{V}$. Hence, it follows from (9.31) and (9.32) that $\dot{D}(u_t, u_t) \le 0$, $t \ge t_0$, for any $u_{t_0} \in \mathcal{W}_2^1(\mathcal{V})$. $\square$

Condition (9.31) implies that for an information (or energy) density distribution $u_t(x)$, $x \in \mathcal{V}$, the information (or energy) flow $\phi(x, u_t(x), \nabla u_t(x))$ at $x \in \mathcal{V}$ is proportional to the information (or energy) density at this point. Note that for a linear information (or energy) flow model where $\phi(x, u_t(x), \nabla u_t(x)) = -k[\nabla u_t(x)]^{\mathrm{T}}$ and $k > 0$ is a conductivity constant, condition (9.31) is automatically satisfied since

$$\nabla^2 u_t(x) \nabla \cdot \phi(x, u_t(x), \nabla u_t(x)) = -k[\nabla^2 u_t(x)]^2 \le 0, \quad x \in \mathcal{V}.$$

Equation (9.23) plays a critical role in (boundary) control design of (9.1) and (9.2). In particular, (9.23), along with Assumptions 9.1 and 9.2, gives a criterion for guaranteeing semistability of (9.1) and (9.2). Next, we discuss boundary semistable control of (9.1) and (9.2) using (9.23). First, we consider *Dirichlet boundary control* [193]. The Dirichlet boundary control problem for (9.1) and (9.2) involves the control law given by (9.2) with

$$u(x,t) = U_{\mathrm{d}}(x,t), \quad x \in \partial\mathcal{V}, \quad t \ge t_0. \qquad (9.33)$$

It follows from (9.23) and Assumption 9.1 that for the Dirichlet boundary control problem, the control input $U_{\mathrm{d}}(x,t)$ should be chosen to satisfy

$$\nabla f(x) \nabla^{\mathrm{T}} U_{\mathrm{d}}(x,t) = 0, \quad x \in \partial\mathcal{V}, \quad t \ge t_0. \qquad (9.34)$$

Next, we consider *Neumann boundary control* [193] for (9.1) and (9.2). The Neumann boundary control problem for (9.1) and (9.2) involves the control law given by (9.2) with

$$\frac{\partial u(x,t)}{\partial \mathbf{n}} = U_{\mathrm{n}}(x,t), \quad x \in \partial\mathcal{V}, \quad t \ge t_0. \qquad (9.35)$$

However, since $\frac{\partial u(x,t)}{\partial \mathbf{n}} = \nabla u_t(x) \cdot \mathbf{n}$, it follows from (9.23) and Assumption 9.1 that $U_{\mathrm{n}}(x,t) = 0$, $x \in \partial \mathcal{V}$, $t \geq t_0$, resulting in a trivial Neumann boundary controller.

Finally, we consider a linear form of (9.1) and (9.2). Specifically, consider the linear (heat) equation given by

$$\frac{\partial u(x,t)}{\partial t} = \nabla^2 u(x,t), \quad x \in \mathcal{V}, \quad t \geq t_0, \quad u(x,t_0) = u_{t_0}(x), \quad x \in \mathcal{V},$$
(9.36)

where $u : \mathbb{R} \times [0,\infty) \to \overline{\mathbb{R}}_+$. It can be easily shown that Assumptions 9.1 and 9.2 hold and that (9.31) holds for (9.36). Now, using the Neumann boundary control law

$$\nabla u(x,t) \cdot \mathbf{n}(x) = 0, \quad x \in \partial \mathcal{V}, \quad t \geq t_0,$$
(9.37)

it follows that all of the equilibrium points of (9.36) are given by $u(x,t) \equiv \alpha \in \mathbb{R}$ [88, p. 346]. Hence, it follows from Corollary 9.1 that the linear equation (9.36) achieves uniform information distributions over $\mathcal{V}$. The boundary condition (9.37) implies that there is no information (heat) flow into or out of $\mathcal{V}$; that is, the boundary $\partial \mathcal{V}$ is insulated.

Finally, we consider the Neumann boundary control law given by

$$U_{\mathrm{n}}(x,t) = -c(u(x,t) - u_{\mathrm{e}}), \quad x \in \partial \mathcal{V}, \quad t \geq t_0,$$
(9.38)

where $c > 0$ and $u_{\mathrm{e}} \geq 0$. This control law is also known as *Newton's law of cooling* in the literature [98, p. 155] and guarantees that, outside $\mathcal{V}$, the information (temperature) $u(x,t)$ is maintained at $u_{\mathrm{e}}$ and the rate of information (heat) flow across the boundary is proportional to $u - u_{\mathrm{e}}$.

**Proposition 9.4.** Consider the linear equation (9.36) with the boundary control (9.38). Then $u(x,t) \equiv u_{\mathrm{e}}$ is an asymptotically stable equilibrium state of (9.36) and (9.38).

**Proof.** Consider the Lyapunov functional candidate

$$V(u_t - u_{\mathrm{e}}) = \frac{1}{2} \int_{\mathcal{V}} (u_t(x) - u_{\mathrm{e}})^2 \mathrm{d}\mathcal{V} = \frac{1}{2} \|u_t - u_{\mathrm{e}}\|_{\mathcal{L}_2}^2.$$

Now, it follows from the Green–Gauss theorem that

$$\dot{V}(u_t - u_{\mathrm{e}}) = \int_{\mathcal{V}} (u(x,t) - u_{\mathrm{e}}) \frac{\partial u(x,t)}{\partial t} \mathrm{d}\mathcal{V}$$
$$= \int_{\mathcal{V}} u(x,t) \nabla^2 u(x,t) \mathrm{d}\mathcal{V} - u_{\mathrm{e}} \int_{\mathcal{V}} \nabla^2 u(x,t) \mathrm{d}\mathcal{V}$$

$$= -\int_{\mathcal{V}} \nabla u(x,t) \nabla^{\mathrm{T}} u(x,t) \mathrm{d}\mathcal{V} + \int_{\partial \mathcal{V}} u(x,t) \nabla u(x,t) \cdot \mathbf{n}(x) \mathrm{d}\mathcal{S}_{\mathcal{V}}$$

$$- u_{\mathrm{e}} \int_{\partial \mathcal{V}} \nabla u(x,t) \cdot \mathbf{n}(x) \mathrm{d}\mathcal{S}_{\mathcal{V}}$$

$$= -D(u_t, u_t) + \int_{\partial \mathcal{V}} (u(x,t) - u_{\mathrm{e}}) \nabla u(x,t) \cdot \mathbf{n}(x) \mathrm{d}\mathcal{S}_{\mathcal{V}}$$

$$= -D(u_t, u_t) - c \int_{\partial \mathcal{V}} (u(x,t) - u_{\mathrm{e}})^2 \mathrm{d}\mathcal{S}_{\mathcal{V}}$$

$$< 0, \quad u_t \in \mathcal{W}_2^0(\mathcal{V}), \quad u_t \neq u_{\mathrm{e}}, \tag{9.39}$$

establishing asymptotic stability of the equilibrium state $u(x,t) \equiv u_{\mathrm{e}}$. $\square$

The control problem addressed by Proposition 9.4 can be viewed as a leader-follower coordination problem [180] for dynamical swarm systems.

## 9.5 Advection-Diffusion Network Models

The nonlinear partial differential equation (9.1) describes a general conservation equation which includes many important swarming models discussed in the literature. See, for example, [232]. In this section, we turn our attention to a specific form of (9.1) involving the *advection-diffusion* model [111, 232] defined over a compact connected set $\mathcal{V} \subset \mathbb{R}^n$ with a smooth boundary $\partial \mathcal{V}$ and volume $\mathrm{vol}\,\mathcal{V}$ given by

$$\frac{\partial \rho(x,t)}{\partial t} = -\nabla \cdot (\rho(x,t) v(x,t)) + \nabla \cdot \left( B(x,t) \nabla^{\mathrm{T}} \rho(x,t) \right), \tag{9.40}$$

$$\rho(x,t_0) = \rho_{t_0}(x), \quad x \in \mathcal{V}, \quad t \geq t_0, \tag{9.41}$$

where $\rho : \mathcal{V} \times [0, \infty) \to \overline{\mathbb{R}}_+$ denotes the density distribution of mobile agents at the point $x = [x_1, \dots, x_n]^{\mathrm{T}} \in \mathcal{V}$ and time instant $t \geq t_0$, $v : \mathcal{V} \times [0, \infty) \to \mathbb{R}^n$ is a density-dependent advection velocity, and $B : \mathcal{V} \times [0, \infty) \to \mathbb{R}^{n \times n}$ is a diffusion operator. Here, we consider the case where $v(x,t)$ is given by

$$v(x,t) = -k \nabla^{\mathrm{T}} \rho(x,t), \quad x \in \mathcal{V}, \quad t \geq t_0, \tag{9.42}$$

where $k \in \mathbb{R}$ and $B(x,t) = \lambda I_n \in \mathbb{R}^{n \times n}$ for all $x \in \mathcal{V}$ and $t \geq t_0$, where $\lambda \in \mathbb{R}$.

**Theorem 9.5.** Consider the dynamical system given by (9.40) and (9.41) with $B(x,t) \equiv \lambda I_n$. Assume that $v(x,t)$ satisfies (9.42). If $k, \lambda \geq 0$ are such that $k^2 + \lambda^2 \neq 0$, then, for every $\alpha \in \overline{\mathbb{R}}_+$, $\rho(x,t) \equiv \alpha$ is a semistable equilibrium state of (9.40) and (9.41) if and only if $\nabla \rho(x,t) \cdot \mathbf{n}(x) = 0$, where $x \in \partial \mathcal{V}$ and $t \geq t_0$. In this case, $\rho(x,t) \to \frac{1}{\mathrm{vol}\mathcal{V}} \int_{\mathcal{V}} \rho_{t_0}(x) \mathrm{d}\mathcal{V}$ as $t \to \infty$ for every initial condition $\rho_{t_0} \in \mathcal{W}_2^1(\mathcal{V})$ and every $x \in \mathcal{V}$; moreover, $\frac{1}{\mathrm{vol}\mathcal{V}} \int_{\mathcal{V}} \rho_{t_0}(x) \mathrm{d}\mathcal{V}$ is a semistable equilibrium state of (9.40) and (9.41).

**Proof.** First, let $k \geq 0$ and $\lambda > 0$. In this case, $\phi(x, \rho(x,t), \nabla\rho(x,t)) = -(k\rho(x,t) + \lambda)\nabla^{\mathrm{T}}\rho(x,t)$, and, hence, Assumptions 9.1 and 9.2 hold. Furthermore,

$$
\begin{aligned}
\nabla^2 \rho_t(x)\nabla \cdot \phi&(x, \rho_t(x), \nabla\rho_t(x)) \\
&= \nabla^2\rho_t(x)\left[-k\nabla\rho_t(x)\nabla^{\mathrm{T}}\rho_t(x) - (k\rho_t(x) + \lambda)\nabla^2\rho_t(x)\right] \\
&= -k[\nabla^2\rho_t(x)]^2 - (k\rho_t(x) + \lambda)[\nabla^2\rho_t(x)]^2 \\
&\leq 0, \quad x \in \mathcal{V},
\end{aligned}
\tag{9.43}
$$

and, hence, (9.31) holds. Now, the result is a direct consequence of Corollary 9.1.

Next, let $k > 0$ and $\lambda = 0$, and assume that $\rho(x,t)\nabla\rho(x,t) \cdot \mathbf{n}(x) = 0$ for $x \in \partial\mathcal{V}$ and $t \geq t_0$. To show Lyapunov stability of $\rho(x,t) \equiv \alpha$, consider the Lyapunov functional (9.24) with $u(x,t)$ replaced by $\rho(x,t)$. Now, it follows from the Green–Gauss theorem that

$$
\begin{aligned}
\dot{V}(\rho_t - \alpha) &= \int_{\mathcal{V}} (\rho(x,t) - \alpha)\frac{\partial\rho(x,t)}{\partial t}\mathrm{d}\mathcal{V} \\
&= -\int_{\mathcal{V}} \rho(x,t)\nabla \cdot (\rho(x,t)v(x,t))\mathrm{d}\mathcal{V} + \alpha\int_{\mathcal{V}} \nabla \cdot (\rho(x,t)v(x,t))\mathrm{d}\mathcal{V} \\
&= \int_{\mathcal{V}} \nabla\rho(x,t)\rho(x,t)v(x,t)\mathrm{d}\mathcal{V} - \int_{\partial\mathcal{V}} \rho(x,t)\rho(x,t)v(x,t) \cdot \mathbf{n}(x)\mathrm{d}\mathcal{S}_{\mathcal{V}} \\
&\quad + \alpha\int_{\partial\mathcal{V}} \rho(x,t)v(x,t) \cdot \mathbf{n}(x)\mathrm{d}\mathcal{S}_{\mathcal{V}} \\
&= -\int_{\mathcal{V}} \rho(x,t)\nabla\rho(x,t)\nabla^{\mathrm{T}}\rho(x,t)\mathrm{d}\mathcal{V} \\
&\leq 0, \quad \rho_t \in \mathcal{W}_2^0(\mathcal{V}),
\end{aligned}
\tag{9.44}
$$

which proves Lyapunov stability of $\rho(x,t) \equiv \alpha$.

To show semistability of $\rho(x,t) \equiv \alpha$, consider the Lyapunov functionals (9.26) and (9.29). Now, it follows from (9.44), with $\alpha = 0$, that $V(\rho_t)$ is a nonincreasing functional of time for all $\rho_{t_0} \in \mathcal{W}_2^0(\mathcal{V})$. Furthermore, it follows from the Green–Gauss theorem that

$$
\begin{aligned}
\frac{1}{2}\dot{D}(\rho_t, \rho_t) &= \int_{\mathcal{V}} \nabla\rho(x,t)\frac{\partial}{\partial t}(\nabla\rho(x,t))^{\mathrm{T}}\mathrm{d}\mathcal{V} \\
&= \int_{\partial\mathcal{V}} \frac{\partial\rho(x,t)}{\partial t}D_{\mathbf{n}(x)}\rho(x,t)\mathrm{d}\mathcal{S}_{\mathcal{V}} - k\int_{\mathcal{V}} [\nabla^2\rho(x,t)]^2\mathrm{d}\mathcal{V}.
\end{aligned}
\tag{9.45}
$$

Next, using similar arguments as in the proof of Corollary 9.1, it can be shown that $D(\rho_t, \rho_t)$ is a nonincreasing functional of time for all $\rho_{t_0} \in \mathcal{W}_2^1(\mathcal{V})$. Furthermore, note that $\mathcal{E}(\rho_t) = V(\rho_t) + D(\rho_t, \rho_t)$. Hence, $\mathcal{E}(\rho_t)$

is a nonincreasing functional of time for all $\rho_{t_0} \in \mathcal{W}_2^1(\mathcal{V})$. The rest of the proof follows as in the proof of Theorem 9.4.

The converse follows as in the proof of Theorem 9.4. $\qquad\square$

## 9.6 Connections between Eulerian and Lagrangian Models for Information Consensus

As shown in Chapter 2, information consensus for a Lagrangian network model involves the dynamical system

$$\dot{x}(t) = -Lx(t), \quad x(0) = x_0, \quad t \geq 0, \tag{9.46}$$

where $x = [x_1, \ldots, x_n]^{\mathrm{T}} \in \mathbb{R}^n$ is the information state and $L \in \mathbb{R}^{n \times n}$ is the *Laplacian* of the underlying communication graph topology of the network [180]. Recall that the entries of a Laplacian matrix $L$ of a directed graph are given by $L_{(i,i)} = \sum_{i=1, i \neq j}^n A_{(i,j)}$, $j = 1, \ldots, n$, and $L_{(i,j)} = -A_{(i,j)}$ for all $i \neq j$, where $A_{(i,j)}$, $i, j = 1, \ldots, n$, are the entries of the weighted adjacency matrix of the directed graph [256]. Consensus is achieved by a group of agents if, for all $x_i(0)$ and $i = 1, \ldots, n$, $\lim_{t \to \infty} x_i(t) \to \alpha$ as $t \to \infty$, where $x_i(t)$ denotes the $i$th component of $x(t)$ and $\alpha \in \mathbb{R}$ depends on the system initial conditions.

Next, we compare our Eulerian framework for information consensus developed in this chapter to the Lagrangian framework for information consensus given by (9.46). Specifically, consider for simplicity the partial differential equation given by (9.36) and (9.37). In this case, (9.36) can be rewritten as

$$\frac{\partial}{\partial t} u(x,t) = -\mathfrak{L}u(x,t), \quad x \in \mathcal{V}, \quad t \geq t_0, \quad u(x,t_0) = u_{t_0}(x), \quad x \in \mathcal{V}, \tag{9.47}$$

where $\mathfrak{L} \triangleq -\nabla^2$ is the Laplacian operator so that (9.47) has the same form as (9.46). Condition (9.37) is a sufficient condition for guaranteeing a uniform information distribution of (9.36). Since $\mathfrak{L}$ is a self-adjoint operator, consider (9.46) with $L = L^{\mathrm{T}}$ and note that, since $L$ has zero row sums, zero is an eigenvalue of $L$ with an associated eigenvector $\mathbf{e} = [1, \ldots, 1]^{\mathrm{T}} \in \mathbb{R}^n$. Next, by Proposition 6.1 of [127], (the lumped parameter versions of) Assumptions 9.1 and 9.2 hold if and only if rank $L = n - 1$. Now, information consensus for (9.46) is immediate by Theorem 6.1 of [127].

**Definition 9.3.** We say that $\lambda$ is an *eigenvalue* of the operator $\mathfrak{L}$ on $\mathcal{V}$ subject to the Neumann boundary condition (9.37) if there exists an *eigenfunction* $w$, not identically equal to zero, solving the boundary value problem

$$\mathfrak{L}w = \lambda w \text{ in } \mathcal{V}, \tag{9.48}$$

$$\frac{\partial w}{\partial \mathbf{n}} = 0 \text{ on } \partial \mathcal{V}. \tag{9.49}$$

Note that it follows from [88, p. 346] that the Neumann boundary value problem

$$\mathfrak{L}u = 0 \text{ in } \mathcal{V}, \tag{9.50}$$

$$\frac{\partial u}{\partial \mathbf{n}} = 0 \text{ on } \partial \mathcal{V} \tag{9.51}$$

has a smooth solution for $u \equiv C \in \mathbb{R}$. Hence, it follows from Definition 9.3 that zero is an eigenvalue of the operator $\mathfrak{L}$ with an associated eigenfunction $w = C$. Thus, $\mathfrak{L}$ plays the same role as $L$. This provides an explicit connection between Lagrangian (discrete) and Eulerian (continuum) network consensus models.

# Chapter Ten

---

# $\mathcal{H}_2$ Optimal Semistable Control for Network Systems

## 10.1 Introduction

Since multiagent systems can involve information laws governed by nodal dynamics and reacting strategies that can be modified to minimize waiting times and optimize system throughput, optimality considerations in network systems are of paramount importance. Optimality considerations are especially relevant for the problem of network caching, wherein decisions can affect future traffic on links, future buffer levels, network delays and congestion, and server loads. In this chapter, we use linear matrix inequalities (LMIs) to develop $\mathcal{H}_2$ optimal semistable controllers for linear dynamical systems. LMIs provide a powerful design framework for linear control problems [46]. Since LMIs lead to convex or quasi-convex optimization problems, they can be solved very efficiently using interior-point algorithms.

Unlike the standard $\mathcal{H}_2$ optimal control problem, a complicating feature of the $\mathcal{H}_2$ optimal semistable stabilization problem is that the closed-loop Lyapunov equation guaranteeing semistability can admit multiple solutions. An interesting feature of the proposed approach, however, is that a least squares solution over all possible semistabilizing solutions corresponds to the $\mathcal{H}_2$ optimal solution. It is shown that this least squares solution can be characterized by an LMI minimization problem. Finally, we apply our results to address the consensus control problems in networks of dynamic agents.

## 10.2 $\mathcal{H}_2$ Semistability Theory

In this section, we establish notation along with several key results on $\mathcal{H}_2$ semistability theory involving the notions of semistability, semicontrollability, and semiobservability. The notation we use in this chapter is consistent with the notation developed in the earlier chapters. Additionally, here we write $(\cdot)^*$ for the complex conjugate transpose, $\|\cdot\|_{\mathrm{F}}$ for the Frobenius matrix norm, $\mathcal{S}^{\perp}$ for the orthogonal complement of a set $\mathcal{S}$, $\det A$ for the determinant of the square matrix $A$, $\mathrm{tr}(\cdot)$ for the trace operator, $\mathbb{E}$ for the expectation operator, and $\mathrm{vec}(\cdot)$ for the column-stacking operator.

The following definition for semistability with respect to $\mathcal{D} \subseteq \mathbb{R}^n$ is needed. For this definition, consider the nonlinear dynamical system given by

$$\dot{x}(t) = f(x(t)), \quad x(0) = x_0, \quad t \geq 0, \tag{10.1}$$

where $x(t) \in \mathcal{D} \subseteq \mathbb{R}^n$, $t \geq 0$, and $f : \mathcal{D} \to \mathbb{R}^q$ is locally Lipschitz continuous on $\mathcal{D}$.

**Definition 10.1.** Let $\mathcal{D} \subseteq \mathbb{R}^n$ be positively invariant under (10.1). The equilibrium solution $x(t) \equiv x_e \in \mathcal{D}$ of (10.1) is *Lyapunov stable with respect to $\mathcal{D}$* if, for every $\varepsilon > 0$, there exists $\delta = \delta(\varepsilon) > 0$ such that if $x_0 \in \mathcal{B}_\delta(x_e) \cap \mathcal{D}$, then $x(t) \in \mathcal{B}_\varepsilon(x_e) \cap \mathcal{D}$, $t \geq 0$. The equilibrium solution $x(t) \equiv x_e \in \mathcal{D}$ of (10.1) is *semistable with respect to $\mathcal{D}$* if it is Lyapunov stable with respect to $\mathcal{D}$ and there exists $\delta > 0$ such that if $x_0 \in \mathcal{B}_\delta(x_e) \cap \mathcal{D}$, then $\lim_{t \to \infty} x(t)$ exists and corresponds to a Lyapunov stable equilibrium point in $\mathcal{D}$. Finally, the system (10.1) is said to be *semistable with respect to $\mathcal{D}$* if every equilibrium point in $\mathcal{D}$ is semistable with respect to $\mathcal{D}$.

Note that if in (10.1) $f(x) = Ax$, where $A \in \mathbb{R}^{n \times n}$, then (10.1) is semistable if and only if $A$ is semistable; that is, $\mathrm{spec}(A) \subset \{s \in \mathbb{C} : \mathrm{Re}\, s < 0\} \cup \{0\}$, and, if $0 \in \mathrm{spec}(A)$, then zero is semisimple. In this case, it can be shown that, for every $x_0 \in \mathbb{R}^n$, $\lim_{t \to \infty} x(t)$ exists or, equivalently, $\lim_{t \to \infty} e^{At}$ exists and is given by $\lim_{t \to \infty} e^{At} = I_n - AA^{\#}$ [26, pp. 437–438].

Next, we present the notions of semicontrollability and semiobservability. For these definitions, let $A \in \mathbb{R}^{n \times n}$, $B \in \mathbb{R}^{n \times m}$, and $C \in \mathbb{R}^{l \times n}$, and consider the linear dynamical system

$$\dot{x}(t) = Ax(t) + Bu(t), \quad x(0) = x_0, \quad t \geq 0, \tag{10.2}$$
$$y(t) = Cx(t), \tag{10.3}$$

with state $x(t) \in \mathbb{R}^n$, input $u(t) \in \mathbb{R}^m$, and output $y(t) \in \mathbb{R}^l$, where $t \geq 0$.

**Definition 10.2.** Let $A \in \mathbb{R}^{n \times n}$ and $B \in \mathbb{R}^{m \times n}$. The pair $(A, B)$ is *semicontrollable* if

$$\left[ \bigcap_{k=1}^{n} \mathcal{N}\left( B^{\mathrm{T}} (A^{k-1})^{\mathrm{T}} \right) \right]^{\perp} = [\mathcal{N}(A^{\mathrm{T}})]^{\perp}, \tag{10.4}$$

where $A^0 \triangleq I_n$.

**Definition 10.3.** Let $A \in \mathbb{R}^{n \times n}$ and $C \in \mathbb{R}^{l \times n}$. The pair $(A, C)$ is *semiobservable* if

$$\bigcap_{k=1}^{n} \mathcal{N}\left(CA^{k-1}\right) = \mathcal{N}(A). \tag{10.5}$$

Semicontrollability and semiobservability are extensions of the classical notions of controllability and observability. In particular, semicontrollability is an extension of null controllability to *equilibrium controllability*, whereas semiobservability is an extension of zero-state observability to *equilibrium observability*. It is important to note here that since Definitions 10.2 and 10.3 are dual, dual results to the semiobservability results that we establish in this section also hold for semicontrollability.

**Definition 10.4.** Let $A \in \mathbb{R}^{n \times n}$, $C \in \mathbb{R}^{l \times n}$, and $K \in \mathbb{R}^{m \times n}$. The pair $(A, C)$ is *semiobservable with respect to $K$* if

$$\mathcal{N}(K) \cap \left(\bigcap_{i=1}^{n} \mathcal{N}\left(CA^{i-1}\right)\right) = \mathcal{N}(K) \cap \mathcal{N}(A). \tag{10.6}$$

The following result shows that semiobservability is unchanged by full state feedback.

**Proposition 10.1.** Let $A \in \mathbb{R}^{n \times n}$, $B \in \mathbb{R}^{n \times m}$, $C \in \mathbb{R}^{l \times n}$, $K \in \mathbb{R}^{m \times n}$, and $R \in \mathbb{R}^{n \times n}$, where $R$ is positive definite. If the pair $(A, C)$ is semiobservable, then the pair $(A + BK, C^{\mathrm{T}}C + K^{\mathrm{T}}RK)$ is semiobservable with respect to $K$.

**Proof.** Note that $\mathcal{N}(C^{\mathrm{T}}C + K^{\mathrm{T}}RK) = \mathcal{N}(C) \cap \mathcal{N}(K)$. Hence,

$$\mathcal{N}(K) \cap \left(\bigcap_{i=1}^{n} \mathcal{N}((C^{\mathrm{T}}C + K^{\mathrm{T}}RK)(A + BK)^{i-1})\right)$$
$$= \bigcap_{i=1}^{n} \mathcal{N}((C^{\mathrm{T}}C + K^{\mathrm{T}}RK)(A + BK)^{i-1})$$
$$= \mathcal{N}(K) \cap \left(\bigcap_{i=1}^{n} \mathcal{N}(CA^{i-1})\right)$$
$$= \mathcal{N}(K) \cap \mathcal{N}(A)$$
$$= \mathcal{N}(K) \cap \mathcal{N}(A + BK), \tag{10.7}$$

which implies that the pair $(A + BK, C^{\mathrm{T}}C + K^{\mathrm{T}}RK)$ is semiobservable with respect to $K$. $\square$

Next, we connect semistability with Lyapunov theory and semiobservability to arrive at a characterization of the $\mathcal{H}_2$ norm of semistable systems. For this result, we consider the linear dynamical system

$$\dot{x}(t) = Ax(t), \quad x(0) = x_0, \quad t \geq 0, \tag{10.8}$$

where $A \in \mathbb{R}^{n \times n}$, with output equation (10.3). Furthermore, for a given semistable system define the $\mathcal{H}_2$ norm of the transfer function $G(s)$ with realization $G(s) \sim [\begin{array}{c|c} A & x_0 \\ \hline C & 0 \end{array}]$ and free response $y(t) = H(t) = Ce^{At}x_0$ by

$$\|G\|_2 \triangleq \left[ \int_0^\infty \|H(t)\|_{\mathrm{F}}^2 \mathrm{d}t \right]^{1/2} = \left[ \frac{1}{2\pi} \int_{-\infty}^\infty \|G(\jmath\omega)\|_{\mathrm{F}}^2 \mathrm{d}\omega \right]^{1/2}. \tag{10.9}$$

The following proposition presents necessary and sufficient conditions for the well-posedness of the $\mathcal{H}_2$ norm of a semistable system.

**Proposition 10.2.** Consider the linear dynamical system (10.8) with output (10.3), and assume that $A$ is semistable. Then the following statements are equivalent.

(i) For every $x_0 \in \mathbb{R}^n$, $\|G\|_2 < \infty$.

(ii) $\int_0^\infty e^{A^{\mathrm{T}}t} Re^{At} \mathrm{d}t < \infty$, where $R \triangleq C^{\mathrm{T}}C$.

(iii) $\mathcal{N}(A) \subset \mathcal{N}(C)$.

**Proof.** The equivalence of (i) and (ii) follows from the fact that

$$\|G\|_2^2 = x_0^{\mathrm{T}} \int_0^\infty e^{A^{\mathrm{T}}t} Re^{At} \mathrm{d}t x_0. \tag{10.10}$$

To show that (ii) implies (iii), note that since $A$ is semistable, it follows that either $A$ is Hurwitz or there exists an invertible matrix $S \in \mathbb{R}^{n \times n}$ such that

$$A = S \begin{bmatrix} J & 0 \\ 0 & 0 \end{bmatrix} S^{-1},$$

where $J \in \mathbb{R}^{r \times r}$, $r = \operatorname{rank} A$, and $J$ is Hurwitz. Now, if $A$ is Hurwitz, then (iii) holds trivially since $\mathcal{N}(A) = \{0\} \subset \mathcal{N}(C)$.

Alternatively, if $A$ is not Hurwitz, then

$$\mathcal{N}(A) = \left\{ x \in \mathbb{R}^n : x = S[0_{1 \times r}, y^{\mathrm{T}}]^{\mathrm{T}}, y \in \mathbb{R}^{n-r} \right\}. \tag{10.11}$$

Now,

$$\int_0^\infty e^{A^{\mathrm{T}}t} R e^{At} \mathrm{d}t = S^{-\mathrm{T}} \int_0^\infty e^{\hat{J}t} \hat{R} e^{\hat{J}t} \mathrm{d}t S$$

$$= S^{-\mathrm{T}} \int_0^\infty \begin{bmatrix} e^{J^{\mathrm{T}}t} \hat{R}_1 e^{Jt} & e^{J^{\mathrm{T}}t} \hat{R}_{12} \\ \hat{R}_{12}^{\mathrm{T}} e^{Jt} & \hat{R}_2 \end{bmatrix} \mathrm{d}t S, \qquad (10.12)$$

where

$$\hat{J} = \begin{bmatrix} J & 0 \\ 0 & 0 \end{bmatrix}, \quad \hat{R} = S^{\mathrm{T}} R S = \begin{bmatrix} \hat{R}_1 & \hat{R}_{12} \\ \hat{R}_{12}^{\mathrm{T}} & \hat{R}_2 \end{bmatrix}. \qquad (10.13)$$

Next, it follows from (10.12) that

$$\int_0^\infty e^{A^{\mathrm{T}}t} R e^{At} \mathrm{d}t < \infty \qquad (10.14)$$

if and only if $\hat{R}_2 = 0$ or, equivalently,

$$[0_{1\times r}, y^{\mathrm{T}}] \hat{R} [0_{1\times r}, y^{\mathrm{T}}]^{\mathrm{T}} = 0, \quad y \in \mathbb{R}^{n-r}, \qquad (10.15)$$

which is further equivalent to $x^{\mathrm{T}} R x = 0$, $x \in \mathcal{N}(A)$. Hence, $\mathcal{N}(A) \subset \mathcal{N}(C)$.

Finally, the proof of (iii) implies that (ii) is immediate by reversing the steps of the proof given above. □

**Theorem 10.1.** Consider the linear dynamical system (10.8). Suppose there exist an $n \times n$ matrix $P \geq 0$ and an $l \times n$ matrix $C$ such that $(A, C)$ is semiobservable and

$$0 = A^{\mathrm{T}} P + P A + R, \qquad (10.16)$$

where $R \triangleq C^{\mathrm{T}} C$. Then (10.8) is semistable with respect to $\mathbb{R}^n$. Furthermore, $\|\|G(s)\|\|_2^2 = (x_0 - x_{\mathrm{e}})^{\mathrm{T}} P (x_0 - x_{\mathrm{e}})$, where $x_{\mathrm{e}} \triangleq x_0 - A A^{\#} x_0$.

**Proof.** The first part of the result is a direct consequence of Proposition 4.1 of [33]. Now, since $A$ is semistable, it follows from (ix) of Proposition 11.7.2 of [26] that $\lim_{t\to\infty} e^{At} = I_q - A A^{\#}$. Next, noting that $A x_{\mathrm{e}} = 0$, (10.8) can be equivalently written as

$$\dot{x}(t) = A(x(t) - x_{\mathrm{e}}), \quad x(0) = x_0, \quad t \geq 0. \qquad (10.17)$$

Hence,

$$\int_0^t (x(s) - x_{\mathrm{e}})^{\mathrm{T}} R (x(s) - x_{\mathrm{e}}) \mathrm{d}s$$

$$= -(x(t) - x_{\mathrm{e}})^{\mathrm{T}} P (x(t) - x_{\mathrm{e}}) + (x_0 - x_{\mathrm{e}})^{\mathrm{T}} P (x_0 - x_{\mathrm{e}}). \qquad (10.18)$$

Now, it follows from the semiobservability of $(A, C)$ that $Rx_{\mathrm{e}} = 0$. Hence, letting $t \to \infty$ and noting that $x(t) \to x_{\mathrm{e}}$ as $t \to \infty$, it follows from (10.18) that

$$\int_0^\infty x^{\mathrm{T}}(t)Rx(t)\mathrm{d}t = (x_0 - x_{\mathrm{e}})^{\mathrm{T}}P(x_0 - x_{\mathrm{e}}). \tag{10.19}$$

Finally, defining the free response of (10.8) by $H(t) \triangleq Cx(t) = Ce^{At}x_0$, $t \geq 0$, and noting that $R = C^{\mathrm{T}}C$, it follows from Parseval's theorem [122, p. 363] that

$$(x_0 - x_{\mathrm{e}})^{\mathrm{T}}P(x_0 - x_{\mathrm{e}}) = \int_0^\infty H^{\mathrm{T}}(t)H(t)\mathrm{d}t = \frac{1}{2\pi}\int_{-\infty}^\infty \|G(\jmath\omega)\|_{\mathrm{F}}^2\mathrm{d}\omega. \tag{10.20}$$

This completes the proof. $\qquad\square$

**Example 10.1.** Consider the linear system (10.8), where $A$ is given by

$$A = \begin{bmatrix} -2 & 1 \\ 2 & -1 \end{bmatrix}, \tag{10.21}$$

and note that $\mathcal{N}(A) = \{(x_1, x_2) \in \mathbb{R}^2 : 2x_1 = x_2 = \alpha, \alpha \in \mathbb{R}\}$. Let $C = [2, -1]$. It is easy to verify that (10.16) holds with

$$P = \begin{bmatrix} 1 & 0 \\ 0 & \frac{1}{2} \end{bmatrix}. \tag{10.22}$$

Furthermore, note that $\mathcal{N}(C) = \mathcal{N}(A)$, and, hence, the pair $(A, C)$ is semiobservable. Now, it follows from Theorem 10.1 that (10.8) is semistable. $\triangle$

Next, we give a necessary and sufficient condition for characterizing semistability using the Lyapunov equation (10.16). Before we state this result, the following lemmas are needed.

**Lemma 10.1.** Consider the linear dynamical system (10.8). If (10.8) is semistable, then, for every $n \times n$ nonnegative definite matrix $R$,

$$\int_0^\infty (x(t) - x_{\mathrm{e}})^{\mathrm{T}}R(x(t) - x_{\mathrm{e}})\mathrm{d}t < \infty, \tag{10.23}$$

where $x_{\mathrm{e}} = (I_n - AA^{\#})x_0$.

**Proof.** Since $A$ is semistable, it follows from the Jordan decomposition that there exists an invertible matrix $S \in \mathbb{C}^{n \times n}$ such that

$$A = S \begin{bmatrix} J & 0 \\ 0 & 0 \end{bmatrix} S^{-1},$$

where $J \in \mathbb{C}^{r \times r}$, $r = \operatorname{rank} A$, and $J$ is asymptotically stable. Let $z(t) \triangleq S^{-1}x(t)$ and $z_e \triangleq S^{-1}x_e$, $t \geq 0$. Then (10.8) becomes

$$\dot{z}(t) = \begin{bmatrix} J & 0 \\ 0 & 0 \end{bmatrix} z(t), \quad z(0) = S^{-1}x_0, \quad t \geq 0, \tag{10.24}$$

which implies that $\lim_{t \to \infty} z_i(t) = 0$, $i = 1, \ldots, r$, and $z_j(t) = z_j(0)$, $j = r+1, \ldots, n$; that is, $z_e = [0, \ldots, 0, z_{r+1}(0), \ldots, z_n(0)]^{\mathrm{T}}$.

Now,

$$\int_0^\infty (x(t) - x_e)^{\mathrm{T}} R(x(t) - x_e)\mathrm{d}t = \int_0^\infty (z(t) - z_e)^* S^* RS(z(t) - z_e)\mathrm{d}t$$

$$= \int_0^\infty \hat{z}^*(t) S^* RS \hat{z}(t)\mathrm{d}t, \tag{10.25}$$

where $\hat{z}(t) \triangleq [z_1(t), \ldots, z_r(t), 0, \ldots, 0]^{\mathrm{T}}$. Since

$$\dot{\hat{z}}(t) = \begin{bmatrix} J & 0 \\ 0 & 0 \end{bmatrix} \hat{z}(t), \quad \hat{z}(0) = \hat{z}_0, \quad t \geq 0, \tag{10.26}$$

and $J$ is asymptotically stable, it follows that

$$\int_0^\infty \hat{z}^*(t) S^* RS \hat{z}(t)\mathrm{d}t < \infty, \tag{10.27}$$

which proves the result. $\qquad \square$

**Lemma 10.2.** Let $A \in \mathbb{R}^{n \times n}$ and $B \in \mathbb{R}^{m \times m}$. If $A$ and $B$ are semistable, then $A \oplus B$ is semistable.

**Proof.** Let $\lambda \in \operatorname{spec}(A)$ and $\mu \in \operatorname{spec}(B)$. Since $A$ and $B$ are both semistable, it follows that $\operatorname{Re} \lambda < 0$ or $\lambda = 0$ and $\operatorname{am}_A(0) = \operatorname{gm}_A(0)$, and $\operatorname{Re} \mu < 0$ or $\mu = 0$ and $\operatorname{am}_B(0) = \operatorname{gm}_B(0)$, where $\operatorname{am}_X(\lambda)$ and $\operatorname{gm}_X(\lambda)$ denote algebraic multiplicity of $\lambda \in \operatorname{spec}(X)$ and geometric multiplicity of $\lambda \in \operatorname{spec}(X)$, respectively. Now, it follows from the fact that $\lambda + \mu \in \operatorname{spec}(A \oplus B)$ that $\operatorname{spec}(A \oplus B) \subset \{z \in \mathbb{C} : \operatorname{Re} z < 0\} \cup \{0\}$. Next, it follows from Fact 7.5.2 of [26] that $\operatorname{gm}_A(0)\operatorname{gm}_B(0) \leq \operatorname{gm}_{A \oplus B}(0) \leq \operatorname{am}_{A \oplus B}(0) = \operatorname{am}_A(0)\operatorname{am}_B(0)$. Since $\operatorname{am}_A(0) = \operatorname{gm}_A(0)$ and $\operatorname{am}_B(0) = \operatorname{gm}_B(0)$, it follows that $\operatorname{gm}_{A \oplus B}(0) = \operatorname{am}_{A \oplus B}(0)$, and, hence, $A \oplus B$ is semistable. $\qquad \square$

**Lemma 10.3.** Let $x \in \mathbb{R}^n$ and $A \in \mathbb{R}^{n \times n}$, and assume that $A$ is semistable. Then $\int_0^\infty e^{At}x\mathrm{d}t$ exists if and only if $x \in \mathcal{R}(A)$. In this case,

$$\int_0^\infty e^{At}x\mathrm{d}t = -A^{\#}x.$$

**Proof.** The proof is similar to the proofs of (vii) and (viii) of Lemma 2.2 of [30] and, hence, is omitted. $\qquad\square$

**Lemma 10.4** (see [33]). Let $A \in \mathbb{R}^{n \times n}$. If there exist an $n \times n$ matrix $P \geq 0$ and an $l \times n$ matrix $C$ such that $(A, C)$ is semiobservable and (10.16) holds, then (i) $\mathcal{N}(P) \subseteq \mathcal{N}(A) \subseteq \mathcal{N}(R)$ and (ii) $\mathcal{N}(A) \cap \mathcal{R}(A) = \{0\}$.

**Theorem 10.2.** Consider the linear dynamical system (10.8). Then (10.8) is semistable if and only if, for every semiobservable pair $(A, C)$, there exists an $n \times n$ matrix $P \geq 0$ such that (10.16) holds. Furthermore, if $(A, C)$ is semiobservable and $P$ satisfies (10.16), then

$$P = \int_0^\infty e^{A^{\mathrm{T}} t} R e^{At} \mathrm{d}t + P_0 \tag{10.28}$$

for some $P_0 = P_0^{\mathrm{T}} \in \mathbb{R}^{n \times n}$ satisfying

$$0 = A^{\mathrm{T}} P_0 + P_0 A \tag{10.29}$$

and

$$P_0 \geq -\int_0^\infty e^{A^{\mathrm{T}} t} R e^{At} \mathrm{d}t. \tag{10.30}$$

In addition, $\min_{P \in \mathcal{P}} \|P\|_{\mathrm{F}}$ has a unique solution $P$ given by

$$P = \int_0^\infty e^{A^{\mathrm{T}} t} R e^{At} \mathrm{d}t, \tag{10.31}$$

where $\mathcal{P}$ denotes the set of all $P$ satisfying (10.16). Finally, (10.8) is semistable if and only if, for every semiobservable pair $(A, C)$, there exists an $n \times n$ matrix $P > 0$ such that (10.16) holds.

**Proof.** Sufficiency for the first implication follows from Theorem 10.1. To show necessity, assume that (10.8) is semistable. Then $\lim_{t \to \infty} x(t) = x_{\mathrm{e}}$, where $x_{\mathrm{e}} = (I_n - AA^{\#})x_0$. For a semiobservable pair $(A, C)$, let

$$P = \int_0^\infty (AA^{\#})^{\mathrm{T}} e^{A^{\mathrm{T}} t} R e^{At} AA^{\#} \mathrm{d}t. \tag{10.32}$$

Then, for $x_0 \in \mathbb{R}^n$,

$$\begin{aligned}
x_0^{\mathrm{T}} P x_0 &= \int_0^\infty x_0^{\mathrm{T}} (AA^{\#})^{\mathrm{T}} e^{A^{\mathrm{T}} t} R e^{At} AA^{\#} x_0 \mathrm{d}t \\
&= \int_0^\infty (x_0 - x_{\mathrm{e}})^{\mathrm{T}} e^{A^{\mathrm{T}} t} R e^{At} (x_0 - x_{\mathrm{e}}) \mathrm{d}t \\
&= \int_0^\infty (x(t) - x_{\mathrm{e}})^{\mathrm{T}} R (x(t) - x_{\mathrm{e}}) \mathrm{d}t,
\end{aligned} \tag{10.33}$$

where we used the fact that $x(t) - x_e = e^{At}(x_0 - x_e)$. It follows from Lemma 10.1 that $P$ is well defined. Since $x_e \in \mathcal{N}(A)$, it follows from (10.5) that $Rx_e = 0$, and, hence,

$$x_0^T P x_0 = \int_0^\infty x^T(t) R x(t) \mathrm{d}t = \int_0^\infty x_0^T e^{A^T t} R e^{At} x_0 \mathrm{d}t, \tag{10.34}$$

which implies that

$$P = \int_0^\infty e^{A^T t} R e^{At} \mathrm{d}t. \tag{10.35}$$

Now, (10.16) is immediate using the fact that $Rx_e = 0$.

Next, since $A$ is semistable, it follows from the above result that there exists an $n \times n$ nonnegative definite matrix $P$ such that (10.16) holds or, equivalently, $(A \oplus A)^T \mathrm{vec}\, P = -\mathrm{vec}\, R$. Hence, $\mathrm{vec}\, R \in \mathcal{R}((A \oplus A)^T)$ and

$$\mathcal{P} = \left\{ P \in \mathbb{R}^{n \times n} : P = -\mathrm{vec}^{-1}\left( ((A \oplus A)^T)^\# \mathrm{vec}\, R \right) + \mathrm{vec}^{-1}(z) \right\}$$

for some $z \in \mathcal{N}((A \oplus A)^T)$. Next, it follows from Lemma 10.2 that $A \oplus A$ is semistable, and, hence, by Lemma 10.3,

$$\mathrm{vec}^{-1}\left( ((A \oplus A)^T)^\# \mathrm{vec}\, R \right) = -\int_0^\infty \mathrm{vec}^{-1}\left( e^{(A \oplus A)^T t} \mathrm{vec}\, R \right) \mathrm{d}t$$

$$= -\int_0^\infty \mathrm{vec}^{-1}\left( e^{A^T t} \otimes e^{A^T t} \right) \mathrm{vec}\, R \mathrm{d}t$$

$$= -\int_0^\infty e^{A^T t} R e^{At} \mathrm{d}t, \tag{10.36}$$

where in (10.36) we used the facts that $(X \otimes Y)^T = X^T \otimes Y^T$, $e^{X \oplus Y} = e^X \otimes e^Y$, and $\mathrm{vec}(XYZ) = (Z^T \otimes X)\mathrm{vec}\, Y$ [26, Chapter 7]. Hence,

$$P = \int_0^\infty e^{A^T t} R e^{At} \mathrm{d}t + \mathrm{vec}^{-1}(z), \tag{10.37}$$

where $\mathrm{vec}^{-1}(z)$ satisfies

$$\mathrm{vec}^{-1}(z) = (\mathrm{vec}^{-1}(z))^T, \quad A^T \mathrm{vec}^{-1}(z) + \mathrm{vec}^{-1}(z)A = 0,$$

and

$$\mathrm{vec}^{-1}(z) \geq -\int_0^\infty e^{A^T t} R e^{At} \mathrm{d}t.$$

If $P$ is such that $\min_{P \in \mathcal{P}} \|P\|_F$ holds, then it follows that $P$ is the unique solution of a least squares minimization problem and is given by

$$P = -\mathrm{vec}^{-1}\left( ((A \oplus A)^T)^\# \mathrm{vec}\, R \right) = \int_0^\infty e^{A^T t} R e^{At} \mathrm{d}t. \tag{10.38}$$

Finally, suppose $(A, C)$ is semiobservable. Then it follows from the first part of the theorem that there exists an $n \times n$ matrix $P \geq 0$ such that (10.16) holds. Since, by Lemma 10.4, $\mathcal{N}(A) \cap \mathcal{R}(A) = \{0\}$, it follows from Lemma 4.14 of [24] that $A$ is group invertible. Thus, let $L \triangleq I_n - AA^{\#}$, and note that $L^2 = L$. Hence, $L$ is the unique $n \times n$ matrix satisfying $\mathcal{N}(L) = \mathcal{R}(A)$, $\mathcal{R}(L) = \mathcal{N}(A)$, and $Lx = x$ for all $x \in \mathcal{N}(A)$. Now, define

$$\hat{P} \triangleq P + L^{\mathrm{T}} L. \tag{10.39}$$

Next, we show that $\hat{P}$ is positive definite.

Consider the function $V(x) = x^{\mathrm{T}} \hat{P} x$, $x \in \mathbb{R}^n$. If $V(x) = 0$ for some $x \in \mathbb{R}^n$, then $Px = 0$ and $Lx = 0$. It follows from (i) of Lemma 10.4 that $x \in \mathcal{N}(A)$, and $Lx = 0$ implies that $x \in \mathcal{R}(A)$. Now, it follows from (ii) of Lemma 10.4 that $x = 0$. Hence, $\hat{P}$ is positive definite. Next, since $LA = A - AA^{\#}A = 0$, it follows that

$$\begin{aligned} A^{\mathrm{T}}\hat{P} + \hat{P}A + R &= A^{\mathrm{T}}P + PA + R + A^{\mathrm{T}}L^{\mathrm{T}}L + L^{\mathrm{T}}LA \\ &= (LA)^{\mathrm{T}}L + L^{\mathrm{T}}LA \\ &= 0. \end{aligned} \tag{10.40}$$

Conversely, if there exists $P > 0$ such that (10.16) holds, consider the function $U(x) = x^{\mathrm{T}} P x$, $x \in \mathbb{R}^n$. Then $\dot{U}(x) = -x^{\mathrm{T}} R x \leq 0$ and $\dot{U}^{-1}(0) = \mathcal{N}(R)$. To obtain the largest invariant set $\mathcal{M}$ contained in $\mathcal{N}(R)$, consider a solution $x(t)$ of (10.8) such that $Cx(t) = 0$ for all $t \geq 0$. On $\mathcal{M}$, it follows that $C\frac{\mathrm{d}^{k-1}}{\mathrm{d}t^{k-1}}x(t) = 0$ for all $t \geq 0$ and $k = 1, \ldots, n$, and, hence, $CA^{k-1}x(t) = 0$ for all $t \geq 0$ and $k = 1, \ldots, n$. Now, it follows from (10.5) that $x(t) \in \mathcal{N}(A)$ for all $t \geq 0$. Thus, $\mathcal{M} \subseteq \mathcal{N}(A)$. Since $\mathcal{N}(A)$ consists of equilibrium points, it follows that $\mathcal{M} = \mathcal{N}(A)$. For $x_{\mathrm{e}} \in \mathcal{N}(A)$, Lyapunov stability of $x_{\mathrm{e}}$ now follows by considering the Lyapunov function $U(x - x_{\mathrm{e}})$. $\square$

**Example 10.2.** Consider the two-agent network consensus problem given by the linear system (10.8), where $A$ is given by [238]

$$A = \begin{bmatrix} -1 & 1 \\ 1 & -1 \end{bmatrix}. \tag{10.41}$$

Note that $\mathcal{N}(A) = \{(x_1, x_2) \in \mathbb{R}^2 : x_1 = x_2 = \alpha, \alpha \in \mathbb{R}\}$. Let $C = [1, -1]$. It is easy to verify that (10.16) holds with

$$P = \frac{1}{2} \begin{bmatrix} 1 & 0 \\ 0 & 1 \end{bmatrix}. \tag{10.42}$$

Furthermore, note that $\mathcal{N}(C) = \mathcal{N}(A)$, and, hence, the pair $(A, C)$ is semiobservable. Now, it follows from Theorem 10.2 that (10.8) is semistable. $\triangle$

Next, we show that the unique solution $P$ given by (10.16) and satisfying $\min_{P \in \mathcal{P}} \|P\|_{\mathrm{F}}$ can be characterized by an LMI minimization problem.

**Theorem 10.3.** Consider the linear dynamical system (10.8) with output (10.3). Assume that $A$ is semistable and $(A, C)$ is semiobservable. Let $P_{\min}$ be the solution to the LMI minimization problem

$$\min \left\{ \operatorname{tr} PV : P \geq 0 \text{ and } A^{\mathrm{T}}P + PA + R \leq 0 \right\}, \qquad (10.43)$$

where $V \in \mathbb{R}^{n \times n}$ and $V \geq 0$. Then

$$\operatorname{tr} P_{\min}V = \operatorname{tr} \int_0^\infty e^{A^{\mathrm{T}}t} R e^{At} \mathrm{d}t V. \qquad (10.44)$$

**Proof.** Let $\hat{P} = \int_0^\infty e^{A^{\mathrm{T}}t} R e^{At} \mathrm{d}t$, and let $P \geq 0$ be such that

$$A^{\mathrm{T}}P + PA + R \leq 0. \qquad (10.45)$$

(Note that $A^{\mathrm{T}}\hat{P} + \hat{P}A + R = 0$, which implies that a $P \geq 0$ satisfying (10.45) exists.) Now, let $W \in \mathbb{R}^{n \times n}$, $W \geq 0$, be such that

$$0 = A^{\mathrm{T}}P + PA + R + W. \qquad (10.46)$$

Next, since $(A, C)$ is semiobservable, it follows that if $x_{\mathrm{e}} \in \mathcal{N}(A)$, then $R x_{\mathrm{e}} = 0$, and, hence, it follows from (10.46) that $W x_{\mathrm{e}} = 0$. Now, using identical arguments as in the proof of Theorem 10.2, it follows that

$$P = \int_0^\infty e^{A^{\mathrm{T}}t}(R + W)e^{At}\mathrm{d}t \geq \int_0^\infty e^{A^{\mathrm{T}}t} R e^{At}\mathrm{d}t = \hat{P}. \qquad (10.47)$$

Finally, since $\hat{P}$ is an element of the feasible set of the optimization problem (10.43), $\operatorname{tr} P_{\min}V = \operatorname{tr} \hat{P}V$. $\qquad \square$

Finally, we provide a dual result to Theorem 10.3 which is necessary for developing feedback controllers guaranteeing closed-loop semistability.

**Theorem 10.4.** Consider the linear dynamical system (10.8) with output (10.3). Assume that $A$ is semistable, and let $V \in \mathbb{R}^{n \times n}$, $V \geq 0$, be such that $(A, V)$ is semicontrollable. Let $Q_{\min}$ be the solution to the LMI minimization problem

$$\min \left\{ \operatorname{tr} QR : Q \geq 0 \text{ and } AQ + QA^{\mathrm{T}} + V \leq 0 \right\}. \qquad (10.48)$$

Then

$$\operatorname{tr} Q_{\min} R = \operatorname{tr} \int_0^\infty e^{A^{\mathrm{T}}t} R e^{At} \mathrm{d}t V = \operatorname{tr} P_{\min} V, \qquad (10.49)$$

where $P_{\min}$ is the solution to the LMI minimization problem given by (10.43).

**Proof.** The proof is a direct consequence of Theorem 10.3 by noting that $(A, V)$ is semicontrollable if and only if $(A^{\mathrm{T}}, V)$ is semiobservable. Now, replacing $A$ with $A^{\mathrm{T}}$ and $R$ with $V$ in Theorem 10.3, it follows that

$$\operatorname{tr} Q_{\min} R = \operatorname{tr} \int_0^\infty e^{At} V e^{A^{\mathrm{T}}t} \mathrm{d}t R = \operatorname{tr} \int_0^\infty e^{A^{\mathrm{T}}t} R e^{At} \mathrm{d}t V = \operatorname{tr} P_{\min} V.$$
$$(10.50)$$

This completes the proof. □

## 10.3 Optimal Semistable Stabilization

In this section, we consider the problem of optimal state feedback control for semistable stabilization of linear dynamical systems. Specifically, we consider the controlled linear system given by

$$\dot{x}(t) = Ax(t) + Bu(t), \quad x(0) = x_0, \quad t \geq 0, \qquad (10.51)$$

where $x(t) \in \mathbb{R}^n$, $t \geq 0$, is the state vector; $u(t) \in \mathbb{R}^m$, $t \geq 0$, is the control input; $A \in \mathbb{R}^{n \times n}$; and $B \in \mathbb{R}^{n \times m}$, with the state feedback controller $u(t) = Kx(t)$, where $K \in \mathbb{R}^{m \times n}$ is such that the closed-loop system given by

$$\dot{x}(t) = (A + BK)x(t), \quad x(0) = x_0, \quad t \geq 0, \qquad (10.52)$$

is semistable and the performance criterion

$$J(K) \triangleq \int_0^\infty \left[ (x(t) - x_{\mathrm{e}})^{\mathrm{T}} R_1 (x(t) - x_{\mathrm{e}}) + (u(t) - u_{\mathrm{e}})^{\mathrm{T}} R_2 (u(t) - u_{\mathrm{e}}) \right] \mathrm{d}t$$
$$(10.53)$$

is minimized, where $R_1 \triangleq E_1^{\mathrm{T}} E_1$, $R_2 \triangleq E_2^{\mathrm{T}} E_2 > 0$, $R_{12} \triangleq E_1^{\mathrm{T}} E_2 = 0$, $u_{\mathrm{e}} = Kx_{\mathrm{e}}$, and $x_{\mathrm{e}} = \lim_{t \to \infty} x(t)$.

The difficulty in addressing the optimal semistable control problem is that the classical optimal control theory cannot be applied to this optimal control problem. More specifically, Examples 2.1–2.3 in [154] show that the optimal semistable control problem proposed here may have a unique solution, infinitely many solutions, or no solution at all, whereas the classical $\mathcal{H}_2$ optimal control problem is characterized by a unique optimal solution.

Hence, it is necessary to develop a new optimal control framework to solve the optimal semistable control problem.

Note that it follows from Lemma 10.1 that if the closed-loop system is semistable, then $J(K)$ is well defined. To develop necessary conditions for the optimal semistable control problem, we assume that $(A, B)$ is semicontrollable, $(A, E_1)$ is semiobservable, and $x_{\mathrm{e}} \in \mathcal{N}(K)$. In this case, it follows from Proposition 10.1 that $(A + BK, R_1 + K^{\mathrm{T}} R_2 K)$ is semiobservable with respect to $K$, and, hence, $(R_1 + K^{\mathrm{T}} R_2 K) x_{\mathrm{e}} = 0$.

Thus,

$$
\begin{aligned}
J(K) &= \int_0^\infty x_0^{\mathrm{T}} e^{\tilde{A}^{\mathrm{T}} t} (R_1 + K^{\mathrm{T}} R_2 K) e^{\tilde{A} t} x_0 \mathrm{d}t \\
&= \mathrm{tr} \int_0^\infty e^{\tilde{A}^{\mathrm{T}} t} (R_1 + K^{\mathrm{T}} R_2 K) e^{\tilde{A} t} x_0 x_0^{\mathrm{T}} \mathrm{d}t \\
&= \mathrm{tr} P_{\mathrm{LS}} V,
\end{aligned}
\tag{10.54}
$$

where we assume that the initial state $x_0$ is a random variable such that $\mathbb{E}[x_0] = 0$ and $\mathbb{E}[x_0 x_0^{\mathrm{T}}] = V$; $\tilde{A} \triangleq A + BK$; and

$$
P_{\mathrm{LS}} \triangleq \int_0^\infty e^{\tilde{A}^{\mathrm{T}} t} (R_1 + K^{\mathrm{T}} R_2 K) e^{\tilde{A} t} \mathrm{d}t
$$

denotes the least squares solution to

$$
0 = \tilde{A}^{\mathrm{T}} P + P \tilde{A} + \tilde{R},
\tag{10.55}
$$

where $\tilde{R} \triangleq R_1 + K^{\mathrm{T}} R_2 K$. Unlike the standard $\mathcal{H}_2$ optimal control problem, $P_{\mathrm{LS}} \geq 0$ is not a unique solution to (10.55).

The following theorem presents an LMI solution to the $\mathcal{H}_2$ optimal semistable control problem.

**Theorem 10.5.** Consider the linear dynamical system (10.51), and assume that $(A, E_1)$ is semiobservable and $(A, V)$ is semicontrollable. Let $Q \in \mathbb{R}^{n \times n}$ and $X \in \mathbb{R}^{m \times n}$ be the solution to the LMI minimization problem

$$
\min_{Q \in \mathbb{R}^{n \times n}, X \in \mathbb{R}^{m \times n}, W \in \mathbb{R}^{p \times p}} \mathrm{tr}\, W
\tag{10.56}
$$

subject to

$$
\begin{bmatrix} Q & (E_1 Q + E_2 X)^{\mathrm{T}} \\ E_1 Q + E_2 X & W \end{bmatrix} > 0,
\tag{10.57}
$$

$$
AQ + BX + QA^{\mathrm{T}} + X^{\mathrm{T}} B^{\mathrm{T}} + V \leq 0.
\tag{10.58}
$$

Then $K = XQ^{-1}$ is a semistabilizing controller for (10.51); that is, $A + BK$ is semistable. Furthermore, $K$ minimizes the $\mathcal{H}_2$ performance criterion $J(K)$ given by (10.53).

**Proof.** Since $K = XQ^{-1}$, it follows from (10.58) that

$$(A + BK)Q + Q(A + BK)^{\mathrm{T}} + V \le 0, \tag{10.59}$$

which, since $(A, V)$ is semicontrollable, implies that $A + BK$ is semistable. Next, note that (10.57) holds if and only if

$$W > (E_1 Q + E_2 X)Q^{-1}(E_1 Q + E_2 X)^{\mathrm{T}}, \tag{10.60}$$

which implies that the minimization problem (10.56)–(10.58) is equivalent to

$$\min \operatorname{tr}(E_1 Q + E_2 X)Q^{-1}(E_1 Q + E_2 X)^{\mathrm{T}} \tag{10.61}$$

subject to

$$AQ + BX + QA^{\mathrm{T}} + X^{\mathrm{T}}B^{\mathrm{T}} + V \le 0, \tag{10.62}$$
$$Q > 0. \tag{10.63}$$

Hence, noting that (10.61)–(10.63) is equivalent to

$$\min \operatorname{tr} Q\tilde{R} \tag{10.64}$$

subject to

$$\tilde{A}Q + Q\tilde{A}^{\mathrm{T}} + V \le 0, \tag{10.65}$$
$$Q > 0, \tag{10.66}$$

the result follows as a direct consequence of Theorems 10.4 and 10.2.    □

If the dimension of the system is very large, then LMIs cannot be implemented in finite time. In this case, instead of seeking a globally optimal solution, we can solve for an approximate suboptimal solution. For the LMI problem developed by Theorem 10.5, a suboptimal solution to this problem can be obtained by using a two-stage optimization process. Specifically, by fixing $Q$ one can design the controller $K$. Then, with $K$ fixed, $Q$ can be obtained. This process continues until convergence or until an acceptable controller is found.

Finally, we note that the framework developed in this chapter is restricted to time-invariant dynamical systems. However, part of the results developed in this chapter can be applied to switched systems. For details, see [163].

## 10.4 Optimal Fixed-Structure Control for Network Consensus

In this section, we use the optimal control framework developed in Section 10.3 to design optimal controllers for multiagent network dynamical systems. Specifically, we use undirected graphs to represent a dynamical network and present solutions to the consensus problem for networks with undirected graph topologies (or information flow) [238]. The notation we use here is identical to the notation established in Chapter 2.

The information flow model we consider is a network dynamical system involving the trajectories of the dynamical network characterized by the multiagent dynamical system $\mathcal{G}$ given by

$$\dot{x}_i(t) = u_i(t), \quad x_i(0) = x_{i0}, \quad t \geq 0, \quad i = 1, \ldots, q, \tag{10.67}$$

$$u_i(t) = \sum_{j=1, j\neq i}^{q} \frac{1}{k_i} \mathcal{A}_{(i,j)}(x_j(t) - x_i(t)), \tag{10.68}$$

where $q \geq 2$; $x_i(t) \in \mathbb{R}$, $t \geq 0$, represents an information state; $u_i(t) \in \mathbb{R}$, $t \geq 0$, represents the control input; $k_i > 0$, $i = 1, \ldots, q$; and $\mathcal{A}_{(i,j)} \geq 0$, $i, j = 1, \ldots, q$, $i \neq j$.

**Assumption 10.1.** The *connectivity matrix* $\mathcal{C} \in \mathbb{R}^{q \times q}$ associated with the multiagent dynamical system $\mathcal{G}$ is defined by

$$\mathcal{C}_{(i,j)} = \begin{cases} 1 & \text{if } (j,i) \in \mathcal{E}, \\ 0, & \text{otherwise,} \end{cases} \quad i \neq j, \quad i, j = 1, \ldots, q, \tag{10.69}$$

and $\mathcal{C}_{(i,i)} = -\sum_{k=1, k\neq i}^{q} \mathcal{C}_{(i,k)}$, $i = j$, $i = 1, \ldots, q$, with $\operatorname{rank} \mathcal{C} = q - 1$ and $\mathcal{C} = \mathcal{C}^{\mathrm{T}}$.

Recall that the negative of the connectivity matrix, that is, $-\mathcal{C}$, is the Laplacian of the graph $\mathfrak{G} = (\mathcal{V}, \mathcal{E}, \mathcal{A})$. Furthermore, note that $\mathcal{C}_{(i,j)} = \mathcal{A}_{(i,j)}$ for all $i, j = 1, \ldots, q$, $i \neq j$. In multiagent coordination [180, 238] and distributed network averaging [315] with a fixed communication topology, we require that $x_{\mathrm{e}} \in \operatorname{span}\{\mathbf{e}\}$, where $\mathbf{e} \in \mathbb{R}^q$. In this section, we consider the design of a fixed-structure consensus protocol for (10.67) and (10.68) such that the closed-loop system is semistable; that is, $\lim_{t\to\infty} x_i(t) = \alpha$, $i = 1, \ldots, q$, $\alpha \in \mathbb{R}$, and (10.53) is minimized.

**Proposition 10.3.** Consider the information flow model (10.67)–(10.68), and assume that Assumption 10.1 holds. Then $\alpha\mathbf{e}$, $\alpha \in \mathbb{R}$, is an equilibrium state of (10.67) and (10.68).

**Proof.** The proof is similar to the proof of Proposition 2.1 and, hence, is omitted. $\qquad\square$

**Proposition 10.4.** Consider the information flow model (10.67) and (10.68), and assume that Assumption 10.1 holds. Then, for every $\alpha \in \mathbb{R}$, $\alpha\mathbf{e}$ is a semistable equilibrium state of (10.67) and (10.68). Furthermore, $x(t) \to \alpha_*\mathbf{e}$ as $t \to \infty$, where $\alpha_* = \sum_{i=1}^{q} k_i x_i(0)/(\sum_{i=1}^{q} k_i)$, and $\alpha_*\mathbf{e}$ is a semistable equilibrium state.

**Proof.** First, note that if Assumption 10.1 holds for (10.67) and (10.68), then it follows from Proposition 10.3 that $\alpha\mathbf{e}$, $\alpha \in \mathbb{R}$, is an equilibrium state of (10.67) and (10.68). To show Lyapunov stability of the equilibrium state $\alpha\mathbf{e}$, consider the Lyapunov function candidate

$$V(x) = \frac{1}{2}(x - \alpha\mathbf{e})^{\mathrm{T}} K (x - \alpha\mathbf{e}), \tag{10.70}$$

where $K \triangleq \mathrm{diag}[k_1, \ldots, k_q] \in \mathbb{R}^{q \times q}$.

Now, using similar arguments as in the proof of Theorem 2.1 and noting that $\mathcal{A}^{\mathrm{T}} = \mathcal{A}$ and $k_i \phi_{ij}(x) = -k_j \phi_{ji}(x)$, $i, j = 1, \ldots, q$, $i \neq j$, it follows that

$$\dot{V}(x) = (x - \alpha\mathbf{e})^{\mathrm{T}} K \dot{x}$$

$$= \sum_{i=1}^{q} x_i \sum_{j=1, j\neq i}^{q} \mathcal{A}_{(i,j)}(x_j - x_i)$$

$$= -\sum_{i=1}^{q} \sum_{j=i+1}^{q-1} \mathcal{A}_{(i,j)}(x_i - x_j)^2$$

$$\leq 0, \quad x \in \mathbb{R}^q, \tag{10.71}$$

which establishes Lyapunov stability of the equilibrium state $\alpha\mathbf{e}$. Next, using similar arguments as in the proof of Theorem 2.1, it can be shown that the largest invariant set $\mathcal{M}$ contained in $\dot{V}^{-1}(0)$ is given by $\mathcal{M} = \{\alpha\mathbf{e}\}$, and, hence, $\alpha\mathbf{e}$ is semistable.

Finally, note that $\mathbf{e}^{\mathrm{T}} K \dot{x}(t) = 0$ for all $t \geq 0$, and, hence,

$$\mathbf{e}^{\mathrm{T}} K x(0) = \lim_{t\to\infty} \mathbf{e}^{\mathrm{T}} K x(t) = \alpha_* \mathbf{e}^{\mathrm{T}} K \mathbf{e}, \tag{10.72}$$

which completes the proof.                                                           $\square$

Since, by Proposition 10.4, the closed-loop system given by (10.67) and (10.68) is semistable, the optimal fixed-structure control problem involves seeking $k_i > 0$, $i = 1, \ldots, q$, such that the cost functional

$$J(K) = \int_0^\infty [(x(t) - \alpha_*\mathbf{e})^{\mathrm{T}} R_1(x(t) - \alpha_*\mathbf{e}) + (u(t) - u_{\mathrm{e}})^{\mathrm{T}} R_2(u(t) - u_{\mathrm{e}})]\mathrm{d}t$$

$$\tag{10.73}$$

is minimized, where $u_e = \alpha_* K^{-1} \mathcal{A}e$, $R_1 = E_1^{\mathrm{T}} E_1 \geq 0$, $R_2 = E_2^{\mathrm{T}} E_2 > 0$, and $R_{12} = E_1^{\mathrm{T}} E_2 = 0$.

The following theorem presents a bilinear matrix inequality (BMI) solution to the fixed-structure optimal semistable control problem for network consensus. For this result, define

$$\mathcal{L} \triangleq \{L \in \mathbb{R}^{q \times q} : L = \mathrm{diag}[\ell_1, \ldots, \ell_q] \in \mathbb{R}^{q \times q}, \ell_i > 0, i = 1, \ldots, q\}.$$

**Theorem 10.6.** Consider the multiagent dynamical system (10.67)–(10.68), and assume that $(\mathcal{A}, E_1)$ is semiobservable and $(\mathcal{A}, V)$ is semicontrollable. Let $Q \in \mathbb{R}^{q \times q}$ and $L \in \mathcal{L}$ be the solution to the BMI minimization problem

$$\min_{Q \in \mathbb{R}^{q \times q}, L \in \mathcal{L}, W \in \mathbb{R}^{p \times p}} \mathrm{tr}\, W \tag{10.74}$$

subject to

$$\begin{bmatrix} Q & (E_1 Q + E_2 L \mathcal{A} Q)^{\mathrm{T}} \\ E_1 Q + E_2 L \mathcal{A} Q & W \end{bmatrix} > 0, \tag{10.75}$$

$$L \mathcal{A} Q + Q \mathcal{A}^{\mathrm{T}} L + V \leq 0. \tag{10.76}$$

Then $u = K^{-1} \mathcal{A} x$ is a semistabilizing controller for (10.67) and $x(t) \to \alpha_* \mathbf{e}$ as $t \to \infty$, where $K^{-1} = L$ and $\alpha_* = \sum_{i=1}^{q} k_i x_i(0) / (\sum_{i=1}^{q} k_i)$. Furthermore, $K$ minimizes the $\mathcal{H}_2$ performance criterion $J(K)$ given by (10.73).

**Proof.** Convergence to the consensus state $\alpha_* \mathbf{e}$ is a direct consequence of Proposition 10.4. The optimality proof is similar to the proof of Theorem 10.5 and, hence, is omitted. $\square$

Because of the diagonal structure of $K$, the optimization problem given in Theorem 10.6 is a BMI. A suboptimal solution to this problem can be obtained by using a two-stage optimization process. Specifically, by fixing $Q$, one can design the controller $K$. Then, with $K$ fixed, $Q$ can be obtained. This process continues until convergence or until an acceptable controller is found.

## Chapter Eleven

# Optimal Control for Linear and Nonlinear Semistabilization

## 11.1 Introduction

In Chapter 10, we developed an $\mathcal{H}_2$ optimal semistable control framework for linear dynamical systems. In this chapter, we address the problem of finding a state feedback nonlinear control law $u = \phi(x)$ that minimizes the performance measure

$$J(x_0, u(\cdot)) \triangleq \int_0^\infty L(x(t), u(t))\mathrm{d}t \tag{11.1}$$

and guarantees semistability of the nonlinear dynamical system

$$\dot{x}(t) = F(x(t), u(t)), \quad x(0) = x_0, \quad t \geq 0, \tag{11.2}$$
$$y(t) = H(x(t), u(t)), \tag{11.3}$$

where, for every $t \geq 0$, $x(t) \in \mathcal{D} \subseteq \mathbb{R}^n$, $\mathcal{D}$ is an open set, $u(t) \in U \subseteq \mathbb{R}^m$, $y(t) \in Y \subseteq \mathbb{R}^l$, $L : \mathcal{D} \times U \to \mathbb{R}$, $F : \mathcal{D} \times U \to \mathbb{R}^n$ is Lipschitz continuous in $x$ and $u$ on $\mathcal{D} \times U$, and $H : \mathcal{D} \times U \to Y$. Specifically, our approach focuses on the role of the Lyapunov function guaranteeing semistability of (11.2) with a feedback control law $u = \phi(x)$, and we provide sufficient conditions for optimality in a form that corresponds to a steady-state version of the Hamilton–Jacobi–Bellman equation.

In addition, we provide sufficient conditions for the existence of a feedback gain $K \in \mathbb{R}^{m \times n}$ such that the state feedback control law $u = Kx$ minimizes the quadratic performance measure

$$J(x_0, u(\cdot)) = \int_0^\infty [(x(t) - x_\mathrm{e})^\mathrm{T} C^\mathrm{T} C(x(t) - x_\mathrm{e}) + (u(t) - u_\mathrm{e})^\mathrm{T} R_2 (u(t) - u_\mathrm{e})]\mathrm{d}t \tag{11.4}$$

and guarantees semistability of the linear dynamical system

$$\dot{x}(t) = Ax(t) + Bu(t), \quad x(0) = x_0, \quad t \geq 0, \tag{11.5}$$
$$y(t) = Cx(t), \tag{11.6}$$

where $u_e \triangleq Kx_e$, $x_e \triangleq \lim_{t\to\infty} x(t)$, $R_2$ is positive definite, $A \in \mathbb{R}^{n\times n}$, $B \in \mathbb{R}^{n\times m}$, and $C \in \mathbb{R}^{l\times n}$. The proposed Riccati equation–based framework for optimal linear semistable stabilization presented in this chapter is different from the framework presented in Chapter 10 using LMIs.

The contents of the chapter are as follows. In Section 11.2, we establish some key results on semistability, semicontrollability, semiobservability, and semistabilization. In Section 11.3, we consider a nonlinear system with a performance functional evaluated over the infinite horizon. The performance functional is then evaluated in terms of a Lyapunov function that guarantees semistability. This result is then specialized to the linear-quadratic case. We then, in Section 11.4, state an optimal control problem and provide sufficient conditions for characterizing an optimal nonlinear feedback controller guaranteeing semistable stabilization.

## 11.2 Mathematical Preliminaries

Consider the nonlinear dynamical system given by

$$\dot{x}(t) = f(x(t)), \quad x(0) = x_0, \quad t \geq 0, \tag{11.7}$$

where, for every $t \geq 0$, $x(t) \in \mathcal{D} \subseteq \mathbb{R}^n$ and $f : \mathcal{D} \to \mathbb{R}^n$ is locally Lipschitz continuous on $\mathcal{D}$. The solution of (11.7) with initial condition $x(0) = x$ defined on $[0, \infty)$ is denoted by $s(\cdot, x)$. The above assumptions imply that the map $s : [0, \infty) \times \mathcal{D} \to \mathcal{D}$ is continuous [144, Thm. 2.1], satisfies the consistency property $s(0, x) = x$, and possesses the semigroup property $s(t, s(\tau, x)) = s(t + \tau, x)$ for all $t, \tau \geq 0$ and $x \in \mathcal{D}$. Given $t \geq 0$ and $x \in \mathcal{D}$, we denote the map $s(t, \cdot) : \mathcal{D} \to \mathcal{D}$ by $s_t$ and the map $s(\cdot, x) : [0, \infty) \to \mathcal{D}$ by $s^x$.

The *orbit* $\mathcal{O}_x$ of a point $x \in \mathcal{D}$ is the set $s^x([0, \infty))$. A set $\mathcal{D}_p \subseteq \mathcal{D}$ is *positively invariant* relative to (11.7) if $s_t(\mathcal{D}_p) \subseteq \mathcal{D}_p$ for all $t \geq 0$ or, equivalently, $\mathcal{D}_p$ contains the orbits of all its points. The set $\mathcal{D}_p$ is *invariant* relative to (11.7) if $s_t(\mathcal{D}_p) = \mathcal{D}_p$ for all $t \geq 0$. The *positive limit set* of $x \in \mathbb{R}^n$ is the set $\omega(x)$ of all subsequential limits of sequences of the form $\{s(t_i, x)\}_{i=0}^{\infty}$, where $\{t_i\}_{i=0}^{\infty}$ is an increasing divergent sequence in $[0, \infty)$. Recall that, for every $x \in \mathbb{R}^n$ that has bounded orbits, $\omega(x)$ is nonempty and compact, and, for every neighborhood $\mathcal{N}$ of $\omega(x)$, there exists $T > 0$ such that $s_t(x) \in \mathcal{N}$ for every $t > T$ [122, Ch. 2]. If $\mathcal{D}_p \subset \mathcal{D}$ is positively invariant and closed, then $\omega(x) \subseteq \mathcal{D}_p$ for all $x \in \mathcal{D}_p$. In addition, $\lim_{t\to\infty} s(t, x)$ exists if and only if $\omega(x)$ is a singleton. Finally, the set of equilibrium points of (11.7) is denoted by $f^{-1}(0) \triangleq \{x \in \mathcal{D} : f(x) = 0\}$.

**Lemma 11.1** (see [122, Prop. 4.7]). Consider the nonlinear dynamical system (11.7), and let $x \in \mathbb{R}^n$. If the positive limit set of (11.7) contains a Lyapunov stable equilibrium point $y$ with respect to $\mathcal{D}$, then $y = \lim_{t \to \infty} s(t, x)$; that is, $\omega(x) = \{y\}$.

Next, we recall the definitions of semicontrollability and semiobservability for linear systems given in Chapter 10, which for convenience we restate here.

**Definition 11.1.** Consider the system given by (11.5). The pair $(A, B)$ is *semicontrollable* if

$$\left[ \bigcap_{k=1}^{n} \mathcal{N}(B^{\mathrm{T}}(A^{k-1})^{\mathrm{T}}) \right]^{\perp} = [\mathcal{N}(A^{\mathrm{T}})]^{\perp}, \tag{11.8}$$

where $A^0 \triangleq I_n$.

In [169], the pair $(A, B)$ is called semicontrollable if and only if

$$\sum_{k=1}^{n} \mathcal{R}(A^{k-1}B) = \mathcal{R}(A), \tag{11.9}$$

where for given sets $\mathcal{S}_1$ and $\mathcal{S}_2$, $\mathcal{S}_1 + \mathcal{S}_2 \triangleq \{x + y : x \in \mathcal{S}_1, y \in \mathcal{S}_2\}$ denotes the Minkowski sum.

**Proposition 11.1.** Consider the dynamical system given by (11.5). Then (11.9) holds if and only if (11.8) holds. Furthermore, (11.8) is equivalent to

$$\mathrm{span}\left\{ \bigcup_{k=1}^{n} \mathcal{R}(A^{k-1}B) \right\} = \mathcal{R}(A). \tag{11.10}$$

**Proof.** First we show that

$$\sum_{i=1}^{n} \mathcal{R}\left(A^{i-1}B\right) = \mathrm{span}\left\{ \bigcup_{i=1}^{n} \mathcal{R}(A^{i-1}B) \right\}.$$

Note that, for every $i \in \{1, \ldots, n\}$, $\mathcal{R}(A^{i-1}B)$ is a subspace of $\mathbb{R}^n$, and, hence, by Fact 2.9.13 of [27, p. 121], the above equality holds. Now, it follows from (11.9) that $(A, B)$ is semicontrollable if and only if (11.10) holds.

Finally, to show that (11.9) is equivalent to (11.8), note that it follows from equation (2.4.14) of [27, p. 103] that $[\mathcal{N}(A^{\mathrm{T}})]^{\perp} = \mathcal{R}(A)$. Hence, by Fact 2.9.16 of [27, p. 121], (11.8) holds if and only if

$$\sum_{i=1}^{n} [\mathcal{N}(B^{\mathrm{T}}(A^{i-1})^{\mathrm{T}})]^{\perp} = \sum_{i=1}^{n} \mathcal{R}(A^{i-1}B) = [\mathcal{N}(A^{\mathrm{T}})]^{\perp} = \mathcal{R}(A).$$

Consequently, (11.10) is equivalent to (11.8).                           □

**Definition 11.2.** Consider the system given by (11.5) and (11.6) with $B = 0$. The pair $(A, C)$ is *semiobservable* if

$$\bigcap_{k=1}^{n} \mathcal{N}(CA^{k-1}) = \mathcal{N}(A). \tag{11.11}$$

The following result is a restatement of Theorem 10.2 and gives necessary and sufficient conditions for semistability of (11.5) and (11.6).

**Theorem 11.1.** Consider the dynamical system $\mathcal{G}$ given by (11.5) with $B = 0$ and output given by (11.6). Then $\mathcal{G}$ is semistable if and only if, for every semiobservable pair $(A, C)$, there exists an $n \times n$ matrix $P = P^{\mathrm{T}} \geq 0$ such that

$$0 = A^{\mathrm{T}} P + PA + C^{\mathrm{T}} C \tag{11.12}$$

is satisfied. Furthermore, if $(A, C)$ is semiobservable and $P$ satisfies (11.12), then

$$P = \int_0^\infty e^{A^{\mathrm{T}} t} C^{\mathrm{T}} C e^{At} \mathrm{d}t + P_0 \tag{11.13}$$

for some $P_0 = P_0^{\mathrm{T}} \in \mathbb{R}^{n \times n}$ satisfying

$$0 = A^{\mathrm{T}} P_0 + P_0 A \tag{11.14}$$

and

$$P_0 \geq -\int_0^\infty e^{A^{\mathrm{T}} t} C^{\mathrm{T}} C e^{At} \, \mathrm{d}t. \tag{11.15}$$

In addition, $\min_{P \in \mathcal{P}} \|P\|_{\mathrm{F}}$ has a unique least squares solution $P$ given by

$$P_{\mathrm{LS}} = \int_0^\infty e^{A^{\mathrm{T}} t} C^{\mathrm{T}} C e^{At} \mathrm{d}t, \tag{11.16}$$

where $\mathcal{P}$ denotes the set of all $P$ satisfying (11.12).

Next, we introduce the notions of *semistabilizability* and *semidetectability* [168] as generalizations of stabilizability and detectability.

**Definition 11.3.** Consider the dynamical system given by (11.5) and (11.6). The pair $(A, B)$ is *semistabilizable* if

$$\mathrm{rank} \begin{bmatrix} B & \jmath\omega I_n - A \end{bmatrix} = n \tag{11.17}$$

for every nonzero $\omega \in \mathbb{R}$. The pair $(A, C)$ is *semidetectable* if

$$\operatorname{rank} \begin{bmatrix} C \\ \jmath\omega I_n - A \end{bmatrix} = n \tag{11.18}$$

for every nonzero $\omega \in \mathbb{R}$.

Note that $(A, C)$ is semidetectable if and only if $(A^{\mathrm{T}}, C^{\mathrm{T}})$ is semistabilizable. Furthermore, it is important to note that semistabilizability and semidetectability are *different* notions from the standard notions of stabilizability and detectability used in linear system theory. Recall that $(A, B)$ is stabilizable if and only if

$$\operatorname{rank} \begin{bmatrix} B & \lambda I_n - A \end{bmatrix} = n$$

for every $\lambda \in \mathbb{C}$ in the closed right half-plane, and $(A, C)$ is detectable if and only if

$$\operatorname{rank} \begin{bmatrix} C \\ \lambda I_n - A \end{bmatrix} = n$$

for every $\lambda \in \mathbb{C}$ in the closed right half-plane. Hence, if $(A, C)$ is detectable, then $(A, C)$ is semidetectable; however, the converse is not true.

A similar remark holds for the notions of controllability and observability. Namely, if $(A, C)$ (respectively, $(A, B)$) is observable (respectively, controllable), then $(A, C)$ (respectively, $(A, B)$) is semidetectable (respectively, semicontrollable); however, the converse is not true. Hence, semidetectability (respectively, semistabilizability) is a weaker notion than both observability and detectability (respectively, controllability and stabilizability). Since (11.17) and (11.18) only concern stabilizability and detectability of the pairs $(A, B)$ and $(A, C)$ on the imaginary axis, we refer to these notions as semistabilizability and semidetectability.

It follows from Facts 2.11.1–2.11.3 of [27, pp. 130–131] that (11.17) and (11.18) are equivalent to

$$\dim[\mathcal{R}(\jmath\omega I_n - A) + \mathcal{R}(B)] = n \tag{11.19}$$

and

$$\mathcal{N}(\jmath\omega I_n - A) \cap \mathcal{N}(C) = \{0\}, \tag{11.20}$$

respectively, where $\dim(\cdot)$ denotes the dimension of a set.

**Example 11.1.** Consider

$$A = \begin{bmatrix} 0 & 0 \\ 0 & 0 \end{bmatrix}, \quad B = \begin{bmatrix} 0 \\ 1 \end{bmatrix}.$$

Clearly, $(A, B)$ is not stabilizable. However, it can be verified using (11.17) that $(A, B)$ is semistabilizable.                                                                         $\triangle$

As in the case of controllability and stabilizability, state feedback control does not destroy semistabilizability and semicontrollability. This is shown in the next lemma.

**Lemma 11.2.** Let $A \in \mathbb{R}^{n \times n}$, $B \in \mathbb{R}^{n \times m}$, and $K \in \mathbb{R}^{m \times n}$. If $(A, B)$ is semistabilizable (respectively, semicontrollable), then $(A+BK, B)$ is semistabilizable (respectively, semicontrollable).

**Proof.** Since $(A, B)$ is semistabilizable, it follows that

$$\text{rank} \begin{bmatrix} B & \jmath\omega I_n - A \end{bmatrix} = n$$

for all nonzero $\omega \in \mathbb{R}$. Hence, using Sylvester's inequality, it follows that

$$
\begin{aligned}
n &= n + (m + n) - (m + n) \\
&= \text{rank} \begin{bmatrix} B & \jmath\omega I_n - A \end{bmatrix} + \text{rank} \begin{bmatrix} I_m & -K \\ 0 & I_n \end{bmatrix} - (m + n) \\
&\leq \text{rank} \left( \begin{bmatrix} B & \jmath\omega I_n - A \end{bmatrix} \begin{bmatrix} I_m & -K \\ 0 & I_n \end{bmatrix} \right) \\
&\leq \text{rank} \begin{bmatrix} B & \jmath\omega I_n - A \end{bmatrix} \\
&= n
\end{aligned}
\tag{11.21}
$$

for all nonzero $\omega \in \mathbb{R}$. Now, since

$$\begin{bmatrix} B & \jmath\omega I_n - A - BK \end{bmatrix} = \begin{bmatrix} B & \jmath\omega I_n - A \end{bmatrix} \begin{bmatrix} I_m & -K \\ 0 & I_n \end{bmatrix},$$

it follows from (11.21) that

$$\text{rank} \begin{bmatrix} B & \jmath\omega I_n - A - BK \end{bmatrix} = n$$

for all nonzero $\omega \in \mathbb{R}$. Thus, $(A + BK, B)$ is semistabilizable.

The proof for semicontrollability follows similarly to the proof of Proposition 10.1.                                                                                              $\square$

Next, using the notions of semistabilizability and semidetectability, we provide a generalization of Theorem 11.1. First, however, the following lemmas are needed.

**Lemma 11.3.** Let $A \in \mathbb{R}^{n \times n}$. Then $A$ is semistable if and only if $\mathcal{N}(A) \cap \mathcal{R}(A) = \{0\}$ and $\text{spec}(A) \subseteq \{\lambda \in \mathbb{C} : \lambda + \lambda^* < 0\} \cup \{0\}$, where $\lambda^*$ denotes the complex conjugate of $\lambda$.

**Proof.** If $A$ is semistable, then it follows from Definition 11.8.1 of [27, p. 727] that $\text{spec}(A) \subseteq \{\lambda \in \mathbb{C} : \lambda + \lambda^* < 0\} \cup \{0\}$ and either $A$ is Hurwitz or there exists an invertible matrix $S \in \mathbb{R}^{n \times n}$ such that

$$A = S \begin{bmatrix} J & 0 \\ 0 & 0 \end{bmatrix} S^{-1},$$

where $J \in \mathbb{R}^{r \times r}$, $r = \text{rank}\, A$, and $J$ is Hurwitz. If $A$ is Hurwitz, then $\mathcal{N}(A) = \{0\} = \mathcal{N}(A) \cap \mathcal{R}(A)$.

Alternatively, if $A$ is not Hurwitz, then $\mathcal{N}(A) = \{S[0_{1 \times r}, y_2^{\mathrm{T}}]^{\mathrm{T}} : y_2 \in \mathbb{R}^{n-r}\}$. In this case, for every $S[0_{1 \times r}, x_2^{\mathrm{T}}]^{\mathrm{T}} \in \mathcal{N}(A) \cap \mathcal{R}(A)$, there exists $z \in \mathbb{R}^n$ such that $S[0_{1 \times r}, x_2^{\mathrm{T}}]^{\mathrm{T}} = Az$. Hence,

$$S[0_{1 \times r}, x_2^{\mathrm{T}}]^{\mathrm{T}} = S \begin{bmatrix} J & 0 \\ 0 & 0 \end{bmatrix} S^{-1} z;$$

that is,

$$\begin{bmatrix} 0 \\ x_2 \end{bmatrix} = \begin{bmatrix} J & 0 \\ 0 & 0 \end{bmatrix} S^{-1} z,$$

which implies that $x_2 = 0$. Thus, $\mathcal{N}(A) \cap \mathcal{R}(A) = \{0\}$.

Conversely, assume that $\mathcal{N}(A) \cap \mathcal{R}(A) = \{0\}$ and $\text{spec}(A) \subseteq \{\lambda \in \mathbb{C} : \lambda + \lambda^* < 0\} \cup \{0\}$. If $A$ is nonsingular, then $A$ is Hurwitz, and, hence, $A$ is semistable. Next, we consider the case where $A$ is singular. Let $x \in \mathcal{N}(A^2)$, and note that it follows from $A^2 x = AAx = 0$ that $Ax \in \mathcal{N}(A)$. Now, noting that $Ax \in \mathcal{R}(A)$, it follows from $\mathcal{N}(A) \cap \mathcal{R}(A) = \{0\}$ that $Ax = 0$, that is, $x \in \mathcal{N}(A)$. Hence, $\mathcal{N}(A^2) \subseteq \mathcal{N}(A)$. However, since $\mathcal{N}(A) \subseteq \mathcal{N}(A^2)$, it follows that $\mathcal{N}(A) = \mathcal{N}(A^2)$. Thus, by Proposition 5.5.8 of [27, p. 323], $0 \in \text{spec}(A)$ is semisimple, and, hence, by Definition 11.8.1 of [27, p. 727], $A$ is semistable. $\square$

**Lemma 11.4.** Let $A \in \mathbb{R}^{n \times n}$ and $C \in \mathbb{R}^{l \times n}$. If $A$ is semistable and $\mathcal{N}(A) \subseteq \mathcal{N}(C)$, then $CL = 0$, where $L$ is given by

$$L \triangleq I_n - AA^{\#}. \tag{11.22}$$

**Proof.** It follows from the semistability of $A$ and Proposition 11.8.1 of [27] that $L$ is well defined. Next, we show that $CLx = 0$ for every $x \in \mathbb{R}^n$. Suppose, *ad absurdum*, that there exists $x \in \mathbb{R}^n$, $x \neq 0$, such that $CLx \neq 0$. Then $Lx \notin \mathcal{N}(C)$. Since $\mathcal{N}(A) \subseteq \mathcal{N}(C)$, it follows that $Lx \notin \mathcal{N}(A)$. However, $ALx = A(I_n - AA^{\#})x = (A - AAA^{\#})x = 0$, which implies that $Lx \in \mathcal{N}(A)$, which is a contradiction. Hence, $CLx = 0$ for every $x \in \mathbb{R}^n$. $\square$

**Theorem 11.2.** Consider the dynamical system $\mathcal{G}$ given by (11.5) with $B = 0$ and output given by (11.6). Then the following statements are equivalent.

(i) $\mathcal{G}$ is semistable.

(ii) $\mathrm{rank}(\jmath\omega I_n - A) = n$ for every nonzero $\omega \in \mathbb{R}$, and there exist a positive integer $p$, a $p \times n$ matrix $E$, and an $n \times n$ matrix $P = P^{\mathrm{T}} \geq 0$ such that

$$0 = A^{\mathrm{T}}P + PA + E^{\mathrm{T}}E. \tag{11.23}$$

In this case,

$$P = \int_0^\infty e^{A^{\mathrm{T}}t}(E^{\mathrm{T}}E + L^{\mathrm{T}}E^{\mathrm{T}}EL)e^{At}\mathrm{d}t + P_0, \tag{11.24}$$

where $L = I_n - AA^{\#}$ and $P_0$ satisfies (11.14) and (11.15).

(iii) For every matrix $C \in \mathbb{R}^{l \times n}$ such that $(A, C)$ is semiobservable, there exists an $n \times n$ matrix $P = P^{\mathrm{T}} \geq 0$ such that (11.12) holds.

(iv) There exist a positive integer $p$, a $p \times n$ matrix $E$, and an $n \times n$ matrix $P = P^{\mathrm{T}} \geq 0$ such that $(A, E)$ is semiobservable and (11.23) holds.

(v) There exist a positive integer $p$, a $p \times n$ matrix $E$, and an $n \times n$ matrix $P = P^{\mathrm{T}} \geq 0$ such that $(A, E)$ is semidetectable and (11.23) holds.

**Proof.** First, note that if $A$ is semistable, then it follows from the definition of semistability that $\jmath\omega \notin \mathrm{spec}(A)$, $\omega \neq 0$. Hence, $\mathrm{rank}(A - \jmath\omega I_n) = n$ for every nonzero $\omega \in \mathbb{R}$.

To prove the existence of a nonnegative definite solution to (11.23), let $E$ be such that $\mathcal{N}(A) \subseteq \mathcal{N}(E)$. For every such pair $(A, E)$, let

$$\hat{P} = \int_0^\infty e^{A^{\mathrm{T}}t}E^{\mathrm{T}}Ee^{At}\mathrm{d}t. \tag{11.25}$$

Now, it follows from Proposition 2.2 of [132] that $\hat{P}$ is well defined. Clearly, $\hat{P} = \hat{P}^{\mathrm{T}} \geq 0$. Since $A$ is semistable, it follows from Lemma 11.3 that $\mathcal{N}(A) \cap \mathcal{R}(A) = \{0\}$, and, hence, $A$ is group invertible [24, p. 119]. Hence, it follows from (11.25) and (11.22) that

$$A^{\mathrm{T}}\hat{P} + \hat{P}A = \int_0^\infty \frac{\mathrm{d}}{\mathrm{d}t}\left(e^{A^{\mathrm{T}}t}E^{\mathrm{T}}Ee^{At}\right)\mathrm{d}t$$
$$= (I_n - AA^{\#})^{\mathrm{T}}E^{\mathrm{T}}E(I_n - AA^{\#}) - E^{\mathrm{T}}E$$
$$= L^{\mathrm{T}}E^{\mathrm{T}}EL - E^{\mathrm{T}}E. \tag{11.26}$$

Next, setting $\hat{P} = P - Z$, where $Z \in \mathbb{R}^{n \times n}$ and $Z = Z^{\mathrm{T}} \geq 0$, it follows from (11.26) that

$$A^{\mathrm{T}}P + PA + E^{\mathrm{T}}E = A^{\mathrm{T}}Z + ZA + L^{\mathrm{T}}E^{\mathrm{T}}EL. \qquad (11.27)$$

Furthermore, it follows from Lemma 2.4 in [132] that $Lx = 0$ for all $x \in \mathcal{N}(A)$, and, hence, the pair $(A, EL)$ is semiobservable since $ELx = 0$ for all $x \in \mathcal{N}(A)$. Consequently, it follows from Theorem 11.1 that

$$Z = \int_0^\infty e^{A^{\mathrm{T}}t} L^{\mathrm{T}} E^{\mathrm{T}} E L e^{At} \mathrm{d}t + P_0, \qquad (11.28)$$

which is a nonnegative definite solution of

$$0 = A^{\mathrm{T}}Z + ZA + L^{\mathrm{T}}E^{\mathrm{T}}EL. \qquad (11.29)$$

Thus, it follows from (11.27) that (11.24) satisfies (11.23), which proves that (i) implies (ii).

Let $V(x) = x^{\mathrm{T}}P_1 x$, where $P_1 \triangleq \hat{P} + L^{\mathrm{T}}L$. If $V(x) = 0$ for some $x \in \mathbb{R}^n$, then $\hat{P}x = 0$ and $Lx = 0$. It follows from (i) of Lemma 2.4 in [132] that $x \in \mathcal{N}(A)$, and $Lx = 0$ implies that $x \in \mathcal{R}(A)$. Now, it follows from (ii) of Lemma 2.4 in [132] that $x = 0$. Hence, $P_1$ is positive definite. Note that $P_1$ satisfies (11.23) since $LA = A - AA^{\#}A = 0$, and, hence,

$$\begin{aligned}
A^{\mathrm{T}}P_1 + P_1 A + E^{\mathrm{T}}E &= A^{\mathrm{T}}\hat{P} + \hat{P}A + E^{\mathrm{T}}E + A^{\mathrm{T}}L^{\mathrm{T}}L + L^{\mathrm{T}}LA \\
&= L^{\mathrm{T}}E^{\mathrm{T}}EL + (LA)^{\mathrm{T}}L + L^{\mathrm{T}}LA \\
&= 0.
\end{aligned}$$

Also note that $\dot{V}(x) = -x^{\mathrm{T}}E^{\mathrm{T}}Ex \leq 0$, $x \in \mathbb{R}^n$, which implies that $A$ is Lyapunov stable. Furthermore, it follows from $\mathrm{rank}(A - \jmath\omega I_n) = n$ for every nonzero $\omega \in \mathbb{R}$ that $\jmath\omega \in \mathrm{spec}(A)$, $\omega \neq 0$. Hence, $A$ is semistable, which proves that (ii) implies (i).

The proof of the equivalence of (i) and (iii) follows from Theorem 2.2 in [132]. Next, we show that (i) is equivalent to (iv). It follows from Theorem 11.1 that (iv) implies (i). Alternatively, if (i) holds, then choose $E$ such that $\mathcal{N}(E) = \mathcal{N}(A)$ (an obvious choice is $E = A$). Since $\mathcal{N}(EA^i) \supseteq \mathcal{N}(A)$ and $\mathcal{N}(EA^{i+1}) \supseteq \mathcal{N}(EA^i)$ for every $i \in \{0, \ldots, n-1\}$, it follows that

$$\mathcal{N}(A) \subseteq \bigcap_{i=1}^n \mathcal{N}(EA^{i-1}) \subseteq \mathcal{N}(E) = \mathcal{N}(A),$$

and, hence,

$$\bigcap_{i=1}^n \mathcal{N}(EA^{i-1}) = \mathcal{N}(A).$$

Thus, $(A, E)$ is semiobservable. Now, using similar arguments as in the proof of the equivalence of (i) and (ii), there exists $P = P^{\mathrm{T}} \geq 0$ such that (11.23) holds, which shows that (i) implies (iv).

Finally, we show the equivalence of (i) and (v). If $A$ is semistable, then $\jmath\omega \notin \mathrm{spec}(A)$, $\omega \neq 0$, and, hence, $\mathrm{rank}(\jmath\omega I_n - A) = n$ for every nonzero $\omega \in \mathbb{R}$. Thus,

$$\mathrm{rank}\begin{bmatrix} E \\ \jmath\omega I_n - A \end{bmatrix} = n$$

for every $E \in \mathbb{R}^{p \times n}$ and every positive integer $p$. The proof of the existence of a positive definite solution to (11.23) follows exactly as in the proof of (i) $\Rightarrow$ (ii). The converse follows using similar arguments as in (ii) $\Rightarrow$ (i) for $A$ Lyapunov stable.

To show that $A$ is semistable, suppose, *ad absurdum*, that $\jmath\omega \in \mathrm{spec}(A)$, where $\omega \in \mathbb{R}$ is nonzero, and let $x \in \mathbb{C}^n$, $x \neq 0$, be an associated eigenvector of $A$. Then it follows from (11.23) that

$$\begin{aligned} -x^* E^{\mathrm{T}} E x &= x^*(A^{\mathrm{T}} P + PA)x \\ &= x^*[(\jmath\omega I_n - A)^* P + P(\jmath\omega I_n - A)]x \\ &= 0. \end{aligned}$$

Hence, $Ex = 0$, and thus

$$\begin{bmatrix} E \\ \jmath\omega I_n - A \end{bmatrix} x = 0,$$

which, since

$$\mathrm{rank}\begin{bmatrix} E \\ \jmath\omega I_n - A \end{bmatrix} = n,$$

implies that $x = 0$, which is a contradiction. Consequently, $\jmath\omega \notin \mathrm{spec}(A)$ for all nonzero $\omega \in \mathbb{R}$. Hence, $\mathrm{spec}(A) \subset \{\lambda \in \mathbb{C} : \mathrm{Re}\,\lambda < 0\} \cup \{0\}$, and, if $0 \in \mathrm{spec}(A)$, then zero is semisimple. Therefore, $A$ is semistable. $\qquad\square$

**Lemma 11.5.** Let $x, y \in \mathbb{R}^n$ be such that $xy^{\mathrm{T}} = yx^{\mathrm{T}} \geq 0$. Then $y = \alpha x$, where $\alpha \geq 0$.

**Proof.** Note that for $x = 0$ or $y = 0$ the inequality is immediate. Next, if $x$ and $y$ are linearly dependent, then it follows from $xy^{\mathrm{T}} = yx^{\mathrm{T}} \geq 0$ that $y = \alpha x$, where $\alpha \geq 0$.

Alternatively, assume, *ad absurdum*, that $x$ and $y$ are linearly independent. In this case, it follows from Proposition 7.1.8 of [27, p. 441] that

$xy^T = yx^T$ if and only if $\text{vec}^{-1}(y \otimes x) = \text{vec}^{-1}(x \otimes y)$, which further implies that $y \otimes x = x \otimes y$. Let $x = [x_1, \ldots, x_n]^T$ and $y = [y_1, \ldots, y_n]^T$. Then it follows from $y \otimes x = x \otimes y$ that $y_i x = x_i y$ for every $i \in \{1, \ldots, n\}$. Since $x$ and $y$ are linearly independent, it follows that $y_i x - x_i y = 0$ for every $i \in \{1, \ldots, n\}$ if and only if $y_i = x_i = 0$ for every $i \in \{1, \ldots, n\}$. This contradicts the assumption that $x$ and $y$ are linearly independent. Now, the assertion follows directly from the first case. $\qquad\square$

**Theorem 11.3.** Consider the dynamical system $\mathcal{G}$ given by (11.5) with $B = 0$ and output given by (11.6). Assume that there exists an $n \times n$ matrix $P = P^T \geq 0$ such that (11.12) holds. Then $\mathcal{G}$ is semistable if and only if the pair $(A, C)$ is semidetectable. Furthermore, if $(A, C)$ is semidetectable and $P$ satisfies (11.12), then

$$P = \int_0^\infty e^{A^T t} C^T C e^{At} \mathrm{d}t + \alpha zz^T, \tag{11.30}$$

where $\alpha \geq 0$, $z \in \mathcal{N}(A^T)$,

$$\alpha zz^T = \int_0^\infty e^{A^T t} L^T C^T C L e^{At} \mathrm{d}t + P_0, \tag{11.31}$$

$L = I_n - AA^\#$, and $P_0$ satisfies (11.14) and (11.15).

**Proof.** The first part of the result is a direct consequence of Theorem 11.2. To prove that $P$ has the form given by (11.30), first note that it follows from (11.12) that $(A \oplus A)^T \text{vec} P = -\text{vec}(C^T C)$. Hence, $\text{vec}(C^T C) \in \mathcal{R}((A \oplus A)^T)$.

Next, it follows from Lemma 3.8 of [132] that $(A \oplus A)^T$ is semistable, and, hence, by Lemma 3.9 of [132],

$$\text{vec}^{-1}\left(((A \oplus A)^T)^\# \text{vec}(C^T C)\right) = -\int_0^\infty \text{vec}^{-1}\left(e^{(A \oplus A)^T t} \text{vec}(C^T C)\right) \mathrm{d}t$$

$$= -\int_0^\infty \text{vec}^{-1}\left(e^{A^T t} \otimes e^{A^T t}\right) \text{vec}(C^T C) \mathrm{d}t$$

$$= -\int_0^\infty e^{A^T t} C^T C e^{At} \mathrm{d}t, \tag{11.32}$$

where in (11.32) we used the facts that $e^{X \oplus Y} = e^X \otimes e^Y$ and $\text{vec}(XYZ) = (Z^T \otimes X)\text{vec} Y$ [27]. Hence, $P = \int_0^\infty e^{A^T t} C^T C e^{At} \mathrm{d}t + \text{vec}^{-1}(w)$, where $w$ satisfies $w \in \mathcal{N}((A \oplus A)^T)$ and $\text{vec}^{-1}(w) = (\text{vec}^{-1}(w))^T \geq 0$. (The nonnegative definiteness of $\text{vec}^{-1}(w)$ is guaranteed by Theorem 4.2(a) of [282].)

Since $(A \oplus A)^{\mathrm{T}}$ is semistable, it follows that a general solution to the equation $(A \oplus A)^{\mathrm{T}} w = 0$ is given by $w = z \otimes y$, where $z, y \in \mathcal{N}(A^{\mathrm{T}})$. Hence, $\mathrm{vec}^{-1}(w) = \mathrm{vec}^{-1}(z \otimes y) = y z^{\mathrm{T}}$, where we used the fact that $z y^{\mathrm{T}} = \mathrm{vec}^{-1}(y \otimes z)$. Furthermore, $z y^{\mathrm{T}} = y z^{\mathrm{T}} \geq 0$. Now, it follows from Lemma 11.5 that $y = \alpha z$, where $\alpha \geq 0$. Finally, (11.31) directly follows from Theorem 11.2 by comparing (11.24) with (11.30) for $C = E$. $\square$

Consider the dynamical system given by (11.5) and (11.6) with $B = 0$. If the pair $(A, C)$ is semiobservable, then $(A, C)$ is semidetectable and, in this case, it follows from Theorems 11.1 and 11.3 that

$$\int_0^\infty e^{A^{\mathrm{T}}t} L^{\mathrm{T}} C^{\mathrm{T}} C L e^{At} \mathrm{d}t = 0.$$

The following theorem is a direct consequence of Theorem 11.1.

**Theorem 11.4.** Consider the closed-loop system $\mathcal{G}$ given by (11.5) and (11.6) with feedback controller $u(t) = Kx(t)$, where $K \in \mathbb{R}^{m \times n}$. Then $\mathcal{G}$ is semistable if and only if for every semicontrollable pair $(A, B)$ and semiobservable pair $(A, C)$ there exists an $n \times n$ matrix $P = P^{\mathrm{T}} \geq 0$ such that

$$0 = \tilde{A}^{\mathrm{T}} P + P \tilde{A} + C^{\mathrm{T}} C + K^{\mathrm{T}} R_2 K, \tag{11.33}$$

where $\tilde{A} \triangleq A + BK$. Furthermore, the least squares solution of (11.33) is given by

$$P_{\mathrm{LS}} \triangleq \int_0^\infty e^{\tilde{A}^{\mathrm{T}}t} (C^{\mathrm{T}} C + K^{\mathrm{T}} R_2 K) e^{\tilde{A}t} \, \mathrm{d}t. \tag{11.34}$$

Finally, in this case (11.4) is given by

$$J(x_0, K) = x_0^{\mathrm{T}} P_{\mathrm{LS}} x_0. \tag{11.35}$$

Next, we give an alternative form of Theorem 11.4 using semidetectability.

**Theorem 11.5.** Consider the closed-loop system $\mathcal{G}$ given by (11.5) and (11.6) with feedback controller $u(t) = Kx(t)$, where $K \in \mathbb{R}^{m \times n}$. Assume that there exists an $n \times n$ matrix $P = P^{\mathrm{T}} \geq 0$ such that (11.33) holds. Then $\mathcal{G}$ is semistable if and only if $(A, C)$ is semidetectable. Furthermore, (11.4) is given by (11.35).

**Proof.** The first assertion is a direct consequence of Theorem 11.3. To show that (11.4) is given by (11.35), it follows from (11.33) that

$$-x^{\mathrm{T}} (\tilde{A}^{\mathrm{T}} P + P \tilde{A}) x = x (C^{\mathrm{T}} C + K^{\mathrm{T}} R_2 K) x, \quad x \in \mathbb{R}^n,$$

and, hence, $\mathcal{N}(\tilde{A}) \subseteq \mathcal{N}(C) \cap \mathcal{N}(R_2 K)$. Thus, for $x_e \in \mathcal{N}(\tilde{A})$, $C x_e = 0$ and $R_2 K x_e = 0$. Now, it follows from (11.4) that

$$
\begin{aligned}
J(x_0, K) &= \int_0^\infty x^{\mathrm{T}}(t)(C^{\mathrm{T}} C + K^{\mathrm{T}} R_2 K) x(t) \mathrm{d}t \\
&= x_0^{\mathrm{T}} \int_0^\infty e^{\tilde{A}^{\mathrm{T}} t}(C^{\mathrm{T}} C + K^{\mathrm{T}} R_2 K) e^{\tilde{A} t}\, \mathrm{d}t\, x_0 \\
&= x_0^{\mathrm{T}} P_{\mathrm{LS}} x_0,
\end{aligned}
$$

which completes the proof.                                                   $\square$

Finally, the following lemma is a restatement of Lemma 10.1.

**Lemma 11.6.** Consider the linear dynamical system $\mathcal{G}$ given by (11.5) and (11.6) with $u \equiv 0$. If $\mathcal{G}$ is semistable, then for every $x_0 \in \mathbb{R}^n$, the performance measure

$$
J(x_0) = \int_0^\infty [(x(t) - x_e)^{\mathrm{T}} C^{\mathrm{T}} C (x(t) - x_e)] \mathrm{d}t, \tag{11.36}
$$

where $x_e = (I - AA^{\#}) x_0$, is finite.

## 11.3 Semistability Analysis of Nonlinear Systems

In this section, we provide connections between Lyapunov functions and nonquadratic cost evaluation. Specifically, we consider the problem of evaluating a nonlinear-nonquadratic cost functional that depends on the solution of the nonlinear dynamical system (11.7). In particular, we show that the nonlinear-nonquadratic cost functional

$$
J(x_0) \triangleq \int_0^\infty L(x(t)) \mathrm{d}t, \tag{11.37}
$$

where $L : \mathcal{D} \to \mathbb{R}$ and $x(t)$, $t \geq 0$, satisfies (11.7), can be evaluated in a convenient form so long as (11.7) is related to an underlying Lyapunov-like function that proves semistability of (11.7).

**Theorem 11.6.** Consider the nonlinear dynamical system $\mathcal{G}$ given by (11.7) with performance functional (11.37), and let $\mathcal{Q}$ be an open neighborhood of $f^{-1}(0)$. Suppose that the solution $x(t)$, $t \geq 0$, of (11.7) is bounded for all $x \in \mathcal{Q}$, and assume that there exists a continuously differentiable function $V : \mathcal{D} \to \mathbb{R}$ such that

$$
V'(x) f(x) \leq 0, \quad x \in \mathcal{Q}, \tag{11.38}
$$

$$
L(x) + V'(x) f(x) = 0, \quad x \in \mathcal{D}. \tag{11.39}
$$

If every point in the largest invariant set $\mathcal{M}$ of $\{x \in \mathcal{Q} : V'(x)f(x) = 0\}$ is Lyapunov stable, then (11.7) is semistable and

$$J(x_0) = V(x_0) - V(x_{\mathrm{e}}), \quad x_0 \in \mathcal{Q}, \tag{11.40}$$

where $x_{\mathrm{e}} = \lim_{t \to \infty} x(t)$.

**Proof.** Let $x(t)$, $t \geq 0$, satisfy (11.7). Then

$$\dot{V}(x(t)) \triangleq \frac{\mathrm{d}}{\mathrm{d}t} V(x(t)) = V'(x(t))f(x(t)), \quad t \geq 0.$$

Hence, it follows from (11.38) that $\dot{V}(x(t)) \leq 0$, $t \geq 0$. Since every solution of (11.7) is bounded, it follows from the hypothesis on $V(\cdot)$ that, for every $x \in \mathcal{Q}$, the positive limit set $\omega(x)$ of (11.7) is nonempty and contained in the largest invariant set $\mathcal{M}$ of $\{x \in \mathcal{Q} : V'(x)f(x) = 0\}$. Since every point in $\mathcal{M}$ is a Lyapunov stable equilibrium point, it follows from Lemma 11.1 that $\omega(x)$ contains a single point for every $x \in \mathcal{Q}$, and $\lim_{t \to \infty} s(t, x)$ exists for every $x \in \mathcal{Q}$. Now, since $\lim_{t \to \infty} s(t, x) \in \mathcal{M}$ is Lyapunov stable for every $x \in \mathcal{Q}$, semistability is immediate. Consequently, $x(t) \to x_{\mathrm{e}}$ as $t \to \infty$ for all initial conditions $x_0 \in \mathcal{Q}$.

Next, since

$$0 = -\dot{V}(x(t)) + V'(x(t))f(x(t)), \quad t \geq 0, \tag{11.41}$$

it follows from (11.39) that

$$L(x(t)) = -\dot{V}(x(t)) + L(x(t)) + V'(x)f(x(t)) = -\dot{V}(x(t)). \tag{11.42}$$

Now, integrating over $[0, t]$ yields

$$\int_0^t L(x(s))\mathrm{d}s = V(x_0) - V(x(t)). \tag{11.43}$$

Letting $t \to \infty$ and noting that $V(x(t)) \to V(x_{\mathrm{e}})$ for all $x_0 \in \mathcal{Q}$ yields (11.40). $\qquad \square$

The following theorem uses Theorem 11.6 to develop an analogous result for linear dynamical systems without the *a priori* assumption of boundedness of solutions. First, however, recall that a continuous function $V : \mathcal{D} \to \mathbb{R}$ is said to be *proper relative to* $\mathcal{D}_{\mathrm{p}} \subseteq \mathcal{D}$ if $V^{-1}(\mathcal{D}_{\mathrm{c}})$ is a relatively compact subset of $\mathcal{D}_{\mathrm{p}}$ for all compact subsets $\mathcal{D}_{\mathrm{c}}$ of $\mathbb{R}$, where $V^{-1}(\cdot)$ denotes the inverse image of $\mathcal{D}_{\mathrm{c}}$.

**Theorem 11.7.** Consider the linear dynamical system $\mathcal{G}$ given by (11.5) and (11.6) with $B = 0$ and with quadratic performance measure (11.36). If

$(A, C)$ is semiobservable, then $\mathcal{G}$ is globally semistable and

$$J(x_0) = x_0^{\mathrm{T}}(AA^{\#})^{\mathrm{T}}PAA^{\#}x_0, \qquad (11.44)$$

where $P = P^{\mathrm{T}} \geq 0$ is a solution to

$$\begin{bmatrix} AA^{\#} \\ I_n \end{bmatrix}^{\mathrm{T}} \begin{bmatrix} P & 0 \\ 0 & -P + P_0 \end{bmatrix} \begin{bmatrix} AA^{\#} \\ I_n \end{bmatrix} = 0 \qquad (11.45)$$

and $P_0$ satisfies (11.14) and (11.15).

**Proof.** Let $f(x) = A(x - x_e)$, $L(x) = (x - x_e)^{\mathrm{T}}C^{\mathrm{T}}C(x - x_e)$, and $\mathcal{Q} = \mathbb{R}^n$, and note that with $V(x) = (x - x_e)^{\mathrm{T}}P(x - x_e)$, where $P = P^{\mathrm{T}} \geq 0$, (11.39) specializes to (11.12), and (11.38) is satisfied for all $x \in \mathbb{R}^n$. Furthermore, note that

$$\begin{aligned} V'(x)f(x) &= V'(x)A(x - x_e) \\ &= (x - x_e)^{\mathrm{T}}(A^{\mathrm{T}}P + PA)(x - x_e) \\ &= -(x - x_e)^{\mathrm{T}}C^{\mathrm{T}}C(x - x_e), \end{aligned}$$

and, hence, $\mathcal{N}(A) \subseteq \mathcal{N}(C)$. In addition, since $(A, C)$ is semiobservable, it follows that $\mathcal{N}(C) \subseteq \mathcal{N}(A)$, and, hence, $\mathcal{N}(C) = \mathcal{N}(A)$. Thus, $\mathcal{N}(A)$ is the largest invariant set of $\{x \in \mathcal{Q} : V'(x)A(x - x_e) = 0\}$.

Next, since $(A, C)$ is semiobservable, it follows from (ii) of Lemma 10.4 that $\mathcal{R}(A) \cap \mathcal{N}(A) = \{0\}$, which implies that $A$ is group invertible [23, p. 119]. Now, let $L = I - AA^{\#}$, and consider the Lyapunov function candidate $\hat{V}(\hat{x}) \triangleq \hat{x}^{\mathrm{T}}(P + L^{\mathrm{T}}L)\hat{x}$, where $\hat{x} \triangleq x - x_e$. If $\hat{V}(\hat{x}) = 0$ for some $\hat{x} \in \mathbb{R}^n$, then $P\hat{x} = 0$ and $L\hat{x} = 0$, and, hence, $\hat{x} \in \mathcal{N}(P)$. Thus, it follows from (11.12) and the semiobservability of $(A, C)$ that $\hat{x} \in \mathcal{N}(A)$. In addition, $\hat{V}(\hat{x}) = 0$ for some $\hat{x} \in \mathbb{R}^n$ implies that $\hat{x} \in \mathcal{N}(L)$, and, hence, $\hat{x} \in \mathcal{R}(A)$. Thus, it follows from Lemma 10.4 that $\hat{V}(\hat{x}) = 0$ only if $\hat{x} = 0$, and, hence, $\hat{V}(\cdot)$ is positive definite and proper relative to $\mathbb{R}^n$.

Next, note that the time derivative of $\hat{V}(\hat{x})$ along the trajectories of (11.5) with $B = 0$ is given by

$$\begin{aligned} \hat{V}'(\hat{x}(t))A\hat{x}(t) &= -\hat{x}^{\mathrm{T}}(t)C^{\mathrm{T}}C\hat{x}(t) + 2\hat{x}^{\mathrm{T}}(t)L^{\mathrm{T}}LA\hat{x}(t) \\ &= -\hat{x}^{\mathrm{T}}(t)C^{\mathrm{T}}C\hat{x}(t) \\ &\leq 0, \quad t \geq 0, \end{aligned}$$

and, hence, $x(t) \equiv x_e$, $t \geq 0$, is Lyapunov stable for every $x_e \in \mathcal{N}(A)$, which implies that every orbit of (11.5) with $B = 0$ is bounded. Therefore, it follows from Theorem 11.6 that $x(t)$, $t \geq 0$, is semistable and, since $V(\cdot)$ and $\hat{V}(\cdot)$ are sign definite and proper relative to $\mathbb{R}^n$, $\mathcal{G}$ is globally semistable.

Since $\mathcal{G}$ is globally semistable, it follows from Lemma 11.6 that the quadratic performance measure (11.36) is finite, and, by (11.40) of Theorem 11.6, it follows that

$$J(x_0) = (x_0 - x_e)^T P(x_0 - x_e) = x_0^T (AA^\#)^T PAA^\# x_0, \qquad (11.46)$$

which proves (11.44).

Finally, note that the performance measure (11.36) can be equivalently written as

$$J(x_0) = x_0^T \int_0^\infty e^{A^T t} C^T C e^{At} dt\, x_0, \qquad (11.47)$$

which, using Theorem 11.1, yields

$$J(x_0) = x_0^T (P - P_0) x_0. \qquad (11.48)$$

Now, (11.45) follows from (11.46) and (11.48).                                    $\square$

Note that (11.45) can be written as

$$P = (AA^\#)^T PAA^\# + P_0. \qquad (11.49)$$

Hence, since $A^\# A = AA^\#$ and $AA^\# A = A$ [27, p. 403], premultiplying and postmultiplying (11.49) by $A^T$ and $A$, respectively, it follows that $A^T P_0 A = 0$, which is implied by (11.14).

**Proposition 11.2.** Consider the linear dynamical system $\mathcal{G}$ given by (11.5) and (11.6) with $B = 0$ and with quadratic performance measure (11.36). If $(A, C)$ is semidetectable and there exists $P = P^T \geq 0$ such that (11.12) holds, then $\mathcal{G}$ is globally semistable and (11.44) holds. In addition, $P$ satisfies

$$\begin{bmatrix} AA^\# \\ I_n \end{bmatrix}^T \begin{bmatrix} P & 0 \\ 0 & -P + \alpha zz^T \end{bmatrix} \begin{bmatrix} AA^\# \\ I_n \end{bmatrix} = 0, \qquad (11.50)$$

where $\alpha \geq 0$ and $z \in \mathcal{N}(A^T)$ satisfies (11.31).

**Proof.** Global semistability of $\mathcal{G}$ is a direct consequence of Theorem 11.2. Next, let $f(x) = A(x - x_e)$, $L(x) = (x - x_e)^T C^T C(x - x_e)$, $\mathcal{Q} = \mathbb{R}^n$, and $V(x) = (x - x_e)^T P(x - x_e)$. Since $\mathcal{G}$ is globally semistable, it follows from Lemma 11.6 that the quadratic performance measure (11.36) is finite, and, by Theorem 11.6, it follows that

$$J(x_0) = (x_0 - x_e)^T P(x_0 - x_e) = x_0^T (AA^\#)^T PAA^\# x_0, \qquad (11.51)$$

which proves (11.44).

Finally, note that the performance measure (11.36) can be equivalently written as

$$J(x_0) = x_0^{\mathrm{T}} \int_0^{\infty} e^{A^{\mathrm{T}} t} C^{\mathrm{T}} C e^{At} \mathrm{d}t \, x_0, \qquad (11.52)$$

which, using Theorem 11.3, yields

$$J(x_0) = x_0^{\mathrm{T}} (P - \alpha z z^{\mathrm{T}}) x_0, \qquad (11.53)$$

where $\alpha \geq 0$ and $z \in \mathcal{N}(A^{\mathrm{T}})$ satisfies (11.31). Now, (11.50) follows from (11.51) and (11.53). □

## 11.4 Optimal Control for Semistabilization

In this section, we use the approach of Theorem 11.6 to obtain a characterization of optimal feedback controllers that guarantee closed-loop semistability. Specifically, sufficient conditions for optimality are given in a form that corresponds to a steady-state version of the Hamilton–Jacobi–Bellman equation. To address the optimal semistabilization problem, we consider the controlled nonlinear dynamical system (11.2) with $u(\cdot)$ restricted to the class of *admissible* controls consisting of measurable functions $u(\cdot)$ such that $u(t) \in U$, $t \geq 0$.

A measurable function $\phi : \mathcal{D} \to U$ satisfying $\phi(x_{\mathrm{e}}) = u_{\mathrm{e}}$, where $x_{\mathrm{e}} \in \mathcal{D}$ is an equilibrium point of (11.2) for some $u_{\mathrm{e}} \in U$, is called a *control law*. If $u(t) = \phi(x(t))$, $t \geq 0$, where $\phi(\cdot)$ is a control law and $x(t)$ satisfies (11.2), then we call $u(\cdot)$ a *feedback control law*. Note that the feedback control law is an admissible control since $\phi(\cdot)$ has values in $U$. Given a control law $\phi(\cdot)$ and a feedback control $u(t) = \phi(x(t))$, $t \geq 0$, the closed-loop system (11.2) is given by

$$\dot{x}(t) = F(x(t), \phi(x)), \quad x(0) = x_0, \quad t \geq 0. \qquad (11.54)$$

For the statement of the main theorem of this section, define the set of convergent controllers $\mathcal{S}(x_0)$ for every initial condition $x_0 \in \mathcal{D}$; that is,

$$\mathcal{S}(x_0) \triangleq \{u(\cdot) : u(\cdot) \text{ is admissible and } x(\cdot) \text{ given by (11.2) is bounded}$$
$$\text{and satisfies } x(t) \to x_{\mathrm{e}} \text{ as } t \to \infty\},$$

where $x_{\mathrm{e}} \in \mathcal{D}$ is an equilibrium point of (11.2) for some $u_{\mathrm{e}} \in U$. Note that restricting our minimization problem to $u(\cdot) \in \mathcal{S}(x_0)$, that is, control inputs corresponding to convergent solutions, can be interpreted as incorporating a semidetectability condition through the cost.

**Theorem 11.8.** Consider the controlled nonlinear dynamical system (11.2) with $u(\cdot) \in \mathcal{S}(x_0)$ and performance measure (11.1), and suppose that there exists a continuously differentiable function $V : \mathcal{D} \to \mathbb{R}$ and a control law $\phi : \mathcal{D} \to U$ such that

$$\phi(x_e) = u_e, \quad (x_e, u_e) \in \mathcal{Q} \times U, \tag{11.55}$$

$$V'(x)F(x, \phi(x)) \leq 0, \quad x \in \mathcal{Q}, \tag{11.56}$$

$$L(x, \phi(x)) + V'(x)F(x, \phi(x)) = 0, \quad x \in \mathcal{D}, \tag{11.57}$$

$$L(x, u) + V'(x)F(x, u) \geq 0, \quad (x, u) \in \mathcal{D} \times U, \tag{11.58}$$

where $\mathcal{Q}$ is an open neighborhood of $F^{-1}(0) \triangleq \{x \in \mathcal{D} : F(x, \phi(x)) = 0\}$. If every point in the largest invariant set $\mathcal{M}$ of $\{x \in \mathcal{Q} : V'(x)F(x, \phi(x)) = 0\}$ is Lyapunov stable, then, with the feedback control $u(\cdot) = \phi(x(\cdot))$, the solution $x(t) = x_e$, $t \geq 0$, of the closed-loop system (11.54) is semistable and

$$J(x_0, \phi(x(\cdot))) = V(x_0) - V(x_e). \tag{11.59}$$

Furthermore, the feedback control $u(\cdot) = \phi(x(\cdot))$ minimizes $J(x_0, u(\cdot))$ in the sense that

$$J(x_0, \phi(x(\cdot))) = \min_{u(\cdot) \in \mathcal{S}(x_0)} J(x_0, u(\cdot)). \tag{11.60}$$

**Proof.** If $u(\cdot) \in \mathcal{S}(x_0)$, then the solution $x(t)$, $t \geq 0$, of (11.2) is bounded for all initial conditions $x_0 \in \mathcal{Q}$. Thus, semistability is a direct consequence of (11.56) and (11.57) by applying Theorem 11.6 to the closed-loop system (11.54). Furthermore, using (11.57), condition (11.59) is a re-statement of (11.40).

To prove (11.60), note that

$$\dot{V}(x(t)) = V'(x(t))F(x(t), u(t)) \tag{11.61}$$

or, equivalently,

$$0 = -\dot{V}(x(t)) + V'(x(t))F(x(t), u(t)). \tag{11.62}$$

Hence,

$$L(x(t), u(t)) = -\dot{V}(x(t)) + L(x(t), u(t)) + V'(x(t))F(x(t), u(t)). \tag{11.63}$$

Now, using (11.58) and (11.59) and the fact that $u(\cdot) \in \mathcal{S}(x_0)$, it follows that

$$J(x_0, u(\cdot)) = \int_0^\infty L(x(t), u(t))\mathrm{d}t$$

$$= \int_0^\infty -\dot{V}(x(t))\,\mathrm{d}t + \int_0^\infty (L(x(t), u(t)) + V'(x)F(x(t), u(t)))\mathrm{d}t$$

$$= V(x_0) - V(x_e) + \int_0^\infty (L(x(t), u(t)) + V'(x)F(x(t), u(t)))\mathrm{d}t$$

$$\geq V(x_0) - V(x_e), \tag{11.64}$$

which yields (11.60). $\qquad\qquad\qquad\qquad\qquad\qquad\qquad\qquad\qquad\qquad$ $\square$

Theorem 11.8 requires that $u(\cdot) \in \mathcal{S}(x_0)$ or, equivalently, the solution of the closed-loop system is bounded for all $x \in \mathcal{Q}$. For asymptotic stabilization, this is automatically satisfied since we additionally require $V(0) = 0$, $V(x) > 0$, $x \in \mathcal{D} \setminus \{0\}$, and $V'(x)F(x, \phi(x)) < 0$, $x \in \mathcal{D}$, in place of (11.56) (see [25, Thm. 3.1] and [122, Thm. 8.2]). This guarantees asymptotic stability of the closed-loop system, and, hence, all closed-loop solutions are bounded.

One can replace the assumption $u(\cdot) \in \mathcal{S}(x_0)$ in Theorem 11.8 with $u(\cdot)$ being simply admissible by not requiring any assumption on the sign definiteness of $V(\cdot)$ but instead supplementing the conditions of Theorem 11.8 by assuming a nontangency condition (see Chapter 4) of the closed-loop vector field to invariant or negatively invariant subsets of the level sets of $V(\cdot)$ containing the system equilibrium. For details, see [36].

Note that Theorem 11.8 guarantees optimality with respect to the set of admissible semistabilizing controllers $\mathcal{S}(x_0)$ with the optimal control law given by the state feedback controller

$$\phi(x) = \underset{u \in \mathcal{S}(x_0)}{\arg\min} \left[ L(x, u) + V'(x)F(x, u) \right], \tag{11.65}$$

which invokes the steady-state Hamilton–Jacobi–Bellman equation and is independent of the initial condition $x_0$. It is important to note that an explicit characterization of $\mathcal{S}(x_0)$ is not required.

Next, we consider the LQR problem for semistabilization; that is, we seek controllers $u(\cdot)$ that minimize (11.4) and guarantee semistability of the linear system given by (11.5) and (11.6). The feedback gain $K$ that minimizes (11.4) and guarantees semistability of (11.5) can be characterized via a solution to an LMI, as shown in Chapter 10. The following result provides a useful alternative for finding the optimal gain $K$ via an algebraic Riccati equation.

**Theorem 11.9.** Consider the linear controlled dynamical system $\mathcal{G}$ given by (11.5) and (11.6) with quadratic performance measure (11.4), assume that the pair $(A, B)$ is semicontrollable and the pair $(A, C)$ is semiobservable, and let $P_{\mathrm{LS}} = P_{\mathrm{LS}}^{\mathrm{T}} \geq 0$ be the least squares solution to the algebraic

Riccati equation

$$0 = A^{\mathrm{T}}P + PA + C^{\mathrm{T}}C - PBR_2^{-1}B^{\mathrm{T}}P. \tag{11.66}$$

Then, with $u = Kx = -R_2^{-1}BP_{\mathrm{LS}}x$, the solution $x(t) = x_{\mathrm{e}}$, $t \geq 0$, to (11.5) is globally semistable;

$$J(x_0, K) = x_0^{\mathrm{T}}\left[\int_0^\infty (\tilde{A}\tilde{A}^\#)^{\mathrm{T}}e^{\tilde{A}^{\mathrm{T}}t}(C^{\mathrm{T}}C + K^{\mathrm{T}}R_2K)e^{\tilde{A}t}\tilde{A}\tilde{A}^\#\,\mathrm{d}t\right]x_0, \tag{11.67}$$

where $\tilde{A} = A + BK$; and

$$J(x_0, K) = \min_{u(\cdot)\in\mathcal{S}(x_0)} J(x_0, u(\cdot)). \tag{11.68}$$

**Proof.** Let

$$F(x, u) = A(x - x_{\mathrm{e}}) + B(u - u_{\mathrm{e}}),$$
$$L(x, u) = (x - x_{\mathrm{e}})^{\mathrm{T}}C^{\mathrm{T}}C(x - x_{\mathrm{e}}) + (u - u_{\mathrm{e}})^{\mathrm{T}}R_2(u - u_{\mathrm{e}}),$$
$$V(x) = (x - x_{\mathrm{e}})^{\mathrm{T}}\hat{P}(x - x_{\mathrm{e}}), \quad \hat{P} = \hat{P}^{\mathrm{T}} \geq 0,$$
$$\mathcal{Q} = \mathcal{D} = \mathbb{R}^n, \quad U = \mathbb{R}^m,$$

and note that (11.57) specializes to

$$(x - x_{\mathrm{e}})^{\mathrm{T}}C^{\mathrm{T}}C(x - x_{\mathrm{e}}) + (u - u_{\mathrm{e}})^{\mathrm{T}}R_2(u - u_{\mathrm{e}})$$
$$+ 2(x - x_{\mathrm{e}})^{\mathrm{T}}\hat{P}[A(x - x_{\mathrm{e}}) + B(u - u_{\mathrm{e}})] = 0. \tag{11.69}$$

Hence,

$$V'(x)F(x, \phi(x)) \leq 0, \quad x \in \mathbb{R}^n.$$

Now, note that

$$L(x, u) + V'(x)F(x, u)$$
$$= L(x, u) + V'(x)F(x, u) - [L(x, \phi(x)) + V'(x)F(x, \phi(x))]$$
$$= [u - \phi(x)]^{\mathrm{T}}R_2[u - \phi(x)]$$
$$\geq 0, \quad x \in \mathbb{R}^n, \tag{11.70}$$

so that conditions (11.56)–(11.58) of Theorem 11.8 are satisfied.

Next, it follows from (11.65) and (11.69) that $u = -R_2^{-1}B^{\mathrm{T}}\hat{P}x = Kx$, and, hence,

$$V'(x)F(x, \phi(x)) = 2(x - x_{\mathrm{e}})^{\mathrm{T}}\hat{P}(A + BK)(x - x_{\mathrm{e}})$$
$$= (x - x_{\mathrm{e}})^{\mathrm{T}}[(A + BK)^{\mathrm{T}}\hat{P} + \hat{P}(A + BK)](x - x_{\mathrm{e}})$$
$$= -(x - x_{\mathrm{e}})^{\mathrm{T}}(C^{\mathrm{T}}C + K^{\mathrm{T}}R_2K)(x - x_{\mathrm{e}}). \tag{11.71}$$

Now, note that (11.33) is equivalent to (11.66) with $K = -R_2^{-1}B^T P$, and, since semiobservability is preserved under full state feedback [132], it follows that if $(A, C)$ is semiobservable, then $(\tilde{A}, \tilde{R})$ is semiobservable, where $\tilde{R} \triangleq C^T C + K^T R_2 K$. Since $(\tilde{A}, \tilde{R})$ is semiobservable, it follows from (ii) of Lemma 10.4 that $\mathcal{R}(\tilde{A}) \cap \mathcal{N}(\tilde{A}) = \{0\}$, which implies that $\tilde{A}$ is group invertible [23, p. 119].

Thus, defining $L = I - \tilde{A}\tilde{A}^{\#}$ and considering the Lyapunov function candidate $\hat{V}(\hat{x}) = \hat{x}^T(\hat{P} + L^T L)\hat{x}$, where $\hat{x} \triangleq x - x_e$, global semistability follows as in the proof of Theorem 11.7. Now, it follows from Theorem 11.4 that the least squares solution $P_{LS}$ of (11.66) is given by (11.34), and, hence, taking $\hat{P} = P_{LS}$, (11.67) directly follows from (11.59). Finally, (11.68) is a restatement of (11.60).     $\square$

It is important to note that unlike Theorem 11.8, in Theorem 11.9 we do not require the assumption that $u(\cdot) \in \mathcal{S}(x_0)$. Rather, Lyapunov stability and, hence, boundedness of solutions of the closed-loop system follow from the hypothesis of the theorem.

**Proposition 11.3.** Consider the controlled linear dynamical system $\mathcal{G}$ given by (11.5) and (11.6) with quadratic performance measure (11.4), assume that the pair $(A, B)$ is semicontrollable and the pair $(A, C)$ is semiobservable, and let $P_{LS} = P_{LS}^T \geq 0$ be the least squares solution to (11.66). Then, with $u = Kx = -R_2^{-1}B^T P_{LS}x$, the equilibrium solution $x(t) \equiv x_e$ to (11.5) is globally semistable and (11.67) holds. Furthermore, (11.68) holds.

**Proof.** Since $(A, B)$ is semicontrollable and $(A, C)$ is semiobservable, the conditions of Theorem 3.7 of [191] are satisfied, and, hence, there exists an $n \times n$ matrix $P = P^T \geq 0$ such that (11.66) holds. Let $P_{LS} = \text{argmin}_{P \in \mathcal{P}} \|P\|_F$ be the least squares solution of (11.66), where $\mathcal{P}$ denotes the set of all $P$ satisfying (11.66). Now, noting that, with $K = -R_2^{-1}B^T P$, (11.33) is equivalent to (11.66), it follows from Theorems 11.4 and 11.9 that (11.5), with $u = -R_2^{-1}B^T P_{LS}x$ and $P_{LS}$ given by (11.34), is globally semistable and

$$J(x_0, K) = x_0^T P_{LS} x_0 \leq J(x_0, u(\cdot)), \qquad (11.72)$$

where $K = -R_2^{-1}B^T P_{LS}$.     $\square$

**Proposition 11.4.** Consider the controlled linear dynamical system $\mathcal{G}$ given by (11.5) and (11.6) with quadratic performance measure (11.4), assume that the pair $(A, B)$ is semistabilizable and the pair $(A, C)$ is semidetectable, and assume that there exists $P = P^T \geq 0$ such that (11.66) holds.

Then, with $u = Kx = -R_2^{-1}B^{\mathrm{T}}Px$, the equilibrium solution $x(t) \equiv x_{\mathrm{e}}$ to (11.5) is globally semistable and

$$J(x_0, K) = x_0^{\mathrm{T}}\left[\int_0^\infty e^{\tilde{A}^{\mathrm{T}}t}(C^{\mathrm{T}}C + K^{\mathrm{T}}R_2K)e^{\tilde{A}t}\,\mathrm{d}t\right]x_0, \tag{11.73}$$

where $\tilde{A} = A + BK$, and

$$\min_{u(\cdot)\in\mathcal{S}(x_0)} J(x_0, u(\cdot)) = J(x_0, K_*) \le J(x_0, K) = 2x_0^{\mathrm{T}}\tilde{A}\tilde{A}^{\#}x_0, \tag{11.74}$$

where $K_* = -R_2^{-1}B^{\mathrm{T}}P_{\mathrm{LS}}$ and $P_{\mathrm{LS}} = P_{\mathrm{LS}}^{\mathrm{T}} \ge 0$ is the least squares solution to (11.66).

**Proof.** Global semistability of (11.5), with $u = -R_2^{-1}B^{\mathrm{T}}Px$, and (11.73) follow directly from Theorem 11.5. To show (11.74), note that it follows from (11.73) and (11.66) that

$$J(x_0, K) = -x_0^{\mathrm{T}}\left[\int_0^\infty e^{\tilde{A}^{\mathrm{T}}t}(\tilde{A}^{\mathrm{T}}P + P\tilde{A})e^{\tilde{A}t}\,\mathrm{d}t\right]x_0$$

$$= -x_0^{\mathrm{T}}\left[e^{\tilde{A}^{\mathrm{T}}t}P\Big|_{t=0}^\infty + Pe^{\tilde{A}t}\Big|_{t=0}^\infty\right]x_0$$

$$= x_0^{\mathrm{T}}\left[(\tilde{A}\tilde{A}^{\#})^{\mathrm{T}}P + P\tilde{A}\tilde{A}^{\#}\right]x_0. \tag{11.75}$$

Since, by Theorem 11.3, $P = P_{\mathrm{LS}} + \alpha zz^{\mathrm{T}}$, where $\alpha \ge 0$ and $z \in \mathcal{N}(\tilde{A}^{\mathrm{T}})$ satisfies (11.31), it follows from (11.75) that

$$J(x_0, K) = x_0^{\mathrm{T}}\left[(\tilde{A}\tilde{A}^{\#})^{\mathrm{T}}P_{\mathrm{LS}} + P_{\mathrm{LS}}\tilde{A}\tilde{A}^{\#}\right]x_0 = 2x_0^{\mathrm{T}}P_{\mathrm{LS}}\tilde{A}\tilde{A}^{\#}x_0.$$

Finally, with

$$F(x, u) = A(x - x_{\mathrm{e}}) + B(u - u_{\mathrm{e}}),$$
$$L(x, u) = (x - x_{\mathrm{e}})^{\mathrm{T}}C^{\mathrm{T}}C(x - x_{\mathrm{e}}) + (u - u_{\mathrm{e}})^{\mathrm{T}}R_2(u - u_{\mathrm{e}}),$$
$$V(x) = (x - x_{\mathrm{e}})^{\mathrm{T}}P_{\mathrm{LS}}(x - x_{\mathrm{e}}),$$
$$\mathcal{Q} = \mathcal{D} = \mathbb{R}^n, \quad U = \mathbb{R}^m,$$

it follows, using similar arguments as in the proof of Theorem 11.9, that $J(x_0, K_*) = \min_{u(\cdot)\in\mathcal{S}(x_0)} J(x_0, u(\cdot))$. Hence, (11.74) holds. $\square$

**Definition 11.4.** A nonnegative definite matrix $P \in \mathbb{R}^{n\times n}$ is a *semistabilizing solution* of (11.66) if $A - BR_2^{-1}B^{\mathrm{T}}P$ is semistable. Furthermore, a semistabilizing solution $P_{\min}$ of (11.66) is the *minimally semistabilizing solution* to (11.66) if $P \ge P_{\min}$ for every semistabilizing solution $P$ to (11.66).

It follows from Definition 11.4 that the least squares solution $P_{\mathrm{LS}}$ to (11.66) is the minimally semistabilizing solution to (11.66). Given the linear dynamical system given by (11.5) and (11.6), if the pair $(A, B)$ is semicontrollable and the pair $(A, C)$ is semiobservable, then it follows from Lemma 11.2 that, for every $K \in \mathbb{R}^{m \times n}$, the pair $(A + BK, B)$ is semicontrollable, and, by Proposition 2.1 in [132], it follows that, for every $R_2 \in \mathbb{R}^{n \times n}$ such that $R_2 = R_2^{\mathrm{T}} > 0$, the pair $(A + BK, C^{\mathrm{T}}C + K^{\mathrm{T}}R_2 K)$ is semiobservable.

Furthermore, if the pair $(A, C)$ is semiobservable, then $(A, C)$ is semidetectable and it follows from Theorems 11.1 and 11.3 that every solution $P = P^{\mathrm{T}} \geq 0$ of (11.33) is given by

$$P = \int_0^\infty e^{\tilde{A}^{\mathrm{T}}t}(C^{\mathrm{T}}C + K^{\mathrm{T}}R_2 K)e^{\tilde{A}t}\mathrm{d}t + zz^{\mathrm{T}}, \qquad (11.76)$$

where $\tilde{A} = A + BK$ and $z \in \mathcal{N}(\tilde{A}^{\mathrm{T}})$. Now, if $K = -R_2^{-1}B^{\mathrm{T}}P$, then (11.33) is equivalent to (11.66), where $P$ can be computed using the Schur decomposition of the Hamiltonian matrix [27, pp. 853–859], and the least squares solution $P_{\mathrm{LS}} = P_{\mathrm{LS}}^{\mathrm{T}} \geq 0$ of (11.66) is given by $P_{\mathrm{LS}} = P - zz^{\mathrm{T}}$, where $z$ is the solution of the optimization problem

$$\min_{z \in \mathbb{R}^n} \|P - zz^{\mathrm{T}}\|_{\mathrm{F}} \qquad (11.77)$$

subject to

$$0 \leq P - zz^{\mathrm{T}}, \qquad (11.78)$$
$$0 = (A^{\mathrm{T}} - PBR_2^{-1}B^{\mathrm{T}})z. \qquad (11.79)$$

One might surmise that Theorem 11.9 and Proposition 11.3 give different values for $J(x_0, K)$. However, note that

$$J(x_0, u(\cdot))$$
$$= \int_0^\infty \left[ (x(t) - x_{\mathrm{e}})^{\mathrm{T}}C^{\mathrm{T}}C(x(t) - x_{\mathrm{e}}) + (u(t) - u_{\mathrm{e}})^{\mathrm{T}}R_2(u(t) - u_{\mathrm{e}}) \right] \mathrm{d}t$$
$$= \int_0^\infty \left[ (x_0 - x_{\mathrm{e}})^{\mathrm{T}}e^{\tilde{A}^{\mathrm{T}}t}(C^{\mathrm{T}}C + K^{\mathrm{T}}R_2 K)e^{\tilde{A}t}(x_0 - x_{\mathrm{e}}) \right] \mathrm{d}t$$
$$= x_0^{\mathrm{T}}\left[ \int_0^\infty (\tilde{A}\tilde{A}^{\#})^{\mathrm{T}}e^{\tilde{A}^{\mathrm{T}}t}(C^{\mathrm{T}}C + K^{\mathrm{T}}R_2 K)e^{\tilde{A}t}\tilde{A}\tilde{A}^{\#}\,\mathrm{d}t\right] x_0, \qquad (11.80)$$

and, since

$$J(x_0, u(\cdot))$$
$$= \int_0^\infty \left[ (x(t) - x_{\mathrm{e}})^{\mathrm{T}}C^{\mathrm{T}}C(x(t) - x_{\mathrm{e}}) + (u(t) - u_{\mathrm{e}})^{\mathrm{T}}R_2(u(t) - u_{\mathrm{e}}) \right] \mathrm{d}t$$

$$= \int_0^\infty x^{\mathrm{T}}(t)(C^{\mathrm{T}}C + K^{\mathrm{T}}R_2 K)x(t)\,\mathrm{d}t, \tag{11.81}$$

it follows that

$$
\begin{aligned}
J(x_0, u(\cdot)) \\
&= \int_0^\infty \left[ (x(t) - x_{\mathrm{e}})^{\mathrm{T}}C^{\mathrm{T}}C(x(t) - x_{\mathrm{e}}) + (u(t) - u_{\mathrm{e}})^{\mathrm{T}}R_2(u(t) - u_{\mathrm{e}}) \right]\,\mathrm{d}t \\
&= \int_0^\infty x^{\mathrm{T}}(t)(C^{\mathrm{T}}C + K^{\mathrm{T}}R_2 K)x(t)\,\mathrm{d}t \\
&= x_0^{\mathrm{T}} \int_0^\infty e^{\tilde{A}^{\mathrm{T}}t}(C^{\mathrm{T}}C + K^{\mathrm{T}}R_2 K)e^{\tilde{A}t}\,\mathrm{d}t\, x_0. \tag{11.82}
\end{aligned}
$$

Hence, (11.67) and (11.73) are equivalent.

Finally, in light of Theorem 11.9 and Lemma 4.3 of [132], the following result is immediate.

**Proposition 11.5.** Consider the linear controlled dynamical system $\mathcal{G}$ given by (11.5) and (11.6). If the pair $(A, B)$ is semicontrollable; the pair $(A, C)$ is semiobservable; and $\mathcal{G}$, with $u = Kx$, is semistable, then

$$P = \int_0^\infty (\tilde{A}\tilde{A}^{\#})^{\mathrm{T}} e^{\tilde{A}^{\mathrm{T}}t}(C^{\mathrm{T}}C + K^{\mathrm{T}}R_2 K)e^{\tilde{A}t}\tilde{A}\tilde{A}^{\#}\,\mathrm{d}t \tag{11.83}$$

satisfies

$$0 = \tilde{A}^{\mathrm{T}}(\tilde{A}^{\mathrm{T}}P + P\tilde{A} + C^{\mathrm{T}}C + K^{\mathrm{T}}R_2 K)\tilde{A} \tag{11.84}$$

or, equivalently, (11.33).

Next, we provide two numerical examples to highlight the optimal semistabilization framework developed in this chapter.

**Example 11.2.** For the first example, we use the optimal semistabilization framework to design consensus controllers for multiagent networks of single integrator systems. To address the consensus problem of $n$ agents exchanging information with collective dynamics given by (11.5) and (11.6), we set the entries $a_{ij}$, $i, j = 1, \ldots, n$, of the system matrix $A$ such that, if agent $j$ receives information from agent $i$, $i \neq j$, then $a_{ij} = 1$; otherwise, $a_{ij} = 0$ and $a_{ii} = -\sum_{j=1, j \neq i}^n a_{ij}$.

Here, we design a control law $u = Kx$ such that (11.5) with $u = Kx$ is semistable; the performance measure (11.4) is minimized in the sense of (11.60); and

$$x_{\mathrm{e}} = \lim_{t \to \infty} x(t) = \alpha \mathbf{e}, \tag{11.85}$$

where $\mathbf{e} = [1, \ldots, 1]^{\mathrm{T}}$ and $\alpha \in \mathbb{R} \backslash \{0\}$ [254]. In order to account for the constraint (11.85), we introduce a terminal steady-state constraint to the performance measure (11.4) so that

$$J(x_0, u(\cdot)) = \lim_{\tau \to \infty} \left\{ \mu^{\mathrm{T}}(x(\tau) - \alpha\mathbf{e}) + \int_0^\tau [(x(t) - x_{\mathrm{e}})^{\mathrm{T}} C^{\mathrm{T}} C(x(t) - x_{\mathrm{e}}) \right.$$
$$\left. + (u(t) - u_{\mathrm{e}})^{\mathrm{T}} R_2(u(t) - u_{\mathrm{e}})]\mathrm{d}t \right\}, \qquad (11.86)$$

where $\mu \in \mathbb{R}^n$, is minimized in the sense of (11.60). This optimization problem is in the form of a Bolza problem [54, Ch. 2], whereas the optimization problems discussed in Section 11.3 are in the form of Lagrange problems.

To account for the terminal consensus constraint, we introduce the additional scalar state $x_{n+1} : [0, \infty) \to \mathbb{R}$ and define $\hat{x} \triangleq [x^{\mathrm{T}}, x_{n+1}]^{\mathrm{T}}$ so that

$$\dot{\hat{x}}(t) = \hat{A}\hat{x}(t) + \hat{B}u(t), \quad \hat{x}(0) = \begin{bmatrix} x_0 \\ \lim_{\tau \to \infty} \dfrac{\mu^{\mathrm{T}}(x(\tau) - \alpha\mathbf{e})}{\tau} \end{bmatrix}, \quad t \geq 0, \quad (11.87)$$

$$y(t) = \hat{C}\hat{x}(t), \qquad (11.88)$$

where

$$\hat{A} \triangleq \begin{bmatrix} A & 0_n \\ 0_n^{\mathrm{T}} & 0 \end{bmatrix}, \qquad \hat{B} \triangleq \begin{bmatrix} B \\ 0_m^{\mathrm{T}} \end{bmatrix}, \qquad \hat{C} \triangleq \begin{bmatrix} C & 0_n \end{bmatrix},$$

and $0_n$ denotes the $n$-dimensional zero vector. In this case, the performance measure (11.86) can be rewritten as

$$J(x_0, u(\cdot)) = \int_0^\infty [(\hat{x}(t) - \hat{x}_{\mathrm{e}})^{\mathrm{T}} \hat{C}^{\mathrm{T}} \hat{C}(\hat{x}(t) - \hat{x}_{\mathrm{e}}) + (u(t) - u_{\mathrm{e}})^{\mathrm{T}} R_2(u(t) - u_{\mathrm{e}})]\mathrm{d}t, \qquad (11.89)$$

where $\hat{x}_{\mathrm{e}}$ is an equilibrium point of (11.87) for some $u_{\mathrm{e}} \in \mathbb{R}^m$.

Note that if the pair $(A, B)$ is semistabilizable and the pair $(A, C)$ is semidetectable, then it follows from Definitions 11.1 and 11.2 that the pair $(\hat{A}, \hat{B})$ is semistabilizable and the pair $(\hat{A}, \hat{C})$ is semidetectable. Hence, it follows from Theorem 11.9 that the solution $\hat{x}(t) = \hat{x}_{\mathrm{e}}$, $t \geq 0$, to (11.87) with $u = K\hat{x}$ and $K = -R_2^{-1}\hat{B}^{\mathrm{T}}\hat{P}_{\mathrm{LS}}$ is globally semistable, where $\hat{P}_{\mathrm{LS}}$ is the least squares solution of

$$0 = \hat{A}^{\mathrm{T}}\hat{P} + \hat{P}\hat{A} + \hat{C}^{\mathrm{T}}\hat{C} - \hat{P}\hat{B}R_2^{-1}\hat{B}^{\mathrm{T}}\hat{P} \qquad (11.90)$$

and (11.67) and (11.68) hold with $\tilde{A} = \hat{A} + \hat{B}K$.

Next, define $\hat{\mu} \triangleq [\mu^{\mathrm{T}}, 0]^{\mathrm{T}}$ and note that if $\hat{x}(t) = [x^{\mathrm{T}}(t), x_{n+1}(t)]^{\mathrm{T}}$, $t \geq 0$, is the solution of (11.87) with $u = K\hat{x}$, then

$$\lim_{\tau \to \infty} \frac{\mu^{\mathrm{T}}(x(\tau) - \alpha\mathbf{e})}{\tau} = \mu^{\mathrm{T}} \lim_{\tau \to \infty} \frac{x(\tau)}{\tau}$$

$$= \hat{\mu}^{\mathrm{T}} \lim_{\tau \to \infty} \frac{e^{\tilde{A}\tau}\hat{x}(0)}{\tau}$$

$$= \hat{\mu}^{\mathrm{T}}\tilde{A} \lim_{\tau \to \infty} e^{\tilde{A}\tau}\hat{x}(0).$$

Now, it follows from Proposition 11.8.1 of [27] that

$$\lim_{\tau \to \infty} \frac{\mu^{\mathrm{T}}(x(\tau) - \alpha\mathbf{e})}{\tau} = \hat{\mu}^{\mathrm{T}}\tilde{A} \lim_{\tau \to \infty} e^{\tilde{A}\tau}\hat{x}(0)$$

$$= \hat{\mu}^{\mathrm{T}}\tilde{A}(I_{n+1} - \tilde{A}\tilde{A}^{\#})\hat{x}(0)$$

$$= \hat{\mu}^{\mathrm{T}}(\tilde{A} - \tilde{A}\tilde{A}\tilde{A}^{\#})\hat{x}(0)$$

$$= 0, \tag{11.91}$$

and, hence, the system given by (11.87) and (11.88) with $u = K\hat{x}$ is equivalent to

$$\dot{\hat{x}}(t) = \tilde{A}\hat{x}(t), \quad \hat{x}(0) = \begin{bmatrix} x_0 \\ 0 \end{bmatrix}, \quad t \geq 0, \tag{11.92}$$

$$y(t) = \hat{C}\hat{x}(t). \tag{11.93}$$

For our simulation, we consider five agents so that

$$A = \begin{bmatrix} -2 & 1 & 1 & 0 & 0 \\ 0 & -1 & 0 & 1 & 0 \\ 1 & 1 & -4 & 1 & 1 \\ 0 & 1 & 1 & -2 & 0 \\ 1 & 1 & 0 & 0 & -2 \end{bmatrix}, \quad B = I_5, \tag{11.94}$$

and

$$C = \begin{bmatrix} 1 & 0 & 0 & 0 & -1 \\ 0 & 1 & 0 & 0 & -1 \\ 0 & 0 & 1 & 0 & -1 \\ 0 & 0 & 0 & 1 & -1 \\ 0 & 0 & 0 & 0 & 0 \end{bmatrix}, \quad R_2 = I_5. \tag{11.95}$$

Note that the pair $(A, B)$ is controllable and, hence, semistabilizable, and the pair $(A, C)$ is semidetectable but not observable. In this case, the least

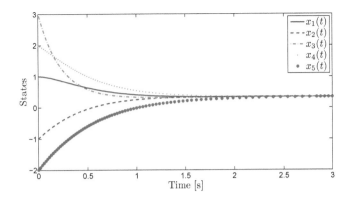

Figure 11.1 State trajectories of the closed-loop system.

squares solution to (11.90) is given by

$$
\hat{P}_{\mathrm{LS}} = \begin{bmatrix}
0.1963 & -0.0513 & 0.0115 & -0.0646 & -0.0919 & 0 \\
-0.0513 & 0.2261 & -0.0082 & 0.0360 & -0.2024 & 0 \\
0.0115 & -0.0082 & 0.1320 & 0.0417 & -0.1770 & 0 \\
-0.0646 & 0.0360 & 0.0417 & 0.2533 & -0.2663 & 0 \\
-0.0919 & -0.2024 & -0.1770 & -0.2663 & 0.7376 & 0 \\
0 & 0 & 0 & 0 & 0 & 0
\end{bmatrix}. \quad (11.96)
$$

For $x_0 = [1, -1, 3, 2, -2]^{\mathrm{T}}$, the trajectories of the closed-loop system are shown in Figure 11.1. $\triangle$

**Example 11.3.** Consider the mechanical system adopted from [122] shown in Figure 11.2 involving an eccentric rotational inertia on a translational oscillator giving rise to nonlinear coupling between the undamped oscillator and the rotational rigid body mode. The oscillator cart of mass $M$ is connected to a fixed support via a linear spring of stiffness $k$. The cart is constrained to one-dimensional motion, and the rotational proof-mass actuator consists of a mass $m$ and mass moment of inertia $I$ located at a distance $e$ from the cart's center of mass.

Letting $q$, $\dot{q}$, $\theta$, $\dot{\theta}$, $u_1$, and $u_2$ denote the translational position and velocity of the cart, the angular position and velocity of the rotational proof mass, and the force acting on the cart and the moment acting on the rotating mass, respectively, the dynamic equations of motion are given by

$$(M + m)\ddot{q}(t) + me\left[\ddot{\theta}(t) \cos \theta(t) - \dot{\theta}^2(t) \sin \theta(t)\right] + kq(t) = u_1(t), \quad (11.97)$$

$$(I + me^2)\ddot{\theta}(t) + me\ddot{q}(t) \cos \theta(t) = u_2(t), \quad (11.98)$$

where $t \geq 0$, $q(0) = q_0$, $\dot{q}(0) = \dot{q}_0$, $\theta(0) = \theta_0$, and $\dot{\theta}(0) = \dot{\theta}_0$.

For this example, we seek a state feedback controller $u = [u_1, u_2]^T = \phi(x)$, where $x = [q, \dot{q}, \dot{\theta}, \theta]^T$, such that the performance measure

$$J(x(0), u(\cdot))$$
$$= \int_0^\infty \left[(x_1(t) - x_e)^T R_1(x_1(t) - x_e) + (u(t) - u_e)^T(u(t) - u_e)\right] dt, \quad (11.99)$$

where

$$R_1 = \frac{1}{4} \begin{bmatrix} 0 & 0 & 0 & 0 \\ 0 & 1 & 0 & 0 \\ 0 & 0 & 1 & 0 \\ 0 & 0 & 0 & 0 \end{bmatrix},$$

is minimized in the sense of (11.60) and (11.97)–(11.98) is semistable.

Next, note that (11.97) and (11.98) with performance measure (11.99) can be cast in the form of (11.2) with performance measure (11.1). In this case, Theorem 11.8 can be applied with $n = 4$, $m = 2$, and

$$L(x, u) = (x - x_e)^T R_1(x - x_e) + (u - u_e)^T(u - u_e) \quad (11.100)$$

to characterize the optimal semistabilizing controllers. The explicit expression of $F(x, u)$ is omitted for brevity. In this case, (11.57) specializes to

$$0 = (x - x_e)^T R_1(x - x_e) + (\phi(x) - u_e)^T(\phi(x) - u_e) + V'(x)F(x, \phi(x)), \quad (11.101)$$

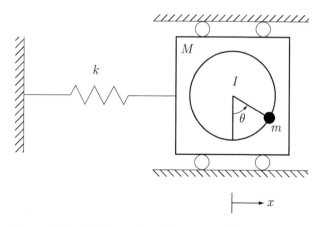

Figure 11.2 Rotational/translational proof-mass actuator.

which implies that

$$V(x) = \frac{1}{2}(x - x_e)^T P(x - x_e)(x - x_e), \quad x \in \mathbb{R}^4, \tag{11.102}$$

where

$$P(x) \triangleq \begin{bmatrix} k & 0 & 0 & 0 \\ 0 & M + m & me\cos\theta & 0 \\ 0 & me\cos\theta & I + me^2 & 0 \\ 0 & 0 & 0 & 0 \end{bmatrix} \tag{11.103}$$

and

$$\phi(x) = -\frac{1}{2}\begin{bmatrix} 0 & 1 & 0 & 0 \\ 0 & 0 & 1 & 0 \end{bmatrix}(x - x_e), \quad x \in \mathbb{R}^4. \tag{11.104}$$

Note that the state feedback control law (11.104) is equivalent to a virtual damper applied to the translational mass $M$ and the rotational mass $m$. In this case, it follows from (11.101) that (11.56) and (11.58) hold since $L(x, u) \geq 0$ and

$$\begin{aligned} L(x, u) &+ V'(x)F(x, u) \\ &= L(x, u) + V'(x)F(x, u) - [L(x, \phi(x)) + V'(x)F(x, \phi(x))] \\ &= [u - \phi(x)]^T[u - \phi(x)] \\ &\geq 0, \quad x \in \mathbb{R}^4. \end{aligned} \tag{11.105}$$

Finally, to show boundedness of solutions of the closed-loop system (11.97)–(11.98) with $u = \phi(x)$ given by (11.104), note that the largest invariant set of $\mathcal{M} = \{x \in \mathbb{R}^4 : V'(x)F(x, \phi(x)) = 0\}$ is $\mathcal{Z} \triangleq \{(0, 0, 0, \theta), \theta \in \mathbb{R}\}$. Now, Lyapunov stability of $x_e = [0, 0, 0, \theta_e]^T \in \mathcal{Z}$ for every $\theta_e \in \mathbb{R}$ follows from Theorem 2 of [175] by noting that $V(0) = 0$, $V(x) \geq 0$, $x \in \mathbb{R}^4$; $V'(x)F(x, \phi(x)) = -\dot{q}^2 - \dot{\theta}^2 \leq 0$, $x \in \mathbb{R}^4$; $[\dot{q}(t), \dot{\theta}(t)]^T = [0, 0]^T$, $t \in \mathbb{R}$, if and only if $[q(t), \theta(t)]^T = [q_e, \theta_e]^T$, $q_e \in \mathbb{R}$; and $x(t) \equiv \hat{x}_e \triangleq [q_e, 0, 0, \theta_e]^T \in \mathcal{M}$, $t < 0$, if and only if $q_e = 0$. Hence, it follows from Theorem 11.8 that the solution $x(t) \equiv x_e$, $t \geq 0$, is semistable.

Let $M = 2\,\text{kg}$, $m = 1\,\text{kg}$, $e = 0.2\,\text{m}$, $k = 10\,\text{N/m}$, $I = 4\,\text{kg}\cdot\text{m}^2$, $q_0 = 1\,\text{m}$, $\dot{q}_0 = 0\,\text{m/s}$, $\theta_0 = \pi/2$, and $\dot{\theta}_0 = 2\,\text{Hz}$. Figure 11.3 shows the state trajectories of the controlled system versus time. Figure 11.4 shows the control signal versus time. Finally,

$$J(x(0), \phi(x(\cdot))) = \frac{1}{2}(x(0) - x_e)^T P(x(0) - x_e)(x(0) - x_e) = 26.16\,\text{N}\cdot\text{m} \tag{11.106}$$

and $\theta_e = 18.5407$.                                                                                     △

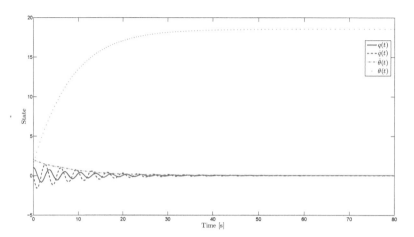

Figure 11.3 Closed-loop system trajectories versus time.

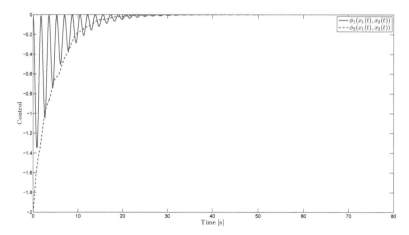

Figure 11.4 Control signal versus time.

*Chapter Twelve*

---

# Optimal Robust Consensus for Network Systems with Imperfect Information

## 12.1 Introduction

Information state equipartition protocols for network systems have been widely used in many engineering system applications, including communication protocol designs for wireless networks [47]; distributed Kalman filtering for sensor fusion [72]; continuous-time Markovian jump processes for stochastic models of chemical reactions [85]; swarm intelligent algorithm design for complex system optimization problems [328]; optimal resource allocation for network systems [171]; consensus, synchronization, and control of complex networks [236]; and system thermodynamics [131], to cite but a few examples. These protocols guarantee that the information state of each agent of a network system sharing information between its neighbors can be synchronized to the same value [167]. A challenging problem for information state equipartitioning is to design proper state equipartition protocols that achieve agreement for different scenarios.

To address this challenge, the authors in [47] developed random gossip algorithms for wireless network systems to achieve information state equipartitioning, where the communication link between every pair of agents is randomly selected. Alternatively, [60] proposes a gossip algorithm using quantized communication. Ergodic theory has emerged as a new tool for information state equipartition design and is related to the consensus problem discussed in [291, 299]. This theory allows for the development of a unified framework for designing deterministic and stochastic information state equipartition protocols and, hence, leads to general results.

Although the primary focus of the information state equipartitioning problem has been on convergence and stability of the information state, performance and optimality properties of these algorithms have also received some attention in the literature [17, 45, 47, 59, 273, 274, 315, 316]. Optimality here refers to the minimization of a given cost functional that captures agent coordination subject to convergence, stability, and network connectivity constraints. To address this problem, the authors in [315] and [47] consider an eigenvalue optimization problem for linear time-invariant and time-varying averaging algorithms, respectively. A least squares minimization problem is proposed in [316] that minimizes a cost associated with the trajectories of the linear averaging algorithm.

The authors in [17] use a static optimization method that treats the state of the neighboring agents as constant when evaluating the minimum value of each individual cost. Reference [273] uses dynamic programming to solve an optimal leader-follower problem, whereas [274] uses a decomposition method to formulate an LQR optimal information state equipartitioning problem into an LMI problem. Similar LQR-type optimization problems are also considered in [45] and [59] with a local cost functional of information interaction type.

While convergence and stability for optimal information state equipartition protocols have been addressed for network systems, robustness to exogenous system disturbances and system uncertainty has remained an open problem. In this chapter, we consider the robustness problem for information state equipartition protocols with respect to imperfect information communication between agents. Specifically, we incorporate *robustness*, *semistability*, and *optimality* for information state equipartitioning in heterogeneous network systems to address a semistable Gaussian linear-quadratic consensus problem motivated by the deterministic optimal semistable control problems addressed in Chapter 10 and in [156, 158], and the optimal distributed agreement [155] and stochastic semistable $\mathcal{H}_2$ control problems addressed in [168, 169].

The optimal semistable control problem can be viewed as an optimal regulation control problem with nondeterministic nonzero set points. In this chapter, we consider a stochastic optimal semistable control problem for addressing optimal information state equipartitioning in heterogeneous network systems under an additive Gaussian white noise disturbance and a random distribution of the system initial conditions. We show that, as in the deterministic optimal semistabilization problem considered in Chapter 10, a distinct feature of the proposed semistable Gaussian linear-quadratic consensus problem is the possibility of nonuniqueness of the solutions, and, hence, it cannot be treated using the standard methods developed for addressing classical linear-quadratic optimal control problems.

Specifically, we present necessary and sufficient conditions for semistability and optimal information state equipartitioning in heterogeneous network systems in the face of additive Gaussian white noise disturbances.

Although the nature of the disturbance or uncertainty is not necessarily known to the network system, its weight matrix information is embedded in the information state equipartition protocol design so that both optimality and robustness can be ensured.

To solve the constrained optimization semistable Gaussian linear-quadratic consensus problem, we first address the existence of optimal solutions to this optimization problem and then develop an iterative numerical algorithm that can be used for solving a general class of nonlinear, nonconvex constrained optimization problems. Finally, we provide sufficient conditions under which global convergence for the proposed algorithm is guaranteed.

## 12.2 Loss of Robustness with Imperfect Information

We start this section by highlighting a serious drawback of network systems sharing imperfect information between agents that was first pointed out in [308]. Consider a network system consisting of $n$ agents whose dynamics are given by

$$\dot{x}_i(t) = u_i(t), \quad x_i(0) = x_{i0}, \quad t \geq 0, \tag{12.1}$$

where $x_i(t) \in \mathbb{R}$ is the system state and $u_i(t) \in \mathbb{R}$ is the system input. The agents communicate through a directed graph $\mathfrak{G} = (\mathcal{V}, \mathcal{E})$ (i.e., the *information topology* of the network), where agent $j$ can send information to agent $i$ only if there exists an edge from $j$ to $i$, and where $\mathcal{V} = \{1, \ldots, n\}$ denotes the set of nodes and $\mathcal{E} = \{(i, j) : i, j \in \mathcal{V}\}$ denotes the set of edges between nodes, which consists of the directed pairs $(i, j)$ from $j$ to $i$. We denote by $\mathcal{N}_i$ the *neighboring set* of agent $i$, which consists of all other agents having edges toward agent $i$, and denote the adjacency matrix of $\mathfrak{G}$ by $\mathcal{A} = [a_{ij}] \in \mathbb{R}^{n \times n}$, whose $(i, j)$th entry is one if agent $i$ has an edge from agent $j$ and zero otherwise.

Recall that a directed graph is *strongly connected* if and only if any two distinct nodes of the directed graph can be connected via a path that follows the direction of the edges of the directed graph. It follows from Corollary 1 of [238] that if the directed graph $\mathfrak{G}$ is strongly connected, then, under the distributed control protocol

$$u_i = \sum_{j \in \mathcal{N}_i} a_{ij}(x_j - x_i), \tag{12.2}$$

all of the agents will asymptotically approach the same value; that is, $x_i(t) \to \bar{x}$ as $t \to \infty$ for some constant $\bar{x}$ which depends on the initial states of the agents. If the information equipartition state $\bar{x}$ is the average of the states, that is, $\bar{x} = \frac{1}{n} \sum_{i=1}^{n} x_i$, then the protocol is called an *average consensus protocol*.

An LQR-type optimal information state equipartition protocol involves the determination of (12.1) with a given communication topology $\mathfrak{G}$ so that (12.2) solves the minimization problem $\min_{\mathfrak{G} \in \mathbb{G}} J(\mathfrak{G})$, where $\mathbb{G}$ denotes an admissible set (normally a strongly connected directed graph) and the cost functional $J(\mathfrak{G})$ is given by (see [45, 59])

$$J(\mathfrak{G}) = \int_0^\infty \left[ \sum_{i=1}^n \sum_{j=1}^{i-1} c_{ij}(x_i(t) - x_j(t))^2 + \sum_{i=1}^n r_i u_i^2(t) \right] dt$$

or (see [132, 274])

$$J(\mathfrak{G}) = \int_0^\infty \left[ \sum_{i=1}^n s_i(x_i(t) - \bar{x})^2 + \sum_{i=1}^n r_i u_i^2(t) \right] dt,$$

where $c_{ij}, r_i, s_i > 0$ are given weights ($c_{ij}$ could be related to $a_{ij}$, as indicated in [59]). An implicit assumption in designing an optimal protocol for (12.2) is that the information that agent $i$ receives from agent $j$ is perfectly known. Here we consider the case where the information that agent $i$ receives from agent $j$ is $x_j + w_j$, where $w_j$ captures information uncertainty. Note that agent $i$ knows its own information state $x_i$ since this information is readily available to agent $i$.

The source of the information uncertainty can come from the physical communication channels between the agents (such as noise in the communication channels) as well as from a quantization error resulting from converting real numbers into finite-bit data (for storing and digital communication). For simplicity of exposition, we assume that the uncertainty $w_i$ is the same for all agent states $x_i$; however, in general, we can have different uncertainties $w_{ij}$ for different communication links.

In the presence of the uncertainty $w_i$, (12.2) becomes

$$u_i = \sum_{j \in \mathcal{N}_i} a_{ij}(x_j + w_j - x_i).$$

The collective dynamics of the network system in this case is given by

$$\dot{x}(t) = -\mathcal{L}x(t) - \mathcal{A}w(t), \quad x(0) = x_0, \quad t \geq 0, \tag{12.3}$$

where $x = [x_1, \ldots, x_n]^{\mathrm{T}}$ is the state, $w = [w_1, \ldots, w_n]^{\mathrm{T}}$ is the system uncertainty, $\mathcal{L} = \mathcal{L}(\mathcal{A})$ is the weighted system Laplacian, and $\mathcal{A}$ is the adjacency matrix of $\mathfrak{G}$. Since $\mathcal{L}$ has a zero eigenvalue, the system (12.3), with $w = 0$, is marginally stable, and, hence, in the presence of a bounded disturbance $w$, the state $x$ can become unbounded.

An alternative way to see this drifting phenomenon is discussed in [156, 157] by considering the dynamics of $\alpha = \sum_{i=1}^{n} x_i = \mathbf{e}^{\mathrm{T}} x$, where $\mathbf{e}$ is the vector whose components are all one. The dynamics for $\alpha$ are then given by

$$\dot{\alpha}(t) = \mathbf{e}^{\mathrm{T}} \left[ -\mathcal{L}x(t) - \mathcal{A}w(t) \right] = -\mathbf{e}^{\mathrm{T}} \mathcal{A}w(t), \quad \alpha(0) = \mathbf{e}^{\mathrm{T}} x(0), \quad t \geq 0,$$

since $\mathbf{e}^{\mathrm{T}} \mathcal{L} = 0$. Therefore, if $\int_0^t \mathbf{e}^{\mathrm{T}} \mathcal{A}w(s)\mathrm{d}s \to \infty$ as $t \to \infty$, then $\alpha(t) \to \infty$ as $t \to \infty$. In particular, if $\mathbf{e}^{\mathrm{T}} \mathcal{A}w$ is a nonzero constant (one can always pick such a nonzero $w$ since $\mathbf{e}^{\mathrm{T}} \mathcal{A}w$ is continuous in $w$), then, no matter how small the disturbance $w$ is, $\alpha(t)$, $t \geq 0$, can grow unbounded, and, thus, $\|x(t)\| \to \infty$ as $t \to \infty$.

The following result shows that for every linear information state equipartition protocol, there always exists an arbitrarily small bounded disturbance signal $w$ that drives $\|x(t)\|$ to infinity as $t \to \infty$. To see this, define the class $\mathcal{H}$ of information state equipartition functions such that if $h \in \mathcal{H}$, then $h(x)$ is unbounded only if $x$ is bounded; note that the average function $h(x) = \frac{1}{n} \sum_{i=1}^{n} x_i$ is one such function.

**Definition 12.1.** A control protocol is called an *information state equipartition* or *consensus protocol with objective function* $h \in \mathcal{H}$ if all the agents asymptotically approach $h(x(0))$ for all initial states $x(0)$ in the following sense:

$$\|x(t) - h(x(0))\mathbf{e}\| \leq \beta(\|x(0) - h(x(0))\mathbf{e}\|, t), \quad t \geq 0, \tag{12.4}$$

for some class $\mathcal{KL}$ function $\beta$.

**Theorem 12.1.** Consider a network system with a given fixed communication topology. Let $g$ be a linear information state equipartition protocol with a given linear objective function $h \in \mathcal{H}$ for $w = 0$. Then, for every $\varepsilon > 0$ and $x(0)$, there exists a function $w$ such that, for $\|w\| < \varepsilon$, $\|x(t)\| \to \infty$ as $t \to \infty$.

**Proof.** First we show that $h$ is time invariant. To see this, note that, for a given continuous $g$, the collective closed-loop system dynamics has the form

$$\dot{x}(t) = f(x(t), w(t)), \quad x(0) = x_0, \quad t \geq 0,$$

where $f$ is a continuous function which depends on the control protocol $g$ and the topology of the network system. For $w = 0$, the network dynamics is autonomous, and, hence, it follows from (12.4) that $x(t) \to h(x(0))$ as $t \to \infty$, which implies that $x(t + s) \to h(x(s))$ as $t \to \infty$ for every $s \geq 0$. Since $x(t + s)$ and $x(t)$ converge to the same limit, $h(x(0)) = h(x(s))$ for all

$s \geq 0$; that is, $h(x(t))$ is time invariant when $w = 0$. Thus,

$$\frac{\mathrm{d}}{\mathrm{d}t}h(x(t)) = \nabla_x h f(x, 0) = 0, \quad x \in \mathbb{R}^n, \tag{12.5}$$

where $\nabla_x$ denotes the nabla operator with respect to $x$.

Since the protocol is linear, in the presence of a disturbance $w$,

$$f(x, w) = f_1(x) + f_2(w)$$

for some continuous functions $f_1$ and $f_2$ such that $f_2(0) = 0$. Now, it follows from (12.5) that $\nabla_x h f_1(x) = 0$, and, hence,

$$\frac{\mathrm{d}}{\mathrm{d}t}h(x(t)) = \nabla_x h f_2(w), \quad x \in \mathbb{R}^n.$$

Since $h$ is linear in $x$, $\nabla_x$ does not depend on $x$, which implies that $\nabla_x h f_2(w)$ is a continuous function of $w$. Thus, for every $\varepsilon > 0$, there exists $\bar{w}$ such that $\|\bar{w}\| < \varepsilon$ and $\nabla_x h f_2(\bar{w}) = c > 0$ for some constant $c$. For the constant signal $w \equiv \bar{w}$, we have $h(t) = ct \to \infty$ as $t \to \infty$, which implies that $\|x(t)\| \to \infty$ as $t \to \infty$ since $h \in \mathcal{H}$. $\qquad \square$

A critical issue with information state equipartitioning in the presence of system uncertainty is that in order to achieve information state equipartitioning via a distributed control protocol, we need the collective objective function $h$ to be time invariant. As shown by Theorem 12.1, information uncertainty, no matter how small, can circumvent time invariance. However, Theorem 12.1 holds for deterministic continuous protocols. If we allow deterministic discontinuous protocols, or randomized protocols, or stochastic protocols driven by standard Brownian motion, then we can achieve approximate consensus [157] or almost sure consensus in the presence of bounded uncertainty.

*Imperfect information* does not necessarily affect the stability of network systems subject to *exogenous reference signals* where the objective is to attain information state equipartitioning to an exogenous signal [99, 239]. Since in this case all the agents are guided toward a common exogenous reference signal, for a linear protocol the uncertainty $w_j$ in the intercommunication among the agents plays a role similar to an additive uncertainty on the reference signal. If one can achieve *input-to-state stability* of the closed-loop system with respect to the reference signal (as in [99, 239]), then the closed-loop system is also input-to-state stable with respect to the system uncertainty. However, for autonomous networks without exogenous reference signals and with information state equipartition protocols, imperfect information turns out to be a serious hurdle in guaranteeing system boundedness.

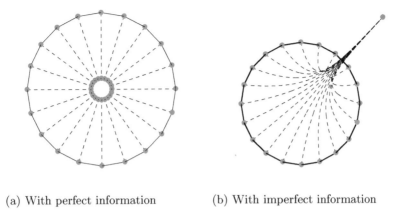

(a) With perfect information          (b) With imperfect information

Figure 12.1 Drifting phenomenon in information state equipartitioning networks; the
points on the larger circles in both the left and right figures are the initial
states, while the points on the smaller circle in the left figure or the aggregated
points in the right figure are the states after a certain time.

Since there is no common exogenous reference, the agents can drift to
infinity in the presence of information uncertainty when using the protocol
(12.2) even though their states may still approach one another. However, as
shown in Figure 12.1, the whole group can drift to infinity. If the quantity of
interest is the agent states themselves and not the values of the states on a
modulus space (e.g., when the agent states are angles evolving in the space
of the modulus $2\pi$, as opposed to length or velocity states which are not
evolving in a modulus space), then the situation of state drifting to infinity
is clearly not desirable.

The problem of average information state equipartitioning in the presence of uncertainty was first considered by [145] for the case where $w_i$ is
a Gaussian disturbance with zero mean. In this case, information state
equipartitioning can still be achieved with probability one. Although almost sure consensus is guaranteed in [145], the optimality of this information state equipartition protocol may not be well defined, as shown in Example 12.1. Alternatively, average information state equipartitioning with
bounded uncertainty was first considered by [18] with a modified information
state equipartition protocol. However, the result of [18] does not improve
on the protocol (12.2) since (12.2) achieves $\varepsilon$-consensus (or approximate
consensus) in the presence of bounded uncertainty.

Neither [145] nor [18] addresses the *drifting phenomenon* shown in
Figure 12.1 when using LQR-type optimal information state equipartition
protocols. This phenomenon is not just peculiar to the average information state equipartition protocol (12.2) but is also manifest in every linear

continuous information state equipartition protocol in the presence of uncertainty. While the information state equipartitioning drifting phenomenon is well recognized within the research community, it has been largely ignored. In this chapter, we develop a general framework that combines a system-theoretic optimal control framework with a numerical optimization problem to address this problem.

## 12.3 Semistable Gaussian Linear-Quadratic Consensus

For standard optimal output feedback, it has been shown in [9, 29] that it is possible to design static output feedback for nonzero set point regulation to guarantee asymptotic stability of the closed-loop system and optimality with respect to a quadratic cost functional. In this section, we formulate our robust and optimal information state equipartitioning problem by presenting a general framework for optimal semistabilization via static output feedback.

Consider the $N$ heterogeneous network systems with exogenous disturbances and measurement noise given by

$$\dot{x}_i(t) = A_i x_i(t) + B_i u_i(t) + D_{i1} w_i(t), \quad x_i(0) = x_{i0}, \quad t \geq 0, \qquad (12.6)$$
$$y_i(t) = x_i(t) + D_{i2} w_i(t), \quad i = 1, \ldots, N, \qquad (12.7)$$

where $x_i(t) \in \mathbb{R}^{n_i}$ is the $i$th system state vector, $u_i(t) \in \mathbb{R}^{m_i}$ is the $i$th system input, $y_i(t) \in \mathbb{R}^{n_i}$ is the $i$th system output, $w_i(t) \in \mathbb{R}^{p_i}$ is a $p_i$-dimensional standard Gaussian white noise process, $D_{i1} w_i(t)$ captures exogenous disturbances, and $D_{i2} w_i(t)$ captures measurement noise. Our aim is to design a distributed control system input $u_i$ of the form

$$u_i = \sum_{j=1}^{N} K_{(i,j)}(\mathfrak{G}) y_j, \quad i = 1, \ldots, N,$$

where $K_{(i,j)}(\mathfrak{G}) \in \mathbb{R}^{m_i \times n_i}$ is a gain associated with the connectivity of nodes $i$ and $j$ for some directed graph $\mathfrak{G}$ and $i, j = 1, \ldots, N$.

Note that we can rewrite (12.6) and (12.7) as

$$\dot{x}(t) = Ax(t) + Bu(t) + D_1 w(t), \quad x(0) = x_0, \quad t \geq 0, \qquad (12.8)$$
$$y(t) = x(t) + D_2 w(t), \qquad (12.9)$$

where $x(t) = [x_1^{\mathrm{T}}(t), \ldots, x_N^{\mathrm{T}}(t)]^{\mathrm{T}} \in \mathbb{R}^n$ is the system state vector, $u(t) = [u_1^{\mathrm{T}}(t), \ldots, u_N^{\mathrm{T}}(t)]^{\mathrm{T}} \in \mathbb{R}^m$ is the system input, $y(t) = [y_1^{\mathrm{T}}(t), \ldots, y_N^{\mathrm{T}}(t)]^{\mathrm{T}} \in \mathbb{R}^n$ is the system output, $w(t) = [w_1^{\mathrm{T}}(t), \ldots, w_N^{\mathrm{T}}(t)]^{\mathrm{T}} \in \mathbb{R}^p$ is a $p$-dimensional standard white noise process, $n = \sum_{i=1}^{N} n_i$, $m = \sum_{i=1}^{N} m_i$, and $p = \sum_{i=1}^{N} p_i$. Here, for simplicity of exposition, we assume that $D_1 D_2^{\mathrm{T}} = 0$.

The aim of the Gaussian linear-quadratic consensus problem is to design a static output feedback control law of the form

$$u(t) = K(\mathfrak{G})y(t) \tag{12.10}$$

such that the following design criteria hold.

(i) The closed-loop system (12.8)–(12.10) is semistable; that is, $\tilde{A} \triangleq A + BK(\mathfrak{G})$ is semistable.

(ii) The performance functional

$$J(K(\mathfrak{G})) = \lim_{t \to \infty} \frac{1}{t} \mathbb{E} \left\{ \int_0^t \left[ (x(s) - x_\infty)^\mathrm{T} R_1 (x(s) - x_\infty) \right. \right.$$
$$\left. \left. + (u(s) - u_\infty)^\mathrm{T} R_2 (u(s) - u_\infty) \right] ds \right\} \tag{12.11}$$

is minimized, where $R_1 \triangleq E_1^\mathrm{T} E_1$, $R_2 \triangleq E_2^\mathrm{T} E_2$, $R_{12} \triangleq E_1^\mathrm{T} E_2 = 0$, $E_1 \in \mathbb{R}^{q \times n}$, $E_2 \in \mathbb{R}^{q \times m}$, $x_\infty = \lim_{t \to \infty} \mathbb{E}[x(t)]$, and $u_\infty = K(\mathfrak{G})x_\infty$.

(iii) $x_\infty = \mathbf{e}c$, where $\mathbf{e} = [1, \dots, 1]^\mathrm{T} \in \mathbb{R}^n$ and $c \in \mathbb{R}$.

Note that the closed-loop system (12.8)–(12.10) is given by

$$\dot{x}(t) = \tilde{A}x(t) + \tilde{D}w(t), \quad x(0) = x_0, \quad t \geq 0, \tag{12.12}$$

where $\tilde{D} \triangleq D_1 + BKD_2$ and, for simplicity of exposition, we write $K$ for $K(\mathfrak{G})$.

Even though the information state equipartitioning problem can be formulated as a Gaussian linear-quadratic consensus problem, it does not guarantee that optimality is always well posed when the minimization problem of $J(K)$ is defined by $\min_K J(K)$, as the following example shows.

**Example 12.1.** Consider the two-agent system with exogenous disturbance given by

$$\dot{x}_1(t) = u_1(t) + w(t), \quad x_1(0) = x_{10}, \quad t \geq 0, \tag{12.13}$$
$$\dot{x}_2(t) = u_2(t) - w(t), \quad x_2(0) = x_{20}, \tag{12.14}$$

where $x_i(t) \in \mathbb{R}$ denotes the state of agent $i$; $u_i(t) \in \mathbb{R}$ denotes the system input of agent $i$, $i = 1, 2$; and $w(t) \in \mathbb{R}$ denotes a standard Gaussian white noise process. Assume that the feedback system inputs are given by $u = Kx$, where $u = [u_1, \ u_2]^\mathrm{T}$ and $x = [x_1, \ x_2]^\mathrm{T}$; that is,

$$u_1 = k_{11}x_1 + k_{12}x_2, \tag{12.15}$$

$$u_2 = k_{21}x_1 + k_{22}x_2, \tag{12.16}$$

where $k_{ij} \in \mathbb{R}$, $i, j = 1, 2$. The closed-loop system is then given by

$$\begin{bmatrix} \dot{x}_1 \\ \dot{x}_2 \end{bmatrix} = \begin{bmatrix} k_{11} & k_{12} \\ k_{21} & k_{22} \end{bmatrix} \begin{bmatrix} x_1 \\ x_2 \end{bmatrix} + \begin{bmatrix} 1 \\ -1 \end{bmatrix} w.$$

For average information state equipartitioning, we require that the limiting state satisfy

$$\lim_{t \to \infty} \mathbb{E}[x_1(t)] = \lim_{t \to \infty} \mathbb{E}[x_2(t)] = 0.5x_{10} + 0.5x_{20}, \tag{12.17}$$

where $\mathbb{E}$ denotes the expectation operator. Thus, we require that $k_{11} + k_{12} = 0$, $k_{21} + k_{22} = 0$, and $k_{11} + k_{22} < 0$ for average information state equipartitioning in the absence of noise. Hence, the closed-loop system is given by

$$\dot{x}(t) = \tilde{A}x(t) + \tilde{D}w(t),$$

where

$$\tilde{A} = \begin{bmatrix} -k_{12} & k_{12} \\ k_{21} & -k_{21} \end{bmatrix}, \quad \tilde{D} = \begin{bmatrix} 1 \\ -1 \end{bmatrix},$$

and $k_{12} + k_{21} > 0$. Note that

$$e^{\tilde{A}t} = \begin{bmatrix} \frac{k_{21}+k_{12}e^{-(k_{12}+k_{21})t}}{k_{12}+k_{21}}, & \frac{k_{12}-k_{12}e^{-(k_{12}+k_{21})t}}{k_{12}+k_{21}} \\ \frac{k_{21}-k_{21}e^{-(k_{12}+k_{21})t}}{k_{12}+k_{21}}, & \frac{k_{12}+k_{21}e^{-(k_{12}+k_{21})t}}{k_{12}+k_{21}} \end{bmatrix}, \tag{12.18}$$

and, hence,

$$\lim_{t \to \infty} \mathbb{E}[x(t)] = \begin{bmatrix} \frac{k_{21}}{k_{12}+k_{21}} & \frac{k_{12}}{k_{12}+k_{21}} \\ \frac{k_{21}}{k_{12}+k_{21}} & \frac{k_{12}}{k_{12}+k_{21}} \end{bmatrix} \mathbb{E}[x(0)]. \tag{12.19}$$

Next, let

$$J(K) = \lim_{t \to \infty} \mathbb{E}[(u_1(t) - u_{e1})^2 + (u_2(t) - u_{e2})^2], \tag{12.20}$$

where $u_{ei} = \lim_{t \to \infty} \mathbb{E}[u_i(t)]$, $i = 1, 2$, and let $\tilde{x}(t) = x(t) - x_\infty$, where $x_\infty = \lim_{t \to \infty} \mathbb{E}[x(t)]$. Note that

$$\dot{\tilde{x}}(t) = \tilde{A}\tilde{x}(t) + \tilde{D}w(t), \quad \tilde{x}(0) = \tilde{x}_0, \quad t \geq 0.$$

To simplify our discussion, we assume that $K$ is symmetric, that is, $k_{12} = k_{21}$. In this case, it follows that

$$J(K) = \lim_{t \to \infty} \mathbb{E}[\tilde{x}^{\mathrm{T}}(t) K^{\mathrm{T}} K \tilde{x}(t)]$$
$$= \mathrm{tr}\,(K^{\mathrm{T}} K \lim_{t \to \infty} \mathbb{E}[\tilde{x}(t)\tilde{x}^{\mathrm{T}}(t)])$$
$$= 2k_{12}.$$

Note that $k_{12} > 0$ and $J(K)$ does not have a minimum value for every $k_{12} = k_{21} \in \mathbb{R}$ satisfying $k_{12} + k_{21} > 0$. Hence, in this case, the Gaussian linear-quadratic consensus problem is not well defined. $\triangle$

If we change the cost function (12.20), then the Gaussian linear-quadratic consensus problem can result in a well-defined problem. However, unlike the standard LQR problem, stabilizability and detectability cannot be used to guarantee the existence of optimal solutions to the semistable Gaussian linear-quadratic consensus problem. Both of these points are highlighted in the following example.

**Example 12.2.** Consider the two-agent system with exogenous disturbance given by (12.13) and (12.14) with a feedback control input given by (12.15) and (12.16). For average information state equipartitioning, we require that (12.17) hold and we assume that $K$ is symmetric, that is, $k_{12} = k_{21}$. If we choose the cost function

$$J(K) = \lim_{t \to \infty} \mathbb{E}[(x_1(t) - x_{e1})^2 + (x_2(t) - x_{e2})^2 + (u_1(t) - u_{e1})^2 + (u_2(t) - u_{e2})^2],$$

where $x_{ei} = \lim_{t \to \infty} \mathbb{E}[x_i(t)]$, $i = 1, 2$, then it follows that

$$J(K) = \lim_{t \to \infty} \mathbb{E}[\tilde{x}^{\mathrm{T}}(t)(I_2 + K^{\mathrm{T}} K)\tilde{x}(t)]$$
$$= \mathrm{tr}((I_2 + K K^{\mathrm{T}}) \lim_{t \to \infty} \mathbb{E}[\tilde{x}(t)\tilde{x}^{\mathrm{T}}(t)])$$
$$= 2k_{12} + \frac{1}{2k_{12}} + \frac{1}{2}\mathrm{tr}\Big(\mathbf{e}\mathbf{e}^{\mathrm{T}} \mathbb{E}[x(0)x^{\mathrm{T}}(0)]\Big).$$

Now, since $2k_{12} + \frac{1}{2k_{12}} \geq 2$, it follows that

$$J(K) \geq 2 + \frac{1}{2}\mathrm{tr}\Big(\mathbf{e}\mathbf{e}^{\mathrm{T}} \mathbb{E}[x(0)x^{\mathrm{T}}(0)]\Big),$$

where equality holds if and only if $k_{12} = k_{21} = 1/2$. Hence, $\min_K J(K)$ exists, and the optimal gain $K_{\min}$ is unique and is given by

$$K_{\min} = \frac{1}{2}\begin{bmatrix} -1 & 1 \\ 1 & -1 \end{bmatrix}.$$

Note that in this case

$$A = 0, \quad B = I_2, \quad D_1 = \begin{bmatrix} 1 \\ -1 \end{bmatrix}, \quad D_2 = 0.$$

Clearly $(A, B)$ is not stabilizable and $(A, D_1)$ is not detectable. This shows that the classical notions of stabilizability and detectability are not sufficient to guarantee the existence of the optimal solutions to semistable Gaussian linear-quadratic consensus problems. $\triangle$

Both Examples 12.1 and 12.2 assume that the feedback signals involve perfect measurements, that is, $y = x$. However, in many network systems, the measurement can be corrupted by sensor noise. In this case, an important question is whether the resulting Gaussian linear-quadratic consensus problem is well defined.

**Example 12.3.** Consider the two-agent system with measurement noise given by

$$\dot{x}_1(t) = u_1(t), \quad x_1(0) = x_{10}, \quad t \geq 0, \tag{12.21}$$

$$\dot{x}_2(t) = u_2(t), \quad x_2(0) = x_{20}, \tag{12.22}$$

$$y_1(t) = x_1(t) + w_1(t), \tag{12.23}$$

$$y_2(t) = x_2(t) + w_2(t), \tag{12.24}$$

where $w_1(t) \in \mathbb{R}$ and $w_2(t) \in \mathbb{R}$ are standard Gaussian white noise processes. Assume that the output feedback control input is given by $u = Ky$, where $u = [u_1, \ u_2]^{\mathrm{T}}$ and $y = [y_1, \ y_2]^{\mathrm{T}}$, that is, $u_1 = k_{11}y_1 + k_{12}y_2$ and $u_2 = k_{21}y_1 + k_{22}y_2$. In this case, the closed-loop system is given by

$$\left[ \begin{array}{c} \dot{x}_1 \\ \dot{x}_2 \end{array} \right] = \left[ \begin{array}{cc} k_{11} & k_{12} \\ k_{21} & k_{22} \end{array} \right] \left[ \begin{array}{c} x_1 \\ x_2 \end{array} \right] + \left[ \begin{array}{cc} k_{11} & k_{12} \\ k_{21} & k_{22} \end{array} \right] w,$$

where $w = [w_1, w_2]^{\mathrm{T}}$.

For average information state equipartitioning, we require that (12.17) hold. Thus, we require that $k_{11} + k_{12} = 0$, $k_{21} + k_{22} = 0$, and $k_{11} + k_{22} < 0$. Hence, $k_{12} + k_{21} > 0$, and the closed-loop system is given by

$$\dot{x}(t) = \tilde{A}x(t) + \tilde{D}w(t), \quad x(0) = x_0, \quad t \geq 0,$$

where $x = [x_1, \ x_2]^{\mathrm{T}}$,

$$\tilde{A} = \left[ \begin{array}{cc} -k_{12} & k_{12} \\ k_{21} & -k_{21} \end{array} \right], \quad \tilde{D} = \left[ \begin{array}{cc} -k_{12} & k_{12} \\ k_{21} & -k_{21} \end{array} \right].$$

In this case, (12.18) and (12.19) hold.

Now, let the cost function be given by (12.20), and note that

$$J(K)$$
$$= \mathrm{tr}(KK^{\mathrm{T}} \lim_{t \to \infty} \mathbb{E}[\tilde{x}(t)\tilde{x}^{\mathrm{T}}(t)])$$

$$= \mathrm{tr} \left( K K^{\mathrm{T}} \frac{1}{k_{12} + k_{21}} \begin{bmatrix} k_{12}^2 & -k_{12}k_{21} \\ -k_{12}k_{21} & k_{21}^2 \end{bmatrix} \right.$$

$$\left. + \frac{1}{(k_{12} + k_{21})^2} \begin{bmatrix} k_{21} & k_{12} \\ k_{21} & k_{12} \end{bmatrix} \mathbb{E}[x(0)x^{\mathrm{T}}(0)] \begin{bmatrix} k_{21} & k_{21} \\ k_{12} & k_{12} \end{bmatrix} \right)$$

$$= \frac{2(k_{12}^2 + k_{21}^2)^2}{k_{12} + k_{21}} + \frac{2(k_{21} - k_{12})^2}{(k_{12} + k_{21})^2} \mathrm{tr} \left( \begin{bmatrix} k_{21}^2 & k_{12}k_{21} \\ k_{21}k_{12} & k_{12}^2 \end{bmatrix} \mathbb{E}[x(0)x^{\mathrm{T}}(0)] \right).$$

Clearly, $\frac{2(k_{12}^2 + k_{21}^2)^2}{k_{12} + k_{21}}$ has no minimum for every $k_{12}, k_{21} \in \mathbb{R}$ satisfying $k_{12} + k_{21} > 0$. Hence, even for an asymmetric $K$, there is no solution to the Gaussian linear-quadratic consensus problem. $\triangle$

Finally, we show that the semistable Gaussian linear-quadratic consensus problem can have multiple solutions.

**Example 12.4.** Consider the two-agent system with randomly distributed initial conditions given by (12.21) and (12.22). Assume that the feedback control inputs are given by $u = Kx$ or, equivalently, (12.15) and (12.16). For average information state equipartitioning, we require that (12.17) hold. In this case, $K$ should satisfy $k_{11} + k_{12} = 0$, $k_{21} + k_{22} = 0$, and $k_{11} + k_{22} < 0$. Thus, (12.18) holds.

Next, let the cost function be given by (12.20) and note that

$$J(K) = \mathrm{tr} \left( K K^{\mathrm{T}} \lim_{t \to \infty} \mathbb{E}[\tilde{x}(t)\tilde{x}^{\mathrm{T}}(t)] \right)$$

$$= \mathrm{tr} \left( K K^{\mathrm{T}} \frac{1}{(k_{12} + k_{21})^2} \begin{bmatrix} k_{21} & k_{12} \\ k_{21} & k_{12} \end{bmatrix} \mathbb{E}[x(0)x^{\mathrm{T}}(0)] \begin{bmatrix} k_{21} & k_{21} \\ k_{12} & k_{12} \end{bmatrix} \right)$$

$$= \frac{2(k_{21} - k_{12})^2}{(k_{12} + k_{21})^2} \mathrm{tr} \left( \begin{bmatrix} k_{21}^2 & k_{12}k_{21} \\ k_{21}k_{12} & k_{12}^2 \end{bmatrix} \mathbb{E}[x(0)x^{\mathrm{T}}(0)] \right)$$

$$= \frac{2(k_{21} - k_{12})^2}{(k_{12} + k_{21})^2} \mathbb{E}[(k_{21}x_1(0) + k_{12}x_2(0))^2].$$

Clearly $J(K) \geq 0$ and $J(K) = 0$ if $k_{12} = k_{21} > 0$. Hence, $\min_K J(K)$ exists and the optimal gain $K_{\min}$ is not unique. In particular, all of the optimal gains can be parameterized by

$$K_{\min} = k_{12} \begin{bmatrix} -1 & 1 \\ 1 & -1 \end{bmatrix},$$

where $k_{12} > 0$ is an arbitrary constant. $\triangle$

## 12.4 Semistabilization and Information State Equipartitioning

In this section, we develop a semistabilization framework for information state equipartitioning. First, however, the following technical lemmas are needed.

**Lemma 12.1.** Let $A \in \mathbb{R}^{n \times n}$, and let $\mathbf{u} \in \mathbb{R}^n$ be a unit vector. Then $\lim_{t \to \infty} e^{At} = \frac{1}{\mathbf{e}^{\mathrm{T}} \mathbf{u}} \mathbf{e} \mathbf{u}^{\mathrm{T}}$ if and only if $\mathbf{e}^{\mathrm{T}} \mathbf{u} \neq 0$, $A$ is semistable, $\mathcal{R}(A) \subseteq \mathcal{N}(\mathbf{e} \mathbf{u}^{\mathrm{T}})$, and $\mathcal{N}(A) \subseteq \mathcal{R}(\mathbf{e} \mathbf{u}^{\mathrm{T}})$.

**Proof.** To prove necessity, first note that by assumption, $A$ is semistable. Furthermore, note that $I_n - AA^{\#} = \frac{1}{\mathbf{e}^{\mathrm{T}} \mathbf{u}} \mathbf{e} \mathbf{u}^{\mathrm{T}}$. Now, it follows from Proposition 6.2.3 of [27, p. 403] that

$$\mathcal{R}(A) = \mathcal{N}(I_n - AA^{\#}) = \mathcal{N}\left(\frac{1}{\mathbf{e}^{\mathrm{T}} \mathbf{u}} \mathbf{e} \mathbf{u}^{\mathrm{T}}\right) = \mathcal{N}(\mathbf{e} \mathbf{u}^{\mathrm{T}})$$

and

$$\mathcal{N}(A) = \mathcal{R}(I_n - AA^{\#}) = \mathcal{R}\left(\frac{1}{\mathbf{e}^{\mathrm{T}} \mathbf{u}} \mathbf{e} \mathbf{u}^{\mathrm{T}}\right) = \mathcal{R}(\mathbf{e} \mathbf{u}^{\mathrm{T}}).$$

Next, to show sufficiency, note that it follows from the semistability of $A$ that

$$\lim_{t \to \infty} e^{At} = I_n - AA^{\#}.$$

Now, it follows from Proposition 6.2.3 of [27, p. 403] that

$$\mathcal{N}(I_n - AA^{\#}) = \mathcal{R}(A) \subseteq \mathcal{N}\left(\frac{1}{\mathbf{e}^{\mathrm{T}} \mathbf{u}} \mathbf{e} \mathbf{u}^{\mathrm{T}}\right)$$

and

$$\mathcal{R}(I_n - AA^{\#}) = \mathcal{N}(A) \subseteq \mathcal{R}\left(\frac{1}{\mathbf{e}^{\mathrm{T}} \mathbf{u}} \mathbf{e} \mathbf{u}^{\mathrm{T}}\right).$$

Note that $I_n - AA^{\#}$ and $\frac{1}{\mathbf{e}^{\mathrm{T}} \mathbf{u}} \mathbf{e} \mathbf{u}^{\mathrm{T}}$ are both idempotent matrices, and, hence, it follows from Lemma 3.4 of [160] that $I_n - AA^{\#} = \frac{1}{\mathbf{e}^{\mathrm{T}} \mathbf{u}} \mathbf{e} \mathbf{u}^{\mathrm{T}}$.  $\square$

Next, we present an eigenvalue-eigenvector structure that provides necessary and sufficient conditions for guaranteeing information state equipartitioning and semistability for linear systems.

**Lemma 12.2.** Let $A \in \mathbb{R}^{n \times n}$, and let $\mathbf{u} \in \mathbb{R}^n$ be a unit vector. Then $\lim_{t \to \infty} e^{At} = \frac{1}{\mathbf{e}^{\mathrm{T}} \mathbf{u}} \mathbf{e} \mathbf{u}^{\mathrm{T}}$ and $\mathbf{e}^{\mathrm{T}} \mathbf{u} \neq 0$ if and only if rank $A = n - 1$, $A\mathbf{e} = 0$, $\mathbf{u}^{\mathrm{T}} A = 0$, and $A$ is semistable.

**Proof.** To show sufficiency, note that, since $A$ is semistable, it follows that

$$\lim_{t\to\infty} e^{At}x = (I_n - AA^{\#})x, \quad x \in \mathbb{R}^n.$$

Next, we show that $\mathcal{R}(A) \subseteq \mathcal{N}(\mathbf{eu}^{\mathrm{T}})$ and $\mathcal{N}(A) \subseteq \mathcal{R}(\mathbf{eu}^{\mathrm{T}})$. Let $v \in \mathcal{R}(A)$. Then $v = Az$ for some $z \in \mathbb{R}^n$. Since $\mathbf{u}^{\mathrm{T}}A = 0$, it follows that $\mathbf{eu}^{\mathrm{T}}v = \mathbf{eu}^{\mathrm{T}}Az = 0$, which implies that $v \in \mathcal{N}(\mathbf{eu}^{\mathrm{T}})$. Hence, $\mathcal{R}(A) \subseteq \mathcal{N}(\mathbf{eu}^{\mathrm{T}})$.

Since $A$ is group invertible, it follows from Corollary 2.3.2 of [27, p. 99] and Fact 3.6.1 of [27, p. 191] that rank $A = n - 1$ implies that

$$\dim \mathcal{N}(A) = n - \dim \mathcal{R}(A) = n - \operatorname{rank} A = 1,$$

where dim $\mathcal{X}$ denotes the dimension of the set $\mathcal{X}$. Hence, if $v \in \mathcal{N}(A)$, then it follows from $\mathbf{u} \in \mathcal{N}(A)$ that $v = c\mathbf{e}$ for some $c \in \mathbb{R}$. Now, for every $y \in \mathbb{R}^n$ such that $\mathbf{u}^{\mathrm{T}}y \neq 0$,

$$v = c\mathbf{e} = \frac{c}{\mathbf{u}^{\mathrm{T}}y}(\mathbf{u}^{\mathrm{T}}y)\mathbf{e} = \frac{c}{\mathbf{u}^{\mathrm{T}}y}\mathbf{eu}^{\mathrm{T}}y.$$

This implies that $v = \mathbf{eu}^{\mathrm{T}}z$ for $z = \frac{c}{\mathbf{u}^{\mathrm{T}}y}y \in \mathbb{R}^n$. Hence, $v \in \mathcal{R}(\mathbf{eu}^{\mathrm{T}})$, and, consequently, $\mathcal{N}(A) \subseteq \mathcal{R}(\mathbf{uv}^{\mathrm{T}})$.

Next, note that $\mathcal{N}(\tilde{A}) = \operatorname{span}\{\mathbf{e}\}$ and $\mathcal{N}(\tilde{A}^{\mathrm{T}}) = \operatorname{span}\{\mathbf{u}\}$. Furthermore, it follows from equation (2.4.14) of [27, p. 103] that $\mathcal{N}(\tilde{A}^{\mathrm{T}})^{\perp} = \mathcal{R}(\tilde{A})$, where $\mathcal{S}^{\perp}$ denotes the orthogonal complement of the set $\mathcal{S}$. If $\mathbf{e}^{\mathrm{T}}\mathbf{u} = 0$, then

$$\mathcal{N}(\tilde{A}) \subseteq \mathcal{N}(\tilde{A}^{\mathrm{T}})^{\perp} = \mathcal{R}(\tilde{A}),$$

and, hence,

$$\mathcal{N}(\tilde{A}) \cap \mathcal{R}(\tilde{A}) = \mathcal{N}(\tilde{A}) = \operatorname{span}\{\mathbf{e}\}.$$

This contradicts the fact that

$$\mathcal{N}(\tilde{A}) \cap \mathcal{R}(\tilde{A}) = \{0\}$$

by (vii) of Fact 3.6.1 of [27, p. 191]. Hence, $\mathbf{e}^{\mathrm{T}}\mathbf{u} \neq 0$. Now, it follows from Lemma 12.1 that

$$\lim_{t\to\infty} e^{At} = \frac{1}{\mathbf{e}^{\mathrm{T}}\mathbf{u}}\mathbf{eu}^{\mathrm{T}}.$$

To show necessity, first note that $\lim_{t\to\infty} e^{At}x$ exists for every $x \in \mathbb{R}^n$ if and only if $\lim_{t\to\infty} e^{At}$ exists. By definition, $A$ is semistable. Since $A$ is semistable, it follows that either $A$ is Hurwitz or there exists an invertible matrix $S \in \mathbb{C}^{n \times n}$ such that

$$A = S \begin{bmatrix} J & 0 \\ 0 & 0 \end{bmatrix} S^{-1},$$

where $J \in \mathbb{C}^{r \times r}$, $r = \operatorname{rank} A$, and $J$ is Hurwitz. Now, if $A$ is Hurwitz, then $\lim_{t \to \infty} e^{At} = 0$, which implies that $c = 0$. This contradicts the assumption that $c \neq 0$. Hence, $A$ is not Hurwitz. If $A = 0$, then $\lim_{t \to \infty} e^{At} = I_n$, which contradicts the assumption of the lemma. Thus, $A \neq 0$.

Now, assume that $A$ is neither Hurwitz nor zero. Let $u_i$, $i = 1, \ldots, n$, be the column vectors of $S$, and let $v_i^{\mathrm{T}}$, $i = 1, \ldots, n$, be the row vectors of $S^{-1}$. Then it follows that

$$\lim_{t \to \infty} e^{At} = \lim_{t \to \infty} S \begin{bmatrix} e^{Jt} & 0 \\ 0 & I_{n-r} \end{bmatrix} S^{-1} = S \begin{bmatrix} 0 & 0 \\ 0 & I_{n-r} \end{bmatrix} S^{-1} = \sum_{i=r}^{n} u_i v_i^{\mathrm{T}}.$$

Clearly, $S$ is a full-rank matrix and

$$\operatorname{rank} \left( \sum_{i=r+1}^{n} u_i v_i^{\mathrm{T}} \right) = \operatorname{rank} \begin{bmatrix} 0 & 0 \\ 0 & I_{n-r} \end{bmatrix} = n - r.$$

However, by assumption, $\mathcal{R}(\sum_{i=r+1}^{n} u_i v_i^{\mathrm{T}}) \subseteq \operatorname{span}\{e\}$. Thus,

$$1 \leq \dim \mathcal{R} \left( \sum_{i=r+1}^{n} u_i v_i^{\mathrm{T}} \right) \leq \dim \operatorname{span}\{e\} = 1;$$

that is,

$$\operatorname{rank} \left( \sum_{i=r+1}^{n} u_i v_i^{\mathrm{T}} \right) = n - r = 1.$$

Hence, $\operatorname{rank} A = r = n - 1$.

Note that each matrix $u_i v_i^{\mathrm{T}}$ is a rank-one matrix. In this case,

$$\sum_{i=r+1}^{n} u_i v_i^{\mathrm{T}} = u_{r+1} v_{r+1}^{\mathrm{T}} = \frac{1}{\sqrt{n}} e u^{\mathrm{T}}.$$

Thus,

$$u_{r+1} = \left( \frac{\mathbf{u}^{\mathrm{T}} v_{r+1}}{\sqrt{n}} \right) \mathbf{e}, \quad v_{r+1} = \left( \frac{\mathbf{e}^{\mathrm{T}} u_{r+1}}{\sqrt{n}} \right) \mathbf{u},$$

which implies that $u_{r+1}$ is a multiple of $\mathbf{e}$ and $v_{r+1}$ is a multiple of $\mathbf{u}$. Consequently, zero is a simple eigenvalue of $A$, and $\mathbf{u}$ and $\mathbf{e}$ are its associated left and right eigenvectors, respectively. Thus, $A\mathbf{e} = 0$ and $\mathbf{u}^{\mathrm{T}}A = 0$, which completes the proof.                                                                                            $\square$

For the Gaussian linear-quadratic consensus problem, it follows from Lemma 12.2 that if rank $A = n - 1$, $A\mathbf{e} = 0$, $\mathbf{u}^{\mathrm{T}}A = 0$, and $A$ is semistable, then $x_\infty = \mathbf{e}c$. Hence, Lemma 12.2 gives necessary and sufficient conditions that characterize criterion (iii) in the Gaussian linear-quadratic consensus design problem. Note that if $\mathbf{u} = \frac{1}{\sqrt{n}}\mathbf{e}$, then the closed-loop system achieves *average information state equipartitioning*. However, in general, $\mathbf{u}$ is not necessarily $\frac{1}{\sqrt{n}}\mathbf{e}$ for our problem. Thus, we consider the general problem of information state equipartitioning instead of the average information state equipartitioning problem.

Next, we connect semistability and optimality to develop optimal consensus protocols. In the remainder of the chapter, we assume that the following assumption holds.

**Assumption 12.1.** The system initial condition $x(0)$ and the system disturbance $w(t)$ are independent for all $t \geq 0$.

As the examples in the previous section show, the standard notions of stabilizability and detectability cannot be used to guarantee optimal semistabilization. Next, we develop conditions that guarantee semistability. First, the following key lemma is needed.

**Lemma 12.3.** Let $f : (-\infty, \infty) \to \mathbb{R}$ be integrable, and let

$$\lim_{t \to \infty} f(t) = \alpha, \quad -\infty \leq \alpha \leq \infty.$$

Then

$$\lim_{t \to \infty} \frac{1}{t} \int_0^t f(s)\mathrm{d}s = \lim_{t \to \infty} f(t).$$

**Proof.** We first consider the case where $-\infty < \alpha < \infty$. It follows from $\lim_{t \to \infty} f(t) = \alpha$ that, for every $\varepsilon > 0$, there exists $\delta > 0$ such that, for all $t > \delta$, $|f(t) - \alpha| < \varepsilon$ or, equivalently, $-\varepsilon + \alpha < f(t) < \varepsilon + \alpha$. Hence, for all $t > \delta$,

$$\frac{1}{t}\int_0^\delta f(s)\mathrm{d}s + \frac{t-\delta}{t}(-\varepsilon + \alpha) < \frac{1}{t}\int_0^\delta f(s)\mathrm{d}s + \frac{1}{t}\int_\delta^t f(s)\mathrm{d}s$$

$$< (\varepsilon + \alpha)\frac{t-\delta}{t} + \frac{1}{t}\int_0^\delta f(s)\mathrm{d}s;$$

that is,

$$\frac{1}{t}\int_0^\delta f(s)\mathrm{d}s - \frac{t-\delta}{t}\varepsilon - \frac{\delta}{t}\alpha < \frac{1}{t}\int_0^t f(s)\mathrm{d}s - \alpha$$

$$< \frac{t-\delta}{t}\varepsilon - \frac{\delta}{t}\alpha + \frac{1}{t}\int_0^\delta f(s)\mathrm{d}s, \quad t > \delta,$$

and, hence,

$$\frac{1}{t}\int_0^\delta f(s)\mathrm{d}s - \varepsilon - \frac{\delta}{t}\alpha < \frac{1}{t}\int_0^t f(s)\mathrm{d}s - \alpha$$

$$< \varepsilon - \frac{\delta}{t}\alpha + \frac{1}{t}\int_0^\delta f(s)\mathrm{d}s, \quad t > \delta.$$

Now, note that

$$\lim_{t\to\infty}\left(-\frac{\delta}{t}\alpha + \frac{1}{t}\int_0^\delta f(s)\mathrm{d}s\right) = 0,$$

and, hence, there exists $\hat{\delta} > 0$ such that, for all $t > \hat{\delta}$,

$$\left|-\frac{\delta}{t}\alpha + \frac{1}{t}\int_0^\delta f(s)\mathrm{d}s\right| < \varepsilon$$

or, equivalently,

$$-\varepsilon < -\frac{\delta}{t}\alpha + \frac{1}{t}\int_0^\delta f(s)\mathrm{d}s < \varepsilon.$$

Thus, for $t > \max\{\delta, \hat{\delta}\}$,

$$-2\varepsilon < \frac{1}{t}\int_0^t f(s)\mathrm{d}s - \alpha < 2\varepsilon.$$

Now, by definition, $\lim_{t\to\infty}\frac{1}{t}\int_0^t f(s)\mathrm{d}s = \alpha$.

Next, consider the case where $\alpha = \infty$, and suppose, *ad absurdum*, that

$$\lim_{t\to\infty}\frac{1}{t}\int_0^t f(s)\mathrm{d}s \neq \infty.$$

Since $\lim_{t\to\infty} f(t) = \infty$, it follows that $\int_0^t f(s)\mathrm{d}s$ must have a lower bound for all $t \geq 0$, and, hence,

$$\liminf_{t\to\infty}\frac{1}{t}\int_0^t f(s)\mathrm{d}s \geq 0;$$

that is, there exists $\delta_1 > 0$ such that

$$\frac{1}{t} \int_0^t f(s)\mathrm{d}s \geq 0, \quad t > \delta_1.$$

Consequently, if

$$\lim_{t \to \infty} \frac{1}{t} \int_0^t f(s)\mathrm{d}s < \infty,$$

then there exists $M \in [0, \infty)$ such that

$$\lim_{t \to \infty} \frac{1}{t} \int_0^t f(s)\mathrm{d}s = M,$$

which implies that

$$\limsup_{t \to \infty} \frac{1}{t} \int_0^t f(s)\mathrm{d}s = M.$$

It follows from $\lim_{t \to \infty} f(t) = \infty$ that there exists $\delta_2 > 0$ such that $f(t) > 2M+1$ for all $t > \delta_2$. Let $\delta = \max\{\delta_1, \delta_2\}$. Then, for every $t > 2\delta+1$,

$$\begin{aligned}
\frac{1}{t} \int_0^t f(s)\mathrm{d}s &= \frac{1}{t} \int_0^\delta f(s)\mathrm{d}s + \frac{1}{t} \int_\delta^t f(s)\mathrm{d}s \\
&= \frac{\delta}{t} \frac{1}{\delta} \int_0^\delta f(s)\mathrm{d}s + \frac{1}{t} \int_\delta^t f(s)\mathrm{d}s \\
&\geq \frac{1}{t} \int_\delta^t f(s)\mathrm{d}s \\
&\geq \frac{t - \delta}{t}(2M + 1) \\
&> \frac{1}{2}(2M + 1) \\
&= M + \frac{1}{2},
\end{aligned}$$

which implies that

$$\limsup_{t \to \infty} \frac{1}{t} \int_0^t f(s)\mathrm{d}s > M.$$

This contradicts the fact that

$$\limsup_{t \to \infty} \frac{1}{t} \int_0^t f(s)\mathrm{d}s = M,$$

and, hence,

$$\lim_{t \to \infty} \frac{1}{t} \int_0^t f(s)\mathrm{d}s = \infty.$$

Finally, for the case where $\alpha = -\infty$, define $g(t) = -f(t)$. Then it follows from

$$\lim_{t \to \infty} f(t) = -\infty$$

that $\lim_{t \to \infty} g(t) = \infty$. Now, using the arguments above, we have

$$\lim_{t \to \infty} \frac{1}{t} \int_0^t g(s) \mathrm{d}s = \infty,$$

which implies that

$$-\lim_{t \to \infty} \frac{1}{t} \int_0^t f(s) \mathrm{d}s = \infty.$$

Hence,

$$\lim_{t \to \infty} \frac{1}{t} \int_0^t f(s) \mathrm{d}s = -\infty,$$

which complete the proof.                                                      □

Suppose $\tilde{A}$ is semistable. Then it follows from (12.11) and Lemma 12.3 that

$$
\begin{aligned}
J(K) &= \lim_{t \to \infty} \mathbb{E}[(x(t) - x_\infty)^\mathrm{T} \tilde{R}(x(t) - x_\infty)] \\
&= \mathrm{tr} \lim_{t \to \infty} \mathbb{E}\left[ (x(t) - x_\infty)(x(t) - x_\infty)^\mathrm{T} \right] \tilde{R}, \quad (12.25)
\end{aligned}
$$

where $\tilde{R} = R_1 + K^\mathrm{T} R_2 K$. Next, it follows from (12.8)–(12.10) that

$$\frac{\mathrm{d}}{\mathrm{d}t} \mathbb{E}[x(t)] = \tilde{A} \mathbb{E}[x(t)], \quad t \geq 0,$$

and, hence, $\mathbb{E}[x(t)] = e^{\tilde{A}t} \mathbb{E}[x(0)]$, which implies that

$$x_\infty = (I_n - \tilde{A} \tilde{A}^\#) \mathbb{E}[x(0)]. \quad (12.26)$$

For the next result, define the covariance matrix $Q(t) = \mathbb{E}[x(t) x^\mathrm{T}(t)] - \mathbb{E}[x(t)]\mathbb{E}[x^\mathrm{T}(t)]$, and note that $Q(t) = Q^\mathrm{T}(t) \geq 0$ for all $t \geq 0$. The following result gives an explicit expression for $Q(t)$, $t \geq 0$, by using the standard stochastic formulation of $\mathcal{H}_2$ theory.

**Proposition 12.1.** Consider the closed-loop system (12.12), and assume that Assumption 12.1 holds. Then $Q(t)$, $t \geq 0$, satisfies the Lyapunov differential equation

$$\dot{Q}(t) = \tilde{A} Q(t) + Q(t) \tilde{A}^\mathrm{T} + D_1 D_1^\mathrm{T} + BKD_2 D_2^\mathrm{T} K^\mathrm{T} B^\mathrm{T}, \quad Q(0) = Q_0, \quad t \geq 0, \quad (12.27)$$

where $Q_0 = \mathbb{E}[x(0)x^{\mathrm{T}}(0)] - \mathbb{E}[x(0)]\mathbb{E}[x^{\mathrm{T}}(0)]$. Furthermore,

$$Q(t) = e^{\tilde{A}t}Q(0)e^{\tilde{A}^{\mathrm{T}}t} + \int_0^t e^{\tilde{A}s}\tilde{D}\tilde{D}^{\mathrm{T}}e^{\tilde{A}^{\mathrm{T}}s}\mathrm{d}s, \qquad (12.28)$$

where $\tilde{D} = D_1 + BKD_2$.

**Proof.** Consider the closed-loop system given by (12.12). Evaluating $\dot{Q}(t)$, $t \geq 0$, yields

$$
\begin{aligned}
\dot{Q}(t) &= \mathbb{E}[\dot{x}(t)x^{\mathrm{T}}(t) + x(t)\dot{x}^{\mathrm{T}}(t)] - \mathbb{E}[\dot{x}(t)]\mathbb{E}[x^{\mathrm{T}}(t)] - \mathbb{E}[x(t)]\mathbb{E}[\dot{x}^{\mathrm{T}}(t)] \\
&= \mathbb{E}[(\tilde{A}x(t) + \tilde{D}w(t))x^{\mathrm{T}}(t) + x(t)(\tilde{A}x(t) + \tilde{D}w(t))^{\mathrm{T}}] \\
&\quad - \tilde{A}\mathbb{E}[x(t)]\mathbb{E}[x^{\mathrm{T}}(t)] - \mathbb{E}[x(t)]\mathbb{E}[x^{\mathrm{T}}(t)]\tilde{A}^{\mathrm{T}} \\
&= \mathbb{E}[\tilde{A}x(t)x^{\mathrm{T}}(t)] - \tilde{A}\mathbb{E}[x(t)]\mathbb{E}[x^{\mathrm{T}}(t)] + \mathbb{E}[x(t)x^{\mathrm{T}}(t)\tilde{A}^{\mathrm{T}}] \\
&\quad - \mathbb{E}[x(t)]\mathbb{E}[x^{\mathrm{T}}(t)]\tilde{A}^{\mathrm{T}} + \mathbb{E}[\tilde{D}w(t)x^{\mathrm{T}}(t) + x(t)w^{\mathrm{T}}(t)\tilde{D}^{\mathrm{T}}] \\
&= \tilde{A}Q(t) + Q(t)\tilde{A}^{\mathrm{T}} + \mathbb{E}\left[\tilde{D}w(t)\left(e^{\tilde{A}t}x_0 + \int_0^t e^{\tilde{A}(t-s)}\tilde{D}w(s)\mathrm{d}s\right)^{\mathrm{T}}\right. \\
&\quad \left. + \left(e^{\tilde{A}t}x_0 + \int_0^t e^{\tilde{A}(t-s)}\tilde{D}w(s)\mathrm{d}s\right)w^{\mathrm{T}}(t)\tilde{D}^{\mathrm{T}}\right] \\
&= \tilde{A}Q(t) + Q(t)\tilde{A}^{\mathrm{T}} + \mathbb{E}\left[\tilde{D}\int_0^t w(t)w^{\mathrm{T}}(s)\tilde{D}^{\mathrm{T}}e^{\tilde{A}^{\mathrm{T}}(t-s)}\mathrm{d}s\right. \\
&\quad \left. + \int_0^t e^{\tilde{A}(t-s)}\tilde{D}w(s)w^{\mathrm{T}}(t)\mathrm{d}s\tilde{D}^{\mathrm{T}}\right] \\
&= \tilde{A}Q(t) + Q(t)\tilde{A}^{\mathrm{T}} + \tilde{D}\int_0^t \delta(t-s)\tilde{D}^{\mathrm{T}}e^{\tilde{A}^{\mathrm{T}}(t-s)}\mathrm{d}s \\
&\quad + \int_0^t e^{\tilde{A}(t-s)}\tilde{D}\delta(s-t)\mathrm{d}s\tilde{D}^{\mathrm{T}} \\
&= \tilde{A}Q(t) + Q(t)\tilde{A}^{\mathrm{T}} + \frac{1}{2}\tilde{D}\tilde{D}^{\mathrm{T}} + \frac{1}{2}\tilde{D}\tilde{D}^{\mathrm{T}} \\
&= \tilde{A}Q(t) + Q(t)\tilde{A}^{\mathrm{T}} + \tilde{D}\tilde{D}^{\mathrm{T}}, \qquad (12.29)
\end{aligned}
$$

where we used the fact that $\mathbb{E}[w(t)w^{\mathrm{T}}(s)] = \delta(t-s)$ and where $\delta(t)$ denotes the symmetric unit impulse centered at $t = 0$ given by

$$\delta(t) \triangleq \lim_{a \to 0} \frac{1}{a\sqrt{\pi}}e^{-t^2/a^2}. \qquad (12.30)$$

Hence, (12.27) holds.

Next, to show (12.28), rewrite (12.27) as

$$\text{vec}\,\dot{Q}(t) = \text{vec}\,\tilde{A}Q(t) + \text{vec}\,Q(t)\tilde{A}^{\mathrm{T}} + \text{vec}\,\tilde{D}\tilde{D}^{\mathrm{T}}$$
$$= (\tilde{A} \oplus \tilde{A})\text{vec}\,Q(t) + \text{vec}\,\tilde{D}\tilde{D}^{\mathrm{T}}.$$

Now, using Lagrange's formula yields

$$\text{vec}\,Q(t) = e^{(\tilde{A}\oplus\tilde{A})t}\text{vec}\,Q(0) + \int_0^t e^{(\tilde{A}\oplus\tilde{A})(t-s)}\text{vec}\,\tilde{D}\tilde{D}^{\mathrm{T}}\mathrm{d}s$$

or, equivalently, by changing the variable of integration,

$$\text{vec}\,Q(t) = e^{(\tilde{A}\oplus\tilde{A})t}\text{vec}\,Q(0) + \int_0^t e^{(\tilde{A}\oplus\tilde{A})s}\text{vec}\,\tilde{D}\tilde{D}^{\mathrm{T}}\mathrm{d}s$$

$$= e^{\tilde{A}t} \otimes e^{\tilde{A}t}\text{vec}\,Q(0) + \int_0^t e^{\tilde{A}s} \otimes e^{\tilde{A}s}\text{vec}\,\tilde{D}\tilde{D}^{\mathrm{T}}\mathrm{d}s$$

$$= \text{vec}\,e^{\tilde{A}t}Q(0)e^{\tilde{A}^{\mathrm{T}}t} + \int_0^t \text{vec}\,e^{\tilde{A}s}\tilde{D}\tilde{D}^{\mathrm{T}}e^{\tilde{A}^{\mathrm{T}}s}\mathrm{d}s,$$

which implies (12.28). $\qquad\square$

Next, we provide necessary and sufficient conditions for ensuring the well-posedness of the integral term $\int_0^t e^{\tilde{A}s}\tilde{D}\tilde{D}^{\mathrm{T}}e^{\tilde{A}^{\mathrm{T}}s}\mathrm{d}s$ as $t \to \infty$ in (12.28).

**Proposition 12.2.** Assume that Assumption 12.1 holds. If $\tilde{A}$ is semistable, then

$$\tilde{Q} = \lim_{t\to\infty} \int_0^t e^{\tilde{A}s}\tilde{D}\tilde{D}^{\mathrm{T}}e^{\tilde{A}^{\mathrm{T}}s}\mathrm{d}s$$

exists if and only if $\mathcal{N}(\tilde{A}^{\mathrm{T}}) \subseteq \mathcal{N}(\tilde{D}^{\mathrm{T}})$. Furthermore, if $\mathcal{N}(\tilde{A}^{\mathrm{T}}) \subseteq \mathcal{N}(\tilde{D}^{\mathrm{T}})$, then $Q$ is given by

$$Q = \lim_{t\to\infty} Q(t) = (I_n - \tilde{A}\tilde{A}^{\#})Q(0)(I_n - \tilde{A}\tilde{A}^{\#})^{\mathrm{T}} + \tilde{Q}. \qquad (12.31)$$

**Proof.** Note that since $\tilde{A}$ is semistable, it follows from $\text{spec}(\tilde{A}) = \text{spec}(\tilde{A}^{\mathrm{T}})$ (see Proposition 4.4.5 of [27, p. 263]) that $\tilde{A}^{\mathrm{T}}$ is semistable. Hence, it follows that either $\tilde{A}^{\mathrm{T}}$ is Hurwitz or there exists an invertible matrix $S \in \mathbb{R}^{n\times n}$ such that

$$\tilde{A}^{\mathrm{T}} = S \begin{bmatrix} J & 0 \\ 0 & 0 \end{bmatrix} S^{-1},$$

where $J \in \mathbb{R}^{r\times r}$, $r = \text{rank}\,\tilde{A}^{\mathrm{T}}$, and $J$ is Hurwitz. Now, if $\tilde{A}^{\mathrm{T}}$ is Hurwitz, then $\mathcal{N}(\tilde{A}^{\mathrm{T}}) = \{0\} \subseteq \mathcal{N}(\tilde{D}^{\mathrm{T}})$.

Alternatively, if $\tilde{A}^{\mathrm{T}}$ is not Hurwitz, then

$$\mathcal{N}(\tilde{A}^{\mathrm{T}}) = \left\{ x \in \mathbb{R}^n : x = S[0_{1 \times r}, y^{\mathrm{T}}]^{\mathrm{T}}, y \in \mathbb{R}^{n-r} \right\}.$$

Now,

$$\int_0^\infty e^{\tilde{A}t} \tilde{D} \tilde{D}^{\mathrm{T}} e^{\tilde{A}^{\mathrm{T}}t} \mathrm{d}t = S^{-\mathrm{T}} \int_0^\infty e^{\hat{J}t} \tilde{D} \tilde{D}^{\mathrm{T}} e^{\hat{J}t} \mathrm{d}t S$$

$$= S^{-\mathrm{T}} \int_0^\infty \left[ \begin{array}{cc} e^{J^{\mathrm{T}}t} \hat{R}_1 e^{Jt} & e^{J^{\mathrm{T}}t} \hat{R}_{12} \\ \hat{R}_{12}^{\mathrm{T}} e^{Jt} & \hat{R}_2 \end{array} \right] \mathrm{d}t S, \qquad (12.32)$$

where

$$\hat{J} = \left[ \begin{array}{cc} J & 0 \\ 0 & 0 \end{array} \right]$$

and

$$\hat{R} = S^{\mathrm{T}} \tilde{D} \tilde{D}^{\mathrm{T}} S = \left[ \begin{array}{cc} \hat{R}_1 & \hat{R}_{12} \\ \hat{R}_{12}^{\mathrm{T}} & \hat{R}_2 \end{array} \right].$$

Next, it follows from (12.32) that

$$\int_0^\infty e^{\tilde{A}t} \tilde{D} \tilde{D}^{\mathrm{T}} e^{\tilde{A}^{\mathrm{T}}t} \mathrm{d}t$$

exists if and only if $\hat{R}_2 = 0$ or, equivalently,

$$[0_{1 \times r}, y^{\mathrm{T}}] \hat{R} [0_{1 \times r}, y^{\mathrm{T}}]^{\mathrm{T}} = 0, \quad y \in \mathbb{R}^{n-r},$$

which is further equivalent to $x^{\mathrm{T}} \tilde{D} \tilde{D}^{\mathrm{T}} x = 0$, $x \in \mathcal{N}(\tilde{A}^{\mathrm{T}})$. Hence, $\mathcal{N}(\tilde{A}^{\mathrm{T}}) \subseteq \mathcal{N}(\tilde{D}^{\mathrm{T}})$. The proof of $\mathcal{N}(\tilde{A}^{\mathrm{T}}) \subseteq \mathcal{N}(\tilde{D}^{\mathrm{T}})$ implying the existence of

$$\int_0^\infty e^{\tilde{A}t} \tilde{D} \tilde{D}^{\mathrm{T}} e^{\tilde{A}^{\mathrm{T}}t} \mathrm{d}t$$

is immediate by reversing the steps given above.

Finally, the second assertion is a direct consequence of $\lim_{t \to \infty} e^{\tilde{A}t} = I_n - \tilde{A} \tilde{A}^{\#}$ and (12.28) in Proposition 12.1. $\qquad \square$

Thus, if Assumption 12.1 holds, $\tilde{A}$ is semistable, and $\mathcal{N}(\tilde{A}^{\mathrm{T}}) \subseteq \mathcal{N}(\tilde{D}^{\mathrm{T}})$, then (12.11) can be rewritten as

$$J(K) = \operatorname{tr} Q \tilde{R} = \operatorname{tr} (I_n - \tilde{A} \tilde{A}^{\#}) Q(0)(I_n - \tilde{A} \tilde{A}^{\#})^{\mathrm{T}} \tilde{R} + \operatorname{tr} \tilde{Q} \tilde{R}. \qquad (12.33)$$

Clearly, if $\mathbb{E}[x(0)x^{\mathrm{T}}(0)] = \mathbb{E}[x(0)]\mathbb{E}[x^{\mathrm{T}}(0)]$, then $Q(0) = 0$, and, hence, $J(K) = \operatorname{tr} \hat{Q} \tilde{R}$. A sufficient condition to guarantee this is the case where $x(0)$ is deterministic. However, here we consider the general case where $x(0)$ is not necessarily deterministic. Without loss of generality, we make the following assumption on $x(0)$.

**Assumption 12.2.** $x(0)$ is a random variable having a covariance $V$; that is, $\mathbb{E}[x(0)x^{\mathrm{T}}(0)] - \mathbb{E}[x(0)]\mathbb{E}[x^{\mathrm{T}}(0)] = V$.

Note that $Q$ in (12.33) has two parts, namely, $(I_n - \tilde{A}\tilde{A}^\#)Q(0)(I_n - \tilde{A}\tilde{A}^\#)^{\mathrm{T}}$ and $\tilde{Q}$. Hence, to minimize $J(K)$, one has to minimize the cost functional involving the terms $\mathrm{tr}\,(I_n - \tilde{A}\tilde{A}^\#)Q(0)(I_n - \tilde{A}\tilde{A}^\#)^{\mathrm{T}}\tilde{R}$ and $\mathrm{tr}\,\tilde{Q}\tilde{R}$ in (12.33) simultaneously. The next result combines the two forms in (12.33) into a compact form.

**Proposition 12.3.** Assume that Assumptions 12.1 and 12.2 hold, and assume that $\mathcal{N}(\tilde{A}^{\mathrm{T}}) \subseteq \mathcal{N}(\tilde{D}^{\mathrm{T}})$. If $\tilde{A}$ is semistable, then $J(K) = \mathrm{tr}\,(W+V)\tilde{R}$, where

$$W = \int_0^\infty e^{\tilde{A}s}[\tilde{A}V + V\tilde{A}^{\mathrm{T}} + \tilde{D}\tilde{D}^{\mathrm{T}}]e^{\tilde{A}^{\mathrm{T}}s}\mathrm{d}s. \tag{12.34}$$

**Proof.** Using the fact that

$$\int_0^\infty \frac{\mathrm{d}}{\mathrm{d}s}(e^{\tilde{A}s}Ve^{\tilde{A}^{\mathrm{T}}s})\mathrm{d}s = e^{\tilde{A}s}Ve^{\tilde{A}^{\mathrm{T}}s}\Big|_0^\infty,$$

it follows that

$$\int_0^\infty e^{\tilde{A}s}[\tilde{A}V + V\tilde{A}^{\mathrm{T}}]e^{\tilde{A}^{\mathrm{T}}s}\mathrm{d}s = (I_n - \tilde{A}\tilde{A}^\#)V(I_n - \tilde{A}\tilde{A}^\#)^{\mathrm{T}} - V. \tag{12.35}$$

The result now is immediate. $\qquad\square$

As shown in Example 12.2, the standard notions of stabilizability and detectability cannot be used to guarantee the existence of solutions to the $\mathcal{H}_2$ optimal semistabilization problem. To guarantee semistabilization of the closed-loop system, we present the new notions of *k-semicontrollability* and *k-semiobservability* by extending the notions of semicontrollability and semiobservability presented in Chapter 10.

**Definition 12.2.** Let $A \in \mathbb{R}^{n \times n}$, $B \in \mathbb{R}^{n \times l}$, and $C \in \mathbb{R}^{l \times n}$. The pair $(A, C)$ is *k-semiobservable* if there exists a nonnegative integer $k$ such that

$$k = \min\left\{l \in \overline{\mathbb{Z}}_+ : \bigcap_{i=1}^n \mathcal{N}(CA^{l+i-1}) = \bigcap_{i=1}^n \mathcal{N}(A^i)\right\}. \tag{12.36}$$

The pair $(A, B)$ is *k-semicontrollable* if there exists a nonnegative integer $k$ such that

$$k = \min \left\{ l \in \overline{\mathbb{Z}}_+ : \sum_{i=1}^{n} \mathcal{R}(A^{l+i-1}B) = \sum_{i=1}^{n} \mathcal{R}(A^i) \right\}. \qquad (12.37)$$

The notion of $k$-semiobservability is a generalization of semiobservability from $k = 0, 1$ to an arbitrary integer $k \geq 0$. For $k \geq 2$, this new notion is weaker than the notions of semiobservability discussed in Chapter 10 and weak semiobservability presented in [156].

**Example 12.5.** Let

$$A = \begin{bmatrix} -1 & 1 \\ 1 & -1 \end{bmatrix}, \quad C = \begin{bmatrix} 1 & -1 \end{bmatrix}.$$

Then $(A, C)$ is 0-semiobservable. However, if $C = \begin{bmatrix} 1 & 0 \end{bmatrix}$, then $(A, C)$ is not 0-semiobservable but is 1-semiobservable. $\triangle$

Note that (12.36) is equivalent to

$$k = \min \left\{ l \in \overline{\mathbb{Z}}_+ : \bigcap_{i=1}^{n} \mathcal{N}(CA^{l+i-1}) = \mathcal{N}(A) \right\},$$

since $\bigcap_{i=1}^{n} \mathcal{N}(A^i) = \mathcal{N}(A)$. Similarly, it follows from (i) and (ii) of Fact 2.9.14 in [27, p. 121] that $\sum_{i=1}^{n} \mathcal{R}(A^i) = \mathcal{R}(A)$. Hence, (12.37) is equivalent to

$$k = \min \left\{ l \in \overline{\mathbb{Z}}_+ : \sum_{i=1}^{n} \mathcal{R}(A^{l+i-1}B) = \mathcal{R}(A) \right\}.$$

The duality between both notions follows directly from both definitions and the properties of null and range spaces (see Chapter 2 in [27]).

**Lemma 12.4.** Let $A \in \mathbb{R}^{n \times n}$ and $C \in \mathbb{R}^{l \times n}$. Then $(A, C)$ is $k$-semiobservable if and only if $(A^{\mathrm{T}}, C^{\mathrm{T}})$ is $k$-semicontrollable. Furthermore, if $(A, C)$ is $k$-semiobservable, then $\mathcal{N}(A) \subseteq \mathcal{N}(CA^k)$.

**Proof.** It follows from Theorem 2.4.3 in [27, p. 103] that $\mathcal{N}(A) = \mathcal{R}(A^{\mathrm{T}})^{\perp}$. Hence,

$$\bigcap_{i=1}^{n} \mathcal{N}(CA^{k+i-1}) = \mathcal{N}(A)$$

if and only if

$$\bigcap_{i=1}^{n} \mathcal{R}((A^{\mathrm{T}})^{k+i-1}C^{\mathrm{T}})^{\perp} = \mathcal{R}(A^{\mathrm{T}})^{\perp}.$$

By Fact 2.9.16 in [27, p. 121],

$$\bigcap_{i=1}^{n} \mathcal{R}((A^{\mathrm{T}})^{k+i-1}C^{\mathrm{T}})^{\perp} = \left[\sum_{i=1}^{n} \mathcal{R}((A^{\mathrm{T}})^{k+i-1}C^{\mathrm{T}})\right]^{\perp},$$

and, hence,

$$\left[\sum_{i=1}^{n} \mathcal{R}((A^{\mathrm{T}})^{k+i-1}C^{\mathrm{T}})\right]^{\perp} = \mathcal{R}(A^{\mathrm{T}})^{\perp}.$$

Now, it follows from (i) and (ii) of Fact 2.9.14 in [27, p. 121] that $(A, C)$ is $k$-semiobservable if and only if

$$\sum_{i=1}^{n} \mathcal{R}((A^{\mathrm{T}})^{k+i-1}C^{\mathrm{T}}) = \mathcal{R}(A^{\mathrm{T}}),$$

and, hence, by definition, $(A^{\mathrm{T}}, C^{\mathrm{T}})$ is $k$-semicontrollable. The second assertion directly follows from Definition 12.2. □

Next, we give a rank characterization for $k$-semiobservability which is much easier to verify in practice.

**Proposition 12.4.** Let $A \in \mathbb{R}^{n \times n}$ and $C \in \mathbb{R}^{l \times n}$. Then $(A, C)$ is $k$-semiobservable if and only if there exists a smallest nonnegative integer $k$ such that

$$\mathrm{rank} \begin{bmatrix} A \\ CA^{k} \\ CA^{k+1} \\ \vdots \\ CA^{k+n-1} \end{bmatrix} = \mathrm{rank} \begin{bmatrix} CA^{k} \\ CA^{k+1} \\ \vdots \\ CA^{k+n-1} \end{bmatrix} = \mathrm{rank}\, A. \tag{12.38}$$

**Proof.** Note that $(A, C)$ is $k$-semiobservable if and only if there exists a smallest nonnegative integer $k$ such that

$$\mathcal{N}\left(\begin{bmatrix} CA^{k} \\ CA^{k+1} \\ \vdots \\ CA^{k+n-1} \end{bmatrix}\right) = \mathcal{N}(A). \tag{12.39}$$

Now, it follows from equation (2.4.13) of [27, p. 103] that (12.39) holds if and only if

$$\mathcal{R}[(CA^k)^{\mathrm{T}}, (CA^{k+1})^{\mathrm{T}}, \ldots, (CA^{k+n-1})^{\mathrm{T}}]^{\perp} = \mathcal{R}(A^{\mathrm{T}})^{\perp}.$$

Note that $\mathcal{R}[(CA^k)^{\mathrm{T}}, (CA^{k+1})^{\mathrm{T}}, \ldots, (CA^{k+n-1})^{\mathrm{T}}]$ and $\mathcal{R}(A^{\mathrm{T}})$ are subspaces. Thus, it follows from Fact 2.9.14 of [27, p. 121] that

$$\mathcal{R}[(CA^k)^{\mathrm{T}}, (CA^{k+1})^{\mathrm{T}}, \ldots, (CA^{k+n-1})^{\mathrm{T}}]^{\perp} = \mathcal{R}(A^{\mathrm{T}})^{\perp}$$

if and only if

$$\mathcal{R}[(CA^k)^{\mathrm{T}}, (CA^{k+1})^{\mathrm{T}}, \ldots, (CA^{k+n-1})^{\mathrm{T}}] = \mathcal{R}(A^{\mathrm{T}}).$$

Now, it follows from Fact 2.11.5 of [27, p. 131] that

$$\mathcal{R}[(CA^k)^{\mathrm{T}}, (CA^{k+1})^{\mathrm{T}}, \ldots, (CA^{k+n-1})^{\mathrm{T}}] = \mathcal{R}(A^{\mathrm{T}})$$

if and only if

$$\begin{aligned}
\mathrm{rank}&[(CA^k)^{\mathrm{T}}, (CA^{k+1})^{\mathrm{T}}, \ldots, (CA^{k+n-1})^{\mathrm{T}}] \\
&= \mathrm{rank}(A^{\mathrm{T}}) \\
&= \mathrm{rank}[A^{\mathrm{T}}, (CA^k)^{\mathrm{T}}, (CA^{k+1})^{\mathrm{T}}, \ldots, (CA^{k+n-1})^{\mathrm{T}}],
\end{aligned}$$

which proves (12.38). $\qquad\square$

The following proposition introduces the *power semistable Lyapunov equation*, which generalizes the results in [132] and [156] from $k = 0$ and $k = 1$, respectively, to an arbitrary nonnegative integer $k$.

**Proposition 12.5.** Let $A \in \mathbb{R}^{n \times n}$. If there exist an $n \times n$ matrix $P = P^{\mathrm{T}} \geq 0$ and an $m \times n$ matrix $C \in \mathbb{R}^{m \times n}$ such that $(A, C)$ is $k$-semiobservable and

$$0 = (A^k)^{\mathrm{T}}(A^{\mathrm{T}}P + PA + R)A^k,$$

where $R = C^{\mathrm{T}}C$, then (i) $\mathcal{N}(PA^k) \subseteq \mathcal{N}(A) \subseteq \mathcal{N}(RA^k)$ and (ii) $\mathcal{N}(A) \cap \mathcal{R}(A) = \{0\}$.

**Proof.** The proof is similar to that of Lemma 4.5 of [156] (for the case where $k = 1$) and, hence, is omitted. $\qquad\square$

Using Proposition 12.5, we have the following necessary and sufficient condition for semistability of the closed-loop system (12.8)–(12.10).

**Theorem 12.2.** Assume that Assumptions 12.1 and 12.2 hold. Then the following statements hold.

(i) $\tilde{A}$ is semistable if and only if, for every $k$-semicontrollable pair $(\tilde{A}, \tilde{D})$, there exists an $n \times n$ matrix $P = P^{\mathrm{T}} \geq -V$ such that

$$0 = \tilde{A}^k[\tilde{A}(P+V) + (P+V)\tilde{A}^{\mathrm{T}} + \tilde{D}\tilde{D}^{\mathrm{T}}](\tilde{A}^k)^{\mathrm{T}}. \qquad (12.40)$$

(ii) $\tilde{A}$ is semistable if and only if, for every $k$-semicontrollable pair $(\tilde{A}, \tilde{D})$, there exists an $n \times n$ matrix $\tilde{P} = \tilde{P}^{\mathrm{T}} > -V$ such that

$$0 = \tilde{A}(\tilde{P}+V) + (\tilde{P}+V)\tilde{A}^{\mathrm{T}} + \tilde{A}^k\tilde{D}\tilde{D}^{\mathrm{T}}(\tilde{A}^k)^{\mathrm{T}}. \qquad (12.41)$$

Furthermore, such a $\tilde{P}$ is not unique.

**Proof.** (i) If $\tilde{A}$ is semistable and $(\tilde{A}, \tilde{D})$ is $k$-semicontrollable, then it follows from Lemma 12.4 that $(\tilde{A}^{\mathrm{T}}, \tilde{D}^{\mathrm{T}})$ is $k$-semiobservable. By Lemma 12.4, $\mathcal{N}(\tilde{A}^{\mathrm{T}}) \subseteq \mathcal{N}(\tilde{D}^{\mathrm{T}}(\tilde{A}^k)^{\mathrm{T}})$. Now, it follows from Proposition 12.2 that

$$W_1 = \int_0^\infty e^{\tilde{A}s}\tilde{A}^k\tilde{D}\tilde{D}^{\mathrm{T}}(\tilde{A}^k)^{\mathrm{T}}e^{\tilde{A}^{\mathrm{T}}s}\mathrm{d}s$$

is well defined. Furthermore, it follows from (12.35) that

$$W_2 = \int_0^\infty e^{\tilde{A}s}[\tilde{A}V + V\tilde{A}^{\mathrm{T}}]e^{\tilde{A}^{\mathrm{T}}s}\mathrm{d}s$$

is well defined. Clearly, $W_2 \geq -V$.

Next, consider

$$P = (\tilde{A}^\#)^k W_1((\tilde{A}^\#)^k)^{\mathrm{T}} + W_2 \geq W_2 \geq -V.$$

It follows from (12.35) that

$$\tilde{A}(W_2 + V) + (W_2 + V)\tilde{A}^{\mathrm{T}} = 0.$$

Hence,

$$\tilde{A}^k(P+V)(\tilde{A}^k)^{\mathrm{T}} = \tilde{A}^k\int_0^\infty e^{\tilde{A}s}\tilde{D}\tilde{D}^{\mathrm{T}}e^{\tilde{A}^{\mathrm{T}}s}\mathrm{d}s(\tilde{A}^k)^{\mathrm{T}} + \tilde{A}^k(W_2 + V)(\tilde{A}^k)^{\mathrm{T}}$$
$$= W_1 + \tilde{A}^k(W_2 + V)(\tilde{A}^k)^{\mathrm{T}}.$$

Thus,

$$\tilde{A}\tilde{A}^k(P+V)(\tilde{A}^k)^{\mathrm{T}} + \tilde{A}^k(P+V)(\tilde{A}^k)^{\mathrm{T}}(\tilde{A})^{\mathrm{T}}$$
$$= \tilde{A}^k\int_0^\infty e^{\tilde{A}s}(\tilde{A}\tilde{D}\tilde{D}^{\mathrm{T}} + \tilde{D}\tilde{D}^{\mathrm{T}}\tilde{A}^{\mathrm{T}})e^{\tilde{A}^{\mathrm{T}}s}\mathrm{d}s(\tilde{A}^k)^{\mathrm{T}}$$

$$= \tilde{A}^k \int_0^\infty \frac{\mathrm{d}}{\mathrm{d}s} (e^{\tilde{A}s} \tilde{D}\tilde{D}^{\mathrm{T}} e^{\tilde{A}^{\mathrm{T}}s}) \mathrm{d}s (\tilde{A}^k)^{\mathrm{T}}$$
$$= \tilde{A}^k (-\tilde{D}\tilde{D}^{\mathrm{T}})(\tilde{A}^k)^{\mathrm{T}},$$

which implies that

$$\tilde{A}\tilde{A}^k (P+V)(\tilde{A}^k)^{\mathrm{T}} + \tilde{A}^k (P+V)(\tilde{A}^k)^{\mathrm{T}}(\tilde{A})^{\mathrm{T}} + \tilde{A}^k \tilde{D}\tilde{D}^{\mathrm{T}}(\tilde{A}^k)^{\mathrm{T}} = 0.$$

Hence, (12.40) holds.

Alternatively, if there exists $P \geq -V$ such that (12.40) holds and $(\tilde{A}, \tilde{D})$ is $k$-semicontrollable, then it follows from Lemma 12.4 that $(\tilde{A}^{\mathrm{T}}, \tilde{D}^{\mathrm{T}})$ is $k$-semiobservable, and, hence, by Proposition 12.5, $\mathcal{N}(\tilde{A}^{\mathrm{T}}) \cap \mathcal{R}(\tilde{A}^{\mathrm{T}}) = \{0\}$. Thus, it follows from [24, p. 119] that $\tilde{A}^{\mathrm{T}}$ is group invertible. Let $L \triangleq I_n - (\tilde{A}^{\mathrm{T}})(\tilde{A}^{\mathrm{T}})^\#$, and note that $L^2 = L$. Hence, $L$ is the unique $n \times n$ matrix satisfying

$$\mathcal{N}(L) = \mathcal{R}(\tilde{A}^{\mathrm{T}}), \quad \mathcal{R}(L) = \mathcal{N}(\tilde{A}^{\mathrm{T}}), \quad Lx = x, \quad x \in \mathcal{N}(\tilde{A}^{\mathrm{T}}).$$

Consider the nonnegative function

$$\mathbb{V}(x) = x^{\mathrm{T}} \tilde{A}^k (P+V)(\tilde{A}^k)^{\mathrm{T}} x + x^{\mathrm{T}} L^{\mathrm{T}} L x.$$

If $\mathbb{V}(x) = 0$ for some $x \in \mathbb{R}^n$, then $(P+V)(\tilde{A}^k)^{\mathrm{T}} x = 0$ and $Lx = 0$. It follows from Proposition 12.5 that $x \in \mathcal{N}(\tilde{A}^{\mathrm{T}})$, whereas $Lx = 0$ implies that $x \in \mathcal{R}(\tilde{A}^{\mathrm{T}})$. Now, it follows from Proposition 12.5 that $x = 0$. Hence, $\mathbb{V}(\cdot)$ is positive definite. Next, since

$$L\tilde{A}^{\mathrm{T}} = \tilde{A}^{\mathrm{T}} - (\tilde{A}^{\mathrm{T}})(\tilde{A}^{\mathrm{T}})^\# \tilde{A}^{\mathrm{T}} = 0,$$

it follows that

$$\dot{\mathbb{V}}(x) = -x^{\mathrm{T}} \tilde{A}^k \tilde{D}\tilde{D}^{\mathrm{T}}(\tilde{A}^k)^{\mathrm{T}} x + x^{\mathrm{T}} \tilde{A} L^{\mathrm{T}} L x + x^{\mathrm{T}} L^{\mathrm{T}} L \tilde{A}^{\mathrm{T}} x$$
$$= -x^{\mathrm{T}} \tilde{A}^k \tilde{D}\tilde{D}^{\mathrm{T}}(\tilde{A}^k)^{\mathrm{T}} x$$
$$\leq 0, \quad x \in \mathbb{R}^n.$$

Note that $\dot{\mathbb{V}}^{-1}(0) = \mathcal{N}(\tilde{D}^{\mathrm{T}}(\tilde{A}^k)^{\mathrm{T}})$.

To find the largest invariant subset $\mathcal{M}$ of $\mathcal{N}(\tilde{D}^{\mathrm{T}}(\tilde{A}^k)^{\mathrm{T}})$, consider the system

$$\dot{x}(t) = \tilde{A}^{\mathrm{T}} x(t), \quad x(0) = x_0, \quad t \geq 0, \tag{12.42}$$

with $\tilde{D}^{\mathrm{T}}(\tilde{A}^k)^{\mathrm{T}} x(t) = 0$ for all $t \geq 0$. Then

$$\tilde{D}^{\mathrm{T}}(\tilde{A}^k)^{\mathrm{T}} \frac{\mathrm{d}^{i-1}}{\mathrm{d}t^{i-1}} x(t) = 0, \quad i = 1, 2, \dots, \quad t \geq 0;$$

that is,

$$\tilde{D}^{\mathrm{T}}(\tilde{A}^k)^{\mathrm{T}}(\tilde{A}^{i-1})^{\mathrm{T}}x(t) = \tilde{D}^{\mathrm{T}}(\tilde{A}^{k+i-1})^{\mathrm{T}}x(t) = 0, \quad i = 1, 2, \ldots, \quad t \geq 0.$$

The definition of $k$-semiobservability now implies that $x(t) \in \mathcal{N}(\tilde{A}^{\mathrm{T}})$ for all $t \geq 0$. Thus, $\mathcal{M} \subseteq \mathcal{N}(\tilde{A}^{\mathrm{T}})$. However, $\mathcal{N}(\tilde{A}^{\mathrm{T}})$ consists of only equilibrium points and, hence, is invariant. Hence, $\mathcal{M} = \mathcal{N}(\tilde{A}^{\mathrm{T}})$.

Now, let $x_{\mathrm{e}} \in \mathcal{N}(\tilde{A}^{\mathrm{T}})$ be an equilibrium point of (12.42), and consider the function $\mathbb{U}(x) = \mathbb{V}(x - x_{\mathrm{e}})$, which is positive definite with respect to $x_{\mathrm{e}}$. Then it follows that

$$\dot{\mathbb{U}}(x) = -(x - x_{\mathrm{e}})^{\mathrm{T}}\tilde{A}^k\tilde{D}\tilde{D}^{\mathrm{T}}(\tilde{A}^k)^{\mathrm{T}}(x - x_{\mathrm{e}}) \leq 0, \quad x \in \mathbb{R}^n,$$

which implies that $x_{\mathrm{e}}$ is Lyapunov stable. Now, it follows from Theorem 3.3 of [33] that $\tilde{A}^{\mathrm{T}}$ is semistable, and, hence, $\tilde{A}$ is semistable.

(ii) Necessity follows from (i). More specifically, it follows from (i) that there exists an $n \times n$ matrix $P = P^{\mathrm{T}} \geq -V$ such that (12.40) holds. Next, consider

$$\tilde{P} = \tilde{A}^k(P + V)(\tilde{A}^k)^{\mathrm{T}} + L^{\mathrm{T}}L - V.$$

Now, using similar arguments as in the proof of (i), it follows that

$$\tilde{P} + V = \tilde{A}^k(P + V)(\tilde{A}^k)^{\mathrm{T}} + L^{\mathrm{T}}L$$

is positive definite. Substituting $\tilde{P}$ into (12.40) yields (12.41). Sufficiency follows using a similar argument as in the proof of Theorem 12.2 by considering

$$\mathbb{V}(x) = x^{\mathrm{T}}(\tilde{P} + V)x = x^{\mathrm{T}}\tilde{A}^k(P + V)(\tilde{A}^k)^{\mathrm{T}}x + x^{\mathrm{T}}L^{\mathrm{T}}Lx$$

and $\mathbb{U}(x) = \mathbb{V}(x - x_{\mathrm{e}})$ for $x_{\mathrm{e}} \in \mathcal{N}(\tilde{A}^{\mathrm{T}})$.

To show the second part of the assertion, for every $\tilde{P} > -V$ satisfying (12.41) and $M \geq 0$, let

$$\overline{P} \triangleq \tilde{P} + L^{\mathrm{T}}ML.$$

Clearly, $\overline{P} \geq \tilde{P} > -V$. It is easy to verify that $\overline{P}$ is a solution to (12.41), and, hence, $\tilde{P}$ is not unique. $\qquad \square$

The following lemma is a generalization of Lemma 11.5 and is needed for the next result.

**Lemma 12.5.** Let $x, y \in \mathbb{R}^n$ be such that $x \neq 0$ and $y \neq 0$. Then $xy^{\mathrm{T}} = y^{\mathrm{T}}x$ if and only if $x$ and $y$ are linearly dependent. Furthermore, $xy^{\mathrm{T}} = y^{\mathrm{T}}x \geq 0$ if and only if $y = \alpha x$, where $\alpha > 0$.

**Proof.** If $x$ and $y$ are linearly dependent, then $xy^{\mathrm{T}} = yx^{\mathrm{T}}$ holds. Conversely, suppose that $x$ and $y$ are linearly independent; then it follows from Proposition 7.1.8 of [27, p. 441] that $xy^{\mathrm{T}} = yx^{\mathrm{T}}$ if and only if $\mathrm{vec}^{-1}(y \otimes x) = \mathrm{vec}^{-1}(x \otimes y)$, where $\mathrm{vec}^{-1}$ denotes the inverse operation of the column-stacking vec operator [27, p. 439], which further implies that $y \otimes x = x \otimes y$. Now, let $x = [x_1, \ldots, x_n]^{\mathrm{T}}$ and $y = [y_1, \ldots, y_n]^{\mathrm{T}}$. Then it follows from $y \otimes x = x \otimes y$ that $x_i x = x_i y$ for every $i = 1, \ldots, n$. Since $x$ and $y$ are linearly independent, it follows that $y_i x - x_i y = 0$ for every $i = 1, \ldots, n$ if and only if $y_i = x_i = 0$ for every $i = 1, \ldots, n$. This contradicts the assumption that $x, y \neq 0$. Hence, $x$ and $y$ are linearly dependent. Similar arguments prove the second assertion.                    $\square$

Using Theorem 12.2, the following result is key in obtaining an equivalent, practically solvable optimization problem.

**Theorem 12.3.** Assume that Assumptions 12.1 and 12.2 hold. Then $\tilde{A}$ is semistable if and only if, for every $k$-semicontrollable pair $(\tilde{A}, \tilde{D})$, there exists an $n \times n$ matrix $\hat{P} = \hat{P}^{\mathrm{T}} > -\tilde{A}^k V (\tilde{A}^k)^{\mathrm{T}}$ such that

$$0 = \tilde{A}\hat{P} + \hat{P}\tilde{A}^{\mathrm{T}} + \tilde{A}^k(\tilde{A}V + V\tilde{A}^{\mathrm{T}} + \tilde{D}\tilde{D}^{\mathrm{T}})(\tilde{A}^k)^{\mathrm{T}}, \qquad (12.43)$$

where $\hat{P}$ is not unique. Furthermore, if $(\tilde{A}, \tilde{D})$ is $k$-semicontrollable and $\hat{P}$ satisfies (12.43), then

$$\hat{P} = \int_0^\infty e^{\tilde{A}t}\tilde{A}^k(\tilde{A}V + V\tilde{A}^{\mathrm{T}} + \tilde{D}\tilde{D}^{\mathrm{T}})(\tilde{A}^k)^{\mathrm{T}}e^{\tilde{A}^{\mathrm{T}}t}\mathrm{d}t + \alpha xx^{\mathrm{T}},$$

$$x \in \mathcal{N}(\tilde{A}), \quad \alpha > 0. \qquad (12.44)$$

**Proof.** The first part of the assertion is a direct consequence of Theorem 12.2 by taking

$$\hat{P} = \tilde{P} + V - \tilde{A}^k V (\tilde{A}^k)^{\mathrm{T}}, \quad \mathbb{V}(x) = x^{\mathrm{T}}(\hat{P} + \tilde{A}^k V (\tilde{A}^k)^{\mathrm{T}})x.$$

To prove that $P$ has the form of (12.44), first note that it follows from Theorem 12.2 that there exists an $n \times n$ matrix $\hat{P} > -\tilde{A}^k V (\tilde{A}^k)^{\mathrm{T}}$ such that (12.43) holds or, equivalently,

$$(\tilde{A} \oplus \tilde{A})\mathrm{vec}\,\hat{P} = -\mathrm{vec}\,\tilde{A}^k(\tilde{A}V + V\tilde{A}^{\mathrm{T}} + \tilde{D}\tilde{D}^{\mathrm{T}})(\tilde{A}^k)^{\mathrm{T}}.$$

Hence,

$$\mathrm{vec}\,\tilde{A}^k(\tilde{A}V + V\tilde{A}^{\mathrm{T}} + \tilde{D}\tilde{D}^{\mathrm{T}})(\tilde{A}^k)^{\mathrm{T}} \in \mathcal{R}(\tilde{A} \oplus \tilde{A}).$$

Next, it follows from Lemma 4.8 of [156] that $\tilde{A} \oplus \tilde{A}$ is semistable, and, hence, by Lemma 4.9 of [156],

$$\mathrm{vec}^{-1}\left((\tilde{A} \oplus \tilde{A})^{\#}\mathrm{vec}\,\tilde{A}^k(\tilde{A}V + V\tilde{A}^{\mathrm{T}} + \tilde{D}\tilde{D}^{\mathrm{T}})(\tilde{A}^k)^{\mathrm{T}}\right)$$

$$= -\int_0^\infty \mathrm{vec}^{-1}\Big(e^{(\tilde{A}\oplus\tilde{A})t}\mathrm{vec}\,\tilde{A}^k(\tilde{A}V + V\tilde{A}^\mathrm{T} + \tilde{D}\tilde{D}^\mathrm{T})(\tilde{A}^k)^\mathrm{T}\Big)\mathrm{d}t$$

$$= -\int_0^\infty \mathrm{vec}^{-1}\Big(e^{\tilde{A}t}\otimes e^{\tilde{A}t}\Big)\mathrm{vec}\,\tilde{A}^k(\tilde{A}V + V\tilde{A}^\mathrm{T} + \tilde{D}\tilde{D}^\mathrm{T})(\tilde{A}^k)^\mathrm{T}\mathrm{d}t$$

$$= -\int_0^\infty e^{\tilde{A}t}\tilde{A}^k(\tilde{A}V + V\tilde{A}^\mathrm{T} + \tilde{D}\tilde{D}^\mathrm{T})(\tilde{A}^k)^\mathrm{T}e^{\tilde{A}^\mathrm{T}t}\mathrm{d}t,$$

where we used the facts that $e^{X\oplus Y} = e^X \otimes e^Y$ and $\mathrm{vec}(XYZ) = (Z^\mathrm{T} \otimes X)\mathrm{vec}\,Y$ [27]. Hence,

$$\hat{P} = \int_0^\infty e^{\tilde{A}t}\tilde{A}^k(\tilde{A}V + V\tilde{A}^\mathrm{T} + \tilde{D}\tilde{D}^\mathrm{T})(\tilde{A}^k)^\mathrm{T}e^{\tilde{A}t}\mathrm{d}t + \mathrm{vec}^{-1}(z),$$

where $z$ satisfies $z \in \mathcal{N}(\tilde{A} \oplus \tilde{A})$ and $\mathrm{vec}^{-1}(z) = (\mathrm{vec}^{-1}(z))^\mathrm{T} \geq 0$ (the nonnegative definiteness of $\mathrm{vec}^{-1}(z)$ is guaranteed by Theorem 4.2(a) of [282]).

Since $(\tilde{A} \oplus \tilde{A})$ is semistable, it follows that a general solution to the equation $(\tilde{A} \oplus \tilde{A})z = 0$ is given by $z = x \otimes y$, where $x, y \in \mathcal{N}(\tilde{A})$. Hence,

$$\mathrm{vec}^{-1}(z) = \mathrm{vec}^{-1}(x \otimes y) = yx^\mathrm{T},$$

where we used the fact that $xy^\mathrm{T} = \mathrm{vec}^{-1}(y \otimes x)$. Furthermore, it follows from $\mathrm{vec}^{-1}(z) = (\mathrm{vec}^{-1}(z))^\mathrm{T} \geq 0$ that $xy^\mathrm{T} = yx^\mathrm{T} \geq 0$. Now, it follows from Lemma 12.5 that $y = \alpha x$, where $\alpha > 0$. $\qquad\square$

Using Theorem 12.3, we have the following optimization-based characterization for the robust and optimal information state equipartitioning problem.

**Theorem 12.4.** Assume that Assumptions 12.1 and 12.2 hold and $(\tilde{A}, \tilde{D})$ is $k$-semicontrollable. Let $S_{\min}$ be a solution to the minimization problem

$$\min\Big\{\mathrm{tr}\,S\tilde{R} : S = S^\mathrm{T} > -\tilde{A}^k V(\tilde{A}^k)^\mathrm{T} \text{ and}$$
$$\tilde{A}S + S\tilde{A}^\mathrm{T} + \tilde{A}^k(\tilde{A}V + V\tilde{A}^\mathrm{T} + \tilde{D}\tilde{D}^\mathrm{T})(\tilde{A}^k)^\mathrm{T} = 0,$$
$$\tilde{A}\mathbf{e} = 0, \quad \mathrm{rank}\,\tilde{A} = n - 1\Big\}. \tag{12.45}$$

Then $\mathrm{tr}\,S_{\min}\tilde{R} = \mathrm{tr}\,\tilde{A}^k W(\tilde{A}^k)^\mathrm{T}\tilde{R}$ and, for every $x \in \mathbb{R}^n$, $\lim_{t\to\infty} e^{\tilde{A}t}x = \mathbf{e}c$ for some $c \in \mathbb{R}$, where $W$ is given as in (12.34).

**Proof.** The first assertion is a direct consequence of Theorem 12.3. (See Theorem 4.2 of [156] for a similar proof.) To show the second assertion, note that it follows from Theorem 12.3 that the existence of solutions to

(12.45) guarantees semistability of $\tilde{A}$. Since $\tilde{A}\mathbf{e} = 0$, it follows that zero is an eigenvalue of $\tilde{A}$. Furthermore, since $\operatorname{rank}\tilde{A}^{\mathrm{T}} = \operatorname{rank}\tilde{A} = n - 1$, it follows that there exists a unique unit vector $\mathbf{u} \in \mathbb{R}^n$ such that $\tilde{A}^{\mathrm{T}}\mathbf{u} = 0$ or, equivalently, $\mathbf{u}^{\mathrm{T}}\tilde{A} = 0$. Now, it follows from Lemma 12.2 that

$$\lim_{t\to\infty} e^{\tilde{A}t} = \frac{1}{\mathbf{e}^{\mathrm{T}}\mathbf{u}}\mathbf{e}\mathbf{u}^{\mathrm{T}},$$

and, hence, $\lim_{t\to\infty} e^{\tilde{A}t}x = \mathbf{e}c$, where $c = \frac{1}{\mathbf{e}^{\mathrm{T}}\mathbf{u}}\mathbf{u}^{\mathrm{T}}x$.                    $\square$

Theorem 12.4 gives a key connection between solving the Gaussian linear-quadratic consensus problem and solving the optimization problem (12.45). To see this, note that it follows from Theorem 12.4 that

$$\operatorname{tr}(S_{\min} + \tilde{A}^k V(\tilde{A}^k)^{\mathrm{T}})\tilde{R} = \operatorname{tr}\tilde{A}^k(W + V)(\tilde{A}^k)^{\mathrm{T}}\tilde{R}.$$

Hence, if $k = 0$, then solving the Gaussian linear-quadratic consensus problem is equivalent to solving the optimization problem (12.45). Alternatively, for every $k = 0, 1, 2, \ldots$, it follows from Proposition 8.4.13 of [27, p. 471] that

$$\lambda_{\min}((\tilde{A}^k)^{\mathrm{T}}\tilde{R}\tilde{A}^k)\operatorname{tr}(W + V) \leq \operatorname{tr}\tilde{A}^k(W + V)(\tilde{A}^k)^{\mathrm{T}}\tilde{R}$$
$$\leq \lambda_{\max}((\tilde{A}^k)^{\mathrm{T}}\tilde{R}\tilde{A}^k)\operatorname{tr}(W + V). \qquad (12.46)$$

Next, it follows from Corollary 8.4.2 of [27, p. 467] that $\lambda_{\min}(\tilde{R})I_n \leq \tilde{R}$, and, hence,

$$\lambda_{\min}(\tilde{R})(\tilde{A}^k)^{\mathrm{T}}\tilde{A}^k \leq (\tilde{A}^k)^{\mathrm{T}}\tilde{R}\tilde{A}^k,$$

which implies that

$$\lambda_{\min}(\tilde{R})\lambda_{\min}((\tilde{A}^k)^{\mathrm{T}}\tilde{A}^k) \leq \lambda_{\min}((\tilde{A}^k)^{\mathrm{T}}\tilde{R}\tilde{A}^k).$$

Similarly,

$$\lambda_{\max}((\tilde{A}^k)^{\mathrm{T}}\tilde{R}\tilde{A}^k) \leq \lambda_{\max}(\tilde{R})\lambda_{\max}((\tilde{A}^k)^{\mathrm{T}}\tilde{A}^k).$$

Hence, by (12.46),

$$\lambda_{\min}(\tilde{R})\lambda_{\min}((\tilde{A}^k)^{\mathrm{T}}\tilde{A}^k)\operatorname{tr}(W + V) \leq \operatorname{tr}\tilde{A}^k(W + V)(\tilde{A}^k)^{\mathrm{T}}\tilde{R}$$
$$\leq \lambda_{\max}(\tilde{R})\lambda_{\max}((\tilde{A}^k)^{\mathrm{T}}\tilde{A}^k)\operatorname{tr}(W + V).$$

In particular, for $k = 0$, we have

$$\lambda_{\min}(\tilde{R})\operatorname{tr}(W + V) \leq \operatorname{tr}(W + V)\tilde{R} \leq \lambda_{\max}(\tilde{R})\operatorname{tr}(W + V).$$

Thus, for $k \geq 1$, if the corresponding values in the global optima for both $\operatorname{tr} \tilde{A}^k(W + V)(\tilde{A}^k)^{\mathrm{T}} \tilde{R}$ and $\operatorname{tr}(W + V)\tilde{R}$ are in the interval

$$\left[ \max_{k \geq 0} \{\lambda_{\min}(\tilde{R})\lambda_{\min}((\tilde{A}^k)^{\mathrm{T}}\tilde{A}^k)\operatorname{tr}(W + V)\}, \right.$$

$$\left. \min_{k \geq 0} \{\lambda_{\min}(\tilde{R})\lambda_{\min}((\tilde{A}^k)^{\mathrm{T}}\tilde{A}^k)\operatorname{tr}(W + V)\} \right],$$

then solving the Gaussian linear-quadratic consensus problem is equivalent to solving the optimization problem (12.45).

The optimization problem (12.45) presented in Theorem 12.4 is still challenging to solve numerically since $K$ is embedded *implicitly* in the cost functional; that is, $S$ is an implicit function of $K$. Next, we exploit the structure of $S$ in (12.45) by means of the technique used in the proof of Theorem 12.3 to further simplify the Gaussian linear-quadratic consensus problem into yet another constrained optimization problem where $K$ appears explicitly in the cost functional. Here, we assume Assumptions 12.1 and 12.2 hold and $(\tilde{A}, \tilde{D})$ is $k$-semicontrollable. In this case, it follows from Theorem 12.3 that $\tilde{A}$ is semistable if and only if there exists $P = P^{\mathrm{T}} > -\tilde{A}^k V(\tilde{A}^k)^{\mathrm{T}}$ such that

$$\tilde{A}P + P\tilde{A}^{\mathrm{T}} + \tilde{A}^k(\tilde{A}V + V\tilde{A}^{\mathrm{T}} + \tilde{D}\tilde{D}^{\mathrm{T}})(\tilde{A}^k)^{\mathrm{T}} = 0. \tag{12.47}$$

Hence, $J(K) = \operatorname{tr}(P_{\mathrm{LS}} + V)\tilde{R}$, where $P_{\mathrm{LS}} = P_{\mathrm{LS}}^{\mathrm{T}} \geq -\tilde{A}^k V(\tilde{A}^k)^{\mathrm{T}}$ is the least squares solution to (12.47); that is, $P_{\mathrm{LS}} = \arg\min_{P \in \mathcal{P}} \|P\|_{\mathrm{F}}$, where $\mathcal{P}$ denotes the set of all solutions $P = P^{\mathrm{T}} \geq -\tilde{A}^k V(\tilde{A}^k)^{\mathrm{T}}$ to (12.47).

Note that (12.47) can be rewritten as

$$\tilde{A}(P + \tilde{A}^k V(\tilde{A}^k)^{\mathrm{T}}) + (P + \tilde{A}^k V(\tilde{A}^k)^{\mathrm{T}})^{\mathrm{T}} \tilde{A}^{\mathrm{T}} = -\tilde{A}^k \tilde{D}\tilde{D}^{\mathrm{T}}(\tilde{A}^k)^{\mathrm{T}},$$

which implies that

$$(\tilde{A} \oplus \tilde{A})\operatorname{vec}(P + \tilde{A}^k V(\tilde{A}^k)^{\mathrm{T}}) = -\operatorname{vec}(\tilde{A}^k \tilde{D}\tilde{D}^{\mathrm{T}}(\tilde{A}^k)^{\mathrm{T}}),$$

and, hence,

$$(\tilde{A} \oplus \tilde{A})\operatorname{vec} P = -\operatorname{vec}(\tilde{A}^k \tilde{D}\tilde{D}^{\mathrm{T}}(\tilde{A}^k)^{\mathrm{T}}) - (\tilde{A} \oplus \tilde{A})\operatorname{vec}(\tilde{A}^k V(\tilde{A}^k)^{\mathrm{T}}).$$

It follows from Theorem 2.6.4 of [27, p. 108] and Proposition 6.1.7 of [27, p. 400] that the set of solutions to this equation is given by

$$\mathcal{N}(\tilde{A} \oplus \tilde{A}) - (\tilde{A} \oplus \tilde{A})^+ \operatorname{vec}(\tilde{A}^k \tilde{D}\tilde{D}^{\mathrm{T}}(\tilde{A}^k)^{\mathrm{T}}) - (\tilde{A} \oplus \tilde{A})^+(\tilde{A} \oplus \tilde{A})\operatorname{vec}(\tilde{A}^k V(\tilde{A}^k)^{\mathrm{T}}),$$

where $A^+$ denotes the Moore–Penrose generalized inverse of $A$ and $\mathcal{N}(A) + y = \{x + y : x \in \mathcal{N}(A)\}$. Using similar arguments as in the proof of Theorem 12.3, it follows that

$$P = \alpha x x^{\mathrm{T}} - \operatorname{vec}^{-1}\left((\tilde{A} \oplus \tilde{A})^+ \operatorname{vec}(\tilde{A}^k \tilde{D}\tilde{D}^{\mathrm{T}}(\tilde{A}^k)^{\mathrm{T}})\right)$$

$$- \text{vec}^{-1} \left( (\tilde{A} \oplus \tilde{A})^{+} (\tilde{A} \oplus \tilde{A}) \text{vec} \, (\tilde{A}^k V (\tilde{A}^k)^{\mathrm{T}}) \right), \qquad (12.48)$$

where $\alpha > 0$ and $x \in \mathcal{N}(\tilde{A})$.

Let

$$f(K) = \text{vec}^{-1} \left( (\tilde{A} \oplus \tilde{A})^{+} \text{vec} \, (\tilde{A}^k \tilde{D} \tilde{D}^{\mathrm{T}} (\tilde{A}^k)^{\mathrm{T}}) \right)$$

and

$$g(K) = \text{vec}^{-1} \left( (\tilde{A} \oplus \tilde{A})^{+} (\tilde{A} \oplus \tilde{A}) \text{vec} \, (\tilde{A}^k V (\tilde{A}^k)^{\mathrm{T}}) \right).$$

Since $P = P^{\mathrm{T}} \geq -\tilde{A}^k V (\tilde{A}^k)^{\mathrm{T}}$, it follows from (12.48) that $K$ satisfies

$$f(K) + g(K) = f^{\mathrm{T}}(K) + g^{\mathrm{T}}(K) \leq \tilde{A}^k V (\tilde{A}^k)^{\mathrm{T}}.$$

Note that in this case, $P_{\mathrm{LS}} = -f(K) - g(K)$. Thus, our stochastic linear-quadratic consensus problem can be converted into the following equivalent constrained optimization problem:

$$\max_{K \in \mathcal{S}} f(K) + g(K) \qquad (12.49)$$

subject to

$$f(K) + g(K) = f^{\mathrm{T}}(K) + g^{\mathrm{T}}(K) \leq (A + BK)^k V (A^{\mathrm{T}} + K^{\mathrm{T}} B^{\mathrm{T}})^k, \quad (12.50)$$
$$(A + BK)\mathbf{e} = 0, \quad \text{rank}(A + BK) = n - 1, \qquad (12.51)$$

and

$$(A + BK, D_1 + BKD_2) \text{ is } k\text{-semicontrollable}, \qquad (12.52)$$

where $\mathcal{S}$ denotes the admissible set for $K$ (e.g., $K$ has a block-structure associated with some directed graph $\mathfrak{G}$). This is not a convex optimization problem due to the rank constraint $\text{rank}(A + BK) = n - 1$. Nevertheless, (12.49) can be solved by using numerical optimization techniques to find a near-optimal or best solution for $K$.

The difficulty in applying numerical optimization techniques to either (12.45) or (12.49) lies in how one can verify (12.51) and (12.52). Here, we present two ways to verify (12.52). First, we consider the case where $D_1 = \tilde{A}$ and $D_2 = 0$. In this case, $\tilde{D} = D_1 + BKD_2 = \tilde{A}$ and the pair $(\tilde{A}, \tilde{D})$ becomes $(\tilde{A}, \tilde{A})$. Then it follows from (12.37) in Definition 12.2 that $(\tilde{A}, \tilde{A})$ is 0-semicontrollable for every $K \in \mathbb{R}^{m \times n}$.

The following result gives a sufficient condition under which $(\tilde{A}, \tilde{D})$ is 0-semicontrollable.

**Proposition 12.6.** Consider the semistable Gaussian linear-quadratic consensus problem defined in Section 12.3. If there exists a full column rank matrix $M \in \mathbb{R}^{n \times p}$ such that

$$\operatorname{rank}(I_n - MD_2^{\mathrm{T}})\operatorname{rank}(B)$$
$$= \operatorname{rank}[(I_n - MD_2^{\mathrm{T}}) \otimes B \quad \operatorname{vec}(D_1 M^{\mathrm{T}}) - \operatorname{vec} A], \qquad (12.53)$$

then there exists $K \in \mathbb{R}^{m \times n}$ such that $(\tilde{A}, \tilde{D})$ is 0-semicontrollable or, equivalently, $(A + BK, D_1 + BKD_2)$ is 0-semicontrollable.

**Proof.** We first show that if there exists a full column rank matrix $M \in \mathbb{R}^{n \times p}$ such that (12.53) holds, then there exists $K \in \mathbb{R}^{m \times n}$ such that $\mathcal{N}(\tilde{A}^{\mathrm{T}}) = \mathcal{N}(\tilde{D}^{\mathrm{T}})$. To see this, note that it follows from Fact 7.4.24 of [27, p. 447] that (12.53) is equivalent to

$$\operatorname{rank}((I_n - MD_2^{\mathrm{T}}) \otimes B) = \operatorname{rank}[(I_n - MD_2^{\mathrm{T}}) \otimes B \quad \operatorname{vec}(D_1 M^{\mathrm{T}}) - \operatorname{vec} A].$$

Now, it follows from Theorem 2.6.4 of [27, p. 108] that

$$((I_n - MD_2^{\mathrm{T}}) \otimes B)x = \operatorname{vec}(D_1 M^{\mathrm{T}}) - \operatorname{vec} A$$

has at least one solution $x \in \mathbb{R}^{mn}$. Let $\operatorname{vec} K = x$. Then

$$((I_n - MD_2^{\mathrm{T}}) \otimes B)\operatorname{vec} K = \operatorname{vec}(D_1 M^{\mathrm{T}}) - \operatorname{vec} A$$

or, equivalently, by Proposition 7.1.9 of [27, p. 441],

$$\operatorname{vec}(BK(I_n - D_2 M^{\mathrm{T}})) = \operatorname{vec}(D_1 M^{\mathrm{T}} - A),$$

and, hence, $BK(I_n - D_2 M^{\mathrm{T}}) = D_1 M^{\mathrm{T}} - A$.

Rearranging this equation yields

$$A + BK = (D_1 + BKD_2)M^{\mathrm{T}},$$

and, consequently, $(A+BK)^{\mathrm{T}} = M(D_1+BKD_2)^{\mathrm{T}}$. Since $M$ has full column rank, it follows from Theorem 2.6.1 of [27, p. 107] that $M$ is left invertible. Now, it follows from Proposition 2.6.3 of [27, p. 107] that $\mathcal{N}(\tilde{A}^{\mathrm{T}}) = \mathcal{N}(\tilde{D}^{\mathrm{T}})$.

Finally, note that since $\mathcal{N}(\tilde{D}^{\mathrm{T}}(\tilde{A}^{\mathrm{T}})^i) \supseteq \mathcal{N}(\tilde{A}^{\mathrm{T}})$ for every positive integer $i$, it follows that

$$\bigcap_{i=1}^{n} \mathcal{N}(\tilde{D}^{\mathrm{T}}(\tilde{A}^{\mathrm{T}})^{i-1}) = \mathcal{N}(\tilde{A}^{\mathrm{T}}),$$

which implies that $(\tilde{A}^{\mathrm{T}}, \tilde{D}^{\mathrm{T}})$ is 0-semiobservable. By duality it follows that (see Lemma 12.4) $(\tilde{A}, \tilde{D})$ is 0-semicontrollable.                              $\square$

## 12.5  Existence of Optimal Solutions for the Gaussian Linear-Quadratic Consensus Problem

In this section, we establish sufficient conditions for the existence of an optimal gain $K$ for the semistable Gaussian linear-quadratic consensus problem. To proceed, define the sets

$$\mathcal{C}_\mathrm{s} \triangleq \{K \in \mathbb{R}^{m \times n} : (\tilde{A}, \tilde{D}) \text{ is } 0\text{-semicontrollable}\},$$
$$\mathcal{K}_\mathrm{s} \triangleq \{K \in \mathbb{R}^{m \times n} : \tilde{A} \text{ is semistable}\}.$$

The next result gives a continuity property for the cost functional for the semistable Gaussian linear-quadratic consensus problem.

**Proposition 12.7.** Assume that Assumption 12.1 holds. If $K \in \mathcal{K}_\mathrm{s} \cap \mathcal{C}_\mathrm{s} \neq \varnothing$, then $J(\cdot)$ is finite and continuous on $\mathcal{K}_\mathrm{s} \cap \mathcal{C}_\mathrm{s}$.

**Proof.** It follows from Proposition 12.2 that if $\tilde{A}$ is semistable, then $Q = \lim_{t \to \infty} Q(t)$ exists if and only if $\mathcal{N}(\tilde{A}^\mathrm{T}) \subseteq \mathcal{N}(\tilde{D}^\mathrm{T})$, where $Q(t)$ satisfies (12.28). Since $K \in \mathcal{C}_\mathrm{s}$ implies $\mathcal{N}(\tilde{A}^\mathrm{T}) \subseteq \mathcal{N}(\tilde{D}^\mathrm{T})$ by Lemma 12.4, it follows from Proposition 12.1 and (12.33) that $J(\cdot)$ is finite. Finally, the continuity of $J(\cdot)$ follows from (12.28). $\qquad\square$

For the main result in this section, recall the definitions of semistabilizability and semidetectability given in Chapter 11.

**Definition 12.3.** Let $A \in \mathbb{R}^{n \times n}$, $B \in \mathbb{R}^{n \times l}$, and $C \in \mathbb{R}^{l \times n}$. The pair $(A, B)$ is *semistabilizable* if

$$\operatorname{rank} \begin{bmatrix} B & \jmath\omega I_n - A \end{bmatrix} = n \tag{12.54}$$

for every nonzero $\omega \in \mathbb{R}$. The pair $(A, C)$ is *semidetectable* if

$$\operatorname{rank} \begin{bmatrix} C \\ \jmath\omega I_n - A \end{bmatrix} = n \tag{12.55}$$

for every nonzero $\omega \in \mathbb{R}$.

It follows from Facts 2.11.1–2.11.3 of [27, pp. 130–131] that (12.54) and (12.55) are equivalent to $\dim[\mathcal{R}(\jmath\omega I_n - A) + \mathcal{R}(B)] = n$ and $\mathcal{N}(\jmath\omega I_n - A) \cap \mathcal{N}(C) = \{0\}$ for every nonzero $\omega \in \mathbb{R}$, respectively.

**Example 12.6.** Consider

$$A = \begin{bmatrix} 0 & 0 \\ 0 & 0 \end{bmatrix}, \quad B = \begin{bmatrix} 0 \\ 1 \end{bmatrix}.$$

Clearly $(A, B)$ is not stabilizable. However, $(A, B)$ is semistabilizable. $\quad\triangle$

It is worth noting that semidetectability is weaker than 0-semiobservability. The following lemma is needed.

**Lemma 12.6.** Let $A \in \mathbb{R}^{n \times n}$ and $C \in \mathbb{R}^{l \times n}$. If $(A, C)$ is 0-semiobservable, then $(A, C)$ is semidetectable.

**Proof.** Suppose, *ad absurdum*, that $(A, C)$ is not semidetectable. Then, by definition, there exists $x \neq 0$ such that $Cx = 0$ and $(A - \jmath\omega I_n)x = 0$ for some $\omega \neq 0$. Hence,

$$CA^{i-1}x = (\jmath\omega)^{i-1}Cx = 0, \quad i = 1, \ldots, n.$$

Since $(A, C)$ is 0-semiobservable, it follows from Definition 12.2 that $x \in \mathcal{N}(A)$, and, hence, $\jmath\omega x = 0$, which implies that $x = 0$, leading to a contradiction. Hence, $(A, C)$ is semidetectable. $\qquad \square$

The converse part of Lemma 12.6 is not true in general as the following example shows.

**Example 12.7.** Consider

$$A = \begin{bmatrix} -1 & 0 \\ 0 & 0 \end{bmatrix}, \quad C = \begin{bmatrix} 0 & 1 \end{bmatrix}.$$

Note that

$$\begin{bmatrix} C \\ \jmath\omega I_n - A \end{bmatrix} = \begin{bmatrix} 0 & 1 \\ \jmath\omega + 1 & 0 \\ 0 & \jmath\omega \end{bmatrix}$$

is a full-rank matrix for every $\omega \in \mathbb{R}$. Hence, $(A, C)$ is semidetectable. However, $\mathcal{N}(C) \cap \mathcal{N}(CA) = \mathcal{N}(\begin{bmatrix} 0 & 1 \end{bmatrix}) \neq \mathcal{N}(A)$. Thus, $(A, C)$ is not 0-semiobservable. $\qquad \triangle$

The next result can be viewed as a partial converse to Proposition 12.7.

**Proposition 12.8.** Assume that Assumptions 12.1 and 12.2 hold. If $J(K) < \infty$, $K \in \mathcal{C}_{\mathrm{s}} \neq \emptyset$, and $\tilde{R} > 0$, then $K \in \mathcal{K}_{\mathrm{s}}$.

**Proof.** By (12.25) and Proposition 12.7,

$$J(K) = \lim_{t \to \infty} \mathrm{tr} \left[ e^{\tilde{A}t} V e^{\tilde{A}^{\mathrm{T}}t} \tilde{R} + \int_0^t e^{\tilde{A}s} \tilde{D}\tilde{D}^{\mathrm{T}} e^{\tilde{A}^{\mathrm{T}}s} \mathrm{d}s \tilde{R} \right].$$

Furthermore, note that

$$e^{\tilde{A}t} V e^{\tilde{A}^{\mathrm{T}}t} + \int_0^t e^{\tilde{A}s} \tilde{D}\tilde{D}^{\mathrm{T}} e^{\tilde{A}^{\mathrm{T}}s} \mathrm{d}s$$

$$= V + \int_0^t \frac{\mathrm{d}}{\mathrm{d}s} \left( e^{\tilde{A}s} V e^{\tilde{A}^T s} \right) \mathrm{d}s + \int_0^t e^{\tilde{A}s} \tilde{D} \tilde{D}^T e^{\tilde{A}^T s} \mathrm{d}s$$

$$= V + \int_0^t e^{\tilde{A}s} \left( \tilde{A} V + V \tilde{A}^T + \tilde{D} \tilde{D}^T \right) e^{\tilde{A}^T s} \mathrm{d}s. \qquad (12.56)$$

Hence, it follows that

$$J(K) = \mathrm{tr} V \tilde{R} + \lim_{t \to \infty} \mathrm{tr} \left[ \int_0^t e^{\tilde{A}s} \left( \tilde{A} V + V \tilde{A}^T + \tilde{D} \tilde{D}^T \right) e^{\tilde{A}^T s} \mathrm{d}s \tilde{R} \right].$$

Thus, $J(K) < \infty$ if and only if

$$\int_0^\infty e^{\tilde{A}s} \left( \tilde{A} V + V \tilde{A}^T + \tilde{D} \tilde{D}^T \right) e^{\tilde{A}^T s} \mathrm{d}s$$

exists.

Next, if

$$P = \int_0^\infty e^{\tilde{A}s} \left( \tilde{A} V + V \tilde{A}^T + \tilde{D} \tilde{D}^T \right) e^{\tilde{A}^T s} \mathrm{d}s$$

exists, then $P = P^T \geq -V$ and

$$\lim_{t \to \infty} e^{\tilde{A}t} \left( \tilde{A} V + V \tilde{A}^T + \tilde{D} \tilde{D}^T \right) e^{\tilde{A}^T t} = 0.$$

In this case,

$$\tilde{A}(P + V) + (P + V)\tilde{A}^T$$

$$= \int_0^\infty \left[ \tilde{A} e^{\tilde{A}s} \left( \tilde{A} V + V \tilde{A}^T + \tilde{D} \tilde{D}^T \right) e^{\tilde{A}^T s} \right.$$

$$\left. + e^{\tilde{A}s} \left( \tilde{A} V + V \tilde{A}^T + \tilde{D} \tilde{D}^T \right) e^{\tilde{A}^T s} \tilde{A}^T \right] \mathrm{d}s + \tilde{A} V + V \tilde{A}^T$$

$$= \int_0^\infty \frac{\mathrm{d}}{\mathrm{d}s} e^{\tilde{A}s} \left( \tilde{A} V + V \tilde{A}^T + \tilde{D} \tilde{D}^T \right) e^{\tilde{A}^T s} \mathrm{d}s + \tilde{A} V + V \tilde{A}^T$$

$$= \lim_{t \to \infty} e^{\tilde{A}t} \left( \tilde{A} V + V \tilde{A}^T + \tilde{D} \tilde{D}^T \right) e^{\tilde{A}^T t} - \left( \tilde{A} V + V \tilde{A}^T + \tilde{D} \tilde{D}^T \right)$$

$$+ \tilde{A} V + V \tilde{A}^T$$

$$= -\tilde{D} \tilde{D}^T. \qquad (12.57)$$

Now, it follows from Lemma 6.3 of [170] that the pair $(\tilde{A}, \tilde{D})$ is completely unstabilizable and

$$\int_0^\infty e^{\tilde{A}s} \tilde{D} \tilde{D}^T e^{\tilde{A}^T s} \mathrm{d}s$$

exists.

Next, consider the pair $(\tilde{A}, \tilde{D})$, and note that it follows from the Kalman decomposition that there exists an invertible matrix $T \in \mathbb{R}^{n \times n}$ such that

$$T\tilde{A}T^{\mathrm{T}} = \begin{bmatrix} \hat{A}_1 & \hat{A}_{12} \\ 0 & \hat{A}_2 \end{bmatrix}, \quad T\tilde{D} = \begin{bmatrix} \hat{D}_1 \\ 0 \end{bmatrix},$$

where $(\hat{A}_1, \hat{D}_1)$ is controllable. Thus,

$$\int_0^\infty e^{\tilde{A}s} \tilde{D}\tilde{D}^{\mathrm{T}} e^{\tilde{A}^{\mathrm{T}}s} \mathrm{d}s = T^{-1} \begin{bmatrix} \int_0^\infty e^{\hat{A}_1 s} \hat{D}_1 \hat{D}_1^{\mathrm{T}} e^{\hat{A}_1^{\mathrm{T}}s} \mathrm{d}s & 0 \\ 0 & 0 \end{bmatrix} (T^{-1})^{\mathrm{T}}$$

exists.

First, we show that $\hat{A}_1$ is asymptotically stable. Suppose, *ad absurdum*, that $\hat{A}_1$ is not asymptotically stable. Let $\lambda \in \mathrm{spec}(\tilde{A})$, where $\mathrm{Re}\,\lambda \geq 0$, and let $x \in \mathbb{C}^r$, $x \neq 0$, satisfy $\hat{A}_1 x = \lambda x$, where $r$ denotes the row dimension of $\hat{A}_1$. Since $(\hat{A}_1, \hat{D}_1)$ is controllable, it follows from Theorem 12.6.18 of [27, p. 815] that

$$\int_0^\infty e^{\hat{A}_1 s} \hat{D}_1 \hat{D}_1^{\mathrm{T}} e^{\hat{A}_1^{\mathrm{T}}s} \mathrm{d}s > 0,$$

and, hence,

$$\alpha \triangleq x^* \int_0^\infty e^{\hat{A}_1 s} \hat{D}_1 \hat{D}_1^{\mathrm{T}} e^{\hat{A}_1^{\mathrm{T}}s} \mathrm{d}s\, x > 0, \quad x \in \mathbb{C}^r.$$

However,

$$\alpha = x^* \int_0^\infty e^{\lambda s} \hat{D}_1 \hat{D}_1^{\mathrm{T}} e^{\bar{\lambda}s} \mathrm{d}s\, x = x^* \hat{D}_1 \hat{D}_1^{\mathrm{T}} x \int_0^\infty e^{2(\mathrm{Re}\,\lambda)s} \mathrm{d}s.$$

Now, by controllability of $(\hat{A}_1, \hat{D}_1)$, we have $x^* \hat{D}_1 \hat{D}_1^{\mathrm{T}} x \neq 0$, $x \in \mathbb{C}^r$. Since $\mathrm{Re}\,\lambda \geq 0$, it follows that

$$\int_0^\infty e^{2(\mathrm{Re}\,\lambda)s} \mathrm{d}s = \infty,$$

which contradicts the fact that $\alpha$ is a positive number. Thus, $\hat{A}_1$ is asymptotically stable.

Next, we show that $\tilde{A}$ is Lyapunov stable. To see this, let $\mu \in \mathrm{spec}(\tilde{A}^{\mathrm{T}})$ and let $z \in \mathbb{C}^n$, $z \neq 0$, satisfy $\tilde{A}^{\mathrm{T}} z = \mu z$. Since $(\tilde{A}, \tilde{D})$ is semicontrollable, it follows from Theorem 12.2 that

$$z^* \tilde{D}\tilde{D}^{\mathrm{T}} z = -z^*(\tilde{A}(\hat{P} + V) + (\hat{P} + V)\tilde{A}^{\mathrm{T}})z = -(2\mathrm{Re}\,\mu)z^*(\hat{P} + V)z$$

for some $\hat{P} = \hat{P}^{\mathrm{T}} > -V$. Hence, $\mathrm{Re}\,\mu \leq 0$. If $\mathrm{Re}\,\mu = 0$, then let $\mu = \jmath\omega$ and $x \in \mathcal{N}((\jmath\omega I_n - \tilde{A}^{\mathrm{T}})^2)$, where $\omega \in \mathbb{R}$. Defining $y = (\jmath\omega I_n - \tilde{A}^{\mathrm{T}})x$, it follows that $(\jmath\omega I_n - \tilde{A}^{\mathrm{T}})y = 0$, and, hence, $\tilde{A}^{\mathrm{T}}y = \jmath\omega y$. Therefore, it follows from Theorem 12.2 that

$$-y^* \tilde{D}\tilde{D}^{\mathrm{T}} y = y^* \left( \tilde{A}(\hat{P}+V) + (\hat{P}+V)\tilde{A}^{\mathrm{T}} \right) y$$
$$= -\jmath\omega y^*(\hat{P}+V) + \jmath\omega y^*(\hat{P}+V)y$$
$$= 0,$$

and, hence, $\tilde{D}^{\mathrm{T}}y = 0$. Thus,

$$0 = x^* \tilde{D}\tilde{D}^{\mathrm{T}} y$$
$$= -x^* \left( \tilde{A}(\hat{P}+V) + (\hat{P}+V)\tilde{A}^{\mathrm{T}} \right) y$$
$$= -x^*(\tilde{A} + \jmath\omega I_n)(\hat{P}+V)y$$
$$= y^*(\hat{P}+V)y.$$

Since $\hat{P}+V > 0$, it follows that $y = 0$; that is, $(\jmath\omega I_n - \tilde{A}^{\mathrm{T}})x = 0$. Therefore, $x \in \mathcal{N}(\jmath\omega I_n - \tilde{A}^{\mathrm{T}})$. Now, it follows from Proposition 5.5.8 of [27, p. 323] that $\jmath\omega$ is semisimple. Thus, $\tilde{A}$ is Lyapunov stable and, consequently, so is $\hat{A}_2$.

Finally, we show that $\jmath\omega \notin \mathrm{spec}(\hat{A}_2)$ for every $\omega \neq 0$. Since $(\tilde{A}, \tilde{D})$ is semistabilizable and $\dim [\mathcal{R}(\jmath\omega I_n - A) + \mathcal{R}(B)] = n$, it follows that

$$\dim \left[ \mathcal{R}(\tilde{D}) + \mathcal{R}(\jmath\omega I_n - \tilde{A}) \right] = n, \quad \omega \in \mathbb{R}\backslash\{0\}.$$

We claim that $(T\tilde{A}T^{-1}, T\tilde{D})$ is semistabilizable. Note that since $T$ is invertible, it follows that $\mathcal{R}(T\tilde{D}) = \mathcal{R}(\tilde{D})$ and

$$\mathcal{R}(\jmath\omega I_n - T\tilde{A}T^{-1}) = \mathcal{R}(\jmath\omega I_n - \tilde{A}), \quad \omega \in \mathbb{R}\backslash\{0\}.$$

Hence,

$$\dim \left[ \mathcal{R}(T\tilde{D}) + \mathcal{R}(\jmath\omega I_n - T\tilde{A}T^{-1}) \right] = \dim \left[ \mathcal{R}(\tilde{D}) + \mathcal{R}(\jmath\omega I_n - \tilde{A}) \right] = n,$$

which implies that $\mathrm{rank} \, [T\tilde{D} \quad \jmath\omega I_n - T\tilde{A}T^{-1}] = n$; that is,

$$\mathrm{rank} \begin{bmatrix} \hat{D}_1 & \jmath\omega I - \hat{A}_1 & -\hat{A}_{12} \\ 0 & 0 & \jmath\omega I - \hat{A}_2 \end{bmatrix} = n.$$

Thus, $\jmath\omega I - \hat{A}_2$ must be a full-rank matrix for $\omega \in \mathbb{R}\backslash\{0\}$, which implies that $\jmath\omega$ is not an eigenvalue of $\hat{A}_2$. Hence, the eigenvalue of $\hat{A}_2$ either is a real/complex eigenvalue with a negative real part or is zero, and if the eigenvalue is zero, then it is semisimple. Therefore, $\hat{A}_2$ is semistable. Since $\tilde{A}$ has the same set of eigenvalues as $\hat{A}_1$ and $\hat{A}_2$, it follows that $\tilde{A}$ is semistable; that is, $K \in \mathcal{S}$. $\qquad\square$

For the next result, define the set $\mathcal{K}_\alpha \triangleq \{K \in \mathcal{C}_s : J(K) \le \alpha\}$, and note that, by Proposition 12.8, $\mathcal{K}_\alpha \subset \mathcal{K}_s$, $\alpha \ge 0$, provided that $\tilde{R} > 0$.

**Proposition 12.9.** Assume that Assumptions 12.1 and 12.2 hold. Furthermore, assume that $R_1 > 0$ and $\mathcal{C}_s \ne \emptyset$. Then there exists $\alpha > 0$ such that $\mathcal{K}_\alpha$ is nonempty and compact relative to $\mathcal{C}_s$.

**Proof.** First, note that by assumption $\mathcal{C}_s$ is nonempty and from Lemma 12.6 that $\mathcal{C}_s \subseteq \mathcal{K}$. Thus, there exists $\alpha > 0$ such that $\mathcal{K}_\alpha$ is nonempty. Define the function $\hat{J} : \mathcal{C}_s \to \mathbb{R}$ by

$$\hat{J}(K) \triangleq \begin{cases} J(K), & K \in \mathcal{K}_{2\alpha}, \\ 2\alpha, & K \notin \mathcal{K}_{2\alpha}. \end{cases} \tag{12.58}$$

Since, by Proposition 12.8, $\mathcal{K}_{2\alpha} \subset \mathcal{S}$, it follows from Proposition 12.7 that $J(\cdot)$ is continuous on $\mathcal{K}_{2\alpha}$. However, $\hat{J}(K) = J(K)$, $K \in \mathcal{K}_{2\alpha}$, and, hence, it follows that $\hat{J}(\cdot)$ is continuous on $\mathcal{K}_{2\alpha}$. Next, note that $J(K) \to 2\alpha$ as $K \to \partial\mathcal{K}_{2\alpha}$. Hence, $\hat{J}(\cdot)$ is continuous on $\mathcal{C}_s$. Thus,

$$\hat{\mathcal{K}}_\alpha \triangleq \{K \in \mathcal{C}_s : \hat{J}(K) \le \alpha\}$$

is closed relative to $\mathcal{C}_s$. Now, $\hat{J}(K) \le \alpha$ implies that $\hat{J}(K) = J(K)$, and, hence, $\hat{\mathcal{K}}_\alpha = \mathcal{K}_\alpha$, which implies that $\mathcal{K}_\alpha$ is closed relative to $\mathcal{C}_s$.

Next, we show that $\mathcal{K}_\alpha$ is bounded relative to $\mathcal{C}_s$. Since $\mathcal{K}_\alpha \subset \mathcal{S}$ and $J(K) = \text{tr}(W + V)\tilde{R}$, it follows that $K \in \mathcal{K}_\alpha$ implies that

$$J(K) = \text{tr}(W + V)\tilde{R},$$

where $W + V \ge 0$ and

$$0 = \tilde{A}(W + V) + (W + V)\tilde{A}^{\text{T}} + \tilde{D}\tilde{D}^{\text{T}}.$$

Since $\text{tr}(W+V)\tilde{R} \le \alpha$ for $K \in \mathcal{K}_\alpha$, it follows from $\lambda_{\min}(W+V) = \sigma_{\min}(W+V)$, $\lambda_{\min}(R_2) = \sigma_{\min}(R_2)$, and Proposition 8.4.13 of [27, p. 471] that

$$\begin{aligned} \alpha &\ge \text{tr}(W + V)\tilde{R} \\ &\ge (\text{tr } \tilde{R})\sigma_{\min}(W + V) \\ &= (\text{tr } R_1 + \text{tr } K^{\text{T}} R_2 K)\sigma_{\min}(W + V) \\ &= (\text{tr } R_1 + \text{tr } KK^{\text{T}} R_2)\sigma_{\min}(W + V) \\ &\ge (\text{tr } R_1 + (\text{tr } KK^{\text{T}})\sigma_{\min}(R_2))\sigma_{\min}(W + V). \end{aligned}$$

Now, using Lemma 6.5 of [170], we obtain

$$(\text{tr } R_1 + (\text{tr } KK^{\text{T}})\sigma_{\min}(R_2))\sigma_{\min}(W + V)$$

$$\geq \frac{(\operatorname{tr} R_1 + \|K\|_{\mathrm{F}}^2 \sigma_{\min}(R_2))\sigma_{\min}(\tilde{D}\tilde{D}^{\mathrm{T}})}{2\sigma_{\max}(\tilde{A})},$$

where $\|K\|_{\mathrm{F}}$ denotes the Frobenius matrix norm of $K$.

Next, noting that

$$\sigma_{\max}(A + BK) \leq \sigma_{\max}(A) + \sigma_{\max}(B)\sigma_{\max}(K), \quad \sigma_{\max}(K) \leq \|K\|_{\mathrm{F}},$$

it follows that

$$\frac{(\operatorname{tr} R_1 + \|K\|_{\mathrm{F}}^2 \sigma_{\min}(R_2))\sigma_{\min}(\tilde{D}\tilde{D}^{\mathrm{T}})}{2\sigma_{\max}(\tilde{A})}$$

$$\geq \frac{[\sigma_{\min}(R_2)\|K\|_{\mathrm{F}}^2 + \operatorname{tr} R_1]\sigma_{\min}(\tilde{D}\tilde{D}^{\mathrm{T}})}{2[\sigma_{\max}(\tilde{A}) + \sigma_{\max}(B)\sigma_{\max}(\tilde{K})]}$$

$$\geq \frac{[\sigma_{\min}(R_2)\|K\|_{\mathrm{F}}^2 + \operatorname{tr} R_1]\sigma_{\min}(V)}{2[\sigma_{\max}(B)\|K\|_{\mathrm{F}} + \sigma_{\max}(\tilde{A})]}$$

$$= \frac{\sigma_{\min}(\tilde{D}\tilde{D}^{\mathrm{T}})\sigma_{\min}(R_2)\|K\|_{\mathrm{F}}^2 + \sigma_{\min}(\tilde{D}\tilde{D}^{\mathrm{T}})\operatorname{tr} R_1}{2[\sigma_{\max}(B)\|K\|_{\mathrm{F}} + \sigma_{\max}(\tilde{A})]}.$$

Combining the inequalities yields

$$\sigma_{\min}(\tilde{D}\tilde{D}^{\mathrm{T}})\sigma_{\min}(R_2)\|K\|_{\mathrm{F}}^2 + \sigma_{\min}(\tilde{D}\tilde{D}^{\mathrm{T}})\operatorname{tr} R_1$$
$$\leq 2\alpha\sigma_{\max}(B)\|K\|_{\mathrm{F}} + 2\alpha\sigma_{\max}(A).$$

Finally, define

$$\beta_1 \triangleq \frac{2\alpha\sigma_{\max}(B)}{\sigma_{\min}(\tilde{D}\tilde{D}^{\mathrm{T}})\sigma_{\min}(R_2)}, \quad \beta_2 \triangleq \frac{\sigma_{\min}(\tilde{D}\tilde{D}^{\mathrm{T}})\operatorname{tr} R_1 - 2\alpha\sigma_{\max}(A)}{\sigma_{\min}(\tilde{D}\tilde{D}^{\mathrm{T}})\sigma_{\min}(R_2)},$$

and choose $\alpha > 0$ to be sufficiently large so that $\beta_1^2 \geq 4\beta_2$. Then it follows that

$$\|K\|_{\mathrm{F}} \leq \frac{1}{2}\beta_1 + \frac{1}{2}\sqrt{\beta_1^2 - 4\beta_2} < \infty,$$

which establishes the boundedness of $\mathcal{K}_\alpha$ relative to $\mathcal{C}_s$. Hence, $\mathcal{K}_\alpha$ is compact relative to $\mathcal{C}_s$. $\square$

Finally, we present an existence result for optimality of a class of Gaussian linear-quadratic consensus problems.

**Theorem 12.5.** Assume that Assumptions 12.1 and 12.2 hold, and assume that $R_1 > 0$ and $\mathcal{K}_s \cap \mathcal{C}_s \neq \emptyset$. Then there exists $K_* \in \mathcal{K}_s \cap \mathcal{C}_s$ such that $J(K_*) \leq J(K)$, $K \in \mathcal{C}_s$.

**Proof.** Since $\mathcal{K}_s \cap \mathcal{C}_s \neq \varnothing$, let $\hat{K} \in \mathcal{K}_s \cap \mathcal{C}_s \neq \varnothing$ and define $\hat{\alpha} \triangleq J(\hat{K})$. Then it follows from $\hat{K} \in \mathcal{K}_{\hat{\alpha}}$ that $\mathcal{K}_{\hat{\alpha}} \neq \varnothing$ and, by Proposition 12.9, $\mathcal{K}_{\hat{\alpha}}$ is compact relative to $\mathcal{C}_s$. Since $\mathcal{K}_{\hat{\alpha}} \subset \mathcal{K}_s \cap \mathcal{C}_s$ by Proposition 12.8, it follows from Proposition 12.7 that $J(\cdot)$ is continuous on $\mathcal{K}_{\hat{\alpha}}$. Hence, there exists $K_* \in \mathcal{K}_{\hat{\alpha}}$ such that $J(K_*) \leq J(K)$, $K \in \mathcal{K}_{\hat{\alpha}}$. Furthermore, since $K \notin \mathcal{K}_{\hat{\alpha}}$ implies that $J(K) > \hat{\alpha}$, it follows that $J(K_*) \leq J(K)$, $K \in \mathcal{C}_s$. $\qquad\square$

## 12.6 A Numerical Algorithm for the Gaussian Linear-Quadratic Consensus Problem

In this section, we develop a globally convergent algorithm based on swarm optimization to efficiently solve (12.45). We choose a swarm optimization technique due to its similar collective dynamics with information state equipartitioning and its merit of not requiring properties such as linearity, differentiability, convexity, separability, or nonexistence of constraints for optimization problems. Furthermore, its efficiency, fewer parameters to adjust, and ability to avoid local maxima make this technique efficacious for solving a large number of nonlinear, nonconvex constrained optimization problems. The efficiency factor is especially important for our problem due to the large parameter search space.

Although many swarm optimization algorithms, such as particle swarm optimization (PSO) [184], ant colony optimization (ACO) [83], and honeybee mating optimization (HBMO) [3], have been developed in the literature, the lack of global convergence guarantees for these algorithms hampers them from any further application to modern cybercritical infrastructure and cyberphysical networks, which require high-fidelity optimizing results for planning, operation, and maintenance. Here, we propose a new PSO-based iterative algorithm and give a theoretical convergence analysis to ensure its global convergence when implemented for a large class of nonlinear, nonconvex constrained optimization problems.

Recall that PSO can be described in vector form as [184]

$$\mathbf{v}_k(\kappa + 1) = a\mathbf{v}_k(\kappa) + b_1 r_1(\mathbf{p}_{1,k} - \mathbf{x}_k(\kappa)) + b_2 r_2(\mathbf{p}_2 - \mathbf{x}_k(\kappa)), \qquad (12.59)$$

$$\mathbf{x}_k(\kappa + 1) = \mathbf{x}_k(\kappa) + \mathbf{v}_k(\kappa + 1), \quad k = 1, \ldots, q, \quad \kappa \in \overline{\mathbb{Z}}_+, \qquad (12.60)$$

where $\mathbf{v}_k(\kappa) \in \mathbb{R}^n$ and $\mathbf{x}_k(\kappa) \in \mathbb{R}^n$ are the velocity and position of particle $k$ at iteration $\kappa$, respectively; $\mathbf{p}_{1,k} \in \mathbb{R}^n$ is the position of the previous best value that particle $k$ obtained at iterate $\kappa$; $\mathbf{p}_2 \in \mathbb{R}^n$ is the position of the global best value that the swarm of particles can achieve at iterate $\kappa$; $a$, $b_1$, and $b_2$ are scalar weight coefficients; and $r_1$ and $r_2$ are two scalar random coefficients which are selected over a uniform distribution in the range $[0, 1]$.

The initial velocity and position for each particle are randomly selected in a prescribed rectangular search region.

For every iteration $\kappa$, the velocity of each particle is updated by the interaction of the current velocity, the previous best position $\mathbf{p}_{1,k}$, and the global position $\mathbf{p}_2$. The position of each particle is updated by using the current position and the newly updated velocity. For every iteration, the previous best position $\mathbf{p}_{1,k}$ and the global best position $\mathbf{p}_2$ are updated using a given objective function.

Here we use an improved form of PSO to solve our constrained optimization problem. The constraints are handled by the methods presented in [79, 202]. More specifically, three rules for updating the global best position $\mathbf{p}_2$ and previous best position $\mathbf{p}_{1,k}$ are as follows.

1. If $\mathbf{x}_1$ and $\mathbf{x}_2$ are both feasible solutions, then choose the one which obtains the best value of the objective function.

2. If $\mathbf{x}_1$ and $\mathbf{x}_2$ are not feasible, then choose the one with the lowest sum of the constraint violation

$$s(\mathbf{x}) = \sum_{i=1}^{m} \max(0, g_i(\mathbf{x})) + \sum_{j=1}^{p} \max(0, (|h_j(\mathbf{x})| - \varepsilon)),$$

where $g_i(\mathbf{x})$ is the $i$th inequality constraint and $h_j(\mathbf{x})$ is the $j$th equality constraint.

3. If one is a feasible solution and the other is an infeasible solution, then choose the feasible solution.

The update formulas given by (12.59) and (12.60) in PSO do not have any guarantee of global convergence for a given optimization problem $\min_{\mathbf{x} \in \mathbb{R}^n} f(\mathbf{x})$. Motivated by semistability analysis of discrete-time switched linear systems [277] and quantized PSO algorithms [213], as well as multiagent coordination optimization [329], we propose the following parametrically quantized particle swarm optimization (PQPSO) algorithm to solve this convergence problem:

$$\mathbf{v}_k(\kappa + 1) = a(\kappa)\mathbf{v}_k(\kappa) + b_1(\kappa)(\mathbf{x}_j(\kappa) - \mathbf{x}_k(\kappa)) + b_2(\kappa)(\mathbf{p}(\kappa) - \mathbf{x}_k(\kappa)),$$
$$\kappa \in \overline{\mathbb{Z}}_+, \qquad (12.61)$$
$$\mathbf{x}_k(\kappa + 1) = \mathbf{x}_k(\kappa) + \mathbf{v}_k(\kappa + 1), \qquad (12.62)$$
$$\mathbf{p}(\kappa + 1) = \begin{cases} \mathbf{p}(\kappa) + b_2(\kappa)(\mathbf{x}_j(\kappa) - \mathbf{p}(\kappa)) & \text{if } \mathbf{p}(\kappa) \notin \mathcal{Z}, \\ \mathbf{x}_j(\kappa) & \text{if } \mathbf{p}(\kappa) \in \mathcal{Z}, \end{cases} \qquad (12.63)$$

where $k = 1, \ldots, q$; the coefficients $a(\kappa)$, $b_1(\kappa)$, and $b_2(\kappa)$ are randomly selected from the finite discrete, quantized sets $\Omega_0 \subseteq (0, \tau_0]$, $\Omega_1 \subseteq (0, \tau_1]$, and $\Omega_2 \subseteq (0, \tau_2]$ at each time step $\kappa$, respectively, where $\tau_0, \tau_1, \tau_2 > 0$ are given; $\mathcal{Z} = \{ \mathbf{y} \in \mathbb{R}^n : f(\mathbf{x}_j) < f(\mathbf{y}) \}$; and $\mathbf{x}_j = \arg \min_{1 \leq k \leq q} f(\mathbf{x}_k)$.

Here we have added the gradient-based adaptation (12.63) for $\mathbf{p}(\kappa)$ to improve the convergence performance of the PSO. The proposed PQPSO algorithm has many potential application areas in power systems [242, 330], electromagnetics [260], resource allocation [171], network security [292, 293], and model predictive control [223, 309], to cite but a few examples.

The basic idea in proving the convergence of the above PQPSO algorithm is to view (12.61)–(12.63) as a discrete linear inclusion

$$Z(\kappa + 1) \in \mathbb{P} Z(\kappa), \quad \kappa \in \overline{\mathbb{Z}}_+, \tag{12.64}$$

where $\mathbb{P}$ is a finite set of matrices and $\mathbb{P}x = \{ Px : P \in \mathbb{P} \}$, and address the semistability property of (12.64) to obtain its convergence information. The following theorem presents a global convergence of the iterative process in PQPSO.

**Theorem 12.6.** Consider the iterative process given by (12.61)–(12.63). Define the (possibly infinite) matrix functions $A^{[j]}(\kappa)$, $A_c^{[j]}(\kappa)$, and $B^{[j]}(\kappa)$ for $j = 1, \ldots, q$ and $\kappa \in \overline{\mathbb{Z}}_+$ as

$$A^{[j]}(\kappa) = \begin{bmatrix} 0_{nq \times nq} & I_{nq} & 0_{nq \times n} \\ b_1(\kappa) W^{[j]} - [b_1(\kappa) + b_2(\kappa)] I_{nq} & -[1-a(\kappa)] I_{nq} & b_2(\kappa) \mathbf{e}_{q \times 1} \otimes I_n \\ b_2(\kappa) E_{n \times nq}^{[j]} & 0_{n \times nq} & -b_2(\kappa) I_n \end{bmatrix}, \tag{12.65}$$

$$A_c^{[j]}(\kappa) = \begin{bmatrix} b_1(\kappa) W^{[j]} - [b_1(\kappa) + b_2(\kappa)] I_{nq} & -[1-a(\kappa)] I_{nq} & b_2(\kappa) \mathbf{e}_{q \times 1} \otimes I_n \\ 0_{nq \times nq} & 0_{nq \times nq} & 0_{nq \times n} \\ 0_{n \times nq} & 0_{n \times nq} & 0_{n \times n} \end{bmatrix}, \tag{12.66}$$

$$B^{[j]}(\kappa) = \begin{bmatrix} 0_{nq \times nq} & I_{nq} & 0_{nq \times n} \\ b_1(\kappa) W^{[j]} - [b_1(\kappa) + b_2(\kappa)] I_{nq} & -[1-a(\kappa)] I_{nq} & b_2(\kappa) \mathbf{e}_{q \times 1} \otimes I_n \\ E_{n \times nq}^{[j]} & 0_{n \times nq} & -I_n \end{bmatrix}, \tag{12.67}$$

where $a(\kappa) \in \Omega_0$, $b_1(\kappa) \in \Omega_1$, and $b_2(\kappa) \in \Omega_2$, and $E_{n \times nq}^{[j]} \in \mathbb{R}^{n \times nq}$ denotes a block matrix whose $j$th block column is $I_n$ and the rest of the block entries are all zero matrices; that is,

$$E_{n \times nq}^{[j]} = [0_{n \times n}, \ldots, 0_{n \times n}, I_n, 0_{n \times n}, \ldots, 0_{n \times n}], \quad j = 1, \ldots, q.$$

Also, $W^{[j]} = (\mathbf{e}_{q \times 1} \otimes I_n) E_{n \times nq}^{[j]}$ for every $j = 1, \ldots, q$, and $\mathbf{e}_{m \times n}$ denotes the $m \times n$ matrix whose entries are all ones. Assume that, for every $j = 1, \ldots, q$ and $\kappa \in \overline{\mathbb{Z}}_+$, the following statements hold.

(i)

$$a(\kappa) < \min\left\{1 + b_1(\kappa) + b_2(\kappa), 2 + b_1(\kappa) + \frac{b_1(\kappa)}{b_2(\kappa)}\right\}$$

and

$$
\begin{aligned}
b_1(\kappa)b_2(\kappa) < \min\Big\{ &(b_1(\kappa) + 2b_2(\kappa) + 1 \\
& - a(\kappa))(b_1(\kappa)b_2(\kappa) + b_1(\kappa) + 2b_2(\kappa) - a(\kappa)b_2(\kappa)), \\
& (b_1(\kappa) + 2b_2(\kappa) + 1 - a(\kappa))(b_1(\kappa)b_2(\kappa) \\
& + b_1(\kappa) + 2b_2(\kappa) + 2b_2^2(\kappa) \\
& - a(\kappa)b_2(\kappa)) - 2b_2^2(\kappa)\Big\}.
\end{aligned}
$$

(ii) $b_2(\kappa) < 2$ and $\lambda_i + \bar{\lambda}_i + \lambda_i\bar{\lambda}_i < 0$ for every $i = 1,\ldots,14$, where $\bar{\lambda}$ denotes the complex conjugate of $\lambda \in \mathbb{C}$; $\lambda_1, \lambda_2, \lambda_3$ satisfy

$$
\begin{aligned}
\lambda^3 &+ (b_1(\kappa) + 2b_2(\kappa) + 1 - a(\kappa))\lambda^2 + (b_1(\kappa)b_2(\kappa) \\
&+ b_1(\kappa) + 2b_2(\kappa) - a(\kappa)b_2(\kappa))\lambda + b_1(\kappa)b_2(\kappa) = 0;
\end{aligned}
$$

$\lambda_4, \lambda_5\, \lambda_6$ satisfy

$$
\begin{aligned}
\lambda^3 &+ (b_1(\kappa) + 2b_2(\kappa) + 1 - a(\kappa))\lambda^2 + (b_1(\kappa)b_2(\kappa) + b_1(\kappa) \\
&+ 2b_2(\kappa) + 2b_2^2(\kappa) - a(\kappa)b_2(\kappa))\lambda + b_1(\kappa)b_2(\kappa) + 2b_2^2(\kappa) = 0;
\end{aligned}
$$

$\lambda_7, \lambda_8$ are given by

$$
\begin{aligned}
\lambda_{7,8} = &-\frac{b_1(\kappa) + b_2(\kappa) + 1 - a(\kappa)}{2} \\
&\pm \left(\frac{(b_1(\kappa) + b_2(\kappa) + 1 - a(\kappa))^2}{4} - (b_1(\kappa) + b_2(\kappa))\right)^{1/2};
\end{aligned}
$$

$\lambda_9, \lambda_{10}$ are given by

$$
\begin{aligned}
\lambda_{9,10} = &-\frac{b_1(\kappa) + b_2(\kappa) + 1 - a(\kappa)}{2} \\
&\pm \left(\frac{(b_1(\kappa) + b_2(\kappa) + 1 - a(\kappa))^2}{4} - b_1(\kappa)\right)^{1/2};
\end{aligned}
$$

$\lambda_{11}, \lambda_{12}$ are given by

$$
\begin{aligned}
\lambda_{11,12} = &-\frac{b_1(\kappa) + b_2(\kappa) + 1 - a(\kappa)}{2} \\
&\pm \left(\frac{(b_1(\kappa) + b_2(\kappa) + 1 - a(\kappa))^2}{4} - (b_1(\kappa) + 2b_2(\kappa))\right)^{1/2};
\end{aligned}
$$

and $\lambda_{13}$, $\lambda_{14}$ are given by

$$\lambda_{13,14} = -\frac{b_1(\kappa) + b_2(\kappa) + 1 - a(\kappa)}{2}$$
$$\pm \left( \frac{(b_1(\kappa) + b_2(\kappa) + 1 - a(\kappa))^2}{4} - (b_1(\kappa) + b_2(\kappa)) \right)^{1/2}.$$

(iii) $\|I_{2nq+n} + A^{[j]}(\kappa) + A_c^{[j]}(\kappa)\| \le 1$ and $\|I_{2nq+n} + B^{[j]}(\kappa) + A_c^{[j]}(t)\| \le 1$.

(iv)

$$\mathcal{N}((A^{[j]}(\kappa) + A_c^{[j]}(\kappa))^{\mathrm{T}}(A^{[j]}(\kappa) + A_c^{[j]}(\kappa))$$
$$+ (A^{[j]}(\kappa) + A_c^{[j]}(\kappa))^{\mathrm{T}} + A^{[j]}(\kappa) + A_c^{[j]}(\kappa))$$
$$= \mathcal{N}((A^{[j]}(\kappa) + A_c^{[j]}(\kappa))^{\mathrm{T}}(A^{[j]}(\kappa) + A_c^{[j]}(\kappa)) + (A^{[j]}(\kappa) + A_c^{[j]}(\kappa))^2)$$

and

$$\mathcal{N}((B^{[j]}(\kappa) + A_c^{[j]}(\kappa))^{\mathrm{T}}(B^{[j]}(\kappa) + A_c^{[j]}(\kappa))$$
$$+ (B^{[j]}(\kappa) + A_c^{[j]}(\kappa))^{\mathrm{T}} + B^{[j]}(\kappa) + A_c^{[j]}(\kappa))$$
$$= \mathcal{N}((B^{[j]}(\kappa) + A_c^{[j]}(\kappa))^{\mathrm{T}}(B^{[j]}(\kappa) + A_c^{[j]}(\kappa)) + (B^{[j]}(\kappa) + A_c^{[j]}(\kappa))^2).$$

Then $\mathbf{x}_k(\kappa) \to \mathbf{p}^{\dagger}$ as $\mathbf{v}_k(\kappa) \to 0$ and $\mathbf{p}(\kappa) \to \mathbf{p}^{\dagger}$ as $\kappa \to \infty$ for every $\mathbf{x}_k(0) \in \mathbb{R}^n$, $\mathbf{v}_k(0) \in \mathbb{R}^n$, and $\mathbf{p}(0) \in \mathbb{R}^n$ and every $k = 1, \ldots, q$, where $\mathbf{p}^{\dagger} \in \mathbb{R}^n$ is some constant vector.

**Proof.** The proof involves several technical lemmas and is given in [174]. $\qquad \square$

Theorem 12.6 also holds for the averaged model of (12.61)–(12.63). Specifically, if we replace $a(\kappa)$, $b_1(\kappa)$, and $b_2(\kappa)$ with $\mathbb{E}[a(\kappa)]$, $\mathbb{E}[b_1(\kappa)]$, and $\mathbb{E}[b_2(\kappa)]$ in Theorem 12.6 (including in Assumptions (i)–(iv)), then the conclusion of Theorem 12.6 becomes $\mathbb{E}[\mathbf{x}_k(\kappa)] \to \mathbf{x}^{\dagger}$ as $\mathbb{E}[\mathbf{v}_k(\kappa)] \to 0$ and $\mathbb{E}[\mathbf{p}(\kappa)] \to \mathbf{x}^{\dagger}$ as $\kappa \to \infty$ for every $\mathbf{x}_k(0) \in \mathbb{R}^n$, $\mathbf{v}_k(0) \in \mathbb{R}^n$, and $\mathbf{p}(0) \in \mathbb{R}^n$ and every $k = 1, \ldots, q$, where $\mathbf{x}^{\dagger} \in \mathbb{R}^n$ is some constant vector.

It is important to note that the authors in [66] and [302] prove that each individual particle converges to a weighted average of its personal best and neighborhood best positions. However, these results differ from Theorem 12.6 in that they need to assume that the initial positions of the particles are within a small neighborhood of their best position. In other words, the convergence results in [66] and [302] only hold locally in the search space. In contrast, the main contribution of Theorem 12.6 is the guarantee of global convergence of PQPSO regardless of the locations of its initial conditions in the search space.

**Example 12.8.** We apply the proposed numerical algorithm to the network system of a 20-node undirected graph topology given by Figure 12.2, adopted from [236]. Hence, $n = 20$. Let $\tau_0 = \tau_1 = \tau_2 = 1$, and let $\Omega_0 = \Omega_1 = \Omega_2 = \{0.01k : k = 1, \ldots, 100\}$. For our numerical simulation, we choose $A = 0$ and $B = I_n$ and let

$$K = \begin{bmatrix} -\sum_{j=1,j\neq 1}^{n} k_{1j} & k_{12} & \cdots & k_{1n} \\ k_{21} & -\sum_{j=1,j\neq 2}^{n} k_{2j} & \cdots & k_{2n} \\ \vdots & \vdots & \ddots & \vdots \\ k_{n1} & k_{n2} & \cdots & -\sum_{j=1,j\neq n}^{n} k_{nj} \end{bmatrix},$$

where $k_{ij} \in \mathbb{R}$, $i, j = 1, \ldots, n$, $i \neq j$, so that $K\mathbf{e} = 0$, $D_1 = K$, $D_2 = 0$, and $V = 0$. Hence, $\tilde{A} = K$, $\tilde{D} = K$, and $(K, K)$ is 0-semicontrollable. Using Theorem 12.4, we can formulate the semistable Gaussian linear-quadratic consensus problem into a constrained optimization problem for $k = 0$ as

$$\min F(K) = \min \mathrm{tr}(S(E_1^{\mathrm{T}} E_1 + K^{\mathrm{T}} E_2^{\mathrm{T}} E_2 K))$$

subject to

$$h_1(K) = \mathrm{rank}(K) - (n-1) = 0, \quad h_2(K) = p = 0,$$
$$h_3(K) = KS + SK^{\mathrm{T}} + KK^{\mathrm{T}} = 0,$$

where $p$ is the indicator of the Cholesky decomposition command $\mathtt{chol}(S) = [C, p]$ in MATLAB used to test the positive definiteness of matrix $S$, $C$ is an upper triangular matrix so that $C^{\mathrm{T}} C = S$, $p = 0$ if $S$ is positive definite, and $p > 0$ if $S$ is not positive definite.

The PQPSO-based numerical algorithm is described in Algorithm 12.1 with $\theta = 10$. The matrices $E_1$ and $E_2$ are given by

$$E_1 = \begin{bmatrix} 1 & 1 & 1 & 1 & 1 & 1 & 1 & 1 & 1 & 1 & 1 & 1 & 1 & 1 & 1 & 1 & 1 & 1 & 1 & 1 \\ -1 & -1 & -1 & -1 & -1 & -1 & -1 & -1 & -1 & -1 & -1 & -1 & -1 & -1 & -1 & -1 & -1 & -1 & -1 & -1 \end{bmatrix}, \tag{12.68}$$

$$E_2 = \begin{bmatrix} 1 & 1 & 1 & 1 & 1 & 1 & 1 & 1 & 1 & 1 & 1 & 1 & 1 & 1 & 1 & 1 & 1 & 1 & 1 & 1 \\ 1 & 1 & 1 & 1 & 1 & 1 & 1 & 1 & 1 & 1 & 1 & 1 & 1 & 1 & 1 & 1 & 1 & 1 & 1 & 1 \end{bmatrix}, \tag{12.69}$$

and the initial condition $x_0$ is given by

$$x_0 = \begin{bmatrix} -1 & 20 & 3 & 2 & 5 & 30 & 10 & 12 & 13 & -5 & 22 & 33 & -3 & 18 & 19 & 15 & -3 & 25 & 24 & 27 \end{bmatrix}^{\mathrm{T}}. \tag{12.70}$$

The time history of trajectories of the closed-loop network system is shown in Figure 12.3.

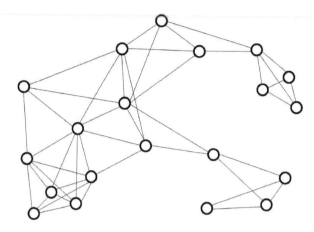

Figure 12.2 Graph topology of the 20-node network system.

---

### Algorithm 12.1

Initialize matrix particles $K$ in the search space with each entry in $K$ being a uniform distribution $U(-\theta, \theta)$, where $\theta > 0$ is a given constant.

**repeat**

    Solve the matrix equation $h_3(K) = 0$ for $S$.

    Use the constrained PQPSO to optimize the problem $\min F(K)$ subject to the constraints $h_1(K) = 0$ and $h_2(K) = 0$.

**until** All of the particles converge to the same position or an exit condition is satisfied.

---

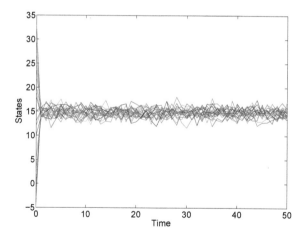

Figure 12.3 Time response of the 20-node network system.

Next, we compare the PSO algorithm with the PQPSO algorithm. The $p$-value, which is the probability of obtaining a test statistic result, that is, at least as extreme as the one that was actually observed in the analysis of variance (ANOVA) test, is 0.5503. This implies that there is no significant difference between the means of the simulation results from both the PSO and the PQPSO as shown in Table 12.1, where the notation $\alpha E + \beta$ denotes $\alpha \times 10^{\beta}$. Since we can guarantee global convergence of the PQPSO, we use the PQPSO algorithm instead of the PSO algorithm to solve the proposed optimization problem.

Finally, we compare our results with the PSO algorithm, the PQPSO algorithm, and three other PSO algorithm variants, namely, the IWPSO [280], CPSO [212], and CLPSO [201] algorithms. We ran the algorithms 20 times and obtained the average for each variant. As can be seen from Table 12.1, the average of the PQPSO algorithm outperforms those of the IWPSO and CLPSO algorithms but does not perform as well as those of the PSO and CPSO algorithms. However, our proposed algorithm guarantees global convergence. $\triangle$

Table 12.1 Comparison of the PSO, PQPSO, IWPSO, CPSO, and CLPSO algorithms.

| Iteration | PSO | PQPSO | IWPSO | CPSO | CLPSO |
|---|---|---|---|---|---|
| 1 | 1.50E+05 | 1.79E+05 | 1.24E+05 | 9.11E+04 | 1.58E+05 |
| 2 | 2.03E+05 | 1.49E+05 | 4.09E+05 | 1.53E+05 | 3.14E+05 |
| 3 | 1.23E+05 | 1.61E+05 | 4.90E+05 | 1.50E+05 | 3.23E+05 |
| 4 | 1.09E+05 | 3.71E+05 | 4.77E+05 | 2.05E+05 | 4.11E+05 |
| 5 | 9.99E+04 | 8.09E+04 | 5.97E+05 | 1.36E+05 | 3.71E+05 |
| 6 | 2.11E+05 | 6.16E+04 | 4.91E+05 | 1.28E+05 | 4.41E+05 |
| 7 | 1.77E+05 | 1.16E+05 | 5.42E+05 | 1.20E+05 | 3.57E+05 |
| 8 | 1.11E+05 | 1.08E+05 | 6.79E+05 | 1.23E+05 | 4.11E+05 |
| 9 | 1.54E+05 | 1.00E+05 | 3.83E+05 | 9.42E+04 | 5.74E+05 |
| 10 | 1.16E+05 | 1.51E+05 | 4.13E+05 | 1.07E+05 | 2.93E+05 |
| 11 | 1.99E+05 | 1.09E+05 | 3.87E+05 | 1.42E+05 | 2.14E+05 |
| 12 | 1.20E+05 | 8.77E+04 | 3.28E+05 | 1.27E+05 | 3.99E+05 |
| 13 | 1.19E+05 | 4.74E+05 | 4.91E+05 | 6.25E+04 | 5.94E+05 |
| 14 | 1.57E+05 | 1.18E+05 | 4.39E+05 | 1.64E+05 | 6.04E+05 |
| 15 | 1.55E+05 | 1.43E+05 | 3.27E+05 | 1.06E+05 | 4.89E+05 |
| 16 | 2.35E+05 | 1.15E+05 | 5.21E+05 | 1.41E+05 | 3.40E+05 |
| 17 | 1.25E+05 | 2.16E+05 | 4.85E+05 | 1.57E+05 | 3.58E+05 |
| 18 | 1.10E+05 | 1.08E+05 | 4.66E+05 | 6.66E+04 | 4.08E+05 |
| 19 | 1.40E+05 | 3.15E+05 | 4.21E+05 | 9.35E+04 | 3.74E+05 |
| 20 | 1.42E+05 | 9.70E+04 | 4.05E+05 | 1.58E+05 | 6.77E+05 |
| Average | 1.48E+05 | 1.63E+05 | 4.44E+05 | 1.27E+05 | 4.06E+05 |

*Chapter Thirteen*

---

# Mitigating the Effects of Sensor Uncertainties in Network Systems

## 13.1 Introduction

Networked multiagent systems (e.g., communication networks, power systems, and process control systems) consist of interacting agents that locally exchange information, energy, or matter [224, 236, 257, 275]. These systems require a resilient distributed control system design architecture for providing high system performance, reliability, and operation in the presence of system uncertainties [76, 150, 300, 321, 323]. An important class of such system uncertainties that can significantly deteriorate achievable closed-loop dynamical system performance is sensor uncertainties, which can arise due to low sensor quality, sensor failure, sensor bias, or detrimental environmental conditions [31, 108, 303, 331]. If relatively cheap sensor suites are used for low-cost, small-scale unmanned vehicle applications, then this can result in inaccurate sensor measurements. Alternatively, sensor measurements can be corrupted by malicious attacks if these dynamical systems are controlled through large-scale, multilayered communication networks, as in the case of cyberphysical systems.

Early approaches that deal with sensor uncertainties focus on classical fault detection, isolation, and recovery schemes (see, e.g., [40, 222]). In these approaches, sensor measurements are compared with an analytical model of the dynamical system by forming a residual signal and analyzing this signal to determine if a fault has occurred. However, in practice it is difficult to identify a single residual signal per failure mode, and, as the number of failure modes increases, this becomes prohibitive. In addition, a common underlying assumption of classical fault detection, isolation, and recovery schemes is that all dynamical system signals remain bounded during the fault detection process, which may not always be a valid assumption.

More recently, the authors of [244] considered the fundamental limitations of attack detection and identification methods for linear systems.

However, their approach is not only computationally expensive but also not linked to the controller design. In [92], adversarial attacks on actuators and sensors are modeled as exogenous disturbances. However, the presented control methodology cannot address situations where more than half of the sensors are compromised and the set of attacked nodes change over time. Finally, the authors in [268] analyze a case where the interactions between networked agents are corrupted by exogenous disturbances. However, their approach is limited to agents having scalar dynamics and is not linked to the controller design in order to mitigate the effect of such sensor uncertainties.

In this chapter, we present a distributed *adaptive control* architecture for networked multiagent systems with undirected communication graph topologies to mitigate the effect of sensor uncertainties. Specifically, we consider multiagent systems having identical high-order, linear dynamics with agent interactions corrupted by unknown exogenous disturbances. We show that the proposed adaptive control architecture guarantees asymptotic stability of the closed-loop dynamical system when the exogenous disturbances are time invariant and uniform ultimate boundedness when the exogenous disturbances are time varying.

## 13.2 Mathematical Preliminaries and Problem Formulation

In this chapter, we use the notation established in Chapter 5. Specifically, we write $\mathcal{A}(\mathfrak{G}) \in \mathbb{R}^{N \times N}$ for the adjacency matrix of a graph $\mathfrak{G}$ defined by (5.1) and $\mathcal{B}(\mathfrak{G}) \in \mathbb{R}^{N \times M}$ for the (node-edge) *incidence matrix* of a graph $\mathfrak{G}$ defined by

$$[\mathcal{B}(\mathfrak{G})]_{ij} \triangleq \begin{cases} 1 & \text{if node } i \text{ is the head of edge } j, \\ -1 & \text{if node } i \text{ is the tail of edge } j, \\ 0, & \text{otherwise,} \end{cases} \tag{13.1}$$

where $M$ is the number of edges, $i$ is an index for the node set, and $j$ is an index for the edge set.

Recall that the graph Laplacian matrix, denoted by $\mathcal{L}(\mathfrak{G}) \in \mathbb{N}^N$, is defined by $\mathcal{L}(\mathfrak{G}) \triangleq \mathcal{D}(\mathfrak{G}) - \mathcal{A}(\mathfrak{G})$ or, equivalently, $\mathcal{L}(\mathfrak{G}) = \mathcal{B}(\mathfrak{G})\mathcal{B}(\mathfrak{G})^{\mathrm{T}}$, and the spectrum of the graph Laplacian of a connected undirected graph $\mathfrak{G}$ can be ordered as $0 = \lambda_1(\mathcal{L}(\mathfrak{G})) < \lambda_2(\mathcal{L}(\mathfrak{G})) \leq \cdots \leq \lambda_N(\mathcal{L}(\mathfrak{G}))$, with $\mathbf{e}_N$ being the eigenvector corresponding to the zero eigenvalue $\lambda_1(\mathcal{L}(\mathfrak{G}))$, $\mathcal{L}(\mathfrak{G})\mathbf{e}_N = \mathbf{0}_N$, and $e^{\mathcal{L}(\mathfrak{G})}\mathbf{e}_N = \mathbf{e}_N$. Finally, we partition the incidence matrix $\mathcal{B}(\mathfrak{G}) = \left[\mathcal{B}_{\mathrm{L}}(\mathfrak{G})^{\mathrm{T}}, \mathcal{B}_{\mathrm{F}}(\mathfrak{G})^{\mathrm{T}}\right]^{\mathrm{T}}$, where $\mathcal{B}_{\mathrm{L}}(\mathfrak{G}) \in \mathbb{R}^{N_{\mathrm{L}} \times M}$; $\mathcal{B}_{\mathrm{F}}(\mathfrak{G}) \in \mathbb{R}^{N_{\mathrm{F}} \times M}$; $N_{\mathrm{L}} + N_{\mathrm{F}} = N$; and $N_{\mathrm{L}}$ and $N_{\mathrm{F}}$, respectively, denote the cardinalities of the leader and follower groups [224].

Furthermore, without loss of generality, we assume that the leader agents are indexed first and the follower agents are indexed last in the graph $\mathfrak{G}$ so that $\mathcal{L}(\mathfrak{G}) = \mathcal{B}(\mathfrak{G})\mathcal{B}(\mathfrak{G})^{\mathrm{T}}$ is given by

$$\mathcal{L}(\mathfrak{G}) = \begin{bmatrix} L(\mathfrak{G}) & G^{\mathrm{T}}(\mathfrak{G}) \\ G(\mathfrak{G}) & F(\mathfrak{G}) \end{bmatrix}, \qquad (13.2)$$

where

$$L(\mathfrak{G}) \triangleq \mathcal{B}_{\mathrm{L}}(\mathfrak{G})\mathcal{B}_{\mathrm{L}}^{\mathrm{T}}(\mathfrak{G}), \quad G(\mathfrak{G}) \triangleq \mathcal{B}_{\mathrm{F}}(\mathfrak{G})\mathcal{B}_{\mathrm{L}}^{\mathrm{T}}(\mathfrak{G}), \quad F(\mathfrak{G}) \triangleq \mathcal{B}_{\mathrm{F}}(\mathfrak{G})\mathcal{B}_{\mathrm{F}}^{\mathrm{T}}(\mathfrak{G}).$$

Note that $F(\mathfrak{G}) \in \mathbb{P}^{N_{\mathrm{F}}}$ for a connected, undirected graph $\mathfrak{G}$ and satisfies $F(\mathfrak{G})\mathbf{e}_{N_{\mathrm{F}}} = -G(\mathfrak{G})\mathbf{e}_{N_{\mathrm{L}}}$. This implies that each row sum of $-F^{-1}(\mathfrak{G})G(\mathfrak{G})$ is equal to one.

Consider the networked multiagent system consisting of $N$ agents with the dynamics of agent $i$, $i \in \{1, \ldots, N\}$, given by

$$\dot{x}_i(t) = Ax_i(t) + Bu_i(t), \quad x_i(0) = x_{i0}, \quad t \geq 0, \qquad (13.3)$$

where $x_i(t) \in \mathbb{R}^n$, $t \geq 0$, is the state vector of agent $i$; $u_i(t) \in \mathbb{R}^m$, $t \geq 0$, is the control input of agent $i$; and $A \in \mathbb{R}^{n \times n}$ and $B \in \mathbb{R}^{n \times m}$ are system matrices. We assume that the pair $(A, B)$ is controllable and the control input $u_i(\cdot)$, $i = 1, \ldots, N$, is restricted to the class of admissible controls consisting of measurable functions such that $u_i(t) \in \mathbb{R}^m$, $t \geq 0$. In addition, we assume that the agents can measure their own state and can locally exchange information via a connected, undirected graph $\mathfrak{G}$ with nodes and edges representing agents and interagent information exchange links, respectively, resulting in a static network topology; that is, the time evolution of the agents does not result in edges appearing or disappearing in the network.

Here, we consider a networked multiagent system, where the agents lie on an agent layer and their local controllers lie on a control layer, as depicted in Figure 13.1. Furthermore, we assume that the graph for the controller structure is the same as the graph for the agents' communication. Specifically, agent $i \in \{1, \ldots, N\}$ sends its state measurement to its corresponding local controller at a given control layer, and this controller sends its control input to agent $i$ lying on the agent layer.

In addition, we assume that the compromised state measurement

$$\tilde{x}_i(t) = x_i(t) + \delta_i(t), \quad i = 1, \ldots, N, \qquad (13.4)$$

is available to the local controller and the neighboring agents of agent $i \in \{1, \ldots, N\}$, where $\tilde{x}_i(t) \in \mathbb{R}^n$, $t \geq 0$, and $\delta_i(t) \in \mathbb{R}^n$, $t \geq 0$, captures sensor uncertainties. In particular, if $\delta_i(\cdot)$ is nonzero, then the state vector $x_i(t)$, $t \geq 0$, of agent $i \in \{1, \ldots, N\}$ is corrupted with a faulty or malicious

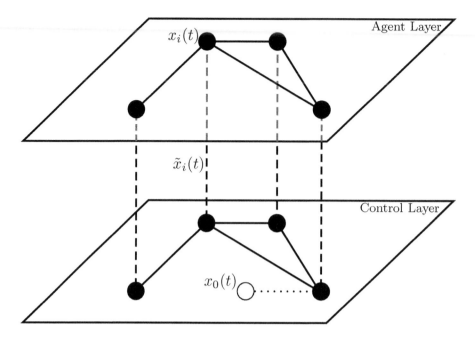

Figure 13.1 A networked multiagent system with agents lying on an agent layer and their
local controllers lying on a control layer.

signal $\delta_i(\cdot)$. Alternatively, if $\delta_i(\cdot)$ is zero, then $\tilde{x}_i(t) = x_i(t)$, $t \geq 0$, and
the uncompromised state measurement is available to the local controller of
agent $i \in \{1, \ldots, N\}$.

Given the two-layer networked multiagent system hierarchy, we are
interested in the problem of asymptotically (or approximately) driving the
state vector of each agent $x_i(t)$, $i = 1, \ldots, N$, $t \geq 0$, to the state vector
of a (virtual) leader $x_0(t) \in \mathbb{R}^n$, $t \geq 0$, that lies on the control layer with
dynamics

$$\dot{x}_0(t) = A_0 x_0(t), \quad x_0(0) = x_{00}, \quad t \geq 0, \tag{13.5}$$

where $A_0 \in \mathbb{R}^{n \times n}$ is Lyapunov stable and is given by $A_0 \triangleq A - BK_0$ with
$K_0 \in \mathbb{R}^{m \times n}$.

For the case where the uncompromised state measurement is avail-
able to the local controller of agent $i \in \{1, \ldots, N\}$, that is, $\delta_i(t) \equiv 0$, the
controller

$$u_i(t) = -K_0 x_i(t) - cK \left[ l_i \big( x_i(t) - x_0(t) \big) + \sum_{i \sim j} \big( x_i(t) - x_j(t) \big) \right] \tag{13.6}$$

guarantees that $\lim_{t \to \infty} x_i(t) = x_0(t)$ for all $i = 1, \ldots, N$, where $l_i = 1$ for a set of $N_L$ agents that have access to the state of the leader $x_0(t)$, $t \geq 0$, and $l_i = 0$ for the remaining $N_F$ agents with $N = N_L + N_F$, and where $K \in \mathbb{R}^{m \times n}$ and $c \in \mathbb{R}_+$ denote an appropriate feedback gain matrix and coupling strength, respectively, such that $A_\xi \triangleq A_0 - \eta_i cBK$ is Hurwitz for all $\eta_i \in \text{spec}(F(\mathfrak{G}))$.

To see this, let $\xi_i(t) \triangleq x_i(t) - x_0(t)$, and note that, using (13.3), (13.5), and (13.6),

$$\dot{\xi}_i(t) = A_0 \xi_i(t) - cBK \left[ l_i \xi_i(t) + \sum_{i \sim j} (\xi_i(t) - \xi_j(t)) \right], \quad \xi_i(0) = \xi_{i0}, \quad t \geq 0,$$

$$(13.7)$$

with $\xi_{i0} \triangleq x_{i0} - x_0$. In addition, defining the augmented state $\xi(t) \triangleq [\xi_1^T(t), \ldots, \xi_N^T(t)]^T$, (13.7) can be written in compact form as

$$\dot{\xi}(t) = \left[ I_N \otimes A_0 - cF(\mathfrak{G}) \otimes BK \right] \xi(t), \quad \xi(0) = \xi_0, \quad t \geq 0. \qquad (13.8)$$

Now, using the results in [93, 219, 323], it can be shown that $I_N \otimes A_0 - cF(\mathfrak{G}) \otimes BK$ is Hurwitz when $A_\xi$ is Hurwitz for all $\eta_i \in \text{spec}(F(\mathfrak{G}))$. Hence, $\lim_{t \to \infty} \xi(t) = 0$; that is, $\lim_{t \to \infty} x_i(t) = x_0(t)$ for all $i = 1, \ldots, N$.

For $\delta(\cdot) \neq 0$, our objective is to design a local controller for each agent $i \in \{1, \ldots, N\}$ of the form

$$u_i(t) = -K_0 \tilde{x}_i(t) - cK \left[ l_i \big( \tilde{x}_i(t) - x_0(t) \big) + \sum_{i \sim j} \big( \tilde{x}_i(t) - \tilde{x}_j(t) \big) \right] + v_i(t),$$

$$(13.9)$$

where $v_i(t) \in \mathbb{R}^m$, $t \geq 0$, is a local corrective signal that suppresses or counteracts the effect of $\delta_i(t)$, $t \geq 0$, to asymptotically (or approximately) recover the ideal system performance (i.e., $\lim_{t \to \infty} x_i(t) = x_0(t)$ for all $i = 1, \ldots, N$) that is achieved when the state vector is available for feedback. Thus, assuming that $A_\xi$ is Hurwitz for all $\eta_i \in \text{spec}(F(\mathfrak{G}))$ implies that there exists an ideal system performance that can be recovered by designing the local corrective signals $v_i(t) \in \mathbb{R}^m$, $t \geq 0$, for each agent $i \in \{1, \ldots, N\}$. Although we consider this specific problem in this chapter, the proposed approach dealing with sensor uncertainties can be used in many other problems that exist in the networked multiagent systems literature [224, 257].

## 13.3 Adaptive Leader Following with Time-Invariant Sensor Uncertainties

In this section, we design the local corrective signal $v_i(t)$, $i = 1, \ldots, N$, $t \geq 0$, in (13.9) to achieve asymptotic adaptive leader following in the presence of time-invariant sensor uncertainties, that is, $\delta_i(t) \equiv \delta_i$, $i = 1, \ldots, N$, $t \geq 0$. For this problem, we propose the corrective signal

$$v_i(t) = K_0 \hat{\delta}_i(t) + cK \left[ l_i \hat{\delta}_i(t) + \sum_{i \sim j} \left( \hat{\delta}_i(t) - \hat{\delta}_j(t) \right) \right], \qquad (13.10)$$

where

$$\dot{\hat{\delta}}_i(t) = -\gamma A^{\mathrm{T}} P \big( \tilde{x}_i(t) - \hat{x}_i(t) - \hat{\delta}_i(t) \big), \quad \hat{\delta}_i(0) = \hat{\delta}_{i0}, \quad t \geq 0; \qquad (13.11)$$

$$\dot{\hat{x}}_i(t) = A_0 \hat{x}_i(t) - cBK \left[ l_i \big( \hat{x}_i(t) - x_0(t) \big) + \sum_{i \sim j} \big( \hat{x}_i(t) - \hat{x}_j(t) \big) \right]$$
$$+ \big( \gamma A^{\mathrm{T}} P + \mu I_n \big) \big( \tilde{x}_i(t) - \hat{x}_i(t) - \hat{\delta}_i(t) \big), \quad \hat{x}_i(0) = \hat{x}_{i0}, \quad t \geq 0; \qquad (13.12)$$

$\hat{\delta}_i(t) \in \mathbb{R}^n$, $t \geq 0$, is the estimate of the sensor uncertainty $\delta_i(t)$; $\hat{x}_i(t) \in \mathbb{R}^n$, $t \geq 0$, is the state estimate of the uncompromised state vector $x_i(t)$; $\gamma \in \mathbb{R}_+$ and $\mu \in \mathbb{R}_+$ are design gains; and $P \in \mathbb{P}^n$ is a solution to the LMI given by

$$I_N \otimes \big( A_0^{\mathrm{T}} P + P A_0 - 2\mu P \big) - cF(\mathfrak{G}) \otimes \big( K^{\mathrm{T}} B^{\mathrm{T}} P + P B K \big) < 0. \qquad (13.13)$$

For the statement of the next result, we note from (13.3) and (13.9) that

$$\dot{x}_i(t) = A x_i(t) - B K_0 x_i(t) - cBK \left[ l_i \big( x_i(t) - x_0(t) \big) + \sum_{i \sim j} \big( x_i(t) - x_j(t) \big) \right]$$
$$- cBK \sum_{i \sim j} (\delta_i - \delta_j) - B \big( c l_i K + K_0 \big) \delta_i + B v_i(t),$$

$$x_i(0) = x_{i0}, \quad t \geq 0. \qquad (13.14)$$

Now, define $x(t) \triangleq [x_1^{\mathrm{T}}(t), \ldots, x_N^{\mathrm{T}}(t)]^{\mathrm{T}}$, $\delta \triangleq [\delta_1^{\mathrm{T}}, \ldots, \delta_N^{\mathrm{T}}]^{\mathrm{T}}$, and $v(t) \triangleq [v_1(t), \ldots, v_N(t)]^{\mathrm{T}}$, and note that (13.14) can be written in compact form as

$$\dot{x}(t) = [I_N \otimes A_0 - cF(\mathfrak{G}) \otimes BK] x(t) + \big( cG(\mathfrak{G}) \otimes BK \big) x_0(t)$$
$$- \big( cF(\mathfrak{G}) \otimes BK + I_N \otimes B K_0 \big) \delta + (I_N \otimes B) v(t),$$
$$x(0) = x_0, \quad t \geq 0. \qquad (13.15)$$

Using (13.4) and (13.15), the dynamics for $\tilde{x}(t)$, $t \geq 0$, with

$$\tilde{x}(t) \triangleq [\tilde{x}_1^{\mathrm{T}}(t), \ldots, \tilde{x}_N^{\mathrm{T}}(t)]^{\mathrm{T}},$$

can also be written in compact form as

$$\dot{\tilde{x}}(t) = [I_N \otimes A_0 - cF(\mathfrak{G}) \otimes BK]\tilde{x}(t) + (cG(\mathfrak{G}) \otimes BK)x_0(t)$$
$$- (I_N \otimes A)\delta + (I_N \otimes B)v(t), \quad \tilde{x}(0) = \tilde{x}_0, \quad t \geq 0. \qquad (13.16)$$

Next, letting $\hat{x}(t) \triangleq [\hat{x}_1^{\mathrm{T}}(t), \ldots, \hat{x}_N^{\mathrm{T}}(t)]^{\mathrm{T}}$, a compact form for the dynamics of $\hat{x}(t)$, $t \geq 0$, is given by

$$\dot{\hat{x}}(t) = [I_N \otimes A_0 - cF(\mathfrak{G}) \otimes BK]\hat{x}(t) + (cG(\mathfrak{G}) \otimes BK)x_0(t)$$
$$- (cF(\mathfrak{G}) \otimes BK + I_N \otimes BK_0)\hat{\delta}(t) + (I_N \otimes B)v(t) + \phi(t),$$
$$\hat{x}(0) = \hat{x}_0, \quad t \geq 0, \qquad (13.17)$$

where $\phi(t) \triangleq [\phi_1^{\mathrm{T}}(t), \ldots, \phi_N^{\mathrm{T}}(t)]^{\mathrm{T}}$, with $\phi_i(t) \triangleq -\dot{\hat{\delta}}_i(t) + \mu e_i(t)$. Finally, define $e_i(t) \triangleq \tilde{x}_i(t) - \hat{x}_i(t) - \hat{\delta}_i(t)$ and $\tilde{\delta}_i(t) \triangleq \delta_i - \hat{\delta}_i(t)$, and note that

$$\dot{e}(t) = (A_r - \mu I_{nN})e(t) - (I_N \otimes A)\tilde{\delta}(t), \quad e(0) = e_0, \quad t \geq 0, \qquad (13.18)$$

$$\dot{\tilde{\delta}}(t) = (I_N \otimes \gamma A^{\mathrm{T}}P)e(t), \quad \tilde{\delta}(0) = \tilde{\delta}_0, \qquad (13.19)$$

where $A_r \triangleq I_N \otimes A_0 - cF(\mathfrak{G}) \otimes BK$, $e(t) \triangleq [e_1^{\mathrm{T}}(t), \ldots, e_N^{\mathrm{T}}(t)]^{\mathrm{T}}$, and $\tilde{\delta}(t) \triangleq [\tilde{\delta}_1^{\mathrm{T}}(t), \ldots, \tilde{\delta}_N^{\mathrm{T}}(t)]^{\mathrm{T}}$.

**Theorem 13.1.** Consider the networked multiagent system consisting of $N$ agents on a connected undirected graph $\mathfrak{G}$, where the dynamics of agent $i \in \{1, \ldots, N\}$ is given by (13.3). In addition, assume that the local controller $u_i(t)$, $i = 1, \ldots, N$, $t \geq 0$, for each agent is given by (13.9), with the corrective signal $v_i(t)$, $i = 1, \ldots, N$, $t \geq 0$, given by (13.10). Moreover, assume that $\delta_i(t) \equiv \delta_i$, $t \geq 0$, and $\det(A) \neq 0$. Then the zero solution $(e(t), \tilde{\delta}(t)) \equiv (0,0)$ of the closed-loop system given by (13.18) and (13.19) is Lyapunov stable and $\lim_{t \to \infty} e(t) = 0$ and $\lim_{t \to \infty} \tilde{\delta}(t) = 0$ for all $(e_0, \tilde{\delta}_0) \in \mathbb{R}^{nN} \times \mathbb{R}^{nN}$.

**Proof.** To show Lyapunov stability of the zero solution $(e(t), \tilde{\delta}(t)) \equiv (0,0)$ of the closed-loop system given by (13.18) and (13.19), consider the Lyapunov function candidate given by

$$V(e, \tilde{\delta}) = \sum_{i=1}^{N}(e_i^{\mathrm{T}}Pe_i + \gamma^{-1}\tilde{\delta}_i^{\mathrm{T}}\tilde{\delta}_i) = e^{\mathrm{T}}(I_N \otimes P)e + \gamma^{-1}\tilde{\delta}^{\mathrm{T}}\tilde{\delta}, \qquad (13.20)$$

where $P$ is a solution to the LMI given by (13.13). Note that $V(0,0) = 0$, $V(e, \tilde{\delta}) > 0$ for all $(e, \tilde{\delta}) \neq (0,0)$, and $V(e, \tilde{\delta})$ is radially unbounded. The time derivative of (13.20) along the trajectories of (13.18) and (13.19) is given by

$$
\dot{V}\big(e(t), \tilde{\delta}(t)\big)
$$
$$
= e^{\mathrm{T}}(t)\left(I_N \otimes (A_0^{\mathrm{T}}P + PA_0 - 2P\mu) - cF(\mathfrak{G}) \otimes (K^{\mathrm{T}}B^{\mathrm{T}}P + PBK)\right)e(t)
$$
$$
\leq 0, \quad t \geq 0. \tag{13.21}
$$

Hence, the zero solution $\big(e(t), \tilde{\delta}(t)\big) \equiv (0,0)$ of the closed-loop system given by (13.18) and (13.19) is Lyapunov stable for all $\big(e_0, \tilde{\delta}_0\big) \in \mathbb{R}^{nN} \times \mathbb{R}^{nN}$.

To show that $\lim_{t\to\infty} e(t) = 0$, note that

$$
\ddot{V}\big(e(t), \tilde{\delta}(t)\big)
$$
$$
= 2e^{\mathrm{T}}(t)\big[I_N \otimes (A_0^{\mathrm{T}}P + PA_0 - 2P\mu) - cF(\mathfrak{G}) \otimes (K^{\mathrm{T}}B^{\mathrm{T}}P + PBK)\big]\dot{e}(t).
$$

Now, it follows from the Lyapunov stability of the zero solution $\big(e(t), \tilde{\delta}(t)\big) \equiv (0,0)$ of (13.18) and (13.19) and the boundedness of $\dot{e}(t), t \geq 0$, that $\ddot{V}\big(e(t), \tilde{\delta}(t)\big)$ is bounded for all $t \geq 0$. Thus, $\dot{V}\big(e(t), \tilde{\delta}(t)\big), t \geq 0$, is uniformly continuous in $t$. Now, it follows from Barbalat's lemma [122] that $\lim_{t\to\infty} \dot{V}\big(e(t), \tilde{\delta}(t)\big) = 0$, and, hence, $\lim_{t\to\infty} e(t) = 0$.

Finally, to show that $\lim_{t\to\infty} \tilde{\delta}(t) = 0$, define $\mathcal{R} \triangleq \{(e, \tilde{\delta}) : \dot{V}(e, \tilde{\delta}) = 0\}$, and let $\mathcal{M}$ be the largest invariant set contained in $\mathcal{R}$. In this case, it follows from (13.18) that $(I_N \otimes A)\tilde{\delta} = 0$, and, hence, $\tilde{\delta} = 0$ since $\det(A) \neq 0$. Thus, $\big(e(t), \tilde{\delta}(t)\big) \to \mathcal{M} = \{(0,0)\}$ as $t \to \infty$. $\qquad\square$

It follows from (13.3), (13.9), and (13.10) that

$$
\dot{x}_i(t) = A_0 x_i(t) - cBK\left[l_i\big(x_i(t) - x_0(t)\big) + \sum_{i\sim j}\big(x_i(t) - x_j(t)\big)\right] - BK_0\tilde{\delta}_i(t)
$$
$$
- cBK\left[l_i\tilde{\delta}_i(t) + \sum_{i\sim j}\big(\tilde{\delta}_i(t) - \tilde{\delta}_j(t)\big)\right], \quad x_i(0) = x_{i0}, \quad t \geq 0, \tag{13.22}
$$

which, using the boundedness of $\tilde{\delta}_i(t)$, $i = 1, \ldots, N$, $t \geq 0$, and the assumption that $A_\xi$ is Hurwitz for all $\eta_i \in \operatorname{spec}(F(\mathfrak{G}))$, implies that $x_i(t)$ is bounded for all $t \geq 0$ and $i \in \{1, \ldots, N\}$. Hence, using (13.4), $\tilde{x}_i(t)$ is bounded for all $t \geq 0$ and $i \in \{1, \ldots, N\}$. Furthermore, since $e_i(t)$, $t \geq 0$; $\tilde{x}_i(t)$, $t \geq 0$; and $\hat{\delta}_i(t)$, $t \geq 0$, are bounded for all $i \in \{1, \ldots, N\}$, it follows that $\hat{x}_i(t)$ is bounded for all $t \geq 0$ and $i \in \{1, \ldots, N\}$.

Since, by Theorem 13.1, $\lim_{t\to\infty}\tilde{\delta}(t) = 0$, it follows from (13.22) that each agent subject to the dynamics given by (13.3) asymptotically recovers the ideal system performance (i.e., $\lim_{t\to\infty} x_i(t) = x_0(t)$ for all $i = 1,\dots,N$), which is a direct consequence of the discussion given in Section 13.2 and the assumption that $A_\xi$ is Hurwitz for all $\eta_i \in \mathrm{spec}(F(\mathfrak{G}))$, $i = 1,\dots,N$. In addition, $\lim_{t\to\infty} e(t) = 0$ and $\lim_{t\to\infty}\tilde{\delta}(t) = 0$ imply that $\lim_{t\to\infty}\big(x(t) - \hat{x}(t)\big) = 0$, which shows that the state estimate $\hat{x}(t)$, $t \geq 0$, converges to the uncompromised state measurement $x(t)$, $t \geq 0$.

Let $Q \in \mathbb{P}^m$; set $K = Q^{-1}B^\mathrm{T}P$; and assume that $A_\xi$ is Hurwitz for all $\eta_i \in \mathrm{spec}(F(\mathfrak{G}))$, $i = 1,\dots,N$, for the given selection of $K$. Substituting $K = Q^{-1}B^\mathrm{T}P$ in (13.13) yields

$$I_N \otimes (A_0^\mathrm{T}P + PA_0 - 2\mu P) - 2cF(\mathfrak{G}) \otimes (PBQ^{-1}B^\mathrm{T}P) < 0. \qquad (13.23)$$

Let $T$ be such that $TF(\mathfrak{G})T^{-1} = J$, where $J$ is the Jordan form of $F(\mathfrak{G})$. Multiplying (13.23) by $T \otimes I_N$ from the left and by $T^{-1} \otimes I_N$ from the right yields

$$(T \otimes I_N)(I_N \otimes (A_0^\mathrm{T}P + PA_0 - 2\mu P))(T^{-1} \otimes I_N)$$
$$- 2(T \otimes I_N)(cF(\mathfrak{G}) \otimes (PBQ^{-1}B^\mathrm{T}P))(T^{-1} \otimes I_N) < 0, \quad (13.24)$$

which can be equivalently written as

$$I_N \otimes (A_0^\mathrm{T}P + PA_0 - 2\mu P) - 2cJ \otimes (PBQ^{-1}B^\mathrm{T}P) < 0. \qquad (13.25)$$

Note that since $F(\mathfrak{G})$ is symmetric, $F(\mathfrak{G})$ is real diagonalizable. Hence, the diagonal form of $J$ allows one to rewrite (13.25) as

$$A_0^\mathrm{T}P + PA_0 - 2\mu P - 2c\eta_i PBQ^{-1}B^\mathrm{T}P < 0, \qquad (13.26)$$

where $\eta_i \in \mathrm{spec}(F(\mathfrak{G}))$, $i = 1,\dots,N$. Now, letting the coupling strength be such that

$$c \geq \frac{1}{\min\{\eta_i\}}, \quad i = 1,\dots,N,$$

it follows from (13.26), using the results in [200], that

$$A_0^\mathrm{T}P + PA_0 - 2\mu P - 2c\eta_i PBQ^{-1}B^\mathrm{T}P$$
$$\leq A_0^\mathrm{T}P + PA_0 - 2\mu P - 2PBQ^{-1}B^\mathrm{T}P, \qquad (13.27)$$

since $-c\eta_i \leq -1, i = 1,\dots,N$, and, hence, one can equivalently solve the LMI

$$(A_0 - \mu I_n)^\mathrm{T}P + P(A_0 - \mu I_n) - 2PBQ^{-1}B^\mathrm{T}P < 0 \qquad (13.28)$$

to obtain $P \in \mathbb{P}^n$ rather than the LMI given by (13.13).

By setting $S \triangleq P^{-1}$ [46], the LMI given by (13.28) can be further simplified to

$$(A_0 - \mu I_n)S + S(A_0 - \mu I_n)^{\mathrm{T}} - 2BQ^{-1}B^{\mathrm{T}} < 0. \tag{13.29}$$

Thus, one can alternatively solve (13.29) for $S \in \mathbb{P}^n$ and then set $P = S^{-1}$ to obtain $P \in \mathbb{P}^n$ for (13.11) and (13.12). Note that since one can solve (13.29) instead of (13.13), the computational complexity does not increase as the number of agents gets larger.

Finally, note that in this chapter we consider the case where each agent is subject to sensor uncertainties. If, however, only a fraction of the agents are subject to sensor uncertainties, then the proposed corrective signal in (13.10) can be applied only to those agents subject to sensor uncertainty and not to all of the agents.

## 13.4 Adaptive Leader Following with Time-Varying Sensor Uncertainties

In this section, we generalize the results of the previous section by designing the local corrective signal $v_i(t)$, $i = 1, \ldots, N$, $t \geq 0$, in (13.9) to achieve approximate adaptive leader following in the presence of time-varying sensor uncertainties $\delta_i(t)$, $i = 1, \ldots, N$, $t \geq 0$. We assume that the time-varying sensor uncertainties are bounded and have bounded time rates of change; that is, $\|\delta_i(t)\|_2 \leq \bar{\delta}, t \geq 0$, and $\|\dot{\delta}_i(t)\|_2 \leq \bar{\dot{\delta}}, t \geq 0$, for all $i = 1, \ldots, N$.

For the statement of our next result, it is necessary to introduce the projection operator [249]. Specifically, let $\phi : \mathbb{R}^n \to \mathbb{R}$ be a continuously differentiable convex function given by

$$\phi(\theta) \triangleq \frac{(\varepsilon_\theta + 1)\theta^{\mathrm{T}}\theta - \theta_{\max}^2}{\varepsilon_\theta \theta_{\max}^2},$$

where $\theta_{\max} \in \mathbb{R}$ is a *projection norm bound* imposed on $\theta \in \mathbb{R}^n$ and $\varepsilon_\theta > 0$ is a *projection tolerance bound*. Then the *projection operator* $\mathrm{Proj} : \mathbb{R}^n \times \mathbb{R}^n \to \mathbb{R}^n$ is defined by

$$\mathrm{Proj}(\theta, y) \triangleq \begin{cases} y & \text{if } \phi(\theta) < 0, \\ y & \text{if } \phi(\theta) \geq 0 \text{ and } \phi'(\theta)y \leq 0, \\ y - \frac{\phi'^{\mathrm{T}}(\theta)\phi'(\theta)y}{\phi'(\theta)\phi'^{\mathrm{T}}(\theta)}\phi(\theta) & \text{if } \phi(\theta) \geq 0 \text{ and } \phi'(\theta)y > 0, \end{cases} \tag{13.30}$$

where $y \in \mathbb{R}^n$ and $\phi'(\theta) \triangleq \frac{\partial \phi(\theta)}{\partial \theta}$. Note that it follows from the definition of the projection operator that $(\theta^* - \theta)^{\mathrm{T}} [\mathrm{Proj}(\theta, y) - y] \geq 0$, $\theta^* \in \mathbb{R}^n$.

Next, for the controller given by (13.9), we use the local corrective signal

$$v_i(t) = K_0 \hat{\delta}_i(t) + cK \left[ l_i \hat{\delta}_i(t) + \sum_{i \sim j} \left( \hat{\delta}_i(t) - \hat{\delta}_j(t) \right) \right], \qquad (13.31)$$

where

$$\dot{\hat{\delta}}_i(t) = \gamma \mathrm{Proj}\left( \hat{\delta}_i(t), -A^{\mathrm{T}} P \left( \tilde{x}_i(t) - \hat{x}_i(t) - \hat{\delta}_i(t) \right) \right), \quad \hat{\delta}_i(0) = \hat{\delta}_{i0}, \quad t \geq 0,$$
$$(13.32)$$

$$\dot{\hat{x}}_i(t) = A_0 \hat{x}_i(t) - cBK \left[ l_i \left( \hat{x}_i(t) - x_0(t) \right) + \sum_{i \sim j} \left( \hat{x}_i(t) - \hat{x}_j(t) \right) \right]$$
$$+ \mu I_n \left( \tilde{x}_i(t) - \hat{x}_i(t) - \hat{\delta}_i(t) \right)$$
$$- \gamma \mathrm{Proj}\left( \hat{\delta}_i(t), -A^{\mathrm{T}} P \left( \tilde{x}_i(t) - \hat{x}_i(t) - \hat{\delta}_i(t) \right) \right), \quad \hat{x}_i(0) = \hat{x}_{i0}, \quad t \geq 0,$$
$$(13.33)$$

and $P \in \mathbb{P}^n$ is the solution to the LMI given by

$$I_N \otimes (A_0^{\mathrm{T}} P + P A_0 - 2\mu P) - cF(\mathfrak{G}) \otimes (K^{\mathrm{T}} B^{\mathrm{T}} P + P B K) < 0. \quad (13.34)$$

For the statement of the next theorem, define

$$-R \triangleq I_N \otimes (A_0^{\mathrm{T}} P + P A_0 - 2\mu P) - cF(\mathfrak{G}) \otimes (K^{\mathrm{T}} B^{\mathrm{T}} P + P B K) < 0,$$
$$(13.35)$$

where $R \in \mathbb{P}^{Nn}$, and note that, using similar arguments as given in the previous section, the dynamics for $e(t) = \tilde{x}(t) - \hat{x}(t) - \hat{\delta}(t)$ and $\tilde{\delta}(t) = \delta(t) - \hat{\delta}(t)$ are given by

$$\dot{e}(t) = (A_r - \mu I_{nN})e(t) - (I_N \otimes A)\tilde{\delta}(t) + \dot{\delta}(t), \quad e(0) = e_0, \quad t \geq 0,$$
$$(13.36)$$
$$\dot{\tilde{\delta}}(t) = \dot{\delta}(t) - \gamma \hat{\delta}_{\mathrm{P}}(t), \quad \hat{\delta}_{\mathrm{P}}(t) \triangleq \left[ \hat{\delta}_{\mathrm{P}1}^{\mathrm{T}}(t), \dots, \hat{\delta}_{\mathrm{P}N}^{\mathrm{T}}(t) \right]^{\mathrm{T}}, \quad \tilde{\delta}(0) = \tilde{\delta}_0, \quad t \geq 0,$$
$$(13.37)$$

where

$$\hat{\delta}_{\mathrm{P}i}(t) \triangleq \mathrm{Proj}\left( \hat{\delta}_i(t), -A^{\mathrm{T}} P e_i(t) \right), \quad i = 1, \dots, N.$$

**Theorem 13.2.** Consider the networked multiagent system consisting of $N$ agents on a connected, undirected graph $\mathfrak{G}$, where the dynamics of

agent $i \in \{1, \ldots, N\}$ is given by (13.3). In addition, assume that the local
controller $u_i(t)$, $i = 1, \ldots, N$, $t \geq 0$, for each agent is given by (13.9), with
the corrective signal $v_i(t)$, $i = 1, \ldots, N$, $t \geq 0$, given by (13.31). Moreover,
assume that the sensor uncertainties are time varying and $\det(A) \neq 0$. Then
the closed-loop system dynamics given by (13.36) and (13.37) are uniformly
bounded for all $(e_0, \tilde{\delta}_0) \in \mathbb{R}^{nN} \times \mathbb{R}^{nN}$ with the ultimate bounds

$$\|e(t)\|_2 \leq \left[ \frac{\lambda_{\max}(P)}{\lambda_{\min}(P)} \eta_1^2 + \frac{1}{\gamma \lambda_{\min}(P)} \eta_2^2 \right]^{\frac{1}{2}}, \quad t \geq T, \qquad (13.38)$$

$$\|\tilde{\delta}(t)\|_2 \leq \left[ \gamma \lambda_{\max}(P) \eta_1^2 + \eta_2^2 \right]^{\frac{1}{2}}, \quad t \geq T, \qquad (13.39)$$

where $\eta_1 \triangleq \frac{1}{\sqrt{d_1}} \left[ \frac{d_2}{2\sqrt{d_1}} + \left( \frac{d_2^2}{4d_1} + d_3 \right)^{\frac{1}{2}} \right]$, $\eta_2 \triangleq \hat{\delta}_{\max} + \bar{\delta}$, $d_1 \triangleq \lambda_{\min}(R)$, $d_2 \triangleq$
$2N\lambda_{\max}(P)\bar{\delta}$, and $d_3 \triangleq 2N\gamma^{-1}\bar{\delta}(\hat{\delta}_{\max} + \bar{\delta})$.

**Proof.** To show uniform boundedness of the system dynamics given
by (13.36) and (13.37), consider the Lyapunov-like function given by (13.20),
where $P$ satisfies (13.34). Note that $V(0, 0) = 0$, $V(e, \tilde{\delta}) > 0$ for all $(e, \tilde{\delta}) \neq$
$(0, 0)$, and $V(e, \tilde{\delta})$ is radially unbounded. The time derivative of (13.20)
along the closed-loop system trajectories of (13.36) and (13.37) is given by

$$\dot{V}(e(t), \tilde{\delta}(t)) = \sum_{i=1}^{N} \left[ 2e_i^{\mathrm{T}}(t) P A_0 e_i(t) - 2e_i^{\mathrm{T}}(t) P B K \sum_{i \sim j} (e_i(t) - e_j(t)) \right.$$
$$- 2e_i^{\mathrm{T}}(t) P A \tilde{\delta}_i(t) + 2e_i^{\mathrm{T}}(t) P \dot{\delta}_i(t)$$
$$- 2e_i^{\mathrm{T}}(t) P B K l_i e_i(t) - 2e_i^{\mathrm{T}}(t) P \mu e_i(t)$$
$$\left. + 2\gamma^{-1}\tilde{\delta}_i^{\mathrm{T}}(t) \left( \dot{\delta}_i(t) - \gamma \mathrm{Proj}(\hat{\delta}_i(t), -A^{\mathrm{T}} P e_i(t)) \right) \right]$$

$$= -e^{\mathrm{T}}(t) R e(t) + \sum_{i=1}^{N} \left[ - 2e_i^{\mathrm{T}}(t) P A \tilde{\delta}_i(t) + 2e_i^{\mathrm{T}}(t) P \dot{\delta}_i(t) \right.$$
$$\left. + 2\gamma^{-1}\tilde{\delta}_i^{\mathrm{T}}(t) \dot{\delta}_i(t) - 2\tilde{\delta}_i^{\mathrm{T}}(t) \mathrm{Proj}(\hat{\delta}_i(t), -A^{\mathrm{T}} P e_i(t)) \right]$$

$$= -e^{\mathrm{T}}(t) R e(t) + \sum_{i=1}^{N} \left[ 2(\hat{\delta}_i(t) - \delta_i(t))^{\mathrm{T}} \right.$$
$$\cdot \left( \mathrm{Proj}(\hat{\delta}_i(t), -A^{\mathrm{T}} P e_i(t)) - ( - A^{\mathrm{T}} P e_i(t)) \right)$$
$$\left. + 2e_i^{\mathrm{T}}(t) P \dot{\delta}_i(t) + 2\gamma^{-1}\tilde{\delta}_i^{\mathrm{T}}(t) \dot{\delta}_i(t) \right]$$

$$\leq -e^{\mathrm{T}}(t)Re(t) + \sum_{i=1}^{N}\left[2e_i^{\mathrm{T}}(t)P\dot{\delta}_i(t) + 2\gamma^{-1}\tilde{\delta}_i^{\mathrm{T}}(t)\dot{\delta}_i(t)\right]$$

$$\leq -d_1\|e(t)\|_2^2 + d_2\|e(t)\|_2 + d_3$$

$$= -\left[\sqrt{d_1}\|e(t)\|_2 - \frac{d_2}{2\sqrt{d_1}}\right]^2 + \frac{d_2^2}{4d_1} + d_3, \quad t \geq 0, \qquad (13.40)$$

and, hence, $\dot{V}\big(e(t),\tilde{\delta}(t)\big) < 0$ outside the compact set

$$\Omega \triangleq \left\{(e,\tilde{\delta}): \|e\|_2 \leq \eta_1 \text{ and } \|\tilde{\delta}\|_2 \leq \eta_2\right\}.$$

This proves the uniform boundedness of the solution $\big(e(t),\tilde{\delta}(t)\big)$ of the system dynamics given by (13.36) and (13.37) for all $\big(e_0,\tilde{\delta}_0\big)\in \mathbb{R}^{nN} \times \mathbb{R}^{nN}$ [186].

To show the ultimate bounds for $e(t)$, $t \geq T$, and $\tilde{\delta}(t)$, $t \geq T$, given by (13.38) and (13.39), respectively, note that

$$\lambda_{\min}(P)\|e(t)\|_2^2 + \gamma^{-1}\|\tilde{\delta}(t)\|_2^2 \leq \lambda_{\max}(P)\eta_1^2 + \gamma^{-1}\eta_2^2, \quad t \geq T,$$

or, equivalently,

$$\lambda_{\min}(P)\|e(t)\|_2^2 \leq \lambda_{\max}(P)\eta_1^2 + \gamma^{-1}\eta_2^2, \quad t \geq T,$$

and

$$\gamma^{-1}\|\tilde{\delta}(t)\|_2^2 \leq \lambda_{\max}(P)\eta_1^2 + \gamma^{-1}\eta_2^2, \quad t \geq T,$$

which proves (13.38) and (13.39). $\qquad\square$

Note that all signals used to construct the local controller $u_i(t)$, $i = 1,\ldots,N$, $t \geq 0$, for each agent given by (13.9) with the local corrective signal $v_i(t)$, $i = 1,\ldots,N$, $t \geq 0$, given by (13.31), (13.32), and (13.33) are bounded.

The projection operator is used in order to guarantee a bounded estimate of the unknown parameter since otherwise one cannot conclude uniform ultimate boundedness from (13.40). The ultimate bounds given by (13.38) and (13.39) characterize the controller design parameters that need to be chosen in order to achieve small excursions of $\|e(t)\|_2$ and $\|\tilde{\delta}(t)\|_2$ for $t \geq T$. This is particularly important for obtaining accurate estimates for $\hat{x}_i(t)$, $i = 1,\ldots,N$, $t \geq T$, and $\hat{\delta}_i(t)$, $i = 1,\ldots,N$, $t \geq T$, as well as suppressing the effect of $\tilde{\delta}_i(t)$, $i = 1,\ldots,N$, $t \geq T$.

To elucidate the effect of the controller design parameters on (13.38) and (13.39), let $A = 1$, $B = 1$, $K = 1$, $A_0 = 0$, $K_0 = 1$, $N = 1$, $c = 1$,

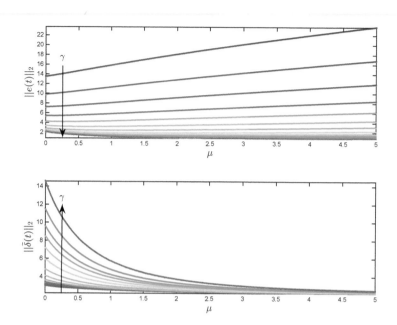

Figure 13.2 Effect of $\mu$ and $\gamma$ on the ultimate bounds given by (13.38) and (13.39) (arrow directions denote the increase of $\gamma$ from 0.1 to 100).

$\hat{\delta}_{\max} = 1$, $\overline{\delta} = 1$, and $\overline{\dot{\delta}} = 2$. In this case, it follows from (13.35) that $P = 0.5(1 + \mu)^{-1}R$, where we set $R = 1$. Figure 13.2 shows the effect of $\mu \in [0, 5]$ and $\gamma \in [0.1, 100]$ on the ultimate bounds given by (13.38) and (13.39). Specifically, as expected, increasing both $\mu$ and $\gamma$ yields smaller ultimate bounds for $\|e(t)\|_2$ and $\|\tilde{\hat{\delta}}(t)\|_2$ for $t \geq T$.

Next, we present two numerical examples to demonstrate the utility and efficacy of the proposed distributed adaptive control architectures for networked multiagent systems to mitigate the effect of time-invariant and time-varying sensor uncertainties.

**Example 13.1.** To illustrate the key ideas presented in Section 13.3, consider a group of $N = 4$ agents subject to the connected, undirected graph $\mathfrak{G}$ given in Figure 13.1, where the dynamics of agent $i$ satisfy

$$\begin{bmatrix} \dot{x}_i^1(t) \\ \dot{x}_i^2(t) \end{bmatrix} = \begin{bmatrix} 0 & 0.3 \\ -2 & 0 \end{bmatrix} \begin{bmatrix} x_i^1(t) \\ x_i^2(t) \end{bmatrix} + \begin{bmatrix} 0 \\ 1 \end{bmatrix} u(t), \quad i = 1, \dots, 4, \quad t \geq 0, \quad (13.41)$$

with initial conditions $x_1(0) = [0, 0]^{\mathrm{T}}$, $x_2(0) = [1, 2]^{\mathrm{T}}$, $x_3(0) = [3, 1]^{\mathrm{T}}$, and $x_4(0) = [2, 2]^{\mathrm{T}}$. Note that $\det(A) \neq 0$, where $A$ is the system matrix of (13.41). For this example, we are interested in the problem of asymptotically driving the state vector of each agent $x_i(t)$, $i = 1, \dots, 4$, $t \geq 0$, to the state vector of a leader $x_0(t)$, $t \geq 0$, having dynamics given by

$$\begin{bmatrix} \dot{x}_0^1(t) \\ \dot{x}_0^2(t) \end{bmatrix} = \begin{bmatrix} 0 & 0.3 \\ -2 & 0 \end{bmatrix} \begin{bmatrix} x_0^1(t) \\ x_0^2(t) \end{bmatrix}, \quad x_0(0) = \begin{bmatrix} 1 \\ 1 \end{bmatrix}, \quad t \geq 0. \tag{13.42}$$

For this problem, $A_0 = A - BK_0$ holds, with $K_0 = [0, 0]$ a direct consequence of (13.41) and (13.42).

To design the proposed local controllers, let $K = Q^{-1}B^{\mathrm{T}}P$, and set $Q = 0.1I_2$; $\mu = 3.2$; and $c = 6 \geq \frac{1}{\min\{\eta_i\}}$, $i = 1, \ldots, 4$, so that the LMI given by (13.29) for $P = S^{-1}$ is satisfied with

$$P = \begin{bmatrix} 0.0539 & 0.0105 \\ 0.0105 & 0.0520 \end{bmatrix}, \tag{13.43}$$

and, hence, $K = [0.1048, \ 0.5199]$. Note with this selection for $K$, $A_\xi$ is Hurwitz for all $\eta_i \in \mathrm{spec}(F(\mathfrak{G}))$, $i = 1, \ldots, 4$. The nominal system performance for the case when the uncompromised state measurement is available to the local controller of agent $i$, $i \in \{1, \ldots, 4\}$, (i.e., $\delta_i(\cdot) = 0$), is shown in Figures 13.3 and 13.4 using (13.6).

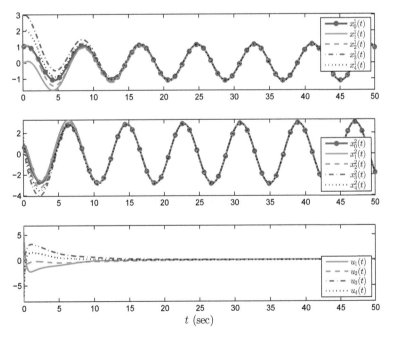

Figure 13.3 Nominal system performance for the group of agents in Example 13.1 with the local controller given by (13.6) (i.e., $v_i(t) \equiv 0$, $i = 1, \ldots, 4$) when the uncompromised state measurement is available for feedback.

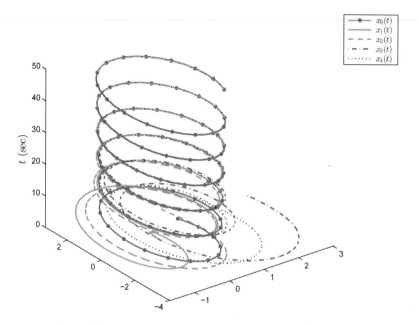

Figure 13.4 System trajectory of each agent in Figure 13.3.

Next, consider a time-invariant sensor uncertainty given by (13.4), with

$$\delta_1 = \begin{bmatrix} 10 \\ 7 \end{bmatrix}, \quad \delta_2 = \begin{bmatrix} 8 \\ 5 \end{bmatrix}, \quad \delta_3 = \begin{bmatrix} 6 \\ 3 \end{bmatrix}, \quad \delta_4 = \begin{bmatrix} 4 \\ 1 \end{bmatrix}. \tag{13.44}$$

The system performance for the case when the compromised state measurement is available to the local controller of agent $i \in \{1, \ldots, 4\}$ (i.e., $\delta_i(\cdot) \neq 0$) is shown in Figures 13.5 and 13.6 using (13.6) (i.e., $v_i(t) \equiv 0$, $i = 1, \ldots, 4$). Now, to illustrate the results of Theorem 13.1, we use the proposed local controller $u_i(t)$, $i = 1, \ldots, 4$, $t \geq 0$, for each agent given by (13.9) and the local corrective signal $v_i(t)$, $i = 1, \ldots, 4$, $t \geq 0$, given by (13.10) with $\gamma = 620$. For this case, the system performance in the presence of time-invariant sensor uncertainties is shown in Figures 13.7 and 13.8.

As expected, the proposed distributed control architecture of Theorem 13.1 allows the state vector of each agent $x_i(t)$, $i = 1, \ldots, 4$, $t \geq 0$, to asymptotically track the state vector of the leader $x_0(t)$, $t \geq 0$. Finally, the time evolutions of the error signals given by $e(t), t \geq 0$, and $\tilde{\delta}(t), t \geq 0$, are shown, respectively, in Figures 13.9 and 13.10 for the closed-loop system with the proposed distributed control architecture. $\triangle$

**Example 13.2.** To illustrate the key ideas presented in this section, consider the same group of agents as in Example 13.1 with the dynamics given by (13.41). For this example, we are interested in the problem of approximately driving the state vector of each agent $x_i(t)$, $i = 1, \ldots, 4$, $t \geq 0$, to the state vector of a leader $x_0(t)$, $t \geq 0$, having the dynamics given by (13.42). Here, once again, we take the same selections for $K$ and $c$ as in Example 13.1 to design the proposed local controllers.

Consider the time-varying sensor uncertainty given by (13.4), with

$$\delta_1(t) = 2 + \sin(0.1t) \begin{bmatrix} 0.1 \\ 0.2 \end{bmatrix}, \quad \delta_2(t) = 1 + \sin(0.15t) \begin{bmatrix} 0.2 \\ 0.4 \end{bmatrix}, \quad (13.45)$$

$$\delta_3(t) = 4 + \sin(0.05t) \begin{bmatrix} 0.3 \\ 0.6 \end{bmatrix}, \quad \delta_4(t) = 4 + \sin(0.05t) \begin{bmatrix} 0.4 \\ 0.8 \end{bmatrix}. \quad (13.46)$$

The system performance for the case when the compromised state measurement is available to the local controller of agent $i \in \{1, \ldots, 4\}$ (i.e., $\delta_i(t) \neq 0$) is shown in Figures 13.11 and 13.12 using (13.6) (i.e., $v_i(t) \equiv 0$, $i = 1, \ldots, 4$). Now, to illustrate the results of Theorem 13.2, we use the proposed local controller $u_i(t)$, $i = 1, \ldots, 4$, $t \geq 0$, for each agent given by (13.9) and the local corrective signal $v_i(t)$, $i = 1, \ldots, 4$, $t \geq 0$, given by (13.31) with $\gamma = 550$, $\mu = 5.2$, and $\hat{\delta}_{\max} = 100$. For this case, the system performance in the presence of time-varying sensor uncertainties is shown in Figures 13.13 and 13.14.

As expected, the proposed distributed control architecture of Theorem 13.2 allows the state vector of each agent $x_i(t)$, $i = 1, \ldots, 4$, $t \geq 0$, to approximately track the state vector of the leader $x_0(t)$, $t \geq 0$. Finally, the time evolutions of the error signals given by (13.36) and (13.37) are, respectively, shown in Figures 13.15 and 13.16. $\triangle$

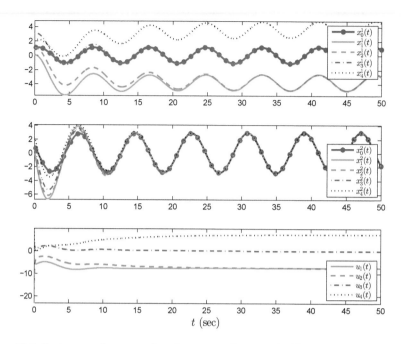

Figure 13.5 System performance for the group of agents in Example 13.1 with the local controller given by (13.6) (i.e., $v_i(t) \equiv 0$, $i = 1, \ldots, 4$) when the compromised state measurement is available for feedback.

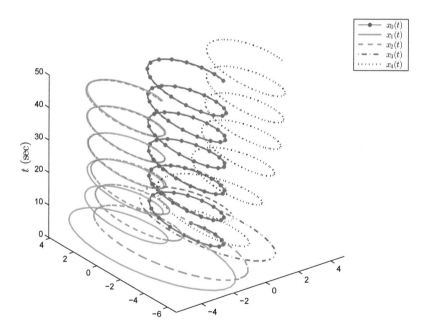

Figure 13.6 System trajectory of each agent in Figure 13.5.

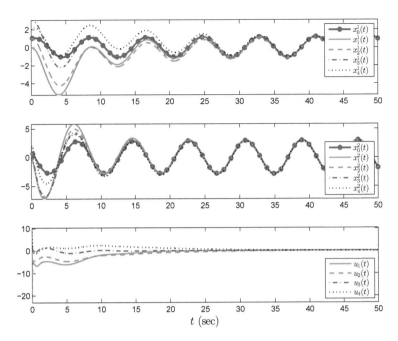

Figure 13.7 System performance for the group of agents in Example 13.1 with the proposed local controller given by (13.9) and the local corrective signal given by (13.10) when the compromised state measurement is available for feedback.

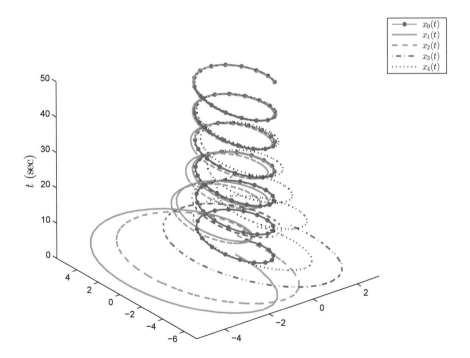

Figure 13.8 System trajectory of each agent in Figure 13.7.

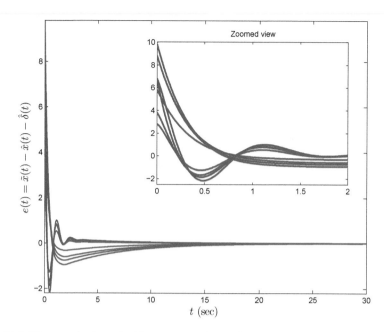

Figure 13.9 Time evolution of $e(t), t \geq 0$, in Example 13.1 with the proposed local controller given by (13.9) and the local corrective signal given by (13.10) when the compromised state measurement is available for feedback.

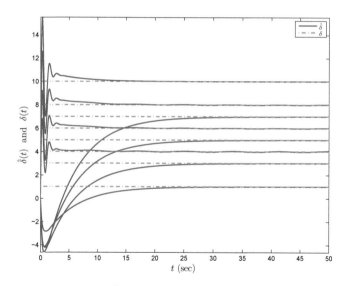

Figure 13.10 Time evolution of $\tilde{\delta}(t), t \geq 0$, in Example 13.1 with the proposed local controller given by (13.9) and the local corrective signal given by (13.10) when the compromised state measurement is available for feedback.

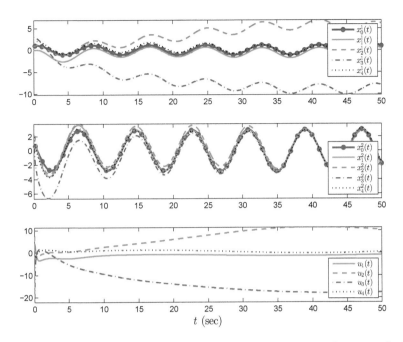

Figure 13.11 System performance for the group of agents in Example 13.2 with the local controller given by (13.6) (i.e., $v_i(t) \equiv 0$, $i = 1, \ldots, 4$) when the compromised state measurement is available for feedback.

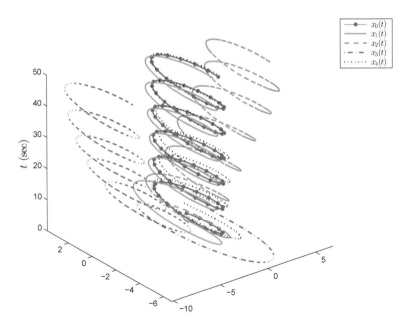

Figure 13.12 System trajectory of each agent in Figure 13.11.

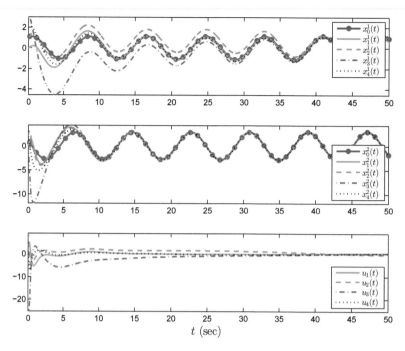

Figure 13.13 System performance for the group of agents in Example 13.2 with the proposed local controller given by (13.9) and the local corrective signal given by (13.31) when the compromised state measurement is available for feedback.

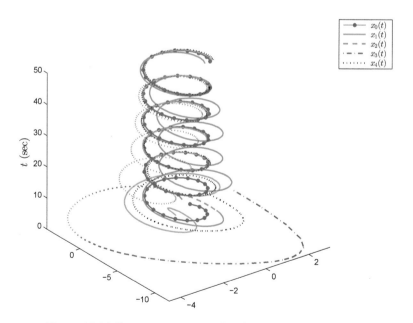

Figure 13.14 System trajectory of each agent in Figure 13.12.

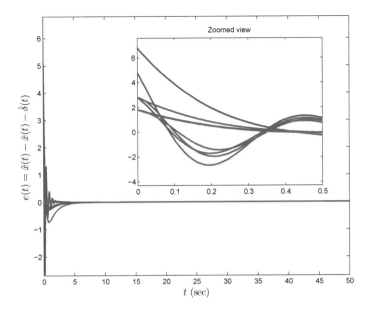

Figure 13.15 Time evolution of (13.36) in Example 13.2 with the proposed local controller given by (13.9) and the local corrective signal given by (13.31) when the compromised state measurement is available for feedback.

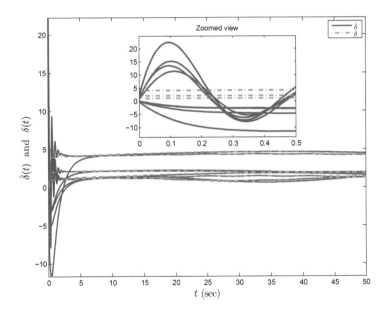

Figure 13.16 Time evolution of (13.37) in Example 13.2 with the proposed local controller given by (13.9) and the local corrective signal given by (13.31) when the compromised state measurement is available for feedback.

*Chapter Fourteen*

# Adaptive Estimation Using Network Identifiers

## 14.1 Introduction

In this chapter, we consider the problem of *adaptive estimation* of a linear system with unknown plant and input matrices. In particular, we propose a novel distributed observer architecture that adaptively identifies the dynamic system matrices using a group of $N$ agents. Each agent generates its own adaptive identifier based on the identifier architecture presented in [230]. Furthermore, it is shown that if the adaptive identifiers have the same structure but do not share information (i.e., are not connected), then there is no guarantee that the $N$ adaptive identifiers will have their estimates converge to the same value without a persistency of excitation condition being imposed. Alternatively, when the update laws for the parameter identifiers are modified to include interagent information exchange, then consensus of both the state and parameter estimates is guaranteed, thus emulating a persistency of excitation condition.

The proposed adaptive identifier architecture includes additional terms in both the state and parameter equations, which effectively penalize the mismatch between all estimates and take the form of nonnegative damping terms that serve to enhance the convergence properties of the state and parameter errors. The adaptive estimation architecture builds on the work of [288] on adaptive consensus control of multiagent systems, with the key difference being that the mismatch between the state and parameter estimates is also penalized, thus accounting for interagent communication constraints.

For nonadaptive estimators, a linear estimator scheme that considers a penalized mismatch of the parameter estimates is proposed in [77, 285]. Alternatively, within the context of distributed Kalman filtering for sensor networks, agreement of the state and parameter estimates, as a measure that is independent of the network topology and wherein the deviations of the parameter estimates are measured from their mean, is considered in [235]. Distributed adaptive control for convergence using consensus learning of sensory information for networked robots is addressed in [271, 272].

The added benefit of the proposed network architecture of the adaptive identifiers, which penalize the mismatch between *both* state and parameter estimates, is the abstract form that the collective error dynamics takes. In particular, the proposed framework allows one to decouple the graph connectivity (i.e., the graph Laplacian) from the stability analysis of the parameter errors by simply replacing a nonnegative damping-like matrix representing the connectivity of the graph topology with another matrix representing a more general interagent connectivity.

## 14.2 Mathematical Preliminaries and Problem Formulation

The graph-theoretic notation and terminology we use in this chapter were introduced in our previous chapters. Specifically, we let $\mathfrak{G} = (\mathcal{V}, \mathcal{E}, \mathcal{A})$ denote a weighted directed graph (or digraph) with the set of nodes (or vertices) $\mathcal{V} = \{1, \ldots, N\}$, the set of edges $\mathcal{E} \subseteq \mathcal{V} \times \mathcal{V}$, and a weighted adjacency matrix $\mathcal{A} \in \mathbb{R}^{N \times N}$. Every edge $\ell \in \mathcal{E}$ corresponds to an ordered pair of vertices $(i, j) \in \mathcal{V} \times \mathcal{V}$, where $i$ and $j$ are the *initial* and *terminal* vertices of the edge $\ell$. In this case, $\ell$ is *incident into* $j$ and *incident out of* $i$. We say that $\mathfrak{G}$ is *strongly* (respectively, *weakly*) *connected* if for every ordered pair of vertices $(i, j)$, $i \neq j$, there exists a *directed* (respectively, *undirected*) *path*, that is, a directed (respectively, undirected) sequence of arcs, leading from $i$ to $j$. The in-neighbors and out-neighbors of node $i$ are, respectively, defined as $\mathcal{N}_{\mathrm{in}}(i) \triangleq \{j \in \mathcal{V} : (j, i) \in \mathcal{E}\}$ and $\mathcal{N}_{\mathrm{out}}(i) \triangleq \{j \in \mathcal{V} : (i, j) \in \mathcal{E}\}$. Furthermore, recall that the graph Laplacian of $\mathfrak{G}$ is defined as $\mathcal{L} \triangleq \Delta - \mathcal{A}$, where $\Delta \triangleq \mathrm{diag}[\deg_{\mathrm{in}}(1), \ldots, \deg_{\mathrm{in}}(N)]$. Finally, for an undirected graph for which the arc set is symmetric, that is, $\mathcal{A} = \mathcal{A}^{\mathrm{T}}$, $\mathcal{N}_{\mathrm{in}}(i) = \mathcal{N}_{\mathrm{out}}(i) \triangleq \mathcal{N}(i)$ and $\deg_{\mathrm{in}}(i) = \deg_{\mathrm{out}}(i) \triangleq \deg(i)$, $i = 1, \ldots, N$.

Recall that the consensus problem involves the system $\mathcal{G}$ given by

$$\dot{x}_i(t) = \sum_{j \in \mathcal{N}_{\mathrm{in}}(i)} \phi_{ij}(x_i(t), x_j(t)), \quad x_i(0) = x_{i0}, \quad t \geq 0, \quad i = 1, \ldots, N,$$

(14.1)

where $\phi_{ij}(\cdot, \cdot)$, $i, j = 1, \ldots, N$, are locally Lipschitz continuous. Here, $x_i(t)$, $t \geq 0$, represents an information state with a distributed consensus algorithm involving neighbor-to-neighbor interaction between agents. In this chapter, we consider distributed consensus estimation algorithms of the form [256]

$$\dot{x}_i(t) = \sum_{j \in \mathcal{N}_{\mathrm{in}}(i)} (x_j(t) - x_i(t)), \quad x_i(0) = x_{i0}, \quad t \geq 0, \quad i = 1, \ldots, N. \quad (14.2)$$

## 14.3 Adaptive Estimation Problem

In this section, we present a brief exposition of the standard centralized adaptive estimation for plant parameter estimation in dynamical systems involving full state information. Specifically, we consider dynamical systems of the form

$$\dot{x}(t) = Ax(t) + Bu(t), \quad x(0) = x_0, \quad t \geq 0, \tag{14.3}$$

where $x(t) \in \mathbb{R}^n$, $t \geq 0$, is the state vector and $u(t) \in \mathbb{R}^m$, $t \geq 0$, is the control input. Here, we assume that the plant and input matrices $A$ and $B$ are unknown and the state $x(t)$ and control input signal $u(t)$ are bounded for all $t \geq 0$. To identify the matrices $A$ and $B$ online, we consider the adaptive observer given by [176]

$$\dot{\hat{x}}(t) = \hat{A}(t)x(t) + \hat{B}(t)u(t) + A_\mathrm{m}(\hat{x}(t) - x(t))$$
$$= A_\mathrm{m}\hat{x}(t) + (\hat{A}(t) - A_\mathrm{m})x(t) + \hat{B}(t)u(t),$$
$$\hat{x}(0) = \hat{x}_0 \neq x_0, \quad t \geq 0, \tag{14.4}$$

where $\hat{x}(t) \in \mathbb{R}^n$, $t \geq 0$, is the observer state; $\hat{A}(t) \in \mathbb{R}^{n \times n}$, $t \geq 0$, is the adaptive estimate of $A$; and $\hat{B}(t) \in \mathbb{R}^{n \times m}$, $t \geq 0$, is the adaptive estimate of $B$. The matrix $A_\mathrm{m} \in \mathbb{R}^{n \times n}$ is a design matrix that is Hurwitz and defines the observer poles.

To establish online estimates for the system matrices $A$ and $B$, define the state and parameter errors, respectively, by $e(t) \triangleq \hat{x}(t) - x(t)$, $t \geq 0$; $\tilde{A}(t) \triangleq \hat{A}(t) - A$, $t \geq 0$; and $\tilde{B}(t) \triangleq \hat{B}(t) - B$, $t \geq 0$. Then the system error dynamics and parameter update dynamics are given by

$$\dot{e}(t) = A_\mathrm{m}e(t) + \tilde{A}(t)x(t) + \tilde{B}(t)u(t), \quad e(0) = e_0, \quad t \geq 0, \tag{14.5}$$

$$\dot{\hat{A}}(t) = -\Gamma_\mathrm{a}Pe(t)x^\mathrm{T}(t), \quad \hat{A}(0) = \hat{A}_0 \neq A, \tag{14.6}$$

$$\dot{\hat{B}}(t) = -\Gamma_\mathrm{b}Pe(t)u^\mathrm{T}(t), \quad \hat{B}(0) = \hat{B}_0 \neq B, \tag{14.7}$$

where $\Gamma_\mathrm{a} \in \mathbb{R}^{n \times n}$ and $\Gamma_\mathrm{b} \in \mathbb{R}^{m \times m}$ are positive definite gain matrices and $P \in \mathbb{R}^{n \times n}$ is a positive definite solution of the Lyapunov equation

$$0 = A_\mathrm{m}^\mathrm{T}P + PA_\mathrm{m} + R, \tag{14.8}$$

where $R \in \mathbb{R}^{n \times n}$ is a given positive definite matrix. Since $A_m$ is Hurwitz, it follows from converse Lyapunov theory [122] that there exists a unique positive definite matrix $P \in \mathbb{R}^{n \times n}$ satisfying (14.8) for a given positive definite matrix $R \in \mathbb{R}^{n \times n}$. The adaptive update laws for $\hat{A}(t)$, $t \geq 0$, and $\hat{B}(t)$, $t \geq 0$, given by (14.6) and (14.7), respectively, can be derived using standard Lyapunov analysis by considering the Lyapunov function candidate

$$V(e, \tilde{A}, \tilde{B}) = e^{\mathrm{T}} P e + \mathrm{tr} \ \tilde{A}^{\mathrm{T}} \Gamma_{\mathrm{a}}^{-1} \tilde{A} + \mathrm{tr} \ \tilde{B}^{\mathrm{T}} \Gamma_{\mathrm{b}}^{-1} \tilde{B}. \qquad (14.9)$$

Note that $V(0,0,0) = 0$ and $V(e, \tilde{A}, \tilde{B}) > 0$ for all $(e, \tilde{A}, \tilde{B}) \neq (0,0,0)$. Now, differentiating (14.9) along the trajectories of (14.5)–(14.7) yields

$$\dot{V}(e(t), \tilde{A}(t), \tilde{B}(t)) = \dot{e}^{\mathrm{T}}(t) P e(t) + e^{\mathrm{T}}(t) P \dot{e}(t) + 2 \mathrm{tr} \ \tilde{A}^{\mathrm{T}}(t) \Gamma_{\mathrm{a}}^{-1} \dot{\tilde{A}}(t)$$
$$+ 2 \mathrm{tr} \ \tilde{B}^{\mathrm{T}}(t) \Gamma_{\mathrm{b}}^{-1} \dot{\tilde{B}}(t)$$
$$= e^{\mathrm{T}}(t)(A_m^{\mathrm{T}} P + P A_m) e(t) + 2 \mathrm{tr} \ \tilde{A}^{\mathrm{T}}(t) P e(t) x^{\mathrm{T}}(t)$$
$$+ 2 \mathrm{tr} \ \tilde{B}^{\mathrm{T}}(t) P e(t) u^{\mathrm{T}}(t) + 2 \mathrm{tr} \ \tilde{A}^{\mathrm{T}}(t) \Gamma_{\mathrm{a}}^{-1} \dot{\tilde{A}}(t)$$
$$+ 2 \mathrm{tr} \ \tilde{B}^{\mathrm{T}}(t) \Gamma_{\mathrm{b}}^{-1} \dot{\tilde{B}}(t). \qquad (14.10)$$

Using the update laws (14.6) and (14.7) in (14.10), it follows that

$$\dot{V}(e(t), \tilde{A}(t), \tilde{B}(t)) = -e^{\mathrm{T}}(t) R e(t) \leq 0, \quad t \geq 0, \qquad (14.11)$$

which guarantees that the error signal $e(t)$, $t \geq 0$, and parameter errors $\tilde{A}(t)$, $t \geq 0$, and $\tilde{B}(t)$, $t \geq 0$, are Lyapunov stable and, hence, are bounded for all $t \geq 0$.

Since $e(t)$, $t \geq 0$; $\hat{A}(t)$, $t \geq 0$; $\hat{B}(t)$, $t \geq 0$; $x(t)$, $t \geq 0$; and $u(t)$, $t \geq 0$, are bounded for all $t \geq 0$, it follows that $\dot{e}(t)$, $t \geq 0$, is bounded, and, hence, $\ddot{V}(e(t), \tilde{A}(t), \tilde{B}(t))$ is bounded for all $t \geq 0$. Now, it follows from Barbalat's lemma [122, p. 221] that $\dot{V}(e(t), \tilde{A}(t), \tilde{B}(t)) \to 0$ as $t \to \infty$, and, hence, $e(t)$ converges to zero asymptotically. Convergence of the adaptive estimates to their true values can be shown when a persistency of excitation condition is imposed [176, 230].

The system error dynamics (14.5) and parameter dynamics (14.6) and (14.7) can be written in operator form as

$$\begin{bmatrix} \dot{e}(t) \\ \dot{\tilde{A}}(t) \\ \dot{\tilde{B}}(t) \end{bmatrix} = \mathfrak{A}(x(t), u(t)) \begin{bmatrix} e(t) \\ \tilde{A}(t) \\ \tilde{B}(t) \end{bmatrix}, \quad \begin{bmatrix} e(0) \\ \tilde{A}(0) \\ \tilde{B}(0) \end{bmatrix} = \begin{bmatrix} e_0 \\ \tilde{A}_0 \\ \tilde{B}_0 \end{bmatrix}, \quad t \geq 0,$$

$$(14.12)$$

where

$$\mathfrak{A}(x(t), u(t)) \triangleq \left[ \begin{array}{c|cc} A_{\mathrm{m}} & (\cdot) x(t) & (\cdot) u(t) \\ \hline -\Gamma_{\mathrm{a}} P(\cdot) x^{\mathrm{T}}(t) & 0 & 0 \\ -\Gamma_{\mathrm{b}} P(\cdot) u^{\mathrm{T}}(t) & 0 & 0 \end{array} \right].$$

The structure given in (14.12) involving the skew-adjoint, state-dependent operator $\mathfrak{A}(\cdot, \cdot)$ is characteristic of adaptive systems [229]. The same structure is observed in the case of distributed adaptive consensus identifiers presented in Section 14.4, in which the operator form involves the same structure as above with additional terms arising due to consensus enforcement. As we see in Section 14.4, this form can be related to the Laplacian of the graph topology of the network.

## 14.4 Adaptive Distributed Observers

In this section, we consider a distributed adaptive observer problem for (14.3). Specifically, we consider $N$ *noninteracting* agents given by

$$\dot{\hat{x}}_i(t) = A_{\mathrm{m}}\hat{x}_i(t) + (\hat{A}_i(t) - A_{\mathrm{m}})x(t) + \hat{B}_i(t)u(t),$$
$$\hat{x}_i(0) = \hat{x}_{i0} \neq x(0), \quad t \geq 0, \tag{14.13}$$

$$\dot{\hat{A}}_i(t) = -\Gamma_{\mathrm{a}i}Pe_i(t)x^{\mathrm{T}}(t), \quad \hat{A}_i(0) = \hat{A}_{i0}, \tag{14.14}$$

$$\dot{\hat{B}}_i(t) = -\Gamma_{\mathrm{b}i}Pe_i(t)u^{\mathrm{T}}(t), \quad \hat{B}_i(0) = \hat{B}_{i0}, \tag{14.15}$$

where, for $i = 1, \ldots, N$, $\hat{x}_i(t) \in \mathbb{R}^n$, $t \geq 0$; $\hat{A}_i(t) \in \mathbb{R}^{n \times n}$, $t \geq 0$; $\hat{B}_i(t) \in \mathbb{R}^{n \times m}$, $t \geq 0$; $e_i(t) \triangleq \hat{x}_i(t) - x(t)$, $t \geq 0$; and $\Gamma_{\mathrm{a}i} \in \mathbb{R}^{n \times n}$ and $\Gamma_{\mathrm{b}i} \in \mathbb{R}^{m \times m}$ are positive definite gain matrices. Here, we can easily replace $A_{\mathrm{m}}$ in (14.13) with $A_{\mathrm{m}i}$, where $A_{\mathrm{m}i}$, $i = 1, \ldots, N$, are Hurwitz design matrices. In this case, the results in the remainder of the chapter hold with minor extensions.

To quantify a measure of disagreement between the state estimates and parameter estimates that is independent of the network topology, we use the deviation of these estimates from the mean defined by

$$\delta_{ie}(t) \triangleq \hat{x}_i(t) - \frac{1}{N}\sum_{j=1}^{N}\hat{x}_j(t) = e_i(t) - \frac{1}{N}\sum_{j=1}^{N}e_j(t), \tag{14.16}$$

$$\delta_{ia}(t) \triangleq \hat{A}_i(t) - \frac{1}{N}\sum_{j=1}^{N}\hat{A}_j(t) = \tilde{A}_i(t) - \frac{1}{N}\sum_{j=1}^{N}\tilde{A}_j(t), \tag{14.17}$$

$$\delta_{ib}(t) \triangleq \hat{B}_i(t) - \frac{1}{N}\sum_{j=1}^{N}\hat{B}_j(t) = \tilde{B}_i(t) - \frac{1}{N}\sum_{j=1}^{N}\tilde{B}_j(t), \tag{14.18}$$

for $i = 1, \ldots, N$. In this case, the pairwise disagreement is defined as

$$\hat{x}_{ij}(t) \triangleq \hat{x}_i(t) - \hat{x}_j(t) = e_{ij}(t) = e_i(t) - e_j(t), \tag{14.19}$$

$$\hat{A}_{ij}(t) \triangleq \hat{A}_i(t) - \hat{A}_j(t) = \tilde{A}_{ij}(t) = \tilde{A}_i(t) - \tilde{A}_j(t), \tag{14.20}$$

$$\hat{B}_{ij}(t) \triangleq \hat{B}_i(t) - \hat{B}_j(t) = \tilde{B}_{ij}(t) = \tilde{B}_i(t) - \tilde{B}_j(t) \tag{14.21}$$

for $i, j = 1, \ldots, N$, $i \neq j$.

Note that the distributed adaptive observers (14.13)–(14.15) can be placed in the form of (14.12). To see this, define

$$
E(t) \triangleq \begin{bmatrix} e_1(t) \\ e_2(t) \\ \vdots \\ e_N(t) \end{bmatrix} \in \mathbb{R}^{nN}, \quad \hat{\mathbb{A}}(t) \triangleq \begin{bmatrix} \hat{A}_1(t) \\ \hat{A}_2(t) \\ \vdots \\ \hat{A}_N(t) \end{bmatrix} \in \mathbb{R}^{nN \times n},
$$

$$
\hat{\mathbb{B}}(t) \triangleq \begin{bmatrix} \hat{B}_1(t) \\ \hat{B}_2(t) \\ \vdots \\ \hat{B}_N(t) \end{bmatrix} \in \mathbb{R}^{nN \times m},
$$

with $\tilde{\mathbb{A}}(t)$ and $\tilde{\mathbb{B}}(t)$ defined analogously, and define $\mathbb{A}_\mathrm{m} \triangleq I_N \otimes A_\mathrm{m}$, $\mathbb{P} \triangleq I_N \otimes P$,

$$
\Gamma_\mathrm{a} \triangleq \begin{bmatrix} \Gamma_{\mathrm{a}1} & 0_{n \times n} & \cdots \\ \vdots & \ddots & \vdots \\ 0_{n \times n} & \cdots & \Gamma_{\mathrm{a}N} \end{bmatrix}, \quad \Gamma_\mathrm{b} \triangleq \begin{bmatrix} \Gamma_{\mathrm{b}1} & 0_{n \times n} & \cdots \\ \vdots & \ddots & \vdots \\ 0_{n \times n} & \cdots & \Gamma_{\mathrm{b}N} \end{bmatrix}.
$$

Then,

$$
\dot{E}(t) = \mathbb{A}_\mathrm{m} E(t) + \tilde{\mathbb{A}}(t)x(t) + \tilde{\mathbb{B}}(t)u(t), \quad E(0) = E_0, \quad t \geq 0, \tag{14.22}
$$

$$
\dot{\tilde{\mathbb{A}}}(t) = -\Gamma_\mathrm{a}\mathbb{P}E(t)x^\mathrm{T}(t), \quad \tilde{\mathbb{A}}(0) = \tilde{\mathbb{A}}_0, \tag{14.23}
$$

$$
\dot{\tilde{\mathbb{B}}}(t) = -\Gamma_\mathrm{b}\mathbb{P}E(t)u^\mathrm{T}(t), \quad \tilde{\mathbb{B}}(0) = \tilde{\mathbb{B}}_0, \tag{14.24}
$$

or, equivalently,

$$
\begin{bmatrix} \dot{E}(t) \\ \dot{\tilde{\mathbb{A}}}(t) \\ \dot{\tilde{\mathbb{B}}}(t) \end{bmatrix} = \tilde{\mathfrak{A}}(x(t), u(t)) \begin{bmatrix} E(t) \\ \tilde{\mathbb{A}}(t) \\ \tilde{\mathbb{B}}(t) \end{bmatrix}, \quad \begin{bmatrix} E(0) \\ \tilde{\mathbb{A}}(0) \\ \tilde{\mathbb{B}}(0) \end{bmatrix} = \begin{bmatrix} E_0 \\ \tilde{\mathbb{A}}_0 \\ \tilde{\mathbb{B}}_0 \end{bmatrix}, \quad t \geq 0,
$$
$$\tag{14.25}$$

where

$$
\tilde{\mathfrak{A}}(x(t), u(t)) \triangleq \left[ \begin{array}{c|cc} \mathbb{A}_\mathrm{m} & (\cdot)x(t) & (\cdot)u(t) \\ \hline -\Gamma_\mathrm{a}\mathbb{P}(\cdot)x^\mathrm{T}(t) & 0 & 0 \\ -\Gamma_\mathrm{b}\mathbb{P}(\cdot)u^\mathrm{T}(t) & 0 & 0 \end{array} \right].
$$

Equation (14.25) is the multiagent identifier version of (14.12) and shows that the distributed adaptive observers (14.13)–(14.15) have identical stability and convergence properties to (14.4), (14.6), and (14.7).

In particular, consider the distributed Lyapunov function candidates for each agent given by

$$V_i(e_i, \tilde{A}_i, \tilde{B}_i) = e_i^{\mathrm{T}} P e_i + \mathrm{tr}\, \tilde{A}_i^{\mathrm{T}} \Gamma_{\mathrm{a}i}^{-1} \tilde{A}_i + \mathrm{tr}\, \tilde{B}_i^{\mathrm{T}} \Gamma_{\mathrm{b}i}^{-1} \tilde{B}_i, \quad i = 1, \ldots, N.$$
(14.26)

Now, the stability of the collective dynamics of (14.13)–(14.15) can be established using the Lyapunov function candidate

$$V(E, \tilde{\mathbb{A}}, \tilde{\mathbb{B}}) = \sum_{i=1}^{N} V_i(e_i, \tilde{A}_i, \tilde{B}_i).$$
(14.27)

Specifically, differentiating (14.27) along the trajectories of (14.13)–(14.15) yields

$$\dot{V}_i(e_i(t), \tilde{A}_i(t), \tilde{B}_i(t)) = -e_i^{\mathrm{T}}(t) R e_i(t), \quad t \geq 0, \quad i = 1, \ldots, N,$$

and, hence,

$$\dot{V}(E(t), \tilde{\mathbb{A}}(t), \tilde{\mathbb{B}}(t)) = -\sum_{i=1}^{N} e_i^{\mathrm{T}}(t) R e_i(t) = -E^{\mathrm{T}}(t) \mathbb{R} E(t) \leq 0, \quad t \geq 0,$$

where $\mathbb{R} \triangleq I_N \otimes R$. Now, similar arguments as given in Section 14.2 can be used to show that $E(t)$ converges to zero asymptotically.

## 14.5 Adaptive Consensus of Distributed Observers over Networks with Undirected Graph Topologies

In this section, we consider a multiagent system in which $N$ agents are utilized to adaptively estimate the plant parameters $A$ and $B$ over a connected undirected network. Each agent provides its own estimate $\hat{A}_i(t)$, $t \geq 0$, and $\hat{B}_i(t)$, $t \geq 0$, $i = 1, \ldots, N$, and strives to arrive at common estimates, that is, reach consensus on the parameter adaptive estimates. The update laws for the parameter identifiers given by (14.13) are modified to include interagent communication with a penalty on the mismatch between the parameter estimates $\hat{A}_i(t)$, $t \geq 0$, and $\hat{B}_i(t)$, $t \geq 0$. Even though the individual adaptive estimates require a condition of persistency of excitation to ensure parameter convergence, the proposed adaptive consensus modification guarantees that all the parameter estimates agree with each other, which emulates a persistency of excitation condition.

**Theorem 14.1.** Consider the dynamical system (14.3) with $A$ and $B$ unknown. Assume that $\mathfrak{G}$ defines a connected undirected graph of $N$ agents implementing the distributed adaptive observers given by

$$\dot{\hat{x}}_i(t) = A_{\mathrm{m}} \hat{x}_i(t) + (\hat{A}_i(t) - A_{\mathrm{m}}) x(t) + \hat{B}_i(t) u(t) - P^{-1} \sum_{j \in \mathcal{N}(i)} (\hat{x}_i(t) - \hat{x}_j(t)),$$

$$\hat{x}_i(0) = \hat{x}_{i0}, \quad t \geq 0, \tag{14.28}$$

$$\dot{\hat{A}}_i(t) = -\Gamma_{ai} P e_i(t) x^{\mathrm{T}}(t) - \Gamma_{ai} \sum_{j \in \mathcal{N}(i)} (\hat{A}_i(t) - \hat{A}_j(t)), \quad \hat{A}_i(0) = \hat{A}_{i0},$$

$$\tag{14.29}$$

$$\dot{\hat{B}}_i(t) = -\Gamma_{bi} P e_i(t) u^{\mathrm{T}}(t) - \Gamma_{bi} \sum_{j \in \mathcal{N}(i)} (\hat{B}_i(t) - \hat{B}_j(t)), \quad \hat{B}_i(0) = \hat{B}_{i0},$$

$$\tag{14.30}$$

where $i = 1, \ldots, N$ and $P$ satisfies (14.8). Then the equilibrium solution $(E(t), \tilde{\mathbb{A}}(t), \tilde{\mathbb{B}}(t))$ of the parameter error system is Lyapunov stable for all $(E_0, \tilde{\mathbb{A}}_0, \tilde{\mathbb{B}}_0) \in \mathbb{R}^{nN} \times \mathbb{R}^{nN \times n} \times \mathbb{R}^{nN \times m}$ and $t \geq 0$, and

$$\lim_{t \to \infty} \hat{x}_{ij}(t) = 0, \quad i, j = 1, \ldots, N,$$

$$\lim_{t \to \infty} \hat{A}_{ij}(t) = 0, \quad i, j = 1, \ldots, N,$$

$$\lim_{t \to \infty} \hat{B}_{ij}(t) = 0, \quad i, j = 1, \ldots, N,$$

$$\lim_{t \to \infty} e_i(t) = 0, \quad i = 1, \ldots, N.$$

**Proof.** Given (14.28)–(14.30), the state and parameter error dynamics are given by

$$\dot{e}_i(t) = A_{\mathrm{m}} e_i(t) + \tilde{A}_i(t) x(t) + \tilde{B}_i(t) u(t) - P^{-1} \sum_{j \in \mathcal{N}(i)} e_{ij}(t),$$

$$e_i(0) = e_{i0}, \quad t \geq 0, \tag{14.31}$$

$$\dot{\tilde{A}}_i(t) = -\Gamma_{ai} P e_i(t) x^{\mathrm{T}}(t) - \Gamma_{ai} \sum_{j \in \mathcal{N}(i)} \tilde{A}_{ij}(t), \quad \tilde{A}_i(0) = \tilde{A}_{i0}, \tag{14.32}$$

$$\dot{\tilde{B}}_i(t) = -\Gamma_{bi} P e_i(t) u^{\mathrm{T}}(t) - \Gamma_{bi} \sum_{j \in \mathcal{N}(i)} \tilde{B}_{ij}(t), \quad \tilde{B}_i(0) = \tilde{B}_{i0}, \tag{14.33}$$

for $i = 1, \ldots, N$. Next, consider the distributed Lyapunov function candidates given by (14.26), and note that the derivatives of $V_i(e_i, \tilde{A}_i, \tilde{B}_i)$, $i = 1, \ldots, N$, along the trajectories of (14.31)–(14.33) are given by

$$\dot{V}_i(e_i(t), \tilde{A}_i(t), \tilde{B}_i(t))$$
$$= e_i^{\mathrm{T}}(t) \left[ A_{\mathrm{m}}^{\mathrm{T}} P + P A_{\mathrm{m}} \right] e_i(t) - 2e_i^{\mathrm{T}}(t) \sum_{j \in \mathcal{N}(i)} e_{ij}(t) + 2e_i^{\mathrm{T}}(t) P \tilde{A}_i(t) x(t)$$

$$+ 2e_i^{\mathrm{T}}(t) P \tilde{B}_i(t) u(t) + 2\mathrm{tr}\left[ \dot{\tilde{A}}_i^{\mathrm{T}}(t) \Gamma_{ai}^{-1} \tilde{A}_i(t) \right] + 2\mathrm{tr}\left[ \dot{\tilde{B}}_i^{\mathrm{T}}(t) \Gamma_{bi}^{-1} \tilde{B}_i(t) \right]$$

$$= -e_i^{\mathrm{T}}(t) R e_i(t) - 2e_i^{\mathrm{T}}(t) \sum_{j \in \mathcal{N}(i)} e_{ij}(t)$$

$$+ 2\mathrm{tr}\left[ \tilde{A}_i^{\mathrm{T}}(t) [P e_i(t) x^{\mathrm{T}}(t) + \Gamma_{ai}^{-1} \dot{\tilde{A}}_i(t)] \right]$$

$$+ 2\mathrm{tr}\Big[\tilde{B}_i^{\mathrm{T}}(t)[Pe_i(t)u^{\mathrm{T}}(t) + \Gamma_{\mathrm{b}i}^{-1}\dot{\tilde{B}}_i(t)]\Big]$$

$$= -e_i^{\mathrm{T}}(t)Re_i(t) - 2e_i^{\mathrm{T}}(t)\sum_{j\in\mathcal{N}(i)}e_{ij}(t) - 2\mathrm{tr}\Big[\tilde{A}_i^{\mathrm{T}}(t)\sum_{j\in\mathcal{N}(i)}\tilde{A}_{ij}(t)\Big]$$

$$- 2\mathrm{tr}\Big[\tilde{B}_i^{\mathrm{T}}(t)\sum_{j\in\mathcal{N}(i)}\tilde{B}_{ij}(t)\Big]. \tag{14.34}$$

Now, using (14.34), it follows from (14.27) that the derivative of $V(E,\tilde{\mathbb{A}},\tilde{\mathbb{B}})$ along the error trajectories of (14.31)–(14.33) is given by

$$\dot{V}(E(t),\tilde{\mathbb{A}}(t),\tilde{\mathbb{B}}(t))$$

$$= \sum_{i=1}^{N}\dot{V}_i(e_i(t),\tilde{A}_i(t),\tilde{B}_i(t))$$

$$= -\sum_{i=1}^{N}e_i^{\mathrm{T}}(t)Re_i(t) - 2\sum_{i=1}^{N}e_i^{\mathrm{T}}(t)\sum_{j\in\mathcal{N}(i)}(e_i(t)-e_j(t))$$

$$- 2\mathrm{tr}\Big[\sum_{i=1}^{N}\tilde{A}_i^{\mathrm{T}}(t)\sum_{j\in\mathcal{N}(i)}(\tilde{A}_i(t)-\tilde{A}_j(t))\Big]$$

$$- 2\mathrm{tr}\Big[\sum_{i=1}^{N}\tilde{B}_i^{\mathrm{T}}(t)\sum_{j\in\mathcal{N}(i)}(\tilde{B}_i(t)-\tilde{B}_j(t))\Big]$$

$$= -\sum_{i=1}^{N}e_i^{\mathrm{T}}(t)Re_i(t) - \sum_{i=1}^{N}\sum_{j\in\mathcal{N}(i)}\|e_{ij}(t)\|_2^2 - \sum_{i=1}^{N}\sum_{j\in\mathcal{N}(i)}\|\tilde{A}_{ij}(t)\|_{\mathrm{F}}^2$$

$$- \sum_{i=1}^{N}\sum_{j\in\mathcal{N}(i)}\|\tilde{B}_{ij}(t)\|_{\mathrm{F}}^2$$

$$\leq 0, \quad t\geq 0, \tag{14.35}$$

where in (14.35) we used the identities

$$2\sum_{i=1}^{N}e_i^{\mathrm{T}}(t)\sum_{j\in\mathcal{N}(i)}(e_i(t)-e_j(t)) = \sum_{i=1}^{N}\sum_{j\in\mathcal{N}(i)}\|e_i(t)-e_j(t)\|_2^2, \tag{14.36}$$

$$2\mathrm{tr}\sum_{i=1}^{N}\tilde{A}_i^{\mathrm{T}}(t)\sum_{j\in\mathcal{N}(i)}(\tilde{A}_i(t)-\tilde{A}_j(t))$$

$$= \sum_{i=1}^{N}\sum_{j\in\mathcal{N}(i)}\mathrm{tr}[\tilde{A}_i(t)-\tilde{A}_j(t)][\tilde{A}_i(t)-\tilde{A}_j(t)]^{\mathrm{T}}, \tag{14.37}$$

and

$$2\mathrm{tr} \sum_{i=1}^{N} \tilde{B}_i^{\mathrm{T}}(t) \sum_{j\in\mathcal{N}(i)} (\tilde{B}_i(t) - \tilde{B}_j(t))$$

$$= \sum_{i=1}^{N} \sum_{j\in\mathcal{N}(i)} \mathrm{tr}[\tilde{B}_i(t) - \tilde{B}_j(t)][\tilde{B}_i(t) - \tilde{B}_j(t)]^{\mathrm{T}}. \qquad (14.38)$$

(Note that (14.36)–(14.38) hold because $\mathfrak{G}$ is an undirected graph, and, hence, $j \in \mathcal{N}(i)$ if and only if $i \in \mathcal{N}(j)$.) Hence, (14.35) implies that the solution $(E(t), \tilde{\mathbb{A}}(t), \tilde{\mathbb{B}}(t))$ of the parameter error system is Lyapunov stable for all $(E_0, \tilde{\mathbb{A}}_0, \tilde{\mathbb{B}}_0) \in \mathbb{R}^{nN} \times \mathbb{R}^{nN\times n} \times \mathbb{R}^{nN\times m}$ and $t \geq 0$.

Next, note that (14.35) implies that

$$V(E(0), \tilde{\mathbb{A}}(0), \tilde{\mathbb{B}}(0))$$

$$\geq V(E(t), \tilde{\mathbb{A}}(t), \tilde{\mathbb{B}}(t)) + \lambda_{\min}(R) \int_0^t \sum_{i=1}^{N} \|e_i(\tau)\|_2^2 \mathrm{d}\tau$$

$$+ \int_0^t \sum_{i=1}^{N} \sum_{j\in\mathcal{N}(i)} \|e_{ij}(\tau)\|_2^2 \mathrm{d}\tau + \int_0^t \sum_{i=1}^{N} \sum_{j\in\mathcal{N}(i)} \|\tilde{A}_{ij}(\tau)\|_{\mathrm{F}}^2 \mathrm{d}\tau$$

$$+ \int_0^t \sum_{i=1}^{N} \sum_{j\in\mathcal{N}(i)} \|\tilde{B}_{ij}(\tau)\|_{\mathrm{F}}^2 \mathrm{d}\tau,$$

and, hence, $E(\cdot) \in \mathcal{L}_2 \cap \mathcal{L}_\infty$ or, equivalently, $e_i(\cdot) \in \mathcal{L}_2 \cap \mathcal{L}_\infty$, $i = 1, \ldots, N$, with $\hat{A}_i(\cdot) \in \mathcal{L}_\infty$, $\hat{B}_i(\cdot) \in \mathcal{L}_\infty$, $e_{ij}(\cdot) \in \mathcal{L}_2$, $\hat{A}_{ij}(\cdot) \in \mathcal{L}_2$, and $\hat{B}_{ij}(\cdot) \in \mathcal{L}_2$, $i, j = 1, \ldots, N$. Furthermore, since $\hat{A}_i(\cdot)$ and $\hat{B}_i(\cdot)$ are bounded, it follows that $\hat{A}_{ij}(\cdot) \in \mathcal{L}_\infty$ and $\hat{B}_{ij}(\cdot) \in \mathcal{L}_\infty$, $i, j = 1, \ldots, N$. Now, since $x(t)$, $t \geq 0$, and $u(t) \geq 0$ are bounded, it follows from (14.31) that $\dot{e}_i(\cdot) \in \mathcal{L}_\infty$, and, hence, by Barbalat's lemma [122, p. 221] (since $e_i(\cdot) \in \mathcal{L}_2 \cap \mathcal{L}_\infty$ and $\dot{e}_i(\cdot) \in \mathcal{L}_\infty$), it follows that $\lim_{t\to\infty} \|e_i(t)\|_2 = 0$, $i = 1, \ldots, N$.

Next, since $e_i(\cdot) \in \mathcal{L}_2 \cap \mathcal{L}_\infty$ and $\dot{e}_i(\cdot) \in \mathcal{L}_\infty$, it follows that $e_{ij}(\cdot) \in \mathcal{L}_2 \cap \mathcal{L}_\infty$ and $\dot{e}_{ij}(\cdot) \in \mathcal{L}_\infty$, and, hence,

$$\lim_{t\to\infty} \|e_{ij}(t)\|_2 = \lim_{t\to\infty} \|\hat{x}_{ij}(t)\|_2 = 0, \quad i, j = 1, \ldots, N.$$

Finally, it follows that $\dot{\hat{A}}_{ij}(\cdot)$ and $\dot{\hat{B}}_{ij}(\cdot)$ are bounded since $x(t)$, $t \geq 0$, and $u(t)$, $t \geq 0$, are bounded, and, thus, by Barbalat's lemma [122, p. 221], $\lim_{t\to\infty} \|\hat{A}_{ij}(t)\|_{\mathrm{F}} = 0$ and $\lim_{t\to\infty} \|\hat{B}_{ij}(t)\|_{\mathrm{F}} = 0$.                                    $\square$

The proposed adaptive consensus distributed observers given in Theorem 14.1 guarantee state and parameter estimate consensus as well as convergence of the pairwise difference of the adaptive estimates. This follows as a direct consequence of the $\mathcal{L}_2$ boundedness of the pairwise disagreement of the parameter estimates.

Next, to examine the dynamic agreement of the parameter and state estimates, we consider the error system (14.31)–(14.33) in a compact form. Specifically, define $\mathbb{J} \triangleq \mathcal{L} \otimes I_n$ so that (14.22)–(14.24) become

$$\dot{E}(t) = \mathbb{A}_{\mathrm{m}}E(t) + \tilde{\mathbb{A}}(t)x(t) + \tilde{\mathbb{B}}(t)u(t) - \mathbb{P}^{-1}\mathbb{J}E(t), \quad E(0) = E_0, \quad t \geq 0, \tag{14.39}$$

$$\dot{\tilde{\mathbb{A}}}(t) = -\Gamma_{\mathrm{a}}\mathbb{P}E(t)x^{\mathrm{T}}(t) - \Gamma_{\mathrm{a}}\mathbb{J}\tilde{\mathbb{A}}(t), \quad \tilde{\mathbb{A}}(0) = \tilde{\mathbb{A}}_0, \tag{14.40}$$

$$\dot{\tilde{\mathbb{B}}}(t) = -\Gamma_{\mathrm{b}}\mathbb{P}E(t)u^{\mathrm{T}}(t) - \Gamma_{\mathrm{b}}\mathbb{J}\tilde{\mathbb{B}}(t), \quad \tilde{\mathbb{B}}(0) = \tilde{\mathbb{B}}_0. \tag{14.41}$$

Equivalently, (14.39)–(14.41) can be rewritten in operator form as

$$\begin{bmatrix} \dot{E}(t) \\ \dot{\tilde{\mathbb{A}}}(t) \\ \dot{\tilde{\mathbb{B}}}(t) \end{bmatrix} = \left( \tilde{\mathfrak{A}}(x(t), u(t)) - \tilde{G}\tilde{J} \right) \begin{bmatrix} E(t) \\ \tilde{\mathbb{A}}(t) \\ \tilde{\mathbb{B}}(t) \end{bmatrix},$$

$$\begin{bmatrix} E(0) \\ \tilde{\mathbb{A}}(0) \\ \tilde{\mathbb{B}}(0) \end{bmatrix} = \begin{bmatrix} E_0 \\ \tilde{\mathbb{A}}_0 \\ \tilde{\mathbb{B}}_0 \end{bmatrix}, \quad t \geq 0, \tag{14.42}$$

where

$$\tilde{G} \triangleq \begin{bmatrix} \mathbb{P}^{-1} & 0 & 0 \\ 0 & \Gamma_{\mathrm{a}} & 0 \\ 0 & 0 & \Gamma_{\mathrm{b}} \end{bmatrix}, \quad \tilde{J} \triangleq \begin{bmatrix} \mathbb{J} & 0 & 0 \\ 0 & \mathbb{J} & 0 \\ 0 & 0 & \mathbb{J} \end{bmatrix}.$$

Note that (14.42) has a similar structure to (14.25), differing only in the additional term $\tilde{J}$, which enforces consensus.

To assess the convergence properties of the deviation of (14.16)–(14.18) from the mean, let

$$\delta_{ie}(t) = e_i(t) - \frac{1}{N}\sum_{j=1}^{N} e_j(t) = \frac{1}{N}\sum_{j \neq i}^{N} e_{ij}(t),$$

with analogous expressions for $\delta_{ia}(t)$ and $\delta_{ib}(t)$. The convergence of the deviation from the mean of $e_i(t)$, $\tilde{A}_i(t)$, and $\tilde{B}_i(t)$, $t \geq 0$, $i = 1, \ldots, N$, can now be established using the fact that the pairwise disagreement of the state and parameter errors converges to zero, and, hence, $\lim_{t \to \infty} \delta_{ie}(t) = 0$, $\lim_{t \to \infty} \delta_{ia}(t) = 0$, and $\lim_{t \to \infty} \delta_{ib}(t) = 0$, $i = 1, \ldots, N$. This implies that

the individual deviations of the adaptive state and parameter estimates from
their mean (static average) converge to zero. This is summarized in the
following proposition.

**Proposition 14.1.** Consider the dynamical system (14.3) with $A$ and
$B$ unknown. Assume that $\mathfrak{G}$ defines a connected undirected graph of $N$
agents implementing the distributed adaptive consensus observers (14.28)–
(14.30), and let the deviations of the state and parameter estimates of the
observers from their mean (static average) be given by

$$\delta_{ie}(t) = \hat{x}_i(t) - \frac{1}{N} \sum_{j=1}^{N} \hat{x}_j(t) = e_i(t) - \frac{1}{N} \sum_{j=1}^{N} e_j(t)$$

$$= \frac{1}{N} \sum_{j \neq i}^{N} \hat{x}_{ij}(t) = \frac{1}{N} \sum_{j \neq i}^{N} e_{ij}(t),$$

$$\delta_{ia}(t) = \hat{A}_i(t) - \frac{1}{N} \sum_{j=1}^{N} \hat{A}_j(t) = \tilde{A}_i(t) - \frac{1}{N} \sum_{j=1}^{N} \tilde{A}_j(t)$$

$$= \frac{1}{N} \sum_{j \neq i}^{N} \hat{A}_{ij}(t) = \frac{1}{N} \sum_{j \neq i}^{N} \tilde{A}_{ij}(t),$$

$$\delta_{ib}(t) = \hat{B}_i(t) - \frac{1}{N} \sum_{j=1}^{N} \hat{B}_j(t) = \tilde{B}_i(t) - \frac{1}{N} \sum_{j=1}^{N} \tilde{B}_j(t)$$

$$= \frac{1}{N} \sum_{j \neq i}^{N} \hat{B}_{ij}(t) = \frac{1}{N} \sum_{j \neq i}^{N} \tilde{B}_{ij}(t).$$

Then, $\lim_{t \to \infty} \|\delta_{ie}(t)\|_2 = 0$, $\lim_{t \to \infty} \|\delta_{ia}(t)\|_2 = 0$, and $\lim_{t \to \infty} \|\delta_{ib}(t)\|_2 = 0$
for $i = 1, \ldots, N$.

**Proof.** The proof is a direct consequence of Theorem 14.1 by not-
ing the pairwise convergences $\lim_{t \to \infty} \hat{x}_{ij}(t) = 0$, $\lim_{t \to \infty} \hat{A}_{ij}(t) = 0$, and
$\lim_{t \to \infty} \hat{B}_{ij}(t) = 0$ and the fact that $N < \infty$.                                          $\square$

Alternatively, one can also consider the deviations from the mean of
the estimates of the neighboring agents. Specifically, defining

$$\gamma_{ie}(t) \triangleq \frac{1}{\deg(i)} \sum_{j \in \mathcal{N}(i)} e_{ij}(t), \quad \gamma_{ia}(t) \triangleq \frac{1}{\deg(i)} \sum_{j \in \mathcal{N}(i)} \tilde{A}_{ij}(t),$$

$$\gamma_{ib}(t) \triangleq \frac{1}{\deg(i)} \sum_{j \in \mathcal{N}(i)} \tilde{B}_{ij}(t),$$

we can relate these expressions to the graph Laplacian of $\mathfrak{G}$. In particular, let

$$\gamma_e(t) = \begin{bmatrix} \gamma_{1e}(t) \\ \gamma_{2e}(t) \\ \vdots \\ \gamma_{Ne}(t) \end{bmatrix}, \quad \gamma_a(t) = \begin{bmatrix} \gamma_{1a}(t) \\ \gamma_{2a}(t) \\ \vdots \\ \gamma_{Na}(t) \end{bmatrix}, \quad \gamma_b(t) = \begin{bmatrix} \gamma_{1b}(t) \\ \gamma_{2b}(t) \\ \vdots \\ \gamma_{Nb}(t) \end{bmatrix},$$

and note that, since

$$\sum_{j \in \mathcal{N}(i)} E_{ij}(t), \quad \sum_{j \in \mathcal{N}(i)} \tilde{A}_{ij}(t), \quad \sum_{j \in \mathcal{N}(i)} \tilde{B}_{ij}(t)$$

correspond to the $i$th block row of $\mathbb{J}E(t)$, $\mathbb{J}\tilde{A}(t)$, and $\mathbb{J}\tilde{B}(t)$, respectively, it follows that

$$\gamma_e(t) = \left(\Delta^{-1} \otimes I_n\right) \mathbb{J}E(t), \quad \gamma_a(t) = \left(\Delta^{-1} \otimes I_n\right) \mathbb{J}\tilde{A}(t),$$
$$\gamma_b(t) = \left(\Delta^{-1} \otimes I_n\right) \mathbb{J}\tilde{B}(t).$$

Now, it follows from Proposition 14.1 that

$$\lim_{t \to \infty} \|\mathbb{J}E(t)\|_{\mathrm{F}} = \lim_{t \to \infty} \|\mathbb{J}\tilde{A}(t)\|_{\mathrm{F}} = \lim_{t \to \infty} \|\mathbb{J}\tilde{B}(t)\|_{\mathrm{F}} = 0.$$

It is important to note, however, that $\lim_{t \to \infty} \|\tilde{A}(t)\|_{\mathrm{F}} = 0$ and $\lim_{t \to \infty} \|\tilde{B}(t)\|_{\mathrm{F}} = 0$ cannot be established unless one imposes the additional condition of persistency of excitation. This demonstrates the benefit of information sharing (i.e., graph connectivity), wherein the absence of $\mathbb{J}$ removes the convergence results on consensus unless a persistency of excitation condition is imposed.

## 14.6 Extensions to Networks with Directed Graph Topologies

In this section, we extend the results of Section 14.5 to adaptive consensus of distributed observers over networks with directed graph topologies.

**Theorem 14.2.** Consider the dynamical system (14.3) with $A$ and $B$ unknown. Assume that $\mathfrak{G}$ defines a weakly connected and balanced directed graph of $N$ agents implementing the distributed adaptive observers given by

$$\dot{\hat{x}}_i(t) = A_{\mathrm{m}}\hat{x}_i(t) + (\hat{A}_i(t) - A_{\mathrm{m}})x(t) + \hat{B}_i(t)u(t)$$
$$- P^{-1} \sum_{j \in \mathcal{N}_{\mathrm{in}}(i)} (\hat{x}_i(t) - \hat{x}_j(t)), \quad \hat{x}_i(0) = \hat{x}_{i0}, \quad t \geq 0, \qquad (14.43)$$

$$\dot{\hat{A}}_i(t) = -\Gamma_{\mathrm{a}i} P e_i(t) x^{\mathrm{T}}(t) - \Gamma_{\mathrm{a}i} \sum_{j \in \mathcal{N}_{\mathrm{in}}(i)} (\hat{A}_i(t) - \hat{A}_j(t)), \quad \hat{A}_i(0) = \hat{A}_{i0},$$

$$\qquad (14.44)$$

$$\dot{\hat{B}}_i(t) = -\Gamma_{\mathrm{b}i} P e_i(t) u^{\mathrm{T}}(t) - \Gamma_{\mathrm{b}i} \sum_{j \in \mathcal{N}_{\mathrm{in}}(i)} (\hat{B}_i(t) - \hat{B}_j(t)), \quad \hat{B}_i(0) = \hat{B}_{i0},$$

$$(14.45)$$

where $i = 1, \ldots, N$ and $P$ satisfies (14.8). Then the equilibrium solution $(E(t), \tilde{\mathbb{A}}(t), \tilde{\mathbb{B}}(t))$ of the parameter error system is Lyapunov stable for all $(E_0, \tilde{\mathbb{A}}_0, \tilde{\mathbb{B}}_0) \in \mathbb{R}^{nN} \times \mathbb{R}^{nN \times n} \times \mathbb{R}^{nN \times m}$ and $t \geq 0$, and $\lim_{t \to \infty} \hat{x}_{ij}(t) = 0$, $\lim_{t \to \infty} \hat{A}_{ij}(t) = 0$, and $\lim_{t \to \infty} \hat{B}_{ij}(t) = 0$ for $i, j = 1, \ldots, N$, and $\lim_{t \to \infty} e_i(t) = 0$ for $i = 1, \ldots, N$.

**Proof.** Given (14.43)–(14.45) the state and parameter error dynamics satisfy

$$\dot{e}_i(t) = A_{\mathrm{m}} e_i(t) + \tilde{A}_i(t) x(t) + \tilde{B}_i(t) u(t) - P^{-1} \sum_{j \in \mathcal{N}_{\mathrm{in}}(i)} e_{ij}(t),$$

$$e_i(0) = e_{i0}, \quad t \geq 0, \qquad (14.46)$$

$$\dot{\tilde{A}}_i(t) = -\Gamma_{\mathrm{a}i} P e_i(t) x^{\mathrm{T}}(t) - \Gamma_{\mathrm{a}i} \sum_{j \in \mathcal{N}_{\mathrm{in}}(i)} \tilde{A}_{ij}(t), \quad \tilde{A}_i(0) = \tilde{A}_{i0}, \qquad (14.47)$$

$$\dot{\tilde{B}}_i(t) = -\Gamma_{\mathrm{b}i} P e_i(t) u^{\mathrm{T}}(t) - \Gamma_{\mathrm{b}i} \sum_{j \in \mathcal{N}_{\mathrm{in}}(i)} \tilde{B}_{ij}(t), \quad \tilde{B}_i(0) = \tilde{B}_{i0}, \qquad (14.48)$$

for $i = 1, \ldots, N$. Next, consider the distributed Lyapunov function candidates given by (14.26), and note that the derivatives of $V_i(e_i, \tilde{A}_i, \tilde{B}_i)$, $i = 1, \ldots, N$, along the trajectories of (14.46)–(14.48) are given by

$$\dot{V}_i(e_i(t), \tilde{A}_i(t), \tilde{B}_i(t))$$

$$= -e_i^{\mathrm{T}}(t) R e_i(t) - 2 e_i^{\mathrm{T}}(t) \sum_{j \in \mathcal{N}_{\mathrm{in}}(i)} e_{ij}(t)$$

$$- 2 \mathrm{tr} \left[ \tilde{A}_i^{\mathrm{T}}(t) \sum_{j \in \mathcal{N}_{\mathrm{in}}(i)} \tilde{A}_{ij}(t) \right] - 2 \mathrm{tr} \left[ \tilde{B}_i^{\mathrm{T}}(t) \sum_{j \in \mathcal{N}_{\mathrm{in}}(i)} \tilde{B}_{ij}(t) \right]. \qquad (14.49)$$

Since $\mathfrak{G}$ is balanced, $\deg_{\mathrm{in}}(i) = \deg_{\mathrm{out}}(i)$, $i = 1, \ldots, N$, and, hence, it follows that

$$\sum_{i=1}^{N} \sum_{j \in \mathcal{N}_{\mathrm{in}}(i)} \|e_i(t) - e_j(t)\|_2^2$$

$$= \sum_{i=1}^{N} \sum_{j \in \mathcal{N}_{\mathrm{in}}(i)} (e_i^{\mathrm{T}}(t) e_i(t) - 2 e_i^{\mathrm{T}}(t) e_j(t) + e_j^{\mathrm{T}}(t) e_j(t))$$

$$= \sum_{i=1}^{N} (\deg_{\mathrm{in}}(i) + \deg_{\mathrm{out}}(i)) e_i^{\mathrm{T}}(t) e_i(t) - 2 \sum_{i=1}^{N} \sum_{j \in \mathcal{N}_{\mathrm{in}}(i)} e_i^{\mathrm{T}}(t) e_j(t)$$

$$= 2\sum_{i=1}^{N} \deg_{\mathrm{in}}(i)e_i^{\mathrm{T}}(t)e_i(t) - 2\sum_{i=1}^{N}\sum_{j\in\mathcal{N}_{\mathrm{in}}(i)} e_i^{\mathrm{T}}(t)e_j(t)$$

$$= 2\sum_{i=1}^{N} e_i^{\mathrm{T}}(t)\sum_{j\in\mathcal{N}_{\mathrm{in}}(i)} e_i(t) - 2\sum_{i=1}^{N}\sum_{j\in\mathcal{N}_{\mathrm{in}}(i)} e_i^{\mathrm{T}}(t)e_j(t)$$

$$= 2\sum_{i=1}^{N} e_i^{\mathrm{T}}(t)\sum_{j\in\mathcal{N}_{\mathrm{in}}(i)} (e_i(t) - e_j(t)). \qquad (14.50)$$

Next, note that since $\mathfrak{G}$ is balanced, the following identities hold:

$$2\mathrm{tr}\sum_{i=1}^{N} \tilde{A}_i^{\mathrm{T}}(t)\sum_{j\in\mathcal{N}_{\mathrm{in}}(i)} (\tilde{A}_i(t) - \tilde{A}_j(t))$$

$$= \sum_{i=1}^{N}\sum_{j\in\mathcal{N}_{\mathrm{in}}(i)} \mathrm{tr}[\tilde{A}_i(t) - \tilde{A}_j(t)][\tilde{A}_i(t) - \tilde{A}_j(t)]^{\mathrm{T}}, \qquad (14.51)$$

$$2\mathrm{tr}\sum_{i=1}^{N} \tilde{B}_i^{\mathrm{T}}(t)\sum_{j\in\mathcal{N}_{\mathrm{in}}(i)} (\tilde{B}_i(t) - \tilde{B}_j(t))$$

$$= \sum_{i=1}^{N}\sum_{j\in\mathcal{N}_{\mathrm{in}}(i)} \mathrm{tr}[\tilde{B}_i(t) - \tilde{B}_j(t)][\tilde{B}_i(t) - \tilde{B}_j(t)]^{\mathrm{T}}. \qquad (14.52)$$

Now, using (14.50)–(14.52), it follows that

$$\dot{V}(E(t), \tilde{\mathbb{A}}(t), \tilde{\mathbb{B}}(t)) = \sum_{i=1}^{N} \dot{V}_i(e_i(t), \tilde{A}_i(t), \tilde{B}_i(t))$$

$$= -\sum_{i=1}^{N} e_i^{\mathrm{T}}(t)Re_i(t) - 2\sum_{i=1}^{N} e_i^{\mathrm{T}}(t)\sum_{j\in\mathcal{N}_{\mathrm{in}}(i)} (e_i(t) - e_j(t))$$

$$- 2\mathrm{tr}\left[\sum_{i=1}^{N} \tilde{A}_i^{\mathrm{T}}(t)\sum_{j\in\mathcal{N}_{\mathrm{in}}(i)} (\tilde{A}_i(t) - \tilde{A}_j(t))\right]$$

$$- 2\mathrm{tr}\left[\sum_{i=1}^{N} \tilde{B}_i^{\mathrm{T}}(t)\sum_{j\in\mathcal{N}_{\mathrm{in}}(i)} (\tilde{B}_i(t) - \tilde{B}_j(t))\right]$$

$$= -\sum_{i=1}^{N} e_i^{\mathrm{T}}(t)Re_i(t) - \sum_{i=1}^{N}\sum_{j\in\mathcal{N}(i)} \|e_{ij}(t)\|_2^2$$

$$-\sum_{i=1}^{N} \sum_{j \in \mathcal{N}_{\text{in}}(i)} \|\tilde{A}_{ij}(t)\|_{\text{F}}^2 - \sum_{i=1}^{N} \sum_{j \in \mathcal{N}_{\text{in}}(i)} \|\tilde{B}_{ij}(t)\|_{\text{F}}^2$$

$$\leq 0, \quad t \geq 0, \tag{14.53}$$

which shows that the solution $(E(t), \tilde{\mathbb{A}}(t), \tilde{\mathbb{B}}(t))$ of the parameter error system is Lyapunov stable for all $(E_0, \tilde{\mathbb{A}}_0, \tilde{\mathbb{B}}_0) \in \mathbb{R}^{nN} \times \mathbb{R}^{nN \times n} \times \mathbb{R}^{nN \times m}$ and $t \geq 0$. The remainder of the proof now follows using identical arguments as in the proof of Theorem 14.1 and, hence, is omitted. $\qquad \square$

It is important to note that an identical result to Proposition 14.1 also holds for the distributed adaptive observers given by (14.43)–(14.45) with a weakly connected and balanced directed graph communication topology. This is immediate from Theorem 14.2 using the fact that the pairwise disagreement of the state and parameter errors converges to zero.

**Example 14.1.** Consider the aircraft dynamical system representing the controlled longitudinal motion of a Boeing 747 airplane linearized at an altitude of 40 kft and a velocity of 774 ft/sec addressed in Example 2.1 given by

$$\dot{x}(t) = \begin{bmatrix} -0.003 & 0.039 & 0 & -0.332 \\ -0.065 & -0.319 & 7.74 & 0 \\ 0.02 & -0.101 & -0.429 & 0 \\ 0 & 0 & 1 & 0 \end{bmatrix} x(t) + \begin{bmatrix} 0.01 \\ -0.18 \\ -1.16 \\ 0 \end{bmatrix} u(t),$$

$$x(0) = 0, \ t \geq 0, \tag{14.54}$$

where $x(t) = [x_1(t), x_2(t), x_3(t), x_4(t)]^{\text{T}}$, $t \geq 0$, is the system state vector, with $x_1(t)$, $t \geq 0$, representing the $x$-body-axis component of the velocity of the aircraft center of mass with respect to the reference axes (in ft/sec); $x_2(t)$, $t \geq 0$, representing the $z$-body-axis component of the velocity of the aircraft center of mass with respect to the reference axes (in ft/sec); $x_3(t)$, $t \geq 0$, representing the $y$-body-axis component of the angular velocity of the aircraft (pitch rate) with respect to the reference axes (in crad/sec); $x_4(t)$, $t \geq 0$, representing the pitch Euler angle of the aircraft body axes with respect to the reference axes (in crad); and $u(t)$, $t \geq 0$, representing the elevator input (in crad). Figure 14.1 shows the system response to a doublet input.

For our first simulation, we assume that the system matrices $A$ and $B$ characterizing (14.54) are unknown and consider the system (14.54) with four agent identifiers defined by a connected undirected graph topology implementing the distributed adaptive observers given by (14.28)–(14.30). Furthermore, we set $A_{\text{m}} = -10I_4$, $\Gamma_{\text{a1}} = \Gamma_{\text{b1}} = I_4$, $\Gamma_{\text{a2}} = \Gamma_{\text{b2}} = 2I_4$, $\Gamma_{\text{a3}} = \Gamma_{\text{b3}} = 4I_4$, and $\Gamma_{\text{a4}} = \Gamma_{\text{b4}} = 8I_4$. Figures 14.2 and 14.3 show the sys-

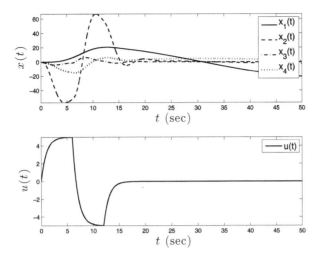

Figure 14.1 System response and doublet input for Boeing 747.

tem response for $e_i(t)$, $t \geq 0$; $\hat{x}_{ij}(t)$, $t \geq 0$; $\hat{A}_{ij}(t)$, $t \geq 0$; and $\hat{B}_{ij}(t)$, $t \geq 0$, $i = 1, \ldots, 4$, which, by Theorem 14.1, guarantees that $\lim_{t \to \infty} e_i(t) = 0$, $\lim_{t \to \infty} \hat{x}_{ij}(t) = 0$, $\lim_{t \to \infty} \hat{A}_{ij}(t) = 0$, and $\lim_{t \to \infty} \hat{B}_{ij}(t) = 0$, $i, j = 1, \ldots, 4$, without imposing a persistency of excitation condition on the input $u(t)$, $t \geq 0$.                                                                                    △

**Example 14.2.** Consider a network of agent identifiers with the strongly connected and balanced directed graph topology shown in Figure 14.4. The parameters used for our simulation are identical to the ones used for the undirected graph topology case. Figures 14.5 and 14.6 show the system response for $e_i(t)$, $t \geq 0$; $\hat{x}_{ij}(t)$, $t \geq 0$; $\hat{A}_{ij}(t)$, $t \geq 0$; and $\hat{B}_{ij}(t)$, $t \geq 0$, $i = 1, \ldots, 4$, which, by Theorem 14.2, guarantees that $\lim_{t \to \infty} e_i(t) = 0$, $\lim_{t \to \infty} \hat{x}_{ij}(t) = 0$, $\lim_{t \to \infty} \hat{A}_{ij}(t) = 0$, and $\lim_{t \to \infty} \hat{B}_{ij}(t) = 0$.

Finally, to assess the efficacy of the proposed approach, we compare our adaptive estimation multiagent network framework with the standard centralized estimator (14.4)–(14.7). Here, we set $\Gamma_a = \Gamma_b = 1$. Figure 14.7 shows the system response for $e(t)$, $t \geq 0$. It can be seen from this figure that the distributed adaptive estimator (Figures 14.2 and 14.5) significantly outperforms the centralized adaptive estimator.                                    △

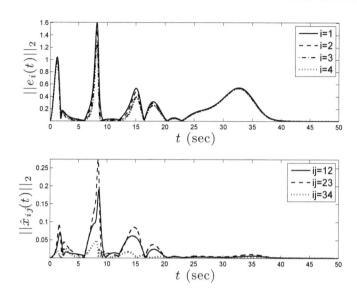

Figure 14.2 State error $\|e_i(t)\|_2$ and $\|\hat{x}_{ij}(t)\|_2$ versus time for the proposed distributed adaptive observers given by (14.28)–(14.30).

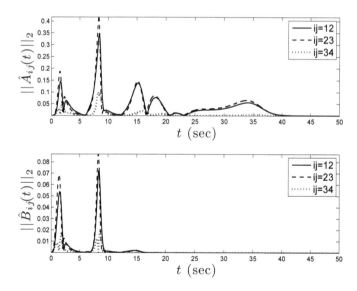

Figure 14.3 Estimated differences $\|\hat{A}_{ij}(t)\|_F$ and $\|\hat{B}_{ij}(t)\|_F$ versus time for the proposed distributed adaptive observers given by (14.28)–(14.30).

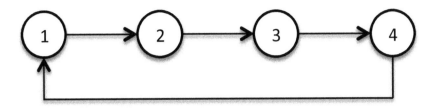

Figure 14.4 Interagent communication graph topology.

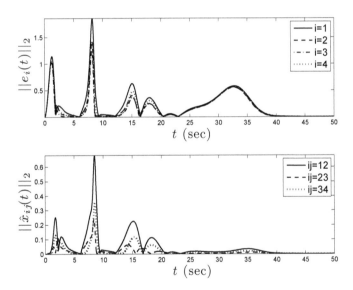

Figure 14.5 State error $\|e_i(t)\|_2$ and $\|\hat{x}_{ij}(t)\|_2$ versus time for the proposed distributed adaptive observers given by (14.43)–(14.45).

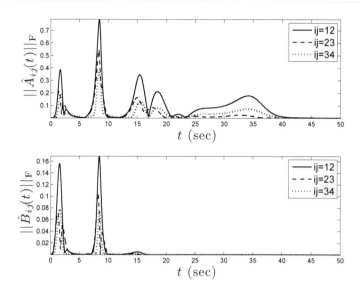

Figure 14.6 Estimated differences $\|\hat{A}_{ij}(t)\|_{\mathrm{F}}$ and $\|\hat{B}_{ij}(t)\|_{\mathrm{F}}$ versus time for the proposed distributed adaptive observers given by (14.43)–(14.45).

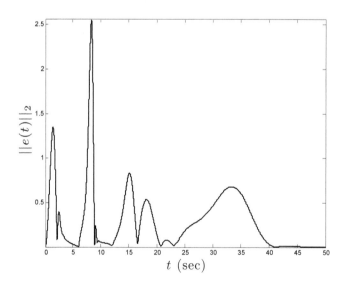

Figure 14.7 State error $\|e(t)\|_2$ versus time for the centralized adaptive observer given by (14.4)–(14.7).

# Chapter Fifteen

## Discrete-Time Network Systems

### 15.1 Introduction

In this chapter, we build on the theories of semistability and finite time semistability for continuous-time dynamical systems developed in Chapters 2 and 4 to develop a rigorous framework for semistability and finite time semistability for discrete-time systems. First, in Section 15.3, we develop the notion of semistability for discrete-time systems and give several alternative equivalent characterizations for semistability. Then, in Section 15.4, we develop the theory of discrete semistability by presenting Lyapunov theorems as well as converse Lyapunov theorems for discrete semistability, which hold with continuous Lyapunov functions whose Lyapunov difference decreases along the dynamical system trajectories and is such that the Lyapunov function satisfies inequalities involving the distance to the set of equilibria.

Next, in Sections 15.5 and 15.6, we develop the notion of finite time semistability and establish finite time semistability theory for discrete-time nonlinear dynamical systems. Specifically, using the existence and uniqueness of solutions for discrete-time nonlinear systems, we define a settling-time function for a finite time semistable system and establish a lower-semicontinuity property of this function. Then we develop sufficient Lyapunov stability theorems for finite time semistability and establish a relationship between finite time convergence and finite time semistability. In addition, we present the first converse Lyapunov theorems for discrete finite time semistability, which are shown to hold for lower semicontinuous Lyapunov functions.

As discussed in Chapter 1, a sizable body of work has emerged in recent years that addresses the distributed consensus problem using the tools of algebraic graph theory (see, e.g., [56, 73, 93, 224, 236, 238, 241, 255, 290, 305, 312] for continuous-time networks and [90, 187, 205, 287, 318, 320, 333] for discrete-time networks). In this monograph as well as in [22, 137, 162, 167], we present an alternative perspective to the distributed consensus problem based on *dynamical thermodynamics* [118, 127], a framework that unifies the foundational disciplines of thermodynamics and dynamical systems theory. As discussed in Chapters 3 and 4, dynamical thermodynamics was developed in [127] and [118] to address the formulation of equilibrium and nonequilibrium thermodynamics in a dynamical systems setting. Dynamical thermodynamics has also been used to apply thermodynamic principles to the analysis and control design of dynamical systems using an energy- and entropy-based hybrid stabilization framework [125, 128, 129].

By generalizing the notions of temperature, energy, and entropy, dynamical thermodynamics is used in [22, 137, 162, 167] to develop a design procedure for distributed consensus controllers that induce networked dynamical systems that emulate a thermodynamic behavior. In particular, for network systems with an undirected communication graph topology, system thermodynamic notions are used to show that every control law protocol of a symmetric Fourier type, with information (or communication) transfer playing the role of energy flow, achieves information consensus [22, 137, 162, 167].

Unlike most of the distributed nonlinear consensus protocols presented in the literature that merely guarantee system convergence, the thermodynamics-based control framework for network systems [22, 137, 162, 167] addresses both convergence and Lyapunov stability. As noted in Chapter 4, from a practical viewpoint it is not sufficient for a nonlinear control protocol to only guarantee that a network converges to a state of consensus since steady-state convergence is not sufficient to guarantee that small perturbations from the limiting state will lead to only small transient excursions from a state of consensus. It is also necessary to guarantee that the equilibrium states representing consensus are Lyapunov stable and, consequently, semistable [36, 122, 162].

In Section 15.7, we use the results of Sections 15.4 and 15.6 to develop consensus protocols for multiagent systems with nonlinear discrete dynamics. Specifically, we use our discrete-time semistability and discrete-time finite time semistability frameworks to design distributed asymptotic and finite time consensus control protocols for nonlinear dynamical networks with bidirectional communication graph topologies. The proposed controller architectures are predicated on the recently developed notion of discrete dynamical thermodynamics [118], resulting in controller architectures involving the exchange of generalized energy state information between agents that guarantee that the closed-loop dynamical network is consistent with basic thermodynamic principles.

## 15.2 Mathematical Preliminaries

Consider the discrete-time nonlinear dynamical system

$$x(k+1) = f(x(k)), \quad x(0) = x_0, \quad k \in \overline{\mathbb{Z}}_+, \tag{15.1}$$

where $x(k) \in \mathcal{D} \subseteq \mathbb{R}^n$, $k \in \overline{\mathbb{Z}}_+$, is the system state vector; $\mathcal{D}$ is an open set, $0 \in \mathcal{D}$; $f : \mathcal{D} \to \mathcal{D}$ is continuous on $\mathcal{D}$; and $\Delta f^{-1}(0) \triangleq \{x \in \mathcal{D} : f(x) = x\}$ is nonempty. We denote the solution to (15.1) with initial condition $x(0) = x_0$ by $s(\cdot, x_0)$ so that the *map* of the dynamical system given by $s : \overline{\mathbb{Z}}_+ \times \mathcal{D} \to \mathcal{D}$ is continuous on $\mathcal{D}$ and satisfies the *consistency* property $s(0, x_0) = x_0$ and the *semigroup* property $s(\kappa, s(k, x_0)) = s(k + \kappa, x_0)$ for all $x_0 \in \mathcal{D}$ and $k, \kappa \in \overline{\mathbb{Z}}_+$. We use the notation $s(k, x_0)$, $k \in \overline{\mathbb{Z}}_+$, and $x(k)$, $k \in \overline{\mathbb{Z}}_+$, interchangeably as the solution of the nonlinear discrete-time dynamical system (15.1) with initial condition $x(0) = x_0$. By a *solution* to (15.1) with initial condition $x(0) = x_0$ we mean a function $x : \overline{\mathbb{Z}}_+ \to \mathcal{D}$ that satisfies (15.1). Given $k \in \overline{\mathbb{Z}}_+$ and $x \in \mathcal{D}$, we denote the map $s(k, \cdot) : \mathcal{D} \to \mathcal{D}$ by $s_k$ and the map $s(\cdot, x) : \overline{\mathbb{Z}}_+ \to \mathcal{D}$ by $s^x$.

If $f(\cdot)$ is continuous, then it follows that $f(s(k-1, \cdot))$ is also continuous since it is constructed as a composition of continuous functions. Hence, $s(k, \cdot)$ is continuous on $\mathcal{D}$. If $f(\cdot)$ is such that $f : \mathbb{R}^n \to \mathbb{R}^n$, then we can construct the *solution sequence* or *discrete trajectory* $x(k) = s(k, x_0)$ to (15.1) iteratively by setting $x(0) = x_0$ and using $f(\cdot)$ to define $x(k)$ recursively by $x(k + 1) = f(x(k))$. This iterative process can be continued indefinitely, and, hence, a solution to (15.1) exists for all $k \geq 0$.

Alternatively, if $f(\cdot)$ is such that $f : \mathcal{D} \to \mathbb{R}^n$, then the solution may cease to exist at some point if $f(\cdot)$ maps $x(k)$ into some point $x(k+1)$ outside the domain of $f(\cdot)$. In this case, the solution sequence $x(k) = s(k, x_0)$ will be defined on the *maximal interval of existence* $\mathcal{I}_{x_0}^+ \subset \overline{\mathbb{Z}}_+$. Note that the solution sequence $x(k)$, $k \in \mathcal{I}_{x_0}^+$, is uniquely defined for every initial condition $x_0 \in \mathcal{D}$ irrespective of whether or not $f(\cdot)$ is a continuous function. That is, any other solution sequence $y(k)$ starting from $x_0$ at $k = 0$ will take exactly the same values as $x(k)$ and can be continued to the same interval as $x(k)$. It is important to note that if $k \in \overline{\mathbb{Z}}_+$, then uniqueness of solutions backward in time need not necessarily hold. This is because that $s(k, x_0) = f^{-1}(s(k + 1, x_0))$, $k \in \overline{\mathbb{Z}}_+$, and there is no guarantee that $f(\cdot)$ is invertible for all $k \in \overline{\mathbb{Z}}_+$. However, if $f : \mathcal{D} \to \mathcal{D}$ is a homeomorphism for all $k \in \overline{\mathbb{Z}}_+$, then the solution sequence is unique for all $k \in \mathbb{Z}$.

In light of the above, the following theorem is immediate.

**Theorem 15.1** (see [122]). Consider the nonlinear dynamical system (15.1), and assume that $f : \mathcal{D} \to \mathcal{D}$. Then, for every $x_0 \in \mathcal{D}$, there exists

$\mathcal{I}_{x_0}^+ \subseteq \overline{\mathbb{Z}}_+$ such that (15.1) has a unique solution $x : \mathcal{I}_{x_0}^+ \to \mathbb{R}^n$. Moreover, if $f(\cdot)$ is continuous, then the solution $s(k, \cdot)$ is continuous for each $k \in \mathcal{I}_{x_0}^+$. If, in addition, $f(\cdot)$ is a homeomorphism of $\mathcal{D}$ onto $\mathbb{R}^n$, then the solution $x : \mathcal{I}_{x_0}^+ \to \mathbb{R}^n$ is unique in all $\mathcal{I}_{x_0} \subseteq \mathbb{Z}$ and $s(k, \cdot)$ is continuous for all $k \in \mathcal{I}_{x_0}$. Finally, if $\mathcal{D} = \mathbb{R}^n$, then $\mathcal{I}_{x_0} = \mathbb{Z}$.

The following definition introduces the notion of class $\mathcal{W}_d$ functions involving *nondecreasing* functions.

**Definition 15.1.** A function $w : \mathbb{R} \to \mathbb{R}$ is of *class* $\mathcal{W}_d$ if $w(z') \le w(z'')$ for all $z', z'' \in \mathbb{R}$ such that $z' \le z''$.

To develop the theory for finite time semistability of discrete autonomous systems, we will require several key results on difference inequalities and the discrete-time comparison principle. Consider the scalar discrete-time nonlinear dynamical system given by

$$z(k+1) = w(z(k)), \quad z(k_0) = z_0, \quad k \in \mathcal{I}_{z_0}, \qquad (15.2)$$

where $z(k) \in \mathcal{Q} \subseteq \mathbb{R}$, $k \in \mathcal{I}_{z_0}$, is the system state vector; $\mathcal{I}_{z_0} \subseteq \mathbb{Z}$ is the maximal interval of existence of a solution $z(k)$ to (15.2); $\mathcal{Q}$ is an open set; $0 \in \mathcal{Q}$; and $w : \mathcal{Q} \to \mathbb{R}$ is a continuous function on $\mathcal{Q}$.

**Theorem 15.2.** Consider the discrete-time nonlinear dynamical system (15.2). Assume that the function $w : \mathcal{Q} \subseteq \mathbb{R} \to \mathbb{R}$ is continuous and $w(\cdot)$ is of class $\mathcal{W}_d$. If there exists a continuous function $V : \mathcal{I}_{z_0} \to \mathcal{Q}$ such that

$$V(k+1) \le w(V(k)), \quad k \in \mathcal{I}_{z_0}, \qquad (15.3)$$

then

$$V(k_0) \le z_0, \quad z_0 \in \mathcal{Q}, \qquad (15.4)$$

implies that

$$V(k) \le z(k), \quad k \in \mathcal{I}_{z_0}, \qquad (15.5)$$

where $z(k)$, $k \in \mathcal{I}_{z_0}$, is the solution to (15.2).

**Proof.** Suppose, *ad absurdum*, that inequality (15.5) does not hold on the entire interval $\mathcal{I}_{z_0}$. Then there exists $\hat{k} \in \mathcal{I}_{z_0}$ such that $V(k) \le z(k)$, $k_0 \le k < \hat{k}$, and

$$V(\hat{k}) > z(\hat{k}). \qquad (15.6)$$

Since $w(\cdot) \in \mathcal{W}_d$, it follows from (15.2), (15.3), and (15.6) that

$$w(z(\hat{k}-1)) = z(\hat{k}) < V(\hat{k}) \le w(V(\hat{k}-1)) \le w(z(\hat{k}-1)), \qquad (15.7)$$

which is a contradiction.                                                □

The following result is a direct corollary of Theorem 15.2.

**Corollary 15.1.** Consider the discrete-time nonlinear dynamical system (15.1). Assume that there exists a continuous function $V : \mathcal{D} \subseteq \mathbb{R}^n \to \mathcal{Q}$ such that

$$V(f(x)) \leq w(V(x)), \quad x \in \mathcal{D}, \tag{15.8}$$

where $w : \mathcal{Q} \subseteq \mathbb{R} \to \mathbb{R}$ is continuous; $w(\cdot) \in \mathcal{W}_{\mathrm{d}}$; and

$$z(k+1) = w(z(k)), \quad z(k_0) = z_0, \quad k \in \mathcal{I}_{z_0}. \tag{15.9}$$

If $\{k_0, \ldots, k_0 + \tau\} \subseteq \mathcal{I}_{x_0} \cap \mathcal{I}_{z_0}$, then

$$V(x_0) \leq z_0, \quad z_0 \in \mathcal{Q}, \tag{15.10}$$

implies that

$$V(x(k)) \leq z(k), \quad k \in \{k_0, \ldots, k_0 + \tau\}. \tag{15.11}$$

**Proof.** For every given $x_0 \in \mathcal{D}$, the solution $x(k)$, $k \in \mathcal{I}_{x_0}$, to (15.1) is well defined. With $\eta(k) \triangleq V(x(k))$, $k \in \mathcal{I}_{x_0}$, it follows from (15.8) that

$$\eta(k+1) \leq w(\eta(k)), \quad k \in \mathcal{I}_{x_0}. \tag{15.12}$$

Moreover, if $\{k_0, \ldots, k_0 + \tau\} \subseteq \mathcal{I}_{x_0} \cap \mathcal{I}_{z_0}$, then it follows from Theorem 15.2 that $V(x_0) = \eta(k_0) \leq z_0$ implies that

$$V(x(k)) = \eta(k) \leq z(k), \quad k \in \{k_0, \ldots, k_0 + \tau\}, \tag{15.13}$$

which establishes the result.                                            □

Note that if the solutions to (15.1) and (15.9) are globally defined for all $x_0 \in \mathbb{R}^n$ and $z_0 \in \mathbb{R}$, then Corollary 15.1 holds for all $k \geq k_0$. For the remainder of the chapter, we assume, without loss of generality, that $k_0 = 0$.

## 15.3 Semistability of Discrete Autonomous Systems

In this and the next section, we develop a stability analysis framework for discrete-time systems having a continuum of equilibria and present necessary and sufficient conditions for *discrete-time semistability*. To develop semistability theory for discrete-time systems, we need some additional notation and definitions.

A set $\mathcal{M} \subset \mathcal{D} \subseteq \mathbb{R}^n$ is a *positively invariant set* with respect to the nonlinear dynamical system (15.1) if $s_k(\mathcal{M}) \subseteq \mathcal{M}$ for all $k \in \overline{\mathbb{Z}}_+$, where

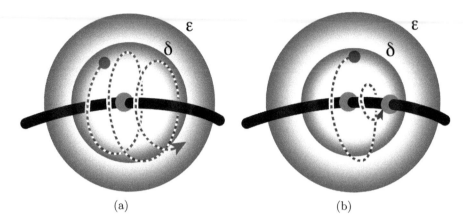

Figure 15.1 (a) Lyapunov stable nonisolated equilibrium point. The perturbed trajectory
need not converge to a new equilibrium. (b) Semistable nonisolated equilibrium point. Semistability guarantees convergence of the perturbed trajectory
to a nearby Lyapunov stable equilibrium point and is a stronger property
than Lyapunov stability.

$s_k(\mathcal{M}) \triangleq \{s_k(x) : x \in \mathcal{M}\}$. A set $\mathcal{M} \subseteq \mathcal{D} \subseteq \mathbb{R}^n$ is an *invariant set* with
respect to the dynamical system (15.1) if $s_k(\mathcal{M}) = \mathcal{M}$ for all $k \in \overline{\mathbb{Z}}_+$. A
point $p \in \mathcal{D}$ is a *limit point* of the trajectory or solution sequence $s(\cdot, x)$
of (15.1) if there exists a monotonic sequence $\{k_n\}_{n=0}^{\infty}$ of positive integers,
with $k_n \to \infty$ as $n \to \infty$, such that $s(k_n, x) \to p$ as $n \to \infty$. The set of
all limit points of $s(k, x)$, $k \in \overline{\mathbb{Z}}_+$, is the *limit set* $\omega(x)$ of $s(\cdot, x)$ of (15.1).
Finally, for $k \geq 0$, we define the *positive orbit* through the point $x_0 \in \mathcal{D}$ as
the motion along the solution sequence $\mathcal{O}_{x_0}^+ \triangleq \{x \in \mathcal{D} : x = s(k, x_0), k \geq 0\}$.

The following definition introduces the notion of semistability for discrete-time systems (see Figure 15.1).

**Definition 15.2.** An equilibrium point $x_e \in \mathcal{D} \subseteq \mathbb{R}^n$ of (15.1) is *Lyapunov stable* if, for all $\varepsilon > 0$, there exists $\delta = \delta(x_e) > 0$ such that if
$x_0 \in \mathcal{B}_\delta(x_e)$, then $x(k) \in \mathcal{B}_\varepsilon(x_e)$, $k \in \overline{\mathbb{Z}}_+$. An equilibrium point $x_e \in \mathcal{D} \subseteq \mathbb{R}^n$
of (15.1) is *semistable* if it is Lyapunov stable and there exists $\delta > 0$ such
that if $x_0 \in \mathcal{B}_\delta(x_e)$, then $\lim_{k\to\infty} s(k, x) = y$, where $y \in \mathcal{D}$ is a Lyapunov
stable equilibrium point of (15.1). An equilibrium point $x_e \in \mathbb{R}^n$ is *globally
semistable* if it is Lyapunov stable and, for every $x_0 \in \mathbb{R}^n$, $\lim_{k\to\infty} x(k) = y$,
where $y \in \mathbb{R}^n$ is Lyapunov stable equilibrium point of (15.1). The system
(15.1) is *semistable* if every equilibrium point of (15.1) is semistable. Finally,
(15.1) is *globally semistable* if every equilibrium point of (15.1) is globally
semistable.

The following proposition gives a sufficient condition for a trajectory
or solution sequence of (15.1) to converge to a limit. For this result, $\mathcal{D}_c \subseteq$

$\mathcal{D} \subseteq \mathbb{R}^n$ denotes a positively invariant set with respect to (15.1) and $s_k(\mathcal{D}_c)$ denotes the image of $\mathcal{D}_c \subset \mathcal{D}$ under the map $s_k : \mathcal{D}_c \to \mathcal{D}$; that is, $s_k(\mathcal{D}_c) \triangleq \{y : y = s_k(x_0) \text{ for some } x(0) = x_0 \in \mathcal{D}_c\}$.

**Proposition 15.1.** Consider the nonlinear discrete-time dynamical system (15.1), and let $x \in \mathcal{D}_c$. If the limit set $\omega(x)$ of (15.1) contains a Lyapunov stable equilibrium point $y$, then $y = \lim_{k \to \infty} s(k, x)$; that is, $\omega(x) = \{y\}$.

**Proof.** Suppose that $y \in \omega(x)$ is Lyapunov stable, and let $\mathcal{N}_\varepsilon \subseteq \mathcal{D}_c$ be an open neighborhood of $y$. Since $y$ is Lyapunov stable, there exists an open neighborhood $\mathcal{N}_\delta \subset \mathcal{D}_c$ of $y$ such that $s_k(\mathcal{N}_\delta) \subseteq \mathcal{N}_\varepsilon$ for every $k \in \mathbb{Z}_+$. Now, since $y \in \omega(x)$, it follows that there exists $\kappa \in \overline{\mathbb{Z}}_+$ such that $s(\kappa, x) \in \mathcal{N}_\delta$. Hence, $s(k + \kappa, x) = s_k(s(\kappa, x)) \in s_k(\mathcal{N}_\delta) \subseteq \mathcal{N}_\varepsilon$ for every $k > 0$. Since $\mathcal{N}_\varepsilon \subseteq \mathcal{D}_c$ is arbitrary, it follows that $y = \lim_{k \to \infty} s(k, x)$. Thus, $\lim_{n \to \infty} s(k_n, x) = y$ for every increasing sequence $\{k_n\}_{n=1}^\infty$, and, hence, $\omega(x) = \{y\}$. $\qquad \square$

Next, we present alternative equivalent characterizations of the semistability of (15.1). For this result, the following definition is required.

**Definition 15.3.** The *domain of semistability* is the set of points $x_0 \in \mathcal{D}$ such that if $x(k)$ is a solution to (15.1) with $x(0) = x_0$, $k \in \overline{\mathbb{Z}}_+$, then $x(k)$ converges to a Lyapunov stable equilibrium point in $\mathcal{D}$.

Note that if (15.1) is semistable, then its domain of semistability contains the set of equilibria in its interior.

**Proposition 15.2.** Consider the nonlinear discrete-time dynamical system $\mathcal{G}$ given by (15.1). Then the following statements are equivalent.

(i) $\mathcal{G}$ is semistable.

(ii) For each $x_e \in \Delta f^{-1}(0)$, there exist class $\mathcal{K}$ and $\mathcal{L}$ functions $\alpha(\cdot)$ and $\beta(\cdot)$, respectively, and $\delta = \delta(x_e) > 0$, such that if $\|x_0 - x_e\| < \delta$, then $\|x(k) - x_e\| \leq \alpha(\|x_0 - x_e\|)$, $k \in \overline{\mathbb{Z}}_+$, and $\text{dist}(x(k), \Delta f^{-1}(0)) \leq \beta(k)$, $k \in \overline{\mathbb{Z}}_+$.

(iii) For each $x_e \in \Delta f^{-1}(0)$, there exist class $\mathcal{K}$ functions $\alpha_1(\cdot)$ and $\alpha_2(\cdot)$, a class $\mathcal{L}$ function $\beta(\cdot)$, and $\delta = \delta(x_e) > 0$ such that if $\|x_0 - x_e\| < \delta$, then $\text{dist}(x(k), \Delta f^{-1}(0)) \leq \alpha_1(\|x(k) - x_e\|)\beta(k) \leq \alpha_2(\|x_0 - x_e\|)\beta(k)$, $k \in \overline{\mathbb{Z}}_+$.

**Proof.** To show that (i) implies (ii), suppose that (15.1) is semistable, and let $x_e \in \Delta f^{-1}(0)$. Since $x_e$ is Lyapunov stable, it follows that there exists $\delta = \delta(x_e) > 0$ and a class $\mathcal{K}$ function $\alpha(\cdot)$ such that if $\|x_0 - x_e\| \leq \delta$, then $\|x(k) - x_e\| \leq \alpha(\|x_0 - x_e\|)$, $k \in \mathbb{Z}_+$. Without loss of generality, we may assume that $\delta$ is such that $\overline{\mathcal{B}_\delta(x_e)}$ is contained in the domain of semistability of (15.1). Hence, for every $x_0 \in \overline{\mathcal{B}_\delta(x_e)}$, $\lim_{k \to \infty} x(k) = x^* \in \Delta f^{-1}(0)$ and, consequently, $\lim_{k \to \infty} \mathrm{dist}(x(k), \Delta f^{-1}(0)) = 0$.

For each $\varepsilon > 0$ and $x_0 \in \overline{\mathcal{B}_\delta(x_e)}$, define $K_{x_0}(\varepsilon)$ to be the infimum of $K$ with the property that $\mathrm{dist}(x(k), \Delta f^{-1}(0)) < \varepsilon$ for all $k \geq K$; that is, $K_{x_0}(\varepsilon) \triangleq \inf\{K : \mathrm{dist}(x(k), \Delta f^{-1}(0)) < \varepsilon, k \geq K\}$. For each $x_0 \in \overline{\mathcal{B}_\delta(x_e)}$, the function $K_{x_0}(\varepsilon)$ is nonnegative and nonincreasing in $\varepsilon$, and $K_{x_0}(\varepsilon) = 0$ for sufficiently large $\varepsilon$.

Next, let $K(\varepsilon) \triangleq \sup\{K_{x_0}(\varepsilon) : x_0 \in \overline{\mathcal{B}_\delta(x_e)}\}$. We claim that $K$ is well defined. To show this, consider $\varepsilon > 0$ and $x_0 \in \overline{\mathcal{B}_\delta(x_e)}$. Since $\mathrm{dist}(s(k, x_0), \Delta f^{-1}(0)) < \varepsilon$ for every $k > K_{x_0}(x_e)$, it follows from the continuity of $s$ that, for every $\varepsilon > 0$, there exists an open neighborhood $\mathcal{U}$ of $x_0$ such that $\mathrm{dist}(s(k, z), \Delta f^{-1}(0)) < \varepsilon$ for every $z \in \mathcal{U}$, $k > K_{x_0}(\varepsilon)$. Hence, $\limsup_{z \to x_0} K_z(\varepsilon) \leq K_{x_0}(\varepsilon)$, implying that the function $x_0 \mapsto K_{x_0}(\varepsilon)$ is upper semicontinuous at the arbitrarily chosen point $x_0$ and hence on $\overline{\mathcal{B}_\delta(x_e)}$. Since an upper semicontinuous function defined on a compact set achieves its supremum, it follows that $K(\varepsilon)$ is well defined. The function $K(\cdot)$ is the pointwise supremum of a collection of nonnegative and nonincreasing functions and, hence, is nonnegative and nonincreasing. Moreover, $K(\varepsilon) = 0$ for every $\varepsilon > \max\{\alpha(\|x_0 - x_e\|) : x_0 \in \overline{\mathcal{B}_\delta(x_e)}\}$.

Let

$$\psi(\varepsilon) \triangleq \frac{2}{\varepsilon} \int_{\varepsilon/2}^{\varepsilon} K(\sigma) \mathrm{d}\sigma + \frac{1}{\varepsilon} \geq K(\varepsilon) + \frac{1}{\varepsilon}.$$

Note that $K(\cdot)$ is measurable since it is upper semicontinuous and, hence, integrable. The function $\psi(\varepsilon)$ is positive, continuous, and strictly decreasing, and $\psi(\varepsilon) \to 0$ as $\varepsilon \to \infty$. Choose $\beta(\cdot) = \psi^{-1}(\cdot)$. Then $\beta(\cdot)$ is positive, continuous, and strictly decreasing, and $\beta(\sigma) \to 0$ as $\sigma \to \infty$. Furthermore, $K(\beta(\sigma)) < \psi(\beta(\sigma)) = \sigma$. Hence, $\mathrm{dist}(x(k), \Delta f^{-1}(0)) \leq \beta(k)$, $k \in \mathbb{Z}_+$.

Next, to show that (ii) implies (iii), suppose that (ii) holds, and let $x_e \in \Delta f^{-1}(0)$. Then $x_e$ is Lyapunov stable. Choosing $x_0$ sufficiently close to $x_e$, it follows from the inequality $\|x(k) - x_e\| \leq \alpha(\|x_0 - x_e\|)$, $k \geq 0$, that trajectories of (15.1) starting sufficiently close to $x_e$ are bounded, and, hence, the positive limit set of (15.1) is nonempty. Since $\lim_{k \to \infty} \mathrm{dist}(x(k), \Delta f^{-1}(0)) = 0$, it follows that the positive limit set is contained in $\Delta f^{-1}(0)$.

Now, since every point in $\Delta f^{-1}(0)$ is Lyapunov stable, it follows from Proposition 15.1 that $\lim_{k \to \infty} x(k) = x^*$, where $x^* \in \Delta f^{-1}(0)$ is Lyapunov stable. If $x^* = x_e$, then it follows using similar arguments as above that there exists a class $\mathcal{L}$ function $\hat{\beta}(\cdot)$ such that

$$\text{dist}(x(k), \Delta f^{-1}(0)) \leq \|x(k) - x_e\| \leq \hat{\beta}(k)$$

for every $x_0$ satisfying $\|x_0 - x_e\| < \delta$ and $k \geq 0$. Hence,

$$\text{dist}(x(k), \Delta f^{-1}(0)) \leq \sqrt{\|x(k) - x_e\|}\sqrt{\hat{\beta}(k)}, \quad k \geq 0.$$

Next, consider the case where $x^* \neq x_e$, and let $\alpha_1(\cdot)$ be a class $\mathcal{K}$ function. In this case, note that

$$\lim_{k \to \infty} \text{dist}(x(k), \Delta f^{-1}(0))/\alpha_1(\|x(k) - x_e\|) = 0,$$

and, hence, it follows using similar arguments as above that there exists a class $\mathcal{L}$ function $\beta(\cdot)$ such that

$$\text{dist}(x(k), \Delta f^{-1}(0)) \leq \alpha_1(\|x(k) - x_e\|)\beta(k), \quad k \geq 0.$$

Finally, note that $\alpha_1 \circ \alpha$ is of class $\mathcal{K}$, and, hence, (iii) follows immediately.

Finally, to show that (iii) implies (i), suppose that (iii) holds, and let $x_e \in \Delta f^{-1}(0)$. Then it follows that $\alpha_1(\|x(k) - x_e\|) \leq \alpha_2(\|x(0) - x_e\|)$, $k \geq 0$; that is, $\|x(k) - x_e\| \leq \alpha(\|x(0) - x_e\|)$, where $k \geq 0$ and $\alpha = \alpha_1^{-1} \circ \alpha_2$ is of class $\mathcal{K}$. Then $x_e$ is Lyapunov stable. Since $x_e$ was chosen arbitrarily, it follows that every equilibrium point is Lyapunov stable. Furthermore, $\lim_{k \to \infty} \text{dist}(x(k), \Delta f^{-1}(0)) = 0$. Choosing $x_0$ sufficiently close to $x_e$, it follows from the inequality $\|x(k) - x_e\| \leq \alpha(\|x_0 - x_e\|)$, $k \geq 0$, that trajectories of (15.1) starting sufficiently close to $x_e$ are bounded, and, hence, the positive limit set of (15.1) is nonempty. Since every point in $\Delta f^{-1}(0)$ is Lyapunov stable, it follows from Proposition 15.1 that $\lim_{k \to \infty} x(k) = x^*$, where $x^* \in \Delta f^{-1}(0)$ is Lyapunov stable. Hence, by definition, (15.1) is semistable. $\qquad\square$

## 15.4 Lyapunov and Converse Lyapunov Theorems for Semistability

In this section, we present Lyapunov and converse Lyapunov theorems for discrete-time semistability. For the results in this section, define $\Delta V(x) \triangleq V(f(x)) - V(x)$ for a given continuous function $V : \mathcal{D} \to \mathbb{R}$.

**Theorem 15.3.** Consider the nonlinear discrete-time dynamical system (15.1). Let $\mathcal{Q} \subseteq \mathbb{R}^n$ be an open neighborhood of $\Delta f^{-1}(0)$, and assume that there exists a continuous function $V : \mathcal{Q} \to \mathbb{R}$ such that

$$\Delta V(x) < 0, \quad x \in \mathcal{Q} \setminus \Delta f^{-1}(0). \tag{15.14}$$

If every equilibrium point of (15.1) is Lyapunov stable, then (15.1) is semistable. Moreover, if $\mathcal{Q} = \mathbb{R}^n$ and $V(x) \to \infty$ as $\|x\| \to \infty$, then (15.1) is globally semistable.

**Proof.** Since every equilibrium point of (15.1) is Lyapunov stable by assumption, for every $z \in \Delta f^{-1}(0)$, there exists an open neighborhood $\mathcal{V}_z$ of $z$ such that $s(\overline{\mathbb{Z}}_+ \times \mathcal{V}_z)$ is bounded and contained in $\mathcal{Q}$. The set $\mathcal{V} \triangleq \cup_{z \in \Delta f^{-1}(0)} \mathcal{V}_z$ is an open neighborhood of $\Delta f^{-1}(0)$ contained in $\mathcal{Q}$. Consider $x \in \mathcal{V}$ so that there exists $z \in \Delta f^{-1}(0)$ such that $x \in \mathcal{V}_z$ and $s(k, x) \in \mathcal{V}_z$, $k \in \overline{\mathbb{Z}}_+$. Since $\mathcal{V}_z$ is bounded, it follows that the positive limit set of $x$ is nonempty and invariant. Furthermore, it follows from (15.14) that $\Delta V(x) \le 0$, $k \in \overline{\mathbb{Z}}_+$, and, hence, it follows from Theorem 13.3 of [118] that $s(k, x) \to \mathcal{M}$ as $k \to \infty$, where $\mathcal{M}$ is the largest invariant set contained in the set $\mathcal{R} = \{y \in \mathcal{V}_z : \Delta V(x) = 0\}$. Note that $\mathcal{R} = \Delta f^{-1}(0)$ is invariant and, hence, $\mathcal{M} = \mathcal{R}$, which implies that $\lim_{k \to \infty} \text{dist}(s(k, x), \Delta f^{-1}(0)) = 0$. Now, since every point in $\Delta f^{-1}(0)$ is Lyapunov stable, it follows from Proposition 15.1 that $\lim_{k \to \infty} s(k, x) = x^*$, where $x^* \in \Delta f^{-1}(0)$ is Lyapunov stable. Hence, by definition, (15.1) is semistable.

Finally, if $\mathcal{Q} = \mathbb{R}^n$ and $V(\cdot)$ is radially unbounded, then global semistability follows using standard arguments.                                                              $\square$

Next, we present a slightly more general theorem for semistability wherein we do not assume that all points in $\Delta V^{-1}(0) \triangleq \{x \in \mathcal{Q} : V(f(x)) = V(x)\}$ are Lyapunov stable but rather we assume that all points in the largest invariant subset of $\Delta V^{-1}(0)$ are Lyapunov stable.

**Theorem 15.4.** Consider the nonlinear discrete-time dynamical system (15.1), and let $\mathcal{Q}$ be an open neighborhood of $\Delta f^{-1}(0)$. Suppose that the positive orbit $\mathcal{O}_x$ of (15.1) is bounded for all $x \in \mathcal{Q}$, and assume that there exists a continuous function $V : \mathcal{Q} \to \mathbb{R}$ such that

$$\Delta V(x) \le 0, \quad x \in \mathcal{Q}. \tag{15.15}$$

If every point in the largest invariant subset $\mathcal{M}$ of $\{x \in \mathcal{Q} : \Delta V(x) = 0\}$ is Lyapunov stable, then (15.1) is semistable. Moreover, if $\mathcal{Q} = \mathbb{R}^n$ and $V(x) \to \infty$ as $\|x\| \to \infty$, then (15.1) is globally semistable.

**Proof.** Since every solution of (15.1) is bounded, it follows from the hypotheses on $V(\cdot)$ that, for every $x \in \mathcal{Q}$, the positive limit set $\omega(x)$ of (15.1) is nonempty and contained in the largest invariant subset $\mathcal{M}$ of $\{x \in \mathcal{Q} : \Delta V(x) = 0\}$. Since every point in $\mathcal{M}$ is a Lyapunov stable equilibrium, it follows from Proposition 15.1 that $\omega(x)$ contains a single point for every $x \in \mathcal{Q}$ and $\lim_{k \to \infty} s(k, x)$ exists for every $x \in \mathcal{Q}$. Now, since $\lim_{k \to \infty} s(k, x) \in \mathcal{M}$ is Lyapunov stable for every $x \in \mathcal{Q}$, it follows from the definition of semistability that every equilibrium point in $\mathcal{M}$ is semistable.

Finally, if $\mathcal{Q} = \mathbb{R}^n$ and $V(\cdot)$ is radially unbounded, then global semistability follows using standard arguments.                                      $\square$

Next, we provide a converse Lyapunov theorem for semistability.

**Theorem 15.5.** Consider the nonlinear discrete-time dynamical system (15.1). Suppose that (15.1) is discrete-time semistable with domain of semistability $\mathcal{D}_0$. Then there exist a continuous nonnegative function $V : \mathcal{D}_0 \to \overline{\mathbb{R}}_+$ and a class $\mathcal{K}$ function $\alpha(\cdot)$ such that the following conditions hold.

(i) $V(x) = 0$, $x \in \Delta f^{-1}(0)$.

(ii) $V(x) \geq \alpha(\mathrm{dist}(x, \Delta f^{-1}(0)))$, $x \in \mathcal{D}_0$.

(iii) $\Delta V(x) < 0$, $x \in \mathcal{D}_0 \backslash \Delta f^{-1}(0)$.

**Proof.** Define the function $V : \mathcal{D}_0 \to \overline{\mathbb{R}}_+$ by

$$V(x) \triangleq \sup_{k \geq 0} \left\{ \frac{1 + 2k}{1 + k} \mathrm{dist}(s(k, x), \Delta f^{-1}(0)) \right\}, \quad x \in \mathcal{D}_0. \tag{15.16}$$

Note that $V(\cdot)$ is well defined since (15.1) is semistable. Clearly, (i) holds. Furthermore, since $V(x) \geq \mathrm{dist}(x, \Delta f^{-1}(0))$, $x \in \mathcal{D}_0$, it follows that (ii) holds.

To show that $V(\cdot)$ is continuous on $\mathcal{D}_0 \backslash \Delta f^{-1}(0)$, define $K : \mathcal{D}_0 \backslash \Delta f^{-1}(0) \to \mathbb{Z}_+$ by

$$K(z) \triangleq \inf\{h : \mathrm{dist}(s(k, z), \Delta f^{-1}(0)) < \mathrm{dist}(z, \Delta f^{-1}(0))/2$$
$$\text{for all } k \geq h > 0\},$$

and write

$$\mathcal{W}_\varepsilon \triangleq \{x \in \mathcal{D}_0 : \mathrm{dist}(s(k, x), \Delta f^{-1}(0)) < \varepsilon, k \in \mathbb{Z}_+\}. \tag{15.17}$$

Note that $\mathcal{W}_\varepsilon \supset \Delta f^{-1}(0)$ is open and positively invariant and contains an open neighborhood of $\Delta f^{-1}(0)$. Consider $z \in \mathcal{D}_0 \backslash \Delta f^{-1}(0)$, and define $\lambda \triangleq \mathrm{dist}(z, \Delta f^{-1}(0)) > 0$. Then it follows from the semistability of (15.1) that there exists $h > 0$ such that $s(h, z) \in \mathcal{W}_{\varepsilon/2}$. Consequently, $s(h+k, z) \in \mathcal{W}_{\varepsilon/2}$ for all $k \in \mathbb{Z}_+$, and, hence, it follows that $K(z)$ is well defined. Since $\mathcal{W}_{\varepsilon/2}$ is open, there exists a neighborhood $\mathcal{B}_\sigma(s(K(z), z)) \subset \mathcal{W}_{\varepsilon/2}$. Hence, $\mathcal{N} \triangleq s_{-K(z)}(\mathcal{B}_\sigma(s(K(z), z)))$ is a neighborhood of $z$ and $\mathcal{N} \subset \mathcal{D}_0$.

Next, choose $\eta > 0$ such that $\eta < \lambda/2$ and $\mathcal{D}_\eta(z) \subset \mathcal{N}$. Then, for every $k > K(z)$ and $y \in \mathcal{B}_\eta(z)$,

$$\frac{1+2k}{1+k}\mathrm{dist}(s(k,y),\Delta f^{-1}(0)) \le 2\mathrm{dist}(s(k,y),\Delta f^{-1}(0))$$

$$\le \lambda.$$

Therefore, for every $y \in \mathcal{B}_\eta(z)$,

$$\begin{aligned}
V(z) - V(y) &= \sup_{k\ge 0}\left\{\frac{1+2k}{1+k}\mathrm{dist}(s(k,z),\Delta f^{-1}(0))\right\} \\
&\quad - \sup_{k\ge 0}\left\{\frac{1+2k}{1+k}\mathrm{dist}(s(k,y),\Delta f^{-1}(0))\right\} \\
&= \sup_{0\le k\le K(z)}\left\{\frac{1+2k}{1+k}\mathrm{dist}(s(k,z),\Delta f^{-1}(0))\right\} \\
&\quad - \sup_{0\le k\le K(z)}\left\{\frac{1+2k}{1+k}\mathrm{dist}(s(k,y),\Delta f^{-1}(0))\right\}. \quad (15.18)
\end{aligned}$$

Hence,

$$\begin{aligned}
|V(z) - V(y)| &\le \sup_{0\le k\le K(z)}\left|\frac{1+2k}{1+k}\left(\mathrm{dist}(s(k,z),\Delta f^{-1}(0))\right.\right. \\
&\qquad\qquad \left.\left. -\mathrm{dist}(s(k,y),\Delta f^{-1}(0))\right)\right| \\
&\le 2\sup_{0\le k\le K(z)}\left|\mathrm{dist}(s(k,z),\Delta f^{-1}(0)) - \mathrm{dist}(s(k,y),\Delta f^{-1}(0))\right| \\
&\le 2\sup_{0\le k\le K(z)}\mathrm{dist}(s(k,z),s(k,y)), \\
&\qquad z \in \mathcal{D}_0\setminus\Delta f^{-1}(0), \quad y \in \mathcal{B}_\eta(z). \quad (15.19)
\end{aligned}$$

Now, it follows from continuous dependence of solutions $s(\cdot,\cdot)$ on system initial conditions and (15.19) that $V(\cdot)$ is continuous on $\mathcal{D}_0\setminus\Delta f^{-1}(0)$.

To show that $V(\cdot)$ is continuous on $\Delta f^{-1}(0)$, consider $x_\mathrm{e} \in \Delta f^{-1}(0)$, and let $\{x_n\}_{n=1}^\infty$ be a sequence in $\mathcal{D}_0\setminus\Delta f^{-1}(0)$ that converges to $x_\mathrm{e}$. Since $x_\mathrm{e}$ is Lyapunov stable, it follows that $x(k) \equiv x_\mathrm{e}$ is the unique solution to (15.1) with $x_0 = x_\mathrm{e}$. By continuous dependence of solutions $s(\cdot,\cdot)$ on system initial conditions, $s(k,x_n) \to s(k,x_\mathrm{e}) = x_\mathrm{e}$ as $n \to \infty$, $k \in \mathbb{Z}_+$.

Let $\varepsilon > 0$, and note that it follows from (ii) of Proposition 15.2 that there exists $\delta = \delta(x_\mathrm{e}) > 0$ such that, for every solution of (15.1) in $\mathcal{B}_\delta(x_\mathrm{e})$, there exists $\hat{K} = \hat{K}(x_\mathrm{e},\varepsilon) > 0$ such that $s_k(\mathcal{B}_\delta(x_\mathrm{e})) \subset \mathcal{W}_\varepsilon$ for all $k \ge \hat{K}$.

Next, note that there exists a positive integer $N_1$ such that $x_n \in \mathcal{B}_\delta(x_e)$ for all $n \geq N_1$. Now, it follows from (15.16) that

$$V(x_n) \leq 2 \sup_{0 \leq k \leq \hat{K}} \text{dist}(s(k, x_n), \Delta f^{-1}(0)) + 2\varepsilon, \quad n \geq N_1. \qquad (15.20)$$

Next, it follows from Lemma 3.1 of Chapter I of [141] that $s(\cdot, x_n)$ converges to $s(\cdot, x_e)$ uniformly on $[0, \hat{K}]$. Hence,

$$\lim_{n \to \infty} \sup_{0 \leq k \leq \hat{K}} \text{dist}\left(s(k, x_n), \Delta f^{-1}(0)\right)$$

$$= \sup_{0 \leq k \leq \hat{K}} \text{dist}\left(\lim_{n \to \infty} s(k, x_n), \Delta f^{-1}(0)\right)$$

$$= \sup_{0 \leq k \leq \hat{K}} \text{dist}(x_e, \Delta f^{-1}(0))$$

$$= 0, \qquad (15.21)$$

which implies that there exists a positive integer $N_2 = N_2(x_e, \varepsilon) \geq N_1$ such that

$$\sup_{0 \leq k \leq \hat{K}} \text{dist}(s(k, x_n), \Delta f^{-1}(0)) < \varepsilon, \quad n \geq N_2. \qquad (15.22)$$

Combining (15.16) with (15.22) yields $V(x_n) < 4\varepsilon$ for all $n \geq N_2$, which implies that $\lim_{n \to \infty} V(x_n) = 0 = V(x_e)$.

Finally, we show that $V(x(k))$ is strictly decreasing along the solution of (15.1) on $\mathcal{D} \backslash \Delta f^{-1}(0)$. Now, note that it follows from the definition of $K(\cdot)$ that the supremum in the definition of $V(s(1, x))$ is reached at some time $\hat{k}$ such that $0 \leq \hat{k} < K(x)$. Hence,

$$V(s(1, x)) = \text{dist}(s(\hat{k} + 1, x), \Delta f^{-1}(0)) \frac{1 + 2\hat{k}}{1 + \hat{k}}$$

$$= \text{dist}(s(\hat{k} + 1, x), \Delta f^{-1}(0)) \frac{1 + 2\hat{k} + 2}{1 + \hat{k} + 1} \left[1 - \frac{1}{(1 + 2\hat{k} + 2)(1 + \hat{k})}\right]$$

$$\leq V(x) \left[1 - \frac{1}{2(1 + K(x))^2}\right], \qquad (15.23)$$

which implies that

$$\Delta V(x) \leq -\frac{1}{2} V(x)(1 + K(x))^{-2} < 0, \quad x \in \mathcal{D}_0 \backslash \Delta f^{-1}(0), \qquad (15.24)$$

and, hence, (iii) holds. $\qquad \qquad \square$

## 15.5 Finite Time Semistability of Discrete Autonomous Systems

In this and the next section, we extend the results of Sections 15.3 and 15.4 to address *finite time semistability* of discrete-time nonlinear dynamical systems. The notion of finite time semistability involves finite time convergence along with Lyapunov stability, as detailed in the next definition.

**Definition 15.4.** Consider the nonlinear dynamical system (15.1). An equilibrium point $x_e \in \Delta f^{-1}(0)$ of (15.1) is *finite time semistable* if there exist an open neighborhood $\mathcal{N} \subseteq \mathcal{D}$ of $x_e$ and a function $K : \mathcal{N} \backslash \Delta f^{-1}(0) \rightarrow \mathbb{Z}_+$, called the *settling-time function*, such that the following statements hold.

(i) *Finite time convergence.* For every $x \in \mathcal{N} \setminus \Delta f^{-1}(0)$, $s^x(k) \in \mathcal{N} \setminus \Delta f^{-1}(0)$ is defined on $k \in \{0, \ldots, K(x) - 1\}$, and $s^x(k)$, $k \geq K(x)$, is contained in $\mathcal{N} \cap \Delta f^{-1}(0)$.

(ii) *Semistability.* $x_e \in \Delta f^{-1}(0)$ is semistable.

An equilibrium point $x_e \in \Delta f^{-1}(0)$ of (15.1) is *globally finite time semistable* if it is finite time semistable with $\mathcal{N} = \mathcal{D} = \mathbb{R}^n$. The system (15.1) is said to be *finite time semistable* if every equilibrium point in $\Delta f^{-1}(0)$ is finite time semistable. Finally, (15.1) is said to be *globally finite time semistable* if every equilibrium point in $\Delta f^{-1}(0)$ is globally finite time semistable.

Note that the definition of finite time convergence in Definition 15.4 is simpler than the corresponding definition in the case of continuous-time systems given in Chapter 4 and [34]. In the case of continuous-time systems, the usual sufficient conditions for existence and uniqueness of solutions necessarily fail to hold at a finite time stable equilibrium. Since discrete-time systems possess existence and uniqueness of solutions without any additional assumptions on $f(\cdot)$, the definition of finite time convergence can be stated in a simpler manner than in the case of continuous-time systems.

It is easy to see from Definition 15.4 that

$$K(x) = \min\{k \in \mathbb{Z}_+ : f(s(k, x)) = s(k, x)\}, \quad x \in \mathcal{N}. \tag{15.25}$$

In particular, $K(x_e) = 0$ for any equilibrium point $x_e \in \Delta f^{-1}(0)$ of (15.1).

The following definition is needed for the next result.

**Definition 15.5.** Let $\mathcal{D} \subseteq \mathbb{R}^n$, $g : \mathcal{D} \rightarrow \mathbb{R}$, and $x \in \mathcal{D}$. The function $g$ is *lower semicontinuous at* $x \in \mathcal{D}$ if for every sequence $\{x_n\}_{n=0}^{\infty} \subset \mathcal{D}$ such that $\lim_{n \to \infty} x_n = x$, $g(x) \leq \liminf_{n \to \infty} g(x_n)$. The function $g$ is *lower* *semicontinuous on* $\mathcal{D}$ if $g$ is lower semicontinuous at every point $x \in \mathcal{D}$.

The next proposition shows that if the settling-time function of a finite time semistable system is lower semicontinuous at each $x_e \in \mathcal{N} \cap \Delta f^{-1}(0)$, then it is lower semicontinuous on $\mathcal{N}$.

**Proposition 15.3.** Consider the nonlinear dynamical system (15.1). Assume that every equilibrium point $x_e \in \Delta f^{-1}(0)$ of (15.1) is finite time semistable, let $\mathcal{N} \subseteq \mathcal{D}$ be as in Definition 15.4, and let $K : \mathcal{N} \to \overline{\mathbb{Z}}_+$ be the settling-time function. Then $K(\cdot)$ is lower semicontinuous on $\mathcal{N}$.

**Proof.** Let $y \in \mathcal{N}$, consider the sequence $\{y_n\}_{n=0}^\infty$ in $\mathcal{N}$ converging to $y$, and let $\tau^- = \liminf_{n\to\infty} K(y_n)$. Let $\{y_m^-\}_{m=0}^\infty$ be a subsequence of $\{y_n\}_{n=0}^\infty$ such that $K(y_m^-) \to \tau^-$ as $m \to \infty$. Since $K$ only takes integer values, it follows that there exists $M > 0$ such that $K(y_m^-) = \tau^-$ for all $m > M$. Since $s(k, \cdot)$ is continuous for each $k$, and since $K(y_m^-) = \tau^{-1}$ for $m > M$, it follows that $s(K(y_m^-), y_m^-) \to s(\tau^-, y)$ as $m \to \infty$. Now, it follows from (15.25) that $s(K(y_m^-), y_m^-) \in \Delta f^{-1}(0)$ for each $m$. Since the set $\Delta f^{-1}(0)$ is closed, we conclude that $s(\tau^-, y) \in \Delta f^{-1}(0)$. Equation (15.25) now implies that

$$K(y) \leq \tau^- = \liminf_{n\to\infty} K(y_n), \qquad (15.26)$$

which implies that $K(\cdot)$ is lower semicontinuous at $y$. Since $y \in \mathcal{N}$ was chosen arbitrarily, the assertion follows. $\square$

In the case of continuous-time systems, it is known that the settling-time function $T(\cdot)$ of a finite time stable equilibrium is continuous in the domain of convergence if and only if it is continuous at the equilibrium (see Proposition 2.4 of [34] and Lemma 4.1 of [167]). In the case of discrete-time systems, the integer-valued function $K(\cdot)$ is continuous at a point only if it is locally constant. Thus, if $K(\cdot)$ is continuous at an equilibrium point $x_e$, then $x_e$ necessarily has to lie in the interior of $\Delta f^{-1}(0)$. On the other hand, the set of equilibrium points is closed. Hence, $K(\cdot)$ can be continuous at all equilibrium points only in the uninteresting case where the set of equilibria is either empty or the whole state space.

## 15.6 Lyapunov and Converse Lyapunov Theorems for Finite Time Semistability

In this section, we present necessary and sufficient conditions for finite time semistability. For these results, we assume that $f : \mathcal{D} \to \mathcal{D}$ and, for every $x_0 \in \mathcal{D}$, that (15.1) is forward complete. The first result establishes a relationship between finite time convergence and finite time semistability.

**Theorem 15.6.** Consider the nonlinear discrete-time dynamical system (15.1). Assume that there exist a continuous nonnegative function $V : \mathcal{D} \to \overline{\mathbb{R}}_+$ such that $\Delta V^{-1}(0) = \Delta f^{-1}(0)$; an open neighborhood $\mathcal{Q} \subseteq \mathcal{D}$ such that $\mathcal{Q} \cap \Delta f^{-1}(0)$ is nonempty; and

$$V(f(x)) \leq w(V(x)), \quad x \in \mathcal{Q} \setminus \Delta f^{-1}(0), \tag{15.27}$$

where $w : \mathbb{R} \to \mathbb{R}$ is continuous; $w(0) = 0$; and

$$z(k+1) = w(z(k)), \quad z(0) = z_0, \quad k \geq 0. \tag{15.28}$$

If (15.28) is finite time convergent to the origin for $\overline{\mathbb{R}}_+$ and every point in $\mathcal{Q} \cap \Delta f^{-1}(0)$ is a Lyapunov stable equilibrium point of (15.1), then every equilibrium point $x_e \in \mathcal{Q} \cap \Delta f^{-1}(0)$ of (15.1) is finite time semistable. Moreover, the settling-time function of (15.1) is lower semicontinuous on an open neighborhood of $\mathcal{Q} \cap \Delta f^{-1}(0)$. If, in addition, $\mathcal{Q} = \mathcal{D}$, then (15.1) is finite time semistable. Finally, if $\mathcal{D} = \mathbb{R}^n$, $V(\cdot)$ is radially unbounded, and (15.27) holds on $\mathbb{R}^n$, then every equilibrium point $x_e \in \Delta f^{-1}(0)$ of (15.1) is globally finite time semistable.

**Proof.** Consider $x_e \in \mathcal{Q} \cap \Delta f^{-1}(0)$. Since $x(k) \equiv x_e$ is Lyapunov stable, it follows that there exists an open positively invariant set $\mathcal{V} \subseteq \mathcal{Q}$ such that $x_e \in \mathcal{V}$. Next, it follows from (15.27) that

$$V(s(k+1, x)) \leq w(V(s(k, x))), \quad x \in \mathcal{V}, \quad k \in \overline{\mathbb{Z}}_+. \tag{15.29}$$

Now, it follows from Corollary 15.1 that

$$V(s(k, x)) \leq \psi(k, V(x_0)), \quad x \in \mathcal{V}, \quad k \in \overline{\mathbb{Z}}_+, \tag{15.30}$$

where $\psi : \overline{\mathbb{Z}}_+ \times \mathbb{R} \to \mathbb{R}$ is the global semiflow of (15.28). Since (15.28) is finite time convergent to the origin for $\overline{\mathbb{R}}_+$, it follows from (15.30) and the nonnegativity of $V(\cdot)$ that

$$V(s(k, x)) = 0, \quad k \geq \hat{K}(V(x)), \quad x \in \mathcal{V}, \tag{15.31}$$

where $\hat{K}(\cdot)$ denotes the settling-time function of (15.28).

Next, since $s(0, x) = x$, $s(k, \cdot)$ is continuous, and $\Delta V(s(k, x)) = 0$ is equivalent to $f(s(k, x)) = s(k, x)$ on $\mathcal{V}$, it follows that $\min\{k \in \overline{\mathbb{Z}}_+ : f(s(k, x)) = s(k, x)\} > 0$, $x \in \mathcal{V} \setminus \Delta f^{-1}(0)$. Furthermore, it follows from (15.31) that $\min\{k \in \overline{\mathbb{Z}}_+ : f(s(k, x)) = s(k, x)\} < \infty$, $x \in \mathcal{V}$. Now, define $K : \mathcal{V} \setminus \Delta f^{-1}(0) \to \mathbb{Z}_+$ by using (15.25). Then it follows that every point in $\mathcal{V} \cap \Delta f^{-1}(0)$ is finite time semistable, and, by Proposition 15.3, $K$ is a lower semicontinuous settling-time function on $\mathcal{V}$. Furthermore, it follows from (15.31) that $K(x) \leq \hat{K}(V(x))$, $x \in \mathcal{V}$.

Moreover, if $\mathcal{Q} = \mathcal{D}$, then $\mathcal{Q}$ is positively invariant by the fact that (15.1) with $f : \mathcal{D} \to \mathcal{D}$ is forward complete with unique solutions, and, hence, the preceding arguments hold with $\mathcal{V} = \mathcal{Q} = \mathcal{D}$. Finally, if $\mathcal{D} = \mathbb{R}^n$ and $V(\cdot)$ is radially unbounded, then global finite time semistability follows using identical arguments. $\qquad\square$

The following definition and lemma are needed for the next results of the chapter.

**Definition 15.6.** A continuous function $w : \mathbb{R} \to \mathbb{R}$ is a *generalized deadzone function* if (i) $|w(z)| < |z|$, $z \in \mathbb{R}$, and (ii) there exists $\varepsilon > 0$ such that $w(z) = 0$ for all $z \in \mathcal{B}_\varepsilon(0)$.

**Lemma 15.1.** Consider the scalar nonlinear discrete-time dynamical system (15.2) with $\mathcal{Q} = \mathbb{R}$, $k_0 = 0$, and $\mathcal{I}_{z_0} = \mathbb{Z}$. Then the zero solution $z(k) \equiv 0$ to (15.2) is a globally finite time stable equilibrium point of (15.2) if and only if $w : \mathbb{R} \to \mathbb{R}$ is a generalized deadzone function.

**Proof.** To show sufficiency, suppose that $w : \mathbb{R} \to \mathbb{R}$ is a generalized deadzone function. Then the zero solution $z(k) \equiv 0$ to (15.2) is an equilibrium point of (15.2). Let $|z(0)| > 0$, and consider the solution sequence $\{z(k)\}_{k=0}^\infty$ generated by (15.2). Suppose, *ad absurdum*, that $|z(k)| \geq \varepsilon$, $k \in \mathbb{Z}$. Since $|w(z)| < |z|$, $z \in \mathbb{R}$, it follows that $|z(k+1)| < |z(k)|$. Thus, since $|z(k)|$, $k \in \mathbb{Z}_+$, is a decreasing sequence that is bounded from below, there exists $z^* > 0$ such that $|z(k)| \to z^* \geq \varepsilon > 0$ as $k \to \infty$. Now, since $w$ is continuous, it follows that

$$w(z^*) = w\left(\min_{k\to\infty} |z(k)|\right) = \lim_{k\to\infty} w(|z(k)|) = z^*,$$

which is a contradiction. Hence, there exists $k$ such that $|z(k)| < \varepsilon$, and, hence, $z(k+1) = 0$. Thus, the zero solution $z(k) \equiv 0$ to (15.2) is globally finite time convergent. Lyapunov stability now follows trivially since $w(z) = 0$, $z \in \mathcal{B}_\varepsilon(0)$.

Conversely, to show necessity, suppose that the zero solution $z(k) \equiv 0$ to (15.2) is a globally finite time stable equilibrium point of (15.2). Let $z(0) \in \mathbb{R}$, consider the solution sequence $\{z(k)\}_{k=0}^\infty$ generated by (15.2), and let $\kappa = \min\{k : z(k) = 0\} - 1$. It follows from finite time stability that $\kappa < \infty$. By the definition of $\kappa$, it follows that $z(\kappa + 1) = 0$, while $z(\kappa) \neq 0$. Now, since $|w(z)| < |z|$, $z \in \mathbb{R}$, it also follows that $w(z) = 0$ for all $z \in \mathcal{B}_{|z(\kappa)|}(0)$. Hence, there exists $\varepsilon = |z(\kappa)| > 0$ such that $w(z) = 0$, $z \in \mathcal{B}_\varepsilon(0)$. $\qquad\square$

Next, using Lemma 15.1, we present two concrete forms for $w(\cdot)$ in

Theorem 15.6 for establishing finite time semistability. For the statement of the next results, we write $\lceil \alpha \rceil$ for the ceiling function denoting the smallest integer greater than or equal to $\alpha$.

**Corollary 15.2.** Consider the nonlinear discrete-time dynamical system (15.1). Assume that there exist a continuous nonnegative function $V : \mathcal{D} \to \overline{\mathbb{R}}_+$ such that $\Delta V^{-1}(0) = \Delta f^{-1}(0)$, real numbers $\alpha \in (0,1)$ and $c > 0$, and an open neighborhood $\mathcal{Q} \subseteq \mathcal{D}$ such that $\mathcal{Q} \cap \Delta f^{-1}(0)$ is nonempty, and assume that

$$\Delta V(x) \leq -c \min \left\{ \frac{V(x)}{c}, V(x)^\alpha \right\}, \quad x \in \mathcal{Q} \setminus \Delta f^{-1}(0). \tag{15.32}$$

If every equilibrium point $x_e \in \mathcal{Q} \cap \Delta f^{-1}(0)$ is a Lyapunov stable equilibrium point of (15.1), then every equilibrium point $x_e \in \mathcal{Q} \cap \Delta f^{-1}(0)$ of (15.1) is finite time semistable. Moreover, there exist an open neighborhood $\mathcal{N}$ of $\mathcal{Q} \cap \Delta f^{-1}(0)$ and a settling-time function $K : \mathcal{N} \to \overline{\mathbb{Z}}_+$ such that either

$$K(x_0) \leq \left\lceil \log_{[1-cV(x_0)^{\alpha-1}]} \frac{c^{\frac{1}{1-\alpha}}}{V(x_0)} \right\rceil + 1, \quad x_0 \in \mathcal{N}, \quad V(x_0) > c^{\frac{1}{1-\alpha}}, \tag{15.33}$$

or

$$K(x_0) = 1, \quad x_0 \in \mathcal{N} \setminus \Delta f^{-1}(0), \quad V(x_0) \leq c^{\frac{1}{1-\alpha}}, \tag{15.34}$$

where $K(\cdot)$ is lower semicontinuous on $\mathcal{N}$. If, in addition, $\mathcal{Q} = \mathcal{D}$, then (15.1) is finite time semistable. Finally, if $\mathcal{D} = \mathbb{R}^n$, $V(\cdot)$ is radially unbounded, and (15.32) holds on $\mathbb{R}^n$, then every equilibrium point $x_e \in \Delta f^{-1}(0)$ of (15.1) is globally finite time semistable.

**Proof.** Consider the scalar discrete-time nonlinear dynamical system given by

$$z(k+1) = z(k) - c\,\text{sign}(z(k)) \min \left\{ \frac{|z(k)|}{c}, |z(k)|^\alpha \right\}, \quad z(0) = z_0, \quad k \geq 0, \tag{15.35}$$

where $z(k) \in \mathbb{R}$, $k \in \overline{\mathbb{Z}}_+$, $\text{sign}(z) \triangleq z/|z|$, $z \neq 0$, $\text{sign}(0) \triangleq 0$, $\alpha \in (0,1)$, and $c > 0$. Note that the right-hand side of (15.35) is a generalized deadzone function, and, hence, by Lemma 15.1, the zero solution $z(k) \equiv 0$ to (15.35) is globally finite time stable. Furthermore, note that if $|z(k)| \leq c^{\frac{1}{1-\alpha}}$, $k \in \overline{\mathbb{Z}}_+$, then $z(k+1) = 0$, and if $|z(k)| > c^{\frac{1}{1-\alpha}}$, $k \in \overline{\mathbb{Z}}_+$, then

$$|z(k)| = \left| z(k-1) \left( 1 - c|z(k-1)|^{\alpha-1} \right) \right| < |z(k-1)|, \quad k \in \overline{\mathbb{Z}}_+. \tag{15.36}$$

Since $\alpha \in (0,1)$, then $|z(k)|^{\alpha-1} > |z(k-1)|^{\alpha-1}$, $k \in \overline{\mathbb{Z}}_+$, and

$$1 - c|z(k)|^{\alpha-1} < 1 - c|z(k-1)|^{\alpha-1}, \quad k \in \overline{\mathbb{Z}}_+. \tag{15.37}$$

Next, it follows from (15.36), (15.37), and $|z(k)| > c^{\frac{1}{1-\alpha}}$, $k \in \overline{\mathbb{Z}}_+$, that

$$|z(k)| = |z(k-1)| \left(1 - c\,|z(k-1)|^{\alpha-1}\right)$$

$$\vdots$$

$$= |z_0| \left(1 - c\,|z_0|^{\alpha-1}\right) \cdots \left(1 - c\,|z(k-1)|^{\alpha-1}\right)$$

$$< |z_0| \left(1 - c\,|z_0|^{\alpha-1}\right)^k, \quad k \in \overline{\mathbb{Z}}_+. \tag{15.38}$$

Now, if $|z_0(1 - c|z_0|^{\alpha-1})^k| \le c^{\frac{1}{1-\alpha}}$, $k \in \overline{\mathbb{Z}}_+$, then $|z(k)| < c^{\frac{1}{1-\alpha}}$, $k \in \overline{\mathbb{Z}}_+$, which implies that $z(k+1) = 0$ for

$$k \ge \log_{[1-c|z_0|^{\alpha-1}]} \frac{c^{\frac{1}{1-\alpha}}}{|z_0|}, \quad |z_0| > c^{\frac{1}{1-\alpha}},$$

and, hence, (i) of Definition 15.4 is satisfied with $\mathcal{N} = \mathcal{D} = \mathbb{R}$ and with the settling-time function $\hat{K}(z_0)$ given by either

$$\hat{K}(z_0) \le \left\lceil \log_{[1-c|z_0|^{\alpha-1}]} \frac{c^{\frac{1}{1-\alpha}}}{|z_0|} \right\rceil + 1, \quad |z_0| > c^{\frac{1}{1-\alpha}}, \tag{15.39}$$

or

$$\hat{K}(z_0) = 1, \quad |z_0| \le c^{\frac{1}{1-\alpha}}, \quad z_0 \ne 0. \tag{15.40}$$

Next, with $z = V(x)$ and

$$w(z) = w(V(x)) = V(x) - c \min\left\{\frac{V(x)}{c}, V(x)^\alpha\right\},$$

it follows from Corollary 15.1 and (15.39) that

$$\hat{K}(V(x_0)) \le \left\lceil \log_{[1-cV(x_0)^{\alpha-1}]} \frac{c^{\frac{1}{1-\alpha}}}{V(x_0)} \right\rceil + 1, \quad x_0 \in \mathcal{B}_\delta(x_e), \quad V(x_0) > c^{\frac{1}{1-\alpha}}. \tag{15.41}$$

Hence, it follows from (15.31) that $K(x) \le \hat{K}(V(x))$, and, hence,

$$x(k) \in \mathcal{Q} \cap \Delta f^{-1}(0), \quad k \ge \left\lceil \log_{[1-cV(x_0)^{\alpha-1}]} \frac{c^{\frac{1}{1-\alpha}}}{V(x_0)} \right\rceil + 1,$$

$$x_0 \in \mathcal{B}_\delta(x_e), \quad V(x_0) > c^{\frac{1}{1-\alpha}}, \tag{15.42}$$

which implies finite time convergence of the trajectories of (15.1) for all $x_0 \in \mathcal{B}_\delta(x_e)$ such that $V(x_0) > c^{\frac{1}{1-\alpha}}$. Alternatively, if $V(x_0) \le c^{\frac{1}{1-\alpha}}$, then it follows from (15.40) that the equilibrium point $x(k) \equiv x_\mathrm{e}$ is finite time convergent with settling-time function $K(x_0) = 1$.

Now, since every point in $\mathcal{Q} \cap \Delta f^{-1}(0)$ is a Lyapunov stable equilibrium point of (15.1), it follows from Theorem 15.6 that every equilibrium point $x_\mathrm{e} \in \mathcal{Q} \cap \Delta f^{-1}(0)$ of (15.1) with $\mathcal{N} \triangleq \mathcal{B}_\delta(x_\mathrm{e})$ is finite time semistable. The remainder of the proof now follows as in the proof of Theorem 15.6. $\qquad\square$

**Example 15.1.** Consider the discrete-time collective dynamics of two agents on $\mathbb{R}^2$ described by

$$x_i(k+1) = x_i(k) + u_i(k), \quad x_i(0) = x_{i0}, \quad k \in \overline{\mathbb{Z}}_+, \quad i = 1, 2, \quad (15.43)$$

where, for $k \in \overline{\mathbb{Z}}_+$, $x_1(k), x_2(k) \in \mathbb{R}$, and $u_1(k)$ and $u_2(k)$ are given by

$$
\begin{aligned}
u_1(k) = &-c\operatorname{sign}\left(x_1(k) - x_2(k)\right) \\
&\cdot \min\left\{\frac{|x_1(k) - x_2(k)|}{2c}, \left[\frac{|x_1(k) - x_2(k)|}{2}\right]^\alpha\right\},
\end{aligned} \quad (15.44)
$$

$$
\begin{aligned}
u_2(k) = &-c\operatorname{sign}\left(x_2(k) - x_1(k)\right) \\
&\cdot \min\left\{\frac{|x_2(k) - x_1(k)|}{2c}, \left[\frac{|x_2(k) - x_1(k)|}{2}\right]^\alpha\right\},
\end{aligned} \quad (15.45)
$$

where $\alpha \in (0, 1)$ and $c \in \mathbb{R}_+$.

First, note that $\mathcal{M} = \left\{x \in \mathbb{R}^2 : x = \beta\mathbf{e}_2, \beta \in \mathbb{R}\right\}$, where $\mathbf{e}_2 \triangleq [1\ 1]^\mathrm{T}$, is the set of equilibria for (15.43) with (15.44) and (15.45), $u_1(k) = -u_2(k)$, and $x_1(k) + x_2(k) = 2\beta$, $k \in \overline{\mathbb{Z}}_+$. Now, consider the Lyapunov function candidate $V(x) = (x - \beta\mathbf{e}_2)^\mathrm{T}(x - \beta\mathbf{e}_2)$. Note that $V(x) > 0$, $x \in \mathbb{R}^2$, $x \neq \beta\mathbf{e}_2$, and $V(x) = 0$ if and only if $x_1 = x_2$. Furthermore, note that $V(\cdot)$ is radially unbounded, and, since $x_1(k) + x_2(k) = 2\beta$, $k \in \overline{\mathbb{Z}}_+$, it follows that $|x_1 - x_2|^2 = 2V(x)$.

Next, note that

$$
\begin{aligned}
\Delta V(x) &= (x_1 + u_1 - \beta)^2 + (x_2 + u_2 - \beta)^2 - (x_1 - \beta)^2 + (x_2 - \beta)^2 \\
&= u_1(2x_1 + u_1 - 2\beta) + u_2(2x_2 + u_2 - 2\beta) \\
&= 2u_1(x_1 - x_2 + u_1) \\
&= -2c\min\left\{\frac{|x_1 - x_2|}{2c}, \left[\frac{|x_1 - x_2|}{2}\right]^\alpha\right\} \\
&\quad \cdot \left[|x_1 - x_2| - c\min\left\{\frac{|x_1 - x_2|}{2c}, \left[\frac{|x_1 - x_2|}{2}\right]^\alpha\right\}\right] \\
&\leq -2c^2\left[\min\left\{\frac{|x_1 - x_2|}{2c}, \left[\frac{|x_1 - x_2|}{2}\right]^\alpha\right\}\right]^2
\end{aligned}
$$

$$= -2c^2 \min \left\{ \frac{|x_1 - x_2|^2}{4c^2}, \left[ \frac{|x_1 - x_2|}{2} \right]^{2\alpha} \right\}$$

$$= -2c^2 \min \left\{ \frac{V(x)}{2c^2}, \left[ \frac{V(x)}{2} \right]^{\alpha} \right\}$$

$$\leq 0, \quad x \in \mathbb{R}^2, \tag{15.46}$$

and, hence, it follows from Corollary 15.2 that (15.43), with control inputs (15.44) and (15.45), is finite time semistable with settling-time function given by either

$$K(x_0) \leq \left\lceil \log_{[1-2^{(1-\alpha)}c^2 V(x_0)^{\alpha-1}]} \frac{2c^{\frac{2}{1-\alpha}}}{V(x_0)} \right\rceil + 1, \quad V(x_0) > 2c^{\frac{1}{1-\alpha}}, \tag{15.47}$$

or

$$K(x_0) = 1, \quad V(x_0) \leq 2c^{\frac{1}{1-\alpha}}. \tag{15.48}$$

The system trajectory and control profile of (15.43), with control inputs (15.44) and (15.45), for the initial conditions $x_0 = [0 \quad 20]^{\mathrm{T}}$, $c = 2$, and $\alpha = 0.5$, are shown in Figure 15.2. The guaranteed settling-time function is given by $K(x_0) = 5$, whereas the achieved finite time convergence step is 4. $\triangle$

**Corollary 15.3.** Consider the nonlinear discrete-time dynamical system (15.1). Assume that there exist a continuous nonnegative function $V : \mathcal{D} \to \overline{\mathbb{R}}_+$ such that $\Delta V^{-1}(0) = \Delta f^{-1}(0)$; a real number $c > 0$; an open neighborhood $\mathcal{Q} \subseteq \mathcal{D}$ such that $\mathcal{Q} \cap \Delta f^{-1}(0)$ is nonempty; and

$$\Delta V(x) \leq -\min \{V(x), c\}, \quad x \in \mathcal{Q} \setminus \Delta f^{-1}(0). \tag{15.49}$$

If every equilibrium point $x_e \in \mathcal{Q} \cap \Delta f^{-1}(0)$ is a Lyapunov stable equilibrium point of (15.1), then every equilibrium point $x_e \in \mathcal{Q} \cap \Delta f^{-1}(0)$ of (15.1) is finite time semistable. Moreover, there exist an open neighborhood $\mathcal{N}$ of $\mathcal{Q} \cap \Delta f^{-1}(0)$ and a settling-time function $K : \mathcal{N} \to \overline{\mathbb{Z}}_+$ such that

$$K(x_0) \leq \left\lceil \frac{V(x_0)}{c} \right\rceil, \quad x_0 \in \mathcal{N}, \tag{15.50}$$

where $K(\cdot)$ is lower semicontinuous on $\mathcal{N}$. If, in addition, $\mathcal{Q} = \mathcal{D}$, then (15.1) is finite time semistable. Finally, if $\mathcal{D} = \mathbb{R}^n$, $V(\cdot)$ is radially unbounded, and (15.49) holds on $\mathbb{R}^n$, then every equilibrium point $x_e \in \Delta f^{-1}(0)$ of (15.1) is globally finite time semistable.

**Proof.** Consider the scalar discrete-time nonlinear dynamical system given by

$$z(k + 1) = z(k) - \text{sign}(z(k)) \min\{|z(k)|, c\}, \quad z(0) = z_0, \quad k \geq 0, \tag{15.51}$$

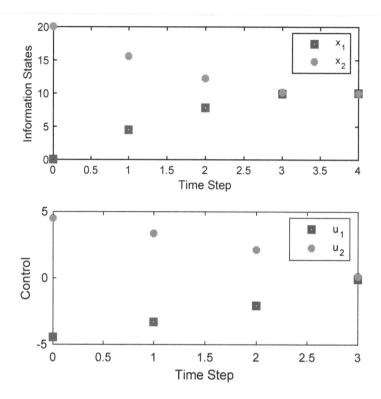

Figure 15.2 Information states and control inputs versus time of (15.43) with control
inputs (15.44) and (15.45). The two agents achieve finite time consensus in 4
sec.

where $z(k) \in \mathbb{R}$, $k \in \overline{\mathbb{Z}}_+$, and $c > 0$. Note that the right-hand side of (15.51)
is a generalized deadzone function, and, hence, by Lemma 15.1, the zero
solution $z(k) \equiv 0$ to (15.51) is globally finite time stable. Furthermore, note
that if $|z(k)| \leq c$, $k \in \overline{\mathbb{Z}}_+$, then $z(k + 1) = z(k) - \text{sign}(z(k))\,|z(k)| = 0$,
and, hence, (i) of Definition 15.4 is satisfied, with $\mathcal{N} = \mathcal{D} = \mathbb{R}$ and with the
settling-time function $\hat{K}(z_0)$ given by

$$\hat{K}(z_0) = \left\lceil \frac{|z_0|}{c} \right\rceil. \tag{15.52}$$

Next, using identical arguments as in the proof of Corollary 15.2 with
$z = V(x)$ and $w(z) = w(V(x)) = V(x) - \min\{V(x), c\}$, it follows that the
equilibrium point $x_e \in \mathcal{Q} \cap \Delta f^{-1}(0)$ of (15.1) is finite time semistable with
settling-time function

$$K(x_0) \leq \left\lceil \frac{V(x_0)}{c} \right\rceil, \quad x_0 \in \mathcal{N}. \tag{15.53}$$

The remainder of the proof now follows as in the proof of Theorem 15.6. $\square$

**Example 15.2.** Consider the two-agent system given by (15.43) with control inputs $u_1(k)$ and $u_2(k)$ given by

$$u_1(k) = -\text{sign}\,(x_1(k) - x_2(k)) \min \left\{ \frac{|x_1(k) - x_2(k)|}{2}, c \right\}, \tag{15.54}$$

$$u_2(k) = -\text{sign}\,(x_2(k) - x_1(k)) \min \left\{ \frac{|x_2(k) - x_1(k)|}{2}, c \right\}, \tag{15.55}$$

where $c \in \mathbb{R}_+$. First, note that $\mathcal{M} = \{x \in \mathbb{R}^2 : x = \beta e_2, \beta \in \mathbb{R}\}$ is the set of equilibria for (15.43), with (15.44) and (15.45), $u_1(k) = -u_2(k)$, and $x_1(k) + x_2(k) = 2\beta$, $k \in \overline{\mathbb{Z}}_+$. Now, consider the Lyapunov function candidate $V(x) = (x - \beta e_2)^{\mathrm{T}} (x - \beta e_2)$. Note that $V(x) > 0$, $x \in \mathbb{R}^2$, $x \neq \beta e_2$, and $V(x) = 0$ if and only if $x_1 = x_2$. Furthermore, note that $V(\cdot)$ is radially unbounded, and, since $x_1(k) + x_2(k) = 2\beta$, $k \in \overline{\mathbb{Z}}_+$, it follows that $|x_1 - x_2|^2 = 2V(x)$.

Next, note that

$$\begin{aligned}
\Delta V(x) &= (x_1 + u_1 - \alpha)^2 + (x_2 + u_2 - \alpha)^2 - (x_1 - \alpha)^2 + (x_2 - \alpha)^2 \\
&= 2u_1[x_1 - x_2 + u_1] \\
&= -2\min \left\{ \frac{|x_1 - x_2|}{2}, c \right\} \left[ |x_1 - x_2| - \min \left\{ \frac{|x_1 - x_2|}{2}, c \right\} \right] \\
&\leq -2 \left[ \min \left\{ \frac{|x_1 - x_2|}{2}, c \right\} \right]^2 \\
&= -2\min \left\{ \frac{|x_1 - x_2|^2}{4}, c^2 \right\} \\
&= -2\min \left\{ \frac{V(x)}{2}, c^2 \right\} \\
&\leq 0, \quad x \in \mathbb{R}^2, \tag{15.56}
\end{aligned}$$

and, hence, it follows from Corollary 15.3 that (15.43), with control inputs (15.54) and (15.55), is finite time semistable with settling-time function

$$K(x_0) \leq \left\lceil \frac{V(x_0)}{2c^2} \right\rceil. \tag{15.57}$$

The system trajectory and control profile of (15.43) with control inputs (15.54) and (15.55) for the initial conditions $x_0 = [0 \quad 20]^{\mathrm{T}}$ and $c = 3$ are shown in Figure 15.3. The guaranteed settling-time function is given by $K(x_0) = 12$, whereas the achieved finite time convergence step is 4. $\triangle$

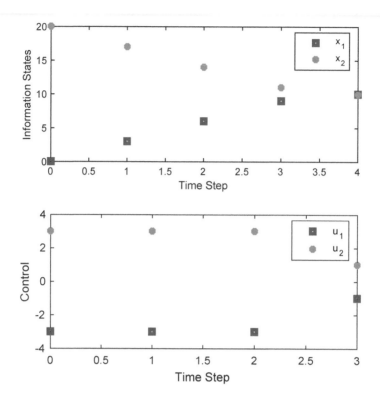

Figure 15.3 Information states and control inputs versus time of (15.43) with control
            inputs (15.54) and (15.55). The two agents achieve finite time consensus in 4
            sec.

Finally, we present partial converse theorems to Theorem 15.6 and
Corollaries 15.2 and 15.3.

**Theorem 15.7.** Consider the nonlinear discrete-time dynamical system (15.1), and let $\mathcal{N}$ be as in Definition 15.4. If every equilibrium point
$x_e \in \mathcal{N} \cap \Delta f^{-1}(0)$ of (15.1) is finite time semistable, then there exist a
nonnegative lower semicontinuous function $V : \mathcal{N} \to \overline{\mathbb{R}}_+$ and a continuous
function $w : \mathbb{R} \to \mathbb{R}$ such that $V(f(x)) \leq w(V(x))$, $x \in \mathcal{N}$, where $w(\cdot)$ is of
class $\mathcal{W}_d$, and the zero solution $z(k) \equiv 0$ to (15.28) is finite time convergent.

**Proof.** First, note that it follows from Proposition 15.3 that the settling-
time function $K : \mathcal{N} \to \overline{\mathbb{Z}}_+$ is lower semicontinuous on $\mathcal{N}$. Next, define
$V : \mathcal{N} \to \overline{\mathbb{R}}_+$ by

$$V(x) \triangleq \sup_{k \geq 0} \frac{1 + bk}{1 + ak} \left[ K(s(k, x)) \right]^{\beta}, \qquad (15.58)$$

where $\beta > 2$, $\beta \in \mathbb{Z}_+$, and $b > a > 0$. Note that $V(\cdot)$ is lower semicontinuous and nonnegative, and, by $s(K(x) + k, x) \in \mathcal{N} \cap \Delta f^{-1}(0)$ for all $x \in \mathcal{N}$ and $k \in \mathbb{Z}_+$, $\Delta V(x) = 0$, $x \in \mathcal{N} \cap \Delta f^{-1}(0)$. Now, note that it follows from the definition of $K(\cdot)$ that the supremum in the definition of $V(s(1, x))$ is reached at some time $\hat{k}$ such that $0 \leq \hat{k} \leq K(x)$. If $\hat{k} < K(x)$, then

$$
\begin{aligned}
V(s(1, x)) &= \frac{1 + b\hat{k}}{1 + a\hat{k}} \left[ K(s(\hat{k} + 1, x)) \right]^{\beta} \\
&= \left[ 1 - \frac{b - a}{(1 + b\hat{k} + b)(1 + a\hat{k})} \right] \frac{1 + b\hat{k} + b}{1 + a\hat{k} + a} \left[ K(s(\hat{k} + 1, x)) \right]^{\beta} \\
&\leq \left[ 1 - \frac{a(b - a)}{b \left[ 1 + aK(x) \right]^2} \right] V(x), \quad K(x) > \hat{k}. \tag{15.59}
\end{aligned}
$$

Alternatively, if $\hat{k} = K(x)$, then $V(s(1, x)) = 0$, which implies that $x \in \mathcal{N} \cap \Delta f^{-1}(0)$.

Next, if $x \notin \mathcal{N} \cap \Delta f^{-1}(0)$, then

$$
V(x) = \sup_{k \geq 0} \frac{1 + bk}{1 + ak} [K(s(k, x))]^{\beta} \geq [K(x)]^{\beta} \geq 1, \tag{15.60}
$$

and, hence, $[1 + aK(x)]^{\beta} \leq (1 + a)^{\beta} V(x)$. Now, (15.59) yields

$$
\begin{aligned}
V(f(x)) - V(x) &\leq -\frac{a(b - a)}{b} V(x) \left( 1 + K(x) \right)^{-2} \\
&\leq -\frac{a(b - a)}{b} V(x) \left[ (1 + a)^{\beta} V(x) \right]^{-\frac{2}{\beta}} \\
&= -\frac{a(b - a)}{b(1 + a)^2} V(x)^{\frac{\beta - 2}{\beta}} \\
&\leq -\frac{a(b - a)}{b(1 + a)^2} \min \left\{ \frac{b(1 + a)^2}{a(b - a)} V(x), V(x)^{\frac{\beta - 2}{\beta}} \right\}. \tag{15.61}
\end{aligned}
$$

Now, letting $\alpha = \frac{\beta - 2}{\beta} \in (0, 1)$ and $c = \frac{a(b - a)}{b(1 + a)^2} > 0$, (15.61) becomes

$$
V(f(x)) \leq V(x) - c \min \left\{ \frac{V(x)}{c}, V(x)^{\alpha} \right\}. \tag{15.62}
$$

Finally, using Lemma 15.1, it follows that the zero solution $z(k) \equiv 0$ to (15.28), with the class $\mathcal{W}_d$ function $w(z) = z - c \operatorname{sign}(z) \min\{\frac{|z|}{c}, |z|^{\alpha}\}$, is finite time convergent. $\square$

**Theorem 15.8.** Consider the nonlinear discrete-time dynamical system (15.1), let $\alpha \in (0, 1)$, and let $\mathcal{N}$ be as in Definition 15.4. If every equilibrium point $x_e \in \mathcal{N} \cap \Delta f^{-1}(0)$ of (15.1) is finite time semistable, then there exist a nonnegative lower semicontinuous function $V : \mathcal{N} \to \overline{\mathbb{R}}_+$ and real numbers $\alpha \in (0, 1)$ and $c > 0$ such that

$$V(f(x)) \leq V(x) - c \min \left\{ \frac{V(x)}{c}, V(x)^\alpha \right\}, \quad x \in \mathcal{N}. \tag{15.63}$$

**Proof.** The proof is identical to that of Theorem 15.7 with class $\mathcal{W}_{\mathrm{d}}$ function $w : \mathbb{R} \to \mathbb{R}$ given by $w(z) = z - c \operatorname{sign}(z) \min\{\frac{|z|}{c}, |z|^\alpha\}$. $\qquad \square$

**Theorem 15.9.** Consider the nonlinear discrete-time dynamical system (15.1), and let $\mathcal{N}$ be as in Definition 15.4. If every equilibrium point $x_\mathrm{e} \in \mathcal{N} \cap \Delta f^{-1}(0)$ of (15.1) is finite time semistable, then there exist a nonnegative lower semicontinuous function $V : \mathcal{N} \to \mathbb{R}$ and a real number $c > 0$ such that

$$V(f(x)) \leq V(x) - \min \left\{ V(x), c \right\}, \quad x \in \mathcal{N}. \tag{15.64}$$

**Proof.** First, note that Proposition 15.3 implies that the settling-time function $K : \mathcal{N} \to \overline{\mathbb{Z}}_+$ is lower semicontinuous on $\mathcal{N}$. Next, define $V : \mathcal{N} \to \overline{\mathbb{R}}_+$ by $V(x) \triangleq cK(x)$, where $c > 0$. Note that $V(\cdot)$ is lower semicontinuous and nonnegative, and, by $s(K(x) + k, x) \in \mathcal{N} \cap \Delta f^{-1}(0)$ for all $x \in \mathcal{N}$ and $k \in \mathbb{Z}_+$, $\Delta V(x) = 0$, $x \in \mathcal{N} \cap \Delta f^{-1}(0)$. Now, since every equilibrium point $x_\mathrm{e} \in \mathcal{N} \cap \Delta f^{-1}(0)$ of (15.1) is finite time semistable and $K(s(1, x)) = K(x) - 1$, it follows that

$$V(f(x)) = cK(s(1, x)) = c(K(x) - 1) = V(x) - c \tag{15.65}$$

for $x \notin \mathcal{N} \cap \Delta f^{-1}(0)$, and, hence,

$$V(f(x)) - V(x) = -c \leq -\min\{V(x), c\}. \tag{15.66}$$

Finally, using Lemma 15.1, it follows that the zero solution $z(k) \equiv 0$ to (15.28) with the class $\mathcal{W}_{\mathrm{d}}$ function $w(z) = z - \operatorname{sign}(z) \min\{z, c\}$ is finite time convergent. $\qquad \square$

## 15.7 A Thermodynamic-Based Architecture for Asymptotic Network Consensus

In this section, we develop a thermodynamically motivated information consensus framework for discrete-time nonlinear network systems to achieve semistability and state equipartition. The consensus problem we address in this section is the discrete-time analogue of the continuous-time problem addressed in Chapters 2 and 4. As in Chapter 4, we use graph-theoretic notions to represent a dynamical network and present solutions to the consensus problem for networks with undirected communication graph topologies (or information flows). The notation we use here is identical to the notation established in Chapter 4.

Consider the $q$ discrete-time dynamical agents $\mathcal{G}_i$ with dynamics given by

$$x_i(k+1) = x_i(k) + u_i(k), \quad x_i(0) = x_{i0}, \quad k \in \overline{\mathbb{Z}}_+, \tag{15.67}$$

where, for each agent $i \in \{1, \ldots, q\}$, $x_i(k) \in \mathbb{R}$, $k \in \overline{\mathbb{Z}}_+$, denotes the information state of agent $\mathcal{G}_i$ and $u_i(k) \in \mathbb{R}$, $k \in \overline{\mathbb{Z}}_+$, denotes the information control input of agent $\mathcal{G}_i$. The nonlinear consensus protocol is given by

$$u_i(k) = \sum_{j=1, j \neq i}^{q} \phi_{ij}\left(x_i(k), x_j(k)\right), \quad i = 1, \ldots, q, \tag{15.68}$$

where $\phi_{ij}(\cdot, \cdot)$, $i, j = 1, \ldots, q$, are continuous functions characterizing the information exchange between agents $\mathcal{G}_j$ and $\mathcal{G}_i$.

In this case, the closed-loop system (15.67) and (15.68) is given by

$$x_i(k+1) = x_i(k) + \sum_{j=1, j \neq i}^{q} \phi_{ij}\left(x_i(k), x_j(k)\right), \quad x_i(0) = 0,$$
$$k \in \overline{\mathbb{Z}}_+, \quad i = 1, \ldots, q, \tag{15.69}$$

or, equivalently, in vector form

$$x(k+1) = f(x(k)), \quad x(0) = x_0, \quad k \in \overline{\mathbb{Z}}_+, \tag{15.70}$$

where $x(k) \triangleq [x_1(k), \ldots, x_q(k)]^{\mathrm{T}}$ and $f = [f_1, \ldots, f_q]^{\mathrm{T}} : \mathcal{D} \subseteq \mathbb{R}^q \to \mathbb{R}^q$ is such that

$$f_i(x(k)) = x_i(k) + \sum_{j=1, j \neq i}^{q} \phi_{ij}\left(x_i(k), x_j(k)\right), \quad i = 1, \ldots, q. \tag{15.71}$$

Note that $\mathcal{G}$ given by (15.70) describes an interconnected network where information states are updated using a distributed controller involving neighbor-to-neighbor interaction between agents.

Although our results can be directly extended to the case where (15.67) and (15.68) describe the dynamics of an aggregate multiagent system with an aggregate state vector $x(k) = \left[x_1^{\mathrm{T}}(k), \ldots, x_q^{\mathrm{T}}(k)\right]^{\mathrm{T}} \in \mathbb{R}^{Nq}$, where $x_i(k) \in \mathbb{R}^N$ and $u_i(k) \in \mathbb{R}^N$, $i = 1, \ldots, q$, by using Kronecker calculus, for simplicity of exposition we focus on individual agent states evolving in $\mathbb{R}$ (i.e., $N = 1$).

To ensure a thermodynamically consistent information flow model, we make the following assumptions on the information flow functions $\phi_{ij}(\cdot, \cdot)$,

$i = 1, \ldots, q$. These assumptions are the discrete-time analogue of Assumptions 4.1 and 4.2.

**Assumption 15.1.** The connectivity matrix $\mathcal{C} \in \mathbb{R}^{q \times q}$ associated with the multiagent dynamical system $\mathcal{G}$ given by (15.70) is defined by

$$\mathcal{C}_{(i,j)} \triangleq \begin{cases} 0 & \text{if } \phi_{ij}(x_i, x_j) \equiv 0, \\ 1, & \text{otherwise,} \end{cases} \quad i \neq j, \quad i, j = 1, \ldots, q,$$

and $\mathcal{C}_{(i,i)} \triangleq -\sum_{m=1, m \neq i}^{q} \mathcal{C}_{(i,m)}$, $i = 1, \ldots, q$, with $\operatorname{rank} \mathcal{C} = q - 1$, and for $\mathcal{C}_{(i,j)} = 1$, $i \neq j$, $\phi_{ij}(x_i, x_j) = 0$ if and only if $x_i = x_j$.

**Assumption 15.2.** For $i, j = 1, \ldots, q$, $(x_i - x_j) \phi_{ij}(x_i, x_j) \leq 0$, $x_i, x_j \in \mathbb{R}$.

**Assumption 15.3.** For $i, j = 1, \ldots, q$, $\frac{\Delta x_i - \Delta x_j}{x_i - x_j} \geq -1$, $x_i \neq x_j$, where $\Delta x_m(k) \triangleq x_m(k+1) - x_m(k)$.

The condition that $\phi(x_i, x_j) = 0$ if and only if $x_i = x_j$, $i \neq j$, implies that agents $\mathcal{G}_i$ and $\mathcal{G}_j$ are *connected* and, hence, can share information; alternatively, $\phi_{ij}(x_i, x_j) \equiv 0$ implies that agents $\mathcal{G}_i$ and $\mathcal{G}_j$ are disconnected and, hence, cannot share information.

Assumption 15.1 implies that if the information or energies in the connected agents $\mathcal{G}_i$ and $\mathcal{G}_j$ are equal, then information or energy exchange between these agents is not possible. This statement is reminiscent of the *zeroth law of discrete thermodynamics*, which postulates that temperature equality is a necessary and sufficient condition for thermal equilibrium. Furthermore, if $\mathcal{C} = \mathcal{C}^{\mathrm{T}}$ and $\operatorname{rank} \mathcal{C} = q - 1$, then it follows that the connectivity matrix $\mathcal{C}$ is irreducible, which implies that for any pair of agents $\mathcal{G}_i$ and $\mathcal{G}_j$, $i \neq j$, of $\mathcal{G}$ there exists a sequence of information connectors (information arcs) of $\mathcal{G}$ that connect agents $\mathcal{G}_i$ and $\mathcal{G}_j$.

Assumption 15.2 implies that energy or information flows from more energetic or information-rich agents to less energetic or information-poor agents and is reminiscent of the *second law of discrete thermodynamics*, which states that heat (i.e., energy in transition) must flow in the direction of lower temperatures. Finally, Assumption 15.3 implies that the energy or information difference between any consecutive time instants is monotonic for any pair of connected agents $\mathcal{G}_i$ and $\mathcal{G}_j$, $i \neq j$; that is,

$$[x_i(k+1) - x_j(k+1)] [x_i(k) - x_j(k)] \geq 0$$

for all $x_i(k) \neq x_j(k)$, $k \geq 0$, $i, j = 1, \ldots, q$. It is important to note here that both finite time consensus controllers in Examples 15.1 and 15.2 satisfy Assumptions 15.1–15.3 and, hence, satisfy basic thermodynamic principles. For further details on Assumptions 15.1–15.3, see [118].

**Theorem 15.10.** Consider the discrete-time multiagent dynamical system (15.69) or, equivalently, (15.70). Assume that Assumptions 15.1–15.3 hold and $\phi_{ij}(x_i, x_j) = -\phi_{ji}(x_j, x_i)$ for all $i, j = 1, \ldots, q$, $i \neq j$. Then, for every $\alpha \in \mathbb{R}$, $\alpha \mathbf{e}$ is a globally semistable equilibrium state of (15.70). Furthermore, $x(k) \rightarrow \frac{1}{q} \mathbf{e} \mathbf{e}^{\mathrm{T}} x(0)$ as $k \rightarrow \infty$, and $\frac{1}{q} \mathbf{e} \mathbf{e}^{\mathrm{T}} x(0)$ is a globally semistable equilibrium state.

**Proof.** Here we write $x(k) \rightarrow \mathcal{M}$ as $k \rightarrow \infty$ to denote that $x(k)$ approaches the set $\mathcal{M}$; that is, for every $\varepsilon > 0$ there exists $K > 0$ such that $\mathrm{dist}(x(k), \mathcal{M}) < \varepsilon$ for all $k > K$. First, we show that $\mathcal{M} = \{x \in \mathbb{R}^q : x = \alpha \mathbf{e}, \alpha \in \mathbb{R}\}$ is the set of equilibria of (15.70). To see this, note that it follows from Assumption 15.1 that for $\mathcal{C}_{(i,j)} = 1$, $i \neq j$, $\phi_{ij}(x_i, x_j) = 0$ if and only if $x_i = x_j$, and, hence, $\alpha \mathbf{e}$ is an equilibrium state of (15.70) for every $\alpha \in \mathbb{R}$.

To show Lyapunov stability of the equilibrium state $\alpha \mathbf{e}$, consider the Lyapunov function candidate $V(x) = \frac{1}{2}(x - \alpha \mathbf{e})^{\mathrm{T}}(x - \alpha \mathbf{e})$. Note that $V(x) > 0$, $x \in \mathbb{R}^q$, $x \neq \alpha \mathbf{e}$, and $V(x) = 0$ if and only if $x = \alpha \mathbf{e}$. Furthermore, $V(\cdot)$ is radially unbounded. Now, since $\phi_{ij}(x_i, x_j) = -\phi_{ji}(x_j, x_i)$ for all $i, j = 1, \ldots, q$, $i \neq j$, and $\mathbf{e}^{\mathrm{T}} x(k+1) = \mathbf{e}^{\mathrm{T}} x(k)$, $k \in \overline{\mathbb{Z}}_+$, it follows from Assumptions 15.2 and 15.3 that

$$
\begin{aligned}
\Delta V(x(k)) &= V(x(k+1)) - V(x(k)) \\
&= \frac{1}{2}[x(k+1) - \alpha \mathbf{e}]^{\mathrm{T}}[x(k+1) - \alpha \mathbf{e}] - \frac{1}{2}[x(k) - \alpha \mathbf{e}]^{\mathrm{T}}[x(k) - \alpha \mathbf{e}] \\
&= \frac{1}{2}x^{\mathrm{T}}(k+1)f(x(k)) - \frac{1}{2}x^{\mathrm{T}}(k)x(k) \\
&= \sum_{i=1}^{q} \sum_{j=1, j \neq i}^{q} x_i(k+1)\phi_{ij}(x_i(k), x_j(k)) \\
&\quad - \frac{1}{2}\sum_{i=1}^{q} \left[ \sum_{j=1, j \neq i}^{q} \phi_{ij}(x_i(k), x_j(k)) \right]^2 \\
&= \sum_{i=1}^{q-1} \sum_{j=i+1}^{q} [x_i(k+1) - x_j(k+1)]\phi_{ij}(x_i(k), x_j(k)) \\
&\quad - \frac{1}{2}\sum_{i=1}^{q} \left[ \sum_{j=1, j \neq i}^{q} \phi_{ij}(x_i(k), x_j(k)) \right]^2
\end{aligned}
$$

$$\leq 0, \quad x(k) \in \mathbb{R}^q, \quad k \in \overline{\mathbb{Z}}_+, \tag{15.72}$$

which, using Theorem 13.2 of [122], establishes Lyapunov stability of the equilibrium state $\alpha\mathbf{e}$.

To show that $\alpha\mathbf{e}$ is semistable, note that

$$\begin{aligned}
\Delta V(x(k)) &= \frac{1}{2}x^{\mathrm{T}}(k+1)x(k+1) - \frac{1}{2}x^{\mathrm{T}}(k)x(k) \\
&= \sum_{i=1}^{q}\sum_{j=1,j\neq i}^{q} x_i(k)\phi_{ij}(x_i(k),x_j(k)) \\
&\quad + \frac{1}{2}\sum_{i=1}^{q}\left[\sum_{j=1,j\neq i}^{q}\phi_{ij}(x_i(k),x_j(k))\right]^2 \\
&\geq \sum_{i=1}^{q-1}\sum_{j=i+1}^{q}\left[x_i(k)-x_j(k)\right]\phi_{ij}(x_i(k),x_j(k)) \\
&= \sum_{i=1}^{q-1}\sum_{j\in\mathcal{K}_i}\left[x_i(k)-x_j(k)\right]\phi_{ij}(x_i(k),x_j(k)), \tag{15.73}
\end{aligned}$$

where $\mathcal{K}_i \triangleq \mathcal{N}_i \setminus \{1,\ldots,i-1\}$ and

$$\mathcal{N}_i \triangleq \{j \in \{1,\ldots,q\} : \phi_{ij}(x_i,x_j) = 0 \text{ if and only if } x_i = x_j\}, \quad i = 1,\ldots,q.$$

Now, note that $\Delta V(x) = 0$ if and only if $(x_i - x_j)\phi_{ij}(x_i,x_j) = 0$, $i = 1,\ldots,q$, $j \in \mathcal{K}_i$.

To see this, first assume that $(x_i - x_j)\phi_{ij}(x_i,x_j) = 0$, $i = 1,\ldots,q$, $j \in \mathcal{K}_i$. Then, it follows from (15.73) that $\Delta V(x) \geq 0$, $x \in \mathbb{R}^q$. However, it follows from (15.72) that $\Delta V(x) \leq 0$, $x \in \mathbb{R}^q$, and, hence, $\Delta V(x) = 0$. Conversely, assume that $\Delta V(x) = 0$. In this case, it follows from Assumption 15.2 and (15.72) that $[x_i(k+1) - x_j(k+1)]\phi_{ij}(x_i(k),x_j(k)) = 0$ and $\sum_{j=1,j\neq i}^{q}\phi_{ij}(x_i(k),x_j(k)) = 0$, $k \in \overline{\mathbb{Z}}_+$, $i,j = 1,\ldots,q$, $i \neq j$. Now, since

$$\begin{aligned}
&[x_i(k+1) - x_j(k+1)]\phi_{ij}(x_i(k),x_j(k)) \\
&= [x_i(k)-x_j(k)]\phi_{ij}(x_i(k),x_j(k)) \\
&\quad + \left[\sum_{h=1,h\neq i}^{q}\phi_{ih}(x_i(k),x_h(k)) - \sum_{l=1,l\neq j}^{q}\phi_{jl}(x_j(k),x_l(k))\right]\phi_{ij}(x_i(k),x_j(k)) \\
&= [x_i(k)-x_j(k)]\phi_{ij}(x_i(k),x_j(k)), \quad k \in \overline{\mathbb{Z}}_+, \quad i,j = 1,\ldots,q, \quad i \neq j, \tag{15.74}
\end{aligned}$$

it follows that $(x_i - x_j)\phi_{ij}(x_i,x_j) = 0$, $i = 1,\ldots,q$, $j \in \mathcal{K}_i$.

Finally, to show that $x(k) \to \frac{1}{q}\mathbf{e}\mathbf{e}^{\mathrm{T}}x(0)$ as $k \to \infty$ and $\frac{1}{q}\mathbf{e}\mathbf{e}^{\mathrm{T}}x(0)$ is a globally semistable equilibrium state, let

$$\mathcal{R} \triangleq \{x \in \mathbb{R}^q : \Delta V(x) = 0\}$$
$$= \{x \in \mathbb{R}^q : [x_i(k) - x_j(k)]\, \phi_{ij}(x_i(k), x_j(k)) = 0,$$
$$i = 1, \ldots, q, \ j \in \mathcal{K}_i\}.$$

Now, it follows from Assumption 15.1 that the communication graph topology of $\mathcal{G}$ is strongly connected, which implies that $\mathcal{R} = \{x \in \mathbb{R}^q : x_1 = x_2 = \cdots = x_q\}$. Since $\mathcal{R}$ consists of the equilibrium states of (15.70), it follows that the largest invariant set $\mathcal{M}$ contained in $\mathcal{R}$ is given by $\mathcal{M} = \mathcal{R}$. Hence, it follows from Theorem 15.4 that for every initial condition $x(0) \in \mathbb{R}^q$, $x(k) \to \mathcal{M}$ as $k \to \infty$, and, hence, $\alpha\mathbf{e}$ is a semistable equilibrium state of (15.70). Moreover, since $\mathbf{e}^{\mathrm{T}}x(k) = \mathbf{e}^{\mathrm{T}}x(0)$ for all $k \in \overline{\mathbb{Z}}_+$, it follows that $x(k) \to \frac{1}{q}\mathbf{e}\mathbf{e}^{\mathrm{T}}x(0)$ as $k \to \infty$. Hence, with $\alpha = \frac{1}{q}\mathbf{e}^{\mathrm{T}}x(0)$, $\alpha\mathbf{e} = \frac{1}{q}\mathbf{e}\mathbf{e}^{\mathrm{T}}x(0)$ is a globally semistable equilibrium state of (15.70). $\square$

Note that in the special case of an *all-to-all connected* communication graph topology, that is, every node of $\mathfrak{G}$ is connected to every other node of $\mathfrak{G}$, Assumption 15.3 can always be satisfied. Specifically, consider the consensus protocol given by

$$u_i(k) = \sum_{j=1, j \neq i}^{q} \phi_{ij}\left(x_i(k), x_j(k)\right)$$
$$= \sum_{j=1, j \neq i}^{q} \mathcal{C}_{(i,j)}\left[\sigma(x_j(k)) - \sigma(x_i(k))\right], \quad i = 1, \ldots, q, \quad (15.75)$$

where $\sigma : \mathbb{R} \to \mathbb{R}$ is such that $\sigma(z) = \beta z$, with $\beta > 0$, and assume that $\mathcal{C}_{(i,j)} = 1$ for all $i, j = 1, \ldots, q$, $i \neq j$. In this case, if $\beta \leq 1/q$, then Assumption 15.3 holds. To see this, first note that if $x_i(k) > x_j(k)$, $i \neq j$, $i, j = 1, \ldots, q$, $k \geq 0$, then

$$\Delta x_i(k) - \Delta x_j(k) = u_i(x(k)) - u_j(x(k))$$
$$= \sum_{h=1}^{q}\left[\sigma(x_h(k)) - \sigma(x_i(k)) - \sigma(x_h(k)) + \sigma(x_j(k))\right]$$
$$= -q\beta\left(x_i(k) - x_j(k)\right)$$
$$\geq -\left(x_i(k) - x_j(k)\right), \quad i \neq j, \quad i, j = 1, \ldots, q, \quad (15.76)$$

and, hence, Assumption 15.3 holds. Alternatively, if $x_i(k) - x_j(k) < 0$, $i \neq j$, $i = 1, \ldots, q$, $k \geq 0$, then analogously it can be shown that Assumption 15.3 holds.

Next, we provide explicit connections of the proposed thermodynamic-based consensus control architecture developed in this section to the recently developed notion of discrete thermodynamics [118]. To develop these connections, the following definition of entropy is needed.

**Definition 15.7.** For the distributed discrete-time consensus protocol $\mathcal{G}$ given by (15.70), a function $\mathcal{S} : \mathbb{R}^q \to \mathbb{R}$ satisfying

$$\mathcal{S}(x(k_2)) \geq \mathcal{S}(x(k_1)), \quad k_2 \geq k_1 \geq 0, \tag{15.77}$$

is called an *entropy* of $\mathcal{G}$.

The next theorem gives an explicit expression for the entropy function of the closed-loop, discrete-time multiagent dynamical system $\mathcal{G}$ given by (15.70).

**Theorem 15.11.** Consider the closed-loop, discrete-time multiagent dynamical system $\mathcal{G}$ given by (15.70), and assume that Assumptions 15.2 and 15.3 hold. Then the function $\mathcal{S} : \mathbb{R}^q \to \mathbb{R}$ given by

$$\mathcal{S}(x) = \mathbf{e}^{\mathrm{T}} \log_e(c\,\mathbf{e} + x) - q \log_e c, \quad x \in \mathbb{R}^q, \tag{15.78}$$

where $\log_e(c\,\mathbf{e} + x)$ denotes the vector natural logarithm given by $[\log_e(c + x_1), \ldots, \log_e(c + x_q)]^{\mathrm{T}}$ and $c > \|x\|_\infty$, is an entropy function of $\mathcal{G}$.

**Proof.** Since $\phi_{ij}(x_i, x_j) = -\phi_{ji}(x_j, x_i)$, $i \neq j$, $i, j = 1, \ldots, q$, and $c > \|x\|_\infty$, it follows that

$$\Delta \mathcal{S}(x(k)) = \sum_{i=1}^{q} \log_e \left[ 1 + \frac{\Delta x_i(k)}{c + x_i(k)} \right]$$

$$\geq \sum_{i=1}^{q} \left[ \frac{\Delta x_i(k)}{c + x_i(k)} \right] \left[ 1 + \frac{\Delta x_i(k)}{c + x_i(k)} \right]^{-1}$$

$$= \sum_{i=1}^{q} \frac{\Delta x_i(k)}{c + x_i(k) + \Delta x_i(k)}$$

$$= \sum_{i=1}^{q} \frac{\Delta x_i(k)}{c + x_i(k+1)}$$

$$= \sum_{i=1}^{q} \sum_{j=1, j\neq i}^{q} \frac{\phi_{ij}(x_i(k), x_j(k))}{c + x_i(k+1)}$$

$$= \sum_{i=1}^{q-1} \sum_{j=i+1}^{q} \left[ \frac{\phi_{ij}(x_i(k), x_j(k))}{c + x_i(k+1)} - \frac{\phi_{ij}(x_i(k), x_j(k))}{c + x_j(k+1)} \right]$$

$$= \sum_{i=1}^{q-1} \sum_{j=1, j\neq i}^{q} \frac{\phi_{ij}(x_i(k), x_j(k))[x_j(k+1) - x_i(k+1)]}{[c + x_i(k+1)][c + x_j(k+1)]}$$

$$\geq 0, \quad k \in \overline{\mathbb{Z}}_+, \tag{15.79}$$

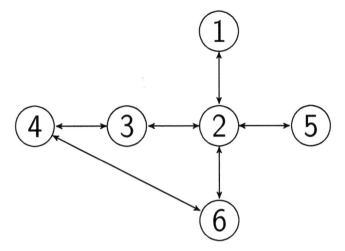

Figure 15.4 Communication graph topology for the six mobile agents.

where in (15.79) we use the fact that $\log_e(1 + x) \geq \frac{x}{x+1}$, $x > -1$. Now, summing (15.79) over $\{k_1, \ldots, k_2 - 1\}$ yields (15.77). $\qquad\square$

Note that it follows from (15.79) that the entropy function given by (15.78) satisfies (15.77) as an equality for an equilibrium (equipartitioned) process and as a strict inequality for a nonequilibrium (nonequipartitioned) process. The entropy expression given by (15.78) is identical in form to the Boltzmann entropy for statistical thermodynamics and the Shannon entropy characterizing the amount of information [118]. In addition, note that $\mathcal{S}(x)$ given by (15.78) achieves a maximum when all of the information states $x_i$, $i = 1, \ldots, q$, are equal [118]. Inequality (15.77) is a generalization of Clausius's inequality for equilibrium and nonequilibrium thermodynamics as well as reversible and irreversible thermodynamics as applied to adiabatically isolated discrete-time thermodynamic systems. For details, see [118].

**Example 15.3.** Consider a network of six dynamical agents $\mathcal{G}$ with the weakly connected, undirected communication graph topology shown in Figure 15.4, with dynamics given by (15.69). The corresponding connectivity matrix is given by

$$\mathcal{C} = \begin{bmatrix} -1 & 1 & 0 & 0 & 0 & 0 \\ 1 & -4 & 1 & 0 & 1 & 1 \\ 0 & 1 & -2 & 1 & 0 & 0 \\ 0 & 0 & 1 & -2 & 0 & 1 \\ 0 & 1 & 0 & 0 & -1 & 0 \\ 0 & 1 & 0 & 1 & 0 & -2 \end{bmatrix}.$$

Note that rank $\mathcal{C} = 5$.

Figure 15.5 shows the information states and control inputs for the six agents versus time with the linear consensus protocol

$$\phi_{ij}(x_i, x_j) = -\frac{1}{6}(x_i - x_j), \quad i, j = 1, \dots, 6,$$

and initial condition $x_0 = [0, 50, 40, -30, -20, -10]^{\mathrm{T}}$. Figure 15.6 shows the information states and control inputs for the six agents versus time with the nonlinear consensus protocol

$$\phi_{ij}(x_i, x_j) = -\frac{1}{2}(\mathrm{sign}(x_i)|x_i|^{0.8} - \mathrm{sign}(x_j)|x_j|^{0.8}), \quad i, j = 1, \dots, 6,$$

and initial condition $x_0 = [0, 50, 40, -30, -20, -10]^{\mathrm{T}}$. Note that for both information flow functions $\phi_{ij}(x_i, x_j)$, $i, j = 1, \dots, 6$, $i \neq j$, considered, Assumptions 15.1–15.3 hold, and by Theorem 15.10 the information states converge to $x_{\mathrm{e}} = \frac{1}{6}\mathbf{e}_6\mathbf{e}_6^{\mathrm{T}}x_0 = 5\mathbf{e}_6$. Finally, Figure 15.7 shows the total agent entropies versus time for both control protocols with $c = \|x\|_\infty + 1$.                    $\triangle$

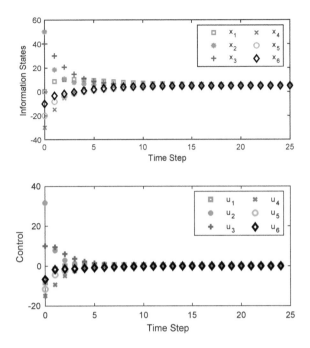

Figure 15.5 Information states and control inputs versus time for the linear consensus protocol.

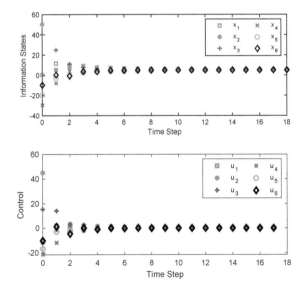

Figure 15.6 Information states and control inputs versus time for the nonlinear consensus protocol. Note that the nonlinear consensus control protocol achieves significantly improved convergence over the linear protocol.

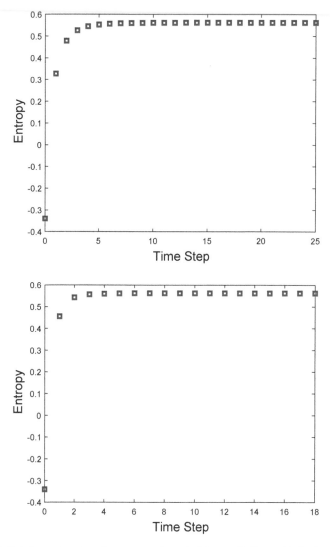

Figure 15.7 Total agent entropies versus time (with $c = \|x\|_\infty + 1$) for both control protocols; linear on the top and nonlinear on the bottom.

## Chapter Sixteen

# $\mathcal{H}_2$ Optimal Semistable Control for Discrete-Time Network Systems

## 16.1 Introduction

In this chapter, we extend the results of Chapter 10 to discrete-time systems. As in the continuous-time case, a complicating feature of the discrete-time $\mathcal{H}_2$ optimal semistable stabilization problem is that the closed-loop Lyapunov equation guaranteeing semistability can admit multiple solutions. However, as in the continuous-time case, a least squares solution over all possible semistabilizing solutions corresponds to the $\mathcal{H}_2$ optimal solution. It is shown that this least squares solution can be characterized by an LMI minimization problem.

## 16.2 Discrete-Time $\mathcal{H}_2$ Semistability Theory

The following definition for semistability with respect to $\mathcal{D} \subseteq \mathbb{R}^n$ for a discrete-time dynamical system is needed. For this definition, consider the nonlinear dynamical system given by

$$x(k+1) = f(x(k)), \quad x(0) = x_0, \quad k \in \overline{\mathbb{Z}}_+, \tag{16.1}$$

where $x(k) \in \mathcal{D} \subseteq \mathbb{R}^n$, $k \in \overline{\mathbb{Z}}_+$, and $f : \mathcal{D} \subseteq \mathbb{R}^n \to \mathbb{R}^n$ is continuous.

**Definition 16.1.** Let $\mathcal{D} \subseteq \mathbb{R}^n$ be positively invariant under (16.1). The equilibrium solution $x(k) \equiv x_{\mathrm{e}} \in \mathcal{D}$ of (16.1) is *Lyapunov stable with respect to $\mathcal{D}$* if, for every $\varepsilon > 0$, there exists $\delta = \delta(\varepsilon) > 0$ such that if $x_0 \in \mathcal{B}_\delta(x_{\mathrm{e}}) \cap \mathcal{D}$, then $x(k) \in \mathcal{B}_\varepsilon(x_{\mathrm{e}}) \cap \mathcal{D}$, $k \in \overline{\mathbb{Z}}_+$. The equilibrium solution $x(k) \equiv x_{\mathrm{e}} \in \mathcal{D}$ of (16.1) is *semistable with respect to $\mathcal{D}$* if it is Lyapunov stable with respect to $\mathcal{D}$ and there exists $\delta > 0$ such that if $x_0 \in \mathcal{B}_\delta(x_{\mathrm{e}}) \cap \mathcal{D}$, then $\lim_{k \to \infty} x(k)$ exists and corresponds to a Lyapunov stable equilibrium point in $\mathcal{D}$. Finally, the system (16.1) is said to be *semistable with respect to $\mathcal{D}$* if every equilibrium point in $\mathcal{D}$ is semistable with respect to $\mathcal{D}$.

**Proposition 16.1.** Let $\mathcal{D}_c \subset \mathbb{R}^n$ be a compact invariant set with respect to (16.1). Suppose that there exists a continuous function $V : \mathcal{D}_c \to \mathbb{R}$ such that $V(f(x)) - V(x) \leq 0$, $x \in \mathcal{D}_c$. Let $\mathcal{R} \triangleq \{x \in \mathcal{D}_c : V(f(x)) = V(x)\}$, and let $\mathcal{M}$ denote the largest invariant set contained in $\mathcal{R}$. If every element in $\mathcal{M}$ is a Lyapunov stable equilibrium point with respect to $\mathcal{D}_c$, then (16.1) is semistable with respect to $\mathcal{D}_c$.

**Proof.** Since every solution of (16.1) is bounded, it follows from the hypotheses on $V(\cdot)$ that, for every $x \in \mathcal{D}_c$, the positive limit set $\omega(x)$ of (16.1) is nonempty and contained in the largest invariant subset $\mathcal{M}$ of $\mathcal{R}$. Since every point in $\mathcal{M}$ is a Lyapunov stable equilibrium point, it follows that every point in $\omega(x)$ is a Lyapunov stable equilibrium point.

Next, let $z \in \omega(x)$, and let $\mathcal{U}_\varepsilon$ be an open neighborhood of $z$. By Lyapunov stability of $z$, it follows that there exists an open subset $\mathcal{U}_\delta$ containing $z$ such that $s_k(\mathcal{U}_\delta) \subseteq \mathcal{U}_\varepsilon$ for every $k \geq k_0$. Since $z \in \omega(x)$, it follows that there exists $h \geq 0$ such that $s(h, x) \in \mathcal{U}_\delta$. Thus, $s(k + h, x) = s_k(s(h, x)) \in s_k(\mathcal{U}_\delta) \subseteq \mathcal{U}_\varepsilon$ for every $k > k_0$. Hence, since $\mathcal{U}_\varepsilon$ was chosen arbitrarily, it follows that $z = \lim_{k \to \infty} s(k, x)$. Now, it follows that $\lim_{i \to \infty} s(k_i, x) \to z$ for every divergent sequence $\{k_i\}$, and, hence, $\omega(x) = \{z\}$. Finally, since $\lim_{k \to \infty} s(k, x) \in \mathcal{M}$ is Lyapunov stable for every $x \in \mathcal{D}_c$, it follows from the definition of semistability that every equilibrium point in $\mathcal{M}$ is semistable. $\qquad\square$

Note that if in (16.1) $f(x) = Ax$, where $A \in \mathbb{R}^{n \times n}$, then (16.1) is semistable with respect to $\mathbb{R}^n$ if and only if $A$ is *semistable*; that is, $\mathrm{spec}(A) \subset \{s \in \mathbb{C} : |s| < 1\} \cup \{1\}$, and, if $1 \in \mathrm{spec}(A)$, then 1 is semisimple. In this case, it can be shown that for every $x_0 \in \mathbb{R}^n$, $\lim_{k \to \infty} x(k)$ exists or, equivalently, $\lim_{k \to \infty} A^k$ exists and is given by $\lim_{k \to \infty} A^k = I_n - (I_n - A)(I_n - A)^\#$ [26, 134].

Next, we present the notions of semicontrollability and semiobservability for discrete-time systems. For these definitions, let $A \in \mathbb{R}^{n \times n}$, $B \in \mathbb{R}^{n \times m}$, and $C \in \mathbb{R}^{l \times n}$, and consider the linear dynamical system

$$x(k+1) = Ax(k) + Bu(k), \quad x(0) = x_0, \quad k \in \overline{\mathbb{Z}}_+, \tag{16.2}$$
$$y(k) = Cx(k), \tag{16.3}$$

with state $x(k) \in \mathbb{R}^n$, input $u(k) \in \mathbb{R}^m$, and output $y(k) \in \mathbb{R}^l$, where $k \in \overline{\mathbb{Z}}_+$.

**Definition 16.2.** Let $A \in \mathbb{R}^{n \times n}$ and $B \in \mathbb{R}^{m \times n}$. The pair $(A, B)$ is *semicontrollable* if

$$\left[\bigcap_{i=1}^{n} \mathcal{N}\left(B^{\mathrm{T}}(A^{\mathrm{T}} - I_n)^{i-1}\right)\right]^{\perp} = \left[\mathcal{N}(A^{\mathrm{T}} - I_n)\right]^{\perp}, \tag{16.4}$$

where $(A^{\mathrm{T}} - I_n)^0 \triangleq I_n$.

**Definition 16.3.** Let $A \in \mathbb{R}^{n \times n}$ and $C \in \mathbb{R}^{l \times n}$. The pair $(A, C)$ is *semiobservable* if

$$\bigcap_{i=1}^{n} \mathcal{N}\left(C(A - I_n)^{i-1}\right) = \mathcal{N}(A - I_n). \tag{16.5}$$

As noted in Chapter 10, semicontrollability and semiobservability are extensions of controllability and observability. In particular, semicontrollability is an extension of null controllability to *equilibrium controllability*, whereas semiobservability is an extension of zero-state observability to *equilibrium observability*. It is important to note here that since Definitions 16.2 and 16.3 are dual, dual results to the semiobservability results that we establish in this section also hold for semicontrollability.

**Definition 16.4.** Let $A \in \mathbb{R}^{n \times n}$, $C \in \mathbb{R}^{l \times n}$, and $K \in \mathbb{R}^{m \times n}$. The pair $(A, C)$ is *semiobservable with respect to $K$* if

$$\mathcal{N}(K) \cap \left(\bigcap_{i=1}^{n} \mathcal{N}\left(C(A - I_n)^{i-1}\right)\right) = \mathcal{N}(K) \cap \mathcal{N}(A - I_n). \tag{16.6}$$

The following result shows that semiobservability is unchanged by full state feedback.

**Proposition 16.2.** Let $A \in \mathbb{R}^{n \times n}$, $B \in \mathbb{R}^{n \times m}$, $C \in \mathbb{R}^{l \times n}$, $K \in \mathbb{R}^{m \times n}$, and $R \in \mathbb{R}^{n \times n}$, where $R$ is positive definite. If the pair $(A, C)$ is semiobservable, then the pair $(A + BK, C^{\mathrm{T}}C + K^{\mathrm{T}}RK)$ is semiobservable with respect to $K$.

**Proof.** Note that $\mathcal{N}(C^{\mathrm{T}}C + K^{\mathrm{T}}RK) = \mathcal{N}(C) \cap \mathcal{N}(K)$. Hence,

$$\mathcal{N}(K) \cap \left(\bigcap_{i=1}^{n} \mathcal{N}((C^{\mathrm{T}}C + K^{\mathrm{T}}RK)(A - I_n + BK)^{i-1})\right)$$

$$= \bigcap_{i=1}^{n} \mathcal{N}((C^{\mathrm{T}}C + K^{\mathrm{T}}RK)(A - I_n + BK)^{i-1})$$

$$= \mathcal{N}(K) \cap \left( \bigcap_{i=1}^{n} \mathcal{N}(C(A - I_n)^{i-1}) \right)$$

$$= \mathcal{N}(K) \cap \mathcal{N}(A - I_n)$$

$$= \mathcal{N}(K) \cap \mathcal{N}(A - I_n + BK), \tag{16.7}$$

which implies that the pair $(A + BK, C^T C + K^T R K)$ is semiobservable with respect to $K$. $\qquad\square$

Next, we connect semistability with Lyapunov theory and semiobservability to arrive at a characterization of the $\mathcal{H}_2$ norm of semistable systems. For this result, we consider the linear dynamical system

$$x(k+1) = Ax(k), \quad x(0) = x_0, \quad k \in \overline{\mathbb{Z}}_+, \tag{16.8}$$

where $A \in \mathbb{R}^{n \times n}$, with output equation (16.3). Furthermore, for a given semistable system, define the $\mathcal{H}_2$ norm of the transfer function $G(z)$ with realization

$$G(z) \sim \left[ \begin{array}{c|c} A & x_0 \\ \hline C & 0 \end{array} \right]$$

and free response $y(k) = H(k) = CA^k x_0$ by

$$\|G\|_2 = \left[ \sum_{k=0}^{\infty} \|H(k)\|_F^2 \right]^{1/2} = \left[ \frac{1}{2\pi} \int_{-\pi}^{\pi} \|G(e^{j\theta})\|_F^2 \mathrm{d}\theta \right]^{1/2}. \tag{16.9}$$

The following proposition presents necessary and sufficient conditions for the well-posedness of the $\mathcal{H}_2$ norm of a semistable system.

**Proposition 16.3.** Consider the linear dynamical system (16.8) with output (16.3), and assume that $A$ is semistable. Then the following statements are equivalent.

(i) For every $x_0 \in \mathbb{R}^n$, $\|G\|_2 < \infty$.

(ii) $\sum_{k=0}^{\infty} (A^k)^T R A^k < \infty$, where $R = C^T C$.

(iii) $\mathcal{N}(A - I_n) \subset \mathcal{N}(C)$.

**Proof.** The equivalence of (i) and (ii) follows from the fact that

$$\|G\|_2^2 = \sum_{k=0}^{\infty} x_0^T (A^k)^T R A^k x_0. \tag{16.10}$$

To show that (ii) implies (iii), note that since $A$ is semistable it follows that either $\rho(A) < 1$ or there exists an invertible matrix $S \in \mathbb{R}^{n \times n}$ such that

$$A = S \begin{bmatrix} J & 0 \\ 0 & I_{n-r} \end{bmatrix} S^{-1},$$

where $J \in \mathbb{R}^{r \times r}$, $r = \operatorname{rank} A$, and $\rho(J) < 1$. Now, if $\rho(A) < 1$, then (iii) holds trivially since $\mathcal{N}(A - I_n) = \{0\} \subset \mathcal{N}(C)$.

Alternatively, if $1 \in \operatorname{spec}(A)$, then

$$\mathcal{N}(A - I_n) = \left\{ x \in \mathbb{R}^n : x = S[0_{1 \times r}, y^{\mathrm{T}}]^{\mathrm{T}}, y \in \mathbb{R}^{n-r} \right\}. \tag{16.11}$$

Now,

$$\sum_{k=0}^{\infty} (A^k)^{\mathrm{T}} R A^k = S^{-\mathrm{T}} \sum_{k=0}^{\infty} \begin{bmatrix} (J^k)^{\mathrm{T}} & 0 \\ 0 & I_{n-r} \end{bmatrix} \hat{R} \begin{bmatrix} J^k & 0 \\ 0 & I_{n-r} \end{bmatrix} S$$

$$= S^{-\mathrm{T}} \sum_{k=0}^{\infty} \begin{bmatrix} (J^k)^{\mathrm{T}} \hat{R}_1 J^k & (J^k)^{\mathrm{T}} \hat{R}_{12} \\ \hat{R}_{12}^{\mathrm{T}} J^k & \hat{R}_2 \end{bmatrix} S, \tag{16.12}$$

where

$$\hat{R} = S^{\mathrm{T}} R S = \begin{bmatrix} \hat{R}_1 & \hat{R}_{12} \\ \hat{R}_{12}^{\mathrm{T}} & \hat{R}_2 \end{bmatrix}. \tag{16.13}$$

Next, it follows from (16.12) that

$$\sum_{k=0}^{\infty} (A^k)^{\mathrm{T}} R A^k < \infty \tag{16.14}$$

if and only if $\hat{R}_2 = 0$ or, equivalently,

$$[0_{1 \times r}, y^{\mathrm{T}}] \hat{R} [0_{1 \times r}, y^{\mathrm{T}}]^{\mathrm{T}} = 0, \quad y \in \mathbb{R}^{n-r}, \tag{16.15}$$

which is further equivalent to $x^{\mathrm{T}} R x = 0$, $x \in \mathcal{N}(A - I_n)$. Hence, $\mathcal{N}(A - I_n) \subset \mathcal{N}(C)$.

Finally, the proof of (iii) implies (ii) is immediate by reversing the steps of the proof given above. $\qquad\square$

**Lemma 16.1.** Let $A \in \mathbb{R}^{n \times n}$. If there exist an $n \times n$ matrix $P \geq 0$ and an $l \times n$ matrix $C$ such that $(A, C)$ is semiobservable and

$$P = A^{\mathrm{T}} P A + R, \tag{16.16}$$

where $R \triangleq C^{\mathrm{T}} C$, then (i) $\mathcal{N}(P) \subseteq \mathcal{N}(A - I_n) \subseteq \mathcal{N}(R)$ and (ii) $\mathcal{N}(A - I_n) \cap \mathcal{R}(A - I_n) = \{0\}$.

**Proof.** (i) If $(A - I_n)x = 0$, then (16.16) implies that $x^{\mathrm{T}}Rx = x^{\mathrm{T}}(P - A^{\mathrm{T}}PA)x = 0$ and, hence, that $Rx = 0$. Thus, $\mathcal{N}(A - I_n) \subseteq \mathcal{N}(R)$. If $Px = 0$, then

$$0 \le x^{\mathrm{T}}Rx = x^{\mathrm{T}}(P - A^{\mathrm{T}}PA)x = -x^{\mathrm{T}}A^{\mathrm{T}}PAx \le 0, \tag{16.17}$$

and, hence, $x^{\mathrm{T}}Rx = 0$ or, equivalently, $Rx = 0$. Thus, $\mathcal{N}(P) \subseteq \mathcal{N}(R)$.

Next, let $x \in \mathcal{N}(P) \subseteq \mathcal{N}(R)$. If $(A - I_n)^k x \in \mathcal{N}(P) \subseteq \mathcal{N}(R)$ for some $k \ge 0$, then

$$\begin{aligned}
0 &= x^{\mathrm{T}}(A^{\mathrm{T}} - I_n)^k R(A - I_n)^k x \\
&= x^{\mathrm{T}}(A^{\mathrm{T}} - I_n)^k (P - A^{\mathrm{T}}PA)(A - I_n)^k x \\
&= -x^{\mathrm{T}}(A^{\mathrm{T}} - I_n)^k A^{\mathrm{T}}PA(A - I_n)^k x \\
&= -x^{\mathrm{T}}(A^{\mathrm{T}} - I_n)^{k+1} P(A - I_n)^{k+1} x, \tag{16.18}
\end{aligned}$$

and, hence, $P(A - I_n)^{k+1}x = 0$, which implies that $(A - I_n)^{k+1}x \in \mathcal{N}(P) \subseteq \mathcal{N}(R)$. Since $(A - I_n)^k x \in \mathcal{N}(P) \subseteq \mathcal{N}(R)$ for $k = 0$, it follows by induction that $x$ is contained in the null space of the left-hand side of (16.5). Equation (16.5) now implies that $x \in \mathcal{N}(A - I_n)$. Thus, $\mathcal{N}(P) \subseteq \mathcal{N}(A - I_n) \subseteq \mathcal{N}(R)$.

(ii) Consider $x \in \mathcal{N}(A - I_n) \cap \mathcal{R}(A - I_n)$. Then $(A - I_n)x = 0$, and there exists $z \in \mathbb{R}^n$ such that $x = (A - I_n)z$. Now, it follows from (i) that $Rx = R(A - I_n)z = 0$. Thus,

$$0 = z^{\mathrm{T}}Rx = z^{\mathrm{T}}(P - A^{\mathrm{T}}PA)x = -z^{\mathrm{T}}(A - I_n)^{\mathrm{T}}Px = -x^{\mathrm{T}}Px, \tag{16.19}$$

and, hence, $Px = 0$. Finally,

$$\begin{aligned}
z^{\mathrm{T}}Rz &= z^{\mathrm{T}}(P - A^{\mathrm{T}}PA)z \\
&= z^{\mathrm{T}}Pz - (x + z)^{\mathrm{T}}P(x + z) \\
&= -x^{\mathrm{T}}Px - x^{\mathrm{T}}Pz - z^{\mathrm{T}}Px \\
&= 0,
\end{aligned}$$

and, hence, $Rz = 0$. This implies that $z$ is contained in the null space of the left-hand side of (16.5). Hence, by (16.5), $(A - I_n)z = x = 0$, as required. $\square$

**Theorem 16.1.** Consider the linear dynamical system (16.8). Suppose that there exist an $n \times n$ matrix $P \ge 0$ and a matrix $C \in \mathbb{R}^{l \times n}$ such that $(A, C)$ is semiobservable and (16.16) holds. Then (16.8) is semistable with respect to $\mathbb{R}^n$. Furthermore, $\|\|G(z)\|\|_2^2 = (x_0 - x_{\mathrm{e}})^{\mathrm{T}}P(x_0 - x_{\mathrm{e}})$, where $x_{\mathrm{e}} \triangleq x_0 - (A - I_n)(A - I_n)^{\#}x_0$.

**Proof.** Since, by Lemma 16.1, $\mathcal{N}(A - I_n) \cap \mathcal{R}(A - I_n) = \{0\}$, it follows from Lemma 4.14 of [24] that $A - I_n$ is group invertible. Let $L \triangleq I_n -$

$(A - I_n)(A - I_n)^{\#}$, and note that $L^2 = L$. Hence, $L$ is the unique $n \times n$ matrix satisfying $\mathcal{N}(L) = \mathcal{R}(A - I_n)$, $\mathcal{R}(L) = \mathcal{N}(A - I_n)$, and $Lx = x$ for all $x \in \mathcal{N}(A - I_n)$.

Consider the nonnegative function

$$V(x) = x^{\mathrm{T}} P x + x^{\mathrm{T}} L^{\mathrm{T}} L x. \tag{16.20}$$

If $V(x) = 0$ for some $x \in \mathbb{R}^n$, then $Px = 0$ and $Lx = 0$. It follows from (i) of Lemma 16.1 that $x \in \mathcal{N}(A - I_n)$, while $Lx = 0$ implies that $x \in \mathcal{R}(A - I_n)$. Now, it follows from (ii) of Lemma 16.1 that $x = 0$. Hence, $V(\cdot)$ is positive definite. Next, since $L(A - I_n) = A - I_n - (A - I_n)(A - I_n)^{\#}(A - I_n) = 0$, it follows that

$$\begin{aligned}
\Delta V(x) &= -x^{\mathrm{T}} R x + x^{\mathrm{T}} (A - I_n)^{\mathrm{T}} L^{\mathrm{T}} L (A - I_n) x \\
&\quad + x^{\mathrm{T}} (A - I_n)^{\mathrm{T}} L^{\mathrm{T}} L x + x^{\mathrm{T}} L^{\mathrm{T}} L (A - I_n) x \\
&= -x^{\mathrm{T}} R x \\
&\leq 0, \quad x \in \mathbb{R}^n.
\end{aligned} \tag{16.21}$$

Note that $\Delta V^{-1}(0) = \mathcal{N}(R)$.

To find the largest invariant subset $\mathcal{M}$ of $\mathcal{N}(R)$, consider a solution $y$ of (16.8) such that $Cx(k) = 0$ for all $k \in \overline{\mathbb{Z}}_+$. Then $Cx(k+1) - Cx(k) = 0$; that is, $C(A - I_n)x(k) = 0$. Similarly, $C(A - I_n)x(k+1) - C(A - I_n)x(k) = C(A - I_n)^2 x(k) = 0$, and so on. This implies that $C(A - I_n)^i x(k) = 0$ for all $k \in \overline{\mathbb{Z}}_+$ and $i = 1, 2, \ldots$. Equation (16.5) now implies that $x(k) \in \mathcal{N}(A - I_n)$ for all $k \in \overline{\mathbb{Z}}_+$. Thus, $\mathcal{M} \subseteq \mathcal{N}(A - I_n)$. However, $\mathcal{N}(A - I_n)$ consists of only equilibrium points and, hence, is invariant. Hence, $\mathcal{M} = \mathcal{N}(A - I_n)$.

Now, let $x_{\mathrm{e}} \in \mathcal{N}(A - I_n)$ be an equilibrium point of (16.8), and consider the function $U(x) = V(x - x_{\mathrm{e}})$, which is positive definite with respect to $x_{\mathrm{e}}$. Then it follows that $\Delta U(x) = -(x - x_{\mathrm{e}})^{\mathrm{T}} R(x - x_{\mathrm{e}}) \leq 0$, $x \in \mathbb{R}^n$. Thus, it follows that $x_{\mathrm{e}}$ is Lyapunov stable, and, hence, by Proposition 16.1, (16.8) is semistable.

Next, since $A$ is semistable, it follows from (vi) of Proposition 11.9.2 of [26] that $\lim_{k \to \infty} A^k = I_n - (A - I_n)(A - I_n)^{\#}$. Now, noting that $Ax_{\mathrm{e}} = x_{\mathrm{e}}$, (16.8) can be equivalently written as

$$x(k+1) - x_{\mathrm{e}} = A(x(k) - x_{\mathrm{e}}), \quad x(0) = x_0, \quad k \in \overline{\mathbb{Z}}_+. \tag{16.22}$$

Hence,

$$\sum_{k=0}^{N} (x(k) - x_{\mathrm{e}})^{\mathrm{T}} R(x(k) - x_{\mathrm{e}})$$

$$= -(x(N) - x_{\mathrm{e}})^{\mathrm{T}} P(x(N) - x_{\mathrm{e}}) + (x_0 - x_{\mathrm{e}})^{\mathrm{T}} P(x_0 - x_{\mathrm{e}}). \quad (16.23)$$

Now, it follows from the semiobservability of $(A, C)$ that $Rx_{\mathrm{e}} = 0$. Hence, letting $N \to \infty$ and noting that $x(k) \to x_{\mathrm{e}}$ as $k \to \infty$, it follows from (16.23) that

$$\sum_{k=0}^{\infty} x^{\mathrm{T}}(k) R x(k) = (x_0 - x_{\mathrm{e}})^{\mathrm{T}} P(x_0 - x_{\mathrm{e}}). \quad (16.24)$$

Finally, defining the free response of (16.8) by $H(k) \triangleq Cx(k) = CA^k x_0$, $k \in \overline{\mathbb{Z}}_+$, and noting that $R = C^{\mathrm{T}}C$, it follows from Parseval's theorem [122, p. 840] that

$$(x_0 - x_{\mathrm{e}})^{\mathrm{T}} P(x_0 - x_{\mathrm{e}}) = \sum_{k=0}^{\infty} H^{\mathrm{T}}(k) H(k) = \frac{1}{2\pi} \int_{-\pi}^{\pi} \|G(e^{\jmath\theta})\|_{\mathrm{F}}^2 \mathrm{d}\theta. \quad (16.25)$$

This completes the proof.                                                                                    $\square$

Next, we give a necessary and sufficient condition for characterizing semistability using the Lyapunov equation (16.16). Before we state this result, the following lemmas are needed.

**Lemma 16.2.** Consider the linear dynamical system (16.8). If (16.8) is semistable, then, for every $n \times n$ nonnegative definite matrix $R$,

$$\sum_{k=0}^{\infty} (x(k) - x_{\mathrm{e}})^{\mathrm{T}} R(x(k) - x_{\mathrm{e}}) < \infty, \quad (16.26)$$

where $x_{\mathrm{e}} = [I_n - (A - I_n)(A - I_n)^{\#}]x_0$.

**Proof.** Since $A$ is semistable, it follows from the Jordan decomposition that there exists an invertible matrix $S \in \mathbb{C}^{n \times n}$ such that

$$A = S \begin{bmatrix} J & 0 \\ 0 & I_{n-r} \end{bmatrix} S^{-1},$$

where $J \in \mathbb{C}^{r \times r}$, $r = \mathrm{rank}\, A$, and $\rho(J) < 1$. Let $z(k) \triangleq S^{-1}x(k)$ and $z_{\mathrm{e}} \triangleq S^{-1}x_{\mathrm{e}}$, $k \in \overline{\mathbb{Z}}_+$. Then (16.8) becomes

$$z(k+1) = \begin{bmatrix} J & 0 \\ 0 & I_{n-r} \end{bmatrix} z(k), \quad z(0) = S^{-1}x_0, \quad k \in \overline{\mathbb{Z}}_+, \quad (16.27)$$

which implies that $\lim_{k \to \infty} z_i(k) = 0$, $i = 1, \ldots, r$, and $z_j(k) = z_j(0)$, $j = r+1, \ldots, n$; that is, $z_{\mathrm{e}} = [0, \ldots, 0, z_{r+1}(0), \ldots, z_n(0)]^{\mathrm{T}}$.

Now,

$$\sum_{k=0}^{\infty}(x(k)-x_{\mathrm{e}})^{\mathrm{T}}R(x(k)-x_{\mathrm{e}}) = \sum_{k=0}^{\infty}(z(k)-z_{\mathrm{e}})^{*}S^{*}RS(z(k)-z_{\mathrm{e}})$$

$$= \sum_{k=0}^{\infty}\hat{z}^{*}(k)S^{*}RS\hat{z}(k), \qquad (16.28)$$

where $\hat{z}(k) \triangleq [z_1(k),\ldots,z_r(k),0,\ldots,0]^{\mathrm{T}}$. Since

$$\hat{z}(k+1) = \begin{bmatrix} J & 0 \\ 0 & 0 \end{bmatrix}\hat{z}(k) \qquad (16.29)$$

and $\rho(J) < 1$, it follows that

$$\sum_{k=0}^{\infty}\hat{z}^{*}(k)S^{*}RS\hat{z}(k) < \infty, \qquad (16.30)$$

which proves the result.  $\square$

**Lemma 16.3.** Let $A \in \mathbb{R}^{n \times n}$ and $B \in \mathbb{R}^{m \times m}$. If $A$ and $B$ are semistable, then $A \otimes B$ is semistable.

**Proof.** Let $\lambda \in \mathrm{spec}(A)$ and $\mu \in \mathrm{spec}(B)$. Since $A$ and $B$ are both semistable, it follows that $|\lambda| < 1$ or $\lambda = 1$ and $\mathrm{am}_A(1) = \mathrm{gm}_A(1)$, and $|\mu| < 1$ or $\mu = 1$ and $\mathrm{am}_B(1) = \mathrm{gm}_B(1)$, where $\mathrm{am}_X(\lambda)$ and $\mathrm{gm}_X(\lambda)$ denote the algebraic multiplicity of $\lambda \in \mathrm{spec}(X)$ and the geometric multiplicity of $\lambda \in \mathrm{spec}(X)$, respectively. Then it follows from the fact that $\lambda\mu \in \mathrm{spec}(A \otimes B)$ that $\mathrm{spec}(A \otimes B) \subset \{z \in \mathbb{C} : |z| < 1\} \cup \{1\}$. Next, it follows from Fact 7.4.12 of [26] that $\mathrm{gm}_A(1)\mathrm{gm}_B(1) \le \mathrm{gm}_{A\otimes B}(1) \le \mathrm{am}_{A\otimes B}(1) = \mathrm{am}_A(1)\mathrm{am}_B(1)$. Since $\mathrm{am}_A(1) = \mathrm{gm}_A(1)$ and $\mathrm{am}_B(1) = \mathrm{gm}_B(1)$, it follows that $\mathrm{gm}_{A\otimes B}(1) = \mathrm{am}_{A\otimes B}(1)$, and, hence, $A \otimes B$ is semistable.  $\square$

**Lemma 16.4.** Let $x \in \mathbb{R}^n$ and $A \in \mathbb{R}^{n \times n}$, and assume that $A$ is semistable. Then $\sum_{k=0}^{\infty}A^k x$ exists if and only if $x \in \mathcal{R}(A - I_n)$. In this case,

$$\sum_{k=0}^{\infty}A^k x = -(A - I_n)^{\#}x.$$

**Proof.** The proof is similar to the proofs of (viii) and (ix) of Lemma 5.2 of [134] and, hence, is omitted.  $\square$

**Theorem 16.2.** Consider the linear dynamical system (16.8). Then (16.8) is semistable if and only if for every semiobservable pair $(A, C)$ there exists an $n \times n$ matrix $P \geq 0$ such that (16.16) holds. Furthermore, if $(A, C)$ is semiobservable and $P$ satisfies (16.16), then

$$P = \sum_{k=0}^{\infty} (A^k)^{\mathrm{T}} R A^k + P_0 \tag{16.31}$$

for some $P_0 = P_0^{\mathrm{T}} \in \mathbb{R}^{n \times n}$ satisfying

$$A^{\mathrm{T}} P_0 A = 0 \tag{16.32}$$

and

$$P_0 \geq -\sum_{k=0}^{\infty} (A^k)^{\mathrm{T}} R A^k. \tag{16.33}$$

In addition, $\min_{P \in \mathcal{P}} \|P\|_{\mathrm{F}}$ has a unique solution $P$ given by

$$P = \sum_{k=0}^{\infty} (A^k)^{\mathrm{T}} R A^k, \tag{16.34}$$

where $\mathcal{P}$ denotes the set of all $P$ satisfying (16.16). Finally, (16.8) is semistable if and only if for every semiobservable pair $(A, C)$ there exists an $n \times n$ matrix $P > 0$ such that (16.16) holds.

**Proof.** Sufficiency for the first implication follows from Theorem 16.1. To show necessity, assume that (16.8) is semistable. Then $\lim_{k \to \infty} x(k) = x_{\mathrm{e}}$, where $x_{\mathrm{e}} = [I_n - (A - I_n)(A - I_n)^{\#}] x_0$. For a semiobservable pair $(A, C)$, let

$$P = \sum_{k=0}^{\infty} ((I_n - A)(I_n - A)^{\#})^{\mathrm{T}} (A^k)^{\mathrm{T}} R A^k (I_n - A)(I_n - A)^{\#}. \tag{16.35}$$

Then, for $x_0 \in \mathbb{R}^n$,

$$\begin{aligned}
x_0^{\mathrm{T}} P x_0 &= \sum_{k=0}^{\infty} x_0^{\mathrm{T}} ((I_n - A)(I_n - A)^{\#})^{\mathrm{T}} (A^k)^{\mathrm{T}} R A^k (I_n - A)(I_n - A)^{\#} x_0 \\
&= \sum_{k=0}^{\infty} (x_0 - x_{\mathrm{e}})^{\mathrm{T}} (A^k)^{\mathrm{T}} R A^k (x_0 - x_{\mathrm{e}}) \\
&= \sum_{k=0}^{\infty} (x(k) - x_{\mathrm{e}})^{\mathrm{T}} R(x(k) - x_{\mathrm{e}}),
\end{aligned} \tag{16.36}$$

where we used the fact that $x(k) - x_e = A^k(x_0 - x_e)$. It follows from Lemma 16.2 that $P$ is well defined. Since $x_e \in \mathcal{N}(A - I_n)$, it follows from (16.5) that $R x_e = 0$, and, hence,

$$x_0^T P x_0 = \sum_{k=0}^{\infty} x^T(k) R x(k) = \sum_{k=0}^{\infty} x_0^T (A^k)^T R A^k x_0, \quad x_0 \in \mathbb{R}^n, \qquad (16.37)$$

which implies that

$$P = \sum_{k=0}^{\infty} (A^k)^T R A^k. \qquad (16.38)$$

Now, (16.16) is immediate using the fact that $R x_e = 0$.

Next, since $A$ is semistable, it follows from the above result that there exists an $n \times n$ nonnegative definite matrix $P$ such that (16.16) holds or, equivalently, $\operatorname{vec} P = (A \otimes A)^T \operatorname{vec} P + \operatorname{vec} R$; that is, $(I_{q^2} - (A \otimes A)^T) \operatorname{vec} P = \operatorname{vec} R$. Hence, $\operatorname{vec} R \in \mathcal{R}(I_{q^2} - (A \otimes A)^T)$, and

$$\mathcal{P} = \left\{ P \in \mathbb{R}^{n \times n} : P = \operatorname{vec}^{-1} \left( (I_{q^2} - (A \otimes A)^T)^{\#} \operatorname{vec} R \right) + \operatorname{vec}^{-1}(z) \right\}$$

for some $z \in \mathcal{N}(I_{q^2} - (A \otimes A)^T)$.

Next, it follows from Lemma 16.3 that $A \otimes A$ is semistable, and, hence, by Lemma 16.4,

$$\operatorname{vec}^{-1} \left( (I_{q^2} - (A \otimes A)^T)^{\#} \operatorname{vec} R \right) = \sum_{k=0}^{\infty} \operatorname{vec}^{-1} (((A \otimes A)^T)^k \operatorname{vec} R)$$

$$= \sum_{k=0}^{\infty} \operatorname{vec}^{-1} (((A^k)^T \otimes (A^k)^T) \operatorname{vec} R)$$

$$= \sum_{k=0}^{\infty} (A^k)^T R A^k, \qquad (16.39)$$

where we used the facts that $(X \otimes Y)^T = X^T \otimes Y^T$, $(X \otimes Y)(Z \otimes W) = XZ \otimes YW$, and $\operatorname{vec}(XYZ) = (Z^T \otimes X) \operatorname{vec} Y$ [26, Ch. 7]. Hence,

$$P = \sum_{k=0}^{\infty} (A^k)^T R A^k + \operatorname{vec}^{-1}(z), \qquad (16.40)$$

where $\operatorname{vec}^{-1}(z)$ satisfies $\operatorname{vec}^{-1}(z) = (\operatorname{vec}^{-1}(z))^T$, $A^T \operatorname{vec}^{-1}(z) A = 0$, and $\operatorname{vec}^{-1}(z) \geq -\sum_{k=0}^{\infty} (A^k)^T R A^k$. If $P$ is such that $\min_{P \in \mathcal{P}} \|P\|_F$ holds, then it follows that $P$ is the unique solution of a least squares minimization problem and is given by

$$P = \mathrm{vec}^{-1}((I_{q^2} - (A \otimes A)^{\mathrm{T}})^{\#}\mathrm{vec}\,R) = \sum_{k=0}^{\infty}(A^k)^{\mathrm{T}}RA^k. \qquad (16.41)$$

Finally, suppose that $(A, C)$ is semiobservable. Then it follows from the first part of the theorem that there exists an $n \times n$ matrix $P \geq 0$ such that (16.16) holds. Let $\hat{P} \triangleq P + L^{\mathrm{T}}L$, where $L = I_n - (A - I_n)(A - I_n)^{\#}$. Then, using similar arguments as in the proof of Theorem 16.1, it can be shown that $\hat{P} > 0$ and satisfies (16.16). Conversely, if there exists $P > 0$ such that (16.16) holds, consider the function $V(x) = x^{\mathrm{T}}Px$. Using similar arguments as in the proof of Theorem 16.1, it can be shown that the largest invariant subset $\mathcal{M}$ of $\mathcal{N}(R)$ is given by $\mathcal{M} = \mathcal{N}(A - I_n)$. For $x_{\mathrm{e}} \in \mathcal{N}(A - I_n)$, Lyapunov stability of $x_{\mathrm{e}}$ now follows by considering the Lyapunov function $V(x - x_{\mathrm{e}})$. $\qquad\square$

Next, we show that the unique solution $P$ given by (16.16) and satisfying $\min_{P \in \mathcal{P}} \|P\|_{\mathrm{F}}$ can be characterized by an LMI minimization problem.

**Theorem 16.3.** Consider the linear dynamical system (16.8) with output (16.3). Assume that $A$ is semistable and $(A, C)$ is semiobservable. Let $P_{\min}$ be the solution to the LMI minimization problem

$$\min\left\{\mathrm{tr}\,PV : P \geq 0 \text{ and } A^{\mathrm{T}}PA + R - P \leq 0\right\}, \qquad (16.42)$$

where $V \in \mathbb{R}^{n \times n}$ and $V \geq 0$. Then

$$\mathrm{tr}\,P_{\min}V = \mathrm{tr}\sum_{k=0}^{\infty}(A^k)^{\mathrm{T}}RA^kV. \qquad (16.43)$$

**Proof.** Let $\hat{P} = \sum_{k=0}^{\infty}(A^k)^{\mathrm{T}}RA^k$, and let $P \geq 0$ be such that

$$A^{\mathrm{T}}PA + R - P \leq 0. \qquad (16.44)$$

(Note that $A^{\mathrm{T}}\hat{P}A + R = \hat{P}$, which implies that a $P \geq 0$ satisfying (16.44) exists.) Now, let $W \in \mathbb{R}^{n \times n}$ and $W \geq 0$ be such that

$$P = A^{\mathrm{T}}PA + R + W. \qquad (16.45)$$

Next, since $(A, C)$ is semiobservable, it follows that if $x_{\mathrm{e}} \in \mathcal{N}(A - I_n)$, then $Rx_{\mathrm{e}} = 0$, and, hence, it follows from (16.45) that $Wx_{\mathrm{e}} = 0$. Now, using identical arguments as in the proof of Theorem 16.2, it follows that

$$P = \sum_{k=0}^{\infty}(A^k)^{\mathrm{T}}(R + W)A^k$$

$$\geq \sum_{k=0}^{\infty} (A^k)^{\mathrm{T}} R A^k$$
$$= \hat{P}. \tag{16.46}$$

Finally, since $\hat{P}$ is an element of the feasible set of the optimization problem (16.42), $\operatorname{tr} P_{\min} V = \operatorname{tr} \hat{P} V$. $\qquad\square$

Finally, we provide a dual result to Theorem 16.3 which is necessary for developing feedback controllers guaranteeing closed-loop semistability.

**Theorem 16.4.** Consider the linear dynamical system (16.8) with output (16.3). Assume that $A$ is semistable, and let $V \in \mathbb{R}^{n \times n}$ and $V \geq 0$ be such that $(A, V)$ is semicontrollable. Let $Q_{\min}$ be the solution to the LMI minimization problem

$$\min \left\{ \operatorname{tr} QR : Q \geq 0 \text{ and } AQA^{\mathrm{T}} + V - Q \leq 0 \right\}. \tag{16.47}$$

Then

$$\operatorname{tr} Q_{\min} R = \operatorname{tr} \sum_{k=0}^{\infty} (A^k)^{\mathrm{T}} R A^k V = \operatorname{tr} P_{\min} V, \tag{16.48}$$

where $P_{\min}$ is the solution to the LMI minimization problem given by (16.42).

**Proof.** The proof is a direct consequence of Theorem 16.3 by noting that $(A, V)$ is semicontrollable if and only if $(A^{\mathrm{T}}, V)$ is semiobservable. Now, replacing $A$ with $A^{\mathrm{T}}$ and $R$ with $V$ in Theorem 16.3, it follows that

$$\operatorname{tr} Q_{\min} R = \operatorname{tr} \sum_{k=0}^{\infty} (A^k)^{\mathrm{T}} V A^k R$$
$$= \operatorname{tr} \sum_{k=0}^{\infty} (A^k)^{\mathrm{T}} R A^k V$$
$$= \operatorname{tr} P_{\min} V. \tag{16.49}$$

This completes the proof. $\qquad\square$

## 16.3 Optimal Semistable Stabilization

In this section, we consider the problem of optimal state feedback control for semistable stabilization of linear dynamical systems. Specifically, we consider the discrete-time controlled linear system given by

$$x(k+1) = Ax(k) + Bu(k), \quad x(0) = x_0, \quad k \in \overline{\mathbb{Z}}_+, \tag{16.50}$$

where $x(k) \in \mathbb{R}^n$, $k \in \overline{\mathbb{Z}}_+$, is the state vector; $u(k) \in \mathbb{R}^m$, $k \in \overline{\mathbb{Z}}_+$, is the control input; $A \in \mathbb{R}^{n \times n}$; and $B \in \mathbb{R}^{n \times m}$, with the state feedback controller $u(k) = Kx(k)$, where $K \in \mathbb{R}^{m \times n}$ is such that the closed-loop system given by

$$x(k+1) = (A + BK)x(k), \quad x(0) = x_0, \quad k \in \overline{\mathbb{Z}}_+, \tag{16.51}$$

is semistable and the performance criterion

$$J(K) \triangleq \sum_{k=0}^{\infty} \left[ (x(k) - x_{\mathrm{e}})^{\mathrm{T}} R_1 (x(k) - x_{\mathrm{e}}) + (u(k) - u_{\mathrm{e}})^{\mathrm{T}} R_2 (u(k) - u_{\mathrm{e}}) \right]$$
$$\tag{16.52}$$

is minimized, where $R_1 \triangleq E_1^{\mathrm{T}} E_1$, $R_2 \triangleq E_2^{\mathrm{T}} E_2 > 0$, $R_{12} \triangleq E_1^{\mathrm{T}} E_2 = 0$, $u_{\mathrm{e}} = K x_{\mathrm{e}}$, and $x_{\mathrm{e}} = \lim_{k \to \infty} x(k)$.

Note that it follows from Lemma 16.2 that if the closed-loop system is semistable, then $J(K)$ is well defined. To develop necessary conditions for the optimal semistable control problem, we assume that $(A, B)$ is semicontrollable, $(A, E_1)$ is semiobservable, and $x_{\mathrm{e}} \in \mathcal{N}(K)$. In this case, it follows from Proposition 16.2 that $(A + BK, R_1 + K^{\mathrm{T}} R_2 K)$ is semiobservable with respect to $K$, and, hence, $(R_1 + K^{\mathrm{T}} R_2 K)x_{\mathrm{e}} = 0$. Thus,

$$J(K) = \sum_{k=0}^{\infty} x_0^{\mathrm{T}} (\tilde{A}^k)^{\mathrm{T}} (R_1 + K^{\mathrm{T}} R_2 K) \tilde{A}^k x_0$$
$$= \mathrm{tr} \sum_{k=0}^{\infty} (\tilde{A}^k)^{\mathrm{T}} \tilde{R} \tilde{A}^k V$$
$$= \mathrm{tr} \, P_{\mathrm{LS}} V, \tag{16.53}$$

where we assume that the initial state $x_0 \in \mathbb{R}^n$ is a random variable such that $\mathbb{E}[x_0] = 0$ and $\mathbb{E}[x_0 x_0^{\mathrm{T}}] = V$, $\tilde{A} \triangleq A + BK$, $\tilde{R} \triangleq R_1 + K^{\mathrm{T}} R_2 K$, and $P_{\mathrm{LS}} \triangleq \mathrm{tr} \sum_{k=0}^{\infty} (\tilde{A}^k)^{\mathrm{T}} \tilde{R} \tilde{A}^k$ denotes the least squares solution to

$$P = \tilde{A}^{\mathrm{T}} P \tilde{A} + \tilde{R}. \tag{16.54}$$

Unlike the standard $\mathcal{H}_2$ optimal control problem, $P_{\mathrm{LS}} \geq 0$ is not a unique solution to (16.54).

The following theorem presents an LMI solution to the $\mathcal{H}_2$ optimal semistable control problem.

**Theorem 16.5.** Consider the linear dynamical system (16.50), and assume that $(A, E_1)$ is semiobservable and $(A, V)$ is semicontrollable. Let $Q \in \mathbb{R}^{n \times n}$ and $X \in \mathbb{R}^{m \times n}$ be the solution to the LMI minimization problem

$$\min_{Q \in \mathbb{R}^{n \times n}, X \in \mathbb{R}^{m \times n}, W \in \mathbb{R}^{p \times p}} \mathrm{tr} \, W \tag{16.55}$$

subject to

$$\begin{bmatrix} Q & (E_1Q + E_2X)^{\mathrm{T}} \\ E_1Q + E_2X & W \end{bmatrix} > 0, \tag{16.56}$$

$$\begin{bmatrix} V - Q & (AQ + BX)^{\mathrm{T}} \\ AQ + BX & -Q \end{bmatrix} \leq 0. \tag{16.57}$$

Then $K = XQ^{-1}$ is a semistabilizing controller for (16.50); that is, $A + BK$ is semistable. Furthermore, $K$ minimizes the $\mathcal{H}_2$ performance criterion $J(K)$ given by (16.52).

**Proof.** Since $K = XQ^{-1}$, it follows from (16.57) using Schur complements that

$$(A + BK)Q(A + BK)^{\mathrm{T}} + V - Q \leq 0, \tag{16.58}$$

which, since $(A, V)$ is semicontrollable, implies that $A + BK$ is semistable. Next, note that (16.56) holds if and only if

$$W > (E_1Q + E_2X)Q^{-1}(E_1Q + E_2X)^{\mathrm{T}}, \tag{16.59}$$

which implies that the minimization problem (16.55)–(16.57) is equivalent to

$$\min \operatorname{tr}(E_1Q + E_2X)Q^{-1}(E_1Q + E_2X)^{\mathrm{T}} \tag{16.60}$$

subject to

$$AQA^{\mathrm{T}} + AX^{\mathrm{T}}B^{\mathrm{T}} + BXA^{\mathrm{T}} + BXQ^{-1}X^{\mathrm{T}}B^{\mathrm{T}} + V - Q \leq 0, \tag{16.61}$$

$$Q > 0. \tag{16.62}$$

Hence, noting that (16.60)–(16.62) is equivalent to

$$\min \operatorname{tr} Q\tilde{R} \tag{16.63}$$

subject to

$$\tilde{A}Q\tilde{A}^{\mathrm{T}} + V - Q \leq 0, \tag{16.64}$$

$$Q > 0, \tag{16.65}$$

the result follows as a direct consequence of Theorems 16.4 and 16.2.  □

## 16.4 Information Flow Models

In the remainder of this chapter, we use the optimal control framework developed in Section 16.3 to design optimal controllers for multiagent network dynamical systems. Specifically, we use undirected and directed graphs to represent a dynamical network and present solutions to the consensus problem for networks with both graph topologies (or information flows) [238]. Here we use the identical graph-theoretic notation as that introduced in Chapters 2 and 4.

The information flow model is a network dynamical system involving the trajectories of the dynamical network characterized by the multiagent dynamical system $\mathcal{G}$ given by

$$x_i(k+1) = x_i(k) + \sum_{j=1, j \neq i}^{q} \phi_{ij}(x(k)), \quad x_i(0) = x_{i0}, \quad k \in \overline{\mathbb{Z}}_+,$$

$$i = 1, \ldots, q, \qquad (16.66)$$

where $q \geq 2$, or, in vector form

$$x(k+1) = f(x(k)), \quad x(0) = x_0, \quad k \in \overline{\mathbb{Z}}_+, \qquad (16.67)$$

where $x(k) \triangleq [x_1(k), \ldots, x_q(k)]^{\mathrm{T}} \in \mathbb{R}^q$, $k \in \overline{\mathbb{Z}}_+$, represents the information state vector; $\phi_{ij} : \mathbb{R}^q \to \mathbb{R}$ is continuous, $i, j = 1, \ldots, q$, $i \neq j$, and represents the information flow from the $j$th agent to the $i$th agent; and $f = [f_1, \ldots, f_q]^{\mathrm{T}} : \mathbb{R}^q \to \mathbb{R}^q$ is such that $f_i(x) = x_i + \mathcal{I}_i(x)$, where for each $i \in \{1, \ldots, q\}$, $\mathcal{I}_i(x) \triangleq \sum_{j=1, j \neq i}^{q} \phi_{ij}(x)$. This nonlinear model is proposed in [118] and [127] and is further discussed in Chapter 15.

**Assumption 16.1.** The connectivity matrix $\mathcal{C} \in \mathbb{R}^{q \times q}$ associated with the multiagent dynamical system $\mathcal{G}$ given by (16.66) is defined by

$$\mathcal{C}_{(i,j)} = \begin{cases} 0 & \text{if } \phi_{ij}(x) \equiv 0, \\ 1, & \text{otherwise,} \end{cases} \quad i \neq j, \quad i, j = 1, \ldots, q, \qquad (16.68)$$

and $\mathcal{C}_{(i,i)} = -\sum_{k=1, k \neq i}^{q} \mathcal{C}_{(i,k)}$, $i = j$, $i = 1, \ldots, q$, with $\operatorname{rank} \mathcal{C} = q - 1$, and for $\mathcal{C}_{(i,j)} = 1$, $i \neq j$, $\phi_{ij}(x) = 0$ if and only if $x_i = x_j$.

**Assumption 16.2.** For $i, j = 1, \ldots, q$, $(x_i - x_j)\phi_{ij}(x) \leq 0$, $x \in \mathbb{R}^q$.

**Assumption 16.3.** For $i, j = 1, \ldots, q$, $|\phi_{ij}(x)| \leq \lambda_{ij}|x_i - x_j|$, $\lambda_{ij} > 0$, $x \in \mathbb{R}^q$.

Assumptions 16.1 and 16.2 are a restatement of Assumptions 15.1 and 15.2. Assumption 16.3 implies that the amount of transferred information flow does not exceed an amount proportional to the information difference between agents $\mathcal{G}_i$ and $\mathcal{G}_j$. This assumption is an information flow constraint similar to the capacity limitation constraint of communication channels in information theory. In particular, this constraint implies that the communication channel for information exchange between two agents cannot exceed an amount of information proportional to the available transferred information between two agents. For further details on Assumptions 16.1–16.3, see [118, 127, 134].

## 16.5 Semistability of Information Flow Models

In this section, we extend the results of Section 15.7.

**Proposition 16.4.** Consider the information flow model (16.67), and assume that Assumptions 16.1 and 16.2 hold. Then $\mathcal{I}_i(x) = 0$ for all $i = 1, \ldots, q$ if and only if $x_1 = \cdots = x_q$. Furthermore, $\alpha \mathbf{e}$, $\alpha \in \mathbb{R}$, is an equilibrium state of (16.67).

**Proof.** The proof of the first conclusion is similar to the proof of Proposition 13.1 of [238] and, hence, is omitted. The second conclusion is a direct consequence of the first conclusion. □

The following lemmas involving graph-theoretic notions are needed for the main result of this section. For the statement of the next result, let $|\mathcal{V}|$ denote the cardinality of the set $\mathcal{V}$.

**Lemma 16.5.** Assume that $\mathfrak{G}$ is an undirected strongly connected graph with $n$ nodes and value $z_i \in \mathbb{R}$ for $i = 1, \ldots, n$. Furthermore, assume that, for each node $i$, the set of nodes of its neighbors is given by $\mathcal{V}_{n_i} = \{i_1, \ldots, i_{n_i}\}$, where $n_i = |\mathcal{V}_{n_i}|$. If for each node $i$, $z_{i_1} = \cdots = z_{i_{n_i}}$ and, for some $m \in \{1, \ldots, n\}$ and some $m_j \in \mathcal{V}_{n_m}$, $z_m = z_{m_j}$, then $z_1 = \cdots = z_n$.

**Proof.** The result is trivial for the cases where $n = 2$ and $n = 3$. Consider the case where $n \geq 4$. Let $m$, $m \in \{1, \ldots, n\}$, be the node satisfying $z_m = z_{m_j}$ for some $m_j \in \mathcal{V}_{n_m}$. If $|\mathcal{V}_{n_m}| = 1$, then we consider the node $m_j$. Since $\mathfrak{G}$ is strongly connected and $n \geq 4$, it follows that $\mathcal{V}_{m_j} \neq \emptyset$. Hence, for every neighbor $s \in \mathcal{V}_{m_j}$, $z_s = z_{m_j} = z_m$. Choose a neighbor $s \in \mathcal{V}_{m_j}$ such that $|\mathcal{V}_s| \geq 2$ (this is possible since $\mathfrak{G}$ is strongly connected). Then, by connectivity, it follows that, for every node $k \in \mathcal{V} \backslash \{s, m_j, m\}$, $z_k = z_{m_j} = z_m$ or $z_k = z_s = z_m$, and, hence, the conclusion follows.

Otherwise, if $|\mathcal{V}_{n_m}| \geq 2$, then choose a neighbor $m_j \in \mathcal{V}_{n_m}$ such that $|\mathcal{V}_{m_j}| \geq 2$ (this is possible since $\mathfrak{G}$ is strongly connected). Then, by connectivity, it follows that, for every node $k \in \mathcal{V} \backslash \{m, m_j\}$, $z_k = z_m$ or $z_k = z_{m_j}$, and, hence, the conclusion follows. □

For the next result, recall that a *cycle* of the graph $\mathfrak{G}$ is a connected graph where every vertex has exactly two neighbors [102] and an *odd cycle* of the graph $\mathfrak{G}$ is a cycle of $\mathfrak{G}$ with an odd number of edges [81, p. 14].

**Lemma 16.6.** Assume that $\mathfrak{G}$ is an undirected strongly connected graph with $n$ nodes and value $z_i \in \mathbb{R}$ for $i = 1, \ldots, n$. Furthermore, assume that for each node $i$, the set of nodes of its neighbors is given by $\mathcal{V}_{n_i} = \{i_1, \ldots, i_{n_i}\}$, where $n_i = |\mathcal{V}_{n_i}|$. If $\mathfrak{G}$ contains an odd cycle and for each $i$, $z_{i_1} = \cdots = z_{i_{n_i}}$, then $z_1 = \cdots = z_n$.

**Proof.** Since $\mathfrak{G}$ contains a cycle of length $m$, where $3 \le m \le n$ is odd, without loss of generality, let $1, \ldots, m$ be the nodes of the cycle. Then, by connectivity, $z_1 = z_3 = \cdots = z_m = z_2 = z_4 = \cdots = z_{m-1}$, which implies that there exists a node $i$ such that $z_i = z_{i_m}$, where $i_m \in \mathcal{V}_{n_i}$. Thus, it follows from Lemma 16.5 that $z_1 = \cdots = z_n$. $\qquad\square$

Next, we present the main stability result of this section for information flow models. Note that although general stability results have been developed in [228] and [6], the conditions of those results are restrictive. Specifically, in [228] it is always required that, for each $i \in \{1, \ldots, q\}$, the right-hand side $f_i(x)$ of (16.67) is contained in the relative *interior* of the convex hull of $x_i$ and its neighbors $x_j$. Although [6] extends the results of [228] to the case where the linear combination of $x_i$ and its neighbors $x_j$ is not necessarily convex, the results still need several technical assumptions.

In the following result, we present improved results for semistability of (16.67). For this result, we define *in-neighbors* of the $i$th agent to be those agents whose information can be received by the $i$th agent.

**Theorem 16.6.** Consider the information flow model (16.67), and assume that Assumptions 16.1–16.3 hold. For $i = 1, \ldots, q \ge 2$, let $n_i \ge 1$ be the number of neighbors of the $i$th agent in the case where $\mathfrak{G}$ is a graph, and let $n_i \ge 1$ be the number of in-neighbors of the $i$th agent in the case where $\mathfrak{G}$ is a digraph. Then the following statements hold.

(i) If $p_i \phi_{ij}(x) = -p_j \phi_{ji}(x)$ and $\lambda_{ij} < \frac{2p_j}{n_i p_j + n_j p_i}$ for all $i, j = 1, \ldots, q$, $i \ne j$, $p_i > 0$, then, for every $\alpha \in \mathbb{R}$, $\alpha \mathbf{e}$ is a semistable equilibrium state of (16.67). Furthermore, $x(k) \to \alpha_* \mathbf{e}$ as $k \to \infty$, where $\alpha_* = \sum_{i=1}^{q} p_i x_i(0) / (\sum_{i=1}^{q} p_i)$.

(ii) If $p_i \phi_{ij}(x) = -p_j \phi_{ji}(x)$, $\frac{n_i}{p_i} = \frac{n_j}{p_j}$, and $\lambda_{ij} \le \frac{2p_j}{n_i p_j + n_j p_i}$ for all $i, j = 1, \ldots, q$, $i \ne j$, $p_i > 0$, and $\lambda_{lm} < \frac{2p_m}{n_l p_l + n_m p_m}$ for some $l, m \in \{1, \ldots, q\}$ and $\mathcal{C}_{(l,m)} = 1$, $l \ne m$, then, for every $\alpha \in \mathbb{R}$, $\alpha \mathbf{e}$ is a semistable equilibrium state of (16.67). Furthermore, $x(k) \to \alpha_* \mathbf{e}$ as $k \to \infty$.

(iii) If $\mathfrak{G}$ contains an odd cycle, $p_i\phi_{ij}(x) = -p_j\phi_{ji}(x)$, $\frac{n_i}{p_i} = \frac{n_j}{p_j}$, and $\lambda_{ij} \leq \frac{2p_j}{n_ip_j+n_jp_i}$ for all $i, j = 1, \ldots, q$, $i \neq j$, $p_i > 0$, then, for every $\alpha \in \mathbb{R}$, $\alpha\mathbf{e}$ is a semistable equilibrium state of (16.67). Furthermore, $x(k) \to \alpha_*\mathbf{e}$ as $k \to \infty$.

(iv) Let $\phi_{ij}(x) = \phi_{ij}(x_i, x_j) = \frac{1}{p_i}\mathcal{A}_{(i,j)}(x_j - x_i)$ for all $i, j = 1, \ldots, q$, $i \neq j$. Assume that $\mathcal{C}^\mathrm{T}\mathbf{e} = 0$ and $p_i \geq n_i^+$, $i = 1, \ldots, q$. Furthermore, assume that $p_r > n_r^+$ for some $r \in \{1, \ldots, q\}$ such that $\mathcal{A}_{(r,j)} = 1$. Then, for every $\alpha \in \mathbb{R}$, $\alpha\mathbf{e}$ is a semistable equilibrium state of (16.67). Furthermore, $x(k) \to \alpha_*\mathbf{e}$ as $k \to \infty$.

**Proof.** First note that it follows from Lemma 16.4 that $\alpha\mathbf{e} \in \mathbb{R}^q$, $\alpha \in \mathbb{R}$, is an equilibrium state of (16.67).

(i) To show Lyapunov stability of the equilibrium state $\alpha\mathbf{e}$, consider the Lyapunov function candidate given by

$$V(x) = (x - \alpha\mathbf{e})^\mathrm{T}P(x - \alpha\mathbf{e}), \qquad (16.69)$$

where $P \triangleq \mathrm{diag}[p_1, \ldots, p_q]$. Now, since $p_i\phi_{ij}(x) = -p_j\phi_{ji}(x)$, $x \in \mathbb{R}^q$, $i \neq j$, $i, j = 1, \ldots, q$, and $\mathbf{e}^\mathrm{T}Px(k+1) = \mathbf{e}^\mathrm{T}Px(k)$, $k \in \mathbb{Z}_+$, it follows from Assumptions 16.2 and 16.3 that

$$\Delta V(x(k))$$

$$= 2\sum_{i=1}^{q}\sum_{j=1,j\neq i}^{q} x_i(k)p_i\phi_{ij}(x(k)) + \sum_{i=1}^{q}p_i\left[\sum_{j=1,j\neq i}^{q}\phi_{ij}(x(k))\right]^2$$

$$= 2\sum_{i=1}^{q}\sum_{j\in\mathcal{K}_i}(x_i(k) - x_j(k))p_i\phi_{ij}(x(k)) + \sum_{i=1}^{q}\frac{1}{p_i}\left[\sum_{j\in\mathcal{N}_i}p_i\phi_{ij}(x(k))\right]^2$$

$$\leq 2\sum_{i=1}^{q}\sum_{j\in\mathcal{K}_i}(x_i(k) - x_j(k))p_i\phi_{ij}(x(k)) + \sum_{i=1}^{q}\sum_{j\in\mathcal{N}_i}\frac{1}{p_i}n_ip_i^2\phi_{ij}^2(x(k))$$

$$= 2\sum_{i=1}^{q}\sum_{j\in\mathcal{K}_i}(x_i(k) - x_j(k))p_i\phi_{ij}(x(k)) + \sum_{i=1}^{q}\sum_{j\in\mathcal{K}_i}\left(\frac{n_i}{p_i} + \frac{n_j}{p_j}\right)p_i^2\phi_{ij}^2(x(k))$$

$$= \sum_{i=1}^{q}\sum_{j\in\mathcal{K}_i}2p_i\left[\left(\frac{n_i}{p_i} + \frac{n_j}{p_j}\right)\frac{p_i}{2}|\phi_{ij}(x(k))|^2 - |(x_i(k) - x_j(k))\phi_{ij}(x(k))|\right]$$

$$\leq \sum_{i=1}^{q}\sum_{j\in\mathcal{K}_i}2p_i\left[\left(\frac{n_i}{p_i} + \frac{n_j}{p_j}\right)\frac{p_i}{2}\lambda_{ij} - 1\right]|(x_i(k) - x_j(k))\phi_{ij}(x(k))|$$

$$\leq 0, \quad k \in \overline{\mathbb{Z}}_+, \qquad (16.70)$$

where $\mathcal{K}_i \triangleq \mathcal{N}_i \setminus \bigcup_{l=1}^{i-1} \{l\}$ and

$$\mathcal{N}_i \triangleq \{j \in \{1, \ldots, q\} : \phi_{ij}(x) = 0 \text{ if and only if } x_i = x_j\}, \quad i = 1, \ldots, q,$$

which establishes Lyapunov stability of the equilibrium state $\alpha \mathbf{e}$.

To show that $\alpha \mathbf{e}$ is semistable, note that

$$\Delta V(x(k)) \geq 2 \sum_{i=1}^{q} \sum_{j \in \mathcal{K}_i} (x_i(k) - x_j(k)) p_i \phi_{ij}(x(k)), \quad k \in \overline{\mathbb{Z}}_+. \quad (16.71)$$

Next, we show that $\Delta V(x) = 0$ if and only if $(x_i - x_j)\phi_{ij}(x) = 0$, $i = 1, \ldots, q$, $j \in \mathcal{K}_i$. First, assume that $(x_i - x_j)\phi_{ij}(x) = 0$, $i = 1, \ldots, q$, $j \in \mathcal{K}_i$. Then it follows from (16.71) that $\Delta V(x) \geq 0$. However, it follows from (16.70) that $\Delta V(x) \leq 0$, and, hence, $\Delta V(x) = 0$. Conversely, assume that $\Delta V(x) = 0$. In this case, note that

$$\Delta V(x) \leq \sum_{i=1}^{q} \sum_{j \in \mathcal{K}_i} 2p_i \left[ \left( \frac{n_i}{p_i} + \frac{n_j}{p_j} \right) \frac{p_i}{2} \lambda_{ij} - 1 \right] |(x_i(t) - x_j(t))\phi_{ij}(x(t))|$$

$$\leq 0, \quad (16.72)$$

and, since $\left( \frac{n_i}{p_i} + \frac{n_j}{p_j} \right) \frac{p_i}{2} \lambda_{ij} - 1 < 0$, it follows that $(x_i - x_j)\phi_{ij}(x) = 0$, $i = 1, \ldots, q$, $j \in \mathcal{K}_i$.

Let

$$\mathcal{R} \triangleq \{x \in \mathbb{R}^q : \Delta V(x) = 0\}$$
$$= \{x \in \mathbb{R}^q : (x_i - x_j)\phi_{ij}(x) = 0, i = 1, \ldots, q, j \in \mathcal{K}_i\}.$$

Now, by Assumption 16.1, the directed graph associated with the connectivity matrix $\mathcal{C}$ for the multiagent dynamical system (16.67) is strongly connected, which implies that $\mathcal{R} = \{x \in \mathbb{R}^q : x_1 = \cdots = x_q\}$. Since the set $\mathcal{R}$ consists of the equilibrium states of (16.67), it follows that the largest invariant set $\mathcal{M}$ contained in $\mathcal{R}$ is given by $\mathcal{M} = \mathcal{R}$. Hence, it follows from Proposition 16.1 that $\alpha \mathbf{e}$ is a semistable equilibrium state of (16.67). To show that $x(k) \to \alpha_* \mathbf{e}$ as $k \to \infty$, note that since $\mathbf{p}^\mathrm{T} x(k) = \mathbf{p}^\mathrm{T} x(0)$ and $x(k) \to \mathcal{M}$ as $k \to \infty$, where $\mathbf{p} \triangleq [p_1, \ldots, p_q]^\mathrm{T} \in \mathbb{R}^q$, it follows that $x(k) \to \alpha_* \mathbf{e}$ as $k \to \infty$.

(ii) Using similar arguments as in (i), it can be shown that $\alpha \mathbf{e}$ is Lyapunov stable. To show semistability of $\alpha \mathbf{e}$, let $\mathcal{R} \triangleq \{x \in \mathbb{R}^q : \Delta V(x) = 0\}$, where $V(\cdot)$ is given by (16.69). In this case, it follows from (16.70) that

$$\mathcal{R} = (\mathcal{R}_1 \cup \mathcal{R}_2) \cap \mathcal{R}_3 = (\mathcal{R}_1 \cap \mathcal{R}_3) \cup (\mathcal{R}_2 \cap \mathcal{R}_3), \quad (16.73)$$

where

$$\mathcal{R}_1 \triangleq \{x \in \mathbb{R}^q : \phi_{ij}(x) = 0, \ i = 1, \ldots, q, \ j \in \mathcal{K}_i\},$$

$$\mathcal{R}_2 \triangleq \left\{x \in \mathbb{R}^q : \left(\frac{n_i}{p_i} + \frac{n_j}{p_j}\right) p_i \phi_{ij}(x) = 2(x_j - x_i), \ i = 1, \ldots, q, \ j \in \mathcal{K}_i\right\},$$

and

$$\mathcal{R}_3 \triangleq \{x \in \mathbb{R}^q : \phi_{ij}(x) = \phi_{ik}(x), \ i = 1, \ldots, q, \ j \in \mathcal{N}_i, \ k \in \mathcal{N}_i \backslash \{j\}\}.$$

If $\phi_{ij}(x) = 0$, then $x_i = x_j$, $i = 1, \ldots, q$, $j \in \mathcal{K}_i$. Now, by Assumption 16.1, the directed graph associated with the connectivity matrix $\mathcal{C}$ for the multiagent dynamical system (16.67) is strongly connected, which implies that $x_1 = \cdots = x_q$. Hence, $\mathcal{R}_1 \cap \mathcal{R}_3 = \{x \in \mathbb{R}^q : x_1 = \cdots = x_q\}$.

Next, we consider the case where $\left(\frac{n_i}{p_i} + \frac{n_j}{p_j}\right) p_i \phi_{ij}(x) = 2(x_j - x_i)$ and $x \in \mathcal{R}_3$, $i = 1, \ldots, q$, $j \in \mathcal{K}_i$. Since $p_i \phi_{ij}(x) = -p_j \phi_{ji}(x)$, it follows that

$$\left(\frac{n_j}{p_j} + \frac{n_i}{p_i}\right) p_j \phi_{ji}(x) = 2(x_i - x_j), \quad i = 1, \ldots, q, \quad j \in \mathcal{K}_i.$$

Hence,

$$\left(\frac{n_i}{p_i} + \frac{n_j}{p_j}\right) p_i \phi_{ij}(x) = 2(x_j - x_i), \quad i = 1, \ldots, q, \quad j \in \mathcal{N}_i.$$

Since $\frac{n_i}{p_i} = \frac{n_j}{p_j}$, it follows that $\phi_{ij}(x) = \frac{1}{n_i}(x_j - x_i)$, $i = 1, \ldots, q$, $j \in \mathcal{N}_i$. Furthermore, since $\phi_{ij}(x) = \phi_{ik}(x)$, it follows that $x_j = x_k$, $i = 1, \ldots, q$, $j, k \in \mathcal{N}_i$, $j \neq k$. Note that since $p_i \phi_{ij}(x) = -p_j \phi_{ji}(x)$, $\mathfrak{G}$ is an undirected graph. Thus, $\mathcal{A} = \mathcal{A}^{\mathrm{T}}$, and, hence, $\mathfrak{G}$ is strongly connected.

Now, it follows from (16.70) that, for $x \in \mathcal{R}_2 \cap \mathcal{R}_3$, $(x_l - x_m)\phi_{lm}(x) = 0$, which implies that $x_l = x_m$. Hence, it follows from Lemma 16.5 that $x_1 = \cdots = x_q$, $\mathcal{R}_2 \cap \mathcal{R}_3 = \{x \in \mathbb{R}^q : x_1 = \cdots = x_q\}$. Therefore, $\mathcal{R} = \{x \in \mathbb{R}^q : x_1 = \cdots = x_q\}$. Now, since the set $\mathcal{R}$ consists of the equilibrium states of (16.67), it follows that the largest invariant set $\mathcal{M}$ contained in $\mathcal{R}$ is the set of equilibria of (16.67). Hence, it follows from Proposition 16.1 that $\alpha \mathbf{e}$ is a semistable equilibrium state of (16.67). To show that $x(k) \to \alpha_* \mathbf{e}$ as $k \to \infty$, note that, since $\mathbf{p}^{\mathrm{T}} x(k) = \mathbf{p}^{\mathrm{T}} x(0)$ and $x(k) \to \mathcal{M}$ as $k \to \infty$, it follows that $x(k) \to \alpha_* \mathbf{e}$ as $k \to \infty$.

(iii) Using similar arguments as in (i), it can be shown that $\alpha \mathbf{e}$ is Lyapunov stable. Furthermore, using similar arguments as in (ii), it follows that, for $x \in \mathcal{R}_2 \cap \mathcal{R}_3$, $x_j = x_k$, $j, k \in \mathcal{N}_i$, $i = 1, \ldots, q$, $j \neq k$. Now, it follows from Lemma 16.6 that $x_1 = \cdots = x_q$. Hence, $\mathcal{R} = \{x \in \mathbb{R}^q : x_1 = \cdots = x_q\}$. The rest of the proof follows as in the proof of (i).

(iv) Let $W \triangleq I_q + P^{-1}\mathcal{A}$. First, we show that $W$ is irreducible. Note that $W$ is a *stochastic matrix* [148, p. 526]. Furthermore, since

$$W - I_q = \begin{bmatrix} \frac{1}{p_1}\mathcal{A}_{(1,1)} & \frac{1}{p_1}\mathcal{A}_{(1,2)} & \cdots & \frac{1}{p_1}\mathcal{A}_{(1,q)} \\ \frac{1}{p_2}\mathcal{A}_{(2,1)} & \frac{1}{p_2}\mathcal{A}_{(2,2)} & \cdots & \frac{1}{p_2}\mathcal{A}_{(2,q)} \\ \vdots & \vdots & \ddots & \vdots \\ \frac{1}{p_q}\mathcal{A}_{(q,1)} & \frac{1}{p_q}\mathcal{A}_{(q,2)} & \cdots & \frac{1}{p_q}\mathcal{A}_{(q,q)} \end{bmatrix}, \tag{16.74}$$

it follows that $\mathrm{rank}(W - I_q) = \mathrm{rank}\,\mathcal{C} = q - 1$. Since $\mathcal{C}^{\mathrm{T}}\mathbf{e} = 0$, it follows that $(W^{\mathrm{T}} - I_q)\mathbf{p} = 0$. Now, it follows from [24, p. 52] that $W$ is irreducible.

Next, note that $|\lambda_i| \le \|W\| = 1$, $i = 1, \ldots, q$; $\lambda_i \in \mathrm{spec}(W)$; and $\|W\|$ is an induced norm of $W$. Then $\rho(W) = 1$. It follows from Theorem 13.1.4 of [24] that $\rho(W) = 1$ is a simple eigenvalue.

Next, we show that $W$ is a *primitive matrix* [148, p. 516]. Since $p_i \ge n_i^+$ for all $i \in \{1, \ldots, q\}$ and $p_r > n_r^+$ for some $r \in \{1, \ldots, q\}$, it follows that

$$\mathrm{tr}\, W = \sum_{i=1}^{q} 1 + \frac{1}{p_i}\mathcal{A}_{(i,i)} \ge 1 + \frac{1}{p_r}\mathcal{A}_{(r,r)} > 0.$$

Then it follows from Corollary 2.28 of [24] that $W$ is primitive.

Now, it follows from Theorem 2 of [123] that $W$ is semistable, and, hence,

$$\lim_{k \to \infty} W^k = I_q - (W - I_q)(W - I_q)^{\#}.$$

Next, it follows from (vi) of Lemma 5.2 of [134] that

$$\mathcal{N}(W - I_q) = \mathcal{R}(I_q - (W - I_q)(W - I_q)^{\#}).$$

Since $(W - I_q)\mathbf{e} = 0$ and $\mathrm{rank}(W - I_q) = q - 1$, it follows that $\mathcal{N}(W - I_q) = \{\alpha\mathbf{e}\}$, where $\alpha \in \mathbb{R}$, and, hence, $\mathcal{R}(I_q - (W - I_q)(W - I_q)^{\#}) = \{\alpha\mathbf{e}\}$, which implies that $\lim_{k \to \infty} x(k) = \lim_{k \to \infty} W^k x(0) = \alpha\mathbf{e}$. To show that $x(k) \to \alpha_*\mathbf{e}$ as $k \to \infty$, note that since $\mathbf{p}^{\mathrm{T}}x(k) = \mathbf{p}^{\mathrm{T}}x(0)$ and $x(k) \to \mathcal{M}$ as $k \to \infty$, it follows that $x(k) \to \alpha_*\mathbf{e}$ as $k \to \infty$. $\square$

**Example 16.1.** To illustrate some of the results of Theorem 16.6, consider the linear dynamical system

$$x_1(k+1) = \frac{1}{2}(x_2(k) + x_3(k)), \quad x_1(0) = x_{10}, \quad k \in \overline{\mathbb{Z}}_+, \tag{16.75}$$

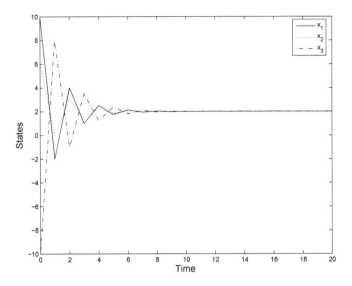

Figure 16.1 Trajectories versus time for (16.75)–(16.77).

$$x_2(k+1) = \frac{1}{2}(x_3(k) + x_1(k)), \quad x_2(0) = x_{20}, \tag{16.76}$$

$$x_3(k+1) = \frac{1}{2}(x_1(k) + x_2(k)), \quad x_3(0) = x_{30}. \tag{16.77}$$

Note that the system (16.75)–(16.77) is an information flow model of the form given by (16.67) and it follows from (iii) of Theorem 16.6 that consensus and semistability of (16.75)–(16.77) are guaranteed. Figure 16.1 shows the trajectories of (16.75)–(16.77) versus time. Note that it is not easy to use the methods in [228] and [6] to prove semistability and consensus for (16.75)–(16.77). However, using Theorem 16.6, this is straightforward.     △

## 16.6 Optimal Fixed-Structure Control of Network Consensus

In multiagent coordination [180, 238] and distributed network averaging [315] with a fixed communication topology, we require that $x_e \in \mathrm{span}\{\mathbf{e}\}$. In this section, we consider the design of a fixed-structure consensus protocol for (16.67) such that the closed-loop system is semistable, kernel $(f) = \mathrm{span}\{\mathbf{e}\}$, and (16.52) is minimized. Here, we consider the consensus protocol (16.67) given by

$$x_i(k+1) = u_i(k), \quad x_i(0) = x_{i0}, \quad k \in \overline{\mathbb{Z}}_+, \tag{16.78}$$

$$u_i(k) = x_i(k) + \sum_{j=1, j \neq i}^{q} \phi_{ij}(x(k)), \tag{16.79}$$

$$\phi_{ij}(x(k)) = \frac{1}{k_i} \mathcal{A}_{(i,j)}(x_j(k) - x_i(k)), \quad i, j = 1, \dots, q, \quad i \neq j, \tag{16.80}$$

where $k_i > n_i^+$, $i = 1, \ldots, q$, and $\mathcal{C}$ satisfies Assumption 16.1 and the conditions of Theorem 16.6. Note that, for (16.78)–(16.80), Assumptions 16.2 and 16.3 are automatically satisfied. Since, by Theorem 16.6, the closed-loop system given by (16.67) is semistable, the optimal fixed-structure control problem involves seeking $k_i$, $k_i > n_i^+$, $i = 1, \ldots, q$, such that the cost functional

$$J(K) = \sum_{k=0}^{\infty} \left[ (x(k) - \alpha^* \mathbf{e})^{\mathrm{T}} R_1 (x(k) - \alpha^* \mathbf{e}) \right.$$
$$\left. + (u(k) - u_e)^{\mathrm{T}} R_2 (u(k) - u_e) \right] \tag{16.81}$$

is minimized, where $u_e = \alpha_* K^{-1} \mathcal{A} \mathbf{e}$, $R_1 = E_1^{\mathrm{T}} E_1 \geq 0$, $R_2 = E_2^{\mathrm{T}} E_2 > 0$, and $R_{12} = E_1^{\mathrm{T}} E_2 = 0$.

The following theorem presents a BMI solution to the fixed-structure optimal semistable control problem for network consensus. For this result, define

$$\mathcal{L} \triangleq \left\{ L \in \mathbb{R}^{q \times q} : L = \mathrm{diag}[\ell_1, \ldots, \ell_q] \in \mathbb{R}^{q \times q}, \ell_i > n_i^+, i = 1, \ldots, q \right\}.$$

**Theorem 16.7.** Consider the consensus protocol (16.78)–(16.80), and assume that $(I_q + \mathcal{A}, E_1)$ is semiobservable and $(I_q + \mathcal{A}, V)$ is semicontrollable. Let $Q \in \mathbb{R}^{q \times q}$ and $L \in \mathcal{L}$ be the solution to the BMI minimization problem

$$\min_{Q \in \mathbb{R}^{q \times q}, L \in \mathcal{L}, W \in \mathbb{R}^{p \times p}} \mathrm{tr}\, W \tag{16.82}$$

subject to

$$\begin{bmatrix} Q & (E_1 Q + E_2 Q + E_2 L \mathcal{A} Q)^{\mathrm{T}} \\ E_1 Q + E_2 Q + E_2 L \mathcal{A} Q & W \end{bmatrix} > 0, \tag{16.83}$$

$$\begin{bmatrix} V - Q & (E_1 Q + E_2 Q + E_2 L \mathcal{A} Q)^{\mathrm{T}} \\ E_1 Q + E_2 Q + E_2 L \mathcal{A} Q & -Q \end{bmatrix} \leq 0. \tag{16.84}$$

Then $u = (I_q + K^{-1} \mathcal{A})x$ is a semistabilizing controller for (16.78), and $x(k) \to \alpha_* \mathbf{e}$ as $k \to \infty$, where $K^{-1} = L$ and $\alpha_* = \sum_{i=1}^{q} k_i x_i(0) / (\sum_{i=1}^{q} k_i)$. Furthermore, $K$ minimizes the $\mathcal{H}_2$ performance criterion $J(K)$ given by (16.81).

**Proof.** Convergence to the consensus state $\alpha_* \mathbf{e}$ is a direct consequence of Theorem 16.6. The optimality proof is similar to the proof of Theorem 16.5 and, hence, is omitted. $\square$

Due to the diagonal structure of $K$, the optimization problem given in Theorem 16.7 is a BMI. A suboptimal solution to this problem can be obtained by using a two-stage optimization process. Specifically, by fixing $Q$, one can design the controller $K$. Then, with $K$ fixed, $Q$ can be obtained. This process continues until convergence or until an acceptable controller is found.

## Chapter Seventeen

# Optimal Network Resource Allocation

## 17.1 Introduction

Balanced coordination for damage mitigation and resource allocation in network systems enhances rapid dissemination of network resources and self-healing of the network, which leads to significant reduction of threats imposed on system network attacks. These can include sensor networks, cooperative satellite systems, and power systems that provide inviting targets to adversaries. Here, by *balanced coordination* we mean that damage compensation is proportional to the severity of the damage caused by adversarial attacks within the network.

To achieve balanced coordination, it is critical to develop new methods for addressing balanced cooperative tasks of network resources and damages with the objective of coordinated action and decision-making. In particular, appropriate sensory and cogitative capabilities, such as adaptation, learning, and agreement on the network levels, need to be developed. As a consequence, designing specific algorithms to achieve balanced coordination tasks in a network-based environment is an important problem for network systems [67].

Apart from the motivation for countering catastrophic network failures and mitigating damage in a geospatial network [159], balanced coordination algorithm design problems for coordinated resource allocation in network systems also arise in a wide range of other emerging applications, such as multiagent and consensus control [180, 273], agent mining [2, 52], multiagent coordination for optimization [39, 329], sensor network design [214], randomized gossip algorithms [47], distributed computation [155, 301, 315], PageRank algorithms for online search engines [177, 278], and target search problems [182], to cite but a few examples. In many of these problems, a unique defining feature is that a group of quantities of interest need to achieve agreement on the final distribution of these quantities, resulting in a Nash-type equilibrium state that is dependent on the system initial condition of these quantities [160].

An example of such a problem is the optimal semistable control problem discussed in Chapters 10–12 and [160], in which the optimal control law is designed so that the limiting state of the closed-loop control system is determined not only by system dynamics but also by the initial state, leading to the notion of semistabilization. However, as shown in Chapters 10–12 and 16, such a problem cannot be solved by means of classical optimal control theory.

In this chapter, we develop a balanced iterative algorithm design framework for network resource allocation for mitigating damage in a network system. Specifically, we consider a network characterized by a strongly connected directed graph $\mathfrak{G} = (\mathcal{V}, \mathcal{E})$ consisting of the set of nodes $\mathcal{V} = \{1, \ldots, n\}$ and the set of edges $\mathcal{E} \subseteq \mathcal{V} \times \mathcal{V}$, where each edge $(i, j) \in \mathcal{E}$ is an ordered pair of distinct nodes. The connectivity matrix $\mathcal{C}$ of $\mathfrak{G}$ in this chapter is defined by $\mathcal{C}_{(i,j)} = 1$ if $(i, j) \in \mathcal{E}$, $\mathcal{C}_{(i,j)} = 0$ if $(i, j) \notin \mathcal{E}$ and $i \neq j$, and $\mathcal{C}_{(i,i)} = 1$, where $\mathcal{C}_{(i,j)}$ represents the $(i, j)$th entry of $\mathcal{C}$, $i, j = 1, \ldots, n$. We denote the quantity of interest of node $i \in \{1, \ldots, n\}$ at time $k$ by a scalar value $x_i(k) \in \mathbb{R}$, which denotes the information state of node $i$ at time $k$. For distributed computation, this value represents the updated data of an iterative algorithm at a particular iterative step, whereas, for resource allocation and damage mitigation, this value represents the available resources in node $i$ to be allocated.

Each node $i$ holds an initial value on the network $x_i(0) \in \mathbb{R}$. The network permits unidirectional communication between two nodes if and only if they are neighbors. We are interested in rearranging the information state of each node of a network to some equilibrium distribution, via a distributed iterative algorithm, so that all of the information states are asymptotically distributed to some pattern that is a function of $x_i(0)$, $i = 1, \ldots, n$. Furthermore, when we model resource allocation problems, we need to account for uncertain disturbances in the balanced iterative algorithm design.

The damage mitigation problem considered in this chapter is defined as a reduction of adversarial attack effectiveness. To effectively model the damage mitigation problem under adversarial attacks, we design a non-fixed-structure *mobile sensor network* for detecting damage caused by adversarial threats. Specifically, we model a network as a strongly connected, non-fixed-structure directed graph in peer-to-peer networks. The nodes in the graph represent critical sites in the network that may be attacked by an adversary. By a non-fixed-structure graph we mean that if there does not exist a link between two nodes in the network, then the mobile sensors in these two nodes can set up a wireless communication link between them so that the graph structure is not necessarily fixed to the actual network system.

One scenario depicting the dynamics of damage mitigation is as follows. Let $x_i(k)$ denote the number of mobile sensors that node $i$ has at time $k$. These mobile sensors can carry equipment and recovery resources to detect possible attacks at node $i$ and evolve along the graph network to mitigate damage based on a given algorithm. At the initial time, the number of mobile sensors at node $i$ is given by $x_i(0)$. At the $k$th time instant, node $i$ communicates with the neighboring node $j$ to ascertain how many mobile sensors both nodes have. Then, to reduce the danger posed by an attack, both nodes relocate their mobile sensors so that the number of mobile sensors at each node is proportional to the severity of risks and damage incurred by an attack. The question is, then, how can we design the evolution process for $x_i(k)$ at node $i$ by means of its neighbor-to-neighbor information so that all the resources can be redistributed as fast as possible in the network?

## 17.2  Resource Allocation Problem Formulation

Consider the iterative algorithm design for the coordinated resource allocation network problem given by

$$x_i(k+1) = a_{ii}x_i(k) - \sum_{j=1,j\neq i}^{n} \mathcal{C}_{(j,i)}a_{ji}[x_i(k) + d_i w_i(k)]$$

$$+ \sum_{j=1,j\neq i}^{n} \mathcal{C}_{(i,j)}a_{ij}[x_j(k) + d_j w_j(k)], \quad x_i(0) = x_{i0},$$

$$k \in \overline{\mathbb{Z}}_+, \qquad (17.1)$$

or, in vector form,

$$x(k+1) = Ex(k) + (E - \Lambda)Dw(k)$$
$$= \Lambda x(k) + (E - \Lambda)x(k) + (E - \Lambda)Dw(k), \quad x(0) = x_0, \quad k \in \overline{\mathbb{Z}}_+,$$
$$(17.2)$$

where $x(k) = [x_1(k), \ldots, x_n(k)]^{\mathrm{T}} \in \mathbb{R}^n$ denotes the information state vector; $w(k) = [w_1(k), \ldots, w_n(k)]^{\mathrm{T}} \in \mathbb{R}^n$ denotes a standard discrete-time Gaussian white noise process; $\Lambda = \mathrm{diag}\,[a_{11}, \ldots, a_{nn}] \in \mathbb{R}^{n\times n}$; $D = \mathrm{diag}\,[d_1, \ldots, d_n] \in \mathbb{R}^{n\times n}$;

$$E_{(i,j)} = \begin{cases} a_{ii} - \sum_{l=1,l\neq i}^{n} \mathcal{C}_{(l,i)}a_{li}, & i = j, \\ \mathcal{C}_{(i,j)}a_{ij}, & i \neq j, \end{cases} \quad i,j = 1, \ldots, n,$$

where $a_{ij}$ are design parameters such that $a_{ij} \geq 0$ and $\sum_{l=1,l\neq i}^{n} a_{li} \leq a_{ii}$; and $d_i$ represents a weighting on the white noise disturbance $w_i(\cdot)$. This network system can, for example, represent a stochastic compartmental model [124] involving a mass balance equation, where $x_i$ denotes the mass (and hence a nonnegative quantity) of the $i$th subsystem or compartment of the compartmental system. Note that due to the time-averaging mass balance principle, the maximum amount of expected value of the mass that can be transported cannot exceed the expected value of the mass in a compartment. Thus, it follows that $a_{ii} \geq \sum_{l=1, l \neq i}^{n} a_{li}$, which captures this physical time-averaging constraint. The term $x_i(k) + d_i w_i(k)$ represents imperfect information in the state at node $i$ due to sensor noise and/or model uncertainty.

This model can alternatively represent a class of gradient-like stochastic optimization algorithms [301] and consensus algorithms [316] arising in distributed computing. The resource allocation design problem involves finding an appropriate matrix $E = [E_{(i,j)}]$ such that the expected values of the trajectories of (17.2) converge and a given performance criterion $J = J(E, x(\cdot))$ subject to (17.2) is optimized.

To elucidate the idea of compartmental modeling for complex network systems, we use an example of a heating, ventilation, and air conditioning (HVAC) network [94]. HVAC systems in large buildings represent complex dynamical systems composed of interconnected thermal, electromechanical, and thermofluidic subsystems. Each subsystem is represented by a zone that is characterized by the volume of air, the external air supply with regulated temperature and supply rate, and the return air from the zone with regulated temperature and exhaust rate. Typically, the central plant in the building supplies chilled and hot water or steam to the cooling and heating coils in the air-handling units of each zone. The air-handling unit further supplies air to the zones, with the air temperature regulated by a heating or cooling coil temperature. The rate of air supply can also be regulated by supply fans. Multiple zones in a building can be represented by the interconnected dynamical system shown in Figure 17.1, composed of $q$ interconnected zones, where $\mathcal{G}_i$ denotes the $i$th zone.

Let $E_{si}$ denote the energy of the $i$th zone, where the amount of energy in each zone is related to the air temperature and the volume of the zone. Furthermore, let $s_i$ denote the external energy supplied to the $i$th zone; $\sigma_{ij} : \overline{\mathbb{R}}_+^q \to \overline{\mathbb{R}}_+$, $i \neq j$, $i, j = 1, \ldots, q$, denote the instantaneous energy flow from the $j$th zone to the $i$th zone, where $\overline{\mathbb{R}}_+^q$ denotes the set of $q$-dimensional vectors whose components are nonnegative; and $\sigma_{ii} : \overline{\mathbb{R}}_+^q \to \overline{\mathbb{R}}_+$, $i = 1, \ldots, q$, denote the instantaneous energy dissipated or lost from the $i$th zone. Hence, an energy balance equation for the $i$th zone yields

$$E_{si}(k+1) = E_{si}(k) + \sum_{j=1, j \neq i}^{q} [\sigma_{ij}(E_s(k)) - \sigma_{ji}(E_s(k))] - \sigma_{ii}(E_s(k)) + s_i(k),$$

$$E_{si}(0) = E_{si0}, \quad k \in \overline{\mathbb{Z}}_+, \qquad (17.3)$$

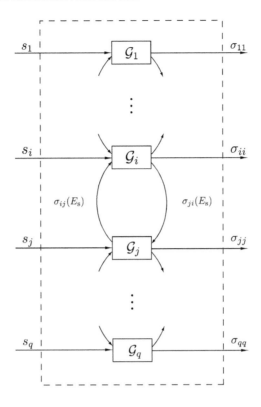

Figure 17.1 Compartmental model for an HVAC system.

where $E_{\mathrm{s}} = [E_{\mathrm{s}1}, \ldots, E_{\mathrm{s}q}]^{\mathrm{T}}$.

Note that (17.3) yields a conservation of energy equation and implies that the energy stored in the $i$th zone is equal to the external energy supplied to the $i$th zone plus the energy gained by the $i$th zone from all other zones due to zone coupling minus the energy dissipated or lost from the $i$th zone. Equivalently, (17.3) can be rewritten in vector form as

$$E_{\mathrm{s}}(k + 1) = \phi(E_{\mathrm{s}}(k)) - d(E_{\mathrm{s}}(k)) + S(k), \quad E_{\mathrm{s}}(0) = E_{\mathrm{s}0}, \quad k \in \overline{\mathbb{Z}}_{+}, \quad (17.4)$$

which yields an *energy balance* equation, where $S(k) = [s_1(k), \ldots, s_q(k)]^{\mathrm{T}}$; $d(E_{\mathrm{s}}) = [\sigma_{11}(E_{\mathrm{s}}), \ldots, \sigma_{qq}(E_{\mathrm{s}})]^{\mathrm{T}}$; and $\phi = [\phi_1, \ldots, \phi_q]^{\mathrm{T}} : \overline{\mathbb{R}}_{+}^{q} \to \mathbb{R}^{q}$ is such that $\phi_i(z) = \sum_{j=1, \, j \neq i}^{q}[\sigma_{ij}(z) - \sigma_{ji}(z)]$, $z \in \overline{\mathbb{R}}_{+}^{q}$. Note that the dynamical system (17.4) conforms with the structure of the iterative algorithm given by (17.1), with $S(\cdot)$ being a vector of disturbance inputs.

The main idea of designing the unknowns $a_{ij}$ in (17.1) lies in the observation that the iterative algorithm (17.2) can be rewritten as the control design problem

$$x(k+1) = \Lambda x(k) + u(k), \quad x(0) = x_0, \quad k \in \overline{\mathbb{Z}}_+, \qquad (17.5)$$

$$u(k) = (E - \Lambda)x(k) + (E - \Lambda)Dw(k), \qquad (17.6)$$

where $u(k)$ represents the state feedback controller corrupted by sensor noise. The optimal control problem then involves the design of the matrix $E$ such that the closed-loop system converges to a given state and a given performance functional is minimized.

It is important to note that the closed-loop system is not required to be asymptotically stable, as is the case for a classical optimal control problem. This is because resource allocation algorithms require only convergence and not asymptotic stability. More specifically, asymptotic stability is not an appropriate notion in this case since resource allocation algorithms need to relocate resources in a certain (possibly unknown) pattern among a network and should not dissipate all of the resources from the network. Therefore, classical optimal control theory cannot be directly applied to solve these types of problems.

To design such an iterative algorithm for resource allocation, we first develop a general optimization-based framework to address an optimal semistable control problem. This framework allows for the transformation of the original control problem, which is challenging to solve numerically, into an equivalent optimization problem that is easier to solve in practice. Then we apply numerical methods to solve the proposed optimization problem that allow for the codesign of the original iterative algorithm.

## 17.3 A Control-Theoretic Approach to Balanced Resource Allocation

In this chapter, we address the *optimal control* design problem for the iterative algorithm (17.2). By optimality, we mean that the iterative algorithm (17.2) minimizes a given cost functional which represents an optimal criterion for (17.2). Here we choose a quadratic cost functional to measure the error between the current state $x(k)$ and the final distribution $\lim_{k\to\infty} x(k)$. In order to solve this optimal control design problem, we first consider a generalization of (17.5) and (17.6) involving a stochastic optimal semistable control problem by extending the results of Chapter 12 to the discrete-time setting.

Specifically, consider the discrete-time linear controlled system driven by a stochastic disturbance given by

$$x(k+1) = Ax(k) + Bu(k) + D_1 w(k), \quad x(0) = x_0, \quad k \in \overline{\mathbb{Z}}_+, \qquad (17.7)$$

where $x(k) = [x_1(k), \ldots, x_n(k)] \in \mathbb{R}^n$ is the system state vector, $u(k) = [u_1(k), \ldots, u_m(k)] \in \mathbb{R}^m$ is the control input, and $w(k) = [w_1(k), \ldots, w_q(k)] \in \mathbb{R}^q$ is a $q$-dimensional standard Gaussian white noise process. Here we assume that $x(0)$ is a random variable. Note that the system (17.7) can be viewed as a generalization of (17.5) for which $A = \Lambda$, $B = I_n$, and $D_1 = 0$.

The following definition is needed for our problem formulation.

**Definition 17.1.** Let $A \in \mathbb{R}^{n \times n}$. $A$ is called *discrete-time semistable* if $\mathrm{spec}(A) \subseteq \{s \in \mathbb{C} : |s| < 1\} \cup \{1\}$, and if $1 \in \mathrm{spec}(A)$, then $1$ is semisimple.

The control design problem involves the state feedback controller given by

$$u(k) = K[x(k) + D_2 w(k)], \tag{17.8}$$

such that the following design criteria are satisfied.

(i) The closed-loop system (17.7) and (17.8) with $w(k) \equiv 0$ is discrete-time semistable; that is, $\tilde{A} \triangleq A + BK$ is discrete-time semistable.

(ii) The performance functional

$$J(K) = \lim_{N \to \infty} \frac{1}{N+1} \mathbb{E} \left\{ \sum_{k=0}^{N} \left[ (x(k) - x_\infty)^{\mathrm{T}} R_1 (x(k) - x_\infty) \right. \right.$$
$$\left. \left. + (u(k) - u_\infty)^{\mathrm{T}} R_2 (u(k) - u_\infty) \right] \right\} \tag{17.9}$$

is minimized, where $R_1 \triangleq E_1^{\mathrm{T}} E_1$, $R_2 \triangleq E_2^{\mathrm{T}} E_2$, $R_{12} \triangleq E_1^{\mathrm{T}} E_2 = 0$, $E_1 \in \mathbb{R}^{r \times n}$, $E_2 \in \mathbb{R}^{r \times m}$, $x_\infty = \lim_{k \to \infty} \mathbb{E}[x(k)]$, and $u_\infty = K x_\infty$, where $\mathbb{E}$ denotes the expectation operator.

(iii) $x_\infty = \frac{1}{\mathbf{u}^{\mathrm{T}} \mathbf{v}} \mathbf{u} \mathbf{v}^{\mathrm{T}} \mathbb{E}[x(0)]$ for some unit vectors $\mathbf{u}, \mathbf{v} \in \mathbb{R}^n$ satisfying $\mathbf{u}^{\mathrm{T}} \mathbf{v} \neq 0$.

Note that the closed-loop system (17.7) and (17.8) is given by

$$x(k+1) = \tilde{A} x(k) + \tilde{D} w(k), \quad x(0) = x_0, \quad k \in \overline{\mathbb{Z}}_+, \tag{17.10}$$

where $\tilde{D} \triangleq D_1 + BK D_2$.

The term $x(k) + D_2 w(k)$ in (17.8) captures the fact that the feedback signal transmission is noisy due to measurement noise, and the term $D_1 w(k)$ in (17.7) captures external disturbances to the plant. Clearly, (17.8) can be viewed as a generalization of (17.6) for which $K = E - \Lambda$ and $D_2 = D$.

Recall from [27, p. 735] that a matrix $A \in \mathbb{R}^{n \times n}$ is discrete-time semistable if and only if $\lim_{k \to \infty} A^k$ exists. Furthermore, if $A$ is discrete-time semistable, then the index of $I_n - A$ is zero or one, and thus $I_n - A$ is group invertible. The *group inverse* $(I_n - A)^\#$ of $I_n - A$ is a special case of the Drazin inverse $(I_n - A)^{\mathrm{D}}$ in the case in which $I_n - A$ has index zero or one [27, p. 323]. In this case, $\lim_{k \to \infty} A^k = I_n - (I_n - A)(I_n - A)^\#$ [27, p. 735]. Hence, the semistability requirement for the closed-loop control system can be interpreted as the convergence requirement for linear iterative algorithms.

The performance cost functional (17.9) represents a trade-off between the fastest convergence rate term $(x(k) - x_\infty)^{\mathrm{T}} R_1 (x(k) - x_\infty)$ and the control energy penalty term $(u(k) - u_\infty)^{\mathrm{T}} R_2 (u(k) - u_\infty)$. The cost functional in (ii) thus captures a trade-off between state convergence and control effort injected into the network system [155]. Although the fastest convergence rate is a good metric for optimal design, one would have to employ significant control effort to achieve this under certain constraints that arise due to limited resources. Thus, we need to balance both quantities in an optimal design.

The following lemma is needed.

**Lemma 17.1.** Let $\{x_n\}_{n=0}^\infty$ be a sequence such that $\lim_{n \to \infty} x_n = a$, where $-\infty \le a \le \infty$. Then

$$\lim_{n \to \infty} \frac{1}{n+1} \sum_{i=0}^n x_i = a.$$

**Proof.** If $-\infty < a < \infty$, then, for every $\varepsilon > 0$, there exists a positive integer $N_1$ such that $|x_n - a| < \varepsilon$ for all $n > N_1$. In this case, for every $n > N_1$,

$$\left| \frac{1}{n+1} \sum_{i=0}^n x_i - a \right| \le \sum_{i=0}^n \frac{1}{n+1} |x_i - a|$$

$$= \sum_{i=0}^{N_1} \frac{1}{n+1} |x_i - a| + \sum_{i=N_1+1}^n \frac{1}{n+1} |x_i - a|$$

$$< \frac{N_1+1}{n+1} M + \frac{n - N_1}{n+1} \varepsilon$$

$$< \frac{N_1+1}{n+1} M + \varepsilon,$$

where $M = \max_{0 \le i \le N_1} \{|x_i - a|\}$. Note that $\lim_{n \to \infty} \frac{N_1+1}{n+1} M = 0$. Then it follows that, for every $\varepsilon > 0$, there exists a positive integer $N_2$ such that $\frac{N_1+1}{n+1} M < \varepsilon$ for all $n > N_2$. Thus, for every $\varepsilon > 0$, there exists a positive integer $N = \max\{N_1, N_2\}$ such that $|\frac{1}{n+1} \sum_{i=0}^{n} x_i - a| < 2\varepsilon$ for all $n > N$. Hence, $\lim_{n \to \infty} \frac{1}{n+1} \sum_{i=0}^{n} x_i = a$.

Alternatively, consider the case where $a = \infty$. To prove that

$$\lim_{n \to \infty} \frac{1}{n+1} \sum_{i=0}^{n} x_i = \infty,$$

suppose, *ad absurdum*, that $a \neq \infty$. Since $\lim_{n \to \infty} x_n = \infty$, it follows that the sequence $\{\sum_{i=0}^{n} x_i\}_{n=0}^{\infty}$ must have a lower bound, and, hence,

$$\liminf_{n \to \infty} \frac{1}{n+1} \sum_{i=0}^{n} x_i \geq 0;$$

that is, there exists a positive integer $N_1$ such that $\frac{1}{n+1} \sum_{i=0}^{n} x_i \geq 0$ for all $n > N_1$. Consequently, if $\lim_{n \to \infty} \frac{1}{n+1} \sum_{i=0}^{n} x_i < \infty$, then there exists $M \in [0, \infty)$ such that $\lim_{n \to \infty} \frac{1}{n+1} \sum_{i=0}^{n} x_i = M$, which implies that $\limsup_{n \to \infty} \frac{1}{n+1} \sum_{i=0}^{n} x_i = M$. However, it follows from $\lim_{n \to \infty} x_n = \infty$ that there exists a positive integer $N_2$ such that $x_n > 2M+1$ for all $n > N_2$.

Let $N = \max\{N_1, N_2\}$. Then, for every $n > 2N + 1$,

$$\frac{1}{n+1} \sum_{i=0}^{n} x_i = \frac{1}{n+1} \sum_{i=0}^{N} x_i + \frac{1}{n+1} \sum_{i=N+1}^{n} x_i$$

$$= \frac{N+1}{n+1} \frac{1}{N+1} \sum_{i=0}^{N} x_i + \frac{1}{n+1} \sum_{i=N+1}^{n} x_i$$

$$\geq \frac{1}{n+1} \sum_{i=N+1}^{n} x_i$$

$$\geq \frac{n-N}{n+1}(2M+1) > \frac{1}{2}(2M+1)$$

$$= M + \frac{1}{2},$$

which implies that $\limsup_{n \to \infty} \frac{1}{n+1} \sum_{i=0}^{n} x_i > M$, contradicting the fact that $\limsup_{n \to \infty} \frac{1}{n+1} \sum_{i=0}^{n} x_i = M$. Thus, $\lim_{n \to \infty} \frac{1}{n+1} \sum_{i=0}^{n} x_i = \infty$.

Finally, for the case where $a = -\infty$, define $y_n = -x_n$. Then it follows from $\lim_{n \to \infty} x_n = -\infty$ that $\lim_{n \to \infty} y_n = \infty$. Hence, using similar arguments as given above, $\lim_{n \to \infty} \frac{1}{n+1} \sum_{i=0}^{n} y_i = \infty$; i.e., $-\lim_{n \to \infty} \frac{1}{n+1} \sum_{i=0}^{n} x_i = \infty$, and, hence, $\lim_{n \to \infty} \frac{1}{n+1} \sum_{i=0}^{n} x_i = -\infty$. $\square$

It follows from Lemma 17.1 that

$$J(K) = \lim_{k \to \infty} \mathbb{E}\Big[(x(k) - x_\infty)^{\mathrm{T}} R_1(x(k) - x_\infty) + (u(k) - u_\infty)^{\mathrm{T}} R_2(u(k) - u_\infty)\Big].$$

Hence, to obtain an optimal design for $K$ in (17.8), the expected state error $\mathbb{E}[(x(k) - x_\infty)^{\mathrm{T}} R_1(x(k) - x_\infty)]$ and the expected control input error $\mathbb{E}[(u(k) - u_\infty)^{\mathrm{T}} R_2(u(k) - u_\infty)]$ need to be minimized simultaneously. It is important to note that $x_\infty$ is not a fixed equilibrium point for the closed-loop system but rather depends on $K$ and $x(0)$, as required by (i) and (iii). As shown in Chapter 12, this problem cannot be converted into a classical LQR problem by shifting the equilibrium point to the origin.

Next, if we view the semistable control problem as a dynamic game, then the requirement that $x_\infty = \frac{1}{\mathbf{u}^{\mathrm{T}}\mathbf{v}}\mathbf{u}\mathbf{v}^{\mathrm{T}}\mathbb{E}[x(0)]$ represents the desired *correlated equilibria* [276] for cooperative games, which connects both control theory and game theory. For coordinated resource allocation, the matrix $\frac{1}{\mathbf{u}^{\mathrm{T}}\mathbf{v}}\mathbf{u}\mathbf{v}^{\mathrm{T}}$ represents the desired final portion of the total resource vector $\mathbb{E}[x(0)]$. Thus, we can disseminate the total resource to every node in the network by weighing the priority for each node.

Let $\mathbf{u} = [u_1, \ldots, u_q]^{\mathrm{T}}$ and $\mathbf{v} = [v_1, \ldots, v_q]^{\mathrm{T}}$ be such that $\mathbf{u}^{\mathrm{T}}\mathbf{v} \neq 0$, and note that

$$\frac{1}{\mathbf{u}^{\mathrm{T}}\mathbf{v}}\mathbf{u}\mathbf{v}^{\mathrm{T}} = \frac{1}{\sum_{i=1}^{q} u_i v_i} \begin{bmatrix} u_1 v_1 & u_1 v_2 & \cdots & u_1 v_q \\ u_2 v_1 & u_2 v_2 & \cdots & u_2 v_q \\ \vdots & \vdots & \ddots & \vdots \\ u_q v_1 & u_q v_2 & \cdots & u_q v_q \end{bmatrix}.$$

The term $\frac{1}{\mathbf{u}^{\mathrm{T}}\mathbf{v}}\mathbf{u}\mathbf{v}^{\mathrm{T}}$ contains a *correlated* steady-state resource distribution characterized by $u_i v_j/(\sum_{i=1}^{q} u_i v_i)$, which is reminiscent of the *Nash equilibrium* arising in game theory. Hence, our control problem can be cast as a game-theoretic problem involving a cooperative network in which every node exchanges resource information with its neighbors that seek a trade-off for its resource. Finally, note that if $\mathbf{u} = \frac{1}{\sqrt{q}}[1, \ldots, 1]^{\mathrm{T}}$, then this problem collapses to a consensus control problem [238, 274] over networks.

Since the proposed semistable control problem is similar to the stochastic version of the LQR problem, one may surmise that this controller can be designed by using the Kalman decomposition of $(A, B)$. That is, if $(A, B)$ is not stabilizable, then we can decompose $(A, B)$ so that we have a controllable subspace and an uncontrollable subspace. For the controllable subspace, we can design an LQR controller so that the closed-loop subsystem is asymptotically stable. For the uncontrollable subspace, the subsystem is either asymptotically stable or neutrally stable. It is clear that in this case the steady state $x_\infty$ is determined by the transformation matrix employed in the Kalman decomposition.

However, such a method has a major deficiency in that the transformation will destroy the network structure given in (17.1). Furthermore, this technique does not yield a suitable transformation matrix that guarantees that $x_\infty = \frac{1}{\mathbf{u}^T \mathbf{v}} \mathbf{u} \mathbf{v}^T \mathbb{E}[x(0)]$ for a given $\mathbf{u}$ and $\mathbf{v}$. This is because the transformation matrix depends only on $(A, B)$ and not on $x_\infty$. Finally, the standard LQR problem requires that $(A, B)$ be stabilizable. However, as shown in Chapters 10–12 and 16, for the proposed semistable control problem, this condition is not the correct condition that guarantees well-posedness for this problem.

## 17.4 Semistable Linear-Quadratic Control Theory

In this section, we develop several technical results to convert the optimal control problem proposed in Section 17.3 into an equivalent optimization problem. We first need the following assumption.

**Assumption 17.1.** The system initial condition $x(0)$ and the system disturbance $w(k)$ are independent for all $k \in \overline{\mathbb{Z}}_+$.

Suppose $\tilde{A}$ is discrete-time semistable; that is, criterion (i) in Section 17.3 holds. Then it follows from (17.9) and Lemma 17.1 that

$$
\begin{aligned}
J(K) \\
&= \lim_{k \to \infty} \mathbb{E}\Big[(x(k) - x_\infty)^T R_1 (x(k) - x_\infty) + (u(k) - u_\infty)^T R_2 (u(k) - u_\infty)\Big] \\
&= \lim_{k \to \infty} \mathbb{E}\Big[(x(k) - x_\infty)^T (R_1 + K^T R_2 K)(x(k) - x_\infty)\Big] \\
&= \lim_{k \to \infty} \mathbb{E}\Big[x^T(k) \tilde{R} x(k)\Big] - x_\infty^T \tilde{R} x_\infty \\
&= \lim_{k \to \infty} \operatorname{tr} \mathbb{E}\Big[x(k) x^T(k)\Big] \tilde{R} - \operatorname{tr}\Big(x_\infty x_\infty^T\Big) \tilde{R} \\
&= \lim_{k \to \infty} \operatorname{tr} \mathbb{E}\Big[x(k) x^T(k)\Big] \tilde{R} - \lim_{k \to \infty} \operatorname{tr}\Big[\mathbb{E}[x(k)] \mathbb{E}[x^T(k)]\Big] \tilde{R} \\
&= \lim_{k \to \infty} \operatorname{tr} Q(k) \tilde{R},
\end{aligned}
\tag{17.11}
$$

where $\tilde{R} = R_1 + K^T R_2 K$ and $Q(k) = \mathbb{E}[x(k) x^T(k)] - \mathbb{E}[x(k)] \mathbb{E}[x^T(k)]$ is the covariance matrix for $x(k)$.

The following result gives a recursive formula for computing $Q(k)$.

**Proposition 17.1.** Consider the closed-loop system (17.10), and assume that Assumption 17.1 holds. Then $Q(k)$ satisfies the Lyapunov difference equation

$$Q(k+1) = \tilde{A}Q(k)\tilde{A}^{\mathrm{T}} + \tilde{D}\tilde{D}^{\mathrm{T}}, \quad Q(0) = Q_0, \quad k \in \overline{\mathbb{Z}}_+, \tag{17.12}$$

where $Q_0 = \mathbb{E}\left[x(0)x^{\mathrm{T}}(0)\right] - \mathbb{E}[x(0)]\mathbb{E}[x^{\mathrm{T}}(0)]$. Furthermore, (17.12) is equivalent to

$$Q(k+1) = \tilde{A}^{k+1}Q(0)(\tilde{A}^{k+1})^{\mathrm{T}} + \sum_{i=0}^{k} \tilde{A}^i \tilde{D}\tilde{D}^{\mathrm{T}}(\tilde{A}^i)^{\mathrm{T}}. \tag{17.13}$$

**Proof.** Consider the closed-loop system given by (17.10), and note that

$$x(k) = \tilde{A}^k x(0) + \sum_{i=0}^{k-1} \tilde{A}^{k-1-i}\tilde{D}w(i).$$

Now, it follows that

$$\begin{aligned}
Q(k+1) &= \mathbb{E}\left[x(k+1)x^{\mathrm{T}}(k+1)\right] - \mathbb{E}[x(k+1)]\mathbb{E}[x^{\mathrm{T}}(k+1)] \\
&= \mathbb{E}\left[(\tilde{A}x(k) + \tilde{D}w(k))(\tilde{A}x(k) + \tilde{D}w(k))^{\mathrm{T}}\right] - \tilde{A}\mathbb{E}[x(k)]\mathbb{E}[x^{\mathrm{T}}(k)]\tilde{A}^{\mathrm{T}} \\
&= \tilde{A}\mathbb{E}[x(k)x^{\mathrm{T}}(k)]\tilde{A}^{\mathrm{T}} + \mathbb{E}[\tilde{A}x(k)w^{\mathrm{T}}(k)\tilde{D}^{\mathrm{T}} \\
&\quad + \tilde{D}w(k)x^{\mathrm{T}}(k)\tilde{A}^{\mathrm{T}}] + \tilde{D}\mathbb{E}[w(k)w^{\mathrm{T}}(k)]\tilde{D}^{\mathrm{T}} - \tilde{A}\mathbb{E}[x(k)]\mathbb{E}[x^{\mathrm{T}}(k)]\tilde{A}^{\mathrm{T}} \\
&= \tilde{A}Q(k)\tilde{A}^{\mathrm{T}} + \tilde{D}\tilde{D}^{\mathrm{T}} \\
&\quad + \mathbb{E}\left[\tilde{A}^{k+1}x(0)w^{\mathrm{T}}(k)\tilde{D}^{\mathrm{T}} + \sum_{i=0}^{k-1}\tilde{A}^{k-i}\tilde{D}w(i)w^{\mathrm{T}}(k)\tilde{D}^{\mathrm{T}} \right. \\
&\quad \left. + \tilde{D}w(k)x^{\mathrm{T}}(0)(\tilde{A}^{\mathrm{T}})^{k+1} + \sum_{i=0}^{k-1}\tilde{D}w(k)w^{\mathrm{T}}(i)\tilde{D}^{\mathrm{T}}(\tilde{A}^{\mathrm{T}})^{k-i}\right] \\
&= \tilde{A}Q(k)\tilde{A}^{\mathrm{T}} + \tilde{D}\tilde{D}^{\mathrm{T}},
\end{aligned}$$

where we used the facts that

$$\mathbb{E}[x(0)w^{\mathrm{T}}(k)] = 0, \quad \mathbb{E}[w(k)x^{\mathrm{T}}(0)] = 0, \quad \mathbb{E}[w(i)w^{\mathrm{T}}(j)] = 0, \quad i \neq j,$$
$$\mathbb{E}[w(k)w^{\mathrm{T}}(k)] = I_n.$$

Finally, recursively using (17.12) yields (17.13). $\qquad\square$

The following result is motivated by Proposition 16.3.

**Lemma 17.2.** Assume that Assumption 17.1 holds. If $\tilde{A}$ is discrete-time semistable, then $\lim_{k\to\infty}\sum_{i=0}^{k}\tilde{A}^i\tilde{D}\tilde{D}^{\mathrm{T}}(\tilde{A}^i)^{\mathrm{T}}$ exists if and only if $\mathcal{N}(I_n - \tilde{A}^{\mathrm{T}}) \subseteq \mathcal{N}(\tilde{D}^{\mathrm{T}})$.

**Proof.** First, it follows from $\operatorname{spec}(\tilde{A}) = \operatorname{spec}(\tilde{A}^{\mathrm{T}})$ (see Proposition 4.4.5 of [27, p. 263]) that if $\tilde{A}$ is discrete-time semistable, then $\tilde{A}^{\mathrm{T}}$ is discrete-time semistable. In this case, it follows that either $\rho(\tilde{A}^{\mathrm{T}}) < 1$ or there exists an invertible matrix $S \in \mathbb{R}^{n \times n}$ such that

$$\tilde{A}^{\mathrm{T}} = S \begin{bmatrix} J & 0 \\ 0 & I_{n-r} \end{bmatrix} S^{-1},$$

where $J \in \mathbb{R}^{r \times r}$, $r = \operatorname{rank}(\tilde{A}^{\mathrm{T}})$, and $\rho(J) < 1$. Now, if $\rho(\tilde{A}^{\mathrm{T}}) < 1$, then $\mathcal{N}(\tilde{A}^{\mathrm{T}} - I_n) = \{0\} \subseteq \mathcal{N}(\tilde{D}^{\mathrm{T}})$.

Alternatively, if $1 \in \operatorname{spec}(\tilde{A}^{\mathrm{T}})$, then

$$\mathcal{N}(\tilde{A}^{\mathrm{T}} - I_n) = \left\{ x \in \mathbb{R}^n : x = S[0_{1 \times r}, y^{\mathrm{T}}]^{\mathrm{T}}, y \in \mathbb{R}^{n-r} \right\}.$$

Now,

$$\sum_{k=0}^{\infty} \tilde{A}^k \tilde{D} \tilde{D}^{\mathrm{T}} (\tilde{A}^{\mathrm{T}})^k = (S^{-1})^{\mathrm{T}} \sum_{k=0}^{\infty} \begin{bmatrix} (J^k)^{\mathrm{T}} & 0 \\ 0 & I_{n-r} \end{bmatrix} \hat{R} \begin{bmatrix} J^k & 0 \\ 0 & I_{n-r} \end{bmatrix} S$$

$$= (S^{-1})^{\mathrm{T}} \sum_{k=0}^{\infty} \begin{bmatrix} (J^k)^{\mathrm{T}} \hat{R}_1 J^k & (J^k)^{\mathrm{T}} \hat{R}_{12} \\ \hat{R}_{12}^{\mathrm{T}} J^k & \hat{R}_2 \end{bmatrix} S, \quad (17.14)$$

where

$$\hat{R} = S^{\mathrm{T}} \tilde{D} \tilde{D}^{\mathrm{T}} S = \begin{bmatrix} \hat{R}_1 & \hat{R}_{12} \\ \hat{R}_{12}^{\mathrm{T}} & \hat{R}_2 \end{bmatrix}.$$

Next, it follows from (17.14) that $\sum_{k=0}^{\infty} \tilde{A}^k \tilde{D} \tilde{D}^{\mathrm{T}} (\tilde{A}^{\mathrm{T}})^k$ exists if and only if $\hat{R}_2 = 0$ or, equivalently,

$$[0_{1 \times r}, y^{\mathrm{T}}] \hat{R} [0_{1 \times r}, y^{\mathrm{T}}]^{\mathrm{T}} = 0, \quad y \in \mathbb{R}^{n-r},$$

which is further equivalent to $x^{\mathrm{T}} \tilde{D} \tilde{D}^{\mathrm{T}} x = 0$, $x \in \mathcal{N}(\tilde{A}^{\mathrm{T}} - I_n)$. Hence, $\mathcal{N}(\tilde{A}^{\mathrm{T}} - I_n) \subseteq \mathcal{N}(\tilde{D}^{\mathrm{T}})$.

Finally, the converse is immediate by reversing the steps given above. $\square$

**Proposition 17.2.** Assume that Assumption 17.1 holds. If $\tilde{A}$ is discrete-time semistable, then $Q = \lim_{k \to \infty} Q(k)$ exists if and only if $\mathcal{N}(I_n - \tilde{A}^{\mathrm{T}}) \subseteq \mathcal{N}(\tilde{D}^{\mathrm{T}})$. Furthermore, if $\mathcal{N}(I_n - \tilde{A}^{\mathrm{T}}) \subseteq \mathcal{N}(\tilde{D}^{\mathrm{T}})$, then $Q$ is given by

$$Q = \hat{Q} + [I_n - (I_n - \tilde{A})(I_n - \tilde{A})^\#] Q(0) [I_n - (I_n - \tilde{A})(I_n - \tilde{A})^\#]^{\mathrm{T}}, \quad (17.15)$$

where $Q(0) = \mathbb{E}[x(0)x^{\mathrm{T}}(0)] - \mathbb{E}[x(0)]\mathbb{E}[x^{\mathrm{T}}(0)]$ and

$$\hat{Q} = \lim_{k \to \infty} \sum_{i=0}^{k} \tilde{A}^i \tilde{D} \tilde{D}^{\mathrm{T}} (\tilde{A}^i)^{\mathrm{T}}.$$

**Proof.** Since $\tilde{A}$ is discrete-time semistable, it follows that $\lim_{k\to\infty} \tilde{A}^{k+1}$ exists. Hence, by (17.13) in Proposition 17.1,

$$\lim_{k\to\infty} Q(k) = \lim_{k\to\infty} Q(k+1)$$

exists if and only if $\lim_{k\to\infty} \sum_{i=0}^{k} \tilde{A}^i \tilde{D} \tilde{D}^{\mathrm{T}} (\tilde{A}^i)^{\mathrm{T}}$ exists. Now, it follows from Lemma 17.2 that $\lim_{k\to\infty} Q(k)$ exists if and only if $\mathcal{N}(I_n - \tilde{A}^{\mathrm{T}}) \subseteq \mathcal{N}(\tilde{D}^{\mathrm{T}})$.

To prove the second assertion, note that it follows from (17.13) in Proposition 17.1 and

$$\lim_{k\to\infty} \tilde{A}^k = \lim_{k\to\infty} \tilde{A}^{k+1} = I_n - (I_n - \tilde{A})(I_n - \tilde{A})^{\#}$$

that

$$\begin{aligned}
Q &= \lim_{k\to\infty} Q(k) \\
&= \lim_{k\to\infty} Q(k+1) \\
&= \lim_{k\to\infty} \tilde{A}^{k+1} Q(0)(\tilde{A}^{k+1})^{\mathrm{T}} + \lim_{k\to\infty} \sum_{i=0}^{k} \tilde{A}^i \tilde{D} \tilde{D}^{\mathrm{T}} (\tilde{A}^i)^{\mathrm{T}} \\
&= [I_n - (I_n - \tilde{A})(I_n - \tilde{A})^{\#}] Q(0) [I_n - (I_n - \tilde{A})(I_n - \tilde{A})^{\#}]^{\mathrm{T}} + \hat{Q},
\end{aligned}$$

which completes the proof. $\qquad\square$

It follows from Proposition 17.2 that

$$\begin{aligned}
&\operatorname{tr} Q\tilde{R} \\
&= \operatorname{tr}[I_n - (I_n - \tilde{A})(I_n - \tilde{A})^{\#}] Q(0)[I_n - (I_n - \tilde{A})(I_n - \tilde{A})^{\#}]^{\mathrm{T}} \tilde{R} + \operatorname{tr} \hat{Q}\tilde{R},
\end{aligned}$$

(17.16)

which implies that $J(K)$ has a finite form. The proposed semistable LQR control design problem satisfying criteria (i) and (ii) in Section 17.3 involves an optimization problem with the objective function $\operatorname{tr} Q\tilde{R}$ subject to the discrete-time semistability of $\tilde{A}$. Note that we have not yet enforced criterion (iii) in the controller design. To incorporate criterion (iii) into our design, we need some additional constraints on $\tilde{A}$.

**Lemma 17.3.** Let $A \in \mathbb{R}^{n \times n}$ and let $\mathbf{u}, \mathbf{v} \in \mathbb{R}^n$ be unit vectors. Then $\lim_{k \to \infty} A^k = \frac{1}{\mathbf{u}^T \mathbf{v}} \mathbf{u} \mathbf{v}^T$ and $\mathbf{u}^T \mathbf{v} \neq 0$ if and only if $\operatorname{rank}(A - I_n) = n - 1$, $A\mathbf{u} = \mathbf{u}$, $\mathbf{v}^T A = \mathbf{v}^T$, and $A$ is discrete-time semistable.

**Proof.** To show sufficiency, note that if $A$ is discrete-time semistable, then it follows from (vi) of Proposition 11.10.2 of [27, p. 735] that

$$\lim_{k \to \infty} A^k x = [I_n - (I_n - A)(I_n - A)^{\#}]x, \quad x \in \mathbb{R}^n.$$

Next, we show that $\mathcal{R}(I_n - A) \subseteq \mathcal{N}(\mathbf{u}\mathbf{v}^T)$ and $\mathcal{N}(I_n - A) \subseteq \mathcal{R}(\mathbf{u}\mathbf{v}^T)$. Let $v \in \mathcal{R}(I_n - A)$. Then $v = z - Az$ for some $z \in \mathbb{R}^n$. Since $\mathbf{v}^T A = \mathbf{v}^T$, it follows that $\mathbf{u}\mathbf{v}^T v = \mathbf{u}\mathbf{v}^T z - \mathbf{u}\mathbf{v}^T A z = \mathbf{u}\mathbf{v}^T z - \mathbf{u}\mathbf{v}^T z = 0$, which implies that $v \in \mathcal{N}(\mathbf{u}\mathbf{v}^T)$. Hence, $\mathcal{R}(I_n - A) \subseteq \mathcal{N}(\mathbf{u}\mathbf{v}^T)$. Note that it follows from Lemma 3.2 of [160] that

$$\mathcal{N}(I_n - (I_n - A)(I_n - A)^{\#}) = \mathcal{R}(I_n - A).$$

Thus,

$$\mathcal{N}(I_n - (I_n - A)(I_n - A)^{\#}) \subseteq \mathcal{N}(\mathbf{u}\mathbf{v}^T).$$

Since $I_n - A$ is group invertible, it follows from Corollary 2.3.2 of [27, p. 99] and Fact 3.6.1 of [27, p. 191] that $\operatorname{rank}(I_n - A) = n - 1$ implies that

$$\dim \mathcal{N}(I_n - A) = n - \dim \mathcal{R}(I_n - A) = n - \operatorname{rank}(I_n - A) = 1,$$

where $\dim \mathcal{X}$ denotes the dimension of the set $\mathcal{X}$. Hence, if $v \in \mathcal{N}(I_n - A)$, then it follows from $\mathbf{u} \in \mathcal{N}(I_n - A)$ that $v = c\mathbf{u}$ for some $c \in \mathbb{R}$. For every $y \in \mathbb{R}^n$ satisfying $\mathbf{v}^T y \neq 0$, we have $v = c\mathbf{u} = \frac{c}{\mathbf{v}^T y}(\mathbf{v}^T y)\mathbf{u} = \frac{c}{\mathbf{v}^T y}\mathbf{u}\mathbf{v}^T y$. This implies that $v = \mathbf{u}\mathbf{v}^T z$ for $z = \frac{c}{\mathbf{v}^T y} y \in \mathbb{R}^n$. Hence, $v \in \mathcal{R}(\mathbf{u}\mathbf{v}^T)$, and, consequently, $\mathcal{N}(I_n - A) \subseteq \mathcal{R}(\mathbf{u}\mathbf{v}^T)$. Note that it follows from Lemma 3.2 of [160] that

$$\mathcal{R}(I_n - (I_n - A)(I_n - A)^{\#}) = \mathcal{N}(I_n - A).$$

Thus,

$$\mathcal{R}(I_n - (I_n - A)(I_n - A)^{\#}) \subseteq \mathcal{R}(\mathbf{u}\mathbf{v}^T).$$

Next, note that $\mathcal{N}(I_n - A) = \operatorname{span}\{\mathbf{u}\}$ and $\mathcal{N}(I_n - A^T) = \operatorname{span}\{\mathbf{v}\}$. Furthermore, it follows from equation (2.4.14) of [27, p. 103] that $\mathcal{N}(I_n - A^T)^{\perp} = \mathcal{R}(I_n - A)$, where $\mathcal{S}^{\perp}$ denotes the orthogonal complement of the set $\mathcal{S}$. If $\mathbf{u}^T \mathbf{v} = 0$, then it follows that

$$\mathcal{N}(I_n - A) \subseteq \mathcal{N}(I_n - A^T)^{\perp} = \mathcal{R}(I_n - A),$$

and, hence,

$$\mathcal{N}(I_n - A) \cap \mathcal{R}(I_n - A) = \mathcal{N}(I_n - A) = \text{span}\{\mathbf{u}\}.$$

This contradicts the fact that $\mathcal{N}(I_n - A) \cap \mathcal{R}(I_n - A) = \{0\}$ by (vii) of Fact 3.6.1 of [27, p. 191]. Hence, $\mathbf{u}^T\mathbf{v} \neq 0$.

Since $\mathbf{u}^T\mathbf{v} \neq 0$, it follows that

$$\mathcal{R}(I_n - (I_n - A)(I_n - A)^\#) \subseteq \mathcal{R}(\mathbf{uv}^T) = \mathcal{R}\left(\frac{1}{\mathbf{u}^T\mathbf{v}}\mathbf{uv}^T\right)$$

and

$$\mathcal{N}(I_n - (I_n - A)(I_n - A)^\#) \subseteq \mathcal{N}(\mathbf{uv}^T) = \mathcal{N}\left(\frac{1}{\mathbf{u}^T\mathbf{v}}\mathbf{uv}^T\right).$$

Note that since both $I_n - (I_n - A)(I_n - A)^\#$ and $\frac{1}{\mathbf{u}^T\mathbf{v}}\mathbf{uv}^T$ are idempotent matrices, it follows from Proposition 3.1 of [160] that $I_n - (I_n - A)(I_n - A)^\# = \frac{1}{\mathbf{u}^T\mathbf{v}}\mathbf{uv}^T$. Thus, $\lim_{k\to\infty} A^k = \frac{1}{\mathbf{u}^T\mathbf{v}}\mathbf{uv}^T$.

To show necessity, first note that $\lim_{k\to\infty} A^k x$ exists for every $x \in \mathbb{R}^n$ if and only if $\lim_{k\to\infty} A^k$ exists. By definition, $A$ is discrete-time semistable. Since $A$ is discrete-time semistable, it follows from [164] that either $A$ is Schur, that is, $\rho(A) < 1$, or there exists an invertible matrix $S \in \mathbb{C}^{n \times n}$ such that

$$A = S \begin{bmatrix} J & 0 \\ 0 & I_{n-r} \end{bmatrix} S^{-1},$$

where $J \in \mathbb{C}^{r \times r}$, $r = \text{rank}(A - I_n)$, and $J$ is Schur. Now, if $A$ is Schur, then $\lim_{k\to\infty} A^k = 0$, which implies that $\mathbf{u} = 0$ or $\mathbf{v} = 0$. This contradicts the assumption that both $\mathbf{u}$ and $\mathbf{v}$ are unit vectors. Hence, $A$ is not Schur. If $A = I_n$, then $\lim_{k\to\infty} A^k = I_n$, which contradicts the assumption of the lemma. Thus, $A \neq I_n$.

Now, assume that $A$ is neither Schur nor $I_n$. Let $u_i$, $i = 1, \ldots, n$, be the column vectors of $S$ and $v_i^T$, $i = 1, \ldots, n$, be the row vectors of $S^{-1}$. Then it follows that

$$\lim_{k\to\infty} A^k = \lim_{k\to\infty} S \begin{bmatrix} J^k & 0 \\ 0 & I_{n-r} \end{bmatrix} S^{-1} = S \begin{bmatrix} 0 & 0 \\ 0 & I_{n-r} \end{bmatrix} S^{-1} = \sum_{i=r+1}^{n} u_i v_i^T.$$

Clearly $S$ is a full-rank matrix and

$$\text{rank}\left(\sum_{i=r+1}^{n} u_i v_i^T\right) = \text{rank} \begin{bmatrix} 0 & 0 \\ 0 & I_{n-r} \end{bmatrix} = n - r.$$

However, by assumption, $\mathcal{R}(\sum_{i=r+1}^{n} u_i v_i^{\mathrm{T}}) \subseteq \mathrm{span}\{\mathbf{u}\}$, and, hence,

$$1 \leq \dim \mathcal{R}\left(\sum_{i=r+1}^{n} u_i v_i^{\mathrm{T}}\right) \leq \dim \mathrm{span}\{\mathbf{u}\} = 1;$$

that is, $\mathrm{rank}(\sum_{i=r+1}^{n} u_i v_i^{\mathrm{T}}) = n - r = 1$. Hence, $\mathrm{rank}\,(A - I_n) = r = n - 1$. Note that each matrix $u_i v_i^{\mathrm{T}}$ is a rank-one matrix. In this case, $\sum_{i=r+1}^{n} u_i v_i^{\mathrm{T}} = u_{r+1} v_{r+1}^{\mathrm{T}} = \mathbf{u}\mathbf{v}^{\mathrm{T}}$. Thus, $u_{r+1} = \mathbf{u}\mathbf{v}^{\mathrm{T}} v_{r+1}$ and $v_{r+1} = \mathbf{v}\mathbf{u}^{\mathrm{T}} u_{r+1}$, which implies that $u_{r+1}$ is a multiple of $\mathbf{u}$ and $v_{r+1}$ is a multiple of $\mathbf{v}$. Consequently, 1 is a simple eigenvalue of $A$, and $\mathbf{v}$ and $\mathbf{u}$ are its associated left and right eigenvectors, respectively. Thus, $A\mathbf{u} = \mathbf{u}$ and $\mathbf{v}^{\mathrm{T}} A = \mathbf{v}^{\mathrm{T}}$, which completes the proof. $\qquad\square$

It follows from Proposition 17.2 and Lemma 17.3 that the semistable control design problem satisfying criteria (i)–(iii) can be cast as the following optimization problem:

$$\min_{K} \mathrm{tr}\, Q\tilde{R} \qquad (17.17)$$

subject to

$$\tilde{A} \text{ is discrete-time semistable}, \quad \mathrm{rank}(\tilde{A} - I_n) = n - 1, \qquad (17.18)$$

$$\tilde{A}\mathbf{u} = \mathbf{u}, \quad \mathbf{v}^{\mathrm{T}} \tilde{A} = \mathbf{v}^{\mathrm{T}}, \qquad (17.19)$$

$$\mathcal{N}(I_n - \tilde{A}^{\mathrm{T}}) \subseteq \mathcal{N}(\tilde{D}^{\mathrm{T}}). \qquad (17.20)$$

Note that this optimization problem is challenging to solve numerically because (i) $Q$ involves the complex operation $(I_n - \tilde{A})^{\#}$, (ii) the null space condition (17.20) is challenging to check in practice, and (iii) the condition of discrete-time semistability cannot be cast as an equality or inequality constraint. One possible way to address (iii) is to compute the eigenvalues of $\tilde{A}$ when minimizing (17.17). However, the existing numerical algorithms for computing *all* of the eigenvalues for a large sparse matrix (e.g., the Arnoldi and Lanczos algorithms) can take too much time to converge and require a lot of memory to store the data when executing (17.17). Thus, this method is untenable for real-time implementation. In the next section, we address these issues by introducing a new Lyapunov equation that characterizes discrete-time semistability and further simplifies (17.17)–(17.20).

## 17.5 Characterization of Semistability via Semicontrollability

To characterize the discrete-time semistability constraint (17.19), we first need to extend our notions of semistabilizability and semidetectability introduced in Chapter 16 for discrete-time linear systems.

**Definition 17.2.** Let $A \in \mathbb{R}^{n \times n}$ and $C \in \mathbb{R}^{l \times n}$. The pair $(A, C)$ is *discrete-time k-semiobservable* if there exists a nonnegative integer $k$ such that

$$k = \min \left\{ l \in \overline{\mathbb{Z}}_+ : \bigcap_{i=1}^{n} \mathcal{N}\left( C(I_n - A)^{l+i-1} \right) = \mathcal{N}(I_n - A) \right\}.$$

**Definition 17.3.** Let $A \in \mathbb{R}^{n \times n}$ and $B \in \mathbb{R}^{n \times l}$. The pair $(A, B)$ is *discrete-time k-semicontrollable* if there exists a nonnegative integer $k$ such that

$$k = \min \left\{ l \in \overline{\mathbb{Z}}_+ : \sum_{i=1}^{n} \mathcal{R}\left( (I_n - A)^{l+i-1}B \right) = \mathcal{R}(I_n - A) \right\}.$$

The notion of $k$-semiobservability is a generalization of semiobservability from $k = 0, 1$ to an arbitrary integer $k \geq 0$. For $k \geq 2$, this new notion is weaker than the notions of semiobservability introduced in Chapter 16 and weak semiobservability introduced in [155, 156].

Next, we present necessary and sufficient conditions for guaranteeing discrete-time semistability of $\tilde{A}$ by using a Lyapunov-like equation. First, however, we require the following assumption and several key lemmas and propositions.

**Assumption 17.2.** $x(0)$ is a random variable having a covariance matrix $V$; that is, $\mathbb{E}[x(0)x^{\mathrm{T}}(0)] - \mathbb{E}[x(0)]\mathbb{E}[x^{\mathrm{T}}(0)] = V$.

First, note that the duality between $k$-semicontrollability and $k$-semiobservability follows from the following lemma.

**Lemma 17.4.** Let $\mathcal{S}_1, \mathcal{S}_2 \subseteq \mathbb{R}^n$ be subspaces. Then $\mathcal{S}_1 = \mathcal{S}_2$ if and only if $\mathcal{S}_1^{\perp} = \mathcal{S}_2^{\perp}$.

**Proof.** It follows from (i) and (ii) of Fact 2.9.14 in [27, p. 121] that $\mathcal{S}_1 \subseteq \mathcal{S}_2$ if and only if $\mathcal{S}_2^{\perp} \subseteq \mathcal{S}_1^{\perp}$. The result is now immediate.  $\square$

**Lemma 17.5.** Let $A \in \mathbb{R}^{n \times n}$ and $C \in \mathbb{R}^{l \times n}$. Then $(A, C)$ is discrete-time $k$-semiobservable if and only if $(A^{\mathrm{T}}, C^{\mathrm{T}})$ is discrete-time $k$-semicontrollable.

**Proof.** It follows from Theorem 2.4.3 in [27, p. 103] that $\mathcal{N}(A) =$

$\mathcal{R}(A^{\mathrm{T}})^{\perp}$. Hence, if

$$\bigcap_{i=1}^{n} \mathcal{N}(C(I_n - A)^{k+i-1}) = \mathcal{N}(I_n - A),$$

then

$$\bigcap_{i=1}^{n} \mathcal{R}((I_n - A^{\mathrm{T}})^{k+i-1}C^{\mathrm{T}})^{\perp} = \mathcal{R}(I_n - A^{\mathrm{T}})^{\perp}.$$

By Fact 2.9.16 in [27, p. 121],

$$\bigcap_{i=1}^{n} \mathcal{R}((I_n - A^{\mathrm{T}})^{k+i-1}C^{\mathrm{T}})^{\perp} = \left[\sum_{i=1}^{n} \mathcal{R}((I_n - A^{\mathrm{T}})^{k+i-1}C^{\mathrm{T}})\right]^{\perp},$$

and, hence,

$$\left[\sum_{i=1}^{n} \mathcal{R}((I_n - A^{\mathrm{T}})^{k+i-1}C^{\mathrm{T}})\right]^{\perp} = \mathcal{R}(I_n - A^{\mathrm{T}})^{\perp}.$$

Now, it follows from Lemma 17.4 that

$$\sum_{i=1}^{n} \mathcal{R}((I_n - A^{\mathrm{T}})^{k+i-1}C^{\mathrm{T}}) = \mathcal{R}(I_n - A^{\mathrm{T}}),$$

and, hence, by definition, $(A^{\mathrm{T}}, C^{\mathrm{T}})$ is discrete-time $k$-semicontrollable. The proof of the converse follows by reversing the steps given above. $\square$

The next result connects the notion of discrete-time $k$-semiobservability with a Lyapunov-type equation.

**Proposition 17.3.** Let $A \in \mathbb{R}^{n \times n}$. If there exist an $n \times n$ matrix $P = P^{\mathrm{T}} \geq 0$ and an $m \times n$ matrix $C \in \mathbb{R}^{m \times n}$ such that $(A, C)$ is discrete-time $k$-semiobservable and

$$0 = ((I_n - A)^k)^{\mathrm{T}}(A^{\mathrm{T}}PA + R - P)(I_n - A)^k, \qquad (17.21)$$

where $R = C^{\mathrm{T}}C$, then (i) $\mathcal{N}(P(I_n - A)^k) \subseteq \mathcal{N}(I_n - A) \subseteq \mathcal{N}(R(I_n - A)^k)$ and (ii) $\mathcal{N}(I_n - A) \cap \mathcal{R}(I_n - A) = \{0\}$.

**Proof.** Note that $(A, C)$ is discrete-time $k$-semiobservable if and only if

$$\bigcap_{i=1}^{n} \mathcal{N}(C(A - I_n)^{k+i-1}) = \mathcal{N}(A - I_n).$$

We claim that

$$\mathcal{N}(((I_n - A)^m)^{\mathrm{T}} R(I_n - A)^{m+j}) = \mathcal{N}(R(I_n - A)^{m+j})$$

for every $m = 0, 1, \ldots$ and every $j = 0, 1, \ldots$. To see this, let

$$x \in \mathcal{N}(((I_n - A)^m)^{\mathrm{T}} R(I_n - A)^{m+j}),$$

that is, $((I_n - A)^m)^{\mathrm{T}} R(I_n - A)^{m+j} x = 0$, and note that

$$((I_n - A)^m)^{\mathrm{T}} R(I_n - A)^{m+j} = ((I_n - A)^m)^{\mathrm{T}} R(I_n - A)^m (I_n - A)^j.$$

Define $y = (I_n - A)^j x$, and note that $((I_n - A)^m)^{\mathrm{T}} R(I_n - A)^m y = 0$. Since $R$ is nonnegative definite, it follows that $((I_n - A)^m)^{\mathrm{T}} R(I_n - A)^m y = 0$ if and only if $R(I_n - A)^m y = 0$, which implies that $R(I_n - A)^{m+j} x = 0$. Hence,

$$\mathcal{N}(((I_n - A)^m)^{\mathrm{T}} R(I_n - A)^{m+j}) \subseteq \mathcal{N}(R(I_n - A)^{m+j})$$

for every $m = 0, 1, \ldots$ and every $j = 0, 1, \ldots$. Clearly,

$$\mathcal{N}(R(I_n - A)^{m+j}) \subseteq \mathcal{N}(((I_n - A)^m)^{\mathrm{T}} R(I_n - A)^{m+j}).$$

Thus, $\mathcal{N}(((I_n - A)^m)^{\mathrm{T}} R(I_n - A)^{m+j}) = \mathcal{N}(R(I_n - A)^{m+j})$ for every $m = 0, 1, \ldots$ and every $j = 0, 1, \ldots$.

Now, letting $m = k$ and $j = i - 1$, it follows that

$$\bigcap_{i=1}^{n} \mathcal{N}(((I_n - A)^k)^{\mathrm{T}} R(I_n - A)^{k+i-1}) = \bigcap_{i=1}^{n} \mathcal{N}(R(I_n - A)^{k+i-1})$$

$$= \mathcal{N}(I_n - A). \qquad (17.22)$$

Let $\tilde{P} \triangleq ((I_n - A)^k)^{\mathrm{T}} P(I_n - A)^k$ and $\tilde{Q} \triangleq ((I_n - A)^k)^{\mathrm{T}} R(I_n - A)^k$. Then (17.21) becomes $\tilde{P} = A^{\mathrm{T}} \tilde{P} A + \tilde{Q}$ and (17.22) becomes $\bigcap_{i=1}^{n} \mathcal{N}(\tilde{Q}(I_n - A)^{i-1}) = \mathcal{N}(I_n - A)$. Now, the result follows directly from Lemma 16.1 by noting that $\mathcal{N}(\tilde{P}) = \mathcal{N}(P(I_n - A)^k)$ and $\mathcal{N}(\tilde{Q}) = \mathcal{N}(R(I_n - A)^k)$.   $\square$

The following result refines the form of $J(K)$ by combining the two separate terms in (17.16) into one compact form.

**Proposition 17.4.** Assume that Assumptions 17.1 and 17.2 hold, and assume that $\mathcal{N}(I_n - \tilde{A}^{\mathrm{T}}) \subseteq \mathcal{N}(\tilde{D}^{\mathrm{T}})$. If $\tilde{A}$ is discrete-time semistable, then $J(K) = \mathrm{tr}\,(W + V)\tilde{R}$, where $W$ is given by

$$W = \sum_{i=0}^{\infty} \tilde{A}^i [\tilde{A} V \tilde{A}^{\mathrm{T}} - V + \tilde{D} \tilde{D}^{\mathrm{T}}](\tilde{A}^i)^{\mathrm{T}}. \qquad (17.23)$$

**Proof.** Using the fact that

$$\sum_{i=0}^{k} \tilde{A}^{i+1} V (\tilde{A}^{i+1})^{\mathrm{T}} - \sum_{i=0}^{k} \tilde{A}^{i} V (\tilde{A}^{i})^{\mathrm{T}} = \tilde{A}^{k+1} V (\tilde{A}^{k+1})^{\mathrm{T}} - V,$$

it follows that

$$\sum_{i=0}^{\infty} \tilde{A}^{i} (\tilde{A} V \tilde{A}^{\mathrm{T}} - V)(\tilde{A}^{i})^{\mathrm{T}}$$

$$= [I_n - (I_n - \tilde{A})(I_n - \tilde{A})^{\#}] V [I_n - (I_n - \tilde{A})(I_n - \tilde{A})^{\#}]^{\mathrm{T}} - V. \quad (17.24)$$

The result is now immediate.                                          □

Finally, the following proposition is needed for the proof of Theorem 17.1 below.

**Proposition 17.5.** Assume that Assumptions 17.1 and 17.2 hold. Then $\tilde{A}$ is discrete-time semistable if and only if, for every discrete-time $k$-semicontrollable pair $(\tilde{A}, \tilde{D})$, there exists an $n \times n$ matrix $P = P^{\mathrm{T}} \geq -(I_n - \tilde{A})V(I_n - \tilde{A})^{\mathrm{T}}$ such that

$$0 = (I_n - \tilde{A})^{k} [\tilde{A}(P + (I_n - \tilde{A})V(I_n - \tilde{A})^{\mathrm{T}})\tilde{A}^{\mathrm{T}} + \tilde{D}\tilde{D}^{\mathrm{T}}$$

$$- P - (I_n - \tilde{A})V(I_n - \tilde{A})^{\mathrm{T}}]((I_n - \tilde{A})^{k})^{\mathrm{T}}. \quad (17.25)$$

**Proof.** If $\tilde{A}$ is discrete-time semistable and $(\tilde{A}, \tilde{D})$ is discrete-time $k$-semicontrollable, then it follows from Lemma 17.5 that $(\tilde{A}^{\mathrm{T}}, \tilde{D}^{\mathrm{T}})$ is discrete-time $k$-semiobservable, which implies that

$$\mathcal{N}(I_n - \tilde{A}^{\mathrm{T}}) \subseteq \mathcal{N}(\tilde{D}^{\mathrm{T}}((I_n - \tilde{A})^{k})^{\mathrm{T}}).$$

Now, it follows from Proposition 17.2 that

$$W_1 = \sum_{i=0}^{\infty} \tilde{A}^{i} (I_n - \tilde{A})^{k} \tilde{D}\tilde{D}^{\mathrm{T}}((I_n - \tilde{A})^{k})^{\mathrm{T}} (\tilde{A}^{i})^{\mathrm{T}}$$

is well defined. Furthermore, it follows from (17.24) that

$$W_2 = \sum_{i=0}^{\infty} \tilde{A}^{i} [\tilde{A} V \tilde{A}^{\mathrm{T}} - V](\tilde{A}^{\mathrm{T}})^{i}$$

is well defined. Clearly, $-V \leq W_2 \leq 0$.

Next, consider

$$P = ((I_n - \tilde{A})^{k})^{\#} W_1 (((I_n - \tilde{A})^{k})^{\#})^{\mathrm{T}} + (I_n - \tilde{A})W_2(I_n - \tilde{A})^{\mathrm{T}}$$

$$\geq (I_n - \tilde{A})W_2(I_n - \tilde{A})^{\mathrm{T}}$$
$$\geq -(I_n - \tilde{A})V(I_n - \tilde{A})^{\mathrm{T}}.$$

It follows from (17.24) and $(I_n - \tilde{A})[I_n - (I_n - \tilde{A})(I_n - \tilde{A})^{\#}] = 0$ that

$$(I_n - \tilde{A})(W_2 + V)(I_n - \tilde{A})^{\mathrm{T}} = 0.$$

Hence,

$$(I_n - \tilde{A})^k (P + (I_n - \tilde{A})V(I_n - \tilde{A})^{\mathrm{T}})((I_n - \tilde{A})^k)^{\mathrm{T}}$$
$$= (I_n - \tilde{A})^k \sum_{i=0}^{\infty} \tilde{A}^i \tilde{D}\tilde{D}^{\mathrm{T}}(\tilde{A}^i)^{\mathrm{T}}((I_n - \tilde{A})^k)^{\mathrm{T}}$$
$$+ (I_n - \tilde{A})^{k+1}(W_2 + V)((I_n - \tilde{A})^{k+1})^{\mathrm{T}}$$
$$= W_1 + (I_n - \tilde{A})^k (I_n - \tilde{A})(W_2 + V)(I_n - \tilde{A})^{\mathrm{T}}((I_n - \tilde{A})^k)^{\mathrm{T}}$$
$$= W_1.$$

Thus,

$$\tilde{A}(I_n - \tilde{A})^k (P + (I_n - \tilde{A})V(I_n - \tilde{A})^{\mathrm{T}})((I_n - \tilde{A})^k)^{\mathrm{T}}(\tilde{A})^{\mathrm{T}}$$
$$- (I_n - \tilde{A})^k (P + (I_n - \tilde{A})V(I_n - \tilde{A})^{\mathrm{T}})((I_n - \tilde{A})^k)^{\mathrm{T}}$$
$$= \tilde{A}W_1\tilde{A}^{\mathrm{T}} - W_1$$
$$= -(I_n - \tilde{A})^k \tilde{D}\tilde{D}^{\mathrm{T}}((I_n - \tilde{A})^k)^{\mathrm{T}},$$

which implies (17.25).

Alternatively, if there exists $P \geq -(I_n - \tilde{A})V(I_n - \tilde{A})^{\mathrm{T}}$ such that (17.25) holds and $(\tilde{A}, \tilde{D})$ is discrete-time $k$-semicontrollable, then it follows from Lemma 17.5 that $(\tilde{A}^{\mathrm{T}}, \tilde{D}^{\mathrm{T}})$ is $k$-semiobservable, and, hence, by Proposition 17.3, $\mathcal{N}(I_n - \tilde{A}^{\mathrm{T}}) \cap \mathcal{R}(I_n - \tilde{A}^{\mathrm{T}}) = \{0\}$. Thus, it follows from [24, p. 119] that $(I_n - \tilde{A})^{\mathrm{T}}$ is group invertible. Let $L \triangleq I_n - ((I_n - \tilde{A})^{\mathrm{T}})((I_n - \tilde{A})^{\mathrm{T}})^{\#}$, and note that $L^2 = L$. Hence, $L$ is the unique $n \times n$ matrix satisfying

$$\mathcal{N}(L) = \mathcal{R}(I_n - \tilde{A}^{\mathrm{T}}), \quad \mathcal{R}(L) = \mathcal{N}(I_n - \tilde{A}^{\mathrm{T}}), \quad Lx = x, \quad x \in \mathcal{N}(I_n - \tilde{A}^{\mathrm{T}}).$$

Now, consider the nonnegative function

$$\mathbb{V}(x) = x^{\mathrm{T}}(I_n - \tilde{A})^k (P + (I_n - \tilde{A})V(I_n - \tilde{A})^{\mathrm{T}})((I_n - \tilde{A})^k)^{\mathrm{T}}x + x^{\mathrm{T}}L^{\mathrm{T}}Lx,$$

and note that if $\mathbb{V}(x) = 0$, then

$$(P + (I_n - \tilde{A})V(I_n - \tilde{A})^{\mathrm{T}})((I_n - \tilde{A})^k)^{\mathrm{T}}x = 0, \quad Lx = 0.$$

By Proposition 17.3,

$$(P + (I_n - \tilde{A})V(I_n - \tilde{A})^{\mathrm{T}})((I_n - \tilde{A})^k)^{\mathrm{T}}x = 0$$

implies that $x \in \mathcal{N}(I_n - \tilde{A}^{\mathrm{T}})$, and, hence, $0 = Lx = x$, which implies that $\mathbb{V}(x)$ is positive definite. Hence,

$$(I_n - \tilde{A})^k (P + (I_n - \tilde{A})V(I_n - \tilde{A})^{\mathrm{T}})((I_n - \tilde{A})^k)^{\mathrm{T}} + L^{\mathrm{T}}L$$

is a positive definite matrix. Now, using similar arguments as in the proof of Theorem 3.1 of [158], it follows that $\tilde{A}^{\mathrm{T}}$ is discrete-time semistable, and, hence, $\tilde{A}$ is discrete-time semistable. $\qquad\square$

**Theorem 17.1.** Assume that Assumptions 17.1 and 17.2 hold. Then $\tilde{A}$ is discrete-time semistable if and only if, for every discrete-time $k$-semicontrollable pair $(\tilde{A}, \tilde{D})$, there exists an $n \times n$ matrix

$$\tilde{P} = \tilde{P}^{\mathrm{T}} > -(I_n - \tilde{A})^{k+1}V((I_n - \tilde{A})^{k+1})^{\mathrm{T}}$$

such that

$$\begin{aligned}
\tilde{P} &+ (I_n - \tilde{A})^{k+1}V((I_n - \tilde{A})^{k+1})^{\mathrm{T}} \\
&= \tilde{A}[\tilde{P} + (I_n - \tilde{A})^{k+1}V((I_n - \tilde{A})^{k+1})^{\mathrm{T}}]\tilde{A}^{\mathrm{T}} \\
&\quad + (I_n - \tilde{A})^k \tilde{D}\tilde{D}^{\mathrm{T}}((I_n - \tilde{A})^k)^{\mathrm{T}}.
\end{aligned} \tag{17.26}$$

Furthermore, if (17.26) holds, then

$$\begin{aligned}
\operatorname{tr}\,(\tilde{P} &+ (I_n - \tilde{A})^{k+1}V((I_n - \tilde{A})^{k+1})^{\mathrm{T}})(I_n - \tilde{A})^{\mathrm{T}}\tilde{R}(I_n - \tilde{A}) \\
&\geq \operatorname{tr}\,(W + V)((I_n - \tilde{A})^{k+1})^{\mathrm{T}}\tilde{R}(I_n - \tilde{A})^{k+1} \\
&= \operatorname{tr}\,Q((I_n - \tilde{A})^{k+1})^{\mathrm{T}}\tilde{R}(I_n - \tilde{A})^{k+1},
\end{aligned} \tag{17.27}$$

where $Q$ is given by (17.15) and $W$ is given by

$$W = \sum_{i=0}^{\infty} \tilde{A}^i [\tilde{A}V\tilde{A}^{\mathrm{T}} - V + \tilde{D}\tilde{D}^{\mathrm{T}}](\tilde{A}^i)^{\mathrm{T}}. \tag{17.28}$$

**Proof.** The first assertion is a direct consequence of Proposition 17.5 by taking

$$\begin{aligned}
\tilde{P} &= (I_n - \tilde{A})^k(P + (I_n - \tilde{A})V(I_n - \tilde{A})^{\mathrm{T}})((I_n - \tilde{A})^k)^{\mathrm{T}} \\
&\quad + L^{\mathrm{T}}L - (I_n - \tilde{A})^{k+1}V((I_n - \tilde{A})^{k+1})^{\mathrm{T}}
\end{aligned}$$

and noting that $L = L\tilde{A}^{\mathrm{T}}$.

To prove the second assertion, it follows, using similar arguments as in the proof of Theorem 3.1 in [158], that if there exists an $n \times n$ matrix $\tilde{P} > -(I_n - \tilde{A})^{k+1}V((I_n - \tilde{A})^{k+1})^{\mathrm{T}}$ such that (17.26) holds, then

$$\tilde{P} + (I_n - \tilde{A})^{k+1}V((I_n - \tilde{A})^{k+1})^{\mathrm{T}}$$

$$= \sum_{i=0}^{\infty} \tilde{A}^i (I_n - \tilde{A})^k \tilde{D}\tilde{D}^{\mathrm{T}}((I_n - \tilde{A})^k)^{\mathrm{T}}(\tilde{A}^i)^{\mathrm{T}} + \alpha x x^{\mathrm{T}},$$

where $\alpha > 0$ and $x \in \mathcal{N}(I_n - \tilde{A})$. Next, using Proposition 17.4, note that

$$W = W_2 + \sum_{i=0}^{\infty} \tilde{A}^i \tilde{D}\tilde{D}(\tilde{A}^{\mathrm{T}})^i.$$

Thus,

$$W + V = W_2 + V + \sum_{i=0}^{\infty} \tilde{A}^i \tilde{D}\tilde{D}(\tilde{A}^{\mathrm{T}})^i.$$

Consequently,

$$
\begin{aligned}
(I_n - \tilde{A})^{k+1} &(W + V)((I_n - \tilde{A})^{k+1})^{\mathrm{T}} \\
&= (I_n - \tilde{A})^k (I_n - \tilde{A})(W_2 + V)(I_n - \tilde{A})^{\mathrm{T}}((I_n - \tilde{A})^k)^{\mathrm{T}} \\
&\quad + (I_n - \tilde{A}) \sum_{i=0}^{\infty} \tilde{A}^i (I_n - \tilde{A})^k \tilde{D}\tilde{D}^{\mathrm{T}}((I_n - \tilde{A})^k)^{\mathrm{T}}(\tilde{A}^i)^{\mathrm{T}}(I_n - \tilde{A})^{\mathrm{T}} \\
&= (I_n - \tilde{A}) \sum_{i=0}^{\infty} \tilde{A}^i (I_n - \tilde{A})^k \tilde{D}\tilde{D}^{\mathrm{T}}((I_n - \tilde{A})^k)^{\mathrm{T}}(\tilde{A}^i)^{\mathrm{T}}(I_n - \tilde{A})^{\mathrm{T}} \\
&= (I_n - \tilde{A})(\tilde{P} + (I_n - \tilde{A})^{k+1}V((I_n - \tilde{A})^{k+1})^{\mathrm{T}} - \alpha x x^{\mathrm{T}})(I_n - \tilde{A})^{\mathrm{T}},
\end{aligned}
$$

and, hence,

$$
\begin{aligned}
\operatorname{tr}(\tilde{P} &+ (I_n - \tilde{A})^{k+1}V((I_n - \tilde{A})^{k+1})^{\mathrm{T}})(I_n - \tilde{A})^{\mathrm{T}}\tilde{R}(I_n - \tilde{A}) \\
&= \operatorname{tr}(I_n - \tilde{A})(\tilde{P} + (I_n - \tilde{A})^{k+1}V((I_n - \tilde{A})^{k+1})^{\mathrm{T}})(I_n - \tilde{A})^{\mathrm{T}}\tilde{R} \\
&= \operatorname{tr}(I_n - \tilde{A})^{k+1}(W + V)((I_n - \tilde{A})^{k+1})^{\mathrm{T}}\tilde{R} + \alpha x^{\mathrm{T}}(I_n - \tilde{A})^{\mathrm{T}}\tilde{R}(I_n - \tilde{A})x \\
&\geq \operatorname{tr}(I_n - \tilde{A})^{k+1}(W + V)((I_n - \tilde{A})^{k+1})^{\mathrm{T}}\tilde{R} \\
&= \operatorname{tr}(W + V)((I_n - \tilde{A})^{k+1})^{\mathrm{T}}\tilde{R}(I_n - \tilde{A})^{k+1},
\end{aligned}
$$

where we used the fact that $\operatorname{tr} BA = \operatorname{tr} AB$, with equality holding if and only if

$$x \in \mathcal{N}(\tilde{R}(I_n - \tilde{A})) \cap \mathcal{N}(I_n - \tilde{A}^{\mathrm{T}}).$$

Finally, it follows from (17.15) and

$$
\begin{aligned}
(I_n - \tilde{A})^{k+1}&[I_n - (I_n - \tilde{A})(I_n - \tilde{A})^{\#}]Q(0) \\
&\cdot [I_n - (I_n - \tilde{A})(I_n - \tilde{A})^{\#}]^{\mathrm{T}}((I_n - \tilde{A})^{k+1})^{\mathrm{T}} = 0
\end{aligned}
$$

that

$$\operatorname{tr} Q((I_n - \tilde{A})^{k+1})^{\mathrm{T}} \tilde{R}(I_n - \tilde{A})^{k+1}$$
$$= \operatorname{tr}(I_n - \tilde{A})^{k+1} Q((I_n - \tilde{A})^{k+1})^{\mathrm{T}} \tilde{R}$$
$$= \operatorname{tr}(I_n - \tilde{A})^{k+1} \sum_{i=0}^{\infty} \tilde{A}^i \tilde{D} \tilde{D}(\tilde{A}^{\mathrm{T}})^i ((I_n - \tilde{A})^{k+1})^{\mathrm{T}} \tilde{R}$$
$$= \operatorname{tr}(I_n - \tilde{A})^{k+1} (W + V)((I_n - \tilde{A})^{k+1})^{\mathrm{T}} \tilde{R},$$

which completes the proof. □

It follows from Theorem 17.1 that discrete-time semistability can be equivalently characterized by the Lyapunov equation (17.26). Hence, the difficulty of handling the discrete-time semistability condition (17.19) for our optimization problem can be mitigated. Furthermore, by Proposition 17.4, (17.27) implies that an optimal solution to (17.17) is also an optimal solution to

$$\min_{K} \operatorname{tr}(\tilde{P} + (I_n - \tilde{A})^{k+1} V((I_n - \tilde{A})^{k+1})^{\mathrm{T}})(I_n - \tilde{A})^{\mathrm{T}} \tilde{R}(I_n - \tilde{A})$$

under some mild conditions. Note that, as shown in the proof of Theorem 17.1, (17.20) is implicitly embedded in (17.26). Thus, Theorem 17.1 serves as a theoretical foundation for converting the optimization problem (17.17) into another equivalent optimization problem, which can be solved by using a fast-convergence swarm algorithm.

## 17.6 Optimization-Based Control Design

In this section, we present an optimization problem that can be used to solve the semistable LQR control problem introduced in Section 17.3. To state this result, let $\mathbf{u} \in \mathbb{R}^n$ be a given unit vector.

**Theorem 17.2.** Assume that Assumptions 17.1 and 17.2 hold, and assume that $(\tilde{A}, \tilde{D})$ is discrete-time $k$-semicontrollable. Let $S_{\min}$ be a solution to the minimization problem

$$\min \Big\{ \operatorname{tr} S(I_n - \tilde{A})^{\mathrm{T}} \tilde{R}(I_n - \tilde{A}) :$$
$$S = S^{\mathrm{T}} > 0, \quad S = \tilde{A} S \tilde{A}^{\mathrm{T}} + (I_n - \tilde{A})^k \tilde{D} \tilde{D}^{\mathrm{T}}((I_n - \tilde{A})^k)^{\mathrm{T}},$$
$$\operatorname{rank}(\tilde{A} - I_n) = n - 1, \quad \tilde{A}\mathbf{u} = \mathbf{u} \Big\}. \tag{17.29}$$

Then

$$\operatorname{tr} S_{\min}(I_n - \tilde{A})^{\mathrm{T}} \tilde{R}(I_n - \tilde{A}) = \operatorname{tr}(W + V)((I_n - \tilde{A})^{k+1})^{\mathrm{T}} \tilde{R}(I_n - \tilde{A})^{k+1}$$

$$= \operatorname{tr} Q((I_n - \tilde{A})^{k+1})^{\mathrm{T}} \tilde{R}(I_n - \tilde{A})^{k+1},$$

and there exists a unique unit vector $\mathbf{v} \in \mathbb{R}^n$ such that $\lim_{k \to \infty} \tilde{A}^k = \frac{1}{\mathbf{u}^{\mathrm{T}} \mathbf{v}} \mathbf{u} \mathbf{v}^{\mathrm{T}}$.

**Proof.** The first assertion is a direct consequence of Theorem 17.1. To show the second assertion, note that it follows from Theorem 17.1 that the existence of solutions to (17.29) guarantees discrete-time semistability of $\tilde{A}$. Since $\tilde{A}\mathbf{u} = \mathbf{u}$, it follows that 1 is an eigenvalue of $\tilde{A}$. Now, since $\det(I_n - \tilde{A}) = \det(I_n - \tilde{A}^{\mathrm{T}})$, it follows that 1 is also an eigenvalue of $\tilde{A}^{\mathrm{T}}$. Furthermore, since

$$\operatorname{rank}(I_n - \tilde{A}^{\mathrm{T}}) = \operatorname{rank}(I_n - \tilde{A}) = n - 1,$$

it follows that $\dim \mathcal{N}(I_n - \tilde{A}^{\mathrm{T}}) = 1$. This implies that there exists a unique unit vector $\mathbf{v} \in \mathbb{R}^n$ such that $\tilde{A}^{\mathrm{T}}\mathbf{v} = \mathbf{v}$ or, equivalently, $\mathbf{v}^{\mathrm{T}}\tilde{A} = \mathbf{v}^{\mathrm{T}}$. Now, it follows from Lemma 17.3 that $\lim_{k \to \infty} \tilde{A}^k = \frac{1}{\mathbf{u}^{\mathrm{T}} \mathbf{v}} \mathbf{u} \mathbf{v}^{\mathrm{T}}$.                  □

Although the minimization problem (17.29) has a different cost functional from (17.17), Theorem 17.2 gives an upper bound between the optimal solutions to (17.29) and the optimal solutions to (17.17). To see this, note that it follows from Proposition 8.4.13 of [27, p. 471] that

$$\lambda_{\min}(\tilde{R}) \operatorname{tr} Q \le \operatorname{tr} Q\tilde{R} \le \lambda_{\max}(\tilde{R}) \operatorname{tr} Q$$

and

$$\lambda_{\min}(((I_n - \tilde{A})^{k+1})^{\mathrm{T}}(I_n - \tilde{A})^{k+1}) \lambda_{\min}(\tilde{R}) \operatorname{tr} Q$$
$$\le \operatorname{tr} Q((I_n - \tilde{A})^{k+1})^{\mathrm{T}} \tilde{R}(I_n - \tilde{A})^{k+1}$$
$$\le \lambda_{\max}(((I_n - \tilde{A})^{k+1})^{\mathrm{T}}(I_n - \tilde{A})^{k+1}) \lambda_{\max}(\tilde{R}) \operatorname{tr} Q.$$

Now, for all possible discrete-time semistable $\tilde{A} \ne I_n$, it follows that the smallest lower bound for $\lambda_{\min}(((I_n - \tilde{A})^{k+1})^{\mathrm{T}}(I_n - \tilde{A})^{k+1})$ is 0 and the supremum for $\lambda_{\max}(((I_n - \tilde{A})^{k+1})^{\mathrm{T}}(I_n - \tilde{A})^{k+1})$ is 2. Hence,

$$0 \le \operatorname{tr} Q((I_n - \tilde{A})^{k+1})^{\mathrm{T}} \tilde{R}(I_n - \tilde{A})^{k+1} \le 2\lambda_{\max}(\tilde{R}) \operatorname{tr} Q.$$

Clearly, $[\lambda_{\min}(\tilde{R}) \operatorname{tr} Q, \lambda_{\max}(\tilde{R}) \operatorname{tr} Q] \subseteq [0, 2\lambda_{\max}(\tilde{R}) \operatorname{tr} Q]$. This implies that solving (17.29) gives an upper and a lower bound to the optimal solutions of (17.17). Furthermore, if both optimization problems have a common global optimum, then the optimal solutions to (17.29) are equivalent to the solutions of the proposed semistable LQR control problem.

Next, we apply Theorem 17.2 to solve the original coordinated resource allocation algorithm design given by (17.2). In this case, we have $A = \Lambda$,

$B = I_n$, $K = E - \Lambda$, $D_1 = 0$, and $D_2 = D$ in the closed-loop system (17.7) and (17.8). Furthermore, $\tilde{A} = E$ and $\tilde{D} = (E-\Lambda)D$. Hence, if $D$ is invertible and $\Lambda = I_n$, then $(E, (E - I_n)D)$ is discrete-time 0-semicontrollable. Now, the optimization problem in Theorem 17.2 reduces to the minimization of

$$\operatorname{tr} S(E - I_n)^{\mathrm{T}}[R_1 + (E - I_n)^{\mathrm{T}}R_2(E - I_n)](E - I_n)$$

for $E$ subject to

$$S = S^{\mathrm{T}} > 0, \quad S = ESE^{\mathrm{T}} + (E - I_n)DD^{\mathrm{T}}((E - I_n)^{\mathrm{T}}),$$
$$\operatorname{rank}(E - I_n) = n - 1, \quad E\mathbf{u} = \mathbf{u}.$$

Finally, if we further assume that $\mathbf{e}^{\mathrm{T}}E = \mathbf{e}^{\mathrm{T}}$, where $\mathbf{e} = [1, \ldots, 1]^{\mathrm{T}}$, and the optimization problem has a solution, then $\lim_{k \to \infty} x(k) = \frac{1}{\mathbf{e}^{\mathrm{T}}\mathbf{u}}(\sum_{i=1}^{n} x_i(0))\mathbf{u}$ for (17.2).

**Corollary 17.1.** Consider the coordinated resource allocation algorithm given by (17.2) or, equivalently, (17.5) and (17.6). Assume that Assumptions 17.1 and 17.2 hold, and assume that $D$ is invertible and $\Lambda = I_n$. Then solving the minimization problem

$$\min_{E}\Big\{ \operatorname{tr} S(E - I_n)^{\mathrm{T}}[R_1 + (E - I_n)^{\mathrm{T}}R_2(E - I_n)](E - I_n) :$$
$$S = S^{\mathrm{T}} > 0, \quad S = ESE^{\mathrm{T}} + (E - I_n)DD^{\mathrm{T}}((E - I_n)^{\mathrm{T}}),$$
$$\operatorname{rank}(E - I_n) = n - 1, \quad \mathbf{v}^{\mathrm{T}}E = \mathbf{v}^{\mathrm{T}}, \quad E\mathbf{u} = \mathbf{u}\Big\} \tag{17.30}$$

gives an optimal solution to the semistable LQR problem with criteria (i)–(iii) given in Section 17.3.

**Proof.** It follows from the assumptions on $D$ and $\Lambda$ that $(E, (E-I_n)D)$ is discrete-time 0-semicontrollable and $\mathcal{N}(I_n - E^{\mathrm{T}}) \subseteq \mathcal{N}(D^{\mathrm{T}}(I_n - E^{\mathrm{T}}))$. Now, the result is a direct consequence of Theorem 17.2, Proposition 17.2, and Lemma 17.3. $\square$

As pointed out in [42, 44], the state feedback stabilization problem involving bound constraints on the entries of the gain matrix $K$ and checking for the existence of a stable matrix in an interval family of matrices is NP-hard. Since the semistable LQR problem involves the design of a state feedback controller as well as checking for the semistability of $\tilde{A}$, the equivalent optimization problem presented in this section is an NP-hard problem. Thus, it cannot be solved by polynomial-time algorithms.

Thus, the proposed optimization problem and its variants are unlikely to allow for *efficient* algorithmic solutions. Furthermore, the rank conditions $\operatorname{rank}(\tilde{A} - I_n) = n - 1$ in (17.29) and $\operatorname{rank}(E - I_n) = n - 1$ in (17.30) are *not* convex in terms of $\tilde{A}$ and $E$, respectively. Hence, in general, (17.29) and (17.30) are *nonconvex* optimization problems, and the matrix constraints in (17.29) and (17.30) involve nonlinear matrix equations as functions of the controller gain $K$. This results in it being a challenging task to seek efficient numerical solvers.

## 17.7 A Numerical Algorithm for Optimal Resource Allocation

Since the optimization problem (17.30) is equivalent to the optimal design problem of balanced coordination algorithms proposed in Section 17.3, one can possibly use approximate computational methods such as the interior-point method or the subgradient method to solve the optimization problem (17.30) over a large-scale graph. However, for problems with more than a few thousand edges, this optimization problem is beyond the computational capabilities of current interior-point semidefinite programming solvers, whereas subgradient methods do not have a simple stopping criterion that guarantees a certain level of suboptimality. Hence, we need to consider alternative algorithms for solving the semistable LQR problem.

Stochastic optimization methods [284] have attracted much attention in recent years since these methods do not require properties such as linearity, differentiability, convexity, separability, or nonexistence of constraints. To solve the nonconvex optimization problem (17.29) and (17.30) efficiently, we resort to a new class of randomized swarm algorithms known as multiagent coordination optimization (MCO) [327, 329]. We choose MCO for several reasons. First, these algorithms show significant efficiency for many benchmark optimization problems compared to the standard PSO and other optimization approaches [172, 326, 327, 328, 329, 330]. Moreover, the MCO algorithms do not require that the optimization problem be differentiable, as is required for classical optimization methods such as gradient descent. Thus, MCO can also be used for optimization problems that are partially irregular, noisy, and changing over time.

Furthermore, we choose an MCO algorithm for its numerical efficiency, fewer parameters to adjust, and ability to avoid local optima. The efficiency factor is especially important for our problem due to the large parameter search space. Additionally, MCO algorithms can be easily implemented in parallel to fully utilize the high-performance, large-scale parallel computing capabilities of multicore computers, which can significantly improve computational efficiency. Preliminary studies on parallel implementation of MCO algorithms have been reported in [172, 327]. Finally, unlike many other swarm intelligence algorithms, whose global convergence guarantees remain open, the global convergence for MCO problems has been solved in [327, 329], and, hence, one can generate a simple stopping criterion for running the MCO algorithm iteratively.

The MCO algorithm has some similarities to PSO [184] but has a clear difference from PSO in that it consists of update laws motivated by hybrid consensus protocols for multiagent systems. The algorithm starts with a set of random solutions for agents which can communicate with each other. The agents then move through the solution space based on an evaluation of their cost functional and neighbor-to-neighbor protocol rules.

More specifically, the velocity update formula for a standard PSO is given by

$$\mathbf{v}_{k+1}^i = \mathbf{v}_k^i + b_1 r_1(\mathbf{g}_{\text{loc},k} - \mathbf{x}_k^i) + b_2 r_2(p - \mathbf{x}_k^i), \quad i = 1, \ldots, q, \qquad (17.31)$$

whereas the velocity update formula for the MCO algorithm is given by

$$\mathbf{v}_{k+1}^i = \mathbf{v}_k^i + \eta \sum_{j \in \mathcal{N}_i} (\mathbf{v}_k^j - \mathbf{v}_k^i) + \mu \sum_{j \in \mathcal{N}_i} (\mathbf{x}_k^j - \mathbf{x}_k^i) + \kappa(p - \mathbf{x}_k^i),$$
$$i = 1, \ldots, q, \qquad (17.32)$$

where $b_1$ and $b_2$ are constant parameters; $r_1$, $r_2$, $\eta$, $\mu$, and $\kappa$ are random numbers in the closed interval [0,1]; $\mathbf{g}_{\text{loc},i}$ is the local best solution found by particle $i$ at step $k$; $p$ is the global best solution achieved by particle $i$ at time $k$; and $\mathcal{N}_i$ represents the set of neighbors of particle $i$ which can share state information with particle $i$. The two cooperative terms $\mu \sum_{j \in \mathcal{N}_i}(\mathbf{x}_k^j - \mathbf{x}_k^i)$ and $\eta \sum_{j \in \mathcal{N}_i}(\mathbf{v}_k^j - \mathbf{v}_k^i)$ in (17.32) are embedded in the velocity update formula for MCO algorithms. This allows the communication topology to share the position and velocity information from different neighbors to enhance the convergence rate of the algorithm [173].

The position update formula for both the PSO and the MCO algorithm is given by

$$\mathbf{x}_{k+1}^i = \mathbf{x}_k^i + \mathbf{v}_{k+1}^i.$$

As the algorithm propagates, agents will accelerate toward individual agents with better cost functional values based on both the position and the velocity update formula. Since $E$ in (17.2) is a weighting matrix for a particular graph topology and the MCO algorithm optimizes vector-based problems, certain transformations between matrices and vectors are needed.

The first transformation is given by the function $v = \texttt{M2V}(M)$, which is shown in Algorithm 17.1, transforming a matrix $M$ into the vector $v$. Here $\texttt{rand}(\cdot)$ denotes the $\texttt{rand}$ function in MATLAB, where $\texttt{rand}(N)$ returns an $N \times N$ matrix containing pseudorandom values drawn from a standard uniform distribution on the open interval (0,1) and $\texttt{rand}(M, N)$ returns an

---

**Algorithm 17.1** Function $v = \texttt{M2V}(M)$

---

**Step 1.** Initialize the vector $v = \texttt{rand}(\sum_{i=1}^{n-1}(n-i), 1)$.
**Step 2.**
$v(1, n-1) = M(1, 2:n)$
$v(n, 2n-3) = M(2, 3:n)$
$\cdots$

**return** $v$

---

$M \times N$ matrix. If we restrict $M$ to be symmetric, then only the upper triangular part of the matrix is needed. The other two restrictions on $M$ are that the sum of each row of the matrix $M$ is one and all the entries of the weighting matrix are in the range $[0, 1]$. To satisfy both conditions, the diagonal entries of $M$ are assigned the difference between one and the sum of all the other entries in the same row. Hence, the upper triangular part of $M$ modulo its diagonal entries is transformed into a vector.

The second transformation is given by the function $M = \texttt{V2M}(v)$, which is shown in Algorithm 17.2, transforming a vector $v$ to the matrix $M$. This function is the inverse of the function $v = \texttt{M2V}(M)$. The matrix $M$ generated by this function satisfies the symmetry property, and the sum of each row and column of the matrix $M$ is one. Moreover, during this process the graph topology of the network does not change. Finally, the MCO algorithm is shown in Algorithm 17.3, with the objective function $f(v) = \texttt{obj}(v)$ shown in Algorithm 17.4. A detailed discussion on the MCO algorithm can be found in [172, 326, 327, 329].

The procedure of using these numerical algorithms to find the best solution to (17.29) can now be summarized as follows.

(i) Transform the connectivity matrix $\mathcal{C}$ into a vector using $v = \texttt{M2V}(\mathcal{C})$.

(ii) Use the MCO algorithm to optimize $f(v) = \texttt{obj}(v)$ and find the best vector solution $p$.

(iii) Transform the vector $p$ to the matrix $E$ by using $E = \texttt{V2M}(p)$, where $E$ is the best matrix for the optimization problem.

**Example 17.1.** To demonstrate the effectiveness of the proposed MCO algorithm, we solve (17.30) and compare the results obtained by the standard PSO with a variation of PSO called the center PSO (CPSO) [212]. The CPSO algorithm involves a center particle that is incorporated into a linearly decreasing weight PSO that achieves faster convergence than the standard

**Algorithm 17.2** Function $M = \text{V2M}(v)$

**Step 1.** Initialize the matrix $M = \text{rand}(n)$.
**Step 2.**
$M(1,:) = v(1 : n - 1)$
$M(2,:) = v(n : 2n - 2)$
$\ldots$

**Step 3.**
**for** $i = 1, \ldots, n$ **do**
  **for** $j = i + 1, \ldots, n$ **do**
    $M(j,i) = M(i,j) \leftarrow \mathcal{C}_{(i,j)} M(i,j)$
  **end for**
  $M(i,i) = 0$
**end for**
**Step 4.**
**for** $i = 1, \ldots, n$ **do**
  $M(i,i) \leftarrow e_i$, where $[e_1, \ldots, e_n]^{\text{T}} = (I_n - M)\mathbf{e}$
**end for**
**return** $M$

PSO algorithm. We chose this algorithm because the velocity update formula for the CPSO algorithm involves neighbor position information, which is similar to the proposed MCO algorithm. Both undirected and directed graph cases are considered. Specifically, we consider designing (17.2) by using its equivalent optimization formulation (17.30).

For the undirected graph case, that is, $\mathfrak{G}$ is an undirected graph, the parameters used in the simulation are $n = 10$, $\varepsilon = 10^{-7}$, $\Lambda = I_{10}$, $D = I_{10}$, $R_1 = 8 \times I_{10} \times 10^5$, $R_2 = 5 \times I_{10} \times 10^5$, $V = 0$,

$$x(0) = [100, 200, 300, 400, 500, 100, 200, 300, 400, 500]^{\text{T}},$$

$\underline{v} = 0$, $\overline{v} = 1$, and $\mathbf{u} = \frac{1}{\sqrt{10}}[1, \ldots, 1]^{\text{T}} \in \mathbb{R}^{10}$. Furthermore, we require $E = E^{\text{T}}$, $E\mathbf{u} = \mathbf{u}$, and $\text{rank}(E - I_{10}) = 9$. Finally, we generate a randomly chosen connectivity matrix $\mathcal{C} = \mathcal{C}^{\text{T}}$ associated with a graph topology $\mathfrak{G}$ given by

$$\mathcal{C} = \begin{bmatrix} 1 & 1 & 1 & 1 & 1 & 0 & 1 & 1 & 1 & 1 \\ 1 & 1 & 1 & 0 & 1 & 1 & 0 & 1 & 1 & 1 \\ 1 & 1 & 1 & 1 & 1 & 1 & 1 & 1 & 1 & 1 \\ 1 & 0 & 1 & 1 & 1 & 1 & 1 & 0 & 1 & 1 \\ 1 & 1 & 1 & 1 & 1 & 0 & 1 & 1 & 1 & 1 \\ 0 & 1 & 1 & 1 & 0 & 1 & 1 & 0 & 1 & 0 \\ 1 & 0 & 1 & 1 & 1 & 1 & 1 & 0 & 1 & 1 \\ 1 & 1 & 1 & 0 & 1 & 0 & 0 & 1 & 1 & 0 \\ 1 & 1 & 1 & 1 & 1 & 1 & 1 & 1 & 1 & 1 \\ 1 & 1 & 1 & 1 & 1 & 0 & 1 & 0 & 1 & 1 \end{bmatrix}.$$

---

**Algorithm 17.3** MCO Function for Resource Allocation $M = \text{AMCO}(\mathcal{C})$

---

**Step 1.** Transform the upper triangular part (excluding the diagonal entries) $\mathcal{C}$ into a vector form $\ell_{\text{int}}$ using $\ell_{\text{int}} = \text{M2V}(\mathcal{C})$.

**Step 2.**

**for** each agent $i = 1, \ldots, q$ **do**

    Initialize the agent's position with a uniformly distributed random vector by $x_i \sim \lambda \ell_{\text{int}}$, where $\lambda \sim U(0, \frac{1}{n})$, $n$ is the number of nodes for $\mathfrak{G}$, and $U$ denotes a uniform distribution.

    Initialize the agent's velocity: $v_i \sim U(\underline{v}, \overline{v})$, where $\underline{v}$ and $\overline{v}$ are the lower and upper bounds of the search speed.

    Update the agent's best-known position to its initial position: $p_i \leftarrow x_i$;

    Call $f(p) = \text{obj}(p)$.

    If $f(p_i) < f(p)$, then update the multiagent network's best-known position: $p \leftarrow p_i$.

**end for**

**repeat**

    $k \leftarrow k + 1$;

    **for** each agent $i = 1, \ldots, q$ **do**

        Choose random parameters: $\eta, \mu, \kappa \sim U(0, 1)$.

        Update the agent's velocity: $v_i \leftarrow v_i + \eta \sum_{j \in \mathcal{N}_i} (v_j - v_i) + \mu \sum_{j \in \mathcal{N}_i} (x_j - x_i) + \kappa(p - x_i)$.

        Update the agent's position: $x_i \leftarrow x_i + v_i$.

        **for** $f(x_i) < f(p_i)$ **do**

            Update the agent's best-known position: $p_i \leftarrow x_i$.

            If $f(p_i) < f(p)$, then update the multiagent network's best-known position: $p \leftarrow p_i$.

        **end for**

    **end for**

**until** $k$ is large enough or the change of $f$ is small.

**Step 3.** Generate the matrix $M$ from $p$ by using $M = \text{V2M}(p)$.

**return** $M$

---

---

**Algorithm 17.4** Objective Function $f(p) = \text{obj}(p)$

---

**Step 1.** Generate the matrix $H$ from $H = \text{V2M}(p)$.

**Step 2.**

**repeat**

   $N \leftarrow N + 1$

   Calculate $S_N$ by solving $S_N = \sum_{i=0}^{N} H^i (I_n - H)^k \tilde{D} \tilde{D}^{\text{T}} ((I_n - H)^k)^{\text{T}} (H^i)^{\text{T}}$.

**until** $\|S_N - S_{N+1}\| \leq \varepsilon$.

Update $S \leftarrow S_N$.

**Step 3.** $f(p) = \text{tr}\, S(I_n - H)^{\text{T}} \tilde{R}(I_n - H)$.

**return** $f(p)$

---

In this case, $E_{(i,i)} \neq 0$ for every $i = 1, \ldots, 10$; $E_{(i,j)} \neq 0$ if $C_{(i,j)} = 1$; and $E_{(i,j)} = 0$ if $C_{(i,j)} = 0$, $i, j = 1, \ldots, 10$, $i \neq j$.

Similarly, for the directed graph case, that is, $\mathfrak{G}$ is a directed graph, the connectivity matrix $C$ associated with a randomly directed graph topology $\mathfrak{G}$ is given by

$$
C = \begin{bmatrix}
1 & 1 & 0 & 0 & 0 & 0 & 0 & 0 & 0 & 0 \\
0 & 1 & 1 & 0 & 0 & 0 & 0 & 0 & 0 & 0 \\
0 & 0 & 1 & 1 & 0 & 0 & 0 & 0 & 0 & 0 \\
0 & 0 & 0 & 1 & 1 & 0 & 0 & 0 & 0 & 0 \\
0 & 0 & 0 & 0 & 1 & 1 & 0 & 0 & 0 & 0 \\
0 & 0 & 0 & 0 & 0 & 1 & 1 & 0 & 0 & 0 \\
0 & 0 & 0 & 0 & 0 & 0 & 1 & 1 & 0 & 0 \\
0 & 0 & 0 & 0 & 0 & 0 & 0 & 1 & 1 & 0 \\
0 & 0 & 0 & 0 & 0 & 0 & 0 & 0 & 1 & 1 \\
1 & 0 & 0 & 0 & 0 & 0 & 0 & 0 & 0 & 1
\end{bmatrix}.
$$

We let $\mathbf{u} = [0.2, 0.2, 0.2, 0.2, 0.2, 0.4, 0.4, 0.4, 0.4, 0.4]^{\text{T}}$, which implies that the final portion of the resource converges to two subgroups. We choose the same parameters as in the undirected graph case except for the initial states, which are taken as

$$
x_0 = [111, 0.2, 3, 1.4, 2.5, 3.1, 42, 3.9, 4.2, 5.3]^{\text{T}}.
$$

Note that for both cases the pair $(E, E - I_{10})$ is discrete-time 0-semicontrollable. Moreover, the positive definite solution $S = E^{\text{T}} S E + (I_{10} - E)(I_{10} - E)^{\text{T}}$ will guarantee the discrete-time semistability of $E$.

We solve the equivalent optimization problem (17.30) in Corollary 17.1 for both cases via the MCO, PSO, and CPSO algorithms, respectively. Each of the three algorithms is run 30 times with randomized initial positions

Table 17.1 Comparison between the PSO, CPSO, and MCO algorithms for the undirected graph case.

|         | PSO | CPSO | MCO |
|---------|---------|---------|---------|
| Min     | 7.6370E4 | 7.3817E4 | 3.6777E4 |
| Max     | 1.1408E5 | 1.1408E5 | 8.6920E4 |
| Average | 1.0094E5 | 1.0054E5 | 6.2507E4 |
| Median  | 1.0610E5 | 1.0737E5 | 7.0353E4 |
| Std     | 8.6988E3 | 1.0543E4 | 1.2896E4 |

Table 17.2 Comparison between the PSO, CPSO, and MCO algorithms for the directed graph case.

|         | PSO | CPSO | MCO |
|---------|---------|---------|---------|
| Min     | 4.2835E2 | 5.7172E2 | 5.7070E-1 |
| Max     | 2.6196E3 | 2.5851E3 | 8.7007E2 |
| Average | 1.4820E3 | 1.5446E3 | 1.3738E2 |
| Median  | 1.7450E3 | 1.7889E3 | 1.8741E2 |
| Std     | 6.1515E2 | 4.6558E2 | 1.6174E2 |

and velocities as well as 30 particles with 500 iterations. The minimum, maximum, average, median, and standard deviation (Std) values for the best values obtained by the three algorithms for the undirected graph case are shown in Table 17.1, whereas Table 17.2 shows these results for the directed graph case.

From the simulation, we can conclude that the MCO algorithm exhibits better performance in all the maximum, minimum, average, median, and Std values for the directed graph case, and all the maximum, minimum, average, and median values for the undirected graph case. Moreover, the averaging convergence rates are shown in Figure 17.2 for the undirected graph case and Figure 17.3 for the directed graph case. Here we take the average value of each algorithm's objective function values at the same iteration for 30 executions to display the average convergence performance of each algorithm. It can be seen from the figures that the MCO algorithm is superior to the PSO and CPSO algorithms.

To see if there is any significant difference between the results obtained by each algorithm, we run the Friedman test [68]. The $p$ value we obtained for the undirected graph case is 2.1880E-11, whereas the $p$ value for the directed graph case is 1.228E-11, which indicates that at least one of the algorithms is significantly different from the other algorithms.

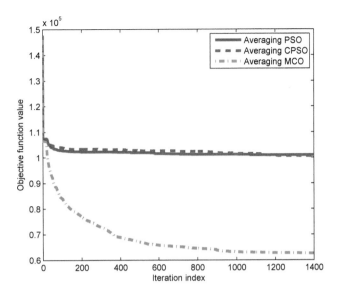

Figure 17.2 Convergence comparison between the PSO, CPSO, and MCO algorithms for the undirected graph case.

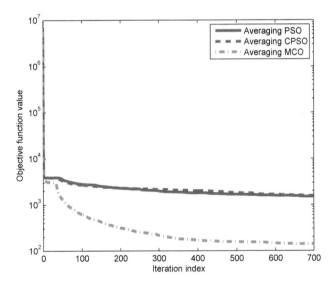

Figure 17.3 Convergence comparison between the PSO, CPSO, and MCO algorithms for the directed graph case.

Moreover, the Wilcoxon unsigned rank test [68] is used for pairwise comparison. The adjusted results for the undirected graph case using the Benjamini–Hochberg method [68] are 5.2032E-6, 2.6016E-6, and 8.8575E-5 for $X_{\mathrm{pso}} - X_{\mathrm{mco}}$, $X_{\mathrm{cpso}} - X_{\mathrm{mco}}$, and $X_{\mathrm{pso}} - X_{\mathrm{cpso}}$, respectively, where $X_{\mathrm{pso}}$,

$X_{\mathrm{cpso}}$, and $X_{\mathrm{mco}}$ are the 30 best results obtained by PSO, CPSO, and MCO after running the algorithms 30 times, respectively. Likewise, the adjusted results for the directed graph case using the Benjamini–Hochberg method [68] are 5.2032E-6, 2.6016E-6, and 8.8575E-5 for $X_{\mathrm{pso}} - X_{\mathrm{mco}}$, $X_{\mathrm{cpso}} - X_{\mathrm{mco}}$, and $X_{\mathrm{pso}} - X_{\mathrm{cpso}}$, respectively.

For the undirected graph case, the best solution obtained by the MCO algorithm is given by

$$
E_{\mathrm{best}}^{UD} = \begin{bmatrix}
0.2476 & 0.0554 & 0.1266 & 0.0362 & 0.0460 & 0 & 0.1078 & 0.1389 & 0.0814 & 0.1601 \\
0.0554 & 0.2405 & 0.1899 & 0 & 0.0686 & 0.1005 & 0 & 0.0901 & 0.0930 & 0.1620 \\
0.1266 & 0.1899 & 0.0242 & 0.0492 & 0.1093 & 0.0714 & 0.1095 & 0.0918 & 0.1093 & 0.1188 \\
0.0362 & 0 & 0.0492 & 0.4356 & 0.1251 & 0.0204 & 0.0817 & 0 & 0.1436 & 0.1082 \\
0.0460 & 0.0686 & 0.1093 & 0.1251 & 0.2756 & 0 & 0.0655 & 0.1175 & 0.0918 & 0.1006 \\
0 & 0.1005 & 0.0714 & 0.0204 & 0 & 0.6983 & 0.0101 & 0 & 0.0993 & 0 \\
0.1078 & 0 & 0.1095 & 0.0817 & 0.0655 & 0.0101 & 0.5189 & 0 & 0.0795 & 0.0270 \\
0.1389 & 0.0901 & 0.0918 & 0 & 0.1175 & 0 & 0 & 0.5125 & 0.0492 & 0 \\
0.0814 & 0.0930 & 0.1093 & 0.1436 & 0.0918 & 0.0993 & 0.0795 & 0.0492 & 0.1745 & 0.0784 \\
0.1601 & 0.1620 & 0.1188 & 0.1082 & 0.1006 & 0 & 0.0270 & 0 & 0.0784 & 0.2449
\end{bmatrix} \tag{17.33}
$$

and is graphically illustrated in Figure 17.4. For the directed graph case, the best solution is given by

$$
E_{\mathrm{best}}^{D} = \begin{bmatrix}
0.9989 & 0.0011 & 0 & 0 & 0 & 0 & 0 & 0 & 0 & 0 \\
0 & 0.9989 & 0.0011 & 0 & 0 & 0 & 0 & 0 & 0 & 0 \\
0 & 0 & 0.9988 & 0.0012 & 0 & 0 & 0 & 0 & 0 & 0 \\
0 & 0 & 0 & 0.9967 & 0.0033 & 0 & 0 & 0 & 0 & 0 \\
0 & 0 & 0 & 0 & 0.9920 & 0.0040 & 0 & 0 & 0 & 0 \\
0 & 0 & 0 & 0 & 0 & 0.9889 & 0.0111 & 0 & 0 & 0 \\
0 & 0 & 0 & 0 & 0 & 0 & 0.9869 & 0.0131 & 0 & 0 \\
0 & 0 & 0 & 0 & 0 & 0 & 0 & 0.9978 & 0.0022 & 0 \\
0 & 0 & 0 & 0 & 0 & 0 & 0 & 0 & 0.9966 & 0.0034 \\
0.0010 & 0 & 0 & 0 & 0 & 0 & 0 & 0 & 0 & 0.9995
\end{bmatrix}, \tag{17.34}
$$

and is graphically illustrated in Figure 17.5. The time history of state trajectories for $E_{\mathrm{best}}^{UD}$ is shown in Figure 17.6, whereas the time history of state trajectories for $E_{\mathrm{best}}^{D}$ is shown in Figure 17.7.

Clearly, the best algorithm matrix (17.33) guarantees that the state $x(k)$ achieves a uniform network resource distribution in expectation. This result is reminiscent of state consensus or agreement from the multiagent control perspective [238] and state synchronization from the neural network perspective [325]. Alternatively, compared with the undirected graph case, the number of communication links between every pair of nodes is less than in the directed graph case. Hence, the state $x(k)$ takes a higher convergence time than that of the undirected graph case. Nevertheless, the matrix (17.34) still guarantees that the state $x(k)$ achieves a nonuniform network resource distribution in expectation.                                                                     △

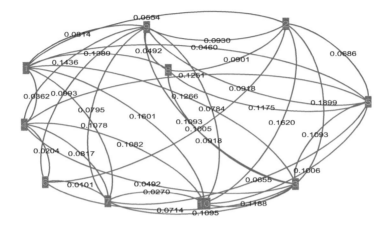

Figure 17.4 Graphical representation of the best solution obtained by the MCO algorithm for the undirected graph case.

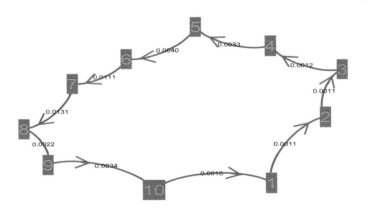

Figure 17.5 Graphical representation of the best solution obtained by the MCO algorithm for the directed graph case.

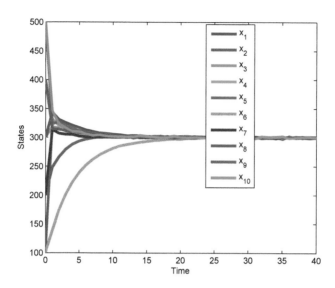

Figure 17.6 State trajectories versus time under the best solution obtained by the MCO
algorithm for the undirected graph case.

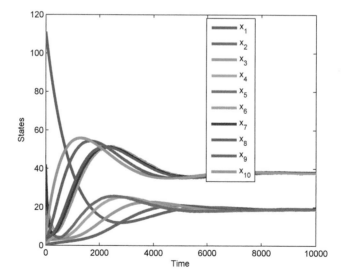

Figure 17.7 State trajectories versus time under the best solution obtained by the MCO
algorithm for the directed graph case.

*Chapter Eighteen*

---

# Approximate Consensus for Network Systems with Inaccurate Sensor Measurements

## 18.1 Introduction

Modern military and national command and control infrastructure capabilities involve large-scale, multilayered network systems placing stringent demands on controller design and implementation of increasing complexity. In numerous large-scale network system applications, agents can detect the location of the neighboring agents only approximately. This problem can arise in network defense systems involving low sensor quality, sensor failure, or detrimental environmental conditions resulting from a large-scale catastrophic event. This problem also arises in many robotics applications with inaccurate sensor data as well as low-cost, small-sized unmanned vehicles with relatively cheap sensors. In such a setting, it is desirable that the agents reach consensus approximately.

In this chapter, we consider a multiagent consensus problem in which agents possess sensors with limited accuracy. Specifically, we consider a group of agents with a connected and undirected communication graph topology and develop consensus control protocols for continuous-time and discrete-time network systems that guarantee that the agents reach an approximate consensus state and converge to a set centered at the centroid of the agents' initial locations. This set is shown to be time varying in the sense that only the differences between agent positions are, in the limit, small.

For discrete-time network systems, we also use difference inclusions and set-valued analysis to describe the inaccurate sensor measurement problem formulation. Set-valued analysis was used in Chapter 6 for consensus control. In [228], the author uses set-valued Lyapunov functions to study convergence of multiagent dynamical systems. The approach involves constructing set-valued Lyapunov functions from convex sets that depend on the agent states. In [6, 215, 228], the authors address the stability of each equilibrium point in the sense that the system solutions approach an equilibrium from a neighborhood of equilibria. Reference [215] considers barycentric coordinate maps, whereas [228] and [6] consider difference equations and difference inclusions, respectively.

Necessary and sufficient conditions for semistability for multiagent consensus problems using set-valued Lyapunov analysis are presented in [103]. The authors in [314] consider an asynchronous rendezvous problem using set-valued consensus theory. Specifically, a design strategy for multiagent consensus is developed by requiring two consecutive way-points to be included within a minimum convex region covering the two associated anticipated way-point sets.

The proposed set-valued consensus protocol builds on the framework of [103, 104, 228] to develop approximate consensus protocols for multiagent systems with uncertain interagent measurements. Specifically, the proposed protocol algorithm modifies the set-valued consensus update maps of the agents by assuming that the locations of all agents, including the agents calculating the update map, are within a ball of radius $r$. However, since the update sets of our design protocol do not satisfy a strict convexity assumption, our results go beyond the results of [228] by employing a set-valued invariance principle.

This chapter can be viewed as a contribution to the literature addressing multiagent systems in the presence of adversarial attacks. Specifically, the authors in [1, 89, 198, 210, 218, 250, 291, 313, 332] utilize stochastic tools and methods to study the behavior of multiagent systems in the presence of communication noise, transmission delays, and packet losses, whereas the authors in [7, 76, 78, 199, 321, 323, 324] utilize nonlinear and adaptive system theory to study the behavior of multiagent systems in the presence of agent and graph topology uncertainties. The results in this chapter addressing multiagent systems with inaccurate sensor measurements add to this literature by utilizing complementary analysis methods, including Lyapunov theory, difference inclusions, and set-valued analysis.

## 18.2  Notation, Definitions, and Mathematical Preliminaries

In this chapter, we use the Minkowski sum for summation of sets, with an analogous definition for set subtraction. Namely, for the sets $\mathcal{X}, \mathcal{Y} \subset \mathbb{R}^n$, $\mathcal{X} + \mathcal{Y}$ and $\mathcal{X} - \mathcal{Y}$ denote, respectively, the set of all vectors $z \in \mathbb{R}^n$ such that $z = x + y$ and $z = x - y$, where $x \in \mathcal{X}$ and $y \in \mathcal{Y}$. We also use the graph-theoretic notation established in Chapters 2 and 14. Furthermore, we define the graph Laplacian and *Perron matrix* of $\mathfrak{G}$ as $\mathcal{L} \triangleq \Delta - \mathcal{A}$ and

$\mathcal{P} \triangleq I - \varepsilon\mathcal{L}$, respectively, where $\varepsilon > 0$ and $\Delta \triangleq \mathrm{diag}[\deg_{\mathrm{in}}(1), \ldots, \deg_{\mathrm{in}}(N)]$; see Section 14.2. Finally, we denote the value of the node $i \in \{1, \ldots, N\}$ at time $t$ (respectively, time step $k$) by $x_i(t) \in \mathbb{R}^n$ (respectively, $x_i(k) \in \mathbb{R}^n$).

As shown in the earlier chapters, the consensus problem involves the network system $\mathcal{G}$ given by

$$\dot{x}_i(t) = u_i(t), \quad x_i(0) = x_{i0}, \quad t \geq 0, \quad i = 1, \ldots, N, \tag{18.1}$$

$$u_i(t) = \sum_{j \in \mathcal{N}_{\mathrm{in}}(i)} (x_j(t) - x_i(t)), \quad i = 1, \ldots, N. \tag{18.2}$$

Here, $x_i(t)$, $t \geq 0$, represents an *information state* and $u_i(t)$, $t \geq 0$, represents an *information control input* with a distributed consensus algorithm involving neighbor-to-neighbor interaction between agents.

In this chapter, we consider the continuous-time distributed consensus algorithm (18.1) and (18.2) resulting in closed-loop systems of the form [256]

$$\dot{x}_i(t) = \sum_{j \in \mathcal{N}_{\mathrm{in}}(i)} (x_j(t) - x_i(t)), \quad x_i(0) = x_{i0}, \quad t \geq 0, \quad i = 1, \ldots, N, \tag{18.3}$$

as well as a discrete-time distributed consensus algorithm resulting in closed-loop systems of the form [224]

$$x_i(k+1) = x_i(k) + \varepsilon \sum_{j \in \mathcal{N}_{\mathrm{in}}(i)} (x_j(k) - x_i(k)), \quad x_i(0) = x_{i0},$$

$$k \in \overline{\mathbb{Z}}_+, \quad i = 1, \ldots, N, \tag{18.4}$$

where $\varepsilon > 0$. Even though in this chapter we limit our attention to a network involving a chain of integrator multiagent dynamical systems $\mathcal{G}$, the proposed framework can be readily extended to designing low-level feedback consensus controllers for a network involving high-order, complex multiagent dynamical systems $\mathcal{G}$.

## 18.3 Consensus Control Problem with Uncertain Interagent Location Measurements

In this chapter, we consider a multiagent network in which $N$ agents with a connected, undirected, and time-invariant communication graph topology reach an approximate consensus state, and we use the terms *agent state* and *agent location* interchangeably. Here we do not consider time delays and communication losses between agents. In particular, each agent $i \in \{1, \ldots, N\}$ has a sensor with accuracy $r$; that is, each agent $i$ can detect the location of the other agents with an accuracy of up to a ball of radius

$r$ centered at the actual location of the other agents. The approximate location of agent $i$ as measured by agent $j$ is given by the set

$$\mathcal{X}_i = \{p \in \mathbb{R}^n : \|p - x_i\|_2 \leq r\}, \quad i = 1, \ldots, N.$$

The network consensus problem considered in this chapter involves the design of a control protocol that guarantees approximate system state equipartition; that is, the difference between any two agent states decreases to below a certain threshold that is dependent on the sensor accuracy $r$. Specifically, each agent $i$ uses an update protocol resulting in a closed-loop system similar to (18.3) or (18.4). However, since only approximate information on the location of the other agents is available at any given instant, the update protocol is constructed using approximate location information only.

In particular, for a discrete-time network system, the update protocol for a connected graph has the form

$$x_i(k + 1) \in \mathcal{F}_i(x(k)), \quad x_i(0) = x_{i0}, \quad k \in \overline{\mathbb{Z}}_+, \quad i = 1, \ldots, N, \qquad (18.5)$$

where

$$\mathcal{F}_i(x(k)) \triangleq x_i(k) + \varepsilon \sum_{j \in \mathcal{N}_{\mathrm{in}}(i)} (\mathcal{X}_j(k) - x_i(k));$$

$x \triangleq [x_1^{\mathrm{T}}, \ldots, x_N^{\mathrm{T}}]^{\mathrm{T}}$; and $\mathcal{X}_j - x_i$ denotes the set of all vectors $z \in \mathbb{R}^n$ such that $z = y - x_i$, with $y \in \mathcal{X}_j$. Note that, for the protocol given by (18.4), every agent has information on the exact location of the other agents, whereas for the protocol given by (18.5), only approximate location information on the other agents is available.

To further elucidate the protocol architecture given by (18.5), consider an all-to-all connected network consisting of three agents. In this case, the update protocol for agent 1 is given by

$$\begin{aligned}
x_1(k + 1) &\in \mathcal{F}_1(x(k)) \\
&= x_1(k) + \varepsilon(\mathcal{X}_1(k) - x_1(k) + \mathcal{X}_2(k) - x_1(k) + \mathcal{X}_3(k) - x_1(k)), \\
&\qquad\qquad x_1(0) = x_{10}, \quad k \in \overline{\mathbb{Z}}_+,
\end{aligned}$$

where the sets $\mathcal{X}_2 - x_1$ and $\mathcal{X}_3 - x_1$ are depicted in Figure 18.1; that is, the measurement of the exact locations of agents 2 and 3 is uncertain due to sensor measurement uncertainty or detrimental environmental conditions.

Figure 18.1 Visualization of sets $\mathcal{X}_2 - x_1$ and $\mathcal{X}_3 - x_1$ used in agent 1's update map.

## 18.4 Continuous-Time Consensus with a Connected Graph Topology

In this section, we consider the continuous-time consensus problem over an undirected communication network with a connected graph topology. We assume that only approximate information on the location of neighboring agents is available at any given instant, with the $i$th agent uncertainty satisfying $\|d_i(t)\|_2 \leq r$, $t \geq 0$, for $i = 1, \ldots, N$. Here we assume that the class of uncertainties we consider does not affect the graph topology (see, for example, [324] for a class of uncertainties affecting the communication graph topology).

In particular, we consider the update protocol for agent $i$ given by

$$\dot{x}_i(t) = \sum_{j \in \mathcal{N}(i)} (z_j(t) - z_i(t)), \quad x_i(0) = x_{i0}, \quad t \geq 0, \quad i = 1, \ldots, N, \quad (18.6)$$

where

$$z_j(t) - z_i(t) \triangleq (x_j(t) - d_j(t)) - (x_i(t) - d_i(t)).$$

In this case, it follows from (18.6) that

$$\dot{x}_i(t) = \sum_{j \in \mathcal{N}(i)} (x_j(t) - x_i(t)) + \sum_{j \in \mathcal{N}(i)} (d_i(t) - d_j(t)),$$

$$x_i(0) = x_{i0}, \quad t \geq 0, \quad i = 1, \ldots, N,$$

or, equivalently, in compact form

$$\dot{x}(t) = -\tilde{\mathcal{L}}x(t) + \tilde{\mathcal{L}}d(t), \quad x(0) = x_0, \quad t \geq 0, \quad (18.7)$$

where $\tilde{\mathcal{L}} \triangleq I_n \otimes \mathcal{L} \in \mathbb{R}^{nN \times nN}$; $\mathcal{L} \in \mathbb{R}^{N \times N}$ denotes the graph Laplacian; $\otimes$ denotes the Kronecker product; $x \triangleq [x_1^1, \ldots, x_N^1, \ldots, x_1^n, \ldots, x_N^n]^{\mathrm{T}}$; $d \triangleq [d_1^1, \ldots, d_N^1, \ldots, d_1^n, \ldots, d_N^n]^{\mathrm{T}}$; and $x_i^j$ and $d_i^j$ denote the $j$th components of $x_i$ and $d_i$, respectively.

Although our results can be directly extended to the case of (18.7), for simplicity of exposition, we will focus on individual agent states evolving in $\mathbb{R}$ (i.e., $n = 1$). In this case, (18.7) becomes

$$\dot{x}(t) = -\mathcal{L}x(t) + \mathcal{L}d(t), \quad x(0) = x_0, \quad t \geq 0. \tag{18.8}$$

For the statement of the next result, let $\mathbf{e}_N \triangleq [1, \ldots, 1]^{\mathrm{T}}$ denote the ones vector of order $N$, and let $\overline{x} \triangleq \frac{1}{N}\mathbf{e}_N^{\mathrm{T}}x$. Furthermore, recall that the Laplacian of an undirected connected graph is a symmetric nonnegative definite matrix with a single zero eigenvalue [224]; specifically, the eigenvalues of the graph Laplacian are given by

$$0 = \lambda_{\min}(\mathcal{L}) \triangleq \lambda_1(\mathcal{L}) < \lambda_2(\mathcal{L}) \leq \lambda_3(\mathcal{L}) \leq \cdots \leq \lambda_N(\mathcal{L}) \triangleq \lambda_{\max}(\mathcal{L}).$$

Hence, the Schur decomposition of $-\mathcal{L}$ is given by $-\mathcal{L} = P_\Sigma \Sigma P_\Sigma^{\mathrm{T}}$, where $P_\Sigma \triangleq [p_1, \ldots, p_{N-1}, \frac{1}{\sqrt{N}}\mathbf{e}_N]$, with $p_i \in \mathbb{R}^N$, $i = 1, \ldots, N-1$;

$$\Sigma \triangleq \begin{bmatrix} \Sigma_0 & 0_{(N-1) \times 1} \\ 0_{1 \times (N-1)} & 0 \end{bmatrix};$$

and $\Sigma_0 \in \mathbb{R}^{(N-1) \times (N-1)}$ is Hurwitz.

**Theorem 18.1.** Consider an undirected network of $N$ agents with a connected graph topology given by (18.8). Then

$$\limsup_{t \to \infty} \|x(t) - \mathbf{e}_N \overline{x}(t)\|_2 \leq \frac{\lambda_N(\mathcal{L})\sqrt{N}r}{\lambda_2(\mathcal{L})}.$$

**Proof.** First, define $\delta(t) \triangleq x(t) - \mathbf{e}_N \overline{x}(t)$ and note that

$$\frac{\mathrm{d}}{\mathrm{d}t}\left(\frac{1}{N}\mathbf{e}_N^{\mathrm{T}}x(t)\right) = \frac{1}{N}\mathbf{e}_N^{\mathrm{T}}(-\mathcal{L}x(t) + \mathcal{L}d(t)) = 0_N,$$

where we used the fact that $\mathcal{L}\mathbf{e}_N = 0_N$ and $\mathcal{L} = \mathcal{L}^{\mathrm{T}}$. Hence, $\overline{x}(t) = \frac{1}{N}\mathbf{e}_N^{\mathrm{T}}x(t) = \frac{1}{N}\mathbf{e}_N^{\mathrm{T}}x(0) = \overline{x}$, $t \geq 0$, which shows that the centroid of the network does not change over time in the presence of time-varying interagent measurement uncertainties.

Next, differentiating $\delta(t)$ with respect to time yields

$$\dot{\delta}(t) = \dot{x}(t) - \mathbf{e}_N \dot{\overline{x}}(t)$$

$$= -\mathcal{L}x(t) + \mathcal{L}d(t)$$
$$= -\mathcal{L}\left[\delta(t) + \mathbf{e}_N\overline{x}(t)\right] + \mathcal{L}d(t)$$
$$= -\mathcal{L}\delta(t) + \mathcal{L}d(t), \quad \delta(0) = \delta_0, \quad t \geq 0. \tag{18.9}$$

Introducing the transformation $q(t) \triangleq P_\Sigma^{\mathrm{T}}\delta(t)$, it follows from (18.9) that

$$\dot{q}(t) = P_\Sigma^{\mathrm{T}}\dot{\delta}(t)$$
$$= -P_\Sigma^{\mathrm{T}}\mathcal{L}P_\Sigma P_\Sigma^{\mathrm{T}}\delta(t) + P_\Sigma^{\mathrm{T}}\mathcal{L}P_\Sigma P_\Sigma^{\mathrm{T}}d(t)$$
$$= -P_\Sigma^{\mathrm{T}}\mathcal{L}P_\Sigma q(t) + P_\Sigma^{\mathrm{T}}\mathcal{L}P_\Sigma\overline{d}(t), \quad q(0) = q_0, \quad t \geq 0,$$

where $\overline{d}(t) \triangleq P_\Sigma^{\mathrm{T}}d(t)$, and, hence,

$$\dot{q}(t) = \begin{bmatrix} \Sigma_0 & 0_{(N-1)\times 1} \\ 0_{1\times(N-1)} & 0 \end{bmatrix} \left[q(t) - \overline{d}(t)\right], \quad q(0) = q_0, \quad t \geq 0. \tag{18.10}$$

Now, it follows from (18.10) that

$$\dot{q}_1(t) = \Sigma_0 q_1(t) - \Sigma_0\overline{d}_1(t), \quad q_1(0) = q_{10}, \quad t \geq 0, \tag{18.11}$$
$$\dot{q}_2(t) = 0, \quad q_2(0) = q_{20}, \tag{18.12}$$

where

$$q_1(t) \triangleq \begin{bmatrix} I_{(N-1)\times(N-1)} & 0_{(N-1)\times 1} \end{bmatrix} q(t),$$
$$\overline{d}_1(t) \triangleq \begin{bmatrix} I_{(N-1)\times(N-1)} & 0_{(N-1)\times 1} \end{bmatrix} \overline{d}(t),$$

and $q_2 \in \mathbb{R}$. Furthermore, note that $q_{20} = 0$ since $\mathbf{e}_N^{\mathrm{T}}\delta(t) = \mathbf{e}_N^{\mathrm{T}}x(t) - \frac{1}{N}\mathbf{e}_N^{\mathrm{T}}\mathbf{e}_N\mathbf{e}_N^{\mathrm{T}}x(t) = 0$.

Next, consider the Lyapunov-like function $V : \mathbb{R}^{(N-1)} \to \mathbb{R}$ given by $V(q_1) = q_1^{\mathrm{T}}Sq_1$, where $S = S^{\mathrm{T}} > 0$, $S \in \mathbb{R}^{(N-1)\times(N-1)}$, satisfies

$$0 = \Sigma_0^{\mathrm{T}}S + S\Sigma_0 + Q, \tag{18.13}$$

with $Q = Q^{\mathrm{T}} > 0$ and $Q \in \mathbb{R}^{(N-1)\times(N-1)}$. Now, note that the derivative of $V(q_1)$ along the trajectories of (18.11) is given by

$$\dot{V}(q_1(t))$$
$$= -q_1^{\mathrm{T}}(t)Qq_1(t) - 2q_1^{\mathrm{T}}(t)S\Sigma_0\overline{d}_1(t)$$
$$\leq -\lambda_{\min}(Q)\|q_1(t)\|_2^2 + 2\sigma_{\max}(S\Sigma_0)\sigma_{\max}\left(\begin{bmatrix} I_{(N-1)\times(N-1)} & 0_{(N-1)\times 1} \end{bmatrix}\right)$$
$$\qquad \cdot \sigma_{\max}(P_\Sigma^{\mathrm{T}})\|d(t)\|_2\|q_1(t)\|_2$$
$$\leq -\lambda_{\min}(Q)\|q_1(t)\|_2^2 + 2\sigma_{\max}(S\Sigma_0)\sqrt{N}r\|q_1(t)\|_2$$
$$= -\|q_1(t)\|_2\left[\lambda_{\min}(Q)\|q_1(t)\|_2 - 2\sigma_{\max}(S\Sigma_0)\sqrt{N}r\right], \quad t \geq 0, \tag{18.14}$$

where we used the facts that

$$\sigma_{\max}\left(\left[\begin{array}{cc} I_{(N-1)\times(N-1)} & 0_{(N-1)\times 1} \end{array}\right]\right) = 1, \quad \sigma_{\max}(P_{\Sigma}^{\mathrm{T}}) = 1,$$

and $\|d(t)\|_2 \leq \sqrt{N}r$, $t \geq 0$.

Next, it follows from (18.14) that $\dot{V}(q_1(t)) \leq 0$, $t \geq 0$, for

$$\|q_1(t)\|_2 \geq \frac{2\sigma_{\max}(S\Sigma_0)\sqrt{N}r}{\lambda_{\min}(Q)} \triangleq \beta, \quad t \geq 0,$$

and, hence, $q_1(t)$, $t \geq 0$, is decreasing for $\|q_1(t)\|_2 > \beta$. Moreover, since $\dot{q}_2(t) = 0$, $t \geq 0$, and $q_2(0) = 0$, then $q_2(t) = 0$ for all $t \geq 0$. Hence, it follows from the definition of $q(t)$ and (18.14) that

$$\|\delta(t)\|_2 = \left\|\left[\begin{array}{c} q_1(t) \\ q_2(t) \end{array}\right]\right\|_2 = \|q_1(t)\|_2 \leq \beta$$

as $t \to \infty$. Now, setting $Q = -\Sigma_0$, it follows from (18.13) that $S = \frac{1}{2}I_{(N-1)}$, and, hence, $\|q_1(t)\|_2 = \|x(t) - \mathbf{e}_N\bar{x}\|_2 \leq \beta$, $t \geq 0$, where

$$\beta = \frac{2\sigma_{\max}(\frac{1}{2}\Sigma_0)\sqrt{N}r}{\lambda_{\min}(-\Sigma_0)} = \frac{\lambda_N(\mathcal{L})\sqrt{N}r}{\lambda_2(\mathcal{L})}, \tag{18.15}$$

which completes the proof. $\qquad\square$

It is important to note that if all the sensor uncertainties are identical, that is, $d_i(t) = d_0(t)$ for all $i = 1, \ldots, N$, then it follows from Theorem 4.1 that all agents reach exact agreement since in this case $\mathcal{L}d(t) = \mathcal{L}\mathbf{e}_N d_0(t) = 0$ in (18.8).

Note that since, by Theorem 18.1,

$$\limsup_{t\to\infty} \|x(t) - \mathbf{e}_N\bar{x}(t)\|_2 \leq \frac{\lambda_N(\mathcal{L})\sqrt{N}r}{\lambda_2(\mathcal{L})},$$

it follows that as the number of agents increases, the uncertainty has a prominent effect on the system. It is also important to note that the bound $\frac{\lambda_N(\mathcal{L})\sqrt{N}r}{\lambda_2(\mathcal{L})}$ depends on the ratio of $\lambda_N(\mathcal{L})$ and $\lambda_2(\mathcal{L})$. For example, consider a set of agents on a line graph. In this case, $\limsup_{t\to\infty} \|x(t) - \mathbf{e}_N\bar{x}(t)\|_2 \leq 1.41r$ for $N = 2$, $\limsup_{t\to\infty} \|x(t) - \mathbf{e}_N\bar{x}(t)\|_2 \leq 5.19r$ for $N = 3$, $\limsup_{t\to\infty} \|x(t) - \mathbf{e}_N\bar{x}(t)\|_2 \leq 11.66r$ for $N = 4$, $\limsup_{t\to\infty} \|x(t) - \mathbf{e}_N\bar{x}(t)\|_2 \leq 21.17r$ for $N = 5$, and $\limsup_{t\to\infty} \|x(t) - \mathbf{e}_N\bar{x}(t)\|_2 \leq 34.77r$ for $N = 6$. Now, consider a set of agents on an all-to-all graph. In this case, $\limsup_{t\to\infty} \|x(t) - \mathbf{e}_N\bar{x}(t)\|_2 \leq 1.41r$ for $N = 2$, $\limsup_{t\to\infty} \|x(t) - \mathbf{e}_N\bar{x}(t)\|_2 \leq 1.73r$ for $N = 3$, $\limsup_{t\to\infty} \|x(t) - \mathbf{e}_N\bar{x}(t)\|_2 \leq 2.00r$ for $N = 4$,

$\limsup_{t\to\infty} \|x(t) - \mathbf{e}_N \overline{x}(t)\|_2 \leq 2.23r$ for $N = 5$, and $\limsup_{t\to\infty} \|x(t) - \mathbf{e}_N \overline{x}(t)\|_2 \leq 2.44r$ for $N = 6$.

It is clear from the above examples that $\limsup_{t\to\infty} \|x(t) - \mathbf{e}_N \overline{x}(t)\|_2$ increases with the size of the network. It is also interesting to note that a network designer can introduce additional connectivity between agents to keep the bound on $\limsup_{t\to\infty} \|x(t) - \mathbf{e}_N \overline{x}(t)\|_2$ small as the size of the network is increased. This is clearly demonstrated in the above examples, wherein less conservative bounds are obtained for an all-to-all graph topology, which has a higher degree of connectivity between agents than for agents with a line graph topology.

Next, we apply Theorem 18.1 to an all-to-all connected graph network. Note that, in this case, $\mathcal{L} = NI_N - E_N$, where $E_N \triangleq \mathbf{e}_N \mathbf{e}_N^\mathrm{T}$ denotes the *ones matrix* of order $N \times N$. Since rank $E_N = 1$, $E_N$ has only one nonzero eigenvalue equal to $N$ with corresponding eigenvector $\mathbf{e}_N$. Next, note that

$$\det[\lambda I_N - \mathcal{L}] = \det[\lambda I_N - (NI_N - E_N)] = \det[(\lambda - N)I_N + E_N].$$

Hence, the eigenvalues of $\mathcal{L}$ are the eigenvalues of $-E_N$ shifted by $N$; that is, $\mathrm{spec}(-E_N) = \{0, N, \ldots, N\}$. Now, with $\lambda_2(\mathcal{L}) = \cdots = \lambda_N(\mathcal{L}) = N$, it follows from Theorem 18.1 that $\limsup_{t\to\infty} \|x(t) - \mathbf{e}_N \overline{x}\|_2 \leq \sqrt{N}r$.

Alternatively, we can arrive at the same result directly by considering the update protocol for the $i$th agent given by

$$\dot{x}_i(t) = \frac{1}{N} \sum_{j=1}^{N} [(x_j(t) - d_j(t)) - (x_i(t) - d_i(t))]$$

$$= \overline{x}(t) - x_i(t) - \overline{d}(t) + d_i(t), \quad x_i(0) = x_{i0}, \quad t \geq 0, \quad i = 1, \ldots, N, \tag{18.16}$$

where $\overline{x}(t) \triangleq \frac{1}{N} \sum_{j=1}^{N} x_j(t) \equiv \overline{x}$ and $\overline{d}(t) \triangleq \frac{1}{N} \sum_{j=1}^{N} d_j(t)$. First, note that it can be shown that $\limsup_{t\to\infty} \|x_i(t) - x_j(t)\|_2 \leq 2r$ for every $i, j = 1, \ldots, N$.

To see this, for $i, j = 1, \ldots, N$, it follows from (18.16) that

$$\frac{\mathrm{d}}{\mathrm{d}t}\left(\frac{1}{2}\|x_i(t) - x_j(t)\|_2^2\right)$$

$$= (x_i(t) - x_j(t))^\mathrm{T} \frac{\mathrm{d}}{\mathrm{d}t}(x_i(t) - x_j(t))$$

$$= (x_i(t) - x_j(t))^\mathrm{T}[\overline{x} - x_i(t) - \overline{d}(t) + d_i(t) - (\overline{x} - x_j(t) - \overline{d}(t) + d_j(t))]$$

$$= -\|x_i(t) - x_j(t)\|_2^2 + (x_i(t) - x_j(t))^\mathrm{T}(d_i(t) - d_j(t))$$

$$\leq -\|x_i(t) - x_j(t)\|_2^2 + 2r\|x_i(t) - x_j(t)\|_2,$$

$$x_i(0) - x_j(0) = x_{i0} - x_{j0}, \quad t \geq 0,$$

where the last inequality follows from the fact that

$$\|d_i(t) - d_j(t)\|_2 \leq \|d_i(t)\|_2 + \|d_j(t)\|_2 \leq 2r, \quad t \geq 0.$$

Hence, $\|x_i(t) - x_j(t)\|_2$ is a decreasing function of time as long as $\|x_i(t) - x_j(t)\|_2 > 2r$, $t \geq 0$. Now, it follows that $\|x_i(t) - x_j(t)\|_2 \leq 2r$ as $t \to \infty$ for all $i, j = 1, \ldots, N$.

Next, since $\bar{x}(t) \equiv \bar{x}$, it follows that $\|x_i(t) - \bar{x}\|_2 \leq r$ as $t \to \infty$ for all $i = 1, \ldots, N$. Furthermore, since

$$\|x(t) - \mathbf{e}_N \bar{x}\|_2^2 = \sum_{i=1}^{N} \|x_i(t) - \bar{x}\|_2^2 \leq Nr^2$$

as $t \to \infty$, it follows that $\limsup_{t \to \infty} \|x(t) - \mathbf{e}_N \bar{x}\|_2 \leq \sqrt{N} r$, which is identical to the result obtained by applying Theorem 18.1 directly.

## 18.5 Discrete-Time Consensus with a Connected Graph Topology

In this section, we consider the discrete-time consensus problem over an undirected network with a connected graph topology. Once again, we assume that only approximate information on the location of neighboring agents is available at any given time, with the $i$th agent uncertainty satisfying $\|d_i(k)\|_2 \leq r$, $k \in \overline{\mathbb{Z}}_+$, for $i = 1, \ldots, N$.

In particular, we consider the update protocol for agent $i$ given by

$$x_i(k+1) = x_i(k) + \varepsilon \sum_{j \in \mathcal{N}(i)} (z_j(k) - z_i(k)), \quad x_i(0) = x_{i0},$$

$$k \in \overline{\mathbb{Z}}_+, \quad i = 1, \ldots, N, \qquad (18.17)$$

where

$$z_j(k) - z_i(k) \triangleq (x_j(k) - d_j(k)) - (x_i(k) - d_i(k))$$

and $\varepsilon > 0$. In this case, it follows from (18.17) that

$$x_i(k+1) = x_i(k) + \varepsilon \sum_{j \in \mathcal{N}(i)} (x_j(k) - x_i(k)) + \varepsilon \sum_{j \in \mathcal{N}(i)} (d_i(k) - d_j(k)),$$

$$x_i(0) = x_{i0}, \quad k \in \overline{\mathbb{Z}}_+, \quad i = 1, \ldots, N,$$

or, equivalently, in compact form

$$x(k+1) = \tilde{\mathcal{P}} x(k) + \varepsilon \tilde{\mathcal{L}} d(k), \quad x(0) = x_0, \quad k \in \overline{\mathbb{Z}}_+, \qquad (18.18)$$

where $\tilde{\mathcal{L}} \triangleq I_n \otimes \mathcal{L} \in \mathbb{R}^{nN \times nN}$; $\tilde{\mathcal{P}} \triangleq I_n \otimes \mathcal{P} \in \mathbb{R}^{nN \times nN}$; $\mathcal{L} \in \mathbb{R}^{N \times N}$ denotes the graph Laplacian; $\mathcal{P} \triangleq I_N - \varepsilon \mathcal{L} \in \mathbb{R}^{N \times N}$ denotes the Perron matrix;

$x \triangleq [x_1^1, \ldots, x_N^1, \ldots, x_1^n, \ldots, x_N^n]^{\mathrm{T}}$; $d \triangleq [d_1^1, \ldots, d_N^1, \ldots, d_1^n, \ldots, d_N^n]^{\mathrm{T}}$; and $x_i^j$ and $d_i^j$ denote the $j$th components of $x_i$ and $d_i$, respectively.

Although our results can be directly extended to the case of (18.18), once again, for simplicity of exposition, we will focus on individual agent states evolving in $\mathbb{R}$ (i.e., $n = 1$). In this case, (18.18) becomes

$$x(k+1) = \mathcal{P}x(k) + \varepsilon \mathcal{L}d(k), \quad x(0) = x_0, \quad k \in \overline{\mathbb{Z}}_+. \qquad (18.19)$$

For the statement of the next result, define $\Delta_{\max} \triangleq \max_{i \in \{1, \ldots, N\}} \deg(i)$.

**Theorem 18.2.** Consider an undirected network of $N$ agents with a connected graph topology given by (18.19), and let $\varepsilon \in (0, \frac{1}{\Delta_{\max}})$. Then

$$\limsup_{k \to \infty} \|x(k) - \mathbf{e}_N \overline{x}(k)\|_2 \le \frac{\varepsilon \lambda_{\max}(\mathcal{L})\sqrt{N}r}{1 - \rho\left(\mathcal{P} - \frac{1}{N}\mathbf{e}_N \mathbf{e}_N^{\mathrm{T}}\right)}.$$

**Proof.** First, define $\delta(k) \triangleq x(k) - \mathbf{e}_N \overline{x}(k)$, and note that $\overline{x}(k+1) = \frac{1}{N}\mathbf{e}_N^{\mathrm{T}} x(k+1) = \frac{1}{N}\mathbf{e}_N^{\mathrm{T}}(x(k) + \varepsilon(-\mathcal{L}x(k) + \mathcal{L}d(k))) = \overline{x}(k)$, where we used the fact that $\mathcal{L}\mathbf{e}_N = 0_N$ and $\mathcal{L} = \mathcal{L}^{\mathrm{T}}$. Hence, $\overline{x}(k) = \frac{1}{N}\mathbf{e}_N^{\mathrm{T}} x(k) = \frac{1}{N}\mathbf{e}_N^{\mathrm{T}} x(0) = \overline{x}$, $k \in \overline{\mathbb{Z}}_+$, which shows that the centroid of the network does not change over time in the presence of time-varying interagent measurement uncertainties. Next, evaluating $\delta(k+1)$, $k \in \overline{\mathbb{Z}}_+$, yields

$$\delta(k+1) = x(k+1) - \mathbf{e}_N \overline{x}(k+1)$$
$$= \mathcal{P}x(k) + \varepsilon \mathcal{L}d(k) - \frac{1}{N}\mathbf{e}_N \mathbf{e}_N^{\mathrm{T}}[\mathcal{P}x(k) + \varepsilon \mathcal{L}d(k)]$$
$$= \mathcal{P}\left[x(k) - \frac{1}{N}\mathbf{e}_N \mathbf{e}_N^{\mathrm{T}} x(k)\right] + \left[I - \frac{1}{N}\mathbf{e}_N \mathbf{e}_N^{\mathrm{T}}\right]\varepsilon \mathcal{L}d(k)$$
$$= \left[\mathcal{P} - \frac{1}{N}\mathbf{e}_N \mathbf{e}_N^{\mathrm{T}}\right]\delta(k) + \varepsilon \mathcal{L}d(k), \quad \delta(0) = \delta_0, \quad k \in \overline{\mathbb{Z}}_+. \quad (18.20)$$

Now, considering a Lyapunov-like function $V : \mathbb{R}^{(N-1)} \to \mathbb{R}$ given by $V(\delta) = \|\delta\|_2$ and recalling that the spectral radius $\rho(M) = \|M\|_2$ for an arbitrary symmetric matrix $M$, it follows from (18.20) that

$$V(\delta(k+1)) = \|\delta(k+1)\|_2$$
$$\le \left\|\left(\mathcal{P} - \frac{1}{N}\mathbf{e}_N \mathbf{e}_N^{\mathrm{T}}\right)\delta(k)\right\|_2 + \|\varepsilon \mathcal{L}d(k)\|_2$$
$$\le \rho\left(\mathcal{P} - \frac{1}{N}\mathbf{e}_N \mathbf{e}_N^{\mathrm{T}}\right)\|\delta(k)\|_2 + \varepsilon \lambda_{\max}(\mathcal{L})\sqrt{N}r$$

$$= \left( \rho \left( \mathcal{P} - \frac{1}{N} \mathbf{e}_N \mathbf{e}_N^{\mathrm{T}} \right) + \frac{\varepsilon \lambda_{\max}(\mathcal{L})\sqrt{N}r}{\|\delta(k)\|_2} \right) V(\delta(k)), \quad k \in \overline{\mathbb{Z}}_+.$$

$$(18.21)$$

Hence, it follows from (18.21) that $V(\delta(k+1)) < V(\delta(k))$ for $\rho(\mathcal{P}-\frac{1}{N}\mathbf{e}_N\mathbf{e}_N^{\mathrm{T}})+$ $\frac{\varepsilon\lambda_{\max}(\mathcal{L})\sqrt{N}r}{\|\delta(k)\|_2} < 1$ and $k \in \overline{\mathbb{Z}}_+$. Now, recalling that all of the eigenvalues of the Perron matrix of an undirected connected graph with $\varepsilon \in (0, \frac{1}{\Delta_{\max}})$ are located in the unit circle and only one eigenvalue has an absolute value of one [236], it follows that $\rho(\mathcal{P} - \frac{1}{N}\mathbf{e}_N\mathbf{e}_N^{\mathrm{T}}) < 1$. Hence, it follows from (18.21) that

$$\|\delta(k)\|_2 \leq \frac{\varepsilon \lambda_{\max}(\mathcal{L})\sqrt{N}r}{1 - \rho\left(\mathcal{P} - \frac{1}{N}\mathbf{e}_N\mathbf{e}_N^{\mathrm{T}}\right)}$$

as $k \to \infty$, which completes the proof.                                       $\square$

Note that

$$\det\left[\lambda I_N - \left(\mathcal{P} - \frac{1}{N}E_N\right)\right] = \det\left[\lambda I_N - \left(I_N - \varepsilon\mathcal{L} - \frac{1}{N}E_N\right)\right]$$

$$= \det\left[(\lambda - 1)I_N - \left(-\varepsilon\mathcal{L} - \frac{1}{N}E_N\right)\right].$$

$$(18.22)$$

Now, since $E_N$ has only one nonzero eigenvalue equal to $N$ with the corresponding eigenvector $\mathbf{e}_N$ and $\mathcal{L}$ has only one zero eigenvalue with the corresponding eigenvector $\mathbf{e}_N$, it follows that

$$\mathrm{spec}\left(-\varepsilon\mathcal{L} - \frac{1}{N}E_N\right) = \{-1, -\varepsilon\lambda_2(\mathcal{L}), \ldots, -\varepsilon\lambda_N(\mathcal{L})\}.$$

Thus, it follows from (18.22) that

$$\mathrm{spec}\left(\mathcal{P} - \frac{1}{N}E_N\right) = \{0, (1 - \varepsilon\lambda_2(\mathcal{L})), \ldots, (1 - \varepsilon\lambda_N(\mathcal{L}))\}.$$

Hence, $\rho\left(\mathcal{P} - \frac{1}{N}\mathbf{e}_N\mathbf{e}_N^{\mathrm{T}}\right) = \max\{|(1 - \varepsilon\lambda_2(\mathcal{L}))|, |(1 - \varepsilon\lambda_N(\mathcal{L}))|\}$.

Next, we apply Theorem 18.2 to an all-to-all connected graph network. Note that, in this case, $\mathcal{L} = NI_N - E_N$. Now, recall that $\lambda_2(\mathcal{L}) = \cdots = \lambda_N(\mathcal{L}) = N$, and, hence, for $\varepsilon \in (0, \frac{1}{N})$, it follows from Theorem 18.2 that

$$\limsup_{k \to \infty} \|x(k) - \mathbf{e}_N\overline{x}(k)\|_2 \leq \frac{\varepsilon\lambda_{\max}(\mathcal{L})\sqrt{N}r}{1 - \rho\left(\mathcal{P} - \frac{1}{N}\mathbf{e}_N\mathbf{e}_N^{\mathrm{T}}\right)} = \frac{\varepsilon N\sqrt{N}r}{1 - (1 - \varepsilon N)} = \sqrt{N}r.$$

Alternatively, we can arrive at the same result directly by considering the update protocol for the $i$th agent given by

$$x_i(k+1) \in \alpha \frac{1}{N} \sum_{j=1}^N \mathcal{X}_j(k) + (1-\alpha)x_i(k) = \mathcal{B}_{\alpha r}(\alpha \overline{x}(k)) + (1-\alpha)x_i(k),$$

$$x_i(0) = x_{i0}, \quad k \in \overline{\mathbb{Z}}_+, \quad i = 1, \ldots, N, \tag{18.23}$$

where $\alpha \in (0,1]$ and $\overline{x}(k) \triangleq \frac{1}{N} \sum_{i=1}^N x_i(k) \equiv \overline{x}$. First, note that it can be shown that $\limsup_{k \to \infty} \|x_i(k) - x_j(k)\|_2 \leq 2r$ for every $i, j = 1, \ldots, N$.

To see this, for $i, j = 1, \ldots, N$, it follows from (18.23) that

$$x_i(k+1) - x_j(k+1)$$
$$\in \mathcal{B}_{\alpha r}(\alpha x_{\text{ave}}(k)) - \mathcal{B}_{\alpha r}(\alpha x_{\text{ave}}(k)) + (1-\alpha)(x_i(k) - x_j(k)), \quad k \in \overline{\mathbb{Z}}_+, \tag{18.24}$$

which implies that

$$\|x_i(k+1) - x_j(k+1)\|_2 \leq (1-\alpha)\|x_i(k) - x_j(k)\|_2 + 2r\alpha. \tag{18.25}$$

Hence, since $\|x_i(k+1) - x_j(k+1)\|_2 \leq \|x_i(k) - x_j(k)\|_2$ for $\|x_i(k) - x_j(k)\|_2 \geq 2r$, it follows that $\|x_i(k) - x_j(k)\|_2 \leq 2r$ as $k \to \infty$ for all $i, j = 1, \ldots, N$. Now, using identical arguments as in Section 18.4, it follows that $\limsup_{k \to \infty} \|x(k) - \mathbf{e}_N \overline{x}\|_2 \leq \sqrt{N}r$, which is identical to the result obtained by using Theorem 18.2 directly.

## 18.6  A Set-Valued Analysis Approach to Discrete-Time Consensus

In this section, we present a set-valued approach for the discrete-time consensus protocol considered in Section 18.5. Due to its mathematical generality, set-valued analysis can prove beneficial for generalizing our results to nonlinear network architectures with a dynamic network topology (see Chapter 6). Before presenting the main results of this section, we require some additional notation and definitions.

Specifically, consider the difference inclusion

$$x(k+1) \in \mathcal{F}(x(k)), \quad x(0) = x_0, \quad k \in \overline{\mathbb{Z}}_+, \tag{18.26}$$

where, for every $k \in \overline{\mathbb{Z}}_+$, $x(k) \in \mathbb{R}^n$, $\mathcal{F} : \mathbb{R}^n \to 2^{\mathbb{R}^n}$ is a *set-valued map* that assigns sets to points, and $2^{\mathbb{R}^n}$ denotes the collection of all subsets of $\mathbb{R}^n$. The set-valued map $\mathcal{F}$ has a *nonempty value at $x$* if $\mathcal{F}(x) \neq \emptyset$. It is assumed that $\mathcal{F}$ has nonempty values for every $x \in \mathbb{R}^n$. Hence, maximal solutions to (18.26) are complete, and, consequently, by a *solution* of (18.26) with initial condition $x(0) = x_0$, we mean a function $x : \overline{\mathbb{Z}}_+ \to \mathbb{R}^n$ that satisfies (18.26).

The set-valued map $\mathcal{F} : \mathbb{R}^n \to 2^{\mathbb{R}^n}$ is *outer semicontinuous at $x$* if, for every sequence $\{x_i\}_{i=0}^{\infty}$ such that $\lim_{i\to\infty} x_i = x$, every convergent sequence $\{y_i\}_{i=0}^{\infty}$ with $y_i \in \mathcal{F}(x_i)$ satisfies $\lim_{i\to\infty} y_i \in \mathcal{F}(x)$. $\mathcal{F}$ is *continuous at $x$* if $\mathcal{F}$ is outer semicontinuous at $x$ and, for every $y \in \mathcal{F}(x)$ and every convergent sequence $\{x_i\}_{i=0}^{\infty}$, there exists $y_i \in \mathcal{F}(x_i)$ such that $\lim_{i\to\infty} y_i = y$. $\mathcal{F}(x)$ is *locally bounded at $x$* if there exists a neighborhood $\mathcal{N}$ of $x$ such that $\mathcal{F}(\mathcal{N}) = \cup_{z\in\mathcal{N}}\mathcal{F}(z)$ is bounded. If $\mathcal{F}$ has compact values and is locally bounded at $x$, then $\mathcal{F}$ is *upper semicontinuous at $x$*; that is, for every $\varepsilon > 0$, there exists $\delta > 0$ such that, for all $z \in \mathbb{R}^n$ satisfying $\|z - x\| < \delta$, $\mathcal{F}(z) \subseteq \mathcal{F}(x) + \overline{\mathcal{B}}_{\varepsilon}(0)$, where $\overline{\mathcal{B}}_{\varepsilon}(0)$ denotes the closure of $\mathcal{B}_{\varepsilon}(0)$.

Given the function $\gamma : \overline{\mathbb{Z}}_+ \to \mathbb{R}^n$, the *positive limit set of $\gamma$* is the set $\Omega(\gamma)$ of points $y \in \mathbb{R}^n$ for which there exists an increasing divergent sequence $\{k_n\}_{n=0}^{\infty}$ satisfying $\lim_{n\to\infty} \gamma(k_n) = y$. We denote the positive limit set of a solution $\psi(\cdot)$ of (18.26) by $\Omega(\psi)$. The positive limit set of a bounded solution of (18.26) is nonempty, compact, and weakly forward invariant with respect to (18.26) [261].

The following theorem gives a general set-valued invariance principle using the set-valued analysis tools developed in [103] and is necessary for the main result of this section.

**Theorem 18.3.** Consider the difference inclusion given by (18.26). Assume that $\mathcal{F} : \mathbb{R}^n \to 2^{\mathbb{R}^n}$ is outer semicontinuous and locally bounded, with nonempty values for all $x \in \mathbb{R}^n$. Let $V : \mathbb{R}^n \to 2^{\mathbb{R}^n}$ be a continuous set-valued map, and let $\mathcal{M} \subset \mathbb{R}^n$ be a closed set such that the following statements hold.

(i) $V(\mathcal{F}(x)) \subseteq V(x)$ for every $x \in \mathbb{R}^n$.

(ii) If $V(y) = V(x)$ for some $y \in \mathcal{F}(x)$, then $x \in \mathcal{M}$.

Then every bounded solution $x : \overline{\mathbb{Z}}_+ \to \mathbb{R}^n$ of (18.26) converges to $\mathcal{M}$; that is, $\lim_{k\to\infty} \text{dist}(x(k), \mathcal{M}) = 0$.

**Proof.** It follows from (i) that $V(\psi(k + 1)) \subseteq V(\psi(k))$ for every solution $\psi(k)$, $k \in \overline{\mathbb{Z}}_+$, of (18.26). Thus, the sequence of closed sets $\{V(\psi(k))\}_{k=0}^{\infty}$ is nonincreasing, and, hence, $\lim_{k\to\infty} V(\psi(k)) = \cap_{k=0}^{\infty} V(\psi(k)) \triangleq \mathcal{V}$ [261]. Next, note that, since $\psi(k)$, $k \in \overline{\mathbb{Z}}_+$, is bounded, $\Omega(\psi)$ is nonempty. Now, for all $x \in \Omega(\psi)$, it follows from the definition of $\Omega(\psi)$ and the continuity of $V$ that $V(x) = \mathcal{V}$. Moreover, the outer semicontinuity of $\mathcal{F}$ ensures that $\Omega(\psi)$ is weakly positively (and negatively) invariant. Specifically, for every $x \in \Omega(\psi)$, there exists $y \in \mathcal{F}(x)$ such that $y \in \Omega(\psi)$. Thus, for every $x \in \Omega(\psi)$, there exists $y \in \mathcal{F}(x)$ such that $V(x) = V(y) = \mathcal{V}$, and, hence,

$\Omega(\psi) \subseteq \mathcal{M}$. Finally, since $\mathrm{dist}(\psi(k), \Omega(\psi)) \to 0$ as $k \to 0$, it follows that $\psi(k) \to \mathcal{M}$ as $k \to \infty$.    □

Next, we illustrate Theorem 18.3 by applying it to the network system given by (18.23). The conclusions of the proposition below are weaker than the results obtained in Section 18.5. However, as noted above, the set-valued approach can prove beneficial for nonlinear network architectures, where direct computation relying on a linear structure is not possible, as well as for partial graph connectivity structures with directed information flow.

**Proposition 18.1.** Consider a network of $N$ agents with an all-to-all graph connectivity given by (18.23), and let $x(\cdot)$ be a bounded solution of (18.23). Then $\lim \sup_{k \to \infty} \|x_i(k) - x_j(k)\|_2 \le 4r$ for every $i, j = 1, \ldots, N$.

**Proof.** Let the set-valued map $V : \mathbb{R}^n \to 2^{\mathbb{R}^n}$ be given by

$$V(x) = \mathcal{B}_{\delta_1(x)}(x_{\mathrm{ave}}) \times \cdots \times \mathcal{B}_{\delta_N(x)}(x_{\mathrm{ave}}),$$

where, for every $i \in \{1, \ldots, N\}$,

$$\delta_i(x) = \begin{cases} \|x_i - x_{\mathrm{ave}}\|_2, & \|x_i - x_{\mathrm{ave}}\|_2 \ge 2r, \\ 2r, & \|x_i - x_{\mathrm{ave}}\|_2 \le 2r, \end{cases}$$

and $\times$ denotes the Cartesian product. Note that $V$ is continuous and has closed and bounded values. Next, it can be shown using a similar argument as in Section 18.5 that

$$\begin{aligned} x_i(k+1) &- x_{\mathrm{ave}}(k+1) \\ &\in \mathcal{B}_{\alpha r}(\alpha x_{\mathrm{ave}}(k)) - \mathcal{B}_{\alpha r}(x_{\mathrm{ave}}(k)) + (1 - \alpha)x_i(k), \quad k \in \overline{\mathbb{Z}}_+, \end{aligned}$$

which implies that

$$\|x_i(k+1) - x_{\mathrm{ave}}(k+1)\|_2 \le (1 - \alpha)\|x_i(k) - x_{\mathrm{ave}}(k)\|_2 + 2r\alpha.$$

Hence, the function $\delta_i(\cdot)$ decreases for $\|x_i - x_{\mathrm{ave}}\|_2 > 2r$ and remains constant for $\|x_i - x_{\mathrm{ave}}\|_2 \le 2r$, $i \in \{1, \ldots, N\}$, and, hence, conditions (i) and (ii) of Theorem 18.3 are satisfied. Now, it follows from Theorem 18.3 that every bounded solution $x_i(\cdot)$, $i \in \{1, \ldots, N\}$, converges to $\mathcal{B}_{2r}(x_{\mathrm{ave}})$. Hence, $\|x_i(k) - x_j(k)\|_2 \le 4r$ as $k \to \infty$ for all $i, j = 1, \ldots, N$.    □

Next, we present two illustrative numerical examples to demonstrate the efficacy of the proposed framework.

**Example 18.1.** In this example, we consider a random network of 10 agents with connected, undirected, and time-invariant communication graph

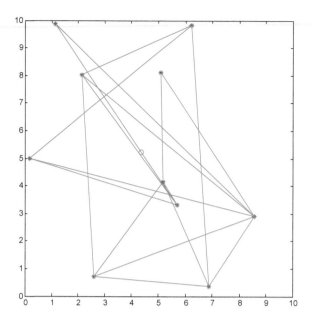

Figure 18.2 Initial network configuration of 10 agents with sensor accuracy of radius $r = 1$.

network topologies and with agent dynamics given by (18.8). Furthermore, we assume that the $i$th agent uncertainty is modeled as a standard white noise process.

Figures 18.2, 18.3, 18.4, and 18.5 show the initial, intermediate, and final network configurations, as well as $\|x(t) - \mathbf{e}_N \bar{x}\|_2$ versus time, of the network of agents when the agents have sensor accuracy of radius one, $\lambda_2(\mathcal{L}) = 1.5568$, and $\lambda_N(\mathcal{L}) = 7.5704$. The circle indicates the location of the initial centroid of the agents. Note that

$$\limsup_{t \to \infty} \|x(t) - \mathbf{e}_N \bar{x}\|_2 \leq \frac{\lambda_N(\mathcal{L}) \sqrt{N} r}{\lambda_2(\mathcal{L})} = 15.3775.$$

Alternatively, Figures 18.6, 18.7, 18.8, and 18.9 show the initial, intermediate, and final network configurations, as well as $\|x(t) - \mathbf{e}_N \bar{x}\|_2$ versus time, of the network of agents when the agents have sensor accuracy of radius one, $\lambda_2(\mathcal{L}) = 0.1172$, and $\lambda_N(\mathcal{L}) = 4.3721$. Once again, the circle indicates the location of the initial centroid of the agents. Note that

$$\limsup_{t \to \infty} \|x(t) - \mathbf{e}_N \bar{x}\|_2 \leq \frac{\lambda_N(\mathcal{L}) \sqrt{N} r}{\lambda_2(\mathcal{L})} = 117.9675.$$

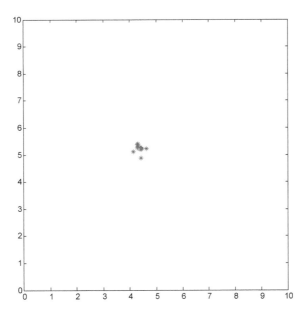

Figure 18.3 Network configuration of 10 agents with sensor accuracy of radius $r = 1$ at $t = 3.5$ sec.

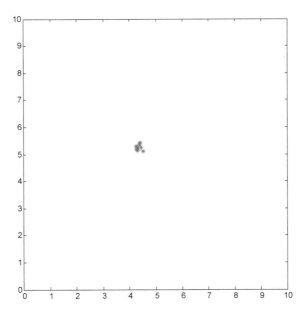

Figure 18.4 Network configuration of 10 agents with sensor accuracy of radius $r = 1$ at $t = 7.5$ sec.

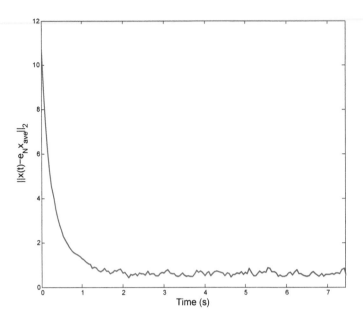

Figure 18.5 Plot of $\|x(t) - \mathbf{e}_N \bar{x}\|_2$ versus time.

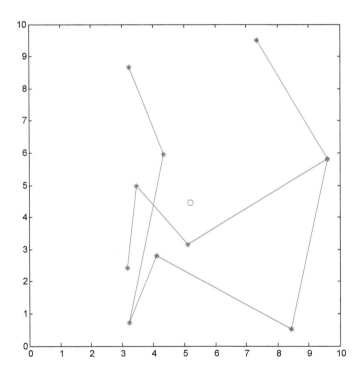

Figure 18.6 Initial network configuration of 10 agents with sensor accuracy of radius $r = 1$.

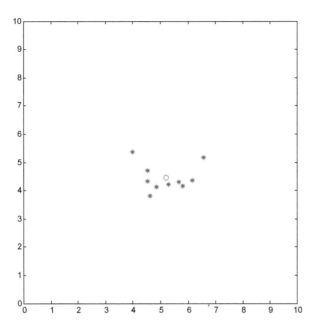

Figure 18.7 Network configuration of 10 agents with sensor accuracy of radius $r = 1$ at $t = 3.5$ sec.

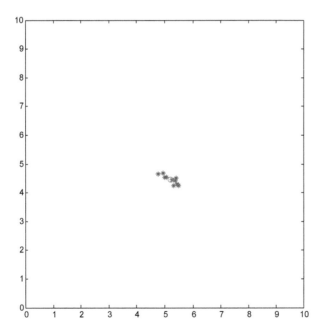

Figure 18.8 Network configuration of 10 agents with sensor accuracy of radius $r = 1$ at $t = 7.5$ sec.

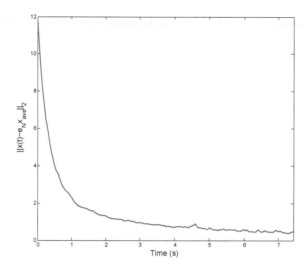

Figure 18.9 Plot of $\|x(t) - \mathbf{e}_N \bar{x}\|_2$ versus time.

Finally, Figures 18.10, 18.11, and 18.12 show the initial, intermediate, and final configurations, respectively, of the network of 10 agents when agents have sensor accuracy of radius 0.5 and the network is all-to-all connected. The simulation shows that the agents reach a consensus set with diameter less than $2r = 1$. The circle indicates a set with diameter one centered at the initial centroid of the agents.                                                  $\triangle$

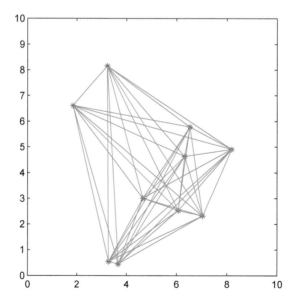

Figure 18.10 Initial network configuration of 10 agents with sensor accuracy of radius $r = 0.5$.

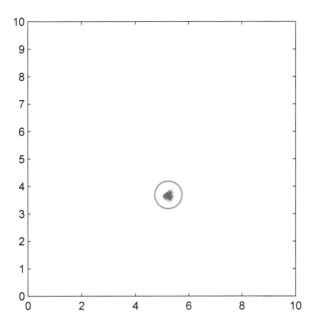

Figure 18.11 Network configuration of 10 agents with sensor accuracy of radius $r = 0.5$ at $t = 3.5$ sec.

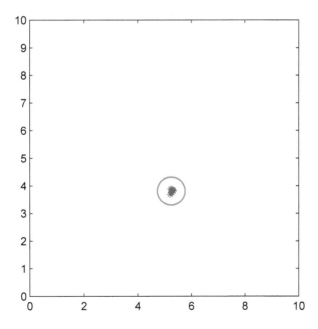

Figure 18.12 Network configuration of 10 agents with sensor accuracy of radius $r = 0.5$ at $t = 7.5$ sec.

**Example 18.2.** In this example, we use the proposed framework for pitch rate consensus of a set of commercial airplanes in the presence of inaccurate sensor measurements, which are modeled as a standard white noise process coupled with sinusoidal time-varying exogenous disturbances. Specifically, consider the multiagent system representing the controlled longitudinal motion of three Boeing 747 airplanes linearized at an altitude of 40 kft and a velocity of 774 ft/sec addressed in Example 2.1 given by

$$\dot{z}_i(t) = A z_i(t) + B \nu_i(t), \quad z_i(0) = z_{i_0}, \quad i = 1,2,3, \quad t \geq 0, \qquad (18.27)$$

where the state

$$z_i(t) = [v_{x_i}(t), v_{z_i}(t), q_i(t), \theta_{e_i}(t)]^{\mathrm{T}} \in \mathbb{R}^4, \quad t \geq 0, \qquad (18.28)$$

of agent $i$, $i = 1,2,3$; control input $\nu_i(t)$; and system matrices $A$ and $B$ are as defined in Example 2.1.

As in Example 2.1, we utilize the two-level hierarchical controller proposed in [322], which is composed of a lower-level controller for command following and a higher-level controller for pitch rate consensus of the three airplanes given by (18.27). To address the lower-level controller design, let $x_i(t)$, $i = 1,2,3$, $t \geq 0$, be a command generated by (18.6) (i.e., the guidance command), and let $s_i(t)$, $i = 1,2,3$, $t \geq 0$, denote the integrator state satisfying

$$\dot{s}_i(t) = E z_i(t) - x_i(t), \quad s_i(0) = s_{i_0}, \quad i = 1,2,3, \quad t \geq 0, \qquad (18.29)$$

where $E = [0,0,1,0]$.

Now, defining the augmented state $\hat{z}_i(t) \triangleq [z_i^{\mathrm{T}}(t), s_i(t)]^{\mathrm{T}}$, (18.27) and (18.29) give

$$\dot{\hat{z}}_i(t) = \hat{A}\hat{z}_i(t) + \hat{B}_1 \nu_i(t) + \hat{B}_2 x_i(t), \quad \hat{z}_i(0) = \hat{z}_{i_0}, \quad i = 1,2,3, \quad t \geq 0, \qquad (18.30)$$

where $\hat{A}$, $\hat{B}_1$, and $\hat{B}_2$ are given by (2.33). Furthermore, let the elevator control input be given by

$$\nu_i(t) = -K\hat{z}_i(t), \quad K = [-0.0157, 0.0831, -4.7557, -0.1400, -9.8603], \quad t \geq 0, \qquad (18.31)$$

which is designed using an optimal LQR.

For the higher-level controller design, we use (18.6) to generate an $x_i(t)$, $t \geq 0$, that has a direct effect on the lower-level controller design to achieve pitch rate consensus. Specifically, the lower-level controller for each agent allows for the tracking of $x_i(t)$, $t \geq 0$, whereas the higher-level controllers allow for the implementation of (18.6). Figures 18.13 and 18.14 present the results for all initial conditions set to zero and $x_1(0) = 10$,

$x_2(0) = 2.5$, and $x_3(0) = 5$. In particular, Figure 18.13 shows that the three airplanes on a line graph achieve approximate pitch rate consensus in the presence of inaccurate sensor measurements with $r = 1$, where the collective behavior of these airplanes satisfies

$$\limsup_{t \to \infty} \|x(t) - \mathbf{e}_N \bar{x}\|_2 \leq \frac{\lambda_N(\mathcal{L})\sqrt{N}r}{\lambda_2(\mathcal{L})} = 5.1962.$$

Figure 18.14 shows a similar collective behavior performance for the airplanes for an all-to-all connected graph with $r = 1$, where

$$\limsup_{t \to \infty} \|x(t) - \mathbf{e}_N \bar{x}\|_2 \leq \frac{\lambda_N(\mathcal{L})\sqrt{N}r}{\lambda_2(\mathcal{L})} = 1.7321$$

holds.                                                                                    △

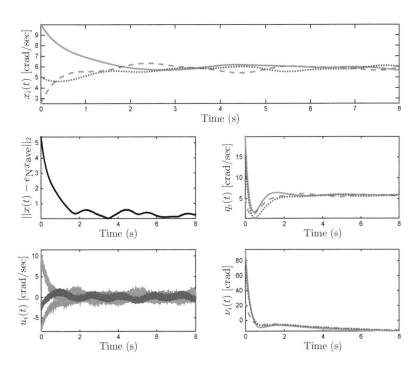

Figure 18.13  Agent guidance state $(x_i(t),\ t \geq 0)$, $\|x(t) - \mathbf{e}_N \bar{x}\|_2$, pitch rate $(q_i(t),\ t \geq 0)$, guidance input $(u_i(t),\ t \geq 0)$, and elevator control $(\nu_i(t),\ t \geq 0)$ responses for the three airplanes on a line graph in the presence of inaccurate sensor measurements (solid, dashed, and dotted lines denote the responses for the first, second, and third airplanes, respectively).

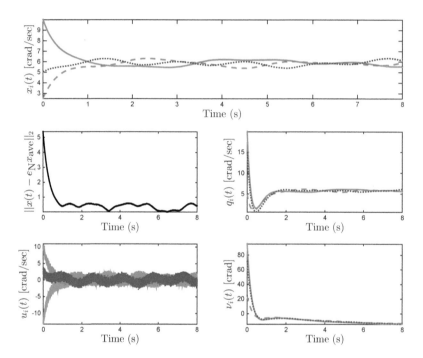

Figure 18.14 Agent guidance state $(x_i(t),\ t \geq 0)$, $\|x(t) - \mathbf{e}_N \overline{x}\|_2$, pitch rate $(q_i(t),\ t \geq 0)$, guidance input $(u_i(t),\ t \geq 0)$, and elevator control $(\nu_i(t),\ t \geq 0)$ responses for the three airplanes on an all-to-all graph in the presence of inaccurate sensor measurements (solid, dashed, and dotted lines denote the responses for the first, second, and third airplanes, respectively).

# A Hybrid Thermodynamic Control Protocol for Network Systems with Intermittent Information

## 19.1 Introduction

In Chapters 2, 3, 4, and 15, we presented an alternative perspective to the consensus control problem that is based on *dynamical thermodynamics* [118, 127]—a framework that unifies the foundational disciplines of thermodynamics and dynamical systems theory. Specifically, by generalizing the notions of temperature, energy, and entropy, dynamical thermodynamics is used in Chapters 2, 3, 4, and 15 to develop a design procedure for distributed consensus controllers that allow networked dynamical systems to emulate thermodynamic behavior.

In particular, for network systems with an undirected and directed communication graph topology, system thermodynamic notions are used to show that every control law protocol of a symmetric Fourier type, with information (or communication) transfer playing the role of energy flow, achieves information consensus [22, 137, 162, 167]. Unlike most of the nonlinear consensus control protocols presented in the literature, which merely guarantee system convergence, a novel feature of the thermodynamic-based consensus control protocols developed in Chapters 2, 3, 4, and 15 is their ability to address both convergence and stability.

Even though the consensus control problem for continuous-time and discrete-time multiagent systems has been extensively studied by numerous researchers, the *hybrid consensus problem* involving controllers that combine a logical switching architecture with continuous dynamics has received far less attention in the literature, with notable exceptions including [112, 113, 136, 151, 153, 161, 181, 208, 209, 248, 252, 317]. In particular, the authors in [136] develop a hybrid control framework for addressing multiagent formation control protocols for general nonlinear dynamical systems using hybrid stabilization of sets, whereas [153, 161] present a hybrid impulsive consensus protocol predicated on a Laplacian disagreement function.

Alternatively, [112, 181, 208, 209] present impulsive consensus algorithms using sampled information, and [113, 151, 252, 317] develop consensus protocols for first- and second-order systems using impulsive information transfer. More recently, [248] developed a hybrid consensus protocol over a network with a directed graph topology in which communication events are triggered stochastically. However, none of the aforementioned references address the problem of hybrid control for network information consensus using a hybrid thermodynamic framework.

In this chapter, we develop a hybrid control framework for semistability and consensus of multiagent systems with intermittent information. Specifically, we use impulsive differential equations [14, 128, 192] to construct a hybrid control architecture for addressing network information consensus wherein communication events are triggered via state-dependent resettings. The proposed controller architecture is predicated on the recently developed notion of hybrid dynamical thermodynamics [118, 119], resulting in a hybrid controller architecture involving the exchange of generalized energy state information between agents that guarantees that the closed-loop dynamical network is semistable and consistent with basic thermodynamic principles. A unique feature of the proposed framework is that the consensus control protocol is hybrid and, hence, can achieve significantly improved convergence in comparison to continuous-time protocols.

## 19.2 Mathematical Preliminaries

In this chapter, we design consensus control protocols wherein information is only available at intermittent instances arising from a hybrid dynamical system $\mathcal{G}$ of the form

$$\dot{x}(t) = f_c(x(t)), \qquad x(0) = x_0, \qquad x(t) \notin \mathcal{Z}, \qquad (19.1)$$
$$\Delta x(t) = f_d(x(t)), \qquad x(t) \in \mathcal{Z}, \qquad (19.2)$$

where, for every $t \geq 0$, $x(t) \in \mathcal{D} \subseteq \mathbb{R}^n$; $\mathcal{D}$ is an open set with $0 \in \mathcal{D}$; $\Delta x(t) \triangleq x(t^+) - x(t)$, where $x(t^+) \triangleq x(t) + f_d(x(t)) = \lim_{\varepsilon \to 0^+} x(t + \varepsilon)$; $f_c : \mathcal{D} \to \mathbb{R}^n$ is Lipschitz continuous and satisfies $f_c(0) = 0$; $f_d : \mathcal{D} \to \mathbb{R}^n$ is continuous; and $\mathcal{Z} \subset \mathcal{D}$ is the *resetting set*. Note that $x_e \in \mathcal{D}$ is an equilibrium point of (19.1) and (19.2) if and only if $f_c(x_e) = 0$ and $f_d(x_e) = 0$. We refer to the differential equation (19.1) as the *continuous-time dynamics*, and we refer to the difference equation (19.2) as the *resetting law*.

A function $x : \mathcal{I}_{x_0} \to \mathcal{D}$ is a *solution* to the impulsive dynamical system (19.1) and (19.2) on the interval $\mathcal{I}_{x_0} \subseteq \mathbb{R}$ with initial condition $x(0) = x_0$, where $\mathcal{I}_{x_0}$ denotes the maximal interval of existence of a solution to (19.1) and (19.2), if $x(\cdot)$ is left-continuous and $x(t)$ satisfies (19.1) and (19.2) for all $t \in \mathcal{I}_{x_0}$. For further discussion on solutions to impulsive differential equations, see [14, 128, 192, 225, 269].

For a particular closed-loop trajectory $x(t)$, we let $t_k \triangleq \tau_k(x_0)$ denote the $k$th instant of time at which $x(t)$ intersects $\mathcal{Z}$, and we call the times $t_k$ the *resetting times*. Thus, the trajectory of the closed-loop system (19.1) and (19.2) from the initial condition $x(0) = x_0$ is given by $\psi(t, x_0)$ for $0 < t \le t_1$, where $\psi(\cdot, \cdot)$ denotes the solution to the continuous-time dynamics (19.1). If the trajectory reaches a state $x_1 \triangleq x(t_1)$ satisfying $x_1 \in \mathcal{Z}$, then the state is instantaneously transferred to $x_1^+ \triangleq x_1 + f_{\mathrm{d}}(x_1)$ according to the resetting law (19.2). The trajectory $x(t)$, $t_1 < t \le t_2$, is then given by $\psi(t - t_1, x_1^+)$, and so on. Our convention here is that the solution $x(t)$ of (19.1) and (19.2) is left-continuous, that is, it is continuous everywhere except at the resetting times $t_k$, and $x_k \triangleq x(t_k) = \lim_{\varepsilon \to 0^+} x(t_k - \varepsilon)$ and $x_k^+ \triangleq x(t_k) + f_{\mathrm{d}}(x(t_k)) = \lim_{\varepsilon \to 0^+} x(t_k + \varepsilon)$ for $k = 1, 2, \ldots$.

To ensure the well-posedness of the solutions to (19.1) and (19.2), we make the following additional assumptions [128].

**Assumption 19.1.** If $x \in \overline{\mathcal{Z}} \backslash \mathcal{Z}$, then there exists $\varepsilon > 0$ such that, for all $0 < \delta < \varepsilon$, $\psi(\delta, x) \notin \mathcal{Z}$.

**Assumption 19.2.** If $x \in \mathcal{Z}$, then $x + f_{\mathrm{d}}(x) \notin \mathcal{Z}$.

Assumption 19.1 ensures that if a trajectory reaches the closure of $\mathcal{Z}$ at a point that does not belong to $\mathcal{Z}$, then the trajectory must be directed away from $\mathcal{Z}$; that is, a trajectory cannot enter $\mathcal{Z}$ through a point that belongs to the closure of $\mathcal{Z}$ but not to $\mathcal{Z}$. Furthermore, Assumption 19.2 ensures that when a trajectory intersects the resetting set $\mathcal{Z}$, it instantaneously exits $\mathcal{Z}$. Finally, we note that if $x_0 \in \mathcal{Z}$, then the system initially resets to $x_0^+ = x_0 + f_{\mathrm{d}}(x_0) \notin \mathcal{Z}$, which serves as the initial condition for the continuous-time dynamics (19.1).

It follows from Assumptions 19.1 and 19.2 that for a particular initial condition, the resetting times $t_k = \tau_k(x_0)$ are distinct and well defined [128]. Since the resetting set $\mathcal{Z}$ is a subset of the state space and is independent of time, impulsive dynamical systems of the form (19.1) and (19.2) are time-invariant systems. These systems are called *state-dependent impulsive dynamical systems* [128]. Since the resetting times are well defined and distinct, and since the solution to (19.1) exists and is unique, it follows that the solution of the impulsive dynamical system (19.1) and (19.2) also exists and is unique over a forward time interval. For details on the existence and uniqueness of solutions of impulsive dynamical systems in forward time, see [14, 192, 269].

Let $x^* \in \mathcal{D}$ satisfy $f_{\mathrm{d}}(x^*) = 0$. Then $x^* \notin \mathcal{Z}$. To see this, suppose that $x^* \in \mathcal{Z}$. Then $x^* + f_{\mathrm{d}}(x^*) = x^* \in \mathcal{Z}$, which contradicts the assumption that if $x \in \mathcal{Z}$, then $x + f_{\mathrm{d}}(x) \notin \mathcal{Z}$. Hence, if $x = 0$ is an equilibrium point of (19.1) and (19.2), then $0 \notin \mathcal{Z}$. Hence, if $x = x^*$ is an equilibrium point of (19.1) and (19.2), then $x^* \notin \mathcal{Z}$.

For the statement of the next result, the following key assumption is needed.

**Assumption 19.3.** Consider the impulsive dynamical system (19.1) and (19.2), and let $s(t, x_0)$, $t \geq 0$, denote the solution to (19.1) and (19.2) with initial condition $x_0$. Then, for every $x_0 \notin \mathcal{Z}$ and every $\varepsilon > 0$ and $t \neq t_k$, there exists $\delta(\varepsilon, x_0, t) > 0$ such that if $\|x_0 - z\| < \delta(\varepsilon, x_0, t)$, $z \in \mathcal{D}$, then $\|s(t, x_0) - s(t, z)\| < \varepsilon$.

Assumption 19.3 is a weakened version of the quasi-continuous dependence assumption given in [128] and is a generalization of the standard continuous dependence property for dynamical systems with continuous flows to dynamical systems with left-continuous flows. Sufficient conditions for guaranteeing that the impulsive dynamical system (19.1) and (19.2) satisfies Assumption 19.3 are given in [125, 128] and Section 21.2.

The next result characterizes impulsive dynamical system limit sets in terms of continuously differentiable functions. In particular, it guarantees that the system trajectories converge to an invariant set contained in a union of level surfaces characterized by the continuous-time dynamics and the resetting system dynamics. For the statement of this result, we assume that $f_{\mathrm{c}}(\cdot)$, $f_{\mathrm{d}}(\cdot)$, and $\mathcal{Z}$ are such that the dynamical system $\mathcal{G}$ given by (19.1) and (19.2) satisfies Assumption 19.3.

**Theorem 19.1** (see [128]). Consider the hybrid nonnegative dynamical system $\mathcal{G}$ given by (19.1) and (19.2), assume that $\mathcal{D}_{\mathrm{c}} \subset \mathcal{D} \subseteq \mathbb{R}^n$ is a compact positively invariant set with respect to (19.1) and (19.2), and assume that there exists a continuously differentiable function $V : \mathcal{D}_{\mathrm{c}} \to \mathbb{R}$ such that

$$V'(x) f_{\mathrm{c}}(x) \leq 0, \quad x \in \mathcal{D}_{\mathrm{c}}, \quad x \notin \mathcal{Z}, \tag{19.3}$$
$$V(x + f_{\mathrm{d}}(x)) \leq V(x), \quad x \in \mathcal{D}_{\mathrm{c}}, \quad x \in \mathcal{Z}. \tag{19.4}$$

Let

$$\mathcal{R} \triangleq \{x \in \mathcal{D}_c : x \notin \mathcal{Z}, \ V'(x)f_c(x) = 0\}$$
$$\cup \ \{x \in \mathcal{D}_c : x \in \mathcal{Z}, \ V(x + f_d(x)) = V(x)\},$$

and let $\mathcal{M}$ denote the largest invariant set contained in $\mathcal{R}$. If $x_0 \in \mathcal{D}_c$, then $x(t) \to \mathcal{M}$ as $t \to \infty$.

## 19.3  A Hybrid Thermodynamic Consensus Control Architecture

In this section, we develop a thermodynamically motivated information consensus framework for multiagent nonlinear systems to achieve semistability and state equipartition. The consensus problem we consider involves a dynamic communication graph with intermittent information over the dynamical network characterized by the multiagent impulsive dynamical systems $\mathcal{G}_i$ given by

$$\dot{x}_i(t) = u_{ci}(t), \quad x_i(0) = x_{i0}, \quad x_i(t) \notin \mathcal{Z}_i, \tag{19.5}$$

$$\Delta x_i(t) = u_{di}(t), \quad x_i(t) \in \mathcal{Z}_i, \quad i = 1, \dots, q, \tag{19.6}$$

where, for every $t \geq 0$, $x_i(t) \in \mathbb{R}$ denotes the information state and $u_{ci}(t)$ and $u_{di}(t) \in \mathbb{R}$, respectively, denote the continuous and discrete information control inputs associated with the local resetting set $\mathcal{Z}_i \subset \mathbb{R}$, $i \in \{1, \dots, q\}$.

The hybrid consensus protocol is given by

$$u_{ci}(t) = \sum_{j=1, i \neq j}^{q} \phi_{cij}(x_i(t), x_j(t)), \tag{19.7}$$

$$u_{di}(t) = \sum_{j=1, i \neq j}^{q} \phi_{dij}(x_i(t), x_j(t)), \tag{19.8}$$

where, for all $i, j = 1, \dots, q$, $i \neq j$, $\phi_{cij}(\cdot, \cdot)$ is locally Lipschitz continuous, $\phi_{dij}(\cdot, \cdot)$ is continuous, $\phi_{cij}(x_i, x_j) = -\phi_{cji}(x_j, x_i)$, and $\phi_{dij}(x_i, x_j) = -\phi_{dji}(x_j, x_i)$. In this case, the closed-loop system (19.5)–(19.8) is given by

$$\dot{x}_i(t) = \sum_{j=1, i \neq j}^{q} \phi_{cij}(x_i(t), x_j(t)), \quad x_i(0) = x_{i0}, \tag{19.9}$$

$$x_i(t) \notin \mathcal{Z}_i, \quad i = 1, \dots, q,$$

$$\Delta x_i(t) = \sum_{j=1, i \neq j}^{q} \phi_{dij}(x_i(t), x_j(t)), \quad x_i(t) \in \mathcal{Z}_i, \tag{19.10}$$

or, equivalently, in vector form,

$$\dot{x}(t) = f_c(x(t)), \qquad x(0) = x_0, \qquad x(t) \notin \mathcal{Z}, \qquad (19.11)$$
$$\Delta x(t) = f_d(x(t)), \qquad x(t) \in \mathcal{Z}, \qquad\qquad\qquad (19.12)$$

where $x(t) \triangleq [x_1(t), \ldots, x_q(t)]^T \in \mathbb{R}^q$, $f_c(x(t)) \triangleq [f_{c1}(x(t)), \ldots, f_{cq}(x(t))]^T \in \mathbb{R}^q$, $f_d(x(t)) \triangleq [f_{d1}(x(t)), \ldots, f_{dq}(x(t))]^T \in \mathbb{R}^q$, and $\mathcal{Z} \triangleq \cup_{i=1}^q \{x \in \mathbb{R}^q : x_i \in \mathcal{Z}_i\}$, with, for $i, j = 1, \ldots, q$,

$$f_{ci}(x(t)) = \sum_{j=1, i \neq j}^q \phi_{cij}(x_i(t), x_j(t)), \qquad (19.13)$$

$$f_{di}(x(t)) = \sum_{j=1, i \neq j}^q \phi_{dij}(x_i(t), x_j(t)). \qquad (19.14)$$

Note that $\mathcal{G}$ given by (19.9) and (19.10) describes an interconnected network where information or communication states are updated using a distributed hybrid controller involving neighbor-to-neighbor interaction between agents. Furthermore, this hybrid control protocol involves a design procedure for consensus with intermittent transmission of information as defined by the local resetting sets $\mathcal{Z}_i$, $i \in \{1, \ldots, q\}$. Although our results can be directly extended to the case where (19.5) and (19.6) describe the hybrid dynamics of an aggregate system with an aggregate state vector $x(t) = [x_1^T(t), \ldots, x_q^T(t)]^T \in \mathbb{R}^{Nq}$, where $x_i(t) \in \mathbb{R}^N$, $u_{ci}(t) \in \mathbb{R}^N$, and $u_{di}(t) \in \mathbb{R}^N$, $i = 1, \ldots, q$, by using Kronecker algebra, for simplicity of exposition we focus on individual agent states evolving in $\mathbb{R}$ (i.e., $N = 1$).

In order to define the global resetting set $\mathcal{Z}$ in terms of the local resetting sets $\mathcal{Z}_i$, $i = 1, \ldots, q$, associated with $\mathcal{G}$, we require some additional notation. Let $\mathcal{O}_i$ denote the set of all agents with information flowing out to the $i$th agent, and let $\mathcal{I}_i$ denote the set of all agents receiving information from the $i$th agent. We define the local resetting sets $\mathcal{Z}_i$ by

$$\mathcal{Z}_i \triangleq \left\{ x_i \in \mathbb{R} : \sum_{j \in \mathcal{O}_i} \phi_{cij}(x_i, x_j)(x_i - x_j) \sum_{j \in \mathcal{I}_i} \phi_{cij}(x_i, x_j)(x_i - x_j) = 0, \right.$$
$$\left. \text{and } x_i \neq x_j, \ j \in \mathcal{O}_i \cup \mathcal{I}_i \right\}, \quad i = 1, \ldots, q, \qquad (19.15)$$

with

$$\mathcal{Z} \triangleq \bigcup_{i=1}^q \{x \in \mathbb{R}^q : x_i \in \mathcal{Z}_i\}. \qquad (19.16)$$

The resetting set (19.15) is proposed in [119] and, as we will see later, is consistent with thermodynamic principles.

To ensure a thermodynamically consistent information flow model, we make the following assumptions on the information flow functions $\phi_{cij}(\cdot, \cdot)$, $i, j = 1, \ldots, q$, between state resettings.

**Assumption 19.4.** The connectivity matrix $\mathcal{C} \in \mathbb{R}^{q \times q}$ associated with the hybrid multiagent dynamical system $\mathcal{G}$ given by (19.11) and (19.12) is defined by

$$\mathcal{C}_{(i,j)} = \begin{cases} 0 & \text{if } \phi_{cij}(x_i(t), x_j(t)) \equiv 0, \\ 1, & \text{otherwise,} \end{cases} \quad i \neq j,$$
$$i, j = 1, \ldots, q, \quad t \geq 0, \quad (19.17)$$

and

$$\mathcal{C}_{(i,i)} = - \sum_{k=1, k \neq i}^{q} \mathcal{C}_{(k,i)}, \quad i = j, \quad i = 1, \ldots, q, \quad (19.18)$$

with rank $\mathcal{C} = q - 1$, and for $\mathcal{C}_{(i,j)} = 1$, $i \neq j$, $\phi_{cij}(x_i(t), x_j(t)) = 0$ if and only if $x_i(t) = x_j(t)$ for all $x(t) \notin \mathcal{Z}$, $t \geq 0$.

**Assumption 19.5.** For $i, j = 1, \ldots, q$, $[x_i(t) - x_j(t)]\phi_{cij}(x_i(t), x_j(t)) \leq 0$, $x(t) \notin \mathcal{Z}$, $t \geq 0$.

Furthermore, in addition to Assumptions 19.4 and 19.5, across resettings the information difference must satisfy the following assumption.

**Assumption 19.6.** For $i, j = 1, \ldots, q$, $[x_i(t_{k+1}) - x_j(t_{k+1})][x_i(t_k) - x_j(t_k)] \geq 0$ for all $x_i(t_k) \neq x_j(t_k)$, $x(t_k) \in \mathcal{Z}$, $k \in \mathbb{Z}_+$.

The condition $\phi_{cij}(x_i(t), x_j(t)) = 0$ if and only if $x_i(t) = x_j(t)$, $i \neq j$, for all $x(t) \notin \mathcal{Z}$ implies that agents $\mathcal{G}_i$ and $\mathcal{G}_j$ are connected and, hence, can share information; alternatively, $\phi_{cij}(x_i(t), x_j(t)) \equiv 0$ implies that agents $\mathcal{G}_i$ and $\mathcal{G}_j$ are disconnected and, hence, cannot share information.

Assumption 19.4 implies that if the information or energies in the connected agents $\mathcal{G}_i$ and $\mathcal{G}_j$ are equal, then information or energy exchange between these agents is not possible. This statement is reminiscent of the *zeroth law of hybrid thermodynamics* [119], which postulates that temperature equality is a necessary and sufficient condition for thermal equilibrium. Furthermore, if $\mathcal{C} = \mathcal{C}^T$ and rank $\mathcal{C} = q - 1$, then it follows that the connectivity matrix $\mathcal{C}$ is irreducible, which implies that, for any pair of agents $\mathcal{G}_i$ and $\mathcal{G}_j$, $i \neq j$, of $\mathcal{G}$, there exists a sequence of information connectors (information or communication arcs) of $\mathcal{G}$ that connect agents $\mathcal{G}_i$ and $\mathcal{G}_j$.

Assumption 19.5 implies that energy or information flows from more energetic or information-rich agents to less energetic or information-poor agents and is reminiscent of the *second law of hybrid thermodynamics* [119], which states that heat (i.e., energy in transition) must flow in the direction of lower temperatures.

Finally, Assumption 19.6 implies that, for any pair of connected agents $\mathcal{G}_i$ and $\mathcal{G}_j$, $i \neq j$, the energy or information difference between consecutive jumps is monotonic.

For the next result, recall that, for addressing the stability of an impulsive dynamical system, the usual stability definitions are valid [128].

**Theorem 19.2.** Consider the closed-loop hybrid multiagent dynamical system $\mathcal{G}$ given by (19.11) and (19.12) with resetting set $\mathcal{Z}$ given by (19.16), and assume Assumptions 19.4, 19.5, and 19.6 hold. Then, for every $\alpha \geq 0$, $\alpha \mathbf{e}$ is a semistable equilibrium state of $\mathcal{G}$. Furthermore, $x(t) \to \frac{1}{q}\mathbf{e}\mathbf{e}^{\mathrm{T}}x(0)$ as $t \to \infty$, and $\frac{1}{q}\mathbf{e}\mathbf{e}^{\mathrm{T}}x(0)$ is a semistable equilibrium state.

**Proof.** It follows from Assumption 19.4 that $\alpha \mathbf{e} \in \mathbb{R}^q, \alpha \in \mathbb{R}$, is an equilibrium state of (19.11) and (19.12). To show Lyapunov stability of the equilibrium state $\alpha \mathbf{e}$, consider the Lyapunov function candidate $V(x) = \frac{1}{2}(x-\alpha\mathbf{e})^{\mathrm{T}}(x-\alpha\mathbf{e})$. Since $\phi_{cij}(x_i(t), x_j(t)) = -\phi_{cji}(x_j(t), x_i(t))$, $x_i(t), x_j(t) \in \mathbb{R}, i \neq j, i,j = 1, \ldots, q$, and $\mathbf{e}^{\mathrm{T}}f_c(x) = 0$, it follows from Assumption 19.5 that

$$
\begin{aligned}
\dot{V}(x(t)) &= (x(t) - \alpha\mathbf{e})^{\mathrm{T}}\dot{x}(t) \\
&= (x(t) - \alpha\mathbf{e})^{\mathrm{T}}u_{\mathrm{c}}(x(t)) \\
&= x^{\mathrm{T}}(t)u_{\mathrm{c}}(x(t)) \\
&= \sum_{i=1}^{q} x_i(t)\left[\sum_{j=1,j\neq i}^{q} \phi_{cij}(x_i(t), x_j(t))\right] \\
&= \sum_{i=1}^{q}\sum_{j=i+1}^{q} [x_i(t) - x_j(t)]\phi_{cij}(x_i(t), x_j(t)) \\
&\leq 0, \quad x(t) \notin \mathcal{Z}.
\end{aligned} \tag{19.19}
$$

Next, it follows from Assumptions 19.5 and 19.6 that

$$
\begin{aligned}
\Delta V(x(t_k)) &= \frac{1}{2}(x(t_k^+) - \alpha\mathbf{e})^{\mathrm{T}}(x(t_k^+) - \alpha\mathbf{e}) \\
&\quad - \frac{1}{2}(x(t_k) - \alpha\mathbf{e})^{\mathrm{T}}(x(t_k) - \alpha\mathbf{e}) \\
&= \sum_{i=1}^{q}\sum_{j=1,j\neq i}^{q} x_i(t_k^+)\phi_{dij}(x_i(t_k), x_j(t_k))
\end{aligned}
$$

$$-\frac{1}{2}\sum_{i=1}^{q}\left[\sum_{j=1,j\neq i}^{q}\phi_{\mathrm{d}ij}(x_i(t_k),x_j(t_k))\right]^2$$

$$=\sum_{i=1}^{q-1}\sum_{j=i+1}^{q}[x_i(t_k^+)-x_j(t_k^+)]\phi_{\mathrm{d}ij}(x_i(t_k),x_j(t_k))$$

$$-\frac{1}{2}\sum_{i=1}^{q}\left[\sum_{j=1,j\neq i}^{q}\phi_{\mathrm{d}ij}(x_i(t_k),x_j(t_k))\right]^2$$

$$\leq 0,\quad x(t_k)\in\mathcal{Z},\tag{19.20}$$

which, by Theorem 2.1 of [128], establishes Lyapunov stability of the equilibrium state $\alpha\mathbf{e}$.

To show that $\alpha\mathbf{e}$ is semistable, note that

$$\dot{V}(x(t))=\sum_{i=1}^{q}\sum_{j=i+1}^{q}[x_i(t)-x_j(t)]\phi_{\mathrm{c}ij}(x_i(t),x_j(t))$$

$$=\sum_{i=1}^{q}\sum_{j\in\mathcal{K}_i}[x_i(t)-x_j(t)]\phi_{\mathrm{c}ij}(x_i(t),x_j(t)),\quad x(t)\notin\mathcal{Z},\tag{19.21}$$

and

$$\Delta V(x(t_k))=\sum_{i=1}^{q}\sum_{j=1,j\neq i}^{q}x_i(t_k)\phi_{\mathrm{d}ij}(x_i(t_k),x_j(t_k))$$

$$+\frac{1}{2}\sum_{i=1}^{q}\left[\sum_{j=1,j\neq i}^{q}\phi_{\mathrm{d}ij}(x_i(t_k),x_j(t_k))\right]^2$$

$$\geq\sum_{i=1}^{q-1}\sum_{j=i+1}^{q}[x_i(t_k)-x_j(t_k)]\phi_{\mathrm{d}ij}(x_i(t_k),x_j(t_k))$$

$$=\sum_{i=1}^{q-1}\sum_{j\in\mathcal{K}_i}[x_i(t_k)-x_j(t_k)]\phi_{\mathrm{d}ij}(x_i(t_k),x_j(t_k)),\quad x(t_k)\in\mathcal{Z},\tag{19.22}$$

where $\mathcal{K}_i=\mathcal{N}_i\backslash\cup_{l=1}^{i-1}\{l\}$ and

$$\mathcal{N}_i=\{j\in\{1,\ldots,q\}:\phi_{\mathrm{c}ij}(x_i,x_j)=0\text{ and }$$
$$\phi_{\mathrm{d}ij}(x_i,x_j)=0\text{ if and only if }x_i=x_j\}.$$

It follows from (19.21) that $\dot{V}(x)=0$, $x\notin\mathcal{Z}$, if and only if

$$(x_i-x_j)\phi_{\mathrm{c}ij}(x_i,x_j)=0,\quad x\notin\mathcal{Z},\quad i=1,\ldots,q,\ j\in\mathcal{K}_i.$$

Next, we show that $\Delta V(x) = 0$, $x \in \mathcal{Z}$, if and only if

$$(x_i - x_j)\phi_{\mathrm{d}ij}(x_i, x_j) = 0, \quad x \in \mathcal{Z}, \quad i = 1, 2, \ldots, q, \ j \in \mathcal{K}_i.$$

Assume that $(x_i - x_j)\phi_{\mathrm{d}ij}(x_i, x_j) = 0$, $i = 1, \ldots, q$, $j \in \mathcal{K}_i$. Then it follows from (19.22) that $\Delta V(x) \geq 0$, $x \in \mathcal{Z}$, and, hence, by (19.20), $\Delta V(x) = 0$, $x \in \mathcal{Z}$. Conversely, assume that $\Delta V(x) = 0$, $x \in \mathcal{Z}$. In this case, it follows from (19.20) that $[x_i(t_k^+) - x_j(t_k^+)]\phi_{\mathrm{d}ij}(x_i(t_k), x_j(t_k)) = 0$ and $\sum_{j=1, j \neq i}^{q} \phi_{\mathrm{d}ij}(x_i(t_k), x_j(t_k)) = 0$, $x(t_k) \in \mathcal{Z}$, $i, j = 1, \ldots, q$, $i \neq j$. Now,

$$
\begin{aligned}
\left[x_i(t_k^+) - x_j(t_k^+)\right] &\phi_{\mathrm{d}ij}(x_i(t_k), x_j(t_k)) \\
&= [x_i(t_k) - x_j(t_k)]\,\phi_{\mathrm{d}ij}(x_i(t_k), x_j(t_k)) \\
&\quad + \left[\sum_{h=1, h \neq i}^{q} \phi_{\mathrm{d}ih}(x_i(t_k), x_h(t_k))\right. \\
&\quad\quad \left. - \sum_{l=1, l \neq j}^{q} \phi_{\mathrm{d}jl}(x_j(t_k), x_l(t_k))\phi_{\mathrm{d}ij}(x_i(t_k), x_j(t_k))\right] \\
&= [x_i(t_k) - x_j(t_k)]\phi_{\mathrm{d}ij}(x_i(t_k), x_j(t_k)), \\
&\qquad\qquad x(t_k) \in \mathcal{Z}, \quad i, j = 1, \ldots, q, \quad i \neq j, \qquad (19.23)
\end{aligned}
$$

and, hence, $(x_i - x_j)\phi_{\mathrm{d}ij}(x_i, x_j) = 0$, $x \in \mathcal{Z}$, $i = 1, \ldots, q$, $j \in \mathcal{K}_i$.

Finally, let

$$
\begin{aligned}
\mathcal{R} &= \left\{x \in \mathbb{R}^q : x \notin \mathcal{Z}, \dot{V}(x) = 0\right\} \cup \left\{x \in \mathbb{R}^q : x \in \mathcal{Z}, \Delta V(x) = 0\right\} \\
&= \left\{x \in \mathbb{R}^q : x \notin \mathcal{Z}, (x_i - x_j)\phi_{\mathrm{c}ij}(x_i, x_j) = 0, \quad i = 1, \ldots, q, \quad j \in \mathcal{K}_i\right\} \\
&\quad \cup \left\{x \in \mathbb{R}^q : x \in \mathcal{Z}, (x_i - x_j)\phi_{\mathrm{d}ij}(x_i, x_j) = 0, \quad i = 1, \ldots, q, \quad j \in \mathcal{K}_i\right\},
\end{aligned}
$$

and note that by Assumption 19.4 the directed graph associated with the connectivity matrix $\mathcal{C}$ for the hybrid multiagent dynamical system $\mathcal{G}$ is strongly connected, which implies that $\mathcal{R} = \{x \in \mathbb{R}^q : x_1 = \cdots = x_q\}$. Since the set $\mathcal{R}$ consists of the equilibrium state of the system, it follows that the largest invariant set $\mathcal{M}$ contained in $\mathcal{R}$ is given by $\mathcal{M} = \mathcal{R}$. Hence, it follows from Theorem 19.1 that for every initial condition $x(0) \in \mathbb{R}^q$, $x(t) \to \mathcal{M}$ as $t \to \infty$, and, hence, $\alpha \mathbf{e}$ is a semistable equilibrium state of the system. Next, note that since $\mathbf{e}^{\mathrm{T}} x(t) = \mathbf{e}^{\mathrm{T}} x(0)$ and $x(t) \to \mathcal{M}$ as $t \to \infty$, it follows that $x(t) \to \frac{1}{q}\mathbf{e}\mathbf{e}^{\mathrm{T}} x(0)$ as $t \to \infty$. Hence, with $\alpha = \frac{1}{q}\mathbf{e}^{\mathrm{T}} x(0)$, $\alpha \mathbf{e} = \frac{1}{q}\mathbf{e}\mathbf{e}^{\mathrm{T}} x(0)$ is a semistable state of the closed-loop hybrid multiagent dynamical system $\mathcal{G}$. $\qquad\square$

## 19.4 Connections to Information Entropy and the Hybrid Second Law of Thermodynamics

In this section, we provide explicit connections of the proposed thermodynamic-based consensus control architecture developed in this section with the recently developed notion of hybrid thermodynamics [119]. To develop these connections, the following definition of hybrid entropy is needed.

**Definition 19.1.** For the distributed hybrid consensus control protocol $\mathcal{G}$ given by (19.11) and (19.12), a function $\mathcal{S} : \mathbb{R}^q \to \mathbb{R}$ satisfying

$$\mathcal{S}(x(T)) \geq \mathcal{S}(x(t_1)), \quad t_1 \leq t_k < T, \quad k \in \mathbb{Z}_+, \tag{19.24}$$

is called an *entropy* function of $\mathcal{G}$.

The next result gives necessary and sufficient conditions for establishing the existence of a hybrid entropy function of $\mathcal{G}$ over an interval $t \in (t_k, t_{k+1}]$ involving the consecutive resetting times $t_k$ and $t_{k+1}$, $k \in \mathbb{Z}_+$.

**Theorem 19.3.** Consider the closed-loop hybrid multiagent dynamical system $\mathcal{G}$ given by (19.11) and (19.12), and assume that Assumptions 19.4, 19.5, and 19.6 hold. Then a function $\mathcal{S} : \mathbb{R}^q \to \mathbb{R}$ is an entropy function of $\mathcal{G}$ if and only if

$$\mathcal{S}(x(\hat{t})) \geq \mathcal{S}(x(t)), \quad t_k < t \leq \hat{t} \leq t_{k+1}, \tag{19.25}$$

$$\mathcal{S}(x(t_k^+)) \geq \mathcal{S}(x(t_k)), \quad k \in \mathbb{Z}_+. \tag{19.26}$$

**Proof.** Let $\mathbb{Z}_{[t_1, T)} \triangleq \{k : t_1 \leq t_k < T\}$, let $k \in \mathbb{Z}_+$, and suppose that $\mathcal{S}(x)$ is an entropy function of $\mathcal{G}$. Then (19.24) holds. Now, since for $t_k < t \leq \hat{t} \leq t_{k+1}$, $\mathbb{Z}_{[t,\hat{t})} = \varnothing$, (19.25) is immediate. Next, note that

$$\mathcal{S}(x(t_k^+)) \geq \mathcal{S}(x(t_k)), \tag{19.27}$$

which, since $\mathbb{Z}_{[t_k, t_k^+)} = k$, implies (19.26).

Conversely, suppose that (19.25) and (19.26) hold, and let $\hat{t} \geq t \geq t_1$ and $\mathbb{Z}_{[t,\hat{t})} = \{i, i+1, \ldots, j\}$. If $\mathbb{Z}_{[t,\hat{t})} = \varnothing$, then it follows from (19.25) that $\mathcal{S}(x)$ is an entropy function of $\mathcal{G}$. Alternatively, if $\mathbb{Z}_{[t,\hat{t})} \neq \varnothing$, it follows from (19.25) and (19.26) that

$$\mathcal{S}(x(\hat{t})) - \mathcal{S}(x(t)) = \mathcal{S}(x(\hat{t})) - \mathcal{S}(x(t_j^+)) + \mathcal{S}(x(t_i^+)) - \mathcal{S}(x(t))$$

$$+ \sum_{m=0}^{j-i-1} [\mathcal{S}(x(t_{j-m}^+)) - \mathcal{S}(x(t_{j-m-1}^+))]$$

$$= \mathcal{S}(x(\hat{t})) - \mathcal{S}(x(t_j^+)) + \mathcal{S}(x(t_i)) - \mathcal{S}(x(t))$$

$$+ \sum_{m=0}^{j-i} [\mathcal{S}(x(t_{j-m}^+)) - \mathcal{S}(x(t_{j-m}))]$$

$$+ \sum_{m=0}^{j-i-1} [\mathcal{S}(x(t_{j-m})) - \mathcal{S}(x(t_{j-m-1}^+))]$$

$$\geq 0, \tag{19.28}$$

which implies that $\mathcal{S}(x)$ is an entropy function of $\mathcal{G}$.                □

The next theorem establishes the existence of a continuously differentiable entropy function for the closed-loop hybrid multiagent dynamical system $\mathcal{G}$ given by (19.11) and (19.12).

**Theorem 19.4.** Consider the closed-loop hybrid multiagent dynamical system $\mathcal{G}$ given by (19.11) and (19.12), and assume that Assumptions 19.5 and 19.6 hold. Then the function $\mathcal{S} : \mathbb{R}^q \to \mathbb{R}$ given by

$$\mathcal{S}(x) = \mathbf{e}^{\mathrm{T}} \mathbf{log}_e(c\mathbf{e} + x) - q \log_e c, \tag{19.29}$$

where $\mathbf{log}_e(c\mathbf{e} + x) = [\log_e(c + x_1), \ldots, \log_e(c + x_q)]^{\mathrm{T}}$ and $c > \|x\|_\infty$, is a continuously differentiable entropy function of $\mathcal{G}$. In addition,

$$\dot{\mathcal{S}}(x(t)) \geq 0, \quad x(t) \notin \mathcal{Z}, \quad t_k < t < t_{k+1}, \tag{19.30}$$
$$\Delta \mathcal{S}(x(t_k)) \geq 0, \quad x(t_k) \in \mathcal{Z}, \quad k \in \mathbb{Z}_+. \tag{19.31}$$

**Proof.** Since, for $x(t) \notin \mathcal{Z}$, $t \in (t_k, t_{k+1}]$, $k \in \mathbb{Z}_+$,

$$\phi_{cij}(x_i, x_j) = -\phi_{cji}(x_i, x_j), \quad i \neq j, \quad i, j = 1, \ldots, q,$$

and $c > \|x\|_\infty$, it follows that

$$
\begin{aligned}
\dot{\mathcal{S}}(x(t)) &= \sum_{i=1}^q \frac{\dot{x}_i(t)}{c + x_i(t)} \\
&= \sum_{i=1}^q \sum_{j=1, j \neq i}^q \frac{\phi_{cij}(x_i(t), x_j(t))}{c + x_i(t)} \\
&= \sum_{i=1}^q \sum_{j=i+1}^q \left( \frac{\phi_{cij}(x_i(t), x_j(t))}{c + x_i(t)} - \frac{\phi_{cij}(x_i(t), x_j(t))}{c + x_j(t)} \right) \\
&= \sum_{i=1}^{q-1} \sum_{j=i+1}^q \frac{\phi_{cij}(x_i(t), x_j(t))[x_j(t) - x_i(t)]}{(c + x_i(t))(c + x_j(t))} \\
&\geq 0, \quad x(t) \notin \mathcal{Z}, \quad t_k < t \leq t_{k+1}. \tag{19.32}
\end{aligned}
$$

Furthermore, since for $x(t_k) \in \mathcal{Z}, k \in \mathbb{Z}_+, \phi_{\mathrm{d}ij}(x_i, x_j) = -\phi_{\mathrm{d}ji}(x_i, x_j), i \neq j, i, j = 1, \ldots, q$, and $c > \|x\|_\infty$, it follows that

$$
\begin{aligned}
\Delta \mathcal{S}(x(t_k)) &= \sum_{i=1}^{q} \log_e \left[ 1 + \frac{\Delta x_i(t_k)}{c + x_i(t_k)} \right] \\
&\geq \sum_{i=1}^{q} \left[ \frac{\Delta x_i(t_k)}{c + x_i(t_k)} \right] \left[ 1 + \frac{\Delta x_i(t_k)}{c + x_i(t_k)} \right]^{-1} \\
&= \sum_{i=1}^{q} \frac{\Delta x_i(t_k)}{c + x_i(t_k) + \Delta x_i(t_k)} \\
&= \sum_{i=1}^{q} \frac{\Delta x_i(t_k)}{c + x_i(t_k^+)} \\
&= \sum_{i=1}^{q} \sum_{i=1, j \neq i}^{q} \frac{\phi_{\mathrm{d}ij}(x_i(t_k), x_j(t_k))}{c + x_i(t_k^+)} \\
&= \sum_{i=1}^{q-1} \sum_{j=i+1}^{q} \left( \frac{\phi_{\mathrm{d}ij}(x_i(t_k), x_j(t_k))}{c + x_i(t_k^+)} - \frac{\phi_{\mathrm{d}ij}(x_i(t_k), x_j(t_k))}{c + x_j(t_k^+)} \right) \\
&= \sum_{i=1}^{q-1} \sum_{j=i+1}^{q} \frac{\phi_{\mathrm{d}ij}(x_i(t_k), x_j(t_k))[x_j(t_k^+) - x_i(t_k^+)]}{(c + x_i(t_k^+))(c + x_j(t_k^+))} \\
&\geq 0, \quad x(t_k) \in \mathcal{Z}, \quad k \in \mathbb{Z}_+, 
\end{aligned}
\tag{19.33}
$$

where in (19.33) we use the fact that $\frac{x}{1+x} < \log_e(1 + x) < x, x > -1, x \neq 0$. The result is now an immediate consequence of Theorem 19.3. □

It follows from (19.32) that (19.15) implies that if the time rate of change of the difference in the input information flow and output information flow between any pair of connected agent entropies is zero and consensus is not reached, then a resetting occurs.

Note that it follows from Theorem 19.3, (19.32), and (19.33) that the entropy function given by (19.29) satisfies (19.24) as an equality for an equilibrium (equipartitioned) process and as a strict inequality for a nonequilibrium (nonequipartitioned) process. The entropy expression given by (19.29) is identical in form to the Boltzmann entropy for statistical thermodynamics [118]. In addition, note that $\mathcal{S}(x)$ given by (19.29) achieves a maximum when all of the information states $x_i, i = 1, \ldots, q$, are equal [118, 127]. Inequality (19.24) is a generalization of Clausius's inequality for equilibrium and nonequilibrium thermodynamics as well as reversible and irreversible thermodynamics as applied to adiabatically isolated hybrid thermodynamic systems involving discontinuous phase transitions. For details, see [119].

Next, we demonstrate the proposed distributed hybrid consensus framework on a set of aircraft achieving pitch rate consensus.

**Example 19.1.** Consider the multiagent system comprising the controlled longitudinal motion of seven Boeing 747 airplanes addressed in Example 2.1 linearized at an altitude of 40 kft and a velocity of 774 ft/sec given by

$$\dot{z}_i(t) = A z_i(t) + B \delta_i(t), \quad z_i(0) = z_{i_0}, \quad i = 1, \dots, 7, \quad t \geq 0, \quad (19.34)$$

where the state $z_i(t) = [v_{x_i}(t), v_{z_i}(t), q_i(t), \theta_{e_i}(t)]^{\mathrm{T}} \in \mathbb{R}^4$, $t \geq 0$, of aircraft $i \in \{1, \dots, 7\}$, control input $\delta_i(t)$, and system matrices $A$ and $B$ are as defined in Example 2.1.

As in Example 2.1, we propose a two-level control hierarchy composed of a lower-level controller for command following and a higher-level hybrid consensus controller for pitch rate consensus with the communication topology shown in Figure 19.1. To address the lower-level controller design, let $x_i(t)$, $i = 1, \dots, 7$, $t \geq 0$, denote an information command generated by (19.9) and (19.10) (i.e., the guidance command), and let $s_i(t)$, $i = 1, \dots, 7$, $t \geq 0$, denote the integrator state satisfying

$$\dot{s}_i(t) = E z_i(t) - x_i(t), \quad s_i(0) = s_{i_0}, \quad i = 1, \dots, 7, \quad t \geq 0, \quad (19.35)$$

where $E = [0, 0, 1, 0]$.

Now, defining the augmented state

$$\hat{z}_i(t) \triangleq [z_i^{\mathrm{T}}(t), s_i(t)]^{\mathrm{T}} \in \mathbb{R}^5,$$

(19.34) and (19.35) give

$$\dot{\hat{z}}_i(t) = \hat{A} \hat{z}_i(t) + \hat{B}_1 \delta_i(t) + \hat{B}_2 x_i(t), \quad \hat{z}_i(0) = \hat{z}_{i_0}, \quad i = 1, \dots, 7, \quad t \geq 0, \quad (19.36)$$

where $\hat{A}$, $\hat{B}_1$, and $\hat{B}_2$ are given by (2.33). Furthermore, let the elevator control input be given by

$$\delta_i(t) = -K \hat{z}_i(t), \quad i = 1, \dots, 7,$$

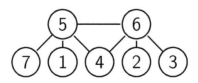

Figure 19.1 Aircraft communication topology.

where

$$K = [-0.0157, 0.0831, -4.7557, -0.1400, -9.8603],$$

which is designed based on an optimal LQR controller.

For the higher-level hybrid consensus controller design, we use (19.9) with functions

$$\phi_{cij}(x_i(t), x_j(t)) = (x_j(t) - x_i(t))^{\frac{1}{3}}, \quad i, j \in \{1, \ldots, 7\}, \quad i \neq j,$$

and (19.10) with

$$\phi_{dij}(x_i(t), x_j(t)) = (x_j(t) - x_i(t))/3, \quad i, j \in \{1, \ldots, 7\}, \quad i \neq j,$$

to generate the information state $x(t)$, $t \geq 0$, that has a direct effect on the lower-level controller design to achieve pitch rate consensus. Here, we used the resetting set $\mathcal{Z}_i$ given by (19.15) and verified Assumptions 19.1–19.3 numerically.

For our simulation, we set $x_1(0) = q_1(0) = 0$, $x_2(0) = q_2(0) = 16$, $x_3(0) = q_3(0) = 8$, $x_4(0) = q_4(0) = 5$, $x_5(0) = q_5(0) = 16$, $x_6(0) = q_6(0) = 19$, and $x_7(0) = q_7(0) = 20$, with all other initial conditions set to zero. Figure 19.2 shows the information command signals and pitch rate of each aircraft versus time for the proposed hybrid thermodynamic control protocol. For comparison purposes, Figure 19.3 shows the information command signals and pitch rate of each aircraft versus time for the continuous-time thermodynamic control protocol proposed in Chapter 2 with

$$\phi_{cij}(x_i(t), x_j(t)) = (x_j(t) - x_i(t))^{\frac{1}{3}}, \quad i, j \in \{1, \ldots, 7\}, \quad i \neq j.$$

Finally, Figure 19.4 shows the total agent entropies versus time for both designs. Since every component of the information state $x(t)$, $t \geq 0$, is nonnegative, here we set $c = 1$ so that $q \log_e c = 0$.

Note that the proposed hybrid consensus protocol gives a much better transient performance than the conventional continuous-time consensus protocol. In addition, even though the closed-loop system with the hybrid protocol resembles a Zeno behavior, this is not the case since a finite number of resettings are executed before consensus is achieved.          △

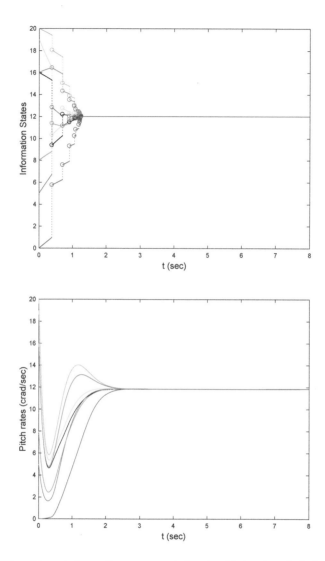

Figure 19.2  Closed-loop information command signal $x_i(t)$ (top) and pitch rate $q_i(t)$ (bottom) trajectories with the proposed higher-level hybrid consensus protocol; $x_1(t), q_1(t)$ in blue; $x_2(t), q_2(t)$ in red; $x_3(t), q_3(t)$ in green; $x_4(t), q_4(t)$ in magenta; $x_5(t), q_5(t)$ in black; $x_6(t), q_6(t)$ in yellow; and $x_7(t), q_7(t)$ in cyan.

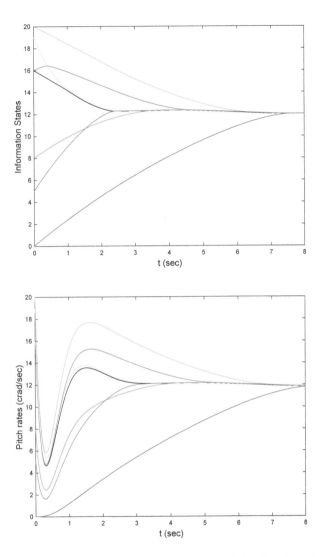

Figure 19.3 Closed-loop information command signal $x_i(t)$ (top) and pitch rate $q_i(t)$ (bottom) trajectories with the higher-level continuous-time consensus protocol given in Example 2.1; $x_1(t), q_1(t)$ in blue; $x_2(t), q_2(t)$ in red; $x_3(t), q_3(t)$ in green; $x_4(t), q_4(t)$ in magenta; $x_5(t), q_5(t)$ in black; $x_6(t), q_6(t)$ in yellow; and $x_7(t), q_7(t)$ in cyan.

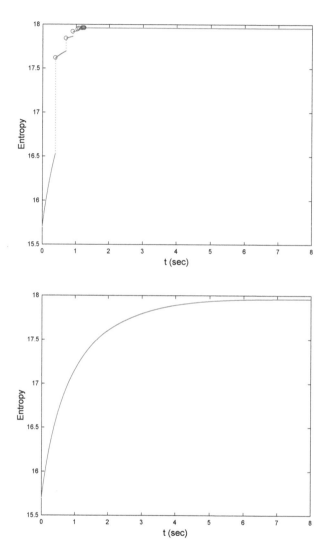

Figure 19.4 Total agent entropies versus time (with $c = 1$ in (19.29)) for both control protocols; hybrid on the top and continuous-time on the bottom.

## Chapter Twenty

---

# Hybrid Control Protocols for Consensus, Parallel Formations, and Collision Avoidance

## 20.1 Introduction

In this chapter, we build on the results of Chapter 19 to develop a hybrid control framework to address fast consensus-seeking problems as well as parallel formation and collision avoidance problems for multiagent dynamical systems. Specifically, we present hybrid distributed controller architectures for multiagent coordination to improve the transient performance of coordination tasks. The proposed controller architectures are predicated on the recently developed hybrid control framework for lossless dynamical systems [127], resulting in hybrid consensus architectures involving the exchange of information between agents.

Another unique feature of our framework is that the proposed hybrid consensus protocols can achieve finite time coordination and, hence, significantly improve the transient performance of the closed-loop system. We use impulsive dynamical systems theory to design hybrid consensus protocols, and, hence, the overall closed-loop dynamics under these controller algorithms achieving consensus possesses discontinuous flows since they combine logical switchings with continuous dynamics, leading to impulsive differential equations [13, 62, 125, 126, 128, 192, 269].

## 20.2 Distributed Nonlinear Control Algorithms for Consensus

As shown in Chapter 2, the consensus problem can be characterized as a dynamical network involving trajectories of a multiagent dynamical system $\mathcal{G}$ given by

$$\dot{x}_i(t) = \sum_{j=1, j \neq i}^{q} \phi_{ij}(x_i(t), x_j(t)), \quad x_i(t_0) = x_{i0}, \quad t \geq 0, \quad i = 1, \ldots, q, \quad (20.1)$$

where $\phi_{ij}(\cdot,\cdot)$, $i,j = 1,\ldots,q$, are locally Lipschitz continuous, or, in vector form,

$$\dot{x}(t) = f(x(t)), \quad x(t_0) = x_0, \quad t \geq 0, \tag{20.2}$$

where $x(t) \triangleq [x_1(t),\ldots,x_q(t)]^{\mathrm{T}}$, $t \geq 0$, and $f = [f_1,\ldots,f_q]^{\mathrm{T}} : \mathcal{D} \to \mathbb{R}^q$ is such that $f_i(x) = \sum_{j=1,j\neq i}^q \phi_{ij}(x_i,x_j)$, where $\mathcal{D} \subseteq \mathbb{R}^q$ is open. Here, $x_i(t)$, $t \geq 0$, represents an information state and $f_i(t) = u_i(t)$ is a distributed consensus algorithm involving neighbor-to-neighbor interaction between agents.

In Chapter 2, we developed a thermodynamically motivated information consensus framework for multiagent dynamical systems that achieves semistability and state equipartition. Specifically, we considered $q$ continuous integrator agents with dynamics

$$\dot{x}_i(t) = u_i(t), \quad x_i(0) = x_{i0}, \quad t \geq 0, \tag{20.3}$$

where, for each $i \in \{1,\ldots,q\}$, $x_i(t) \in \mathbb{R}$ denotes the information state and $u_i(t) \in \mathbb{R}$ denotes the information control input for all $t \geq 0$. The nonlinear consensus protocol in Chapter 2 was given by

$$u_i(t) = \sum_{j=1,j\neq i}^q \phi_{ij}(x_i(t), x_j(t)), \tag{20.4}$$

where $\phi_{ij}(\cdot,\cdot)$, $i,j = 1,\ldots,q$, are locally Lipschitz continuous.

Recall Assumptions 2.1 and 2.2, which are repeated here for convenience.

**Assumption 20.1.** The connectivity matrix $\mathcal{C} \in \mathbb{R}^{q\times q}$ associated with the multiagent dynamical system $\mathcal{G}$ given by (20.3) and (20.4) is defined by

$$\mathcal{C}_{(i,j)} \triangleq \begin{cases} 0 & \text{if } \phi_{ij}(x_i,x_j) \equiv 0, \\ 1, & \text{otherwise,} \end{cases} \quad i \neq j, \ i,j = 1,\ldots,q,$$

$\mathcal{C}_{(i,i)} \triangleq -\sum_{k=1,k\neq i}^q \mathcal{C}_{(i,k)}$, $i = 1,\ldots,q$, with $\mathrm{rank}\,\mathcal{C} = q-1$, and for $\mathcal{C}_{(i,j)} = 1$, $i \neq j$, $\phi_{ij}(x_i,x_j) = 0$ if and only if $x_i = x_j$.

**Assumption 20.2.** For $i,j = 1,\ldots,q$, $(x_i - x_j)\phi_{ij}(x_i,x_j) \leq 0$, $x_i, x_j \in \mathbb{R}$.

Note that (20.4) is a *static* consensus protocol. A natural question regarding (20.3) is, How can we design *dynamic* compensators to achieve fast convergent network consensus? This question is important since it can be used to design fast convergent consensus protocols for multiagent coordination via output feedback. To begin to address this question, we consider

$q$ continuous-time integrator agents given by (20.3) and the dynamic compensators given by

$$\dot{x}_{ci}(t) = \sum_{j=1, j \neq i}^{q} \phi_{ij}(x_{ci}(t), x_{cj}(t)) + \sum_{j=1, j \neq i}^{q} \eta_{ij}(x_i(t), x_j(t)),$$

$$x_{ci}(0) = x_{ci0}, \ t \geq 0, \tag{20.5}$$

$$u_i(t) = - \sum_{j=1, j \neq i}^{q} \mu_{ij}(x_{ci}(t), x_{cj}(t)), \tag{20.6}$$

where $\phi_{ij}(\cdot, \cdot)$, $\eta_{ij}(\cdot, \cdot)$, and $\mu_{ij}(\cdot, \cdot)$, $i, j = 1, \ldots, q$, $i \neq j$, satisfy Assumptions 20.1 and 20.2.

## 20.3 A Hybrid Control Architecture for Consensus

To improve network system performance and, more important, address finite time coordination, we present a *hybrid* consensus protocol that achieves agreement in multiagent dynamical systems using a hybrid dynamic *supervisory control* architecture. Specifically, consider $q$ mobile agents with the dynamics $\mathcal{G}_i$ given by (20.3) and the hybrid dynamic compensators $\mathcal{G}_{ci}$ given by

$$\dot{x}_{ci}(t) = - \sum_{j=1, j \neq i}^{q} \mathcal{C}_{(i,j)}(x_{ci}(t) - x_{cj}(t)) - \sum_{j=1, j \neq i}^{q} \mathcal{C}_{(i,j)}(x_i(t) - x_j(t)),$$

$$(x_i(t), \bar{x}_i(t), x_{ci}(t), \bar{x}_{ci}(t)) \notin \mathcal{Z}_i, \quad x_{ci}(0) = x_{ci0}, \quad t \geq 0, \tag{20.7}$$

$$x_{ci}(t^+) = \underset{x_{ci}(t)}{\arg \min} \sum_{j=1, j \neq i}^{q} \mathcal{C}_{(i,j)} \| x_{ci}(t) - x_{cj}(t) \|_2^2,$$

$$(x_i(t), \bar{x}_i(t), x_{ci}(t), \bar{x}_{ci}(t)) \in \mathcal{Z}_i, \tag{20.8}$$

$$u_i(t) = - \sum_{j=1, j \neq i}^{q} \mathcal{C}_{(j,i)}(x_{cj}(t) - x_{ci}(t)), \tag{20.9}$$

where $x_{ci} \in \mathbb{R}^n$, $\bar{x}_i \triangleq [x_{i_1}^{\mathrm{T}}, \ldots, x_{i_{|\mathcal{K}_i|}}^{\mathrm{T}}]^{\mathrm{T}}$, $\bar{x}_{ci} \triangleq [x_{ci_1}^{\mathrm{T}}, \ldots, x_{ci_{|\mathcal{K}_i|}}^{\mathrm{T}}]^{\mathrm{T}}$, $\mathcal{K}_i \triangleq \{i_1, \ldots, i_{|\mathcal{K}_i|}\}$ denotes the indices of all the other agents which have a communication link with the $i$th agent, $|\mathcal{K}_i|$ denotes the cardinality of the set $\mathcal{K}_i$, and $\mathcal{Z}_i$ is a *resetting set* given by

$$\mathcal{Z}_i \triangleq \Big\{ (x_i, \bar{x}_i, x_{ci}, \bar{x}_{ci}) : \frac{\mathrm{d}}{\mathrm{d}t} \mathcal{L}_i(x_i, \bar{x}_i) = 0 \text{ and}$$

$$\mathcal{L}_i(x_{ci}, \bar{x}_{ci}) > \min_{x_{ci}} \mathcal{L}_i(x_{ci}, \bar{x}_{ci}) \Big\}, \tag{20.10}$$

where

$$\mathcal{L}_i(x_i, \bar{x}_i) \triangleq \sum_{j=1, j\neq i}^{q} \mathcal{C}_{(i,j)} \|x_i - x_j\|_2^2$$

and

$$\mathcal{L}_i(x_{ci}, \bar{x}_{ci}) \triangleq \sum_{j=1, j\neq i}^{q} \mathcal{C}_{(i,j)} \|x_{ci} - x_{cj}\|_2^2, \quad i, j = 1, \ldots, q, \quad i \neq j.$$

Here, we assume that if the $i$th agent has a communication link with the $j$th agent, then they can exchange information via the control inputs $u_i$ and $u_j$, $i, j = 1, \ldots, q$, $i \neq j$. The resetting set $\mathcal{Z}_i$ is thus defined to be the set of all points in the closed-loop state space that correspond to decreasing the *Laplacian disagreement function* $\mathcal{L}_i(x_i, \bar{x}_i)$ [238] of the $i$th subsystem and minimizing the Laplacian disagreement function $\mathcal{L}_i(x_{ci}, \bar{x}_{ci})$ of the $i$th subcontroller. Note that (20.8) is equivalent to a least squares minimization problem with solution

$$x_{ci}(t^+) = \frac{1}{|\mathcal{K}_i|} \sum_{j\in\mathcal{K}_i} x_{cj}(t), \quad (x_i(t), \bar{x}_i(t), x_{ci}(t), \bar{x}_{ci}(t)) \in \mathcal{Z}_i. \quad (20.11)$$

Equation (20.11) is similar in form to the model proposed in [304]. Furthermore, (20.11) implies that the resetting dynamics are such that the state of the controller is reset to the *center of gravity* or average value of all other controller state variables.

It is important to note that our consensus protocol (20.7)–(20.9) is different from the consensus protocols based on nonsmooth gradient flows that are proposed in [69] for two reasons. First, (20.7)–(20.9) is a *dynamic* controller, whereas the controllers proposed in [69] are *static* controllers. Second, (20.7)–(20.9) is a *distributed* protocol, whereas the consensus protocols in [69] are *centralized* protocols. Furthermore, the proposed consensus protocol (20.7)–(20.9) uses a simple linear feedback form plus a logical switching to achieve enhanced performance, as compared with the complicated nonsmooth switched feedback forms of the consensus protocols proposed in the literature.

The closed-loop system $\tilde{\mathcal{G}}$ given by (20.3) and (20.7)–(20.9) gives an *impulsive differential equation* [13, 62, 125, 126, 128, 192, 269] of the form

$$\dot{\tilde{x}}(t) = \tilde{f}_c(\tilde{x}(t)), \quad \tilde{x}(0) = \tilde{x}_0, \quad \tilde{x}(t) \notin \tilde{\mathcal{Z}}, \quad (20.12)$$

$$\Delta\tilde{x}(t) = \tilde{f}_d(\tilde{x}(t)), \quad \tilde{x}(t) \in \tilde{\mathcal{Z}}, \quad (20.13)$$

where $t \geq 0$; $\tilde{x}(t) \in \tilde{\mathcal{D}} \subseteq \mathbb{R}^{2nq}$ denotes the closed-loop state involving the plant state and the compensator state; and $\tilde{f}_c : \tilde{\mathcal{D}} \to \mathbb{R}^{2nq}$ and $\tilde{f}_d : \tilde{\mathcal{D}} \to \mathbb{R}^{2nq}$

denote the closed-loop continuous-time and resetting dynamics, respectively, with $\tilde{f}_{\mathrm{c}}(\tilde{x}_{\mathrm{e}}) = 0$, where $\tilde{x}_{\mathrm{e}} \in \tilde{\mathcal{D}} \backslash \tilde{\mathcal{Z}}$ denotes the closed-loop equilibrium state and $\tilde{\mathcal{Z}} \triangleq \bigcup_{i=1}^{q} \mathcal{Z}_i$. Note that although the closed-loop state vector consists of plant states and controller states, it is clear that only those states associated with the controller are reset. For the closed-loop system (20.12) and (20.13), it follows from the structure of the resetting set (20.10) that Assumptions 19.1 and 19.2 established in Chapter 19 hold. Furthermore, we assume that Assumption 19.3 holds for the closed-loop impulsive dynamical system $\tilde{\mathcal{G}}$.

Sufficient conditions that guarantee that the impulsive dynamical system (20.12) and (20.13) satisfies Assumption 19.3 are given in [128] and Chapter 21. Here, we note that the structure of the resetting set (20.10) is such that if $\tilde{x} \in \tilde{\mathcal{Z}}$, then $\tilde{x} + \tilde{f}_{\mathrm{d}}(\tilde{x}) \in \overline{\tilde{\mathcal{Z}}} \backslash \tilde{\mathcal{Z}}$, and, hence, if $\tilde{x} \in \overline{\tilde{\mathcal{Z}}}$ such that $\tilde{f}_{\mathrm{c}}(\tilde{x})$ is $k$-transversal to (20.12) in the sense of [128] (see also Definition 21.1), then Assumptions 19.1–19.3 are automatically satisfied. For details of this fact, see [128]. Here, we only note that $k$-transversality as introduced in Definition 21.1 guarantees that the sign of the derivative of the Laplacian disagreement function of the $i$th subsystem changes as the closed-loop system trajectory traverses the closure of the resetting set $\tilde{\mathcal{Z}}$ at the intersection with $\overline{\tilde{\mathcal{Z}}}$.

The next definition introduces the notion of semistability for hybrid dynamical systems of the form (20.12) and (20.13).

**Definition 20.1.** Let $\mathcal{D} \subseteq \mathbb{R}^{2nq}$ be an open positively invariant set with respect to (20.12) and (20.13). An equilibrium point $z \in \mathcal{D}$ of (20.12) and (20.13) is *Lyapunov stable* if, for every $\varepsilon > 0$, there exists $\delta = \delta(\varepsilon) > 0$ such that, for every initial condition $x_0 \in \mathcal{B}_\delta(z)$ and every solution $x(t)$ with the initial condition $x(0) = x_0$, $x(t) \in \mathcal{B}_\varepsilon(z)$ for all $t \geq 0$. An equilibrium point $z \in \mathcal{D}$ of (20.12) and (20.13) is *semistable* if $z$ is Lyapunov stable and there exists an open subset $\mathcal{D}_0$ of $\mathcal{D}$ containing $z$ such that, for all initial conditions in $\mathcal{D}_0$, the solutions of (20.12) and (20.13) converge to a Lyapunov stable equilibrium point. The system (20.12) and (20.13) is *semistable* with respect to $\mathcal{D}$ if every solution with initial condition in $\mathcal{D}$ converges to a Lyapunov stable equilibrium. Finally, (20.12) and (20.13) are said to be globally semistable if (20.12) and (20.13) are semistable with respect to $\mathbb{R}^{2nq}$.

Next, we define finite time semistability of (20.12) and (20.13). Let $\mathcal{E}$ denote the set of equilibria for (20.12) and (20.13).

**Definition 20.2.** Let $\mathcal{D} \subseteq \mathbb{R}^{2nq}$ be an open positively invariant set with respect to (20.12) and (20.13). An equilibrium point $x_{\mathrm{e}} \in \mathcal{E}$ of (20.12) and (20.13) is said to be *finite time semistable* if there exist an open neighborhood $\mathcal{U} \subseteq \mathcal{D}$ of $x_{\mathrm{e}}$ and a function $T : \mathcal{U} \backslash \mathcal{E} \to (0, \infty)$, called the *settling-time function*, such that the following statements hold.

(i) For every $x \in \mathcal{U} \backslash \mathcal{E}$ and every solution $\psi(t)$ of (20.12) and (20.13) with $\psi(0) = x$, $\psi(t) \in \mathcal{U} \backslash \mathcal{E}$ for all $t \in [0, T(x))$, and $\lim_{t \to T(x)} \psi(t)$ exists and is contained in $\mathcal{U} \cap \mathcal{E}$.

(ii) $x_{\mathrm{e}}$ is semistable.

An equilibrium point $x_{\mathrm{e}} \in \mathcal{E}$ of (20.12) and (20.13) is said to be *globally finite time semistable* if it is finite time semistable with $\mathcal{D} = \mathcal{U} = \mathbb{R}^{2nq}$. The system (20.12) and (20.13) is said to be *finite time semistable* if every equilibrium point in $\mathcal{E}$ is finite time semistable. Finally, (20.12) and (20.13) are said to be *globally finite time semistable* if every equilibrium point in $\mathcal{E}$ is globally finite time semistable.

Recall that a point $z \in \mathbb{R}^{2nq}$ is a positive limit point of a solution $\tilde{x}(t)$ to (20.12) and (20.13) with $\tilde{x}(0) = \tilde{x}_0$ if there exists a sequence $\{t_n\}_{n=1}^{\infty}$ with $t_n \to \infty$ and $\tilde{x}(t_n) \to z$ as $n \to \infty$. The set $\omega(\tilde{x}_0)$ of all such positive limit points is the positive limit set of $\tilde{x}(0) = \tilde{x}_0$.

**Lemma 20.1.** Consider the system given by (20.12) and (20.13). Assume that Assumptions 19.1–19.3 hold. Furthermore, assume that the solutions of (20.12) and (20.13) are bounded, and let $\tilde{x}(\cdot)$ be a solution of (20.12) and (20.13) with $\tilde{x}(0) = \tilde{x}_0 \in \mathbb{R}^{2nq}$. If $z \in \omega(\tilde{x}_0)$ is a Lyapunov stable equilibrium point, where $\omega(\tilde{x}_0)$ denotes the positive limit set of $\tilde{x}_0$, then $z = \lim_{t \to \infty} \tilde{x}(t)$ and $\omega(\tilde{x}_0) = \{z\}$.

**Proof.** Let $\mathcal{U}_{\varepsilon}$ be an open neighborhood of $z$. By Lyapunov stability of $z$, there exists an open set $\mathcal{U}_{\delta}$ containing $z$ such that $s_t(\mathcal{U}_{\delta}) \subset \mathcal{U}_{\varepsilon}$ for all $t \geq 0$. Since $z \in \omega(\tilde{x}_0)$, it follows that there exists $h \geq 0$ such that $s(h, \tilde{x}_0) \in \mathcal{U}_{\delta}$. Hence, $s(t + h, \tilde{x}_0) = s_t(s(h, \tilde{x}_0)) \in s_t(\mathcal{U}_{\delta}) \subseteq \mathcal{U}_{\varepsilon}$ for all $t > 0$. Thus, $z = \lim_{t \to \infty} s(t, \tilde{x}_0)$. Now, it follows that $s(t_i, \tilde{x}_0) \to z$ for all sequences $t_i \to \infty$, and, hence, $\omega(\tilde{x}_0) = z$. $\qquad\square$

**Theorem 20.1.** Consider the closed-loop system $\tilde{\mathcal{G}}$ given by (20.3) and (20.7)–(20.9) with the resetting set (20.10). Assume that Assumptions 20.1, 20.2, and 19.3 hold. Furthermore, assume that $\mathcal{C}^{\mathrm{T}} \mathbf{e} = 0$. Then the following statements hold.

(i) The state of the closed-loop system $\tilde{\mathcal{G}}$ given by $\tilde{x}(t)$ asymptotically converges to the set $\{x_1 = \cdots = x_q = \alpha, x_{\mathrm{c}1} = \cdots = x_{\mathrm{c}q} = \beta\}$ for all system initial conditions $\tilde{x}_0 \in \mathbb{R}^{2nq}$, where $\alpha \in \mathbb{R}^n$ and $\beta \in \mathbb{R}^n$. In addition, the Laplacian disagreement function of each agent is decreasing and the Laplacian disagreement function of each subcontroller is minimized across resetting events.

(ii) If, in addition,

$$\sum_{i=1}^{q} \|z_i\|_2^2 \geq \sum_{i=1}^{q} \frac{1}{|\mathcal{K}_i|^2} \left\| \sum_{j \in \mathcal{K}_i} z_j \right\|_2^2 \tag{20.14}$$

holds for every $z_i \in \mathbb{R}^n$, $i = 1, \ldots, q$, then, for every $\alpha \in \mathbb{R}^n$ and $\beta \in \mathbb{R}^n$, $(\mathbf{e} \otimes \alpha, \mathbf{e} \otimes \beta)$ is a semistable equilibrium state of $\tilde{\mathcal{G}}$. Furthermore, $(x(t), x_{\mathrm{c}}(t)) \to (\frac{1}{q} \sum_{i=1}^{q} x_i(t_0), \frac{1}{q} \sum_{i=1}^{q} x_{ci}(t_0))$ as $t \to \infty$ and $(\frac{1}{q} \sum_{i=1}^{q} x_i(t_0), \frac{1}{q} \sum_{i=1}^{q} x_{ci}(t_0))$ is a semistable equilibrium state, where $x \triangleq [x_1^{\mathrm{T}}, \ldots, x_q^{\mathrm{T}}]^{\mathrm{T}}$ and $x_{\mathrm{c}} \triangleq [x_{\mathrm{c}1}^{\mathrm{T}}, \ldots, x_{\mathrm{c}q}^{\mathrm{T}}]^{\mathrm{T}}$.

(iii) Finally, if (20.14) holds and there exists a time instant $T > t_0$ such that $\mathcal{L}_i(x_i(T), \bar{x}_i(T)) = 0$; $\frac{\mathrm{d}}{\mathrm{d}t}\mathcal{L}_i(x_i(T), \bar{x}_i(T)) = 0$; $(1/|\mathcal{K}_i|) \sum_{r \in \mathcal{K}_i} x_{cr}(T) = (1/|\mathcal{K}_j|) \sum_{l \in \mathcal{K}_j} x_{cl}(T)$ for all $i, j = 1, \ldots, q$, $i \neq j$; and $\mathcal{L}_i(x_{ci}(T), \bar{x}_{ci}(T)) > 0$ for all $i = 1, \ldots, q$, then, for every $\alpha \in \mathbb{R}^n$ and $\beta \in \mathbb{R}^n$, $(\mathbf{e} \otimes \alpha, \mathbf{e} \otimes \beta)$ is a finite time semistable equilibrium state of $\tilde{\mathcal{G}}$. Furthermore,

$$(x(t), x_{\mathrm{c}}(t)) = \left( \frac{1}{q} \sum_{i=1}^{q} x_i(t_0), \frac{1}{q} \sum_{i=1}^{q} x_{ci}(t_0) \right), \quad t \geq T,$$

and

$$\left( \frac{1}{q} \sum_{i=1}^{q} x_i(t_0), \frac{1}{q} \sum_{i=1}^{q} x_{ci}(t_0) \right)$$

is a finite time semistable equilibrium state.

**Proof.** (i) Consider the nonnegative function

$$V(\tilde{x}) = \frac{1}{4} \sum_{i=1}^{q} \sum_{j=1, j \neq i}^{q} \mathcal{C}_{(i,j)} \|x_i - x_j\|_2^2 + \frac{1}{4} \sum_{i=1}^{q} \sum_{j=1, j \neq i}^{q} \mathcal{C}_{(i,j)} \|x_{ci} - x_{cj}\|_2^2, \tag{20.15}$$

where $\tilde{x} \triangleq [x_1^{\mathrm{T}}, \ldots, x_q^{\mathrm{T}}, x_{\mathrm{c}1}^{\mathrm{T}}, \ldots, x_{\mathrm{c}q}^{\mathrm{T}}]^{\mathrm{T}} \in \mathbb{R}^{2nq}$, and note that

$$\dot{V}(\tilde{x}) = \frac{1}{2} \sum_{i=1}^{q} \sum_{j=1, j \neq i}^{q} (\mathcal{C}_{(i,j)} + \mathcal{C}_{(j,i)})(x_i - x_j)^{\mathrm{T}} \dot{x}_i$$

$$+ \frac{1}{2} \sum_{i=1}^{q} \sum_{j=1, j \neq i}^{q} (\mathcal{C}_{(i,j)} + \mathcal{C}_{(j,i)})(x_{\mathrm{c}i} - x_{\mathrm{c}j})^{\mathrm{T}} \dot{x}_{\mathrm{c}i}$$

$$= - \sum_{i=1}^{q} \left[ \sum_{j=1, j \neq i}^{q} \mathcal{C}_{(i,j)}(x_{\mathrm{c}i} - x_{\mathrm{c}j}) \right]^{\mathrm{T}} \left[ \sum_{j=1, j \neq i}^{q} \mathcal{C}_{(i,j)}(x_{\mathrm{c}i} - x_{\mathrm{c}j}) \right]$$

$$\leq 0, \quad \tilde{x} \notin \tilde{\mathcal{Z}}, \tag{20.16}$$

and

$$\Delta V(\tilde{x}) = \frac{1}{4} \sum_{i=1}^{q} \left[ \min_{x_{\mathrm{c}i}} \sum_{j=1, j \neq i}^{q} \mathcal{C}_{(i,j)} \|x_{\mathrm{c}i} - x_{\mathrm{c}j}\|_2^2 \right]$$

$$- \frac{1}{4} \sum_{i=1}^{q} \sum_{j=1, j \neq i}^{q} \mathcal{C}_{(i,j)} \|x_{\mathrm{c}i} - x_{\mathrm{c}j}\|_2^2$$

$$< 0, \quad \tilde{x} \in \tilde{\mathcal{Z}}. \tag{20.17}$$

Thus, for $i, j = 1, \ldots, q$, $i \neq j$, the set $\tilde{\mathcal{D}}_{\mathrm{c}} \triangleq \{(x_i - x_j, x_{\mathrm{c}i} - x_{\mathrm{c}j}) : V(\tilde{x}) \leq c\}$, where $c > 0$, is a compact positively invariant set.

Let $\mathcal{R} = \{\tilde{x} \in \tilde{\mathcal{D}}_{\mathrm{c}} : \tilde{x} \notin \tilde{\mathcal{Z}}, \dot{V}(\tilde{x}) = 0\} \cup \{\tilde{x} \in \tilde{\mathcal{D}}_{\mathrm{c}} : \tilde{x} \in \tilde{\mathcal{Z}}, \Delta V(\tilde{x}) = 0\}$, and note that

$$\mathcal{R} = \bigcap_{i=1}^{q} \left\{ \tilde{x} \in \tilde{\mathcal{D}}_{\mathrm{c}} : \sum_{j=1, j \neq i}^{q} \mathcal{C}_{(i,j)}(x_{\mathrm{c}i} - x_{\mathrm{c}j}) = 0 \right\}. \tag{20.18}$$

Let $\mathcal{M}$ denote the largest invariant set contained in $\mathcal{R}$. Using similar arguments as in the proof of Theorem 4.9, it follows that $x_i = \alpha_i$ for all $i = 1, \ldots, q$, and $x_{\mathrm{c}1} = \cdots = x_{\mathrm{c}q}$ on $\mathcal{M}$, where $\alpha_i \in \mathbb{R}^n$. Furthermore, since $\sum_{i=1}^{q} \dot{x}_{\mathrm{c}i} = 0$, it follows from Proposition 2.1 that $x_{\mathrm{c}i} = \beta$ for all $i = 1, \ldots, q$, where $\beta \in \mathbb{R}^n$, and, hence, $\sum_{j=1, j \neq i}^{q} \mathcal{C}_{(i,j)}(x_i - x_j) = 0$, which, using Proposition 2.1, implies that $\alpha_i = \alpha$ for all $i = 1, \ldots, q$, where $\alpha \in \mathbb{R}^n$. Now, it follows from Theorem 8.1 of [128] that, for every initial condition $\tilde{x}_0 \in \mathbb{R}^{2nq}$, $\tilde{x}(t)$ converges to the largest invariant set $\mathcal{M}$ contained in the set $\{\tilde{x} \in \mathbb{R}^{2nq} : x_1 = \cdots = x_q = \alpha, x_{\mathrm{c}1} = \cdots = x_{\mathrm{c}q} = \beta\}$.

Finally, the fact that the Laplacian disagreement function of each agent is decreasing and the Laplacian disagreement function of each subcontroller is minimized across resetting events is immediate from (20.8), (20.10), and (20.11).

(ii) For every $\alpha \in \mathbb{R}^n$ and $\beta \in \mathbb{R}^n$, let

$$W(\tilde{x} - \gamma) = \|x - \mathbf{e} \otimes \alpha\|_2^2 + \|x_{\mathrm{c}} - \mathbf{e} \otimes \beta\|_2^2,$$

where $\gamma \triangleq [\mathbf{e}^{\mathrm{T}} \otimes \alpha^{\mathrm{T}}, \mathbf{e}^{\mathrm{T}} \otimes \beta^{\mathrm{T}}]^{\mathrm{T}}$. Then it follows from (20.14) that

$$W(\tilde{x} + \tilde{f}_{\mathrm{d}}(\tilde{x}) - \gamma) \leq W(\tilde{x} - \gamma), \quad \tilde{x} \in \mathbb{R}^{2nq}. \tag{20.19}$$

Furthermore,

$$W'(\tilde{x} - \gamma)\tilde{f}_{\mathrm{c}}(\tilde{x}_{\mathrm{c}}) = -\sum_{i=1}^{q} \sum_{j \in \mathcal{K}_i} \mathcal{C}_{(i,j)}(x_{ci} - x_{cj})^2 \leq 0, \quad \tilde{x} \in \mathbb{R}^{2nq}. \tag{20.20}$$

Hence, it follows that Theorem 1 of [126] that (20.12) and (20.13) are Lyapunov stable; that is, the closed-loop system $\tilde{\mathcal{G}}$ given by (20.3) and (20.7)–(20.9) with the resetting set (20.10) is Lyapunov stable.

Thus, it follows from (i) and Lemma 20.1 that $\lim_{t \to \infty} \tilde{x}(t) = \gamma^*$ and $\gamma^*$ is a Lyapunov stable equilibrium. Hence, by definition, for every $\alpha \in \mathbb{R}^n$ and $\beta \in \mathbb{R}^n$, $(\mathbf{e} \otimes \alpha, \mathbf{e} \otimes \beta)$ is a semistable equilibrium state of $\tilde{\mathcal{G}}$. Finally, since $\sum_{i=1}^{q} \dot{x}_i(t) = 0$ and $\sum_{i=1}^{q} \dot{x}_{ci}(t) = 0$ for all $\tilde{x}(t) \notin \tilde{\mathcal{Z}}$, it follows that $(x(t), x_{\mathrm{c}}(t)) \to (\frac{1}{q} \sum_{i=1}^{q} x_i(t_0), \frac{1}{q} \sum_{i=1}^{q} x_{ci}(t_0))$ as $t \to \infty$.

(iii) It follows from the resetting set (20.10) that if at time instant $t_k$, where $t_k$ denotes the resetting time,

$$\frac{\mathrm{d}}{\mathrm{d}t} \mathcal{L}_i(x_i(t_k), \bar{x}_i(t_k)) = 0$$

and

$$\mathcal{L}_i(x_{ci}(t_k), \bar{x}_{ci}(t_k)) > 0, \quad i = 1, \dots, q,$$

then $\mathcal{L}_i(x_{ci}(t_k^+), \bar{x}_{ci}(t_k^+)) = \min_{x_{ci}} \mathcal{L}_i(x_{ci}(t_k), \bar{x}_{ci}(t_k))$ for all $i = 1, \dots, q$. Since, by assumption, $\mathcal{L}_i(x_i(t_k), \bar{x}_i(t_k)) = 0$ for all $i = 1, \dots, q$, it follows that $\mathcal{L}_i(x_i(t_k^+), \bar{x}_i(t_k^+)) = 0$ for all $i = 1, \dots, q$. In this case, it follows that $x_1(t) = \cdots = x_q(t)$, and, hence, $\dot{x}_{ci}(t) = 0$ for all $t \geq t_k$, $i = 1, \dots, q$. Furthermore, $x_{c1}(t) = \cdots = x_{cq}(t)$, and, hence, $\dot{x}_i(t) = 0$ for all $t \geq t_k$, $i = 1, \dots, q$. Now, we have $x_1(t) = \cdots = x_q(t) = \alpha$ and $x_{c1}(t) = \cdots = x_{cq}(t) = \beta$ for all $t \geq t_k$ and some $\alpha \in \mathbb{R}^n$ and $\beta \in \mathbb{R}^n$.

Since, by (ii), the closed-loop system $\tilde{\mathcal{G}}$ is semistable, it follows from the definition of finite time semistability that, for every $\alpha \in \mathbb{R}^n$ and $\beta \in \mathbb{R}^n$, $(\mathbf{e} \otimes \alpha, \mathbf{e} \otimes \beta)$ is a semistable equilibrium state of $\tilde{\mathcal{G}}$. Finally, since $\sum_{i=1}^{q} \dot{x}_i(t) = 0$ and $\sum_{i=1}^{q} \dot{x}_{ci}(t) = 0$ for all $\tilde{x}(t) \notin \tilde{\mathcal{Z}}$, it follows that $(x(t), x_{\mathrm{c}}(t)) = (\frac{1}{q} \sum_{i=1}^{q} x_i(t_0), \frac{1}{q} \sum_{i=1}^{q} x_{ci}(t_0))$ for all $t \geq t_k$. $\square$

The resetting set $\mathcal{Z}_i$ given by (20.10) is the set of all points in the state space that correspond to the decreasing Laplacian disagreement function

$\mathcal{L}_i(x_i, \bar{x}_i)$, $i = 1, \ldots, q$. Next, we present yet another form of the resetting set $\mathcal{Z}_i$ given by

$$\mathcal{Z}_i \triangleq \left\{ (x_i, \bar{x}_i, x_{ci}, \bar{x}_{ci}) : \frac{\mathrm{d}}{\mathrm{d}t} \mathcal{L}_i(x_{ci}, \bar{x}_{ci}) = 0 \text{ and} \right.$$

$$\left. \mathcal{L}_i(x_{ci}, \bar{x}_{ci}) > \min_{x_{ci}} \mathcal{L}_i(x_{ci}, \bar{x}_{ci}) \right\}. \tag{20.21}$$

**Theorem 20.2.** Consider the closed-loop system $\tilde{\mathcal{G}}$ given by (20.3) and (20.7)–(20.9) with the resetting set (20.21). Assume that Assumptions 20.1, 20.2, and 19.3 hold. Furthermore, assume that $\mathcal{C}^{\mathrm{T}} \mathbf{e} = 0$. Then the following statements hold.

(i) The state of the closed-loop system $\tilde{\mathcal{G}}$ given by $\tilde{x}(t)$ asymptotically converges to the set $\{x_1 = \cdots = x_q = \alpha, x_{c1} = \cdots = x_{cq} = \beta\}$ for all system initial conditions $\tilde{x}_0 \in \mathbb{R}^{2nq}$, where $\alpha \in \mathbb{R}^n$ and $\beta \in \mathbb{R}^n$. In addition, the Laplacian disagreement function of each agent is decreasing, and the Laplacian disagreement function of each subcontroller is minimized across resetting events.

(ii) If, in addition, (20.14) holds, then, for every $\alpha \in \mathbb{R}^n$ and $\beta \in \mathbb{R}^n$, $(\mathbf{e} \otimes \alpha, \mathbf{e} \otimes \beta)$ is a semistable equilibrium state of $\tilde{\mathcal{G}}$. Furthermore, $(x(t), x_{\mathrm{c}}(t)) \to (\frac{1}{q} \sum_{i=1}^{q} x_i(t_0), \frac{1}{q} \sum_{i=1}^{q} x_{ci}(t_0))$ as $t \to \infty$ and

$$\left( \frac{1}{q} \sum_{i=1}^{q} x_i(t_0), \frac{1}{q} \sum_{i=1}^{q} x_{ci}(t_0) \right)$$

is a semistable equilibrium state.

(iii) Finally, if (20.14) holds and there exists a time instant $T > t_0$ such that

$$\mathcal{L}_i(x_i(T), \bar{x}_i(T)) = 0, \quad \frac{\mathrm{d}}{\mathrm{d}t} \mathcal{L}_i(x_{ci}(T), \bar{x}_{ci}(T)) = 0,$$

$$(1/|\mathcal{K}_i|) \sum_{r \in \mathcal{K}_i} x_{cr}(T) = (1/|\mathcal{K}_j|) \sum_{l \in \mathcal{K}_j} x_{cl}(T)$$

for all $i, j = 1, \ldots, q$, $i \neq j$, and $\mathcal{L}_i(x_{ci}(T), \bar{x}_{ci}(T)) > 0$ for all $i = 1, \ldots, q$, then, for every $\alpha \in \mathbb{R}^n$ and $\beta \in \mathbb{R}^n$, $(\mathbf{e} \otimes \alpha, \mathbf{e} \otimes \beta)$ is a finite time semistable equilibrium state of $\tilde{\mathcal{G}}$. Furthermore,

$$(x(t), x_{\mathrm{c}}(t)) = \left( \frac{1}{q} \sum_{i=1}^{q} x_i(t_0), \frac{1}{q} \sum_{i=1}^{q} x_{ci}(t_0) \right), \quad t \geq T,$$

and

$$\left(\frac{1}{q}\sum_{i=1}^{q} x_i(t_0), \frac{1}{q}\sum_{i=1}^{q} x_{ci}(t_0)\right)$$

is a finite time semistable equilibrium state.

**Proof.** The proof is similar to the proof of Theorem 20.1 and, hence, is omitted.                                                                    $\square$

**Example 20.1.** To illustrate the effect of the hybrid consensus protocol (20.7)–(20.9), consider the system with $q = 2$ and $n = 1$ given by

$$\dot{x}_1(t) = x_{c1} - x_{c2}, \quad x_1(0) = x_{10}, \quad t \geq 0, \tag{20.22}$$

$$\dot{x}_2(t) = x_{c2} - x_{c1}, \quad x_2(0) = x_{20}, \tag{20.23}$$

$$\dot{x}_{c1}(t) = x_{c2}(t) - x_{c1}(t) + x_2(t) - x_1(t), \quad x_{c1}(0) = x_{c10},$$
$$(x_1(t), x_2(t), x_{c1}(t), x_{c2}(t)) \notin \mathcal{Z}_1, \tag{20.24}$$

$$x_{c1}(t^+) = \frac{1}{2}[x_{c1}(t) + x_{c2}(t)], \quad (x_1(t), x_2(t), x_{c1}(t), x_{c2}(t)) \in \mathcal{Z}_1, \tag{20.25}$$

$$\dot{x}_{c2}(t) = x_{c1}(t) - x_{c2}(t) + x_1(t) - x_2(t), \quad x_{c2}(0) = x_{c20},$$
$$(x_1(t), x_2(t), x_{c1}(t), x_{c2}(t)) \notin \mathcal{Z}_2, \tag{20.26}$$

$$x_{c2}(t^+) = \frac{1}{2}[x_{c1}(t) + x_{c2}(t)], \quad (x_1(t), x_2(t), x_{c1}(t), x_{c2}(t)) \in \mathcal{Z}_2, \tag{20.27}$$

where $\mathcal{Z}_1$ and $\mathcal{Z}_2$ are resetting sets given by

$$\mathcal{Z}_1 = \{(x_1, x_2, x_{c1}, x_{c2}) : (x_1 - x_2)(x_{c2} - x_{c1}) = 0 \text{ and } x_{c1} \neq x_{c2}\}, \tag{20.28}$$

$$\mathcal{Z}_2 = \{(x_1, x_2, x_{c1}, x_{c2}) : (x_2 - x_1)(x_{c1} - x_{c2}) = 0 \text{ and } x_{c1} \neq x_{c2}\}. \tag{20.29}$$

Figure 20.1 shows the state trajectories of (20.22)–(20.29) with (20.28) and (20.29). We can see from the figure that the hybrid consensus protocol can achieve agreement in finite time, giving us a much better transient performance than the conventional consensus protocols. The plant Laplacian disagreement function is given by $\frac{1}{2}(x_1 - x_2)^2$, the controller Laplacian disagreement function is given by $\frac{1}{2}(x_{c1} - x_{c2})^2$, and the total Laplacian disagreement function is given by $\frac{1}{2}(x_1 - x_2)^2 + \frac{1}{2}(x_{c1} - x_{c2})^2$. Figure 20.2 shows the Laplacian disagreement functions versus time for (20.22)–(20.29).                                                                          $\triangle$

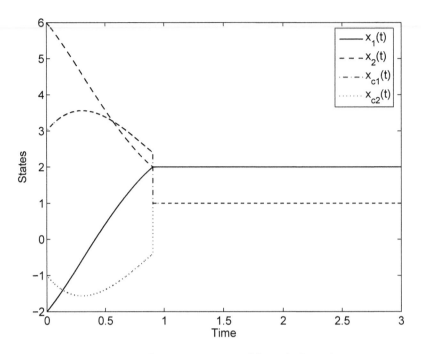

Figure 20.1 State trajectories of (20.22)–(20.29).

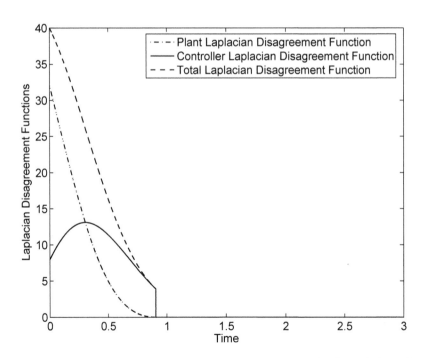

Figure 20.2 Laplacian disagreement functions versus time for (20.22)–(20.29).

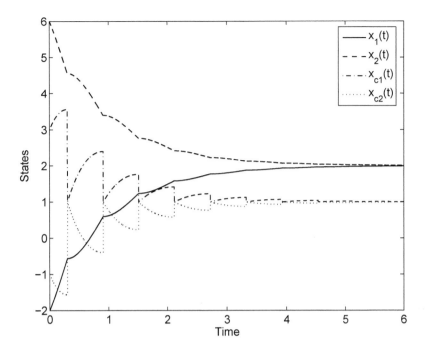

Figure 20.3 State trajectories of (20.22)–(20.27) with (20.30) and (20.31).

**Example 20.2.** Consider the system (20.22)–(20.27) with the resetting sets $\mathcal{Z}_1$ and $\mathcal{Z}_2$ given by

$$\mathcal{Z}_1 = \{(x_1, x_2, x_{c1}, x_{c2}) : (x_{c1} - x_{c2})(x_2 - x_1 + x_{c2} - x_{c1}) = 0$$
$$\text{and } x_{c1} \neq x_{c2}\}, \tag{20.30}$$
$$\mathcal{Z}_2 = \{(x_1, x_2, x_{c1}, x_{c2}) : (x_{c2} - x_{c1})(x_1 - x_2 + x_{c1} - x_{c2}) = 0$$
$$\text{and } x_{c1} \neq x_{c2}\}. \tag{20.31}$$

Figure 20.3 shows the state trajectories of (20.22)–(20.27) with (20.30) and (20.31). Figure 20.4 shows the Laplacian disagreement functions versus time for (20.22)–(20.27) with (20.30) and (20.31). $\triangle$

The above control architecture is based on *state-dependent impulsive dynamical systems* in which the resetting set is defined by a region in the state space that is independent of time. However, knowledge of the Laplacian disagreement function is required, which uses control information between connected agents.

Next, we propose a *time-dependent* hybrid consensus protocol for multiagent dynamical systems whose dynamics are described by (20.3). This hybrid control architecture does not require control information between connected agents. Specifically, consider $q$ mobile agents with the dynamics

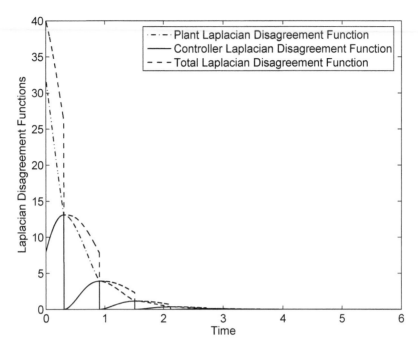

Figure 20.4 Laplacian disagreement functions versus time for (20.22)–(20.27) with (20.30) and (20.31).

$\mathcal{G}_i$ given by (20.3) and the time-dependent hybrid dynamic compensator $\mathcal{G}_{ci}$ given by

$$\dot{x}_{ci}(t) = -\sum_{j=1,j\neq i}^{q} \mathcal{C}_{(i,j)}(x_{ci}(t) - x_{cj}(t)) - \sum_{j=1,j\neq i}^{q} \mathcal{C}_{(i,j)}(x_i(t) - x_j(t)),$$
$$t \neq t_k, \quad x_{ci}(0) = x_{ci0}, \quad t \geq 0, \quad (20.32)$$

$$x_{ci}(t_k^+) = \arg\min_{x_{ci}(t_k)} \sum_{j=1,j\neq i}^{q} \mathcal{C}_{(i,j)}\|x_{ci}(t_k) - x_{cj}(t_k)\|_2^2, \quad t = t_k, \quad (20.33)$$

$$u_i(t) = -\sum_{j=1,j\neq i}^{q} \mathcal{C}_{(j,i)}(x_{cj}(t) - x_{ci}(t)), \quad (20.34)$$

where $t_k$, $k = 1, 2, \ldots$, are prescribed resetting times. Here, we assume that $t_{k+1} - t_k \geq \varepsilon > 0$, $k \in \mathbb{Z}_+$.

**Theorem 20.3.** Consider the closed-loop system $\tilde{\mathcal{G}}$ given by (20.3) and (20.32)–(20.34). Assume that Assumptions 20.1, 20.2, and 19.3 hold. Furthermore, assume that $\mathcal{C}^{\mathrm{T}}\mathbf{e} = 0$ and $t_{k+1} - t_k \geq \varepsilon > 0$, $k \in \mathbb{Z}_+$. Then the following statements hold.

(i) The state of the closed-loop system $\tilde{\mathcal{G}}$ given by $\tilde{x}(t)$ asymptotically converges to the set $\{x_1 = \cdots = x_q = \alpha, x_{c1} = \cdots = x_{cq} = \beta\}$ for all system initial conditions $\tilde{x}_0 \in \mathbb{R}^{2nq}$, where $\alpha \in \mathbb{R}^n$ and $\beta \in \mathbb{R}^n$. In addition, the Laplacian disagreement function of each agent is decreasing, and the Laplacian disagreement function of each subcontroller is minimized across resetting events.

(ii) If, in addition, (20.14) holds, then, for every $\alpha \in \mathbb{R}^n$ and $\beta \in \mathbb{R}^n$, $(\mathbf{e} \otimes \alpha, \mathbf{e} \otimes \beta)$ is a semistable equilibrium state of $\tilde{\mathcal{G}}$. Furthermore, $(x(t), x_c(t)) \to (\frac{1}{q}\sum_{i=1}^q x_i(t_0), \frac{1}{q}\sum_{i=1}^q x_{ci}(t_0))$ as $t \to \infty$, and

$$\left( \frac{1}{q}\sum_{i=1}^q x_i(t_0), \frac{1}{q}\sum_{i=1}^q x_{ci}(t_0) \right)$$

is a semistable equilibrium state.

(iii) Finally, if (20.14) holds, there exists a resetting time $t_k$ such that $\mathcal{L}_i(x_i(t_k), \bar{x}_i(t_k)) = 0$ for all $i = 1, \ldots, q$; and

$$(1/|\mathcal{K}_i|) \sum_{r \in \mathcal{K}_i} x_{cr}(t_k) = (1/|\mathcal{K}_j|) \sum_{l \in \mathcal{K}_j} x_{cl}(t_k)$$

for all $i, j = 1, \ldots, q$, $i \neq j$, then, for every $\alpha \in \mathbb{R}^n$ and $\beta \in \mathbb{R}^n$, $(\mathbf{e} \otimes \alpha, \mathbf{e} \otimes \beta)$ is a finite time semistable equilibrium state of $\tilde{\mathcal{G}}$. Furthermore, $(x(t), x_c(t)) = (\frac{1}{q}\sum_{i=1}^q x_i(t_0), \frac{1}{q}\sum_{i=1}^q x_{ci}(t_0))$, $t \geq t_k$, and $(\frac{1}{q}\sum_{i=1}^q x_i(t_0), \frac{1}{q}\sum_{i=1}^q x_{ci}(t_0))$ is a finite time semistable equilibrium state.

**Proof.** (i) We consider two cases for the resetting set $\mathcal{T} \triangleq \{t_1, t_2, \ldots\}$.

*Case 1.* The set $\mathcal{T}$ is finite. In this case, it follows that for all $t \geq \max_k\{t_k\}$, the closed-loop system $\tilde{\mathcal{G}}$ is given by the time-invariant system (20.3) and (20.32). Now, considering the nonnegative function (20.15) and using similar arguments as in the proof of Theorem 20.1, it follows that $\tilde{x}(t)$ asymptotically converges to the set $\{x_1 = \cdots = x_q = \alpha, x_{c1} = \cdots = x_{cq} = \beta\}$ for all initial conditions $\tilde{x}_0 \in \mathbb{R}^{2nq}$, where $\alpha \in \mathbb{R}^n$ and $\beta \in \mathbb{R}^n$.

*Case 2.* The set $\mathcal{T}$ is infinitely countable. In this case, it follows from the assumption on $t_k$ that $t_k \to \infty$ as $k \to \infty$. Now, using the nonnegative function (20.15), it follows from Theorem 2.5 of [128] that $\tilde{x}(t)$ asymptotically converges to the set $\{x_1 = \cdots = x_q = \alpha, x_{c1} = \cdots = x_{cq} = \beta\}$ for all initial conditions $\tilde{x}_0 \in \mathbb{R}^{2nq}$, where $\alpha \in \mathbb{R}^n$ and $\beta \in \mathbb{R}^n$.

(ii) Using Theorem 1 of [126] for time-dependent impulsive dynamical systems, the proof is similar to the proof of (ii) of Theorem 20.1 and, hence, is omitted.

(iii) The proof is similar to the proof of (iii) of Theorem 20.1 and, hence, is omitted.                                                                  □

**Example 20.3.** To illustrate the time-dependent hybrid consensus protocol (20.32)–(20.34), consider the system with $q = 2$ and $n = 1$ given by

$$\dot{x}_1(t) = x_{c1} - x_{c2}, \quad x_1(0) = x_{10}, \quad t \geq 0, \tag{20.35}$$

$$\dot{x}_2(t) = x_{c2} - x_{c1}, \quad x_2(0) = x_{20}, \tag{20.36}$$

$$\dot{x}_{c1}(t) = x_{c2}(t) - x_{c1}(t) + x_2(t) - x_1(t), \quad x_{c1}(0) = x_{c10}, \quad t \neq t_k, \tag{20.37}$$

$$x_{c1}(t^+) = \frac{1}{2}[x_{c1}(t) + x_{c2}(t)], \quad t = t_k, \tag{20.38}$$

$$\dot{x}_{c2}(t) = x_{c1}(t) - x_{c2}(t) + x_1(t) - x_2(t), \quad x_{c2}(0) = x_{c20}, \quad t \neq t_k, \tag{20.39}$$

$$x_{c2}(t^+) = \frac{1}{2}[x_{c1}(t) + x_{c2}(t)], \quad t = t_k, \tag{20.40}$$

and let $t_k = k$, $k \in \mathbb{Z}_+$. Figure 20.5 shows the state trajectories of (20.35)–(20.40). Figure 20.6 shows the Laplacian disagreement functions versus time for (20.35)–(20.40).                                                        △

## 20.4 Distributed Control for Parallel Formations

In this section, we develop a distributed controller for mobile agents to achieve parallel flocking formations [183, 234, 294, 295]. Here, the model

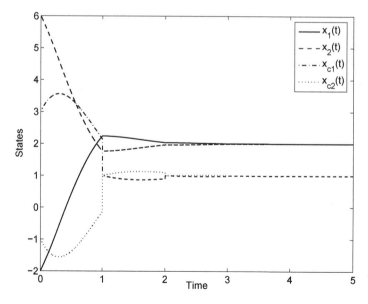

Figure 20.5 State trajectories of (20.35)–(20.40).

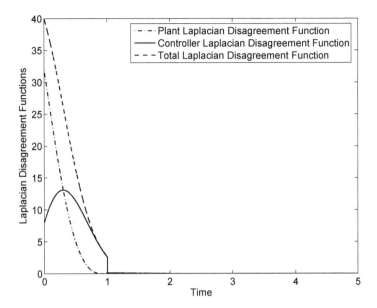

Figure 20.6 Laplacian disagreement functions versus time for (20.35)–(20.40).

considered by (20.3) is extended to double integrator dynamics in order to more naturally model the evolution of physical phenomena.

Specifically, consider $q$ mobile agents given by

$$\ddot{x}_i(t) = u_i(t), \quad x_i(0) = x_{i0}, \quad \dot{x}_i(0) = \dot{x}_{i0}, \quad t \geq 0, \qquad (20.41)$$

where, for $i = 1, \ldots, q$ and $t \geq 0$, $x_i(t) \in \mathbb{R}^n$ is the position vector of the $i$th agent, $\dot{x}_i(t) \in \mathbb{R}^n$ is the velocity vector of the $i$th agent, and $u_i(t) \in \mathbb{R}^n$ is the control input of the $i$th agent. The control aim is to design a hybrid feedback control law so that parallel formation is achieved, wherein the agents $\mathcal{G}_i$ are collectively required to maintain a prescribed geometric shape with $\dot{x}_1 = \cdots = \dot{x}_q = c$, where $c \in \mathbb{R}^n$ is a constant vector, and the relative position between any two mobile agents is asymptotically stabilized to a constant value.

Consider the hybrid dynamic compensator with both $x_i$ and $\dot{x}_i$ transmitted between team agents given by

$$\dot{x}_{ci}(t) = - \sum_{j=1, j \neq i}^{q} \mathcal{C}_{(i,j)}(x_{ci}(t) - x_{cj}(t)) - \sum_{j=1, j \neq i}^{q} \mathcal{C}_{(i,j)}(\dot{x}_i(t) - \dot{x}_j(t)),$$
$$(x_i(t), \bar{x}_i(t), \dot{x}_i(t), \dot{\bar{x}}_i(t), x_{ci}(t), \bar{x}_{ci}(t)) \notin \mathcal{Z}_i,$$
$$x_{ci}(0) = x_{ci0}, \quad t \geq 0, \qquad (20.42)$$

$$x_{ci}(t^+) = \underset{x_{ci}(t)}{\arg\min} \sum_{j=1, j \neq i}^{q} \mathcal{C}_{(i,j)} \|x_{ci}(t) - x_{cj}(t)\|_2^2,$$

$$(x_i(t), \bar{x}_i(t), \dot{x}_i(t), \dot{\bar{x}}_i(t), x_{ci}(t), \bar{x}_{ci}(t)) \in \mathcal{Z}_i, \qquad (20.43)$$

$$u_i(t) = -\sum_{j=1, j \neq i}^{q} \mathcal{C}_{(j,i)}(x_{cj}(t) - x_{ci}(t))$$

$$\qquad\qquad - \sum_{j=1, j \neq i}^{q} \nabla_{x_i}(U_{ij}(x_i(t), x_j(t)) + U_{ji}(x_j(t), x_i(t))), \qquad (20.44)$$

where $x_{ci} \in \mathbb{R}^n$;

$$\bar{x}_i \triangleq [x_{i_1}^T, \ldots, x_{i_{|\mathcal{K}_i|}}^T]^T, \quad \dot{\bar{x}}_i \triangleq [\dot{x}_{i_1}^T, \ldots, \dot{x}_{i_{|\mathcal{K}_i|}}^T]^T, \quad \bar{x}_{ci} \triangleq [x_{ci_1}^T, \ldots, x_{ci_{|\mathcal{K}_i|}}^T]^T;$$

$\mathcal{K}_i \triangleq \{i_1, \ldots, i_{|\mathcal{K}_i|}\}$ denotes the indices of all the other agents which have a communication link with the $i$th agent; $|\mathcal{K}_i|$ denotes the cardinality of the set $\mathcal{K}_i$; $U_{ij}(x_i, x_j)$ is a *generalized potential function* satisfying $U_{ij}(x_i, x_j) \geq 0$; $U_{ij}(x_i, x_j) \equiv 0$ for $\mathcal{C}_{(i,j)} = 0$;

$$\nabla_{x_i} U_{ij}(x_i, x_j) + \nabla_{x_j} U_{ij}(x_i, x_j) = 0; \qquad (20.45)$$

and

$$\sum_{i=1}^{q} \sum_{j=1, j \neq i}^{q} \nabla_{x_i}(U_{ij}(x_i, x_j) + U_{ji}(x_j, x_i)) = 0, \qquad (20.46)$$

where $\nabla_{x_i} U_{ij}(x_i, x_j) \triangleq \left(\frac{\partial U_{ij}}{\partial x_i}(x_i, x_j)\right)^T$ is a gradient vector; and $\mathcal{Z}_i$ is the resetting set given by

$$\mathcal{Z}_i \triangleq \Bigg\{ (x_i, \bar{x}_i, \dot{x}_i, \dot{\bar{x}}_i, x_{ci}, \bar{x}_{ci}) : \frac{\mathrm{d}}{\mathrm{d}t} \Phi_i(\dot{x}_i, \dot{\bar{x}}_i) = 0 \text{ and}$$

$$J_i(x_{ci}, \bar{x}_{ci}) > \min_{x_{ci}} J_i(x_{ci}, \bar{x}_{ci}) \Bigg\}, \qquad (20.47)$$

where $\Phi_i(\dot{x}_i, \dot{\bar{x}}_i) \triangleq \sum_{j=1, j \neq i}^{q} \mathcal{C}_{(i,j)} \|\dot{x}_i - \dot{x}_j\|_2^2$ and

$$J_i(x_{ci}, \bar{x}_{ci}) \triangleq \sum_{j=1, j \neq i}^{q} \mathcal{C}_{(i,j)} \|x_{ci} - x_{cj}\|_2^2, \quad i, j = 1, \ldots, q, \quad i \neq j.$$

**Theorem 20.4.** Consider the closed-loop system $\tilde{\mathcal{G}}$ given by (20.41)–(20.44). Assume that Assumptions 20.1 and 20.2 hold. Furthermore, assume that $\mathcal{C} = \mathcal{C}^T$. Let $\mathcal{D} \subseteq \mathbb{R}^n$ be a compact positively invariant set with respect to the closed-loop system (20.41)–(20.44). Then there exists $\mathcal{D}_0 \subseteq \mathcal{D}$ such that if all the initial conditions are in $\mathcal{D}_0$, then parallel formation is achieved for (20.41) by using the distributed hybrid feedback control law (20.42)–(20.44). In addition, the Laplacian disagreement function of each agent's velocity is decreasing and the Laplacian disagreement function of each subcontroller is minimized across resetting events.

**Proof.** Consider the nonnegative function given by

$$V(\tilde{x}, \dot{\tilde{x}}, \tilde{x}_{\mathrm{c}}) = \frac{1}{4} \sum_{i=1}^{q} \sum_{j=1, j \neq i}^{q} \mathcal{C}_{(i,j)} \|\dot{x}_i - \dot{x}_j\|_2^2 + \sum_{i=1}^{q} \sum_{j=1, j \neq i}^{q} U_{ij}(x_i, x_j)$$

$$+ \frac{1}{4} \sum_{i=1}^{q} \sum_{j=1, j \neq i}^{q} \mathcal{C}_{(i,j)} \|x_{\mathrm{c}i} - x_{\mathrm{c}j}\|_2^2, \tag{20.48}$$

where $\tilde{x} \triangleq [x_1^{\mathrm{T}}, \ldots, x_q^{\mathrm{T}}]^{\mathrm{T}} \in \mathbb{R}^{nq}$ and $\tilde{x}_{\mathrm{c}} \triangleq [x_{\mathrm{c}1}^{\mathrm{T}}, \ldots, x_{\mathrm{c}q}^{\mathrm{T}}]^{\mathrm{T}} \in \mathbb{R}^{nq}$. The derivative of $V(\tilde{x}, \dot{\tilde{x}}, \tilde{x}_{\mathrm{c}})$ along the trajectories of the closed-loop system (20.41)–(20.44) is given by

$$\dot{V}(\tilde{x}, \dot{\tilde{x}}, \tilde{x}_{\mathrm{c}}) = \frac{1}{2} \sum_{i=1}^{q} \sum_{j=1, j \neq i}^{q} (\mathcal{C}_{(i,j)} + \mathcal{C}_{(j,i)})(\dot{x}_i - \dot{x}_j)^{\mathrm{T}} \ddot{x}_i$$

$$+ \sum_{i=1}^{q} \sum_{j=1, j \neq i}^{q} (\dot{x}_i - \dot{x}_j)^{\mathrm{T}} \nabla_{x_i} U_{ij}(x_i, x_j)$$

$$+ \frac{1}{2} \sum_{i=1}^{q} \sum_{j=1, j \neq i}^{q} (\mathcal{C}_{(i,j)} + \mathcal{C}_{(j,i)})(x_{\mathrm{c}i} - x_{\mathrm{c}j})^{\mathrm{T}} \dot{x}_{\mathrm{c}i}$$

$$= - \sum_{i=1}^{q} \left[ \sum_{j=1, j \neq i}^{q} \mathcal{C}_{(i,j)}(x_{\mathrm{c}i} - x_{\mathrm{c}j}) \right]^{\mathrm{T}} \left[ \sum_{j=1, j \neq i}^{q} \mathcal{C}_{(i,j)}(x_{\mathrm{c}i} - x_{\mathrm{c}j}) \right]$$

$$\leq 0, \quad (\tilde{x}, \dot{\tilde{x}}, \tilde{x}_{\mathrm{c}}) \notin \tilde{\mathcal{Z}}, \quad (\tilde{x}, \dot{\tilde{x}}, \tilde{x}_{\mathrm{c}}) \in \mathcal{D}, \tag{20.49}$$

and

$$\Delta V(\tilde{x}, \dot{\tilde{x}}, \tilde{x}_{\mathrm{c}}) = \frac{1}{4} \sum_{i=1}^{q} \left[ \min_{x_{\mathrm{c}i}} \sum_{j=1, j \neq i}^{q} \mathcal{C}_{(i,j)} \|x_{\mathrm{c}i} - x_{\mathrm{c}j}\|_2^2 \right]$$

$$- \frac{1}{4} \sum_{i=1}^{q} \sum_{j=1, j \neq i}^{q} \mathcal{C}_{(i,j)} \|x_{\mathrm{c}i} - x_{\mathrm{c}j}\|_2^2$$

$$< 0, \quad (\tilde{x}, \dot{\tilde{x}}, \tilde{x}_{\mathrm{c}}) \in \tilde{\mathcal{Z}}, \quad (\tilde{x}, \dot{\tilde{x}}, \tilde{x}_{\mathrm{c}}) \in \mathcal{D}. \tag{20.50}$$

Next, let

$$\mathcal{R} \triangleq \left\{ (\tilde{x}, \dot{\tilde{x}}, \tilde{x}_{\mathrm{c}}) \in \mathcal{D} : (\tilde{x}, \dot{\tilde{x}}, \tilde{x}_{\mathrm{c}}) \notin \mathcal{Z}, \dot{V}(\tilde{x}, \dot{\tilde{x}}, \tilde{x}_{\mathrm{c}}) = 0 \right\}$$

$$\cup \{(\tilde{x}, \dot{\tilde{x}}, \tilde{x}_c) \in \mathcal{D} : (\tilde{x}, \dot{\tilde{x}}, \tilde{x}_c) \in \mathcal{Z}, \Delta V(\tilde{x}, \dot{\tilde{x}}, \tilde{x}_c) = 0\},$$

and note that

$$\mathcal{R} = \bigcap_{i=1}^{q} \left\{ (\tilde{x}, \dot{\tilde{x}}, \tilde{x}_c) \in \mathcal{D} : \sum_{j=1, j \neq i}^{q} \mathcal{C}_{(i,j)}(x_{ci} - x_{cj}) = 0 \right\}$$

$$= \{(\tilde{x}, \dot{\tilde{x}}, \tilde{x}_c) \in \mathcal{D} : x_{c1} = \cdots = x_{cq}\}.$$

Furthermore, let $\mathcal{M}$ denote the largest invariant set contained in $\mathcal{R}$. It follows from (20.42) that $\sum_{i=1}^{q} \dot{x}_{ci} = 0$, and, hence, $x_{ci} = \alpha$ on $\mathcal{M}$, where $\alpha \in \mathbb{R}^n$, $i = 1, \ldots, q$. In addition, $\dot{x}_1 = \cdots = \dot{x}_q$ and

$$\ddot{x}_i = - \sum_{j=1, j \neq i}^{q} \nabla_{x_i}(U_{ij}(x_i, x_j) + U_{ji}(x_j, x_i))$$

on $\mathcal{M}$. Since

$$\sum_{i=1}^{q} \sum_{j=1, j \neq i}^{q} \nabla_{x_i}(U_{ij}(x_i, x_j) + U_{ji}(x_j, x_i)) = 0,$$

it follows that $\ddot{x}_i = 0$, and, hence, $\dot{x}_i = \beta$ on $\mathcal{M}$, where $\beta \in \mathbb{R}^n$, $i = 1, \ldots, q$. Now, it follows from Theorem 19.1 (see also Theorem 8.1 of [128]) that, for every initial condition in $\mathcal{D}_0$, $(\tilde{x}(t), \dot{\tilde{x}}(t), \tilde{x}_c(t))$ converges to the largest invariant set $\mathcal{M}$ contained in the set

$$\hat{\mathcal{M}} \triangleq \left\{ (\tilde{x}, \dot{\tilde{x}}, \tilde{x}_c) \in \mathcal{D} : \dot{x}_1 = \cdots = \dot{x}_q = \alpha, x_{c1} = \cdots = x_{cq} = \beta, \right.$$

$$\left. \sum_{j=1, j \neq i}^{q} \nabla_{x_i}(U_{ij}(x_i, x_j) + U_{ji}(x_j, x_i)) = 0, i = 1, \ldots, q \right\}.$$

Finally, the fact that the Laplacian disagreement function of each agent's velocity is decreasing and the Laplacian disagreement function of each subcontroller is minimized across resetting events is immediate from (20.43) and (20.47). □

## 20.5 Collision Avoidance via Hybrid Control

In this section, we develop a hybrid algorithm for collision avoidance between mobile agents. For dynamic models given by (20.41), a potential function approach is used in [298] and [234] to guarantee collision avoidance. However, for kinematic models given by (2.35), the potential function approach developed in [298] and [234] cannot be used. In this section, we develop a hybrid collision avoidance algorithm for the driftless controllable nonholonomic system given by

$$\dot{x}_{\mathrm{p}i}(t) = X_1(x_{\mathrm{p}i}(t))u_{1,i}(t) + \cdots + X_{m+1}(x_{\mathrm{p}i}(t))u_{m+1,i}(t),$$
$$x_{\mathrm{p}i}(0) = x_{\mathrm{p}i0}, \quad t \geq 0, \tag{20.51}$$

where, for each $i \in \{1, \ldots, q\}$ and $t \geq 0$, $x_{\mathrm{p}i}(t) \triangleq [x_i^{\mathrm{T}}(t), z_i^{\mathrm{T}}(t)]^{\mathrm{T}}$; $x_i(t) \in \mathbb{R}^n$; $z_i(t) \in \mathbb{R}^m$; $u_{l,i}(t) \in \mathbb{R}$ denotes the control input; $l = 1, \ldots, m$; $X_1 = [g^{\mathrm{T}}(z_i), 0_{1 \times m}]^{\mathrm{T}} \in \mathbb{R}^{m+n}$; $g : \mathbb{R}^m \to \mathbb{R}^n$ is smooth; and $X_j = \mathbf{e}_{n+j-1}$, $j = 2, \ldots, m+1$, where $\mathbf{e}_{n+j-1} \in \mathbb{R}^{m+n}$ is a vector whose $(n+j-1)$th component is one and remaining components are zero.

Specifically, consider $q$ mobile agents whose dynamics are described by the kinematic model (20.51). We assume that the initial condition for (20.51) satisfies $\|x_{i0} - x_{j0}\|_2 > L$, $i, j = 1, \ldots, q$, $i \neq j$, where $L > 0$ is a constant. Furthermore, we require that the distance between any two connecting agents is strictly greater than or equal to $L$; that is, $\|x_i - x_j\|_2 \geq L$, $\mathcal{C}_{(i,j)} = 1$, $i \neq j$, $i, j = 1, \ldots, q$.

Consider the hybrid controller given by

$$\dot{x}_{\mathrm{c}i}(t) = - \sum_{j=1, j \neq i}^{q} \mathcal{C}_{(i,j)}(x_{\mathrm{c}i}(t) - x_{\mathrm{c}j}(t)) + g^{\mathrm{T}}(z_i(t))\zeta_i(t), \tag{20.52}$$

$$\dot{z}_{\mathrm{c}i}(t) = - \sum_{j=1, j \neq i}^{q} \mathcal{C}_{(i,j)}(z_{\mathrm{c}i}(t) - z_{\mathrm{c}j}(t)) - \sum_{j=1, j \neq i}^{q} \mathcal{C}_{(i,j)}(z_i(t) - z_j(t)),$$
$$(x_i(t), x_j(t), z_i(t), z_j(t), x_{\mathrm{c}i}(t), x_{\mathrm{c}j}(t), z_{\mathrm{c}i}(t), z_{\mathrm{c}j}(t)) \notin \bigcup_{k \in \mathcal{N}_i} \mathcal{Z}_{ik},$$
$$i = 1, \ldots, q, \quad j \in \mathcal{N}_i, \quad x_{\mathrm{c}i}(0) = x_{\mathrm{c}i0}, \quad z_{\mathrm{c}i}(0) = z_{\mathrm{c}i0}, \quad t \geq 0, \tag{20.53}$$

$$x_{\mathrm{c}i}(t^+) = u_{xi}(x_{\mathrm{c}}(t)), \tag{20.54}$$

$$z_{\mathrm{c}i}(t^+) = u_{zi}(z_{\mathrm{c}}(t)), \tag{20.55}$$

$$(x_i(t), x_j(t), z_i(t), z_j(t), x_{\mathrm{c}i}(t), x_{\mathrm{c}j}(t), z_{\mathrm{c}i}(t), z_{\mathrm{c}j}(t)) \in \mathcal{Z}_{ij}, \quad j \in \mathcal{N}_i,$$

$$v_i(t) = -x_{\mathrm{c}i}(t), \tag{20.56}$$

$$\begin{bmatrix} u_{2,i}(t) \\ \vdots \\ u_{m+1,i}(t) \end{bmatrix} = - \sum_{j=1, j \neq i}^{q} \mathcal{C}_{(j,i)}(z_{\mathrm{c}j}(t) - z_{\mathrm{c}i}(t)), \tag{20.57}$$

where $v_i(t) \triangleq u_{1,i}(t) \in \mathbb{R}$; $\mathcal{N}_i \subseteq \{i+1, \ldots, q\}$ is the set containing the maximum number of connecting nodes; and the resetting set $\mathcal{Z}_{ij}$ is given by

$$
\mathcal{Z}_{ij} = \Big\{ (x_i, x_j, z_i, z_j, x_{ci}, x_{cj}, z_{ci}, z_{cj}) : \|x_i - x_j\|_2^2 \leq L^2 \text{ and}
$$
$$
\frac{\mathrm{d}}{\mathrm{d}t} \|x_i - x_j\|_2^2 < 0 \Big\}, \quad i = 1, \ldots, q, \quad j \in \mathcal{N}_i. \quad (20.58)
$$

Our goal is to design $u_{xi}$ and $u_{zi}$ such that for

$$
(x_i, x_j, z_i, z_j, x_{ci}, x_{cj}, z_{ci}, z_{cj}) \in \mathcal{Z}_{ij}, \quad i = 1, \ldots, q, \quad j \in \mathcal{N}_i,
$$

$$
\sum_{i=1}^{q} (x_{ci}^2 + \|z_{ci}\|_2^2) > \sum_{i=1}^{q} (u_{xi}^2(x_c) + \|u_{zi}(z_c)\|_2^2) \quad (20.59)
$$

and

$$
(x_i - x_j)^{\mathrm{T}} [g(z_j)u_{xj}(x_c) - g(z_i)u_{xi}(x_c)] \geq 0. \quad (20.60)
$$

The above constraints lead to the nonlinear optimization problem

$$
\min_{u_{xi} \in \mathbb{R}, u_{zi} \in \mathbb{R}^m} \sum_{i=1}^{q} (u_{xi}^2 + \|u_{zi}\|_2^2) \quad (20.61)
$$

subject to

$$
0 < \varepsilon \leq \frac{\sum_{i=1}^{q} (u_{xi}^2 + \|u_{zi}\|_2^2)}{\sum_{i=1}^{q} (x_{ci}^2 + \|z_{ci}\|_2^2)} \leq \mu < 1, \quad (20.62)
$$
$$
(x_i - x_j)^{\mathrm{T}} [g(z_j)u_{xj}(x_c) - g(z_i)u_{xi}(x_c)] \geq 0, \quad i = 1, \ldots, q, \quad j \in \mathcal{N}_i, \quad (20.63)
$$
$$
\|x_k - x_l\|_2^2 \leq L^2, \quad (20.64)
$$
$$
(x_k - x_l)^{\mathrm{T}} (g(z_l)x_{cl} - g(z_k)x_{ck}) < 0, \quad k \in \{1, \ldots, q\}, \quad l \in \mathcal{N}_k. \quad (20.65)
$$

At each resetting time instant, we can solve this nonlinear optimization problem to obtain $u_{xi}$ and $u_{zi}$. However, since most nonlinear optimization problems are time-consuming and do not usually possess global solutions, a real-time algorithm is needed for implementing the hybrid control architecture (20.52)–(20.57).

Next, we give an explicit formula for $u_{xi}$ and $u_{zi}$ for kinematic nonholonomic systems so that the proposed hybrid control architecture can be easily implemented online. Consider the control strategy

$$
u_{xi}(t) = \rho_i x_{ci}(t), \quad (20.66)
$$

$$u_{zi}(t) = \text{diag}[\eta_{i,1}, \ldots, \eta_{i,m}] z_{ci}(t), \tag{20.67}$$

where $|\rho_i| \leq 1$ and $|\eta_{i,k}| < 1$, $i = 1, \ldots, q$, $k = 1, \ldots, m$, are such that, for all $i = 1, \ldots, q$ and $j \in \mathcal{N}_i$,

$$\rho_j[(x_i - x_j)^{\mathrm{T}} g(z_j) x_{cj}] - \rho_i[(x_i - x_j)^{\mathrm{T}} g(z_i) x_{ci}] \geq 0. \tag{20.68}$$

Furthermore, consider the linear programming problem given by

$$\max_{\rho_i} \sum_{i=1}^{q} \rho_i \tag{20.69}$$

subject to

$$\mathcal{C}_{(i,j)} \rho_j[(x_i - x_j)^{\mathrm{T}} g(z_j) x_{cj}] - \mathcal{C}_{(i,j)} \rho_i[(x_i - x_j)^{\mathrm{T}} g(z_i) x_{ci}] \geq 0, \tag{20.70}$$

$$-1 \leq \rho_i \leq 1, \quad \mathcal{C}_{(i,j)} \in \{0,1\}, \quad i = 1, \ldots, q-1, \quad j = i+1, \ldots, q. \tag{20.71}$$

At each resetting time instant, (20.69) is a standard linear programming problem and can be easily solved using the simplex method. It is clear that (20.59) and (20.60) hold for $(x_i, x_j, z_i, z_j, x_{ci}, x_{cj}, z_{ci}, z_{cj}) \in \mathcal{Z}_{ij}$, $i = 1, \ldots, q$, $j \in \mathcal{N}_i$.

**Theorem 20.5.** Consider the closed-loop system $\tilde{\mathcal{G}}$ given by (20.51), (20.52)–(20.57), (20.66), and (20.67). Let $\mathcal{D}$ be a compact positively invariant set with respect to the closed-loop system $\tilde{\mathcal{G}}$. Assume that Assumptions 19.1–19.3 hold, and assume that $\mathcal{C} = \mathcal{C}^{\mathrm{T}}$. Furthermore, assume that the system initial condition for (20.51) satisfies $\|x_{i0} - x_{j0}\|_2 > L$, $i, j = 1, \ldots, q$, $i \neq j$. Then there exists $\mathcal{D}_0 \subseteq \mathcal{D}$ such that if all the system initial conditions are in $\mathcal{D}_0$, then the state of the closed-loop system $\tilde{\mathcal{G}}$ approaches the largest invariant set $\mathcal{M}$ contained in

$$\hat{\mathcal{M}} \triangleq \bigcap_{i=1}^{q} \left\{ x_{ci} = \alpha, \; z_i = \beta, \; z_{ci} = \gamma, \; \dot{x}_i = g(\beta)\alpha, \; g^{\mathrm{T}}(\beta)\zeta_i = 0 \right\},$$

where $\alpha \in \mathbb{R}$, $\beta \in \mathbb{R}^m$, and $\gamma \in \mathbb{R}^m$. Furthermore, the distance between the $i$th and $j$th agents is greater than or equal to $L$ for all $i = 1, \ldots, q$, $j \in \mathcal{N}_i$.

**Proof.** The proof of convergence is similar to the proof of Theorem 2.5 and, hence, is omitted. Next, we show that if the initial distance between the $i$th agent and the $j$th agent is greater than $L$, $i = 1, \ldots, q$, $j \in \mathcal{N}_i$, then the distance between those two agents is always greater than or equal to $L$ for all time. To see this, let $\{t_k\}_{k \in \mathbb{Z}_+}$ denote the resetting times, and note that it follows from Assumptions 19.1 and 19.2 that the resetting times $t_k$, $k \in \mathbb{Z}_+$, are well defined [62, 126, 128].

First, consider the time interval $[0, t_1)$, and note that for the first resetting time $t_1$, $\|x_i(t_1) - x_j(t_1)\|_2 \le L$ and

$$\frac{\mathrm{d}}{\mathrm{d}t}\|x_i(t) - x_j(t)\|_2 \Big|_{t=t_1} < 0,$$

and, hence, either $\|x_i(t) - x_j(t)\|_2 \ge L$ or

$$\frac{\mathrm{d}}{\mathrm{d}t}\|x_i(t) - x_j(t)\|_2 \ge 0$$

for all $t \in [0, t_1)$, $i = 1, \ldots, q$, $j \in \mathcal{N}_i$. If

$$\frac{\mathrm{d}}{\mathrm{d}t}\|x_i(t) - x_j(t)\|_2 \ge 0$$

for all $t \in [0, t_1)$, then

$$\begin{aligned}
\|x_i(t) - x_j(t)\|_2 &= \|x_{i0} - x_{j0}\|_2 + \int_0^t \frac{\mathrm{d}}{\mathrm{d}s}\|x_i(s) - x_j(s)\|_2 \mathrm{d}s \\
&\ge \|x_{i0} - x_{j0}\|_2 \\
&> L
\end{aligned}$$

for all $t \in [0, t_1)$, $i = 1, \ldots, q$, $j \in \mathcal{N}_i$. In either case, $\|x_i(t) - x_j(t)\|_2 \ge L$ for all $t \in [0, t_1)$, $i = 1, \ldots, q$, $j \in \mathcal{N}_i$.

Next, consider the case where $t \in [t_1, t_2)$. It follows from the continuity of $x(t)$ that

$$\|x_i(t_1) - x_j(t_1)\|_2 = \|x_i(t_1^+) - x_j(t_1^+)\|_2 = L, \quad i = 1, \ldots, q, \quad j \in \mathcal{N}_i.$$

Furthermore, note that the controller (20.52)–(20.57), (20.66), and (20.67) satisfies

$$\frac{\mathrm{d}}{\mathrm{d}t}\|x_i(t) - x_j(t)\|_2 \Big|_{t=t_1^+} \ge 0, \quad i = 1, \ldots, q, \quad j \in \mathcal{N}_i.$$

Hence, it follows that either $\|x_i(t) - x_j(t)\|_2 \ge L$ or

$$\frac{\mathrm{d}}{\mathrm{d}t}\|x_i(t) - x_j(t)\|_2 \ge 0$$

for all $t \in [t_1^+, t_2)$, $i = 1, \ldots, q$, $j \in \mathcal{N}_i$. If $\frac{\mathrm{d}}{\mathrm{d}t}\|x_i(t) - x_j(t)\|_2 \ge 0$ for all $t \in [t_1^+, t_2)$, then

$$\begin{aligned}
\|x_i(t) - x_j(t)\|_2 &= \|x_i(t^+) - x_j(t^+)\|_2 + \int_{t^+}^t \frac{\mathrm{d}}{\mathrm{d}s}\|x_i(s) - x_j(s)\|_2 \mathrm{d}s \\
&\ge \|x_i(t^+) - x_j(t^+)\|_2 \\
&= L
\end{aligned}$$

for all $t \in [0, t_1)$, $i = 1, \ldots, q$, and $j \in \mathcal{N}_i$. In either case, $\|x_i(t) - x_j(t)\|_2 \geq L$ for all $t \in [t_1^+, t_2)$, $i = 1, \ldots, q$, $j \in \mathcal{N}_i$.

Now, repeating the above arguments for each interval $[t_k, t_{k+1})$, $k \geq 2$, yields that $\|x_i(t) - x_j(t)\|_2 \geq L$ for all $t \in [t_k^+, t_{k+1})$, $i = 1, \ldots, q$, $j \in \mathcal{N}_i$. Finally, note that

$$[0, t_1) \cup \left( \bigcup_{k \in \mathbb{Z}_+} [t_k, t_{k+1}) \right) = [0, \infty),$$

which implies that $\|x_i(t) - x_j(t)\|_2 \geq L$ for all $t \geq 0$, $i = 1, \ldots, q$, $j \in \mathcal{N}_i$. $\square$

*Chapter Twenty-One*

---

# Formation Control Protocols for Network Systems via Hybrid Stabilization of Sets

## 21.1 Introduction

Using system-theoretic thermodynamic concepts, an energy- and entropy-based hybrid controller architecture is proposed in [125, 128] as a means for achieving enhanced energy dissipation in lossless and dissipative dynamical systems. These dynamic controllers combine a logical switching architecture with continuous dynamics to guarantee that the system plant energy is strictly decreasing across switchings. The general framework developed in [125] leads to closed-loop systems described by impulsive differential equations [128]. In particular, the authors in [125, 128] construct hybrid dynamic controllers that guarantee that the closed-loop system is consistent with basic thermodynamic principles. Specifically, they establish the existence of an entropy function for the closed-loop system that satisfies a hybrid Clausius-type inequality. Special cases of energy-based and entropy-based hybrid controllers involving state-dependent switching are also developed to show the efficacy of the approach.

As shown in Chapters 19 and 20, hybrid control architectures can be combined with dynamical thermodynamics to design hybrid consensus protocols for multiagent systems wherein a group of agents are required to agree on certain quantities of interest, such as a group of agents having a common leading angle or a shared communication frequency. Convergence and state equipartitioning also arise in numerous complex large-scale dynamical networks that demonstrate a degree of synchronization. System synchronization typically involves coordination of events, allowing a dynamical system to operate in unison and resulting in system self-organization. The onset of synchronization in populations of coupled dynamical networks has been studied for various complex networks, including network models for mathematical biology, statistical physics, kinetic theory, bifurcation theory, and plasma physics [286]. Synchronization of firing neural oscillator populations also appears in the neuroscience literature [53, 165].

Alternatively, in other applications of multiagent systems, groups of agents are required to achieve and maintain a prescribed geometric shape. As shown in Chapter 2, this formation control problem includes flocking [234, 294] and cyclic pursuit [220], wherein parallel and circular formations of vehicles are sought. Since a specified formation of multiagent systems, which can include flocking, cyclic pursuit, rendezvous, or consensus, can be characterized by a hyperplane or manifold in the state space, in this chapter we extend the results of [125, 128] to develop a state-dependent hybrid control framework for addressing multiagent formation control protocols for general nonlinear dynamical systems using hybrid stabilization of sets. The proposed framework involves a novel class of fixed-order, energy-based hybrid dynamic controllers as a means of achieving cooperative control formations.

These dynamic controllers combine a logical switching architecture with continuous dynamics to guarantee that a system-generalized energy function, whose zero level set characterizes a specified system formation, is strictly decreasing across switchings. The general framework leads to hybrid closed-loop systems described by impulsive differential equations and addresses general nonlinear dynamical systems without limiting consensus and formation control protocols to single and double integrator models.

The contents of the chapter are as follows. In Section 21.2, we establish definitions and notation and review some basic results on impulsive differential equations, which provide the mathematical foundation for designing formation control protocols for nonlinear dynamical systems using logic-based hybrid controllers. In Section 21.3, we present a general state-dependent hybrid control framework for stabilization of sets. The main result in this section extends the results of [125] to hybrid stabilization of sets and is not limited to lossless or dissipative dynamical systems. In Section 21.4, we specialize the results of Section 21.3 to linear dynamical systems. We then turn our attention to hybrid control design for parallel and rendezvous formations in Section 21.5 and consensus control of multiagent systems for standard single and double integrator formation models in Section 21.6. In Section 21.7, we use the results of Sections 21.2 and 21.3 to design hybrid controllers for cyclic pursuit.

## 21.2 Hybrid Control and Impulsive Dynamical Systems

In this chapter, we consider continuous-time nonlinear dynamical systems of the form

$$\dot{x}_{\mathrm{p}}(t) = f_{\mathrm{p}}(x_{\mathrm{p}}(t), u(t)), \quad x_{\mathrm{p}}(0) = x_{\mathrm{p}0}, \quad t \geq 0, \qquad (21.1)$$
$$y(t) = h_{\mathrm{p}}(x_{\mathrm{p}}(t)), \qquad (21.2)$$

where $t \geq 0$; $x_{\mathrm{p}}(t) \in \mathcal{D}_{\mathrm{p}} \subseteq \mathbb{R}^{n_{\mathrm{p}}}$; $\mathcal{D}_{\mathrm{p}}$ is an open set; $u(t) \in \mathbb{R}^m$, $f_{\mathrm{p}}$ : $\mathcal{D}_{\mathrm{p}} \times \mathbb{R}^m \to \mathbb{R}^{n_{\mathrm{p}}}$ is smooth (i.e., infinitely differentiable) on $\mathcal{D}_{\mathrm{p}} \times \mathbb{R}^m$; and $h_{\mathrm{p}} : \mathcal{D}_{\mathrm{p}} \to \mathbb{R}^l$ is smooth. Furthermore, we consider hybrid (i.e., resetting) dynamic controllers of the form

$$\dot{x}_{\mathrm{c}}(t) = f_{\mathrm{cc}}(x_{\mathrm{c}}(t), y(t)), \quad x_{\mathrm{c}}(0) = x_{\mathrm{c}0}, \quad (x_{\mathrm{c}}(t), y(t)) \notin \mathcal{Z}_{\mathrm{c}}, \quad (21.3)$$
$$\Delta x_{\mathrm{c}}(t) = f_{\mathrm{dc}}(x_{\mathrm{c}}(t), y(t)), \quad (x_{\mathrm{c}}(t), y(t)) \in \mathcal{Z}_{\mathrm{c}}, \quad (21.4)$$
$$u(t) = h_{\mathrm{cc}}(x_{\mathrm{c}}(t), y(t)), \quad (21.5)$$

where $t \geq 0$; $x_{\mathrm{c}}(t) \in \mathcal{D}_{\mathrm{c}} \subseteq \mathbb{R}^{n_{\mathrm{c}}}$, $\mathcal{D}_{\mathrm{c}}$ is an open set; $\Delta x_{\mathrm{c}}(t) \triangleq x_{\mathrm{c}}(t^+) - x_{\mathrm{c}}(t)$, where $x_{\mathrm{c}}(t^+) \triangleq x_{\mathrm{c}}(t) + f_{\mathrm{dc}}(x_{\mathrm{c}}(t), y(t)) = \lim_{\varepsilon \to 0^+} x_{\mathrm{c}}(t+\varepsilon)$; $(x_{\mathrm{c}}(t), y(t)) \in \mathcal{Z}_{\mathrm{c}}$; $f_{\mathrm{cc}} : \mathcal{D}_{\mathrm{c}} \times \mathbb{R}^l \to \mathbb{R}^{n_{\mathrm{c}}}$ is smooth on $\mathcal{D}_{\mathrm{c}} \times \mathbb{R}^l$; $h_{\mathrm{cc}} : \mathcal{D}_{\mathrm{c}} \times \mathbb{R}^l \to \mathbb{R}^m$ is smooth; $f_{\mathrm{dc}} : \mathcal{D}_{\mathrm{c}} \times \mathbb{R}^l \to \mathbb{R}^{n_{\mathrm{c}}}$ is continuous; and $\mathcal{Z}_{\mathrm{c}} \subset \mathcal{D}_{\mathrm{c}} \times \mathbb{R}^l$ is the *resetting set*. Note that, for generality, we allow the hybrid dynamic controller to be of fixed dimension $n_{\mathrm{c}}$, which may be less than the plant order $n_{\mathrm{p}}$.

The equations of motion for the closed-loop dynamical system (21.1)–(21.5) have the form

$$\dot{x}(t) = f_{\mathrm{c}}(x(t)), \quad x(0) = x_0, \quad x(t) \notin \mathcal{Z}, \quad (21.6)$$
$$\Delta x(t) = f_{\mathrm{d}}(x(t)), \quad x(t) \in \mathcal{Z}, \quad (21.7)$$

where

$$x \triangleq \begin{bmatrix} x_{\mathrm{p}} \\ x_{\mathrm{c}} \end{bmatrix} \in \mathbb{R}^n, \quad f_{\mathrm{c}}(x) \triangleq \begin{bmatrix} f_{\mathrm{p}}(x_{\mathrm{p}}, h_{\mathrm{cc}}(x_{\mathrm{c}}, h_{\mathrm{p}}(x_{\mathrm{p}}))) \\ f_{\mathrm{cc}}(x_{\mathrm{c}}, h_{\mathrm{p}}(x_{\mathrm{p}})) \end{bmatrix},$$
$$f_{\mathrm{d}}(x) \triangleq \begin{bmatrix} 0 \\ f_{\mathrm{dc}}(x_{\mathrm{c}}, h_{\mathrm{p}}(x_{\mathrm{p}})) \end{bmatrix}, \quad (21.8)$$

and $\mathcal{Z} \triangleq \{x \in \mathcal{D} : (x_{\mathrm{c}}, h_{\mathrm{p}}(x_{\mathrm{p}})) \in \mathcal{Z}_{\mathrm{c}}\}$, with $n \triangleq n_{\mathrm{p}} + n_{\mathrm{c}}$ and $\mathcal{D} \triangleq \mathcal{D}_{\mathrm{p}} \times \mathcal{D}_{\mathrm{c}}$. We refer to the differential equation (21.6) as the *continuous-time dynamics*, and we refer to the difference equation (21.7) as the *resetting law*. Note that although the closed-loop state vector consists of plant states and controller states, it is clear from (21.8) that only those states associated with the controller are reset. To ensure the well-posedness of the solutions to (21.6) and (21.7), we make the following additional assumptions, which were discussed in Chapter 19.

**Assumption 21.1.** If $x \in \overline{\mathcal{Z}} \backslash \mathcal{Z}$, then there exists $\varepsilon > 0$ such that, for all $0 < \delta < \varepsilon$, $\psi(\delta, x) \notin \mathcal{Z}$, where $\psi(\cdot, \cdot)$ denotes the solution to the continuous-time dynamics (21.6).

**Assumption 21.2.** If $x \in \mathcal{Z}$, then $x + f_{\mathrm{d}}(x) \notin \mathcal{Z}$.

Recall that a function $x : \mathcal{I}_{x_0} \to \mathcal{D}$ is a *solution* to the impulsive dynamical system (21.6) and (21.7) on the interval $\mathcal{I}_{x_0} \subseteq \mathbb{R}$ with initial condition $x(0) = x_0$, where $\mathcal{I}_{x_0}$ denotes the maximal interval of existence of a solution to (21.6) and (21.7), if $x(\cdot)$ is left-continuous and $x(t)$ satisfies (21.6) and (21.7) for all $t \in \mathcal{I}_{x_0}$. For further discussion on solutions to impulsive differential equations, see [13, 14, 62, 128, 192, 225, 269]. For convenience, we use the notation $s(t, x_0)$ to denote the solution $x(t)$ of (21.6) and (21.7) at time $t \geq 0$ with initial condition $x(0) = x_0$.

For the statement of the next result, we need the following key assumption, stated in Chapter 19 and repeated here for convenience.

**Assumption 21.3.** Consider the impulsive dynamical system (21.6) and (21.7), and let $s(t, x_0)$, $t \geq 0$, denote the solution to (21.6) and (21.7) with initial condition $x_0$. Then, for every $x_0 \notin \mathcal{Z}$ and every $\varepsilon > 0$ and $t \neq t_k$, there exists $\delta(\varepsilon, x_0, t) > 0$ such that if $\|x_0 - z\| < \delta(\varepsilon, x_0, t)$, $z \in \mathcal{D}$, then $\|s(t, x_0) - s(t, z)\| < \varepsilon$.

As discussed in Chapter 19, Assumption 21.3 is a weakened version of the quasi-continuous dependence assumption given in [62, 128] and is a generalization of the standard continuous dependence property for dynamical systems with continuous flows to dynamical systems with left-continuous flows. Specifically, by letting $t \in [0, \infty)$, Assumption 21.3 specializes to the classical continuous dependence of solutions of a given dynamical system on the system's initial conditions $x_0 \in \mathcal{D}$ for every time instant.

It should be noted that the standard continuous dependence property for dynamical systems with continuous flows is defined uniformly in time on compact intervals. Since solutions of impulsive dynamical systems are not continuous in time and solutions are not continuous functions of the system initial conditions, Assumption 21.3, involving pointwise continuous dependence, is needed to apply the hybrid invariance principle developed in [62, 128] to hybrid closed-loop systems. Sufficient conditions that guarantee that the impulsive dynamical system (21.6) and (21.7) satisfies a stronger version of Assumption 21.3 are given in [62, 128] (see also [109]). The following proposition provides a generalization of Proposition 4.1 in [62] for establishing sufficient conditions for guaranteeing that the impulsive dynamical system (21.6) and (21.7) satisfies Assumption 21.3.

**Proposition 21.1** (see [125]). Consider the impulsive dynamical system $\mathcal{G}$ given by (21.6) and (21.7). Assume that Assumptions 21.1 and 21.2 hold; $\tau_1(\cdot)$ is continuous at every $x \notin \overline{\mathcal{Z}}$ such that $0 < \tau_1(x) < \infty$; and, if $x \in \mathcal{Z}$, then $x + f_\mathrm{d}(x) \in \overline{\mathcal{Z}} \backslash \mathcal{Z}$. Furthermore, for every $x \in \overline{\mathcal{Z}} \backslash \mathcal{Z}$ such that $0 < \tau_1(x) < \infty$, assume that the following statements hold.

(i) If a sequence $\{x_i\}_{i=1}^{\infty} \in \mathcal{D}$ is such that $\lim_{i \to \infty} x_i = x$ and $\lim_{i \to \infty} \tau_1(x_i)$

exists, then either both $f_{\mathrm{d}}(x) = 0$ and $\lim_{i\to\infty} \tau_1(x_i) = 0$ or

$$\lim_{i\to\infty} \tau_1(x_i) = \tau_1(x).$$

(ii) If a sequence $\{x_i\}_{i=1}^{\infty} \in \overline{\mathcal{Z}}\backslash\mathcal{Z}$ is such that $\lim_{i\to\infty} x_i = x$ and

$$\lim_{i\to\infty} \tau_1(x_i)$$

exists, then $\lim_{i\to\infty} \tau_1(x_i) = \tau_1(x)$.

Then $\mathcal{G}$ satisfies Assumption 21.3.

The following result provides sufficient conditions for establishing continuity of $\tau_1(\cdot)$ at $x_0 \notin \overline{\mathcal{Z}}$ and *sequential continuity* of $\tau_1(\cdot)$ at $x_0 \in \overline{\mathcal{Z}}\backslash\mathcal{Z}$; that is, $\lim_{i\to\infty} \tau_1(x_i) = \tau_1(x_0)$ for $\{x_i\}_{i=1}^{\infty} \notin \mathcal{Z}$ and $\lim_{i\to\infty} x_i = x_0$. For this result, the following definition is needed. First, however, recall that the *Lie derivative* of a smooth function $\mathcal{X} : \mathcal{D} \to \mathbb{R}$ along the vector field of the continuous-time dynamics $f_{\mathrm{c}}(x)$ is given by

$$L_{f_{\mathrm{c}}}\mathcal{X}(x) \triangleq \frac{\mathrm{d}}{\mathrm{d}t}\mathcal{X}(\psi(t,x))|_{t=0} = \frac{\partial\mathcal{X}(x)}{\partial x}f_{\mathrm{c}}(x),$$

and the *zeroth-* and *higher-order Lie derivatives* are, respectively, defined by $L_{f_{\mathrm{c}}}^0\mathcal{X}(x) \triangleq \mathcal{X}(x)$ and $L_{f_{\mathrm{c}}}^k\mathcal{X}(x) \triangleq L_{f_{\mathrm{c}}}(L_{f_{\mathrm{c}}}^{k-1}\mathcal{X}(x))$, where $k \geq 1$.

**Definition 21.1** (see [125]). Let $\mathcal{Q} \triangleq \{x \in \mathcal{D} : \mathcal{X}(x) = 0\}$, where $\mathcal{X} : \mathcal{D} \to \mathbb{R}$ is an infinitely differentiable function. A point $x \in \mathcal{Q}$ such that $f_{\mathrm{c}}(x) \neq 0$ is *k-transversal* to (21.6) if there exists $k \in \{1, 2, \ldots\}$ such that

$$L_{f_{\mathrm{c}}}^r\mathcal{X}(x) = 0, \quad r = 0, \ldots, 2k-2, \quad L_{f_{\mathrm{c}}}^{2k-1}\mathcal{X}(x) \neq 0. \qquad (21.9)$$

**Proposition 21.2** (see [125]). Consider the impulsive dynamical system (21.6) and (21.7). Let $\mathcal{X} : \mathcal{D} \to \mathbb{R}$ be an infinitely differentiable function such that $\overline{\mathcal{Z}} = \{x \in \mathcal{D} : \mathcal{X}(x) = 0\}$, and assume that every $x \in \overline{\mathcal{Z}}$ is $k$-transversal to (21.6). Then, at every $x_0 \notin \overline{\mathcal{Z}}$ such that $0 < \tau_1(x_0) < \infty$, $\tau_1(\cdot)$ is continuous. Furthermore, if $x_0 \in \overline{\mathcal{Z}}\backslash\mathcal{Z}$ is such that $\tau_1(x_0) \in (0, \infty)$ and either (i) $\{x_i\}_{i=1}^{\infty} \in \overline{\mathcal{Z}}\backslash\mathcal{Z}$ or (ii) $\lim_{i\to\infty} \tau_1(x_i) > 0$, where $\{x_i\}_{i=1}^{\infty} \notin \overline{\mathcal{Z}}$ is such that $\lim_{i\to\infty} x_i = x_0$ and $\lim_{i\to\infty} \tau_1(x_i)$ exists, then $\lim_{i\to\infty} \tau_1(x_i) = \tau_1(x_0)$.

Let $x_0 \notin \mathcal{Z}$ be such that $\lim_{i\to\infty} \tau_1(x_i) \neq \tau_1(x_0)$ for some unbounded sequence $\{x_i\}_{i=1}^{\infty} \notin \mathcal{Z}$ with $\lim_{i\to\infty} x_i = x_0$. Then it follows from Proposition 21.2 that $\lim_{i\to\infty} \tau_1(x_i) = 0$.

The notion of $k$-transversality introduced in Definition 21.1 differs from the well-known notion of transversality [84, 114] involving an orthogonality condition between a vector field and a differentiable submanifold. In the case where $k = 1$, Definition 21.1 coincides with the standard notion of transversality and guarantees that the solution of the closed-loop system (21.6) and (21.7) is not tangent to the closure of the resetting set $\mathcal{Z}$ at the intersection with $\overline{\mathcal{Z}}$ [128]. In general, however, $k$-transversality guarantees that the sign of $\mathcal{X}(x(t))$ changes as the closed-loop system trajectory $x(t)$ traverses the closure of the resetting set $\mathcal{Z}$ at the intersection with $\overline{\mathcal{Z}}$.

Proposition 21.2 is a nontrivial generalization of Proposition 4.2 of [62] and Lemma 3 of [109]. Specifically, Proposition 21.2 establishes the continuity of $\tau_1(\cdot)$ in the case where the resetting set $\mathcal{Z}$ is not a closed set. In addition, the $k$-transversality condition given in Definition 21.1 is also a generalization of the transversality conditions given in [62] and [109] by considering higher-order derivatives of the function $\mathcal{X}(\cdot)$ rather than simply considering the first-order derivative as in [62, 109].

The next result characterizes impulsive dynamical system limit sets in terms of continuously differentiable functions. In particular, we show that the system trajectories of a state-dependent impulsive dynamical system converge to an invariant set contained in a union of level surfaces characterized by the continuous-time system dynamics and the resetting system dynamics. Note that for addressing the stability of sets of an impulsive dynamical system, the usual set stability definitions are valid [122].

Specifically, for a positively invariant set $\mathcal{D}_0 \subset \mathcal{D}$, $\mathcal{D}_0$ is *Lyapunov stable* with respect to (21.6) and (21.7) if and only if, for every open neighborhood $\mathcal{O}_1 \subseteq \mathcal{D}$ of $\mathcal{D}_0$, there exists an open neighborhood $\mathcal{O}_2 \subseteq \mathcal{O}_1$ of $\mathcal{D}_0$ such that $x(t) \in \mathcal{O}_1$, $t \geq 0$, for all $x_0 \in \mathcal{O}_2$. Equivalently, $\mathcal{D}_0$ is Lyapunov stable with respect to (21.6) and (21.7) if and only if, for all $\varepsilon > 0$, there exists $\delta = \delta(\varepsilon) > 0$ such that if $\text{dist}(x_0, \mathcal{D}_0) < \delta$, then $\text{dist}(s(t, x_0), \mathcal{D}_0) < \varepsilon$, $t \geq 0$. $\mathcal{D}_0$ is *attractive* with respect to (21.6) and (21.7) if and only if there exists an open neighborhood $\mathcal{O}_3 \subset \mathcal{D}$ of $\mathcal{D}_0$ such that the omega limit set $\omega(x_0)$ of (21.6) and (21.7) is contained in $\mathcal{D}_0$ for all $x_0 \in \mathcal{O}_3$. $\mathcal{D}_0$ is *asymptotically stable* with respect to (21.6) and (21.7) if and only if $\mathcal{D}_0$ is Lyapunov stable and attractive. Equivalently, $\mathcal{D}_0$ is asymptotically stable with respect to (21.6) and (21.7) if and only if $\mathcal{D}_0$ is Lyapunov stable and there exists $\varepsilon > 0$ such that if $\text{dist}(x_0, \mathcal{D}_0) < \varepsilon$, then $\text{dist}(s(t, x_0), \mathcal{D}_0) \to 0$ as $t \to \infty$. Asymptotic stability is *global* if the previous statement holds for all $x_0 \in \mathbb{R}^n$.

It is important to note here that since state-dependent impulsive dynamical systems are time invariant [128], the notions of asymptotic stability and uniform asymptotic stability with respect to initial times are equivalent. However, unlike continuous-time and discrete-time dynamical systems, wherein asymptotic set stability of autonomous systems is equivalent to the existence of class $\mathcal{K}$ and $\mathcal{L}$ functions $\alpha(\cdot)$ and $\beta(\cdot)$, respectively, such that if $\mathrm{dist}(x_0, \mathcal{D}_0) < \delta$, $\delta > 0$, then $\mathrm{dist}(x(t), \mathcal{D}_0) \le \alpha(\mathrm{dist}(x_0, \mathcal{D}_0))\beta(t)$, $t \ge 0$, this is not generally true for state-dependent impulsive dynamical systems. That is, asymptotic stability might not be uniform with respect to compact sets of initial conditions. If, however, for every compact set the first time-to-impact function $\tau_1(x_0)$ is uniformly bounded with respect to the system initial conditions, then it can be shown that asymptotic stability is uniform with respect to compact sets of initial conditions. For further details on this subtle point, see [107].

**Theorem 21.1.** Consider the impulsive dynamical system (21.6) and (21.7), and assume that Assumptions 21.1–21.3 hold. Assume that $\mathcal{D}_{ci} \subset \mathcal{D}$ is a positively invariant set with respect to (21.6) and (21.7); assume that if $x_0 \in \mathcal{Z}$, then $x_0 + f_d(x_0) \in \overline{\mathcal{Z}} \backslash \mathcal{Z}$; and assume that there exists a continuously differentiable function $V : \mathcal{D}_{ci} \to \mathbb{R}$ such that

$$V'(x)f_c(x) \le 0, \quad x \in \mathcal{D}_{ci}, \quad x \notin \mathcal{Z}, \tag{21.10}$$
$$V(x + f_d(x)) \le V(x), \quad x \in \mathcal{D}_{ci}, \quad x \in \mathcal{Z}. \tag{21.11}$$

Let

$$\mathcal{R} \triangleq \{x \in \mathcal{D}_{ci} : x \notin \mathcal{Z}, V'(x)f_c(x) = 0\}$$
$$\cup \{x \in \mathcal{D}_{ci} : x \in \mathcal{Z}, V(x + f_d(x)) = V(x)\},$$

and let $\mathcal{M}$ denote the largest invariant set contained in $\mathcal{R}$. If $x_0 \in \mathcal{D}_{ci}$, then $x(t) \to \mathcal{M}$ as $t \to \infty$. Furthermore, if $\mathcal{D}_0 \subset \overset{\circ}{\mathcal{D}}_{ci}$; $V(x) = 0$, $x \in \mathcal{D}_0$; $V(x) > 0$, $x_0 \in \mathcal{D}_{ci} \backslash \mathcal{D}_0$; and the set $\mathcal{R}$ contains no invariant set other than the set $\mathcal{D}_0$, then the set $\mathcal{D}_0$ is asymptotically stable with respect to (21.6) and (21.7), and $\mathcal{D}_{ci}$ is a subset of the domain of attraction of (21.6) and (21.7).

**Proof.** The proof of this result is similar to the proof of Corollary 5.1 given in [62] and, hence, is omitted. $\square$

Setting $\mathcal{D} = \mathbb{R}^n$ and requiring $V(x) \to \infty$ as $\|x\| \to \infty$ in Theorem 21.1, it follows that the set $\mathcal{D}_0$ is globally asymptotically stable. A similar remark holds for Theorem 21.2 below.

**Theorem 21.2.** Consider the impulsive dynamical system (21.6) and (21.7), and assume that Assumptions 21.1–21.3 hold. Assume that $\mathcal{D}_{ci} \subset \mathcal{D}$ is a positively invariant set with respect to (21.6) and (21.7) such that $\mathcal{D}_0 \subset \overset{\circ}{\mathcal{D}}_{ci}$; assume that if $x_0 \in \mathcal{Z}$, then $x_0 + f_d(x_0) \in \overline{\mathcal{Z}} \backslash \mathcal{Z}$; and assume that for every $x_0 \in \mathcal{D}_{ci} \backslash \mathcal{D}_0$, there exists $\tau \ge 0$ such that $x(\tau) \in \mathcal{Z}$, where $x(t)$, $t \ge 0$, denotes the solution to (21.6) and (21.7) with the initial condition $x_0$.

Furthermore, assume that there exists a continuously differentiable function $V : \mathcal{D}_{ci} \to \mathbb{R}$ such that $V(x) = 0$, $x \in \mathcal{D}_0$; $V(x) > 0$, $x_0 \in \mathcal{D}_{ci}\backslash\mathcal{D}_0$;

$$V(x + f_{\mathrm{d}}(x)) < V(x), \quad x \in \mathcal{D}_{ci}, \quad x \in \mathcal{Z}; \quad (21.12)$$

and (21.10) is satisfied. Then the set $\mathcal{D}_0 \subset \mathcal{D}_{ci}$ is asymptotically stable with respect to (21.6) and (21.7), and $\mathcal{D}_{ci}$ is a subset of the domain of attraction.

**Proof.** It follows from (21.12) that $\mathcal{R} = \{x \in \mathcal{D}_{ci} : x \notin \mathcal{Z}, V'(x)f_{\mathrm{c}}(x) = 0\}$. Since for every $x_0 \in \mathcal{D}_{ci}\backslash\mathcal{D}_0$, there exists $\tau \geq 0$ such that $x(\tau) \in \mathcal{Z}$, it follows that the largest invariant set contained in $\mathcal{R}$ is $\mathcal{D}_0$. Now, the result is a direct consequence of Theorem 21.1. $\qquad\qquad\square$

## 21.3 Hybrid Stabilization of Sets

In this section, we present a hybrid controller design framework for stabilization of sets. Specifically, we consider nonlinear dynamical systems $\mathcal{G}_{\mathrm{p}}$ of the form given by (21.1) and (21.2). Furthermore, we consider hybrid resetting dynamic controllers $\mathcal{G}_{\mathrm{c}}$ of the form

$$\dot{x}_{\mathrm{c}}(t) = f_{\mathrm{cc}}(x_{\mathrm{c}}(t), y(t)), \quad x_{\mathrm{c}}(0) = x_{\mathrm{c}0}, \quad (x_{\mathrm{c}}(t), y(t)) \notin \mathcal{Z}_{\mathrm{c}}, \quad (21.13)$$
$$\Delta x_{\mathrm{c}}(t) = \eta(y(t)) - x_{\mathrm{c}}(t), \quad (x_{\mathrm{c}}(t), y(t)) \in \mathcal{Z}_{\mathrm{c}}, \quad (21.14)$$
$$y_{\mathrm{c}}(t) = h_{\mathrm{cc}}(x_{\mathrm{c}}(t), y(t)), \quad (21.15)$$

where $x_{\mathrm{c}}(t) \in \mathcal{D}_{\mathrm{c}} \subseteq \mathbb{R}^{n_c}$, $\mathcal{D}_{\mathrm{c}}$ is an open set, $y(t) \in \mathbb{R}^l$, $y_{\mathrm{c}}(t) \in \mathbb{R}^m$, $f_{\mathrm{cc}} : \mathcal{D}_{\mathrm{c}} \times \mathbb{R}^l \to \mathbb{R}^{n_c}$ is smooth on $\mathcal{D}_{\mathrm{c}} \times \mathbb{R}^l$, $\eta : \mathbb{R}^l \to \mathcal{D}_{\mathrm{c}}$ is continuous, and $h_{\mathrm{cc}} : \mathcal{D}_{\mathrm{c}} \times \mathbb{R}^l \to \mathbb{R}^m$ is smooth.

Consider the negative-feedback interconnection of $\mathcal{G}_{\mathrm{p}}$ and $\mathcal{G}_{\mathrm{c}}$ given by $y = u_{\mathrm{c}}$ and $u = -y_{\mathrm{c}}$. In this case, the closed-loop system $\mathcal{G}$ is given by

$$\dot{x}(t) = f_{\mathrm{c}}(x(t)), \quad x(0) = x_0, \quad x(t) \notin \mathcal{Z}, \quad t \geq 0, \quad (21.16)$$
$$\Delta x(t) = f_{\mathrm{d}}(x(t)), \quad x(t) \in \mathcal{Z}, \quad (21.17)$$

where $t \geq 0$, $x(t) \triangleq [x_{\mathrm{p}}^{\mathrm{T}}(t), x_{\mathrm{c}}^{\mathrm{T}}(t)]^{\mathrm{T}}$, $\mathcal{Z} \triangleq \{x \in \mathcal{D} : (x_{\mathrm{c}}, h_{\mathrm{p}}(x_{\mathrm{p}})) \in \mathcal{Z}_{\mathrm{c}}\}$,

$$f_{\mathrm{c}}(x) = \begin{bmatrix} f_{\mathrm{p}}(x_{\mathrm{p}}, -h_{\mathrm{cc}}(x_{\mathrm{c}}, h_{\mathrm{p}}(x_{\mathrm{p}}))) \\ f_{\mathrm{cc}}(x_{\mathrm{c}}, h_{\mathrm{p}}(x_{\mathrm{p}})) \end{bmatrix}, \quad (21.18)$$

$$f_{\mathrm{d}}(x) = \begin{bmatrix} 0 \\ \eta(h_{\mathrm{p}}(x_{\mathrm{p}})) - x_{\mathrm{c}} \end{bmatrix}. \quad (21.19)$$

The objective is to design the hybrid resetting controller (21.13)–(21.15) so that the set $\mathcal{D}_0 = \{(x_{\mathrm{p}}, x_{\mathrm{c}}) \in \mathcal{D}_{\mathrm{p}} \times \mathcal{D}_{\mathrm{c}} : x_{\mathrm{p}} \in \mathcal{D}_{\mathrm{p}0}\}$, where $\mathcal{D}_{\mathrm{p}0} \subset \mathcal{D}_{\mathrm{p}}$, is asymptotically stable with respect to the closed-loop system

(21.16) and (21.17). In order to do this, we associate with the plant a generalized energy function $V_p : \mathcal{D}_p \to \overline{\mathbb{R}}_+$ such that $V_p(x_p) = 0$, $x_p \in \mathcal{D}_{p0}$, and $V_p(x_p) > 0$, $x_p \in \mathcal{D}_p \backslash \mathcal{D}_{p0}$. Furthermore, we associate with the controller a generalized energy function $V_c : \mathcal{D}_c \times \mathbb{R}^l \to \overline{\mathbb{R}}_+$ such that $V_c(x_c, y) \geq 0$, $x_c \in \mathcal{D}_c$, $y \in \mathbb{R}^l$, and $V_c(x_c, y) = 0$ if and only if $x_c = \eta(y)$. Finally, we associate with the closed-loop system the generalized energy function $V(x) \triangleq V_p(x_p) + V_c(x_c, h_p(x_p))$.

Next, we construct the resetting set for the closed-loop system $\mathcal{G}$ in the following way:

$$\mathcal{Z} = \{(x_p, x_c) \in \mathcal{D}_p \times \mathcal{D}_c : L_{f_c} V_c(x_c, h_p(x_p)) = 0 \text{ and } V_c(x_c, h_p(x_p)) > 0\}. \tag{21.20}$$

The resetting set $\mathcal{Z}$ is thus defined to be the set of all points in the closed-loop state space that correspond to the instant when the controller is at the verge of decreasing its generalized energy function $V_c(\cdot)$. If the controller states are reset, then the generalized energy function $V_p(\cdot)$ can never increase after the first resetting event. Furthermore, if the generalized energy function $V(\cdot)$ of the closed-loop system is conserved between resetting events, then a decrease in $V_p(\cdot)$ is accompanied by a corresponding increase in $V_c(\cdot)$. Hence, this approach allows the generalized plant energy to flow to the controller, where it increases the emulated generalized controller energy but does not allow the emulated generalized controller energy to flow back to the plant after the first resetting event.

This energy-dissipating hybrid controller effectively enforces a one-way generalized energy transfer between the plant and the controller after the first resetting event. For practical implementation, knowledge of $x_c$ and $y$ is sufficient to determine whether or not the closed-loop state vector is in the set $\mathcal{Z}$. That is, the full state $x_p$ need not be known in order to determine whether or not the closed-loop state vector is in the set $\mathcal{Z}$, nor is it needed for feedback control between resettings determined by (21.15).

The next theorem gives sufficient conditions for asymptotic stability of the set $\mathcal{D}_0 \subset \mathcal{D}_p \times \mathcal{D}_c$ with respect to the closed-loop system $\mathcal{G}$ using state-dependent hybrid controllers.

**Theorem 21.3.** Consider the closed-loop impulsive dynamical system $\mathcal{G}$ given by (21.16) and (21.17), and assume that $\mathcal{D}_{ci} \subset \mathcal{D}$ is a positively invariant set with respect to $\mathcal{G}$ such that $\mathcal{D}_0 \subset \overset{\circ}{\mathcal{D}}_{ci}$, where $\mathcal{D}_0 = \{(x_p, x_c) \in \mathcal{D}_p \times \mathcal{D}_c : x_p \in \mathcal{D}_{p0}\}$ and $\mathcal{D}_{p0} \subset \mathcal{D}_p$. Assume that there exists a continuously differentiable function $V_p : \mathcal{D}_p \to \overline{\mathbb{R}}_+$ such that $V_p(x_p) = 0$, $x_p \in \mathcal{D}_{p0}$, and $V_p(x_p) > 0$, $x_p \in \mathcal{D}_p \backslash \mathcal{D}_{p0}$, and assume that there exists a smooth (i.e., infinitely differentiable) function $V_c : \mathcal{D}_c \times \mathbb{R}^l \to \overline{\mathbb{R}}_+$ such that $V_c(x_c, y) \geq 0$,

$x_c \in \mathcal{D}_c$, $y \in \mathbb{R}^l$, and $V_c(x_c, y) = 0$ if and only if $x_c = \eta(y)$. Furthermore, assume that every $x_0 \in \overline{\mathcal{Z}}$ is $k$-transversal to (21.16) and

$$\dot{V}_p(x_p(t)) + \dot{V}_c(x_c(t), y(t)) = 0, \quad x(t) \notin \mathcal{Z}, \qquad (21.21)$$

where $y = u_c = h_p(x_p)$ and $\mathcal{Z}$ is given by (21.20). Then the set $\mathcal{D}_0 \subset \mathcal{D}_{ci}$ is asymptotically stable with respect to the closed-loop system $\mathcal{G}$. Finally, if $\mathcal{D}_p = \mathbb{R}^{n_p}$, $\mathcal{D}_c = \mathbb{R}^{n_c}$, and $V(\cdot)$ is radially unbounded, then the set $\mathcal{D}_0 \subset \mathcal{D}_{ci}$ is globally asymptotically stable with respect to $\mathcal{G}$.

**Proof.** First, note that since $V_c(x_c, y) \geq 0$, $x_c \in \mathcal{D}_c$, $y \in \mathbb{R}^l$, it follows that

$$\begin{aligned}
\overline{\mathcal{Z}} &= \{(x_p, x_c) \in \mathcal{D}_p \times \mathcal{D}_c : L_{f_c} V_c(x_c, h_p(x_p)) = 0 \text{ and } V_c(x_c, h_p(x_p)) \geq 0\} \\
&= \{(x_p, x_c) \in \mathcal{D}_p \times \mathcal{D}_c : \mathcal{X}(x) = 0\}, \qquad (21.22)
\end{aligned}$$

where $\mathcal{X}(x) = L_{f_c} V_c(x_c, h_p(x_p))$. Next, we show that if the $k$-transversality condition (21.9) holds, then Assumptions 21.1–21.3 hold and, for every $x_0 \in \mathcal{D}_{ci}$, there exists $\tau \geq 0$ such that $x(\tau) \in \mathcal{Z}$. Note that if $x_0 \in \overline{\mathcal{Z}} \backslash \mathcal{Z}$, that is, $V_c(x_c(0), h_p(x_p(0))) = 0$ and $L_{f_c} V_c(x_c(0), h_p(x_p(0))) = 0$, it follows from the $k$-transversality condition that there exists $\delta > 0$ such that for all $t \in (0, \delta]$, $L_{f_c} V_c(x_c(t), h_p(x_p(t))) \neq 0$.

Hence, since

$$V_c(x_c(t), h_p(x_p(t))) = V_c(x_c(0), h_p(x_p(0))) + t L_{f_c} V_c(x_c(\tau), h_p(x_p(\tau)))$$

for some $\tau \in (0, t]$ and $V_c(x_c, y) \geq 0$, $x_c \in \mathcal{D}_c$, $y \in \mathbb{R}^l$, it follows that $V_c(x_c(t), h_p(x_p(t))) > 0$, $t \in (0, \delta]$, which implies that Assumption 21.1 is satisfied. Furthermore, if $x \in \mathcal{Z}$, then, since $V_c(x_c, y) = 0$ if and only if $x_c = \eta(y)$, it follows from (21.17) that $x + f_d(x) \in \overline{\mathcal{Z}} \backslash \mathcal{Z}$. Hence, Assumption 21.2 holds.

Next, consider the set $\mathcal{M}_\gamma \triangleq \{x \in \mathcal{D}_{ci} : V_c(x_c, h_p(x_p)) = \gamma\}$, where $\gamma \geq 0$. It follows from the $k$-transversality condition that for every $\gamma \geq 0$, $\mathcal{M}_\gamma$ does not contain any nontrivial trajectory of $\mathcal{G}$. To see this, suppose, *ad absurdum*, that there exists a nontrivial trajectory $x(t) \in \mathcal{M}_\gamma$, $t \geq 0$, for some $\gamma \geq 0$. In this case, it follows that

$$\frac{\mathrm{d}^k}{\mathrm{d}t^k} V_c(x_c(t), h_p(x_p(t))) = L_{f_c}^k V_c(x_c(t), h_p(x_p(t))) \equiv 0, \quad k = 1, 2, \dots,$$

which contradicts the $k$-transversality condition.

Next, we show that for every $x_0 \notin \mathcal{Z}$, $x_0 \notin \mathcal{D}_0$, there exists $\tau > 0$ such that $x(\tau) \in \mathcal{Z}$. To see this, suppose, *ad absurdum*, that $x(t) \notin \mathcal{Z}$, $t \geq 0$, which implies that

$$\frac{\mathrm{d}}{\mathrm{d}t} V_{\mathrm{c}}(x_{\mathrm{c}}(t), h_{\mathrm{p}}(x_{\mathrm{p}}(t))) \neq 0, \quad t \geq 0, \tag{21.23}$$

or

$$V_{\mathrm{c}}(x_{\mathrm{c}}(t), h_{\mathrm{p}}(x_{\mathrm{p}}(t))) = 0, \quad t \geq 0. \tag{21.24}$$

If (21.23) holds, then it follows that $V_{\mathrm{c}}(x_{\mathrm{c}}(t), h_{\mathrm{p}}(x_{\mathrm{p}}(t)))$ is a (decreasing or increasing) monotonic function of time. Hence, it follows from the monotone convergence theorem [122, p. 37] that $V_{\mathrm{c}}(x_{\mathrm{c}}(t), h_{\mathrm{p}}(x_{\mathrm{p}}(t))) \to \gamma$ as $t \to \infty$, where $\gamma \geq 0$ is a constant, which implies that the positive limit set of the closed-loop system is contained in $\mathcal{M}_\gamma$ for some $\gamma \geq 0$, which is a contradiction. Similarly, if (21.24) holds, then $\mathcal{M}_0$ contains a nontrivial trajectory of $\mathcal{G}$, also leading to a contradiction. Hence, for every $x_0 \notin \mathcal{Z}$, there exists $\tau > 0$ such that $x(\tau) \in \mathcal{Z}$. Thus, it follows that for every $x_0 \notin \mathcal{Z}$, $0 < \tau_1(x_0) < \infty$.

Now, it follows from Proposition 21.2 that $\tau_1(\cdot)$ is continuous at $x_0 \notin \overline{\mathcal{Z}}$. Furthermore, for all $x_0 \in \overline{\mathcal{Z}} \backslash \mathcal{Z}$ and for every unbounded sequence $\{x_i\}_{i=1}^{\infty} \in \overline{\mathcal{Z}} \backslash \mathcal{Z}$ converging to $x_0 \in \overline{\mathcal{Z}} \backslash \mathcal{Z}$, it follows from the $k$-transversality condition and Proposition 21.2 that $\lim_{i \to \infty} \tau_1(x_i) = \tau_1(x_0)$. Next, let $x_0 \in \overline{\mathcal{Z}} \backslash \mathcal{Z}$, and let $\{x_i\}_{i=1}^{\infty} \in \mathcal{D}_{\mathrm{ci}}$ be such that $\lim_{i \to \infty} x_i = x_0$ and $\lim_{i \to \infty} \tau_1(x_i)$ exists. In this case, it follows from Proposition 21.2 that either $\lim_{i \to \infty} \tau_1(x_i) = 0$ or $\lim_{i \to \infty} \tau_1(x_i) = \tau_1(x_0)$. Furthermore, since $x_0 \in \overline{\mathcal{Z}} \backslash \mathcal{Z}$ corresponds to the case where $V_{\mathrm{c}}(x_{\mathrm{c}0}, h_{\mathrm{p}}(x_{\mathrm{p}0})) = 0$, it follows that $x_{\mathrm{c}0} = \eta(h_{\mathrm{p}}(x_{\mathrm{p}0}))$, and, hence, $f_{\mathrm{d}}(x_0) = 0$. Now, it follows from Proposition 21.1 that Assumption 21.3 holds.

Next, note that if $x_0 \in \mathcal{Z}$ and $x_0 + f_{\mathrm{d}}(x_0) \notin \mathcal{D}_0$, then it follows from the above analysis that there exists $\tau > 0$ such that $x(\tau) \in \mathcal{Z}$. Alternatively, if $x_0 \in \mathcal{Z}$ and $x_0 + f_{\mathrm{d}}(x_0) \in \mathcal{D}_0$, then the solution of the closed-loop system reaches $\mathcal{D}_0$ in finite time, which is a stronger condition than reaching $\mathcal{D}_0$ as $t \to \infty$.

To show that the set $\mathcal{D}_0 \subset \mathcal{D}_{\mathrm{ci}}$ is asymptotically stable, consider the Lyapunov function candidate $V(x) = V_{\mathrm{p}}(x_{\mathrm{p}}) + V_{\mathrm{c}}(x_{\mathrm{c}}, h_{\mathrm{p}}(x_{\mathrm{p}}))$ corresponding to the total generalized energy function. It follows from (21.21) that

$$\dot{V}(x(t)) = 0, \quad x(t) \notin \mathcal{Z}. \tag{21.25}$$

Furthermore, it follows from (21.19) and (21.20) that

$$\begin{aligned}
\Delta V(x(t_k)) &= V_{\mathrm{c}}(x_{\mathrm{c}}(t_k^+), h_{\mathrm{p}}(x_{\mathrm{p}}(t_k^+))) - V_{\mathrm{c}}(x_{\mathrm{c}}(t_k), h_{\mathrm{p}}(x_{\mathrm{p}}(t_k))) \\
&= V_{\mathrm{c}}(\eta(h_{\mathrm{p}}(x_{\mathrm{p}}(t_k))), h_{\mathrm{p}}(x_{\mathrm{p}}(t_k))) - V_{\mathrm{c}}(x_{\mathrm{c}}(t_k), h_{\mathrm{p}}(x_{\mathrm{p}}(t_k))) \\
&= -V_{\mathrm{c}}(x_{\mathrm{c}}(t_k), h_{\mathrm{p}}(x_{\mathrm{p}}(t_k)))
\end{aligned}$$

$$< 0, \quad x(t_k) \in \mathcal{Z}, \quad k \in \overline{\mathbb{Z}}_+. \tag{21.26}$$

Thus, it follows from Theorem 21.2 that the set $\mathcal{D}_0 \subset \mathcal{D}_{\mathrm{ci}}$ is asymptotically stable.

Finally, if $\mathcal{D}_{\mathrm{p}} = \mathbb{R}^{n_{\mathrm{p}}}$, $\mathcal{D}_{\mathrm{c}} = \mathbb{R}^{n_{\mathrm{c}}}$, and $V(\cdot)$ is radially unbounded, then global asymptotic stability is immediate using standard arguments. $\qquad \square$

To demonstrate the utility of Theorem 21.3, let the set $\mathcal{D}_{\mathrm{p}0}$ be given by the zero level set of the function $Q_{\mathrm{p}} : \mathcal{D}_{\mathrm{p}} \to \mathbb{R}^{s_{\mathrm{p}}}$ and let $V_{\mathrm{p}} : \mathcal{D}_{\mathrm{p}} \to \overline{\mathbb{R}}_+$ be given by

$$V_{\mathrm{p}}(x_{\mathrm{p}}) = Q^{\mathrm{T}}(x_{\mathrm{p}}) P Q(x_{\mathrm{p}}), \quad x_{\mathrm{p}} \in \mathcal{D}_{\mathrm{p}}, \tag{21.27}$$

where $P \in \mathbb{R}^{s_{\mathrm{p}} \times s_{\mathrm{p}}}$ and $P > 0$. Furthermore, let $V_{\mathrm{c}} : \mathcal{D}_{\mathrm{c}} \times \mathbb{R}^l \to \overline{\mathbb{R}}_+$ be given by

$$V_{\mathrm{c}}(x_{\mathrm{c}}, h_{\mathrm{p}}(x_{\mathrm{p}})) = (x_{\mathrm{c}} - \eta(h_{\mathrm{p}}(x_{\mathrm{p}})))^{\mathrm{T}} P_{\mathrm{c}} (x_{\mathrm{c}} - \eta(h_{\mathrm{p}}(x_{\mathrm{p}}))), \quad (x_{\mathrm{p}}, x_{\mathrm{c}}) \in \mathcal{D}_{\mathrm{p}} \times \mathcal{D}_{\mathrm{c}}, \tag{21.28}$$

where $P_{\mathrm{c}} \in \mathbb{R}^{n_{\mathrm{c}} \times n_{\mathrm{c}}}$ and $P_{\mathrm{c}} > 0$. In this case, the functions $f_{\mathrm{cc}}(\cdot, \cdot)$, $h_{\mathrm{cc}}(\cdot, \cdot)$, and $\eta(\cdot)$ can be selected using (21.21) in Theorem 21.3. These constructions are shown for the specific problems of consensus and formation control for multiagent systems in the next sections.

## 21.4 Specialization to Linear Dynamical Systems

In this section, we specialize the results of Section 21.3 to the class of linear dynamical systems given by

$$\dot{x}_{\mathrm{p}}(t) = A x_{\mathrm{p}}(t) + B u(t), \quad x_{\mathrm{p}}(0) = x_{\mathrm{p}0}, \quad t \geq 0, \tag{21.29}$$

$$y(t) = C x_{\mathrm{p}}(t), \tag{21.30}$$

where $x_{\mathrm{p}}(t) \in \mathbb{R}^n$, $A \in \mathbb{R}^{n \times n}$, $B \in \mathbb{R}^{n \times m}$, and $C \in \mathbb{R}^{l \times n}$. Here, for simplicity of exposition, we assume that $n_{\mathrm{p}} = n_{\mathrm{c}} = n$ and $C = I_n$. The case where $C \neq I_n$ can be addressed using an identical analysis, as is shown below with $F_2, H_2$, and $M$ in (21.33)–(21.35) replaced by $F_2 C, H_2 C$, and $M C$, respectively.

For the system (21.29) and (21.30), we construct a hybrid feedback controller of the form (21.13)–(21.15) that asymptotically stabilizes the set $\mathcal{D}_0$ given by

$$\mathcal{D}_0 = \{(x_{\mathrm{p}}, x_{\mathrm{c}}) \in \mathcal{D}_{\mathrm{p}} \times \mathcal{D}_{\mathrm{c}} : x_{\mathrm{p}} \in \mathcal{D}_{\mathrm{p}0}\}, \tag{21.31}$$

where

$$\mathcal{D}_{\mathrm{p}0} = \{x_{\mathrm{p}} \in \mathcal{D}_{\mathrm{p}} : T x_{\mathrm{p}} = 0\} \tag{21.32}$$

and $T \in \mathbb{R}^{s_p \times n}$. Specifically, we set

$$f_{cc}(x_c, x_p) = F_1 x_c + F_2 x_p, \tag{21.33}$$

$$h_{cc}(x_c, x_p) = -H_1 x_c - H_2 x_p, \tag{21.34}$$

$$\eta(x_p) = M x_p, \tag{21.35}$$

where $F_1 \in \mathbb{R}^{n \times n}$, $F_2 \in \mathbb{R}^{n \times n}$, $H_1 \in \mathbb{R}^{m \times n}$, $H_2 \in \mathbb{R}^{m \times n}$, and $M \in \mathbb{R}^{n \times n}$. Thus, the closed-loop system (21.29), (21.30), and (21.13)–(21.15) with the negative feedback interconnection $u = -y_c$ is given by

$$\dot{x}_p(t) = (A + BH_2) x_p(t) + BH_1 x_c(t), \quad (x_p(t), x_c(t)) \notin \mathcal{Z}, \tag{21.36}$$

$$\dot{x}_c(t) = F_1 x_c(t) + F_2 x_p(t), \quad (x_p(t), x_c(t)) \notin \mathcal{Z}, \tag{21.37}$$

$$\Delta x_c(t) = M x_p(t) - x_c(t), \quad (x_p(t), x_c(t)) \in \mathcal{Z}, \tag{21.38}$$

where $\mathcal{Z}$ is given by (21.20).

Next, define the generalized energy functions

$$V_p(x_p) = \frac{1}{2} x_p^{\mathrm{T}} T^{\mathrm{T}} T x_p, \quad x_p \in \mathcal{D}_p, \tag{21.39}$$

$$V_c(x_c, x_p) = \frac{1}{2} (x_c - M x_p)^{\mathrm{T}} P_c (x_c - M x_p), \quad (x_p, x_c) \in \mathcal{D}_p \times \mathcal{D}_c, \tag{21.40}$$

where $P_c \in \mathbb{R}^{n \times n}$ and $P_c > 0$. Note that $V_p(x_p) = 0$, $x_p \in \mathcal{D}_{p0}$, and $V_p(x_p) > 0$, $x_p \in \mathcal{D}_p \backslash \mathcal{D}_{p0}$. Furthermore, note that $V_c(x_c, x_p) \geq 0$, $(x_p, x_c) \in \mathcal{D}_p \times \mathcal{D}_c$, and $V_c(x_c, x_p) = 0$ if and only if $x_c = \eta(x_p)$. For the closed-loop system (21.36)–(21.38), condition (21.21) in Theorem 21.3 gives

$$\begin{aligned}
\dot{V}_p&(x_p(t)) + \dot{V}_c(x_c(t), x_p(t)) \\
&= x_p^{\mathrm{T}}(t)(T^{\mathrm{T}} T B H_1 + F_2^{\mathrm{T}} P_c - A^{\mathrm{T}} M^{\mathrm{T}} P_c - H_2^{\mathrm{T}} B^{\mathrm{T}} M^{\mathrm{T}} P_c \\
&\quad - M^{\mathrm{T}} P_c F_1 + M^{\mathrm{T}} P_c M B H_1) x_c(t) \\
&\quad + x_p^{\mathrm{T}}(t)(T^{\mathrm{T}} T A + T^{\mathrm{T}} T B H_2 - M^{\mathrm{T}} P_c F_2 \\
&\quad + M^{\mathrm{T}} P_c M A + M^{\mathrm{T}} P_c M B H_2) x_p(t) \\
&\quad + x_c^{\mathrm{T}}(t)(P_c F_1 - P_c M B H_1) x_c(t) \\
&= 0, \quad (x_p(t), x_c(t)) \notin \mathcal{Z}. \tag{21.41}
\end{aligned}$$

Since $x_p$ and $x_c$ are independent state variables, (21.41) holds if and only if there exist skew-symmetric matrices $A_p \in \mathbb{R}^{n \times n}$ and $A_c \in \mathbb{R}^{n \times n}$ such that

$$\begin{aligned}
T^{\mathrm{T}} T B H_1 &+ F_2^{\mathrm{T}} P_c - A^{\mathrm{T}} M^{\mathrm{T}} P_c - H_2^{\mathrm{T}} B^{\mathrm{T}} M^{\mathrm{T}} P_c \\
&- M^{\mathrm{T}} P_c F_1 + M^{\mathrm{T}} P_c M B H_1 = 0, \tag{21.42}
\end{aligned}$$

$$T^{\mathrm{T}}TA + T^{\mathrm{T}}TBH_2 - M^{\mathrm{T}}P_{\mathrm{c}}F_2 + M^{\mathrm{T}}P_{\mathrm{c}}MA + M^{\mathrm{T}}P_{\mathrm{c}}MBH_2 = A_{\mathrm{p}},$$
$$(21.43)$$

and

$$P_{\mathrm{c}}F_1 - P_{\mathrm{c}}MBH_1 = A_{\mathrm{c}}. \tag{21.44}$$

The skew-symmetric matrices $A_{\mathrm{p}} \in \mathbb{R}^{n \times n}$ and $A_{\mathrm{c}} \in \mathbb{R}^{n \times n}$ are free design parameters. Furthermore, if the matrices $H_1 \in \mathbb{R}^{m \times n}$ and $H_2 \in \mathbb{R}^{m \times n}$ are fixed, then it follows from (21.42)–(21.44) that

$$F_1 = P_{\mathrm{c}}^{-1}A_{\mathrm{c}} + MBH_1, \tag{21.45}$$
$$F_2 = MA + MBH_2 - P_{\mathrm{c}}^{-1}A_{\mathrm{c}}M - P_{\mathrm{c}}^{-1}H_1^{\mathrm{T}}B^{\mathrm{T}}T^{\mathrm{T}}T, \tag{21.46}$$

where $M \in \mathbb{R}^{n \times n}$ satisfies

$$T^{\mathrm{T}}TA + T^{\mathrm{T}}TBH_2 + M^{\mathrm{T}}A_{\mathrm{c}}M + M^{\mathrm{T}}H_1^{\mathrm{T}}B^{\mathrm{T}}T^{\mathrm{T}}T = A_{\mathrm{p}}. \tag{21.47}$$

Note that if $A_{\mathrm{c}}$ is skew symmetric, then $M^{\mathrm{T}}A_{\mathrm{c}}M$ is also skew symmetric. In this case, we can set $A_{\mathrm{p}} = \tilde{A}_{\mathrm{p}} + M^{\mathrm{T}}A_{\mathrm{c}}M$, where $\tilde{A}_{\mathrm{p}} \in \mathbb{R}^{n \times n}$ is an arbitrary skew-symmetric matrix, so that

$$NM = L, \tag{21.48}$$

where

$$N \triangleq T^{\mathrm{T}}TBH_1, \tag{21.49}$$
$$L \triangleq -\tilde{A}_{\mathrm{p}} - A^{\mathrm{T}}T^{\mathrm{T}}T - H_2^{\mathrm{T}}B^{\mathrm{T}}T^{\mathrm{T}}T. \tag{21.50}$$

Recall that a solution $M$ to the matrix equation (21.48) exists if and only if [27, Fact 6.4.43, p. 421]

$$NN^{\dagger}L = L, \tag{21.51}$$

where $N^{\dagger} \in \mathbb{R}^{n \times n}$ is the Moore–Penrose generalized inverse of $N \in \mathbb{R}^{n \times n}$. If (21.51) is satisfied, then every solution to (21.48) is given by

$$M = N^{\dagger}L + Y - N^{\dagger}NY, \tag{21.52}$$

where $Y \in \mathbb{R}^{n \times n}$ is an arbitrary matrix, and, if $Y = 0$, then $\operatorname{tr} M^{\mathrm{T}}M$ is minimized. Thus, the existence of a hybrid controller that asymptotically stabilizes the set $\mathcal{D}_0$ given by (21.31) and (21.32) is characterized by a matrix condition (21.51).

Finally, if

$$T^{\mathrm{T}}TBH_1 \neq 0, \tag{21.53}$$

then the $k$-transversality condition (21.9) is satisfied. To see this, note that (21.53) implies that $\dot{V}_p(x_p) \not\equiv 0$, which, using (21.21), implies that $\dot{V}_c(x_c, x_p) \not\equiv 0$. This shows that the $k$-transversality condition, with $k = 1$, holds for the closed-loop system (21.36)–(21.38). Note that (21.53) is guaranteed by (21.51).

## 21.5 Hybrid Control Design for Parallel and Rendezvous Formations

In this section, we apply the hybrid control framework developed in Section 21.4 to multiagent systems composed of double integrator agents executing various coordinated tasks. First, we consider the parallel formation problem for multiagent systems. Specifically, let $q$ denote the number of mobile agents so that $x_p = [x_{p1}^T, x_{p2}^T]^T \in \mathbb{R}^{2dq}$, where $x_{p1} \in \mathbb{R}^{dq}$ represents a vector of positions, $x_{p2} \triangleq \dot{x}_{p1} \in \mathbb{R}^{dq}$ represents a vector of velocities, $d$ represents the number of degrees of freedom of each agent, and $A$ and $B$ in (21.29) are given by

$$A = \begin{bmatrix} 0_{dq \times dq} & I_{dq} \\ 0_{dq \times dq} & 0_{dq \times dq} \end{bmatrix}, \quad B = \begin{bmatrix} 0_{dq \times dq} \\ I_{dq} \end{bmatrix}. \tag{21.54}$$

The control aim is to design a hybrid feedback control law so that a parallel formation is achieved wherein the agents are collectively required to maintain a prescribed geometric shape with constant velocities and the relative position between any two mobile agents is asymptotically stabilized to a constant value. For this task, we set $d = 2$ and let $q = 5$ so that $x_{p1} = [x_1, y_1, \ldots, x_5, y_5]^T$, where $x_i$, $y_i$, $i = 1, \ldots, 5$, are, respectively, horizontal and vertical coordinates of the $i$th agent. The individual agent dynamics are thus given by

$$\ddot{x}_i(t) = u_{xi}(t), \quad x_i(0) = x_{i0}, \quad t \geq 0, \tag{21.55}$$
$$\ddot{y}_i(t) = u_{yi}(t), \quad y_i(0) = y_{i0}, \tag{21.56}$$

where $i = 1, \ldots, 5$ and $u_{xi}$, $u_{yi}$ are individual control inputs in the horizontal and vertical directions, respectively.

For our hybrid controller design, we set

$$H_1 = [I_{10}, I_{10}], \quad H_2 = [0_{10 \times 10}, H_{22}], \quad P_c = 2I_{20}, \quad \tilde{A}_p = 0,$$

where

$$H_{22} = \begin{bmatrix} -2 & 0 & 1 & 0 & 0 & 0 & 0 & 0 & 1 & 0 \\ 0 & -2 & 0 & 0 & 0 & 0 & 0 & 0 & 0 & 0 \\ 1 & 0 & -2 & 0 & 1 & 0 & 0 & 0 & 0 & 0 \\ 0 & 0 & 0 & -2 & 0 & 0 & 0 & 0 & 0 & 0 \\ 0 & 0 & 1 & 0 & -2 & 0 & 1 & 0 & 0 & 0 \\ 0 & 0 & 0 & 0 & 0 & -2 & 0 & 0 & 0 & 0 \\ 0 & 0 & 0 & 0 & 1 & 0 & -2 & 0 & 1 & 0 \\ 0 & 0 & 0 & 0 & 0 & 0 & 0 & -2 & 0 & 0 \\ 1 & 0 & 0 & 0 & 0 & 0 & 1 & 0 & -2 & 0 \\ 0 & 0 & 0 & 0 & 0 & 0 & 0 & 0 & 0 & -2 \end{bmatrix}, \tag{21.57}$$

and we choose $A_c \in \mathbb{R}^{20 \times 20}$ to be a random skew-symmetric matrix. The specifications of the parallel formation along the $x$ axis with equal distances along the $y$ axis and equal velocities can be characterized by (21.32) with

$$T = \begin{bmatrix} T_1 & 0_{7 \times 10} \\ 0_{9 \times 10} & T_2 \end{bmatrix}, \tag{21.58}$$

where

$$T_1 = \begin{bmatrix} 1 & 0 & -1 & 0 & 0 & 0 & 0 & 0 & 0 & 0 \\ 0 & 0 & 1 & 0 & -1 & 0 & 0 & 0 & 0 & 0 \\ 0 & 0 & 0 & 0 & 1 & 0 & -1 & 0 & 0 & 0 \\ 0 & 0 & 0 & 0 & 0 & 0 & 1 & 0 & -1 & 0 \\ 0 & 1 & 0 & -2 & 0 & 1 & 0 & 0 & 0 & 0 \\ 0 & 0 & 0 & 1 & 0 & -2 & 0 & 1 & 0 & 0 \\ 0 & 0 & 0 & 0 & 0 & 1 & 0 & -2 & 0 & 1 \end{bmatrix}, \tag{21.59}$$

$$T_2 = \begin{bmatrix} 0 & 1 & 0 & 0 & 0 & 0 & 0 & 0 & 0 & 0 \\ 0 & 1 & 0 & -1 & 0 & 0 & 0 & 0 & 0 & 0 \\ 0 & 0 & 0 & 1 & 0 & -1 & 0 & 0 & 0 & 0 \\ 0 & 0 & 0 & 0 & 0 & 1 & 0 & -1 & 0 & 0 \\ 0 & 0 & 0 & 0 & 0 & 0 & 0 & 1 & 0 & -1 \\ 1 & 0 & -1 & 0 & 0 & 0 & 0 & 0 & 0 & 0 \\ 0 & 0 & 1 & 0 & -1 & 0 & 0 & 0 & 0 & 0 \\ 0 & 0 & 0 & 0 & 1 & 0 & -1 & 0 & 0 & 0 \\ 0 & 0 & 0 & 0 & 0 & 0 & 1 & 0 & -1 & 0 \end{bmatrix}. \tag{21.60}$$

In this case, condition (21.51) is verified, and $M \in \mathbb{R}^{20 \times 20}$ is obtained from (21.52) with $Y = 0$. Consequently, the matrices $F_1 \in \mathbb{R}^{20 \times 20}$ and $F_2 \in \mathbb{R}^{20 \times 20}$ are computed using (21.45) and (21.46).

For our simulation, we set $x_p(0) = [0.4, -1, -0.3, 1.3, -1.3, -0.8, -0.1, -0.7, 1.1, 1.9, -0.34, -0.26, 0.2, -0.12, 0.47, 0.47, 0.15, 0.36, -0.1, 0.13]^T$ and $x_c(0) = M x_p(0)$. Figure 21.1 shows the positions of the agents in the plane, whereas Figures 21.2 and 21.3 show the control forces in the $x$ and $y$ directions, respectively, acting on each agent. Figures 21.4 and 21.5 show the agent velocities in the $x$ and $y$ directions, respectively. Finally, Figure 21.6 shows the time history of the generalized energy functions $V_{\mathrm{p}}(x_{\mathrm{p}}(t))$ and $V_{\mathrm{c}}(x_{\mathrm{c}}(t), x_{\mathrm{p}}(t))$, $t \geq 0$.

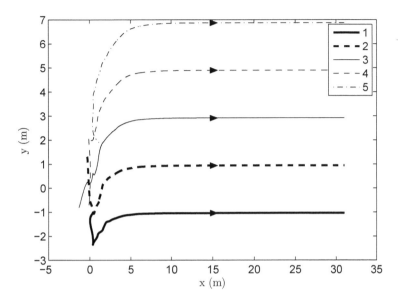

Figure 21.1  Agent positions in the plane.

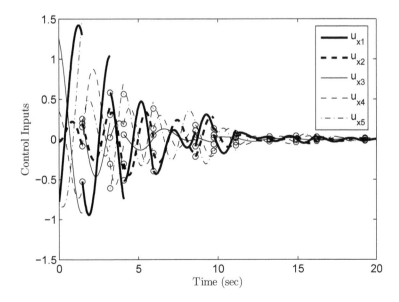

Figure 21.2  Control forces in the $x$ direction.

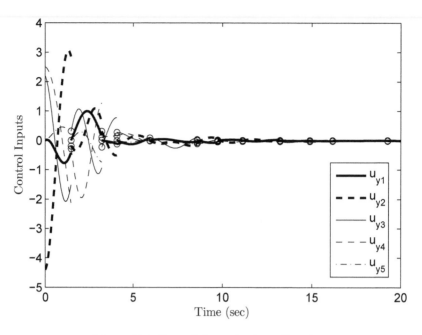

Figure 21.3 Control forces in the $y$ direction.

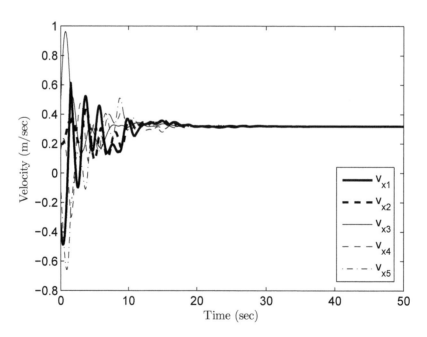

Figure 21.4 Velocities in the $x$ direction.

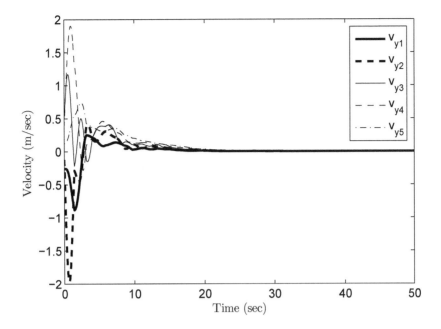

Figure 21.5 Velocities in the $y$ direction.

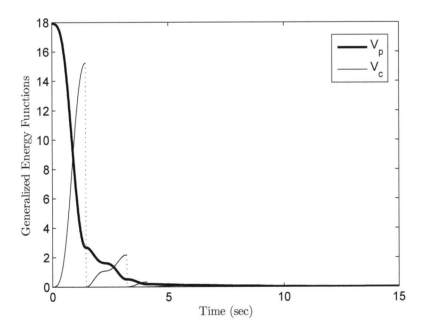

Figure 21.6 Generalized energy functions $V_{\mathrm{p}}(x_{\mathrm{p}}(t))$ and $V_{\mathrm{c}}(x_{\mathrm{c}}(t), x_{\mathrm{p}}(t))$ versus time.

For the next task, we design a hybrid controller (21.13)–(21.15) for the rendezvous problem of planar double integrator agents. Specifically, a cooperative rendezvous task requires that each agent determine the rendezvous time and location through team negotiation. In the following simulation, we consider four agents coming to a square formation with zero terminal velocities. In this case,

$$T = \begin{bmatrix} T_1 & 0_{6\times8} \\ 0_{8\times8} & T_2 \end{bmatrix},$$

where

$$T_1 = \begin{bmatrix} 1 & 0 & -1 & 0 & 1 & 0 & -1 & 0 \\ 0 & 1 & 0 & -1 & 0 & 1 & 0 & -1 \\ 1 & 0 & 1 & 0 & -1 & 0 & -1 & 0 \\ 0 & 1 & 0 & -1 & 0 & -1 & 0 & 1 \\ 1 & -1 & -1 & 0 & 0 & 0 & 0 & 1 \\ 0 & 0 & 0 & 1 & 1 & -1 & -1 & 0 \end{bmatrix}, \tag{21.61}$$

$$T_2 = \begin{bmatrix} 1 & 0 & 0 & 0 & 0 & 0 & 0 & 0 \\ 0 & 1 & 0 & 0 & 0 & 0 & 0 & 0 \\ 1 & 0 & -1 & 0 & 0 & 0 & 0 & 0 \\ 0 & 1 & 0 & -1 & 0 & 0 & 0 & 0 \\ 0 & 0 & 1 & 0 & -1 & 0 & 0 & 0 \\ 0 & 0 & 0 & 1 & 0 & -1 & 0 & 0 \\ 0 & 0 & 0 & 0 & 1 & 0 & -1 & 0 \\ 0 & 0 & 0 & 0 & 0 & 1 & 0 & -1 \end{bmatrix}. \tag{21.62}$$

For our simulation, we set $H_1 = [I_8, I_8]$, $H_2 = [0_{8\times8}, -2I_8]$, $A_c = \tilde{A}_p = 0$, and $P_c = 0.7I_{16}$. As in our previous example, condition (21.51) is verified, and $M \in \mathbb{R}^{16\times16}$ is obtained from (21.52) with $Y = 0$. For the initial conditions $x_p(0) = [10, 10, 1, 5, 2, 4, 9, 1, 0, 0, 0, 0, 0, 0, 0, 0]^T$ and $x_c(0) = Mx_p(0)$, Figure 21.7 shows the positions of the agents in the plane, whereas Figures 21.8 and 21.9 show the control forces in the $x$ and $y$ directions, respectively, acting on each agent. Finally, Figure 21.10 shows the time history of the generalized energy functions $V_p(x_p(t))$ and $V_c(x_c(t), x_p(t))$, $t \geq 0$.

## 21.6 Hybrid Control Design for Consensus in Multiagent Networks

In this section, we specialize the results of Section 21.4 to design hybrid consensus controllers for multiagent networks of single integrator systems. Specifically, the consensus problem involves the design of a dynamic protocol algorithm that guarantees system state equipartition [162, 167]; that is,

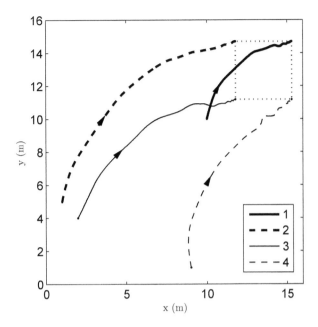

Figure 21.7  Agent positions in the plane.

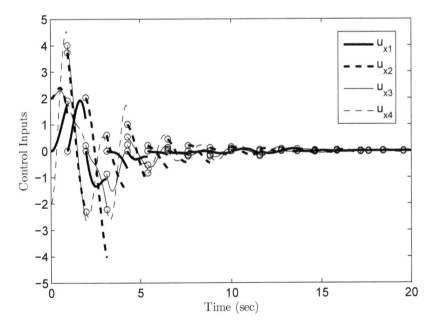

Figure 21.8  Control forces in the $x$ direction.

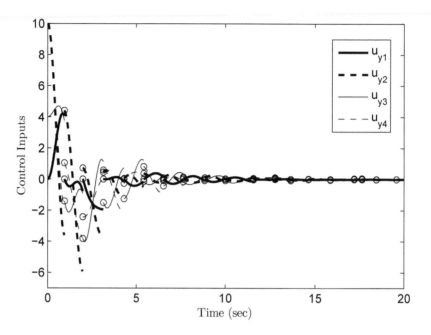

Figure 21.9 Control forces in the $y$ direction.

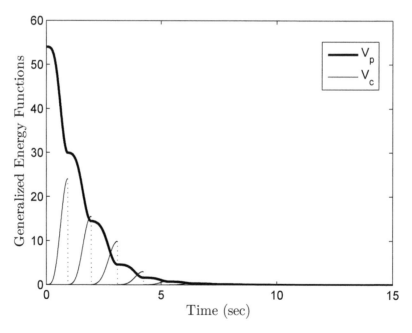

Figure 21.10 Generalized energy functions $V_{\mathrm{p}}(x_{\mathrm{p}}(t))$ and $V_{\mathrm{c}}(x_{\mathrm{c}}(t), x_{\mathrm{p}}(t))$ versus time.

$\lim_{t \to \infty} x_{\mathrm{p}i}(t) = \alpha \in \mathbb{R}$ for $i = 1, \ldots, q$, where $x_{\mathrm{p}i}(t)$ denotes the $i$th component of the system state vector $x_{\mathrm{p}}(t)$. In particular, consider $q$ continuous-time integrator agents with dynamics

$$\dot{x}_{\mathrm{p}i}(t) = u_i(t), \quad x_i(0) = x_{i0}, \quad t \geq 0, \quad i = 1, \ldots, q, \quad (21.63)$$
$$y_i(t) = x_{\mathrm{p}i}(t), \quad\quad\quad\quad\quad\quad\quad (21.64)$$

where, for each $i \in \{1, \ldots, q\}$, $x_{\mathrm{p}i}(t) \in \mathbb{R}$ denotes the information state and $u_i(t) \in \mathbb{R}$ denotes information control input for all $t \geq 0$. In this case, the set $\mathcal{D}_{\mathrm{p}0} = \{x_{\mathrm{p}} \in \mathbb{R}^q : x_{\mathrm{p}1} = \cdots = x_{\mathrm{p}q}\}$, where $x_{\mathrm{p}} \triangleq [x_{\mathrm{p}1}, \ldots, x_{\mathrm{p}q}]^{\mathrm{T}}$, characterizes the state of consensus in the multiagent network.

In the following analysis, we construct a hybrid feedback controller (21.13)–(21.15) that asymptotically stabilizes a more general form of the classical consensus steady state for the multiagent network characterized by

$$\mathcal{D}_0 = \{(x_{\mathrm{p}}, x_{\mathrm{c}}) \in \mathcal{D}_{\mathrm{p}} \times \mathcal{D}_{\mathrm{c}} : x_{\mathrm{p}} \in \mathcal{D}_{\mathrm{p}0}\}, \quad (21.65)$$

where

$$\mathcal{D}_{\mathrm{p}0} = \{x_{\mathrm{p}} \in \mathcal{D}_{\mathrm{p}} : Tx_{\mathrm{p}} = 0\} \quad (21.66)$$

and $T \in \mathbb{R}^{s_{\mathrm{p}} \times q}$. Clearly, $\mathcal{D}_{\mathrm{p}0}$ given by (21.66) characterizes the equipartitioned consensus state of a multiagent network with

$$T = \begin{bmatrix} 1 & -1 & 0 & 0 & \cdots & 0 \\ 0 & 1 & -1 & 0 & \cdots & 0 \\ 0 & 0 & 1 & -1 & \cdots & 0 \\ \vdots & & & & \ddots & \vdots \\ 0 & 0 & 0 & 0 & \cdots & -1 \end{bmatrix} \in \mathbb{R}^{(q-1) \times q}. \quad (21.67)$$

In order to stabilize $\mathcal{D}_0$ given by (21.65), consider the hybrid feedback controller (21.13)–(21.15) with

$$f_{\mathrm{cc}}(x_{\mathrm{c}}, x_{\mathrm{p}}) = F(x_{\mathrm{c}} - x_{\mathrm{p}}), \quad (21.68)$$
$$h_{\mathrm{cc}}(x_{\mathrm{c}}, x_{\mathrm{p}}) = -H(x_{\mathrm{c}} - x_{\mathrm{p}}), \quad (21.69)$$
$$\eta(x_{\mathrm{p}}) = Mx_{\mathrm{p}}, \quad\quad\quad\quad (21.70)$$

where $F \in \mathbb{R}^{q \times q}$, $H \in \mathbb{R}^{q \times q}$, and $M \in \mathbb{R}^{q \times q}$. In this case, equations (21.42)–(21.44) developed for a general class of linear dynamical systems specialize to

$$P_{\mathrm{c}}F - P_{\mathrm{c}}MH = A_{\mathrm{c}}, \quad (21.71)$$
$$-T^{\mathrm{T}}TH + M^{\mathrm{T}}A_{\mathrm{c}} = A_{\mathrm{p}}, \quad (21.72)$$
$$A_{\mathrm{p}} = A_{\mathrm{c}} \quad\quad\quad\quad (21.73)$$

for the system (21.63) and (21.64). Note that (21.71)–(21.73) can be further simplified to give

$$F - MH = P_c^{-1} A_c, \tag{21.74}$$
$$-T^T T H + M^T A_c = A_c. \tag{21.75}$$

Furthermore, note that if $q$ is even, then we can always choose a skew-symmetric matrix $A_c \in \mathbb{R}^{q \times q}$ such that $A_c^{-1}$ exists. In this case, it follows from (21.74) and (21.75) that

$$M = A_c^{-1}(A_c - H^T T^T T), \tag{21.76}$$
$$F = P_c^{-1} A_c + A_c^{-1}(A_c - H^T T^T T)H. \tag{21.77}$$

For the following numerical example, we consider four agents, with the dynamics given by (21.63) and (21.64) and the objective being to stabilize the equipartitioned consensus state with $T \in \mathbb{R}^{3 \times 4}$ given by (21.67). For our design, we set

$$A_c = \begin{bmatrix} 0 & 1 & -1 & 0 \\ -1 & 0 & 1 & -1 \\ 1 & -1 & 0 & 0.5 \\ 0 & 1 & -0.5 & 0 \end{bmatrix}, \tag{21.78}$$

$H = 1.5 I_4$, and $P_c = 0.75 I_4$ so that $M \in \mathbb{R}^{4 \times 4}$ and $F \in \mathbb{R}^{4 \times 4}$ are computed using (21.76) and (21.77). Note that with the above choice of $T \in \mathbb{R}^{3 \times 4}$ and $H \in \mathbb{R}^{4 \times 4}$, condition (21.53) is satisfied.

For the initial conditions $x_p(0) = [0.5, 1, 0.7, 1.2]^T$ and $x_c(0) = Mx_p(0)$, Figure 21.11 shows the system state history versus time, whereas Figure 21.12 shows the control input history versus time. Finally, Figure 21.13 shows the time history of the generalized energy functions $V_p(x_p(t))$ and $V_c(x_c(t), x_p(t))$ versus time. It can be seen from Figure 21.12 that the control inputs $u_i$, $i = 1, \dots, 4$, are discontinuous functions of time.

## 21.7 Hybrid Control Design for Cyclic Pursuit

In this section, we use the results of Section 21.2 to develop a hybrid reset-ting controller of the form (21.3)–(21.5) to achieve circular formations [183] involving cyclic pursuit [87, 161, 220]. The proposed controller has a lead-erless dynamic distributed architecture, which is more robust and exhibits faster convergence than static leader-based or partially leaderless control designs [87, 161, 183, 220].

Consider $q$ mobile autonomous agents in a plane described by the

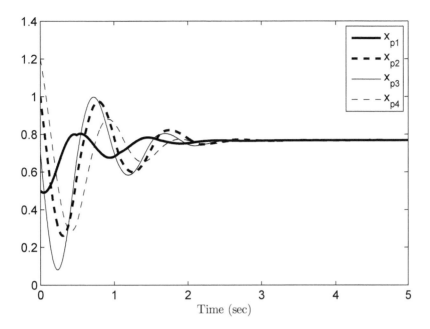

Figure 21.11 Plant states $x_{\mathrm{p}}$ versus time.

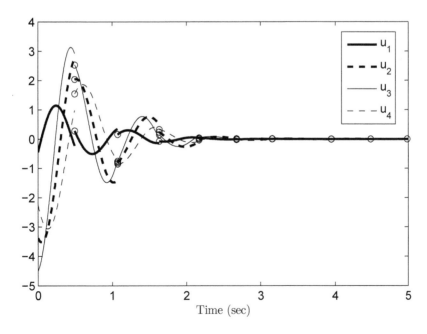

Figure 21.12 Control inputs versus time.

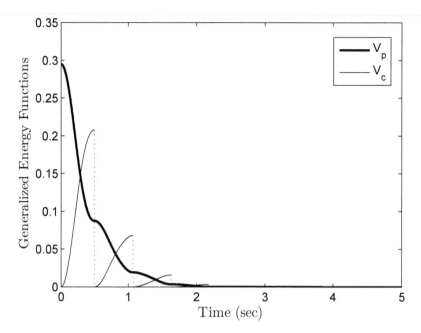

Figure 21.13 Generalized energy functions $V_{\mathrm{p}}$ and $V_{\mathrm{c}}$ versus time.

*unicycle model* given by

$$\dot{x}_i(t) = v_i(t)\cos\theta_i(t), \quad x_i(0) = x_{i0}, \quad t \geq 0, \tag{21.79}$$

$$\dot{y}_i(t) = v_i(t)\sin\theta_i(t), \quad y_i(0) = y_{i0}, \tag{21.80}$$

$$\dot{\theta}_i(t) = \omega_i(t), \quad \theta_i(0) = \theta_{i0}, \tag{21.81}$$

where, for each $i \in \{1, \ldots, q\}$, $[x_i, y_i]^{\mathrm{T}} \in \mathbb{R}^2$ denotes the position vector of the $i$th agent, $\theta_i \in \mathbb{R}$ denotes the orientation of the $i$th agent, $v_i \in \mathbb{R}$ denotes the velocity of the $i$th agent, $\omega_i \in \mathbb{R}$ denotes the angular velocity of the $i$th agent, and $u_i = [v_i, \omega_i]^{\mathrm{T}}$ is the control input of the $i$th agent.

For our result, we assume that the graph $\mathfrak{G}$ of the communication topology for the mobile agents is *undirected* and *strongly connected* [224]. The control aim is to design $u_i$ by means of neighboring information so that a circular formation is achieved; that is, the system state asymptotically converges to an invariant manifold characterized by the following constraints [161]:

$$v_i = c_1, \quad \omega_i = c_2, \quad \sum_{i=1}^{q}\sin\theta_i = 0, \quad \sum_{i=1}^{q}\cos\theta_i = 0, \tag{21.82}$$

$$\sum_{j\in\mathcal{N}_i}[(x_i - x_j)\cos\theta_i + (y_i - y_j)\sin\theta_i] = 0, \quad i = 1, \ldots, q, \tag{21.83}$$

where $c_1 \in \mathbb{R}$, $c_2 \in \mathbb{R}$, and $\mathcal{N}_i$ denotes the set of all neighbors which can communicate with the $i$th agent. Note that if $\theta_i(t) - \theta_j(t) = \frac{2(j-i)}{q}$ for all $i, j = 1, \ldots, q$, then $\sum_{i=1}^q \sin \theta_i = 0$ and $\sum_{i=1}^q \cos \theta_i = 0$.

Define $h_i(x_{\mathrm{p}}) \triangleq \sum_{j \in \mathcal{N}_i}[(x_i - x_j) \cos \theta_i + (y_i - y_j) \sin \theta_i]$, $\tilde{\theta}_i \triangleq \theta_i + \frac{2i}{q}\pi$, $i = 1, \ldots, q$, $h(x_{\mathrm{p}}) \triangleq [h_1(x_{\mathrm{p}}), \ldots, h_q(x_{\mathrm{p}})]^{\mathrm{T}}$, and $\tilde{\theta} \triangleq [\tilde{\theta}_1, \ldots, \tilde{\theta}_q]^{\mathrm{T}}$, where $x_{\mathrm{p}} \triangleq [\tilde{x}^{\mathrm{T}}, \tilde{y}^{\mathrm{T}}, \theta^{\mathrm{T}}]^{\mathrm{T}}$, $\tilde{x} \triangleq [x_1, \ldots, x_q]^{\mathrm{T}}$, $\tilde{y} \triangleq [y_1, \ldots, y_q]^{\mathrm{T}}$, and $\theta \triangleq [\theta_1, \ldots, \theta_q]^{\mathrm{T}}$, and consider the hybrid resetting controller given by

$$\dot{x}_{\mathrm{c}1}(t) = -Lx_{\mathrm{c}1}(t) + h(x_{\mathrm{p}}(t)), \quad x_{\mathrm{c}1}(0) = x_{\mathrm{c}10}, \quad t \geq 0, \quad x(t) \notin \mathcal{Z}, \tag{21.84}$$

$$\dot{x}_{\mathrm{c}2}(t) = -Lx_{\mathrm{c}2}(t) + L\tilde{\theta}(t), \quad x_{\mathrm{c}2}(0) = x_{\mathrm{c}20}, \quad x(t) \notin \mathcal{Z}, \tag{21.85}$$

$$\Delta x_{\mathrm{c}1}(t) = -L^{\dagger}Lx_{\mathrm{c}1}(t), \quad x(t) \in \mathcal{Z}, \tag{21.86}$$

$$\Delta x_{\mathrm{c}2}(t) = -L^{\dagger}Lx_{\mathrm{c}2}(t), \quad x(t) \in \mathcal{Z}, \tag{21.87}$$

$$v(t) = -x_{\mathrm{c}1}(t) - h(x_{\mathrm{p}}(t)), \tag{21.88}$$

$$\omega(t) = -x_{\mathrm{c}2}(t), \tag{21.89}$$

where $L \in \mathbb{R}^{q \times q}$ denotes the Laplacian of the graph $\mathfrak{G}$, $x \triangleq [x_{\mathrm{p}}^{\mathrm{T}}, x_{\mathrm{c}}^{\mathrm{T}}]^{\mathrm{T}}$, $x_{\mathrm{c}} \triangleq [x_{\mathrm{c}1}^{\mathrm{T}}, x_{\mathrm{c}2}^{\mathrm{T}}]^{\mathrm{T}}$, $x_{\mathrm{c}1} \in \mathbb{R}^q$, and $x_{\mathrm{c}2} \in \mathbb{R}^q$. Since, by assumption, $\mathfrak{G}$ is undirected and strongly connected, it follows that $L = L^{\mathrm{T}} \geq 0$ and the rank of $L$ is $q - 1$.

Next, let

$$V_{\mathrm{p}}(x_{\mathrm{p}}) = \frac{1}{2}\tilde{x}^{\mathrm{T}}L\tilde{x} + \frac{1}{2}\tilde{y}^{\mathrm{T}}L\tilde{y} + \frac{1}{2}\tilde{\theta}^{\mathrm{T}}L^2\tilde{\theta} \tag{21.90}$$

and

$$V_{\mathrm{c}}(x_{\mathrm{c}}, x_{\mathrm{p}}) = \frac{1}{2}x_{\mathrm{c}1}^{\mathrm{T}}x_{\mathrm{c}1} + \frac{1}{2}x_{\mathrm{c}2}^{\mathrm{T}}Lx_{\mathrm{c}2}, \tag{21.91}$$

and define the resetting set $\mathcal{Z}$ by

$$\mathcal{Z} = \{x \in \mathbb{R}^{5q} : x_{\mathrm{c}1}^{\mathrm{T}}[-Lx_{\mathrm{c}1} + h(x_{\mathrm{p}})] + x_{\mathrm{c}2}^{\mathrm{T}}L(-Lx_{\mathrm{c}2} + L\tilde{\theta}) = 0, \ x_{\mathrm{c}2}^{\mathrm{T}}Lx_{\mathrm{c}2} > 0\}. \tag{21.92}$$

Furthermore, note that $\mathcal{X}(x) = x_{\mathrm{c}1}^{\mathrm{T}}[-Lx_{\mathrm{c}1} + h(x_{\mathrm{p}})] + x_{\mathrm{c}2}^{\mathrm{T}}L(-Lx_{\mathrm{c}2} + L\tilde{\theta})$ and

$$f_{\mathrm{c}}(x) = \begin{bmatrix} \mathrm{diag}[\cos\theta_1, \ldots, \cos\theta_q][-x_{\mathrm{c}1} - h(x_{\mathrm{p}})] \\ \mathrm{diag}[\sin\theta_1, \ldots, \sin\theta_q][-x_{\mathrm{c}1} - h(x_{\mathrm{p}})] \\ -x_{\mathrm{c}2} \\ -Lx_{\mathrm{c}1} + h \\ -Lx_{\mathrm{c}2} + L\tilde{\theta} \end{bmatrix}, \tag{21.93}$$

where "diag" denotes a diagonal matrix. In addition, note that

$$\frac{\partial h}{\partial \tilde{x}} = \begin{bmatrix} L_{(1,1)} \cos\theta_1 & L_{(1,2)} \cos\theta_1 & \cdots & L_{(1,q)} \cos\theta_1 \\ L_{(2,1)} \cos\theta_2 & L_{(2,2)} \cos\theta_2 & \cdots & L_{(2,q)} \cos\theta_2 \\ \vdots & \vdots & \ddots & \vdots \\ L_{(q,1)} \cos\theta_q & L_{(q,2)} \cos\theta_q & \cdots & L_{(q,q)} \cos\theta_q \end{bmatrix}, \quad (21.94)$$

$$\frac{\partial h}{\partial \tilde{y}} = \begin{bmatrix} L_{(1,1)} \sin\theta_1 & L_{(1,2)} \sin\theta_1 & \cdots & L_{(1,q)} \sin\theta_1 \\ L_{(2,1)} \sin\theta_2 & L_{(2,2)} \sin\theta_2 & \cdots & L_{(2,q)} \sin\theta_2 \\ \vdots & \vdots & \ddots & \vdots \\ L_{(q,1)} \sin\theta_q & L_{(q,2)} \sin\theta_q & \cdots & L_{(q,q)} \sin\theta_q \end{bmatrix}, \quad (21.95)$$

$$\frac{\partial h}{\partial \theta} = \mathrm{diag}[\bar{h}_1(x_{\mathrm{p}}), \ldots, \bar{h}_q(x_{\mathrm{p}})], \quad (21.96)$$

where $L_{(i,j)}$ denotes the $(i,j)$th entry of $L$, $i,j = 1,\ldots,q$, and $\bar{h}_i(x_{\mathrm{p}}) \triangleq \sum_{j\in\mathcal{N}_i}[-(x_i - x_j)\sin\theta_i + (y_i - y_j)\cos\theta_i]$, $i = 1,\ldots,q$.

Then, it follows that

$$\begin{aligned}
L_{f_c}\mathcal{X}(x) \\
&= x_{\mathrm{c1}}^{\mathrm{T}} \frac{\partial h}{\partial \tilde{x}} \mathrm{diag}[\cos\theta_1, \ldots, \cos\theta_q](-x_{\mathrm{c1}} - h) \\
&\quad + x_{\mathrm{c1}}^{\mathrm{T}} \frac{\partial h}{\partial \tilde{y}} \mathrm{diag}[\sin\theta_1, \ldots, \sin\theta_q](-x_{\mathrm{c1}} - h) \\
&\quad - x_{\mathrm{c1}}^{\mathrm{T}} \frac{\partial h}{\partial \theta} x_{\mathrm{c2}} - (Lx_{\mathrm{c2}})^{\mathrm{T}} Lx_{\mathrm{c2}} \\
&\quad + (-Lx_{\mathrm{c1}} + h)^{\mathrm{T}}(-Lx_{\mathrm{c1}} + h) + (-Lx_{\mathrm{c1}})^{\mathrm{T}}(-Lx_{\mathrm{c1}} + h) \\
&\quad - 2(-Lx_{\mathrm{c2}} + L\tilde{\theta})^{\mathrm{T}}(-L)(-Lx_{\mathrm{c2}} + L\tilde{\theta}) + (L\tilde{\theta})^{\mathrm{T}}(-L)(-Lx_{\mathrm{c2}} + L\tilde{\theta}) \\
&= x_{\mathrm{c1}}^{\mathrm{T}} C(\theta)(-x_{\mathrm{c1}} - h) + x_{\mathrm{c1}}^{\mathrm{T}} S(\theta)(-x_{\mathrm{c1}} - h) \\
&\quad - x_{\mathrm{c1}}^{\mathrm{T}} \mathrm{diag}[\bar{h}_1, \ldots, \bar{h}_q] x_{\mathrm{c2}} - (Lx_{\mathrm{c2}})^{\mathrm{T}} Lx_{\mathrm{c2}} \\
&\quad + (-Lx_{\mathrm{c1}} + h)^{\mathrm{T}}(-Lx_{\mathrm{c1}} + h) + (-Lx_{\mathrm{c1}})^{\mathrm{T}}(-Lx_{\mathrm{c1}} + h) \\
&\quad - 2(-Lx_{\mathrm{c2}} + L\tilde{\theta})^{\mathrm{T}}(-L)(-Lx_{\mathrm{c2}} + L\tilde{\theta}) + (L\tilde{\theta})^{\mathrm{T}}(-L)(-Lx_{\mathrm{c2}} + L\tilde{\theta}),
\end{aligned}$$
$$(21.97)$$

where $h = h(x_{\mathrm{p}})$, $\bar{h}_i = \bar{h}_i(x_{\mathrm{p}})$, $i = 1,\ldots,q$, and

$$C(\theta) = \begin{bmatrix} L_{(1,1)} \cos^2\theta_1 & L_{(1,2)} \cos\theta_1 \cos\theta_2 & \cdots & L_{(1,q)} \cos\theta_1 \cos\theta_q \\ L_{(2,1)} \cos\theta_2 \cos\theta_1 & L_{(2,2)} \cos^2\theta_2 & \cdots & L_{(2,q)} \cos\theta_2 \cos\theta_q \\ \vdots & \vdots & \ddots & \vdots \\ L_{(q,1)} \cos\theta_q \cos\theta_1 & L_{(q,2)} \cos\theta_q \cos\theta_2 & \cdots & L_{(q,q)} \cos^2\theta_q \end{bmatrix},$$
$$(21.98)$$

$$S(\theta) = \begin{bmatrix} L_{(1,1)} \sin^2 \theta_1 & L_{(1,2)} \sin \theta_1 \sin \theta_2 & \cdots & L_{(1,q)} \sin \theta_1 \sin \theta_q \\ L_{(2,1)} \sin \theta_2 \sin \theta_1 & L_{(2,2)} \sin^2 \theta_2 & \cdots & L_{(2,q)} \sin \theta_2 \sin \theta_q \\ \vdots & \vdots & \ddots & \vdots \\ L_{(q,1)} \sin \theta_q \sin \theta_1 & L_{(q,2)} \sin \theta_q \sin \theta_2 & \cdots & L_{(q,q)} \sin^2 \theta_q \end{bmatrix}.$$

$$(21.99)$$

Hence, if $L_{f_c} \mathcal{X}(x) \neq 0$ for all $x \in \mathbb{R}^{5q}$ satisfying

$$\begin{bmatrix} x_{c1} & Lx_{c2} \end{bmatrix} \begin{bmatrix} -Lx_{c1} + h \\ -Lx_{c2} + L\tilde{\theta} \end{bmatrix} = 0 \qquad (21.100)$$

and

$$\begin{bmatrix} \mathrm{diag}[\cos\theta_1, \ldots, \cos\theta_q](-x_{c1} - h) \\ \mathrm{diag}[\sin\theta_1, \ldots, \sin\theta_q](-x_{c1} - h) \\ -x_{c2} \\ -Lx_{c1} + h \\ -Lx_{c2} + L\tilde{\theta} \end{bmatrix} \neq 0, \qquad (21.101)$$

then the $k$-transversality condition (21.9) holds with $k = 1$.

Now, it follows from (21.90), (21.91), (21.79)–(21.81), (21.84), (21.85), (21.88), and (21.89) that

$$\begin{aligned} \dot{V}_{\mathrm{p}}(x_{\mathrm{p}}) &+ \dot{V}_{\mathrm{c}}(x_{\mathrm{c}}, x_{\mathrm{p}}) \\ &= \tilde{x}^{\mathrm{T}} L\dot{\tilde{x}} + \tilde{y}^{\mathrm{T}} L\dot{\tilde{y}} + \tilde{\theta} L^{\mathrm{T}} L\dot{\tilde{\theta}} + x_{c1}^{\mathrm{T}} \dot{x}_{c1} + x_{c2}^{\mathrm{T}} L\dot{x}_{c2} \\ &= \tilde{x}^{\mathrm{T}} L\dot{\tilde{x}} + \tilde{y}^{\mathrm{T}} L\dot{\tilde{y}} - \tilde{\theta} L^{\mathrm{T}} Lx_{c1} + x_{c1}^{\mathrm{T}}(-Lx_{c1} + h) \\ &\quad + x_{c2}^{\mathrm{T}} L(-Lx_{c2} + L\tilde{\theta}) \\ &= \tilde{x}^{\mathrm{T}} L\dot{\tilde{x}} + \tilde{y}^{\mathrm{T}} L\dot{\tilde{y}} + x_{c1}^{\mathrm{T}} h - x_{c1}^{\mathrm{T}} Lx_{c1} - x_{c2}^{\mathrm{T}} L^2 x_{c2} \\ &= \tilde{x}^{\mathrm{T}} L\dot{\tilde{x}} + \tilde{y}^{\mathrm{T}} L\dot{\tilde{y}} + (-h - v)^{\mathrm{T}} h - x_{c1}^{\mathrm{T}} Lx_{c1} - x_{c2}^{\mathrm{T}} L^2 x_{c2} \\ &= \tilde{x}^{\mathrm{T}} L\dot{\tilde{x}} + \tilde{y}^{\mathrm{T}} L\dot{\tilde{y}} - v^{\mathrm{T}} h - h^{\mathrm{T}} h - x_{c1}^{\mathrm{T}} Lx_{c1} - x_{c2}^{\mathrm{T}} L^2 x_{c2}, \quad x \notin \mathcal{Z}. \end{aligned}$$

$$(21.102)$$

Note that

$$\frac{1}{2}\tilde{x}^{\mathrm{T}} L\tilde{x} + \frac{1}{2}\tilde{y}^{\mathrm{T}} L\tilde{y} = \frac{1}{4}\sum_{i=1}^{q}\sum_{j\in\mathcal{N}_i}(x_i - x_j)^2 + \frac{1}{4}\sum_{i=1}^{q}\sum_{j\in\mathcal{N}_i}(y_i - y_j)^2,$$

and, hence,

$$\begin{aligned} \tilde{x}^{\mathrm{T}} L\dot{\tilde{x}} &+ \tilde{y}^{\mathrm{T}} L\dot{\tilde{y}} \\ &= \frac{1}{2}\sum_{i=1}^{q}\sum_{j\in\mathcal{N}_i}(x_i - x_j)(\dot{x}_i - \dot{x}_j) + \frac{1}{2}\sum_{i=1}^{q}\sum_{j\in\mathcal{N}_i}(y_i - y_j)(\dot{y}_i - \dot{y}_j) \end{aligned}$$

$$= \frac{1}{2} \sum_{i=1}^{q} \sum_{j \in \mathcal{N}_i} (x_i - x_j)(v_i \cos \theta_i - v_j \cos \theta_j)$$

$$+ \frac{1}{2} \sum_{i=1}^{q} \sum_{j \in \mathcal{N}_i} (y_i - y_j)(v_i \sin \theta_i - v_j \sin \theta_j)$$

$$= \frac{1}{2} \sum_{i=1}^{q} \sum_{j \in \mathcal{N}_i} v_i[(x_i - x_j) \cos \theta_i + (y_i - y_j) \sin \theta_i]$$

$$+ \frac{1}{2} \sum_{i=1}^{q} \sum_{j \in \mathcal{N}_i} v_j[(x_j - x_i) \cos \theta_j + (y_j - y_i) \sin \theta_j]$$

$$= \sum_{i=1}^{q} \sum_{j \in \mathcal{N}_i} v_i[(x_i - x_j) \cos \theta_i + (y_i - y_j) \sin \theta_i]$$

$$= \sum_{i=1}^{q} v_i h_i$$

$$= v^{\mathrm{T}} h, \quad x \notin \mathcal{Z}. \tag{21.103}$$

Hence, combining (21.102) and (21.103) yields

$$\dot{V}_{\mathrm{p}}(x_{\mathrm{p}}) + \dot{V}_{\mathrm{c}}(x_{\mathrm{c}}, x_{\mathrm{p}}) = -h^{\mathrm{T}} h - x_{\mathrm{c}1}^{\mathrm{T}} L x_{\mathrm{c}1} - x_{\mathrm{c}2}^{\mathrm{T}} L^2 x_{\mathrm{c}2} \leq 0, \quad x \notin \mathcal{Z}. \tag{21.104}$$

Alternatively, noting that $L(I_q - L^{\dagger} L) = 0$ and $\|I_q - L^{\dagger} L\|_{\mathrm{F}} \leq 1$, it follows from (21.90), (21.91), (21.86), and (21.87) that

$$\begin{aligned}
\Delta V_{\mathrm{p}}(x_{\mathrm{p}}) &+ \Delta V_{\mathrm{c}}(x_{\mathrm{c}}, x_{\mathrm{p}}) \\
&= \frac{1}{2} x_{\mathrm{c}1}^{\mathrm{T}} (I_q - L^{\dagger} L)^{\mathrm{T}} (I_q - L^{\dagger} L) x_{\mathrm{c}1} - \frac{1}{2} x_{\mathrm{c}1}^{\mathrm{T}} x_{\mathrm{c}1} - \frac{1}{2} x_{\mathrm{c}2}^{\mathrm{T}} L x_{\mathrm{c}2} \\
&\leq -\frac{1}{2} x_{\mathrm{c}2}^{\mathrm{T}} L x_{\mathrm{c}2} \\
&< 0, \quad x \in \mathcal{Z}. \tag{21.105}
\end{aligned}$$

Next, let $\mathcal{R} \triangleq \{x \in \mathcal{D} : \dot{V}_{\mathrm{p}}(x_{\mathrm{p}}) + \dot{V}_{\mathrm{c}}(x_{\mathrm{c}}, x_{\mathrm{p}}) = 0\}$, where $\mathcal{D} \subseteq \mathbb{R}^{5q}$ is positively invariant with respect to (21.79)–(21.81) and (21.84)–(21.89). Then it follows that $\mathcal{R} = \{x \in \mathcal{D} : L x_{\mathrm{c}1} = 0, L x_{\mathrm{c}2} = 0, h = 0\}$. Let $\mathcal{M}$ denote the largest invariant set contained in $\mathcal{R}$, and note that $L\mathbf{e} = L^{\mathrm{T}}\mathbf{e} = 0$ and rank $L = q - 1$, where $\mathbf{e} \triangleq [1, \dots, 1]^{\mathrm{T}} \in \mathbb{R}^q$. Then it follows from (21.85) that $\mathbf{e}^{\mathrm{T}} \dot{x}_{\mathrm{c}2} = 0$. Since on $\mathcal{M}$, $L x_{\mathrm{c}2} = 0$, it follows that $x_{\mathrm{c}2} = c_1 \mathbf{e}$, and, hence, $\omega = -c_1 \mathbf{e}$, where $c_1 \in \mathbb{R}$. Now, it follows from (21.85) that $L\theta = 0$. Thus, it follows that $\tilde{\theta}_i = \tilde{\theta}_j$, and, hence, $\theta_i - \theta_j = \frac{2(j-i)}{q} \pi$ for every $i, j = 1, \dots, q$, which further implies that $\sum_{i=1}^{q} \sin \theta_i = 0$ and $\sum_{i=1}^{q} \cos \theta_i = 0$.

Next, since $Lx_{c1} = 0$ and $h = 0$ on $\mathcal{M}$, it follows from (21.84) that $\dot{x}_{c1} = 0$, and, together with $Lx_{c1} = 0$, we thus have $x_{c1} = c_2\mathbf{e}$, where $c_2 \in \mathbb{R}$. Hence, it follows from (21.88) that on $\mathcal{M}$, $v = -c_2\mathbf{e}$. Finally, it follows from Theorem 21.1 that there exists $\mathcal{D}_0 \subset \mathcal{D}$ such that, for every system initial condition in $\mathcal{D}_0$, $(x(t), y(t), \theta(t), x_c(t), z_c(t)) \to \mathcal{M}$ as $t \to \infty$, which implies convergence to a circular formation characterized by the manifold (21.82) and (21.83).

To show the efficacy of our framework, let $q = 10$, and let

$$
L = \begin{bmatrix}
2 & -1 & 0 & 0 & 0 & 0 & 0 & 0 & 0 & -1 \\
-1 & 2 & -1 & 0 & 0 & 0 & 0 & 0 & 0 & 0 \\
0 & -1 & 2 & -1 & 0 & 0 & 0 & 0 & 0 & 0 \\
0 & 0 & -1 & 2 & -1 & 0 & 0 & 0 & 0 & 0 \\
0 & 0 & 0 & -1 & 2 & -1 & 0 & 0 & 0 & 0 \\
0 & 0 & 0 & 0 & -1 & 2 & -1 & 0 & 0 & 0 \\
0 & 0 & 0 & 0 & 0 & -1 & 2 & -1 & 0 & 0 \\
0 & 0 & 0 & 0 & 0 & 0 & -1 & 2 & -1 & 0 \\
0 & 0 & 0 & 0 & 0 & 0 & 0 & -1 & 2 & -1 \\
-1 & 0 & 0 & 0 & 0 & 0 & 0 & 0 & -1 & 2
\end{bmatrix}. \tag{21.106}
$$

For this system, the transversality condition was verified numerically for $k = 1$. A group of 10 agents is initialized with random initial positions $(\tilde{x}, \tilde{y})$ in the range $[-10, 10] \times [-10, 10]$. The initial states of the hybrid resetting controller (21.84)–(21.89) are chosen randomly within $[-10, 10]$. Figure 21.14 shows that a circular formation is achieved using the hybrid resetting controller (21.84)–(21.89). Figures 21.15–21.17 show the time histories of the velocities and orientations for circular formation design.

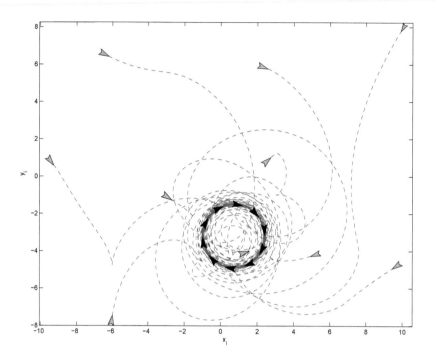

Figure 21.14  Agent positions in the plane.

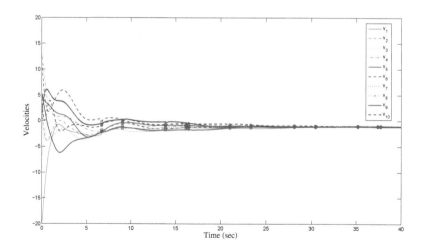

Figure 21.15  Agent velocities versus time.

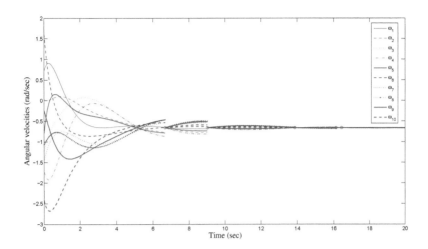

Figure 21.16  Agent angular velocities versus time.

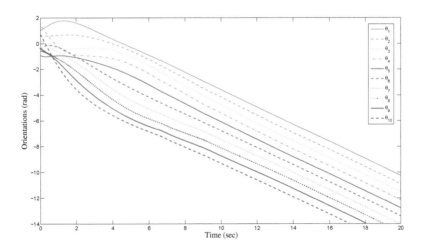

Figure 21.17  Agent orientations versus time.

*Chapter Twenty-Two*

---

# Conclusion

In this monograph, we have developed a dynamical systems and control theory framework for network information systems. These systems are composed of networked interconnected large-scale systems and include air traffic control systems, power and energy grid systems, manufacturing and processing systems, aerospace and transportation systems, communication and information networks, integrative biological systems, biological neural networks, biomolecular and biochemical systems, nervous systems, immune systems, environmental and ecological systems, molecular and chemical reaction systems, economic and financial systems, cellular systems, metabolic systems, and ecosystems, to name but a few examples.

The complexity of large-scale network dynamical systems is due to the natural scale of these systems and often necessitates hierarchical decentralized and distributed architectures for analyzing and controlling these systems due to their high system dimensionality and communication connection constraints. The role of system uncertainty is also critical in the analysis and control design of large-scale interconnected network systems. Information uncertainty and system variations in large-scale network systems may change constantly and unpredictably, resulting in rapid and often catastrophic transitions. To address the stability analysis and control system design of large-scale network systems in the presence of uncertainties, exogenous disturbances, and imperfect system network communication, distributed robust control architectures that combine logical operations with continuous dynamics are needed. These hybrid control architectures provide hierarchical coordination and autonomy by utilizing higher-level reasoning and decision-making.

The underlying intention of this monograph has been to present a general analysis and control design framework for large-scale network systems, with an emphasis on system thermodynamics and dynamical systems theory. It is hoped that this monograph will help stimulate increased interaction between engineers, physicists, computer scientists, information scientists, and dynamical systems and control theorists. The potential for applying and extending this work across disciplines is enormous.

For example, in economic systems, the interaction of raw materials, finished goods, and financial resources can be modeled as large-scale network systems with subsystem interconnections representing various interacting sectors in a dynamic economy. Similarly, network systems, computer networks, and telecommunications systems are amenable to large-scale modeling, with interconnections governed by nodal dynamics and routing strategies that can be controlled to minimize waiting times and optimize system throughput.

Large-scale network system models can also be used to model the interconnecting components of power grid systems with energy flow between regional distribution points subject to control and possible failure. Road, rail, air, and space transportation systems also give rise to large-scale network systems with interconnections subject to failure and real-time modification. Modern HVAC systems in large commercial buildings are characterized by a large number of interconnected zones that require heating, ventilation, and cooling and, thus, also constitute large-scale network systems. In particular, the automated operation of a network of HVAC systems for a regional collection of smart buildings involves a large number of interacting subsystems with multiple zones and components, nonlinear heat transfer models, multiple spatial and temporal timescales, and model uncertainties and disturbances involving changes in weather and solar radiation, varying heat loads, humidity, computers, lab equipment, people, and other latent heat sources.

In all of the aforementioned applications, reliable system analysis and distributed robust control system design, with integrated verification and validation, are essential for providing high network system performance and reconfigurable network system operation in the presence of system uncertainties and system component failures.

# Bibliography

[1] N. Abaid, I. Igel, and M. Porfiri, "On the consensus protocol of conspecific agents," *Linear Algebra and Its Applications*, vol. 437, pp. 221–235, 2012.

[2] G. Acampora, J. M. Cadenas, V. Loia, and E. M. Ballester, "A multi-agent memetic system for human-based knowledge selection," *IEEE Transactions on Systems, Man, and Cybernetics—Part A: Systems and Humans*, vol. 31, pp. 946–960, 2011.

[3] A. Afshar, O. B. Haddad, M. A. Mariño, and B. J. Adams, "Honey-bee mating optimization (HBMO) algorithm for optimal reservoir operation," *Journal of The Franklin Institute*, vol. 344, pp. 452–462, 2007.

[4] M. Andelic and C. M. da Fonseca, "Sufficient conditions for positive definiteness of tridiagonal matrices revisited," *Positivity*, vol. 15, pp. 155–159, 2001.

[5] H. Ando, Y. Oasa, I. Suzuki, and M. Yamashita, "Distributed memoryless point convergence algorithm for mobile robots with limited visibility," *IEEE Transactions on Robotics and Automation*, vol. 15, pp. 818–828, 1999.

[6] D. Angeli and P.-A. Bliman, "Stability of leaderless discrete-time multi-agent systems," *Mathematics of Control, Signals, and Systems*, vol. 18, pp. 293–322, 2006.

[7] E. Arabi, T. Yucelen, and W. M. Haddad, "Mitigating the effects of sensor uncertainties in networked multiagent systems," in *Proceedings of the American Control Conference* (Boston, MA), pp. 5545–5550, 2016.

[8] M. Arcak, "Passivity as a design tool for group coordination," *IEEE Transactions on Automatic Control*, vol. 143, pp. 1380–1390, 2007.

[9] Z. Artstein and A. Leizarowitz, "Tracking periodic signals with the overtaking criterion," *IEEE Transactions on Automatic Control*, vol. 30, pp. 1123–1126, 1985.

[10] J. P. Aubin and A. Cellina, *Differential Inclusions*. Berlin, Germany: Springer-Verlag, 1984.

[11] A. Bacciotti and F. Ceragioli, "Stability and stabilization of discontinuous systems and nonsmooth Lyapunov functions," *ESAIM Control Optimization and Calculus of Variations*, vol. 4, pp. 361–376, 1999.

[12] A. Bacciotti and L. Rosier, *Liapunov Functions and Stability in Control Theory*. London, U.K.: Springer-Verlag, 2001.

[13] D. D. Bainov and P. S. Simeonov, *Systems with Impulse Effect: Stability, Theory and Applications*. Chichester, U.K.: Ellis Horwood, 1989.

[14] D. D. Bainov and P. S. Simeonov, *Impulsive Differential Equations: Asymptotic Properties of the Solutions*. Singapore: World Scientific, 1995.

[15] J. Bang-Jensen and G. Z. Gutin, *Digraphs: Theory, Algorithms and Applications*. New York: Springer, 2002.

[16] A. Banos, J. Carrasco, and A. Barreiro, "Reset times-dependent stability of reset control systems," *IEEE Transactions on Automatic Control*, vol. 56, pp. 217–223, 2011.

[17] D. Bauso, L. Giarre, and R. Pesenti, "Mechanism design for optimal consensus problems," in *Proceedings of the IEEE Conference on Decision and Control* (San Diego, CA), pp. 3381–3386, 2006.

[18] D. Bauso, L. Giarré, and R. Pesenti, "Consensus for networks with unknown but bounded disturbances," *SIAM Journal on Control and Optimization*, vol. 48, pp. 1756–1770, 2009.

[19] O. Beker, C. V. Hollot, and Y. Chait, "Plant with an integrator: An example of reset control overcoming limitations of linear feedback," *IEEE Transactions on Automatic Control*, vol. 46, pp. 1797–1799, 2001.

[20] O. Beker, C. V. Hollot, Y. Chait, and H. Han, "Fundamental properties of reset control systems," *Automatica*, vol. 40, pp. 905–915, 2004.

[21] E. Ben-Jacob, I. Cohen, and H. Levine, "Cooperative self-organization of microorganisms," *Advanced Physics*, vol. 49, pp. 395–554, 2000.

[22] J. M. Berg, D. H. S. Maithripala, Q. Hui, and W. M. Haddad, "Thermodynamics-based control for network systems," *ASME Journal of Dynamic Systems, Measurement and Control*, vol. 135, 051003, 2013.

[23] A. Berman and R. Plemmons, *Nonnegative Matrices in the Mathematical Sciences.* Classics Appl. Math. 9. Philadelphia: SIAM, 1994.

[24] A. Berman and R. J. Plemmons, *Nonnegative Matrices in the Mathematical Sciences.* New York: Academic Press, 1979.

[25] D. S. Bernstein, "Nonquadratic cost and nonlinear feedback control," *International Journal of Robust and Nonlinear Control,* vol. 3, pp. 211–229, 1993.

[26] D. S. Bernstein, *Matrix Mathematics.* Princeton, NJ: Princeton University Press, 2005.

[27] D. S. Bernstein, *Matrix Mathematics,* 2nd ed. Princeton, NJ: Princeton University Press, 2009.

[28] D. S. Bernstein and S. P. Bhat, "Lyapunov stability, semistability, and asymptotic stability of matrix second-order systems," *ASME Journal of Vibration and Acoustics,* vol. 117, pp. 145–153, 1995.

[29] D. S. Bernstein and W. M. Haddad, "Optimal output feedback for nonzero set point regulation," *IEEE Transactions on Automatic Control,* vol. 32, pp. 641–645, 1987.

[30] D. S. Bernstein and D. C. Hyland, "Compartmental modeling and second-moment analysis of state space systems," *SIAM Journal on Matrix Analysis and Applications,* vol. 14, pp. 880–901, 1993.

[31] D. M. Bevly and B. Parkinson, "Cascaded Kalman filters for accurate estimation of multiple biases, dead-reckoning navigation, and full state feedback control of ground vehicles," *IEEE Transactions on Control Systems Technology,* vol. 15, pp. 199–208, 2007.

[32] S. P. Bhat and D. S. Bernstein, "Continuous finite-time stabilization of the translational and rotational double integrators," *IEEE Transactions on Automatic Control,* vol. 43, pp. 678–682, 1998.

[33] S. P. Bhat and D. S. Bernstein, "Lyapunov analysis of semistability," in *Proceedings of the American Control Conference* (San Diego, CA), pp. 1608–1612, 1999.

[34] S. P. Bhat and D. S. Bernstein, "Finite-time stability of continuous autonomous systems," *SIAM Journal on Control and Optimization,* vol. 38, pp. 751–766, 2000.

[35] S. P. Bhat and D. S. Bernstein, "Arc-length-based Lyapunov tests for convergence and stability in systems having a continuum of equilibria," in *Proceedings of the American Control Conference* (Denver, CO), pp. 2961–2966, 2003.

[36] S. P. Bhat and D. S. Bernstein, "Nontangency-based Lyapunov tests for convergence and stability in systems having a continuum of equilibria," *SIAM Journal on Control and Optimization*, vol. 42, pp. 1745–1775, 2003.

[37] S. P. Bhat and D. S. Bernstein, "Geometric homogeneity with applications to finite-time stability," *Mathematics of Control, Signals, and Systems*, vol. 17, pp. 101–127, 2005.

[38] S. P. Bhat and D. S. Bernstein, "Arc-length-based Lyapunov tests for convergence and stability with applications to systems having a continuum of equilibria," *Mathematics of Control, Signals, and Systems*, vol. 22, pp. 155–184, 2010.

[39] P. Bianchi and J. Jakubowicz, "Convergence of a multi-agent projected stochastic gradient algorithm for non-convex optimization," *IEEE Transactions on Automatic Control*, vol. 58, pp. 391–405, 2013.

[40] M. Blanke and J. Schröder, *Diagnosis and Fault-Tolerant Control*. New York: Springer, 2006.

[41] P.-A. Bliman and G. Ferrari-Trecate, "Average consensus problems in networks of agents with delayed communications," in *Proceedings of the IEEE Conference on Decision and Control* (Seville, Spain), pp. 7066–7071, 2005.

[42] V. Blondel and J. N. Tsitsiklis, "NP-hardness of some linear control design problems," *SIAM Journal on Control and Optimization*, vol. 35, pp. 2118–2127, 1997.

[43] V. D. Blondel, J. M. Hendrickx, and J. N. Tsitsiklis, "On the 2R conjecture for multi-agent systems," in *Proceedings of the European Control Conference* (Kos, Greece), pp. 874–881, 2007.

[44] V. D. Blondel and J. N. Tsitsiklis, "A survey of computational complexity results in systems and control," *Automatica*, vol. 36, pp. 1249–1274, 2000.

[45] F. Borrelli and T. Keviczky, "Distributed LQR design for identical dynamically decoupled systems," *IEEE Transactions on Automatic Control*, vol. 53, pp. 1901–1912, 2008.

[46] S. Boyd, L. El Ghaoui, E. Feron, and V. Balakrishnan, *Linear Matrix Inequalities in System and Control Theory*. SIAM Stud. Appl. Math. 15. Philadelphia: SIAM, 1994.

[47] S. Boyd, A. Ghosh, B. Prabhakar, and D. Shah, "Randomized gossip algorithms," *IEEE Transactions on Information Theory*, vol. 52, pp. 2508–2530, 2006.

[48] J. B. Boyling, "An axiomatic approach to classical thermodynamics," *Proceedings of the Royal Society of London. Series A, Mathematical and Physical Sciences*, vol. 329, pp. 35–70, July 1972.

[49] R. M. Brach, *Mechanical Impact Dynamics*. New York: Wiley, 1991.

[50] M. S. Branicky, "Multiple-Lyapunov functions and other analysis tools for switched and hybrid systems," *IEEE Transactions on Automatic Control*, vol. 43, pp. 475–482, 1998.

[51] B. Brogliato, *Nonsmooth Impact Mechanics: Models, Dynamics, and Control*. London, U.K.: Springer-Verlag, 1996.

[52] D. E. Brown and L. F. Gunderson, "Using clustering to discover the preferences of computer criminals," *IEEE Transactions on Systems, Man, and Cybernetics—Part A: Systems and Humans*, vol. 41, pp. 311–318, 2001.

[53] E. Brown, J. Moehlis, and P. Holmes, "On the phase reduction and response dynamics of neural oscillator populations," *Neural Computation*, vol. 16, pp. 673–715, 2004.

[54] A. E. Bryson, *Applied Optimal Control*. New York: Hemisphere, 1975.

[55] A. E. Bryson, *Control of Aircraft and Spacecraft*. Princeton, NJ: Princeton University Press, 1993.

[56] F. Bullo, J. Cortes, and S. Martínez, *Distributed Control of Robotic Networks*. Princeton, NJ: Princeton University Press, 2009.

[57] S. Camazine, J. Deneubourg, N. Franks, J. Sneyd, G. Theraulaz, and E. Bonabeau, *Self-Organization in Biological Systems*. Princeton, NJ: Princeton University Press, 2001.

[58] S. L. Campbell and N. J. Rose, "Singular perturbation of autonomous linear systems," *SIAM Journal on Mathematical Analysis*, vol. 10, pp. 542–551, 1979.

[59] Y. Cao and W. Ren, "Optimal linear-consensus algorithms: An LQR perspective," *IEEE Transactions on Systems, Man, and Cybernetics—Part B: Cybernetics*, vol. 40, pp. 819–830, 2010.

[60] R. Carli, P. Frasca, F. Fagnani, and S. Zampieri, "Gossip consensus algorithms via quantized communication," *Automatica*, vol. 46, pp. 70–80, 2010.

[61] N. L. Carothers, *Real Analysis*. New York: Cambridge University Press, 2000.

[62] V. Chellaboina, S. P. Bhat, and W. M. Haddad, "An invariance principle for nonlinear hybrid and impulsive dynamical systems," *Nonlinear Analysis*, vol. 53, pp. 527–550, 2003.

[63] F. H. Clarke, *Optimization and Nonsmooth Analysis*. New York: Wiley, 1983.

[64] F. H. Clarke, Y. S. Ledyaev, R. J. Stern, and P. R. Wolenski, *Nonsmooth Analysis and Control Theory*. New York: Springer-Verlag, 1998.

[65] J. C. Clegg, "A nonlinear integrator for servomechanisms," *Transactions of the American Institute of Electrical Engineers*, vol. 77, pp. 41–42, 1958.

[66] M. Clerc and J. Kennedy, "The particle swarm-explosion, stability, and convergence in a multidimensional complex space," *IEEE Transactions on Evolutionary Computation*, vol. 6, pp. 58–73, 2002.

[67] Committee on Network Science for Future Army Applications, National Research Council, *Network Science*. Washington, D.C.: The National Academies Press, 2005.

[68] W. J. Conover, *Practical Nonparametric Statistics*. New York: Wiley, 1999.

[69] J. Cortés, "Finite-time convergent gradient flows with applications to network consensus," *Automatica*, vol. 42, pp. 1993–2000, 2006.

[70] J. Cortés, "Discontinuous dynamical systems," *IEEE Control Systems Magazine*, vol. 28, pp. 36–73, 2008.

[71] J. Cortés, "Distributed algorithms for reaching consensus on general functions," *Automatica*, vol. 44, pp. 726–737, 2008.

[72] J. Cortés, "Distributed Kriged Kalman filter for spatial estimation," *IEEE Transactions on Automatic Control*, vol. 54, pp. 2816–2827, 2009.

[73] J. Cortés and F. Bullo, "Coordination and geometric optimization via distributed dynamical systems," *SIAM Journal on Control and Optimization*, vol. 44, pp. 1543–1574, 2005.

[74] J. Cortés, S. Martinez, and F. Bullo, "Robust rendezvous for mobile autonomous agents via proximity graphs in arbitrary dimensions," *IEEE Transactions on Automatic Control*, vol. 51, pp. 1289–1298, 2006.

[75] C. M. Dafermos, *Hyperbolic Conservation Laws in Continuum Physics.* Berlin, Germany: Springer-Verlag, 2000.

[76] A. Das and F. L. Lewis, "Distributed adaptive control for synchronization of unknown nonlinear networked systems," *Automatica*, vol. 46, pp. 2014–2021, 2010.

[77] A. Das and M. Mesbahi, "Distributed linear parameter estimation in sensor networks based on Laplacian dynamics consensus algorithm," in *IEEE Communications Society Conference on Sensor, Mesh, and Ad Hoc Communications and Networks* (Reston, VA), vol. 2, pp. 440–449, 2006.

[78] G. De La Torre and T. Yucelen, "State emulator-based adaptive architectures for resilient networked multiagent systems over directed and time-varying graphs," in *ASME Dynamic Systems and Control Conference* (Houston, TX), American Society of Mechanical Engineers, 2015.

[79] K. Deb, "An efficient constraint handling method for genetic algorithms," *Computer Methods in Applied Mechanics and Engineering*, vol. 186, pp. 311–338, 2007.

[80] J. P. Desai, J. P. Ostrowski, and V. Kumar, "Modeling and control of formations of nonholonomic mobile robots," *IEEE Transactions on Robotics and Automation*, vol. 17, pp. 905–908, 2001.

[81] R. Diestel, *Graph Theory.* New York: Springer-Verlag, 1997.

[82] T. Donchev, V. Ríos, and P. Wolenski, "Strong invariance and one-sided Lipschitz multifunctions," *Nonlinear Analysis*, vol. 60, pp. 849–862, 2005.

[83] M. Dorigo, V. Maniezzo, and A. Colorni, "Ant system: Optimization by a colony of cooperating agents," *IEEE Transactions on Systems, Man, and Cybernetics—Part B: Cybernetics*, vol. 26, pp. 29–41, 1996.

[84] B. A. Dubrovin, A. T. Fomenko, and S. P. Novikov, *Modern Geometry—Methods and Applications: Part II: The Geometry and Topology of Manifolds.* New York: Springer-Verlag, 1985.

[85] B. A. Earnshaw and J. P. Keener, "Global asymptotic stability of solutions of nonautonomous master equations," *SIAM Journal on Applied Dynamical Systems*, vol. 9, pp. 220–237, 2010.

[86] M. Egerstedt and X. Hu, "Formation control with virtual leaders and reduced communications," *IEEE Transactions on Robotics and Automation*, vol. 17, pp. 947–951, 2001.

[87] M. I. El-Hawwary and M. Maggiore, "Distributed circular formation stabilization for dynamic unicycles," *IEEE Transactions on Automatic Control*, vol. 58, pp. 149–162, 2013.

[88] L. C. Evans, *Partial Differential Equations*. Providence, RI: American Mathematical Society, 1998.

[89] F. Fagnani and S. Zampieri, "Average consensus with packet drop communication," *SIAM Journal on Control and Optimization*, vol. 48, pp. 102–133, 2009.

[90] L. Fang and P. J. Antsaklis, "Information consensus of asynchronous discrete-time multi-agent systems," in *Proceedings of the American Control Conference* (Portland, OR), pp. 1883–1888, 2005.

[91] L. Farina and S. Rinaldi, *Positive Linear Systems: Theory and Applications*. New York: John Wiley & Sons, 2000.

[92] H. Fawzi, P. Tabuada, and S. Diggavi, "Secure estimation and control for cyber-physical systems under adversarial attacks," *IEEE Transactions on Automatic Control*, vol. 59, pp. 1454–1467, 2012.

[93] J. A. Fax and R. M. Murray, "Information flow and cooperative control of vehicle formations," *IEEE Transactions on Automatic Control*, vol. 49, pp. 1465–1476, 2004.

[94] C. C. Federspiel and J. E. Seem, "Temperature control in large buildings," in *The Control Handbook* (W. S. Levine, ed.), pp. 1191–1204, Boca Raton, FL: CRC Press in collaboration with IEEE Press, 1996.

[95] C. L. Fefferman, "Existence and smoothness of the Navier-Stokes equations," in *The Millennium Prize Problems* (J. Carlson, A. Jaffe, and A. Wiles, eds.), pp. 57–70, Providence, RI; Oxford, U.K.: American Mathematical Society; Clay Mathematics Institute, 2006.

[96] A. F. Filippov, "Differential equations with discontinuous right-hand side," *American Mathematical Society Translations*, vol. 42, pp. 199–231, 1964.

[97] A. F. Filippov, *Differential Equations with Discontinuous Right-Hand Sides*. Dordrecht, Netherlands: Kluwer, 1988.

[98] G. B. Folland, *Introduction to Partial Differential Equations,* 2nd ed. Princeton, NJ: Princeton University Press, 1995.

[99] R. Freeman, P. Yang, and K. Lynch, "Stability and convergence properties of dynamic average consensus estimators," in *Proceedings of the IEEE Conference on Decision and Control* (San Diego, CA), pp. 338–343, 2006.

[100] A. T. Fuller, "Optimization of some nonlinear control systems by means of Bellman's equation and dimensional analysis," *International Journal of Control*, vol. 3, pp. 359–394, 1966.

[101] V. Gazi and K. M. Passino, "Stability analysis of swarms," *IEEE Transactions on Automatic Control*, vol. 48, pp. 692–697, 2003.

[102] C. Godsil and G. F. Royle, *Algebraic Graph Theory*. New York: Springer-Verlag, 2001.

[103] R. Goebel, "Set-valued Lyapunov functions for difference inclusions," *Automatica*, vol. 47, pp. 127–132, 2011.

[104] R. Goebel, "Robustness of stability through necessary and sufficient Lyapunov-like conditions for systems with a continuum of equilibria," *Systems and Control Letters*, vol. 65, pp. 81–88, 2014.

[105] R. Goebel, R. G. Sanfelice, and A. R. Teel, "Hybrid dynamical systems," *IEEE Control Systems Magazine*, vol. 29, no. 2, pp. 28–93, 2009.

[106] R. Goebel, R. G. Sanfelice, and A. R. Teel, *Hybrid Dynamical Systems: Modeling, Stability, and Robustness*. Princeton, NJ: Princeton University Press, 2012.

[107] R. Goebel and A. R. Teel, "Solutions to hybrid inclusions via set and graphical convergence with stability theory applications," *Automatica*, vol. 42, pp. 573–587, 2006.

[108] H. F. Grip, T. Fossen, T. A. Johansen, and A. Saberi, "Attitude estimation using biased gyro and vector measurements with time-varying reference vectors," *IEEE Transactions on Automatic Control*, vol. 57, pp. 1332–1338, 2012.

[109] J. W. Grizzle, G. Abba, and F. Plestan, "Asymptotically stable walking for biped robots: Analysis via systems with impulse effects," *IEEE Transactions on Automatic Control*, vol. 46, pp. 51–64, 2001.

[110] J. L. Gross and J. Yellen, *Handbook of Graph Theory*. Boca Raton, FL: CRC Press, 2004.

[111] D. Grünbaum and A. Okubo, "Modelling social animal aggregations," in *Frontiers in Mathematical Biology* (S. A. Levin, ed.), pp. 296–325, Berlin, Germany: Springer-Verlag, 1994.

[112] Z.-H. Guan, Z.-W. Liu, G. Feng, and M. Jian, "Impulsive consensus algorithms for second-order multi-agent networks with sampled information," *Automatica*, vol. 48, pp. 1397–1404, 2012.

[113] Z.-H. Guan, B. Hu, M. Chi, D.-X. He, and X.-M. Chen, "Guaranteed performance consensus in second-order multi-agent systems with hybrid impulsive control," *Automatica*, vol. 50, pp. 2415–2418, 2014.

[114] V. Guillemin and A. Pollack, *Differential Topology*. Englewood Cliffs, NJ: Prentice-Hall, 1974.

[115] Y. Guo, Y. Wang, L. Xie, and J. Zheng, "Stability analysis and design of reset systems: Theory and an application," *Automatica*, vol. 45, pp. 492–497, 2009.

[116] V. Gupta, C. Langbort, and R. M. Murray, "On the robustness of distributed algorithms," in *Proceedings of the IEEE Conference on Decision and Control* (San Diego, CA), pp. 3473–3478, 2006.

[117] W. M. Haddad, "Nonlinear differential equations with discontinuous right-hand sides: Filippov solutions, nonsmooth stability and dissipativity theory, and optimal discontinuous feedback control," *Communications in Applied Analysis*, vol. 18, pp. 455–522, 2014.

[118] W. M. Haddad, *A Dynamical Systems Theory of Thermodynamics*. Princeton, NJ: Princeton University Press, 2019.

[119] W. M. Haddad, "Condensed matter physics, hybrid energy and entropy principles, and the hybrid first and second laws of thermodynamics," *Communications in Nonlinear Science and Numerical Simulation*, vol. 83, 105096, 2020.

[120] W. M. Haddad and V. Chellaboina, "Stability theory for nonnegative and compartmental dynamical systems with time delay," *Systems and Control Letters*, vol. 51, pp. 355–361, 2004.

[121] W. M. Haddad and V. Chellaboina, "Stability and dissipativity theory for nonnegative dynamical systems: A unified analysis framework for biological and physiological systems," *Nonlinear Analysis: Real World Applications*, vol. 6, pp. 35–65, 2005.

[122] W. M. Haddad and V. Chellaboina, *Nonlinear Dynamical Systems and Control: A Lyapunov-Based Approach*. Princeton, NJ: Princeton University Press, 2008.

[123] W. M. Haddad, V. Chellaboina, and E. August, "Stability and dissipativity theory for discrete-time nonnegative and compartmental dynamical systems," *International Journal of Control*, vol. 76, pp. 1845–1861, 2003.

[124] W. M. Haddad, V. Chellaboina, and Q. Hui, *Nonnegative and Compartmental Dynamical Systems*. Princeton, NJ: Princeton University Press, 2010.

[125] W. M. Haddad, V. Chellaboina, Q. Hui, and S. G. Nersesov, "Energy-and entropy-based stabilization for lossless dynamical systems via hybrid controllers," *IEEE Transactions on Automatic Control*, vol. 52, pp. 1604–1614, 2007.

[126] W. M. Haddad, V. Chellaboina, and N. A. Kablar, "Nonlinear impulsive dynamical systems. Part I: Stability and dissipativity," *International Journal of Control*, vol. 74, pp. 1631–1658, 2001.

[127] W. M. Haddad, V. Chellaboina, and S. G. Nersesov, *Thermodynamics: A Dynamical Systems Approach*. Princeton, NJ: Princeton University Press, 2005.

[128] W. M. Haddad, V. Chellaboina, and S. G. Nersesov, *Impulsive and Hybrid Dynamical Systems: Stability, Dissipativity, and Control*. Princeton, NJ: Princeton University Press, 2006.

[129] W. M. Haddad, V. Chellaboina, and S. G. Nersesov, "Hybrid decentralized maximum entropy control for large-scale dynamical systems," *Nonlinear Analysis: Hybrid Systems*, vol. 1, pp. 244–263, 2007.

[130] W. M. Haddad, V. Chellaboina, and T. Rajpurohit, "Dissipativity theory for nonnegative and compartmental dynamical systems with time delay," *IEEE Transactions on Automatic Control*, vol. 49, pp. 747–751, 2004.

[131] W. M. Haddad and Q. Hui, "Complexity, robustness, self-organization, swarms, and system thermodynamics," *Nonlinear Analysis: Real World Applications*, vol. 10, pp. 531–543, 2009.

[132] W. M. Haddad, Q. Hui, and V. Chellaboina, "$\mathcal{H}_2$ optimal semistable control for linear dynamical systems: An LMI approach," *Journal of The Franklin Institute*, vol. 348, pp. 2898–2910, 2011.

[133] W. M. Haddad, Q. Hui, V. Chellaboina, and S. G. Nersesov, "Hybrid decentralized maximum entropy control for large-scale dynamical systems," *Nonlinear Analysis: Hybrid Systems*, vol. 1, pp. 244–263, 2007.

[134] W. M. Haddad, Q. Hui, S. G. Nersesov, and V. Chellaboina, "Thermodynamic modeling, energy equipartition, and nonconservation of entropy for discrete-time dynamical systems," *Advances in Difference Equations*, vol. 2005, no. 3, pp. 275–318, 2005.

[135] W. M. Haddad, S. G. Nersesov, and M. Ghasemi, "Consensus control protocols for nonlinear dynamical systems via hybrid stabilization of sets," in *Proceedings of the European Control Conference* (Zurich, Switzerland), pp. 3149–3154, 2013.

[136] W. M. Haddad, S. G. Nersesov, Q. Hui, and M. Ghasemi, "Formation control protocols for nonlinear dynamical systems via hybrid stabilization of sets," *Journal of Dynamic Systems, Measurement, and Control*, vol. 136, 051020, 2014.

[137] W. M. Haddad, T. Rajpurohit, and X. Jin, "Stochastic semistability for nonlinear dynamical systems with application to consensus on networks with communication uncertainty," *IEEE Transactions on Automatic Control*, vol. 65, pp. 2826–2841, 2020.

[138] T. Hagiwara and M. Araki, "Design of a stable feedback controller based on the multirate sampling of the plant output," *IEEE Transactions on Automatic Control*, vol. 33, pp. 812–819, 1988.

[139] V. T. Haimo, "Finite-time controllers," *SIAM Journal on Control and Optimization*, vol. 24, pp. 760–770, 1986.

[140] J. K. Hale, "Dynamical systems and stability," *Journal of Mathematical Analysis and Applications*, vol. 26, pp. 39–59, 1969.

[141] J. K. Hale, *Ordinary Differential Equations*, 2nd ed. New York: Wiley, 1980. Reprinted by Krieger, Malabar, 1991.

[142] J. K. Hale and S. M. V. Lunel, *Introduction to Functional Differential Equations*. New York: Springer-Verlag, 1993.

[143] G. H. Hardy, J. E. Littlewood, and G. Pólya, *Inequalities*, 2nd ed. Cambridge, U.K.: Cambridge University Press, 1952.

[144] P. Hartman, *Ordinary Differential Equations*. Classics Appl. Math. 38. Philadelphia: SIAM, 2002.

[145] Y. Hatano, A. Das, and M. Mesbahi, "Agreement in presence of noise: Pseudogradients on random geometric networks," in *Proceedings of the IEEE Conference on Decision and Control* (Seville, Spain), pp. 6382–6387, 2005.

[146] Y. Hong, "Finite-time stabilization and stabilizability of a class of controllable systems," *Systems and Control Letters*, vol. 46, pp. 231–236, 2002.

[147] Y. Hong, J. Huang, and Y. Xu, "On an output feedback finite-time stabilization problem," *IEEE Transactions on Automatic Control*, vol. 46, pp. 305–309, 2001.

[148] R. A. Horn and R. C. Johnson, *Matrix Analysis*. Cambridge, U.K.: Cambridge University Press, 1985.

[149] I. M. Horowitz, *Synthesis of Feedback Systems*. New York: Academic Press, 1963.

[150] Z.-G. Hou, L. Cheng, and M. Tan, "Decentralized robust adaptive control for the multiagent system consensus problem using neural networks," *IEEE Transactions on Systems, Man, and Cybernetics—Part B: Cybernetics*, vol. 39, pp. 636–647, 2009.

[151] H. Hu, L. Yu, W. Zhang, and H. Song, "Group consensus in multi-agent systems with hybrid protocol," *Journal of The Franklin Institute*, vol. 350, pp. 575–597, 2013.

[152] S. Hu, V. Lakshmikantham, and S. Leela, "Impulsive differential systems and the pulse phenomena," *Journal of Mathematical Analysis and Applications*, vol. 137, pp. 605–612, 1989.

[153] Q. Hui, "Hybrid consensus protocols: An impulsive dynamical system approach," *International Journal of Control*, vol. 83, pp. 1107–1116, 2010.

[154] Q. Hui, "Optimal semistable control for continuous-time coupled systems," in *Proceedings of the American Control Conference* (Baltimore, MD), pp. 6403–6408, 2010.

[155] Q. Hui, "Optimal distributed linear averaging," *Automatica*, vol. 47, pp. 2713–2719, 2011.

[156] Q. Hui, "Optimal semistable control for continuous-time linear systems," *Systems and Control Letters*, vol. 60, pp. 278–284, 2011.

[157] Q. Hui, "Quantized near-consensus via quantized communication links," *International Journal of Control*, vol. 84, pp. 931–946, 2011.

[158] Q. Hui, "Distributed semistable LQR control for discrete-time dynamically coupled systems," *Journal of The Franklin Institute*, vol. 349, pp. 74–92, 2012.

[159] Q. Hui, "Convergence and stability analysis for iterative dynamics with application to compartmental networks: A trajectory distance based Lyapunov approach," *Journal of The Franklin Institute*, vol. 350, pp. 679–697, 2013.

[160] Q. Hui, "Optimal semistable control in *ad hoc* network systems: A sequential two-stage approach," *IEEE Transactions on Automatic Control*, vol. 58, pp. 779–784, 2013.

[161] Q. Hui and W. M. Haddad, "Continuous and hybrid distributed control for multiagent systems: Consensus, flocking, and cyclic pursuit," in *Proceedings of the American Control Conference* (New York), pp. 2576–2581, 2007.

[162] Q. Hui and W. M. Haddad, "Distributed nonlinear control algorithms for network consensus," *Automatica*, vol. 44, pp. 2375–2381, 2008.

[163] Q. Hui and W. M. Haddad, "Semistability of switched dynamical systems, Part I: Linear system theory," *Nonlinear Analysis: Hybrid Systems*, vol. 3, pp. 343–353, 2009.

[164] Q. Hui and W. M. Haddad, "$\mathcal{H}_2$ optimal semistable stabilization for linear discrete-time dynamical systems with applications to network consensus," *International Journal of Control*, vol. 82, pp. 456–469, 2009.

[165] Q. Hui, W. M. Haddad, and J. M. Bailey, "Multistability, bifurcations, and biological neural networks: A synaptic drive firing model for cerebral cortex transition in the induction of general anesthesia," *Nonlinear Analysis: Hybrid Systems*, vol. 5, pp. 554–573, 2011.

[166] Q. Hui, W. M. Haddad, and S. P. Bhat, "Finite-time semistability theory with applications to consensus protocols in dynamical networks," in *Proceedings of the American Control Conference* (New York), pp. 2411–2416, 2007.

[167] Q. Hui, W. M. Haddad, and S. P. Bhat, "Finite-time semistability and consensus for nonlinear dynamical networks," *IEEE Transactions on Automatic Control*, vol. 53, pp. 1887–1900, 2008.

[168] Q. Hui and Z. Liu, "A semistabilizability/semidetectability approach to semistable $\mathcal{H}_2$ and $\mathcal{H}_\infty$ control problems," in *49th Annual Allerton Conference on Communication, Control, and Computing* (Monticello, IL), pp. 566–571, 2011.

[169] Q. Hui and Z. Liu, "Semistability-based robust and optimal control design for network systems," in *Proceedings of the IEEE Conference on Decision and Control* (Maui, HI), pp. 7049–7054, 2012.

[170] Q. Hui and Z. Liu, "Semistability-based robust and optimal control design for network systems," Technical Report CSEL-07-14, Control Center Engineering Laboratory, Department of Mechanical Engineering, Texas Tech University, Lubbock, TX, 2014 [online], http://arxiv.org/abs/1407.6690.

[171] Q. Hui and H. Zhang, "Optimal balanced coordinated network resource allocation using swarm optimization," in *Proceedings of the IEEE Conference on Decision and Control* (Maui, HI), pp. 3936–3941, 2012.

[172] Q. Hui and H. Zhang, "Convergence analysis and parallel computing implementation for the multiagent coordination optimization algorithm," Technical Report CSEL-06-13, Control Center Engineering Laboratory, Department of Mechanical Engineering, Texas Tech University, Lubbock, TX, 2013 [online], http://arxiv.org/abs/1306.0225.

[173] Q. Hui and H. Zhang, "Semistability-based convergence analysis for paracontracting multiagent coordination optimization," Technical Report CSEL-08-13, Control Center Engineering Laboratory, Department of Mechanical Engineering, Texas Tech University, Lubbock, TX, 2013 [online], http://arxiv.org/abs/1308.2930.

[174] Q. Hui, H. Zhang, and Z. Liu, "On robust and optimal imperfect information state equipartitioning for network systems," *Journal of The Franklin Institute*, vol. 352, pp. 3410–3446, 2015.

[175] A. Iggidr, B. Kalitine, and R. Outbib, "Semidefinite Lyapunov functions stability and stabilization," *Mathematics of Control, Signals and Systems*, vol. 9, pp. 95–106, 1996.

[176] P. A. Ioannou and J. Sun, *Robust Adaptive Control*. Englewood Cliffs, NJ: Prentice-Hall, 1995.

[177] H. Ishii and R. Tempo, "Distributed randomized algorithms for the PageRank computation," *IEEE Transactions on Automatic Control*, vol. 55, pp. 1987–2002, 2010.

[178] J. A. Jacquez, *Compartmental Analysis in Biology and Medicine*, 2nd ed. Ann Arbor, MI: University of Michigan Press, 1985.

[179] J. A. Jacquez and C. P. Simon, "Qualitative theory of compartmental systems," *SIAM Review*, vol. 35, pp. 43–79, 1993.

[180] A. Jadbabaie, J. Lin, and A. S. Morse, "Coordination of groups of mobile autonomous agents using nearest neighbor rules," *IEEE Transactions on Automatic Control*, vol. 48, pp. 988–1001, 2003.

[181] N. Jie and L. Zhong, "Sampled-data consensus for high-order multiagent systems under fixed and randomly switching topology," *Discrete Dynamics in Nature and Society*, vol. 2014, 598965, 2014.

[182] M. Jun and D. E. Jeffcoat, "Control theoretic analysis of a target search problem by a team of search vehicles," *International Journal of Control*, vol. 81, pp. 1878–1885, 2008.

[183] E. W. Justh and P. S. Krishnaprasad, "Equilibria and steering laws for planar formations," *Systems and Control Letters*, vol. 52, pp. 25–38, 2004.

[184] J. Kennedy and R. Eberhart, "Particle swarm optimization," in *IEEE International Conference on Neural Networks*, vol. 4, pp. 1942–1946, 1995.

[185] H. K. Khalil, *Nonlinear Systems*. Upper Saddle River, NJ: Prentice-Hall, 1996.

[186] H. K. Khalil, *Nonlinear Systems*, 3rd ed. Upper Saddle River, NJ: Prentice-Hall, 2002.

[187] D. B. Kingston and R. W. Beard, "Discrete-time average-consensus under switching network topologies," in *Proceedings of the American Control Conference* (Minneapolis, MN), pp. 3551–3556, 2006.

[188] D. Kondepudi and I. Prigogine, *Modern Thermodynamics: From Heat Engines to Dissipative Structures*. Chichester, U.K.: John Wiley and Sons, 1998.

[189] N. N. Krasovskii, *Stability of Motion*. Stanford, CA: Stanford University Press, 1963.

[190] J. Kurzweil, "On the inversion of Lyapunov's second theorem on stability of motion," *American Mathematical Society Translations*, vol. 24, pp. 19–77, 1963, translated from *Czechoslovak Mathematical Journal*, vol. 81, pp. 217–259, 455–484, 1956.

[191] H. Kwakernaak and R. Sivan, *Linear Optimal Control Systems*. New York: Wiley, 1972.

[192] V. Lakshmikantham, D. D. Bainov, and P. S. Simeonov, *Theory of Impulsive Differential Equations*. Singapore: World Scientific, 1989.

[193] I. Lasiecka and R. Triggiani, *Control Theory for Partial Differential Equations: Continuous and Approximation Theories*. New York: Cambridge University Press, 2000.

[194] N. E. Leonard and E. Fiorelli, "Virtual leaders, artificial potentials, and coordinated control of groups," in *Proceedings of the IEEE Conference on Decision and Control* (Orlando, FL), pp. 2968–2873, 2001.

[195] A. Leonessa, W. M. Haddad, and V. Chellaboina, *Hierarchical Nonlinear Switching Control Design with Applications to Propulsion Systems*. London, U.K.: Springer-Verlag, 2000.

[196] A. Leonessa, W. M. Haddad, and V. Chellaboina, "Nonlinear system stabilization via hierarchical switching control," *IEEE Transactions on Automatic Control*, vol. 46, pp. 17–28, 2001.

[197] H. Levine, W. Rappel, and I. Cohen, "Self-organization in systems of self-propelled particles," *Physical Review* E, vol. 63, 017101, 2000.

[198] Z. Li, Z. Duan, G. Chen, and L. Huang, "Consensus of multiagent systems and synchronization of complex networks: A unified viewpoint," *IEEE Transactions on Circuits and Systems* I: *Regular Papers*, vol. 57, pp. 213–224, 2010.

[199] Z. Li, Z. Duan, and F. L. Lewis, "Distributed robust consensus control of multi-agent systems with heterogeneous matching uncertainties," *Automatica*, vol. 50, pp. 883–889, 2014.

[200] Z. Li, X. Liu, P. Lin, and W. Ren, "Consensus of linear multi-agent systems with reduced-order observer-based protocols," *Systems and Control Letters*, vol. 60, no. 7, pp. 510–516, 2011.

[201] J. J. Liang, A. K. Qin, P. Suganthan, and S. Baskar, "Comprehensive learning particle swarm optimizer for global optimization of multimodal functions," *IEEE Transactions on Evolutionary Computation*, vol. 10, no. 3, pp. 281–295, 2006.

[202] J. J. Liang, T. P. Runarsson, E. M. Montes, M. Clerc, P. N. Suganthan, C. A. Coello Coello, and K. Deb, "Problem definitions and evaluation criteria for the CEC 2006 special session on constrained real-parameter optimization," Technical Report, School of Electrical and Electronic Engineering, Nanyang Technical University, Singapore, 2006.

[203] J. Lin, A. S. Morse, and B. D. O. Anderson, "The multi-agent rendezvous problem," in *Proceedings of the IEEE Conference on Decision and Control* (Maui, HI), pp. 1508–1513, 2003.

[204] J. Lin, A. S. Morse, and B. D. O. Anderson, "The multi-agent rendezvous problem. Part 1: The synchronous case," *SIAM Journal on Control and Optimization*, vol. 46, pp. 2096–2119, 2007.

[205] P. Lin and Y. Jia, "Consensus of second-order discrete-time multi-agent systems with nonuniform time-delays and dynamically changing topologies," *Automatica*, vol. 45, pp. 2154–2158, 2009.

[206] Z. Lin, M. E. Broucke, and B. A. Francis, "Local control strategies for groups of mobile autonomous agents," *IEEE Transactions on Automatic Control*, vol. 49, no. 4, pp. 622–629, 2004.

[207] Z. Lin, B. Francis, and M. Maggiore, "State agreement for continuous-time coupled nonlinear systems," *SIAM Journal on Control and Optimization*, vol. 46, pp. 288–307, 2007.

[208] H. Liu, G. Xie, and L. Wang, "Necessary and sufficient conditions for solving consensus problems of double-integrator dynamics via sampled control," *International Journal of Robust and Nonlinear Control*, vol. 20, pp. 1706–1722, 2010.

[209] H. Liu, G. Xie, and L. Wang, "Consensus of multi-agent systems with nonlinear dynamics and sampled-data information: A delayed-input approach," *International Journal of Robust and Nonlinear Control*, vol. 23, pp. 602–619, 2013.

[210] J. Liu, H. Zhang, X. Liu, and W.-C. Xie, "Distributed stochastic consensus of multi-agent systems with noisy and delayed measurements," *IET Control Theory and Applications*, vol. 7, pp. 1359–1369, 2013.

[211] Y. Liu, K. M. Passino, and M. M. Polycarpou, "Stability analysis of $m$-dimensional asynchronous swarms with a fixed communication topology," *IEEE Transactions on Automatic Control*, vol. 48, pp. 76–95, 2003.

[212] Y. Liu, Z. Qin, Z. Shi, and J. Lu, "Center particle swarm optimization," *Neurocomputing*, vol. 70, pp. 672–679, 2007.

[213] Z. Liu, S. Deshpande, and Q. Hui, "Quantized particle swarm optimization: An improved algorithm based on group effect and its convergence analysis," in *ASME Dynamic Systems and Control Conference* (Arlington, VA), pp. 321–327, 2011.

[214] Z. Liu, H. Zhang, P. Smith, and Q. Hui, "Hierarchical optimization strategies for sensor network deployment," in *World Automation Congress* (Puerto Vallarta, Mexico), pp. 1028–1041, 2012.

[215] J. Lorenz and D. A. Lorenz, "On conditions for convergence to consensus," *IEEE Transactions on Automatic Control*, vol. 55, pp. 1651–1656, 2010.

[216] J. Lygeros, D. N. Godbole, and S. Sastry, "Verified hybrid controllers for automated vehicles," *IEEE Transactions on Automatic Control*, vol. 43, pp. 522–539, 1998.

[217] N. A. Lynch, *Distributed Algorithms*. San Francisco, CA: Morgan Kaufmann, 1996.

[218] C. Ma, T. Li, and J. Zhang, "Consensus control for leader-following multi-agent systems with measurement noises," *Journal of Systems Science and Complexity*, vol. 23, pp. 35–49, 2010.

[219] C.-Q. Ma and J.-F. Zhang, "Necessary and sufficient conditions for consensusability of linear multi-agent systems," *IEEE Transactions on Automatic Control*, vol. 55, pp. 1263–1268, 2010.

[220] J. A. Marshall, M. E. Broucke, and B. A. Francis, "Formations of vehicles in cyclic pursuit," *IEEE Transactions on Automatic Control*, vol. 49, pp. 1963–1974, 2004.

[221] J. L. Massera, "Contributions to stability theory," *Annals of Mathematics*, vol. 64, pp. 182–206, 1956.

[222] M.-A. Massoumnia, G. C. Verghese, and A. S. Willsky, "Failure detection and identification," *IEEE Transactions on Automatic Control*, vol. 34, pp. 316–321, 1989.

[223] J. Mercieca and S. G. Fabri, "Particle swarm optimization for nonlinear model predictive control," in *5th International Conference on Engineering, Applied Sciences and Technology* (Lisbon, Portugal), pp. 88–93, 2011.

[224] M. Mesbahi and M. Egerstedt, *Graph Theoretic Methods for Multiagent Networks*. Princeton, NJ: Princeton University Press, 2010.

[225] A. N. Michel, K. Wang, and B. Hu, *Qualitative Theory of Dynamical Systems: The Role of Stability Preserving Mappings*. New York: Marcel Dekker, Inc., 2001.

[226] A. Mogilner and L. Edelstein-Keshet, "A non-local model for a swarm," *Journal of Mathematical Biology*, vol. 38, pp. 534–570, 1999.

[227] L. Moreau, "Stability of continuous-time distributed consensus algorithms," in *Proceedings of the IEEE Conference on Decision and Control* (Paradise Island, Bahamas), pp. 3998–4003, 2004.

[228] L. Moreau, "Stability of multiagent systems with time-dependent communication links," *IEEE Transactions on Automatic Control*, vol. 50, pp. 169–182, 2005.

[229] A. P. Morgan and K. S. Narendra, "On the stability of nonautonomous differential equations $\dot{x} = [A + B(t)]x$, with skew symmetric matrix $B(t)$," *SIAM Journal on Control and Optimization*, vol. 15, pp. 163–176, 1977.

[230] K. S. Narendra and A. M. Annaswamy, *Stable Adaptive Systems*. Englewood Cliffs, NJ: Prentice-Hall, 1989.

[231] D. Nesic, L. Zaccrian, and A. R. Teel, "Stability properties of reset systems," *Automatica*, vol. 44, pp. 2019–2026, 2008.

[232] A. Okubo, *Diffusion and Ecological Problems: Mathematical Models.* Berlin, Germany: Springer-Verlag, 1980.

[233] A. Okubo and S. Levin, *Diffusion and Ecological Problems,* 2nd ed. New York: Springer-Verlag, 2001.

[234] R. Olfati-Saber, "Flocking for multi-agent dynamic systems: Algorithms and theory," *IEEE Transactions on Automatic Control,* vol. 51, pp. 401–420, 2006.

[235] R. Olfati-Saber, "Distributed Kalman filtering for sensor networks," in *Proceedings of the IEEE Conference on Decision and Control* (New Orleans, LA), pp. 5492–5498, 2007.

[236] R. Olfati-Saber, J. A. Fax, and R. M. Murray, "Consensus and cooperation in networked multi-agent systems," *Proceedings of the IEEE,* vol. 95, pp. 215–233, 2007.

[237] R. Olfati-Saber and R. M. Murray, "Consensus protocol for networks of dynamic agents," in *Proceedings of the American Control Conference* (Denver, CO), pp. 951–956, 2003.

[238] R. Olfati-Saber and R. M. Murray, "Consensus problems in networks of agents with switching topology and time-delays," *IEEE Transactions on Automatic Control,* vol. 49, pp. 1520–1533, 2004.

[239] R. Olfati-Saber and J. S. Sharmma, "Consensus filters for sensor networks and distributed sensor fusion," in *Proceedings of the IEEE Conference on Decision and Control* (Seville, Spain), pp. 6698–6703, 2005.

[240] B. E. Paden and S. S. Sastry, "A calculus for computing Filippov's differential inclusion with application to the variable structure control of robot manipulators," *IEEE Transactions on Circuits and Systems,* vol. CAS-34, pp. 73–82, 1987.

[241] F. Paganini, J. C. Doyle, and S. H. Low, "Scalable laws for stable network congestion control," in *Proceedings of the IEEE Conference on Decision and Control* (Orlando, FL), pp. 185–190, 2001.

[242] J.-B. Park, K.-S. Lee, J.-R. Shin, and K. Y. Lee, "A particle swarm optimization for economic dispatch with nonsmooth cost functions," *IEEE Transactions on Power Systems,* vol. 20, pp. 34–42, 2005.

[243] J. Parrish and W. Hamner, *Animal Groups in Three Dimensions.* Cambridge, U.K.: Cambridge University Press, 1997.

[244] F. Pasqualetti, F. Dörfler, and F. Bullo, "Attack detection and identification in cyber-physical systems," *IEEE Transactions on Automatic Control,* vol. 58, pp. 2715–2729, 2013.

[245] K. M. Passino, "Biomimicry of bacterial foraging for distributed optimization and control," *IEEE Control Systems Magazine*, vol. 22, pp. 52–67, 2002.

[246] K. M. Passino, A. N. Michel, and P. J. Antsaklis, "Lyapunov stability of a class of discrete event systems," *IEEE Transactions on Automatic Control*, vol. 39, pp. 269–279, 1994.

[247] P. Peleties and R. DeCarlo, "Asymptotic stability of m-switched systems using Lyapunov-like functions," in *Proceedings of the American Control Conference* (Boston, MA), pp. 1679–1684, 1991.

[248] S. Phillips, Y. Li, and R. G. Sanfelice, "A hybrid consensus protocol for pointwise exponential stability with intermittent information," in *Proceedings of the IFAC Symposium on Nonlinear Control Systems* (Monterey, CA), pp. 146–151, 2016.

[249] J.-B. Pomet and L. Praly, "Adaptive nonlinear regulation: Estimation from the Lyapunov equation," *IEEE Transactions on Automatic Control*, vol. 37, pp. 729–740, 1992.

[250] M. Porfiri and D. J. Stilwell, "Consensus seeking over random weighted directed graphs," *IEEE Transactions on Automatic Control*, vol. 52, pp. 1767–1773, 2007.

[251] I. Prigogine, *From Being to Becoming*. New York: Freeman, 1980.

[252] Y. Qian, X. Wu, J. Lu, and J.-A. Lu, "Second-order consensus of multi-agent systems with nonlinear dynamics via impulsive control," *Neurocomputing*, vol. 125, pp. 142–147, 2014.

[253] M. L. Radulescu and F. H. Clarke, "Geometric approximation of proximal normals," *Journal of Convex Analysis*, vol. 4, pp. 373–379, 1997.

[254] W. Ren, "Consensus strategies for cooperative control of vehicle formations," *IET Control Theory and Applications*, vol. 1, pp. 505–512, 2007.

[255] W. Ren and R. W. Beard, "Consensus seeking in multiagent systems under dynamically changing interaction topologies," *IEEE Transactions on Automatic Control*, vol. 50, pp. 655–661, 2005.

[256] W. Ren, R. W. Beard, and E. M. Atkins, "Information consensus and its applications in multi-vehicle cooperative control," *IEEE Control Systems Magazine*, vol. 27, pp. 71–82, 2007.

[257] W. Ren and Y. Cao, *Distributed Coordination of Multi-Agent Networks: Emergent Problems, Models, and Issues*. London, U.K.: Springer-Verlag, 2010.

[258] C. W. Reynolds, "Flocks, herds, and schools: A distributed behavioral model," *Computers and Graphics*, vol. 21, pp. 25–34, 1987.

[259] E. Rimon and D. Koditschek, "Exact robot navigation using artificial potential functions," *IEEE Transactions on Robotics and Automation*, vol. 8, pp. 501–518, 1992.

[260] J. Robinson and Y. Rahmat-Samii, "Particle swarm optimization in electromagnetics," *IEEE Transactions on Antennas and Propagation*, vol. 52, pp. 397–407, 2004.

[261] R. Rockafellar and R. J. B. Wets, *Variational Analysis*. Berlin, Germany: Springer, 1998.

[262] R. T. Rockafellar, *Convex Analysis*. Princeton, NJ: Princeton University Press, 1997.

[263] L. Rosier, "Homogeneous Lyapunov function for homogeneous continuous vector field," *Systems and Control Letters*, vol. 19, pp. 467–473, 1992.

[264] H. L. Royden, *Real Analysis*. New York: Macmillan, 1988.

[265] E. P. Ryan, "Singular optimal controls for second-order saturating systems," *International Journal of Control*, vol. 30, pp. 549–564, 1979.

[266] E. P. Ryan, "Finite-time stabilization of uncertain nonlinear planar systems," *Dynamics and Control*, vol. 1, pp. 83–94, 1991.

[267] E. P. Ryan, "An integral invariance principle for differential inclusions with applications in adaptive control," *SIAM Journal on Control and Optimization*, vol. 36, pp. 960–980, 1998.

[268] T. Sadikhov, W. M. Haddad, R. Goebel, and M. Egerstedt, "Set-valued protocols for almost consensus of multiagent systems with uncertain interagent communication," in *Proceedings of the American Control Conference* (Portland, OR), pp. 4002–4007, 2014.

[269] A. M. Samoilenko and N. A. Perestyuk, *Impulsive Differential Equations*. Singapore: World Scientific, 1995.

[270] W. Sandberg, "On the mathematical foundations of compartmental analysis in biology, medicine and ecology," *IEEE Transactions on Circuits and Systems* I, vol. 25, pp. 273–279, 1978.

[271] M. Schwager, J. J. Slotine, and D. Rus, "Decentralized, adaptive control for coverage with networked robots," in *IEEE International Conference on Robotics and Automation* (Rome, Italy), pp. 3289–3294, 2007.

[272] M. Schwager, J. J. Slotine, and D. Rus, "Consensus learning for distributed coverage control," in *IEEE International Conference on Robotics and Automation* (Pasadena, CA), pp. 1042–1048, 2008.

[273] E. Semsar-Kazerooni and K. Khorasani, "Optimal consensus algorithms for cooperative team of agents subject to partial information," *Automatica*, vol. 44, pp. 2766–2777, 2008.

[274] E. Semsar-Kazerooni and K. Khorasani, "Optimal consensus seeking in a network of multiagent systems: An LMI approach," *IEEE Transactions on Systems, Man, and Cybernetics—Part B: Cybernetics*, vol. 40, pp. 540–547, 2010.

[275] J. S. Shamma, *Cooperative Control of Distributed Multi-Agent Systems*. Hoboken, NJ: John Wiley and Sons, 2007.

[276] J. S. Shamma and G. Arslan, "Dynamic fictitious play, dynamic gradient play, and distributed convergence to Nash equilibria," *IEEE Transactions on Automatic Control*, vol. 50, pp. 312–327, 2005.

[277] J. Shen, J. Hu, and Q. Hui, "Semistability of switched linear systems with applications to sensor networks: A generating function approach," in *Proceedings of the IEEE Conference on Decision and Control* (Orlando, FL), pp. 8044–8049, 2011.

[278] J. Shen, J. Hu, and Q. Hui, "Semistability of switched linear systems with application to PageRank algorithms," *European Journal of Control*, vol. 20, pp. 132–140, 2014.

[279] D. Shevitz and B. Paden, "Lyapunov stability theory of nonsmooth systems," *IEEE Transactions on Automatic Control*, vol. 39, pp. 1910–1914, 1994.

[280] Y. Shi and R. Eberhart, "A modified particle swarm optimizer," in *IEEE International Conference on Evolutionary Computation* (Anchorage, AK), pp. 69–73, 1998.

[281] T. R. Smith, H. Hanssmann, and N. E. Leonard, "Orientation control of multiple underwater vehicles with symmetry-breaking potentials," in *Proceedings of the IEEE Conference on Decision and Control* (Orlando, FL), pp. 4598–4603, 2001.

[282] J. Snyders and M. Zakai, "On nonnegative solutions of the equation $AD + DA' = -C$," *SIAM Journal on Applied Mathematics*, vol. 18, pp. 704–714, 1970.

[283] S. L. Sobolev, "Applications of functional analysis in mathematical physics," in *Translations of Mathematical Monographs*, vol. 7. Providence, RI: American Mathematical Society, 1963.

[284] J. Spall, *Introduction to Stochastic Search and Optimization: Estimation, Simulation, and Control.* New York: Wiley, 2003.

[285] S. Stankovic, M. Stankovic, and D. Stipanovic, "Decentralized parameter estimation by consensus based stochastic approximation," *IEEE Transactions on Automatic Control,* vol. 56, pp. 531–543, 2011.

[286] S. H. Strogatz, "From Kuramoto to Crawford: Exploring the onset of synchronization in populations of coupled oscillators," *Physica D,* vol. 143, pp. 1–20, 2000.

[287] Y. Su and J. Huang, "Two consensus problems for discrete-time multi-agent systems with switching network topology," *Automatica,* vol. 48, pp. 1988–1997, 2012.

[288] K. Sumizaki, L. Liu, and S. Hara, "Adaptive consensus on a class of nonlinear multi-agent dynamical systems," in *SICE Annual Conference* (Taipei, Taiwan), pp. 1141–1145, 2010.

[289] I. Suzuki and M. Yamashita, "Distributed anonymous mobile robots: Formation of geometric patterns," *SIAM Journal on Computing,* vol. 28, pp. 1347–1363, 1999.

[290] D. Swaroop and J. K. Hedrick, "Constant spacing strategies for platooning in automated highway systems," *ASME Journal of Dynamic Systems, Measurement, and Control,* vol. 121, pp. 462–470, 1999.

[291] A. Tahbaz-Salehi and A. Jadbabaie, "A necessary and sufficient condition for consensus over random networks," *IEEE Transactions on Automatic Control,* vol. 53, pp. 791–795, 2008.

[292] Y. Tang, X. Luo, Q. Hui, and R. K. C. Chang, "On generalized low-rate denial-of-quality attack against Internet services," in *17th International Workshop on Quality of Service* (Charleston, SC), pp. 1–5, 2009.

[293] Y. Tang, X. Luo, Q. Hui, and R. K. C. Chang, "Modeling the vulnerability of feedback-control based Internet services to low-rate DoS attacks," *IEEE Transactions on Information Forensics and Security,* vol. 9, pp. 339–353, 2014.

[294] H. G. Tanner, A. Jadbabaie, and G. J. Pappas, "Stable flocking of mobile agents, Part I: Fixed topology," in *Proceedings of the IEEE Conference on Decision and Control* (Maui, HI), pp. 2010–2015, 2003.

[295] H. G. Tanner, A. Jadbabaie, and G. J. Pappas, "Flocking in teams of nonholonomic agents," in *Cooperative Control* (V. Kumar, N. E. Leonard, and A. S. Morse, eds.), pp. 229–239, Berlin, Germany: Springer-Verlag, 2005.

[296] H. G. Tanner, A. Jadbabaie, and G. J. Pappas, "Flocking in fixed and switching networks," *IEEE Transactions on Automatic Control*, vol. 52, pp. 863–868, 2007.

[297] C. N. Taylor, R. W. Beard, and J. Humphreys, "Dynamic input consensus using integrators," in *Proceedings of the American Control Conference* (San Francisco, CA), pp. 3357–3362, 2011.

[298] C. Tomlin, G. J. Pappas, and S. S. Sastry, "Conflict resolution for air traffic management: A study in multiagent hybrid systems," *IEEE Transactions on Automatic Control*, vol. 43, pp. 509–521, 1998.

[299] B. Touri and A. Nedic, "On ergodicity, infinite flow, and consensus in random models," *IEEE Transactions on Automatic Control*, vol. 56, pp. 1593–1605, 2011.

[300] H. L. Trentelman, K. Takaba, and N. Monshizadeh, "Robust synchronization of uncertain linear multi-agent systems," *IEEE Transactions on Automatic Control*, vol. 58, pp. 1511–1523, 2013.

[301] J. N. Tsitsiklis, D. P. Bertsekas, and M. Athans, "Distributed asynchronous deterministic and stochastic gradient optimization algorithms," *IEEE Transactions on Automatic Control*, vol. 31, pp. 803–812, 1986.

[302] F. van den Bergh and A. Engelbrecht, "A study of particle swarm optimization particle trajectories," *Information Sciences*, vol. 176, pp. 937–971, 2006.

[303] A. T. Vemuri, "Sensor bias fault diagnosis in a class of nonlinear systems," *IEEE Transactions on Automatic Control*, vol. 46, pp. 949–954, 2001.

[304] T. Vicsek, A. Czirók, E. Ben-Jacob, I. Cohen, and O. Shochet, "Novel type of phase transition in a system of self-driven particles," *Physical Review Letters*, vol. 75, pp. 1226–1229, 1995.

[305] R. Vidal, O. Shakernia, and S. Sastry, "Formation control of nonholonomic mobile robots with omnidirectional visual servoing and motion segmentation," in *IEEE International Conference on Robotics and Automation* (Taipei, Taiwan), pp. 584–589, 2003.

[306] M. Vidyasagar, *Nonlinear Systems Analysis*, 2nd ed. Classics Appl. Math. 42. Philadelphia: SIAM, 2002.

[307] L. R. Volevich and B. P. Paneyakh, "Certain spaces of generalized functions and embedding theorems," *Russian Mathematics Surveys*, vol. 20, pp. 1–73, 1965.

[308] L. Vu, "A fundamental problem in multiagent networks: Consensus with imperfect information," Technical Report, Department of Aeronautics and Astronomics, University of Washington, Seattle, WA, June 2008.

[309] X. Wang and J. Xiao, "PSO-based model predictive control for nonlinear processes," in *Advances in Natural Computation* (L. Wang, K. Chen, and Y. S. Ong, eds.), vol. 3611, pp. 196–203, Berlin, Germany: Springer-Verlag, 2005.

[310] J. Warga, *Optimal Control of Differential and Functional Equations.* New York: Academic Press, 1972.

[311] F. W. Wilson, "Smoothing derivatives of functions and applications," *Transactions of the American Mathematical Society*, vol. 139, pp. 413–428, 1969.

[312] J. D. Wolfe, D. F. Chichka, and J. L. Speyer, "Decentralized controllers for unmanned aerial vehicle formation flight," in *Proceedings of the AIAA Conference on Guidance, Navigation, and Control* (AIAA-1996-3833, San Diego, CA), 1996.

[313] Z. Wu, L. Peng, L. Xie, and J. Wen, "Stochastic bounded consensus tracking of leader–follower multi-agent systems with measurement noises based on sampled-data with small sampling delay," *Physica A: Statistical Mechanics and Its Applications*, vol. 392, pp. 918–928, 2013.

[314] F. Xiao and L. Wang, "Asynchronous rendezvous analysis via set-valued consensus theory," *SIAM Journal on Control and Optimization*, vol. 50, pp. 196–221, 2012.

[315] L. Xiao and S. Boyd, "Fast linear iterations for distributed averaging," *Systems and Control Letters*, vol. 53, pp. 65–78, 2004.

[316] L. Xiao, S. Boyd, and S. J. Kim, "Distributed average consensus with least-mean-square deviation," *Journal of Parallel Distributed Computing*, vol. 67, pp. 33–46, 2007.

[317] W. Xion, D. W. C. Ho, and J. Cao, "Impulsive consensus of multi-agent directed networks with nonlinear perturbations," *International Journal of Robust and Nonlinear Control*, vol. 22, pp. 1571–1582, 2012.

[318] T. Yang, Z. Meng, D. V. Dimarogonas, and K. H. Johansson, "Global consensus for discrete-time multi-agent systems with input saturation constraints," *Automatica*, vol. 50, pp. 499–506, 2014.

[319] T. Yoshizawa, *Stability Theory by Liapunov's Second Method.* Tokyo, Japan: Mathematical Society of Japan, 1966.

[320] K. You and L. Xie, "Network topology and communication data rate for consensusability of discrete-time multi-agent systems," *IEEE Transactions on Automatic Control*, vol. 56, pp. 2262–2275, 2011.

[321] T. Yucelen and M. Egerstedt, "Control of multiagent systems under persistent disturbances," in *Proceedings of the American Control Conference* (Montreal, Canada), pp. 5264–5269, 2012.

[322] T. Yucelen and W. M. Haddad, "Consensus protocols for networked multi-agent systems with a uniformly continuous quasi-resetting architecture," *International Journal of Control*, vol. 87, pp. 1716–1727, 2014.

[323] T. Yucelen and E. N. Johnson, "Control of multivehicle systems in the presence of uncertain dynamics," *International Journal of Control*, vol. 86, pp. 1540–1553, 2013.

[324] T. Yucelen, J. D. Peterson, and K. L. Moore, "Control of networked multiagent systems with uncertain graph topologies," in *Dynamic Systems and Control Conference* (Columbus, OH), vol. 3, no. 10, 2015.

[325] X. Zeng, Q. Hui, W. M. Haddad, T. Hayakawa, and J. M. Bailey, "Synchronization of biological neural network systems with stochastic perturbations and time delays," *Journal of The Franklin Institute*, vol. 351, pp. 1205–1225, 2014.

[326] H. Zhang and Q. Hui, "Hybrid multiagent swarm optimization: Algorithms, evaluation, and application," in *Proceedings of the IEEE Conference on Decision and Control* (Maui, HI), pp. 5699–5704, 2012.

[327] H. Zhang and Q. Hui, "Convergence analysis and parallel computing implementation for the multiagent coordination optimization algorithm with applications," in *IEEE International Conference on Automation Science and Engineering* (Madison, WI), pp. 837–842, 2013.

[328] H. Zhang and Q. Hui, "Modified hybrid multiagent swarm optimization algorithms for mixed-binary nonlinear programming," in *Hawaii International Conference on System Sciences* (Maui, HI), pp. 1412–1421, 2013.

[329] H. Zhang and Q. Hui, "Multiagent coordination optimization: A control-theoretic perspective of swarm intelligence algorithms," in *IEEE Congress on Evolutionary Computation* (Cancun, Mexico), pp. 3339–3346, 2013.

[330] H. Zhang and Q. Hui, "A new hybrid swarm optimization algorithm for power system vulnerability analysis and sensor network deployment," in *International Joint Conference on Neural Networks* (Dallas, TX), pp. 1221–1228, 2013.

[331] X. Zhang, T. Parisini, and M. M. Polycarpou, "Sensor bias fault isolation in a class of nonlinear systems," *IEEE Transactions on Automatic Control*, vol. 50, pp. 370–376, 2005.

[332] Y. Zhang and Y.-P. Tian, "Consensus of data-sampled multi-agent systems with random communication delay and packet loss," *IEEE Transactions on Automatic Control*, vol. 55, pp. 939–943, 2010.

[333] M. Zhu and S. Martínez, "Discrete-time dynamic average consensus," *Automatica*, vol. 46, pp. 322–329, 2010.

# Index

## A

adaptive control, 344
adaptive distributed
    observers, 371
adaptive estimation, 367
adaptive identifiers, 367
adjacency matrix, 16, 344
advection-diffusion
    model, 238
adversarial attack
    effectiveness, 450
alignment, 1
all-to-all connected
    network, 417
approximate consensus,
    487
asymptotically stable
    equilibrium, 57
  infinite-dimensional
    system, 225
asymptotically stable set,
    560
attractive set, 560
average information state
    equipartitioning, 307

## B

balanced coordination,
    449
balanced graph, 17
Banach space, 199
bilinear matrix inequality
    (BMI), 259
Boltzmann entropy, 32
Bolza problem, 285

## C

ceiling function, 404

centroid convergence, 122
Clarke generalized
    gradient, 136
Clarke upper generalized
    derivative, 136
Clausius inequality, 32
  infinite-dimensional
    system, 228
cohesion, 1
colony optimization, 334
compact embedding, 225
compartmental
    dynamical system,
    205
compartmental matrix,
    198
compartmental models,
    197
compartmental system
    donor controlled, 211
compartmental vector
    field, 205
complete graph, 43
complete strong cycle, 56
complexity, 219
connected agents, 18
connectivity matrix, 16
consensus, 1, 14
consensus problem, 17,
    574
conservation equation,
    221
contingent set, 152
continuous set-valued
    map, 500
continuous
    thermodynamics, 221
continuously
    differentiable entropy

functional, 229
control law, 277
converse Lyapunov
    theorem for
    semistability, 80, 397
cooperative game, 458
cyclic pursuit, 1, 556, 578
cyclic pursuit control, 14

## D

damage mitigation, 450
decentralized control, 13
density distribution, 222
density function, 223
difference inclusion, 488,
    499
differential inclusion, 135
digraph, 16
dilation, 183
direct predecessor, 55
direct successor, 55
directed graph, 16, 17
direction cone, 100
  discontinuous
    dynamical systems,
    143
Dirichlet boundary
    control, 236
Dirichlet integral, 224
disconnected agents, 18
distributed control, 13
domain of semistability,
    81
  discrete-time system,
    393
drifting phenomenon, 297
driftless controllable
    nonholonomic
    system, 35

dynamic equilibrium, 51

# E

edges of a graph, 16
emergent behavior, 220
energy, 50
entropy, 31, 51
   discrete-time system,
    418
   hybrid system, 521
entropy functional, 229
equilibrium
   controllability, 245,
    425
equilibrium observability,
   245, 425
equilibrium point
   hybrid system, 512
   infinite-dimensional
    system, 225
equipartition of energy,
   26
equipartitioning, 14
ergodic theory, 291
essentially nonnegative
   matrix, 198
essentially nonnegative
   vector field, 205
Eulerian swarm model,
   221
existential statement, 15

# F

feedback control law, 277
Filippov set-valued map,
   135
finite time convergent, 88
finite time parallel
   formation, 110, 172
finite time semistability,
   87
   discontinuous system,
    137
   discrete-time system,
    400
finite time semistable, 88
finite time semistable
   equilibrium, 97
   discrete-time system,
    400
   hybrid system, 533

first law of
   thermodynamics, 52
flocking, 1, 556
flocking control, 14
flux function, 222
formation control, 1, 14,
   556

# G

game theory, 458
generalized deadzone
   function, 403
generalized potential
   function, 36, 546
globally finite time
   semistable, 88
gossip algorithms, 291

# H

$\mathcal{H}_2$ norm, 246
   discrete-time system,
    426
$\mathcal{H}_2$ optimal semistable
   control, 243
$\mathcal{H}_2$ optimal semistable
   control problem, 436
Hamilton–Jacobi–
   Bellman equation,
   277, 279
homogeneous body, 49
homogeneous function, 92
honeybee mating
   optimization, 334
hybrid consensus, 531
hybrid consensus
   problem, 511
hybrid consensus
   protocol, 515
hybrid dynamic controller
   557
hybrid dynamical system,
   3, 512
hybrid stabilization of
   sets, 555

# I

impulsive differential
   equation, 3
impulsive dynamical
   system, 513

in-degree, 55
incidence matrix, 344
infinite-dimensional
   system, 225
information consensus, 26
information state, 17
input-to-state stability,
   296
intermittent information,
   512
internal energy, 50
invariant set, 23
   discrete-time system,
    392
   infinite-dimensional
    system, 225
irreducible matrix, 18
isolated system, 51
isotropic body, 49

# K

$k$-semicontrollability, 314
   discrete-time system,
    466
$k$-semiobservability, 314
   discrete-time system,
    466
$k$-transversality, 559
Krasovskii–LaSalle
   theorem, 23
   hybrid systems, 514
Kuramoto model, 191

# L

$\mathcal{L}_\infty$ space, 223
Laplacian disagreement
   function, 532
leader-follower
   coordination
   problem, 238
Lie bracket, 92
Lie derivative, 92, 559
Lie group, 35
linear matrix inequalities
   (LMIs), 243
lower semicontinuous
   function, 401
Lyapunov stable
   equilibrium, 57
   discrete-time system,
    392

hybrid system, 533
infinite-dimensional
    system, 225
Lyapunov stable set, 560
Lyapunov theorem
    semistability, 395

# M

Minkowski sum, 488
mobile sensor network,
    450

# N

Nash equilibrium, 449,
    458
negatively invariant set,
    80
    infinite-dimensional
        system, 225
Nernst's theorem, 58
network resource
    allocation, 449
Neumann boundary
    control, 236
Newton's law of cooling,
    237
nodes of a graph, 16
nondecreasing function,
    390
nonholonomic dynamics,
    35
nonisolated equilibrium
    point, 56
nonnegative function, 198
nonnegative vector field,
    205
nontangent vector field,
    100
    discontinuous systems,
        143
nontangent vector field to
    a set, 100
null space, 15

# O

optimal semistabilization
    problem, 277
optimal state
    equipartitioning
    problem, 322

out-degree, 55
outer semicontinuous
    set-valued map, 500

# P

pairwise symmetry
    condition, 53
parallel formation, 544,
    569
particle swarm
    optimization, 334
path, 17
Perron matrix, 496
positive function, 198
positive orbit, 98
    discrete-time system,
        392
positively invariant set,
    137
    discrete-time system,
        391
    infinite-dimensional
        system, 225
power semistable
    Lyapunov equation,
    317
precompact set, 199
projection, 151
projection norm bound,
    352
projection operator, 352
projection tolerance
    bound, 352
protocol agreement, 26
proximal normal
    direction, 152

# Q

quasi-equilibrium state,
    52

# R

randomized swarm
    algorithm, 476
range space, 15
regular function, 136
rendezvous problem, 1,
    574
resetting law, 512
resetting set, 512

resetting time, 513
resource allocation
    problem, 451
restricted prolongation,
    143
robust consensus, 179

# S

Schwartz distributions,
    223
second law of
    thermodynamics, 18,
    52
sector bound condition,
    53
semicontrollability, 245
    discrete-time system,
        424
semidetectability, 264
semi-Euler vector field, 92
semiobservability, 245
    discrete-time system,
        424
semistability, 2
    discontinuous system,
        136
semistabilizability, 264
semistable equilibrium,
    57
    discrete-time system,
        392
    hybrid system, 533
    infinite-dimensional
        system, 225
semistable LQR (linear
    quadratic regulator)
    control theory, 462
semistable matrix, 424,
    455
semistable with respect
    to a set, 21
separation, 1
sequential continuity, 559
set-valued invariance
    principle, 500
set-valued Lie derivative,
    136
set-valued map, 499
settling-time function, 88
    discrete-time system,
        400

hybrid system, 534
Shannon entropy, 32
simple complete strong
    cycle, 56
simple strong cycle, 56
Sobolev space, 224
solution sequence, 389
spectral abscissa, 15
spectral radius, 15
spectrum, 15
state-dependent
    impulsive dynamical
    system, 513, 541
stochastic optimal
    semistable control,
    292
stochastic optimization
    method, 476
strong cycle, 55
strong path, 55
strongly connected graph,
    17
strongly positively
    invariant, 135

supervisory control, 531
swarm optimization
    algorithm, 334
swarm systems, 3
symmetric Fourier-type
    heat transfer, 53
system thermodynamics,
    49

T

tangent cone, 101, 143
temperature, 51
thermal conductance
    matrix, 56
thermal conductivity, 50
thermal equilibrium, 51
thermally insulated body,
    49
time-dependent hybrid
    protocol, 541
total entropy function, 58
trajectory, 80
    discrete-time system,
    389

U

undirected graph, 17
unicycle model, 43
universal statement, 15
upper right Dini
    derivative, 83
upper semicontinuous
    set-valued map, 500

W

weakly invariant, 135
weakly negatively
    invariant, 135
weakly positively
    invariant, 135
weakly proper function,
    98

Z

zeroth law of
    thermodynamics, 18,
    52